onship

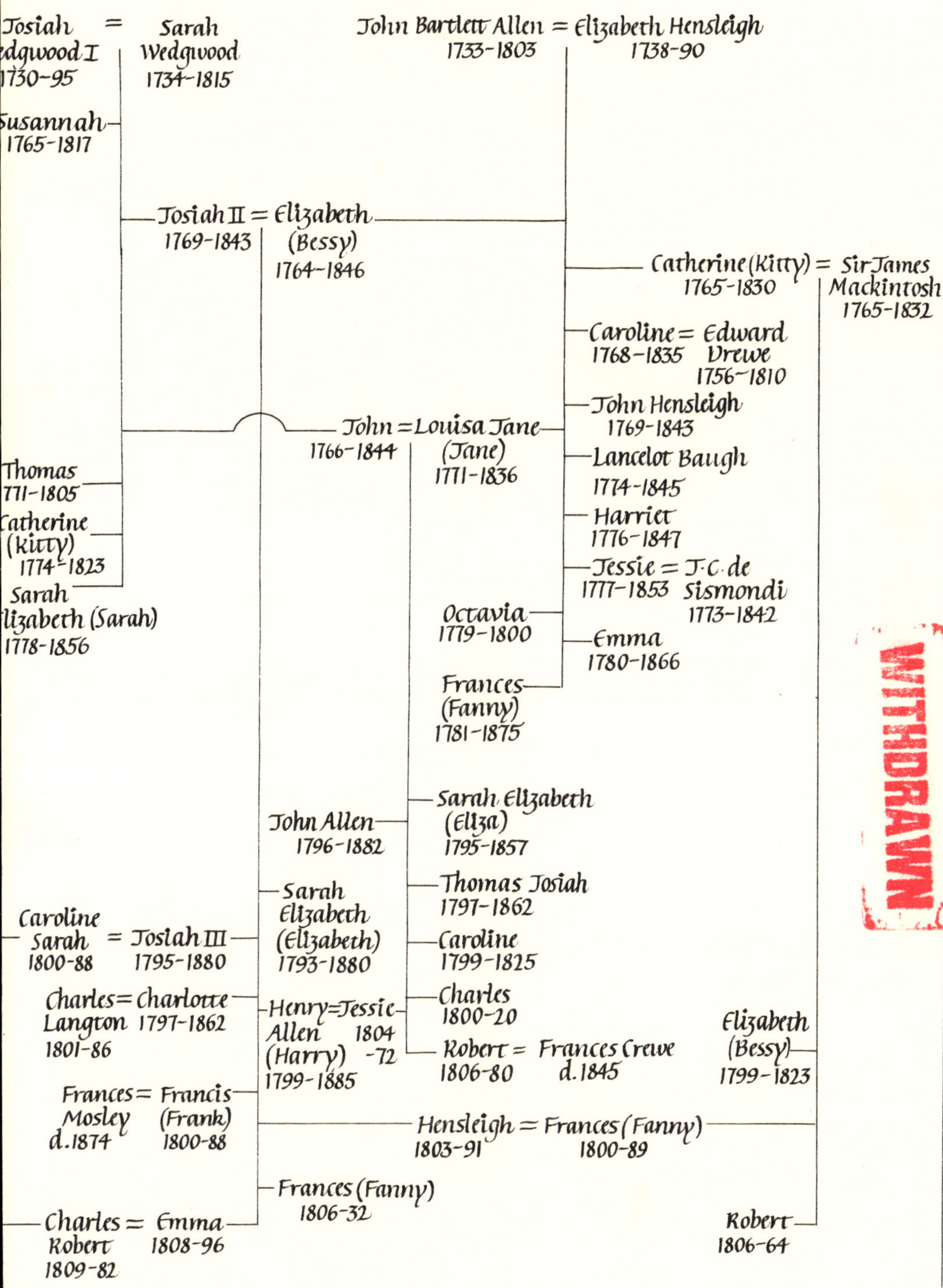

Josiah
dgwood I
1730–95
=
Sarah
Wedgwood
1734–1815
Susannah
1765–1817
John Bartlett Allen = Elizabeth Hensleigh
1733–1803
1738–90
Josiah II = Elizabeth
1769–1843
(Bessy)
1764–1846
Catherine (Kitty) = Sir James
1765–1830
Mackintosh
1765–1832
Caroline = Edward
1768–1835
Drewe
1756–1810
John Hensleigh
1769–1843
John = Louisa Jane
1766–1844
(Jane)
1771–1836
Lancelot Baugh
1774–1845
Harriet
1776–1847
Jessie = J. C. de
1777–1853
Sismondi
1773–1842
Thomas
771–1805
Catherine
(Kitty)
1774–1823
Sarah
lizabeth (Sarah)
1778–1856
Octavia
1779–1800
Emma
1780–1866
Frances
(Fanny)
1781–1875
Sarah Elizabeth
(Eliza)
1795–1857
John Allen
1796–1882
Thomas Josiah
1797–1862
Sarah
Elizabeth
(Elizabeth)
1793–1880
Caroline
Sarah
1800–88
= Josiah III
1795–1880
Caroline
1799–1825
Charles = Charlotte
Langton
1801–86
1797–1862
Henry = Jessie
Allen
(Harry)
1799–1885
1804
–72
Charles
1800–20
Robert = Frances Crewe
1806–80
d. 1845
Elizabeth
(Bessy)
1799–1823
Frances = Francis
Mosley
d. 1874
(Frank)
1800–88
Hensleigh = Frances (Fanny)
1803–91
1800–89
Frances (Fanny)
1806–32
Charles = Emma
Robert
1809–82
1808–96
Robert
1806–64

THE CORRESPONDENCE OF
CHARLES DARWIN

Editors

FREDERICK BURKHARDT DUNCAN M. PORTER

JOY HARVEY MARSHA RICHMOND

Production Manager

SARAH BENTON

Assistant Editors

CHARLOTTE BOWMAN

HEIDI BRADSHAW PERRY O'DONOVAN

Research Associates

ANNE SCHLABACH BURKHARDT

NORA CARROLL STEVENSON

Office Manager

HEDY FRANKS

This edition of the Correspondence of Charles Darwin is sponsored by the American Council of Learned Societies. Its preparation is made possible by the co-operation of the Cambridge University Library and the American Philosophical Society.

Advisory Committees for the edition, appointed by the Council, have the following members:

Support for editing has been received from the National Endowment for the Humanities, the National Science Foundation, the Andrew W. Mellon Foundation, the Alfred P. Sloan Foundation, the Royal Society of London, the British Academy, the Pilgrim Trust, the Stifterverband für die Deutsche Wissenschaft, and the Isaac Newton Trust. The National Endowment's grants (Nos. RE-23166-75-513, RE-27067-77-1359, RE-00082-80-1628, RE-20166-82, RE-20480-85, RE-20764-89, RE-20913-91, and RE-21097-93) were from its Program for Editions; the National Science Foundation's funding of the work was under grants Nos. SOC-75-15840, SES-7912492, SES-8517189, DIR-9020874, and SBR-9020874. Any opinions, findings, conclusions, or recommendations expressed in this publication are those of the editors and do not necessarily reflect the views of the grantors.

Charles Darwin. Taken by his son, William Erasmus Darwin.
(Courtesy of the Archives, Gray Herbarium, Harvard University.)

THE CORRESPONDENCE OF
CHARLES DARWIN

VOLUME 9 1861

CAMBRIDGE
UNIVERSITY PRESS

Published by the Press Syndicate of the University of Cambridge
The Pitt Building, Trumpington Street, Cambridge CB2 1RP
40 West 20th Street, New York, NY 10011–4211, USA
10 Stamford Road, Oakleigh, Melbourne 3166, Australia

First published 1994

Printed in Great Britain at the University Press, Cambridge

British Library cataloguing in publication data available

Library of Congress cataloguing in publication data available

ISBN 0 521 45156 6

CONTENTS

ILLUSTRATIONS

LIST OF LETTERS

The following list is in the order of the entries in the *Calendar of the correspondence of Charles Darwin*. It includes all those letters that are listed in the *Calendar* within the date range covered by this volume of the *Correspondence*. Some letters have been redated since the publication of the *Calendar*, so this list is necessary to enable users of the *Calendar* to locate such letters in the *Correspondence*. Alongside the *Calendar* numbers are the current dates ascribed to every item. In the list, a date printed in italic type indicates that the item has been omitted from the volume entirely.

Letters included in this volume that were not available for inclusion in the *Calendar* appear at the end of the list. These letters are listed in a 'Supplement' to a new edition of the *Calendar* (Cambridge University Press, 1994). The numbers given to the new letters are those assigned in the *Calendar* Supplement; they correspond to the chronological ordering of the original listing with the addition of an alphabetical marker.

1616. [17 July 1861]
2415. 17 February [1861]
2628. See Appendix VI.
2661. 22 January [1861]
2738. Postscript to 3041.
2765. 20 April [1861?]
2868. 16 July [1861]
3037. *Enclosure to 2576.*
3038. [before 5 February 1861]
3039. [before 8 November 1861]
3040. *[after 3 May 1864]*
3041. 3 January [1861]
3042. [before 5 January 1861]
3043. 5 June [1861]
3044. 7 January [1861]
3045. 8 January [1861]
3046. 9 January [1861]
3047. 15 January [1861]
3048. 21 January [1861]
3049. 22 January 1861
3050. 23 [January 1861]
3051. [24 January 1861]
3052. 25 January 1861
3053. 1 February 1861
3054. 2 February [1861]
3055. 3 February [1861]
3056. 3 February [1861]
3057. 4 February [1861]
3058. 4 February 1861
3059. *4 February [1856]. See* Supplement to vol. 7.
3060. 8 [February 1861]
3061. [before 9 February 1861]
3062. 14 February [1861]
3063. 17 February [1861]
3064. 17 February [1861]
3065. 20 [February 1861]
3066. 22 February [1861]
3067. 23 February [1861]
3068. 23 February [1861]
3069. 24 February [1861]
3070. 25 February [1861]
3071. 25 February [1861]
3072. 26 February [1861]
3073. 26–7 February [1861]
3074. [before 6 April 1861]
3075. 1 March [1861]
3076. 2 March [1861]
3077. 3 March 1861
3078. 3 March [1861]
3079. 3 March [1861]
3080. 4 March [1861]
3081. 4 March [1861]
3082. 4 March 1861
3083. 7 March [1861]

INTRODUCTION

By 1861, Darwin had weathered the storm that followed the publication of *Origin of species* and felt cautiously optimistic about the ultimate acceptance of his ideas. As the year opened, he was hard at work on his study of the variations in domestic animals and plants, the first part of the more detailed exposition of the evidence for his theory that he had promised in the preface of *Origin*. But, important as it was to strengthen his case using the data he had compiled over twenty years, he now wished to explore the applications of natural selection that were of most interest to him, tackling problems in natural history, both new and old. These new investigations stemmed from direct botanical observations in the field made during enforced absences from his portfolios of notes at Down House. During the summer of 1860, he had become interested in *Drosera* and other insectivorous plants. His interest in dimorphism, awakened in 1860 following observations of *Primula*, developed into an intensive study of the phenomenon in 1861. Orchids, in particular the astonishing adaptive mechanisms by which cross-pollination is ensured, also became the subject of dedicated research. These studies, initiated as mere pastimes, became full-scale investigations of fundamental features of the natural world—of sexuality, propagation, hybridisation, and other phenomena that, as Darwin said in his *Autobiography*, he had before thought were 'unmeaning variability' (p. 128). Darwin often admitted that he enjoyed working on these fresh projects more than 'sticking' to 'confounded cocks Hens & Ducks'.

The letters in this volume provide a detailed and intimate understanding of Darwin's problem-solving method of work and exude a sense of his confidence in the efficacy of his theory. That is not to say that he was any less active in promoting the acceptance of evolution and natural selection. Indeed, early in the year he acted on two fronts to consolidate and augment support for the views presented in *Origin*. Having learned from his publisher, John Murray, in November 1860 that a new edition of *Origin* was called for, Darwin took the opportunity to include in the third edition corrections and additions that addressed criticisms made by reviewers. He also inserted for his British audience a preface, already published in the first American and German editions, that included a discussion of works pre-dating *Origin* that contained theories of species transmutation, setting in proper relief his own mechanism of natural selection. With this work behind him, Darwin took steps to acquaint those who opposed the theory on religious grounds with the argument that natural selection was not inconsistent with natural theology. He made arrangements with his friend Asa Gray to reprint and distribute in Britain Gray's series of review-essays on this topic previously published in America, telling Gray: 'I believe your pamphlet will do me & Natural Selection, right good service' (letter to Asa Gray, 26–7 Februrary [1861]). Darwin drew up a carefully thought-out list of those individuals, societies, and journals to which to send a copy of the pamphlet (see Appendix III). However, Darwin himself, as his letters to Gray show, remained unconvinced by Gray's suggestion that providence may have played a part in the appearance and selection of variations.

Although sales of *Origin* began to decline later in the year, scientific interest in Darwin's views continued unabated and indeed entered a new phase in which they were subject to detailed scrutiny. The reviews of the third edition, and the comments of naturalists with whom Darwin corresponded, showed that a number of specialists were beginning to examine more carefully the implications of the theory of natural selection for their particular fields. Darwin relished these explorations, whether in support of his views or against them. The review published in the July issue of the *Zoologist* by George Maw, for example, singled out Darwin's explanation of the numerous instances of the 'correlation of growth' of certain organs and characters. Whereas for Darwin such cases exemplified the working of unknown laws of development in conjunction with natural selection, for Maw they remained notable instances of design in nature. Although Darwin, in his subsequent correspondence with Maw, chided him for having mingled science with religion, he nonetheless recommended the review to several friends, most notably his faithful 'barometer' of scientific opinion, Charles Lyell (see letter to Charles Lyell, 20 July [1861]). One reason for Darwin's interest in this piece may have been that Maw touched, albeit negatively, upon what Darwin considered the strongest indications of transmutation, namely the correlative growth of organs during development and the embryological relationships between organisms.

Darwin also found the review by the young geologist Frederick Wollaston Hutton particularly striking. As he told his friend Joseph Dalton Hooker, Hutton was one of the few reviewers who clearly understood that the theory of natural selection could not be 'directly proved' (see second letter to J. D. Hooker, 23 [April 1861]). Darwin continued to stress to his correspondents, as he had done since the publication of *Origin*, that his theory, like the wave theory of light in physics, only gained credence by its ability to explain and harmonise a number of otherwise disparate and unexplained classes of facts.

It is not surprising, then, that Darwin was pleased that the methodology followed in *Origin* was singled out for praise in 1861. He had been disappointed to learn of John Frederick William Herschel's cool initial response to his argument; he was gratified, however, when Herschel, in a newly published text, *Physical geography*, partially endorsed his views, though demurring with regard to man and maintaining that nature offered more evidence of design than Darwin was willing to admit. With the growing confidence of a committed crusader, Darwin wrote to Herschel, on 23 May [1861]:

> You will think me very conceited when I say I feel quite easy about the ultimate success of my views, (with much error, as yet unseen by me, to be

> no doubt eliminated); & I feel this confidence, because I find so many young & middle-aged truly good workers in different branches, either partially or wholly accepting my views, because they find that they can thus group & understand many scattered facts. This has occurred with those who have chiefly or almost exclusively studied morphology, geographical Distribution, systematic Botany, simple geology & palæontology.

Moreover, Darwin found an important philosophical ally in John Stuart Mill. Through Henry Fawcett, a young Cambridge political economist and convert to his theory, Darwin learned of Mill's judgment that the reasoning throughout *Origin* was 'in the most exact accordance with the strict principles of Logic' and that Darwin's methodology was 'the only one proper to such a subject' (letter from Henry Fawcett, 16 July [1861]). Mill in fact included a brief statement to this effect in a footnote to the fifth edition of his *System of logic* (Mill 1862, p. 18 n.). Later in the summer Fawcett himself made Darwin's methodology the subject of a lecture given before the British Association for the Advancement of Science. He subsequently sent Darwin a copy of the manuscript and some newspaper reports of the discussion that ensued wherein there had been criticism of Darwin's hypothesising. Darwin commented in reply: 'How odd it is that anyone should not see that all observation must be for or against some view if it is to be of any service!' (letter to Henry Fawcett, 18 September [1861]).

Some new names were added in 1861 to the list of 'young & middle-aged' naturalists whom Darwin counted among his supporters, including that of one of the 'very best of the younger geologists', Archibald Geikie. Geikie had approved of Darwin's chapter on the imperfection of the geological record (see letter to George Maw, 19 July [1861]). The American palaeontologist, Joseph Leidy, also sent word of his 'approbation' of *Origin*, but of even greater import was the news that he had some interesting facts 'in support of the doctrine of selection' that he planned to report 'at a favourable opportunity' (letter to Joseph Leidy, 4 March [1861]). However, the publication of Leidy's pioneer study of North American fossil vertebrates was interrupted by the outbreak of the Civil War.

Undoubtedly, the news that most excited Darwin was word from Henry Walter Bates, recently returned from an eleven-year scientific expedition in the Amazonian river basin, that he would soon publish empirical evidence in support of natural selection. Like his friend Alfred Russel Wallace, who had independently formulated the hypothesis of natural selection, Bates was a firm believer in the involvement of selection in determining the form and habits of living organisms. He was convinced that the materials he had collected in South America, particularly adaptively marked insects, provided some of the best support to date for the working of selection in nature. As Bates boldly stated: 'I think I have got a glimpse into the laboratory where Nature manufactures her new species' (letter from H. W. Bates, 28 March [1861]).

Bates's clear understanding of the operation of natural selection and its implications for the study of natural history was evident. He told Darwin in his letter of

[1 December] 1861:

> I think the whole tenour of your book teaches ... that the welfare of species is the object of all structures & forms. All species exist by virtue of some endowment enabling them to withstand adverse circumstances.

Bates had investigated cases in several genera of butterflies and other insects, such as the hover-fly *Volucella*, of what he called 'analogical resemblance' or mimicry (referred to today as Batesian mimicry). Applying the notion of selective advantage, Bates explained such cases as illustrations of protective adaptation, with one species assuming the adult coloration of another in order to escape, as he informed Darwin, the 'persecution by insect enemies from which the other set is free' (letter from H. W. Bates, 30 September 1861). As Peter Bowler has commented, mimicry provided natural selection its 'greatest triumph in explaining adaptations' (Bowler 1983, p. 29).

Darwin suggested that Bates write a popular account of his travels, somewhat in the genre of his own *Journal of researches*, believing that it would be a 'very valuable contribution to Nat. History.—' (letter to H. W. Bates, 4 April [1861]). He also advised him that the public appreciated 'a good dose of reasoning', 'new & curious remarks on habits, final causes &c', and 'Monkeys,—our poor cousins.—' (letter to H. W. Bates, 3 December [1861]). Darwin volunteered to read some of the chapters and recommended that Bates offer the manuscript to the publisher, John Murray, with a view to obtaining 'large distribution' for the work (letter to H. W. Bates, 25 September [1861]).

Nevertheless, opposition remained to natural selection as an explanation of organic phenomena that had hitherto been favoured instances of the working of special creation and design. Supported by such authorities as Richard Owen and Louis Agassiz, who provided critical evidence, naturalists like Cuthbert Collingwood and laymen such as the physician, Charles Robert Bree, and the Scottish divine, Gilbert Rorison, published tracts in opposition to Darwin's theory.

The effect of these attacks, however, was mild in comparison with the controversy between Richard Owen and Thomas Henry Huxley concerning the anatomical differences between human and 'anthropoid' brains. Conducted in the *Athenæum*, in the *Natural History Review*, and in other periodicals of the day, the so-called 'hippocampus' debate, although not directly concerned with Darwin's theory, was recognised by most as being of crucial importance to the question of an evolutionary relationship between the human species and the higher apes. Owen had given anatomical lectures to accompany the exhibition in February at the Royal Geographical Society and later at the Royal Institution of Great Britain of the remarkable gorilla specimens brought back from West Africa by Paul Belloni Du Chaillu. Owen, while noting the similarities between the skeletons of apes and humans, maintained that their brain structures were strikingly dissimilar. Huxley pointed out, publicly and acerbicly, that Owen was mistaken in his interpretations and even in his citation of anatomical facts. The clash between the two great authorities in comparative anatomy and physiology provided the scientific and lay communities alike with much material for both serious and satirical commentary. A humorous spoof on the topic, for example, was published in *Punch* (see Appendix VIII). For his part, Darwin, although periodically concerned at the strongly polemical tone of Huxley's attacks, enjoyed Huxley's sparring with Owen. On reading an article by Huxley that criticised Owen's views on the brain, Darwin complimented him on producing 'a complete and awful smasher' (letter to T. H. Huxley, 3 January [1861]). Ever since Owen's highly critical and, as Darwin felt, unfair review of *Origin*, he had nursed a growing animosity toward the man; as he told Huxley at the beginning of the year, he did not wish for '*open* quarrel', but he and Owen would 'never be friends again' (letter to T. H. Huxley, 3 January [1861]).

The relationship between Huxley and Darwin was strengthened in 1861 on a personal as well as on a scientific level. Early in the year, when Darwin learned that Huxley's wife, recuperating from the birth of her fourth child, remained desolate over the death in September 1860 of their first-born, Noel, the Darwins invited Henrietta and the children to Down House. The Darwins' own experience of loss with the death, ten years previously, of their daughter Annie, while Emma too was heavily pregnant, undoubtedly formed a lasting bond between the two families.

In May 1861 Darwin also offered consolation to his friend Hooker whose father-in-law, John Stevens Henslow, died after a brief illness. Darwin expressed his relief that Henslow's sufferings were finally at an end, adding: 'I fully believe a better man never walked this earth' (letter to J. D. Hooker, 18 [May 1861]). Henslow had been a uniquely important figure in Darwin's life. Not only had Henslow taught him the principles of botanical study, but he had also introduced him to geology and guided him, paternally, in other matters of science. Henslow's seminal importance in arranging Darwin's participation in the voyage of the *Beagle* is well known. As late as 1860, Henslow had defended Darwin against criticism from Adam Sedgwick and Richard Owen. Darwin himself was able to recall poignantly other aspects of Henslow's influence and relationship when he was asked to contribute to Leonard Jenyns's *Memoir of the Rev. John Stevens Henslow* (see Appendix X).

Botany, as mentioned earlier, became a major preoccupation of Darwin's during the course of 1861. Projects begun the previous summer as holiday amusements developed into serious scientific pursuits. In his experiments with the two flower forms of *Primula*, he at first formed the hypothesis that dimorphism in the length of stamens was an indication of how dioecious forms might gradually arise, but he became convinced that it was instead an instance of the importance of achieving intercrossing between distinct individuals. He told Gray that such cases could perhaps 'throw some light on Hybridisation' (letter to Asa Gray, 16 September [1861]).

But it was his study of orchids that provided Darwin with the most remarkable examples of adaptations for effecting cross-pollination (see *Autobiography*, p. 133). Having observed orchids sporadically over many years, Darwin became more and more impressed, as he dissected both British orchids and exotic species, with the modifications they exhibited for ensuring cross-pollination by insects. 'I marvel often', he told Hooker, 'as I think over the diversity & perfection of the contrivances.—' (letter of [28 July–10 August 1861]). Later in the year, he went even further, writing to John Lindley on 17 October: 'Orchids have interested me more than almost anything in my life'. Darwin pursued this study with 'half mad', 'boyish delight' throughout the summer of 1861, writing to anyone he thought might be able to provide him with a specimen of a species or tribe that he had not yet examined. Initially intending to publish his results in the journal of the Linnean Society of London, Darwin decided instead to prepare a small volume that would reach a wider public. Having approached John Murray with some hesitation, giving him realistic estimations of the limited popularity of the work, Darwin was surprised when Murray readily agreed to publish the book. Convinced that orchids, primroses, and insectivorous plants (another subject of his botanical investigations) all had 'some direct bearing on the subject of species' (letter to Henry Fawcett, 18 September [1861]), Darwin conceived of the orchid work as a treatise illustrating the operation of natural selection rather than creation (see *Orchids*, pp. 306–7). To Hooker he confided, 'I shall make use of my Orchid little volume in illustrating modification-of-species-doctrine, but I keep very very doubtful whether I am not doing a foolish action in publishing' (letter to J. D. Hooker, 17 November [1861]). Nevertheless, Darwin immediately set to work to prepare the final manuscript, and *Orchids* was published in May of 1862.

The time spent on his botanical investigations and in preparing the results for publication left Darwin little opportunity to pursue his larger study of the variations in animals and plants upon which natural selection could operate. Demonstrating the great range of variation in animals such as domesticated fowls involved a careful osteological comparison of different forms, and on this Darwin made but slow progress. As he told a French correspondent, 'it is such tedious work comparing skeletons—' (letter to Jean Louis Armand de Quatrefage de Bréau, 25 April [1861]). By the end of May, however, he had completed the first drafts of chapters on most of the common domesticated animals, among them horses, rabbits, pigeons, and poultry. As he frequently admitted to friends such as Charles Lyell and interested supporters like Alfred Russel Wallace, he knew he should pursue his work on variation more assiduously; however, his family's long summer holiday in Devon allowed him to postpone the 'tedious' work and indulge in the pleasure of working on fresh and exciting botanical projects.

Although Darwin was no longer directly involved in geological investigations, geology features prominently in the correspondence of 1861. Lyell continued to act as Darwin's adviser and, in turn, relied upon Darwin as a sounding-board for matters arising from his own investigations. Lyell was preparing a major revision of his *Elements of geology*, in which he intended to incorporate the results of recent studies regarding the antiquity of man and a reconsideration of the chronology of major periods of elevation, subsidence, and glaciation in Europe. Through his letters, Lyell involved Darwin in his deliberations. Darwin offered his opinions willingly and soon became more personally involved in the discussions when they turned to a re-evaluation of the 'parallel roads' of Glen Roy, a topic on which he had published a major paper twenty years earlier. Both Lyell and Darwin encouraged the young Scottish geologist Thomas Francis Jamieson to make an excursion to Glen Roy to study the phenomenon. Upon his return, Jamieson reported that, in his view, the evidence overwhelmingly supported the theory that the roads were the shores of a former glacial lake rather than the theory, proposed by Darwin and accepted by Lyell, that they were formed by an arm of the sea. Despite his belief that Jamieson's interpretation still did not encompass certain phenomena that he considered critical, Darwin finally conceded that his paper on the subject had been 'one long gigantic blunder' (letters to Charles Lyell, 6 September [1861], and to T. F. Jamieson, 6 September [1861]). Jamieson's letters to Lyell, which formed part of the fascinating three-way correspondence on the Glen Roy problem, and were passed on to Darwin, are included in Appendix IX.

There were important developments too in Darwin's family life. Over the years he had occasionally expressed worries about the responsibility of finding suitable careers for his five sons. In 1861, an opportunity unexpectedly arose for his oldest son, William Erasmus. Late in May, Darwin's young friend and neighbour, John Lubbock, a partner with his father in the London bank Robarts, Lubbock & Co., told Darwin that he knew of a banker in a town in the south of England who was seeking a partner. William, who was in his final year at Cambridge University, had, with his father's encouragement, been considering a career in the law. But both father and son found the proposal of a banking partnership too good to neglect, despite the prospect of William's having to leave Cambridge prematurely. Although, as Darwin cautioned, banking was 'dull work', William could soon expect an annual income of about £800 and so would 'be at once an almost rich man' (letter to W. E. Darwin, [26 May 1861]). The arrangement would entail, however, providing William with a deposit of £10,000 as guarantee of probity, funds that would have to be advanced from his inheritance.

While the arrangements for William's partnership were being completed during July and August, the Darwin family was in residence at the seaside resort of Torquay. Even though the trip halted a number of Darwin's botanical experiments, the sacrifice was made in the hope that sea air would improve the health of Henrietta Emma Darwin, who had been suffering for well over a year from the after-effects of what was thought to be 'a form of typhus fever' (*Correspondence* vol. 8, letter to J. D. Hooker, 11 May 1860). This hope was realised. By the end of July, Henrietta was strong enough to be able to accompany her mother and cousin on a week-long tour of Dartmoor. The Darwins, however, also gave credit to Joseph Hooker for Henrietta's gradual recuperation; in February Hooker had recommended the external application of cod-liver oil, a remedy only recently introduced to England, as a means to aid her digestion and appetite. Etty's improving health was indeed a cause for rejoicing, for ever since the death from a languishing illness of their eldest daughter, Annie, in 1851 (see *Correspondence* vol. 5), the Darwins had feared that the same fate, perhaps the result of an inherited constitutional weakness, might await their other children.

William travelled to Southampton early in September to take up his new position. He was not expected to begin his new life unaided. Even before the partnership settlement was agreed Darwin began to mobilise his social network in support of his son. On 1 August he wrote to Lyell to ask whether he could suggest any suitable south-coast introductions for William. Through his friend Henry Bonham-Carter, Darwin arranged for an introduction to the Nightingales. 'It is evident that you will have plenty of acquaintances; but I am not sorry that you should know two or three respectable persons on your own account' (letter to W. E. Darwin, 17 [October 1861]). He also wrote a letter introducing William to the director of the ordnance survey, the 'pleasant & distinguished' Henry James.

James was about to organise a Rifle Volunteer Corps in Southampton. The Volunteer Movement was a popular response to the general fear after 1859 of possible French aggression against England stemming from the war of Italian unification. Established under Crown patronage and comprising largely individuals from the professional and commercial classes, the volunteers were strictly local units, and membership carried with it considerable social status. In November, William told his father that he had heard he was to be asked to join, 'but', he complained, 'I have had enough of volunteering.' William held the rank of captain in the Farnborough Rifle Volunteer Corps; however, being based in Southampton, it was not easy for him to comply with the attendance regulations, and he resigned his commission (see letters to W. E. Darwin, 22 October [1861], 27 October [1861], and 15 November [1861]).

It was not France, however, but the United States of America that aroused fears in 1861. By the spring, seven states had seceded from the Union, and in April, the attack by southern forces on Fort Sumter, a Federal base in South Carolina, set off the American Civil War. In Britain, these events were keenly followed. 'I never knew the newspapers so profoundly interesting', Darwin told Asa Gray on 5 June. Despite the British government's declared neutral stance, Darwin told Gray: 'I have not seen or heard a soul who is not with the North'. Darwin and Gray both unreservedly supported the northern cause. 'Of all my English correspondents', Gray wrote on 31 December, 'you are the only one . . . whose views and sentiments are perfectly satisfactory to me'.

For Darwin, the primary issue was slavery, and his letters to Gray are testaments to the depth of his anti-slavery convictions. He told Gray in his letter of 5 June [1861]:

> Some few, & I am one, even wish to God, though at a cost of millions of lives, that the North would proclaim a crusade against Slavery. In the long run, a million horrid deaths would be amply repaid in the cause of humanity. . . . Great God how I sh^{d} like to see that greatest curse on Earth Slavery abolished.

Towards the end of the year, Britain was drawn into a closer involvement in the conflict. In November, two Confederate envoys bound for Europe boarded a

British mail packet, the *Trent*, at Havana. Shortly after leaving port, the British vessel was intercepted by a United States Navy gunboat, the *San Jacinto*. Under threat of force the two envoys were removed. The British government demanded an apology from the United States government and issued an ultimatum for the release of the envoys. A tense diplomatic situation developed, and at the close of the year the two nations stood on the brink of war. Darwin wrote, almost disbelievingly, to Gray on 11 December: 'What a thing it is, that when you receive this we may be at war'.

Despite these alarming political events, life at Down House continued to be relatively tranquil. During December, while everyone awaited news of Washington's response, Darwin was actively engaged in preparing his volume on orchids. He had good reason to be content and confident about his scientific work. He was attracting a number of young and dedicated followers who were convinced by his doctrines and interested in pursuing 'evolutionary' studies in their own particular areas of expertise. His old and established friends—Hooker, Gray, Huxley, and Lyell—continued to support his doctrines, publicly as well as privately, showing only slightly more caution than the younger generation. Even religiously motivated opposition to his views was dying down, in large measure owing to the increasing internal turmoil in the Anglican Church over the liberal theological tract *Essays and reviews* (see Appendix VI). Earlier in the year, Darwin had urged those 'on **our** side' to broadcast their views and hasten the wider acceptance of the theories set forth in *Origin* (see letter to P. L. Sclater, 12 [March 1861]). By the end of 1861, the evidence that his views were indeed gaining wider acceptance increased his conviction that evolution and natural selection held the key to future advances in all areas of natural history.

ACKNOWLEDGMENTS

The editors are grateful to Mr George Pember Darwin for permission to publish the Darwin letters and manuscripts. They also thank the Syndics of the Cambridge University Library and other owners of the manuscript letters who have generously made them available.

The work for this edition has been supported by grants from the National Endowment for the Humanities (NEH) and the National Science Foundation (NSF). The Alfred P. Sloan Foundation and the Andrew W. Mellon Foundation provided grants to match NEH funding, and the Mellon Foundation awarded grants to Cambridge University that made it possible to put the entire Darwin correspondence into machine-readable form. Research and editorial work have also been supported by grants from the Royal Society of London, the British Academy, and the Pilgrim Trust. The Royal Society has helped meet the costs of publication. The Stifterverband für die Deutsche Wissenschaft has provided funds to edit Darwin's correspondence with German naturalists.

The Cambridge University Library and the American Philosophical Society have generously made working space and many services available to the editorial staff.

Since the project began in 1975, we have been fortunate in benefiting from the interest, experience, and practical help of many people in many places, and the editors hope that they have adequately expressed their thanks to them individually as the work proceeded. There are some, however, who have helped over so long a period that it would be ungracious not to thank them personally in the work that has profited so much from their co-operation.

Over the years, Kathy Fuller, formerly assistant director of the Division of Research Grants of NEH, and Ronald J. Overmann, program director for History and Philosophy of Science of NSF, have given far more time and attention to the project than their formal duties required. Margaret Scrymser of NEH has also provided valuable direction. The editors also appreciate the interest and encouragement they received from John E. Sawyer, former president of the Andrew W. Mellon Foundation, and his colleague, James Morris. The editors would like to acknowledge with gratitude the assistance rendered to the project by the late Shepard Stone and by Frank Rhodes.

Without the expert help of John L. Dawson of the Literary and Linguistic Computing Centre of Cambridge University, the computerisation of the correspondence would not have been possible, for the work on both the *Calendar* and the *Correspondence* required the solution of many novel technical problems.

The Cambridge University computer and that of the Cambridge University Library were used for initial work on this edition and for proof production.

W. D. S. Motherwell, former head of automation at the Library, arranged for the installation of terminals in the project's office in the Manuscripts Department and solved the operational problems that arose. More recently, Paul Ayris and Chris Sendall have assisted us in accessing the library's laser printer. Simon Buck has offered invaluable assistance in preparing automatic typesetting procedures and has acted as typesetting advisor to the project.

The camera-ready copy for this book was produced at the Oxford University Computing Service on a Monotype Prism Imagesetter. We are grateful to Stephen Miller and to the Oxford operations staff for their advice and assistance.

The late Sir Hedley Atkins and Philip Titheradge, formerly curators of the Darwin Museum at Down House, Downe, Kent, welcomed the editors on numerous visits and responded most generously to frequent requests for information and for material from the collections at Down. The current curator, Solene Morris of the British Museum (Natural History), has enabled our close relationship with Down to continue. Professor Richard Darwin Keynes kindly made available Emma Darwin's diaries and other Darwin family material in his possession. Ursula Mommens has also provided letters and other materials that belonged to her grandfather, Francis Darwin. The help of Peter J. Gautrey, who, until 1989, directed the Darwin Archive at the Cambridge University Library, has been of special importance. His great knowledge of Darwiniana has always been readily shared with the editors, as with countless other Darwin scholars.

Libraries all over the world have given help that was literally indispensable by making available photocopies of Darwin correspondence and other manuscripts in their collections. The institutions and individuals that furnished copies of letters for this volume can be found in the List of provenances. The editors are extremely indebted to them all. We are also grateful to the many people who have transmitted information regarding the whereabouts of particular letters and, in many cases, have generously provided copies of such letters for the project's use.

The editors make daily use of the incomparable facilities of the Cambridge University Library and have benefited greatly from its services and from the help and expertise of its staff, particularly the staff of the Manuscripts Department. We are especially grateful to the University Librarian, Frederick W. Ratcliffe, and the Under-librarian of Manuscripts, Patrick Zutshi, for their generous support. Other members of the library's staff who have frequently responded to the editors' requests are: Gerry Bye, Roger Fairclough, David Hall, Elisabeth Leedham-Green, Vicki Newman, William Noblett, Jayne Ringrose, Godfrey Waller, and Cynthia Webster. Natasha Charles, Sophie Clark, Paul Emmines, Michelle Gower, Darrell McMurray, Dan Mansfield, and Jonathan Ringer, all of the Manuscripts reading room, have patiently dealt with the editors' often complex requirements.

At the American Philosophical Society Library, a splendid collection of Darwiniana and works in the history of science has been continuously available to the editorial staff since the inception of the project. Whitfield J. Bell Jr, secretary of the society until 1983, serves on the United States Advisory Committee for the project and has done his utmost to further its work. The editors have also benefited from the co-operation of Edward Carter II, Roy C. Goodman, Carl F. Miller, Elizabeth Carroll-Horrocks, and Bertram Dodelin, all of the APS Library.

Douglas W. Bryant, formerly of the Harvard University Library; Rodney Dennis, Jennie Rathbun, and Susan Halpert of the Houghton Library; Constance Carter of the Science Division of the Library of Congress; Lothian Lynas of the New York Botanical Garden Library; Judith Warnement and Jean Cargill of the Gray Herbarium of Harvard; and the staff members of the New York Public Library have all been exceptionally helpful in providing material from their great collections.

In Britain, the editors have often received assistance from Mary Sampson, archivist of the Royal Society; Rex E. R. Banks and John Thackray of the Natural History Museum Library; Gina Douglas, librarian of the Linnean Society of London; and Sylvia FitzGerald, Cheryl Piggott, and Leonora Thompson of the Library of the Royal Botanic Gardens, Kew. We would also like to thank Anthony M. Carr, Miss Williams, and Mrs Ion of the Local Studies Library, Shrewsbury; James B. Lawson of Shrewsbury School Library; Richard G. Williams, Mrs Felton, and Anne Barrett (college archivist) of Imperial College of Science, Technology and Medicine; Christine Fyfe, archivist of Keele University; and Dr and Mrs Courtney of Christ's College, Cambridge. Virginia Murray of John Murray (publishers) Ltd has graciously provided the editors with copies of the Darwin correspondence in the Murray archive and with related information. Leonard G. Wilson kindly provided the editors with photocopies of manuscript material in the possession of the family of Sir Charles Lyell.

Among the others who have advised and assisted the editors in their work are Julius Held, Garry J. Tee, Hugh Torrens, Thomas Junker, and Morton N. Cohen. We are also pleased to acknowledge the invaluable support of Desmond King-Hele and Geoffrey Lloyd of the British Advisory Committee.

Continued thanks are due to William Montgomery, former associate editor with this project, for his assistance in editing the letters in the collection of the American Philosophical Society. We gratefully acknowledge the valuable assistance of Stefanie Cookson, Shelly Innes, Sarah Lavelle, Robert Lindsey, and Leslie Nye. Thanks are also extended to Jean Macqueen who provided the index to the volume.

We are grateful for the years of service, scholarship, and outstanding editorial skill provided by Marsha Richmond who recently returned to the United States. We wish her every success in all she undertakes in the future.

LIST OF PROVENANCES

The following list gives the locations of the original versions of the letters printed in this volume. The editors are grateful to all the institutions and individuals listed for allowing access to the letters in their care.

Academy of Natural Sciences of Philadelphia, PA, USA
American Philosophical Society, Philadelphia, PA, USA
Auckland Public Library, Auckland, New Zealand
Walter J. Benjamin Autographs (dealers)
British Library, The Manuscript Collections, Great Russell Street, London, England
British Museum, Department of Western Asiatic Antiquities, London, England
Cambridge University Library, Manuscripts department, Cambridge, England
Christ's College Library, Cambridge, England
Cleveland Health Sciences Library, Cleveland, OH, USA
Ralph Colp Jr (private collection)
Cottage Gardener (publication)
CUL. *See* Cambridge University Library
DAR. *See* Cambridge University Library
Dawsons of Pall Mall (dealers)
Down House, Downe, Kent, England
Duke University, Durham, NC, USA
Edinburgh University Library, Edinburgh, Scotland
The Field (publication)
Gardeners' Chronicle and Agricultural Gazette (publication)
Grant 1861 (publication)
Gray Herbarium of Harvard University, Cambridge, MA, USA
Imperial College of Science, Technology and Medicine, South Kensington, London, England
Kinnordy MS. *See* Lady Lyell
Leeds University Library, Leeds, England
Linnean Society of London, Burlington House, London, England
Liverpool Public Library, Liverpool, England
LL (publication)
James Lowe Autographs (dealers)
Lady Lyell (private collection)
Mitchell Library, State Library of New South Wales, Sydney, Australia
Moray District Record Office, Tolbooth, Forres, Scotland
John Murray (publishers), London, England

National Library of Scotland, Edinburgh, Scotland
New York Botanical Garden Library, Bronx, New York, NY, USA
Oxford University Museum, Oxford, England
R. D. Pyrah (private collection)
Royal Botanic Gardens, Kew, Richmond, Surrey, England
Royal College of Physicians, Regent's Park, London, England
Royal Horticultural Society, Vincent Square, London, England
Royal Society, Carlton House Terrace, London, England
Elizabeth Rütimeyer (private collection)
Scripps College Library, Claremont, CA, USA
Barton L. Smith (private collection)
Sotheby (dealers)
Stephen 1885 (publication)
J. A. Sulivan (private collection)
Victoria and Albert Museum, South Kensington, London, England
Wellcome Institute for the History of Medicine Library, London, England
Jeffries Wyman Jr (private collection)

A NOTE ON EDITORIAL POLICY

The first and chief objective of this edition is to provide complete and authoritative texts of Darwin's correspondence. In so far as it is possible, the letters have been dated, arranged in chronological order, and the recipients or senders identified. Darwin seldom wrote the full date on his letters and, unless the addressee was well known to him, usually wrote only 'Dear Sir' or 'Dear Madam'. After the adoption of adhesive postage stamps in the 1840s, the separate covers that came into use with them were usually not preserved, and thus the dates and the names of many recipients of Darwin's letters have had to be derived from other evidence. The notes made by Francis Darwin on letters sent to him for his editions of his father's correspondence have been helpful, as have matching letters in the correspondence, but many dates and recipients have had to be deduced from the subject-matter or references in the letters themselves. These tasks, together with deciphering the handwriting of Darwin and others, have presented the most troublesome editorial problems.

Whenever possible, transcriptions have been made from manuscripts. If the manuscript was inaccessible but a photocopy or other facsimile version was available, that version has been used as the source. In many cases, the editors have had recourse to Francis Darwin's large collection of copies of letters, compiled in the 1880s. Other copies, published letters, or drafts have been transcribed when they provide texts that are otherwise unavailable.

The method of transcription employed in this edition is adapted from that described by Fredson Bowers in 'Transcription of manuscripts: the record of variants', *Studies in Bibliography* 29 (1976): 212–64. This system is based on accepted principles of modern textual editing and has been widely adopted in literary editions.

The case for using the principles and techniques of this form of textual editing for historical and non-literary documents, both in manuscript and print, has been forcefully argued by G. Thomas Tanselle in 'The editing of historical documents', *Studies in Bibliography* 31 (1978): 1–56. The editors of the *Correspondence* followed Dr Tanselle in his conclusion that a 'scholarly edition of letters or journals should not contain a text which had editorially been corrected, made consistent, or otherwise smoothed out' (p. 48), but they have not wholly subscribed to the statement made earlier in the article in which he says, 'In the case of notebooks, diaries, letters and the like, whatever state they are in constitutes their finished form, and the question of whether the writer "intended" something else is irrelevant' (p. 47). The editors have preserved the spelling, punctuation, and grammar of the original, but they have found it impossible to set aside entirely the question of authorial intent. One obvious reason is that in reading Darwin's writing, there must necessarily be reliance upon both context and intent. Even when Darwin's general intent is clear, there are cases in which alternative readings are, or may be, possible, and therefore the transcription decided upon must to some extent be conjectural. When the editors were uncertain of their transcription, the doubtful text has been enclosed in italic square brackets.

A major editorial decision was to adopt the so-called 'clear-text' method of transcription, which so far as possible keeps the text free of brackets recording deletions, insertions, and other alterations in the places at which they occur. Darwin's changes are, however, recorded in the back matter of the volume, under 'Manuscript alterations and comments', in notes keyed to the printed text by paragraph and line number. All lines above the first paragraph of the letter (i.e., date, address, or salutation) are referred to as paragraph 'o'. Separate paragraph numbers are used for subscriptions and postscripts. These notes enable the reader who wishes to do so to reconstruct the manuscript versions of Darwin's letters while furnishing printed versions that are uninterrupted by editorial interpolations. The Manuscript alterations and comments record all alterations made by Darwin in his letters and any editorial amendments made in transcription. For copies of Darwin letters included in the correspondence no attempt has been made to record systematically all alterations to the text, but ambiguous passages are noted. The editors believe it would be impracticable to attempt to go further without reliable information about the texts of the original versions of the letters involved. Letters to Darwin have been transcribed without recording any of the writers' alterations unless they reflect significant changes in substance or impede the sense; in such cases footnotes will bring them to the reader's attention.

Misspellings have been preserved, even when it is clear that they were unintentional as, for instance, 'lawer' for 'lawyer'. Such errors often indicate excitement or haste and may exhibit, over a series of letters, a habit of carelessness in writing to a particular correspondent or about a particular subject.

Capital letters have also been transcribed as they occur except in certain cases, such as 'm', 'k', and 'c', which are frequently written somewhat larger than others as initial letters of words. In these cases an attempt has been made to follow the normal practice of the writers.

In some instances that are not misspellings in a strict sense, editorial corrections have been made. In his early manuscripts and letters Darwin consistently wrote 'bl' so that it looks like 'lb' as in 'albe' for 'able', 'talbe' for 'table'. Because the form of the letters is so consistent in different words, the editors consider that this is most unlikely to be a misspelling but must be explained simply as a peculiarity of Darwin's handwriting. Consequently, the affected words have been transcribed as normally spelled and no record of any alteration is given in the textual apparatus. Elsewhere, though, there are misformed letters that the editors have recorded because they do, or could, affect the meaning of the word in which they appear. The main example is the occasional inadvertent crossing of 'l'. When the editors are satisfied that the intended letter was 'l' and not 't', as, for example, in 'stippers' or 'istand', then 'l' has been transcribed, but the actual form of the word in the manuscript has been given in the alteration notes.

If copies of letters provide the only source, the editors have frequently retained corrections made to the text when it is clear that they were based upon comparison with the original. Francis Darwin's corrections of misreadings by copyists have usually been followed; corrections to the text that appear to be editorial alterations have not been retained.

Editorial interpolations in the text are in square brackets. Italic square brackets enclose conjectured readings and descriptions of illegible passages. To avoid confusion, in the few instances in which Darwin himself used square brackets, they have been altered by the editors to parentheses with the change recorded in the alteration notes. In letters to Darwin, square brackets have been changed to parentheses silently.

Material that is irrecoverable because the manuscript has been torn or damaged is indicated by angle brackets; any text supplied within them is obviously the responsibility of the editors. Occasionally, the editors are able to supply missing sections of text by reference to transcripts or photocopies of manuscript material made before the damage occurred.

Words and passages that have been underlined for emphasis are printed in italics in accordance with conventional practice. When the author of a letter has indicated greater emphasis by underlining a word or passage two or more times, then the text is printed in bold type.

Paragraphs are often not clearly indicated in the letters. Darwin and others sometimes marked a change of subject by leaving a somewhat larger space than usual between sentences; sometimes Darwin employed a longer dash. In these cases, and in very long stretches of text when the subject is clearly changed, a new paragraph is started by the editors without notice. The start of letters, valedictions, and postscripts are also treated as new paragraphs regardless of whether they appear as such in the manuscript. Special manuscript devices delimiting sections or paragraphs, e.g., blank spaces left between sections of text and lines drawn across the page, are treated as normal paragraph indicators and are not specially marked or recorded unless their omission leaves the text unclear.

Additions to a letter that run over into the margins, or are continued at its head or foot, are transcribed at the point in the text at which the editors believe they were intended to be read. The placement of such an addition is only recorded in a footnote if it seems to the editors to have some significance or if the position at which it should be transcribed is unclear. Enclosures are transcribed following the letter.

The hand-drawn illustrations and diagrams that occur in some letters are reproduced as faithfully as possible and are usually positioned as they were in the original text. In some cases, however, it has been necessary to reduce the size of a diagram or enhance an outline for clarity; any such alterations are recorded in footnotes. The location of diagrams within a letter is sometimes changed for typesetting reasons. Tables have been reproduced as close to the original format as possible given typesetting constraints.

Occasionally punctuation marking the end of a clause or sentence is not present in the manuscript, but the author has made his or her intention clear by allowing, for example, extra space or a line break to function as punctuation. In such cases, the editors have inserted an extra space following the sentence or clause to set it off from the following text.

Some Darwin letters and an occasional letter to Darwin are known only from entries in the catalogues of book and manuscript dealers or mentions in other published sources. Whatever information these sources provide about the content of such letters has been reproduced without change.

For every Darwin letter, the text that is available to the editors is always given in full. Some other items, however, are not printed in their entirety. Some memorandums and other documents that are not letters but are relevant to the correspondence have been summarised, as have those letters to Darwin that the editors consider can be presented adequately in shortened form. In a few cases the editors have included letters that are not to or from Darwin if they relate significantly to the published correspondence. Letters that can only be given a wide date range, in some instances spanning several decades, will be printed together in one of the concluding volumes.

The format in which the transcriptions are printed in the *Correspondence* is as follows:

1. *Order of letters*. The letters are arranged in chronological sequence. A letter that can be dated only approximately is placed at the earliest date on which the editors believe it could have been written. The basis of a date supplied by the editors is given in a footnote unless it is derived from a postmark, watermark, or endorsement as recorded in the physical description of the letter (see section 4, below). Letters with the same date, or with a range of dates commencing with that date, are printed in the alphabetical order of their senders or recipients unless their contents dictate a clear alternative order. Letters dated only to a year or a range of years precede letters that are dated to a particular month or range of months, and these, in turn, precede those that are dated to a particular day or to a range of dates commencing with a particular day.

2. *Headline*. This gives the name of the sender or recipient of the letter and its date. The date is given in a standard form, but those elements not taken directly from the letter text are supplied in square brackets.

3. *The letter text*. The transcribed text follows as closely as possible the layout of the source, although no attempt is made to produce a type-facsimile of the manuscript: word-spacing and line-division in the running text are not adhered to. Similarly, the typography of printed sources is not replicated. Dates and addresses given by authors are transcribed as they appear, except that if both the date and address are at the head of the letter they are always printed on separate lines with the address

first, regardless of the exact manuscript order. If no address is given on a letter by Darwin, the editors have supplied one, when able to do so, in square brackets at the head of the letter. Similarly, if Darwin was writing from an address different from the one given on the letter, his actual location is given in brackets. Addresses on printed stationery are transcribed in italics. Addresses, dates, and valedictions have been run into single lines to save space, but the positions of line-breaks in the original are marked by vertical bars.

4. *Physical description*. All letters are complete and in the hand of the sender unless otherwise indicated. If a letter was written by an amanuensis, or exists only as a draft or a copy, or is incomplete, or is in some other way unusual, then the editors provide the information needed to complete the description. Postmarks, endorsements, and watermarks are recorded only when they are evidence for the date or address of the letter.

5. *Source*. The final line provides the provenance of the text. Some sources are given in abbreviated form (e.g., DAR 140: 18) but are listed in full in the List of provenances unless the source is a published work. Letters in private collections are also indicated. References to published works are given in author–date or short-title form, with full titles and publication details supplied in the Bibliography at the end of the volume.

6. *Darwin's annotations*. Darwin frequently made notes in the margins of the letters he received, scored significant passages, and crossed through details that were of no further interest to him. These annotations are transcribed or described following the letter text. They are keyed to the letter text by paragraph and line numbers. Most notes are short, but occasionally they run from a paragraph to several pages, and are sometimes written on separate sheets appended to the letter. Extended notes relating to a letter are transcribed whenever practicable following the annotations as 'CD notes'.

Quotations from Darwin manuscripts in footnotes and elsewhere, and the text of his annotations and notes on letters, are transcribed in 'descriptive' style. In this method the alterations in the text are recorded in brackets at the places in which they occur. For example:

'See Daubeny ['vol. 1' *del*] for *descriptions of volcanoes in [*interl*] S.A.' *ink*

means that Darwin originally wrote in ink 'See Daubeny vol. 1 for S.A.' and then deleted 'vol. 1' and inserted 'descriptions of volcanoes in' after 'for'. The asterisk before 'descriptions' marks the beginning of the interlined phrase, which ends at the bracket. The asterisk is used when the alteration applies to more than the immediately preceding word. The final text can be read simply by skipping the material in brackets. Descriptive style is also used in the Manuscript alterations and comments.

Volumes of the *Correspondence* are published in chronological order. Each volume is self-contained, having its own index, bibliography, and biographical register. A chronology covering the period of each volume is supplied, and Appendixes give additional material as appropriate to assist the understanding of the correspondence. A cumulative index is planned once the edition is complete. References are supplied for all persons, publications, and subjects mentioned, even though some repetition of material in earlier volumes is involved.

If the name of a person mentioned in a letter is incomplete or incorrectly spelled, the full, correct form is given in a footnote. Brief biographies of persons mentioned in a volume, and dates of each correspondent's letters to and from Darwin, are given in the Biographical register and index to correspondents.

Short titles are used for references to Darwin's books and to collections of his letters (e.g., *Descent*, *Collected papers*, *LL*). They are also used for some standard reference works (e.g., *Alum. Cantab.*, *Wellesley index*, *DNB*). For all other works, author–date references are used. The full titles of all the books referred to are given in the Bibliography.

Occasional supplements will be published containing letters that have come to light or have been redated since the relevant volumes of the *Correspondence* appeared. The first such supplement was included in Volume 7 and covered the years from 1821 to 1857.

The editors use the abbreviation 'CD' for Charles Darwin throughout the footnotes. A list of all abbreviations used by the editors in this volume is given on p. xxxvi.

The Wedgwood and Darwin

Robert Waring Darwin 1766–1848 = Susannah Wedgwood 1765–1817

- Marianne 1798–1858 = Henry Parker 1788–1856
  - Robert b.1825
  - Henry 1827–92
  - Francis 1829–71
  - Charles b.1831
  - Mary Susan 1836–93
- Caroline Sarah 1800–88 = Josiah III (Jos) 1795–1880
  - Sophy Marianne 1838–9
  - Katherine Elizabeth Sophy (Sophy) 1842–1911
  - Margaret Susan 1843–1937
  - Lucy Caroline 1846–1919
- Susan Elizabeth 1803–66
- Erasmus Alvey 1804–81
- Charles Robert 1809–82 = Emma 1808–96
  - William Erasmus 1839–1914
  - Anne Elizabeth 1841–51
  - Mary Eleanor Sept.–Oct. 1842
  - Henrietta Emma (Etty) 1843–1927
  - George Howard 1845–1912
  - Elizabeth 1847–1926
  - Francis (Frank) 1848–1925
  - Leonard 1850–1943
  - Horace 1851–1928
  - Charles Waring 1856–8
- Emily Catherine (Catherine) 1810–66

amilies up to 1861

- Josiah Wedgwood II 1769-1843 = Elizabeth (Bessy) Allen 1764-1846
  - Sarah Elizabeth (Elizabeth) 1793-1880
  - Mary Anne 1796-8
  - Charlotte 1797-1862 = Charles Langton 1801-86
    - Edmund 1841-75
  - Henry Allen (Harry) 1799-1885 = Jessie Wedgwood 1804-72
    - Louisa Frances 1834-1903
    - Caroline b. 1836
    - John Darwin 1840-70
    - Anne Jane 1841-77
    - Arthur 1843-1900
    - Rowland 1847-1921
  - Francis 1800-88 = Frances Mosley d. 1874
    - Godfrey 1833-1905
    - Amy 1835-1910
    - Cicely Mary 1837-1917
    - Clement 1840-89
    - Lawrence 1844-1913
    - stance Rose 1846-1903
    - Mabel Frances 1852-1930
  - Hensleigh 1803-91 = Frances (Fanny) Mackintosh 1800-89
    - Frances Julia (Snow) 1833-1913
    - James Mackintosh (Bro) 1834-64
    - Ernest Hensleigh 1838-98
    - Katherine Euphemia (Effie) 1839-1931
    - Alfred Allen 1842-92
    - Hope Elizabeth 1844-1934
  - Frances (Fanny) 1806-32

ABBREVIATIONS

LS — letter in hand of amanuensis, signed by sender
LS(A) — letter in hand of amanuensis with additions by author
Mem — memorandum
(S) — signed with sender's name by amanuensis

CD — Charles Darwin
CUL — Cambridge University Library
DAR — Darwin Archive, Cambridge University Library
del — deleted
illeg — illegible
interl — interlined
underl — underlined

TRANSCRIPTION CONVENTIONS

[some text] — 'some text' is an editorial insertion
*[*some text*]* — 'some text' is the conjectured reading of an ambiguous word or passage
[some text] — 'some text' is a description of a word or passage that cannot be transcribed, e.g., '*3 words illeg*'
⟨ ⟩ — word(s) destroyed
⟨some text⟩ — 'some text' is a suggested reading for a destroyed word or passage
⟨*some text*⟩ — 'some text' is a description of a destroyed word or passage, e.g., '*3 lines excised*'

THE CORRESPONDENCE OF CHARLES DARWIN 1861

To Thomas Henry Huxley 3 January [1861][1]

Down Bromley Kent
Jan 3d.

My dear Huxley

I have just finished Nor I of N. H. Review & must congratulate you, as chiefly concerned, on its excellence.[2] The whole seems to me admirable,—so admirable that it is impossible that other numbers shd. be so good, but it would be foolish to expect it. I am rather a croaker & I do rather fear that the merit of the articles will be above the run of common readers & subscribers.

I have been much interested by your Brain article.[3] What a complete & awful smasher (& done like a "buttered angel") it is for Owen! What a canting humbug he is to have left out that sentence in the Lecture before the orthodox Cambridge Dons.[4] I like Lubbock's paper *very* much: how well he writes.[5] MacDonnels, of course, pleases me greatly.—[6] But I am very curious to know who wrote the Protozoa article:[7] I shall hear, if it be not a secret, from Lubbock. It strikes me as very good, & by Jove how Owen is shown up—"this great & sound reasoner".[8]

By the way this reminds me of a passage which I have just observed in Owen's address at Leeds,[9] which a clever Reviewer might turn into good fun. He defines (p. xc) & further on amplifies his definition that Creation means "a process he knows not what". And in previous sentence he says facts shake his confidence that the Apteryx in N. Zealand & Red Grouse in England are "distinct creations". So that he has no confidence that these birds were produced by "processes he knows not what".— What miserable inconsistencies & rubbish this truckling to opposite opinions leads the great generaliser!

Farewell, I heartily rejoice in the clear merit of this number.— I hope Mrs. Huxley goes on well.—[10] Etty keeps much the same, but has not got up to the same pitch, as when you were here.—[11]

Farewell | C. Darwin

P.S.[12] In my little historical sketch of opinion on Species, I have picked out the foregoing sentences and his axiom of ordained becoming &c.;[13] and if the reader has any acuteness, I shall thus take some revenge;[14] but I shall make no comments;—I am not bold enough and do not want to come to *open* quarrel. But we shall never be friends again. What an admirable résume of Botanical Papers; I suppose by Oliver.—[15] What labour!

Imperial College of Science, Technology and Medicine Archives (Huxley 5: 155, 372–6)

[1] The year is provided by the reference to the *Natural History Review* (see n. 2, below).

[2] The first number of the new series of the *Natural History Review: a quarterly journal of biological science*, under Huxley's general editorship, appeared in January 1861. For CD's doubts about whether Huxley should take on this responsibility, see *Correspondence* vol. 8, letter to T. H. Huxley, 20 July [1860]. CD's annotated copies of the journal are in the Darwin Library–CUL.

[3] T. H. Huxley 1861a. The article was a response to Richard Owen's view, published in R. Owen 1857 and 1859, that the brain provided certain anatomical characters by which man could be classified in a sub-class separate from the anthropoid apes. CD had criticised Owen's classification when it was first published in 1857 (see *Correspondence* vol. 6, letters to J. D. Dana, 5 April [1857], and to J. D. Hooker, 5 July [1857]).

[4] CD refers to Owen's Rede lecture delivered at Cambridge University in May 1859. The lecture was the first in a new series of annual lectures to be delivered by a man of eminence in science or literature. Owen's lecture on the classification of the Mammalia (R. Owen 1859) was but a slight modification of R. Owen 1857. In his article, Huxley cited a quotation from the earlier paper, in which Owen stated that it was difficult for an anatomist to distinguish between *Homo* and *Pithecus*; Huxley then pointed out that the statement had been omitted from R. Owen 1859 (see T. H. Huxley 1861a, p. 69).

[5] John Lubbock, one of several co-editors of the *Natural History Review*, contributed a paper published in the first issue on the anatomy and natural history of the entozoan *Sphaerularia bombi*, a worm parasitic on humble-bees (Lubbock 1861a).

[6] Robert M'Donnell's paper (M'Donnell 1861) announced his discovery of the electrical organs in skates. M'Donnell stated that he had been led to search for these organs after considering CD's remarks in *Origin* about the 'unity of type resulting from community of descent' (M'Donnell 1861, p. 58):

> Considering, therefore, that on the theory of Mr. Darwin it was in the highest degree improbable that the electric organs of the Torpedo were totally absent in the Skates, I undertook a careful search, with the view of following out their homologies . . . I have thus been led to make out the bodies which I conceive to be the true homologues of the Torpedo's wondrous organs

CD had learned of M'Donnell's work in advance of publication (see *Correspondence* vol. 8, letters to T. H. Huxley, 16 November [1860] and 22 November [1860]). In his copy of this issue of the *Natural History Review*, CD noted: 'M'Donnell on Electric Organs— gradation, good'.

[7] CD refers to an anonymous review of five recent works on various groups of Protozoa (*Natural History Review* n.s. 1 (1861): 34–43). The author may have been Huxley himself judging from the number of references to his work and the strident attack on Owen's views.

[8] Among the books reviewed (see n. 7, above) was R. Owen 1860a, in which Owen defined the fundamental attributes of the plant and animal kingdoms. On the basis of this definition, he established a third kingdom of 'Protozoa', which embraced all those organisms exhibiting 'common organic characters' that could not, using the specified criteria, be assigned to either of the other two kingdoms. The reviewer criticised Owen's criteria for distinguishing between animal and vegetable modes of existence and suggested that there were indeed sound scientific reasons for assuming the animal nature of true Protozoa (*Natural History Review* n.s. 1 (1861): 41–3).

[9] CD refers to Owen's presidential address at the 1858 meeting of the British Association for the Advancement of Science held in Leeds (R. Owen 1858). See nn. 13 and 14, below.

[10] Henrietta Anne Huxley was recovering from the birth, on 11 December 1860, of the Huxleys' son Leonard (L. Huxley ed. 1900, 1: 216).

[11] Huxley visited Down on 14 December 1860, as related in a letter of 19 December 1860 to Joseph Dalton Hooker: 'I was at Down on Saturday and saw Darwin. He seems fairly well, and his daughter was up and looks better than I expected to see her.—' (L. Huxley ed. 1900, 1: 223). Henrietta Emma Darwin had suffered from poor health throughout 1860. See *Correspondence* vol. 8.

[12] The postscript, although on a separate leaf, is clearly related to the topics discussed in the letter (see nn. 14 and 15, below). It is included among the copies of Huxley's letters made by Francis Darwin in DAR 145.

13 CD refers to Owen's published reference to the 'axiom of the continuous operation of the ordained becoming of living things.' Owen first used this phrase in his presidential address to the British Association meeting at Leeds (R. Owen 1858, p. li). The phrase was also used in Owen's unsigned review of *Origin* ([R. Owen] 1860b, p. 500).

14 CD included a reference to passages in R. Owen 1858 (see n. 9, above) in 'An historical sketch of the recent progress of opinion on the origin of species', included in the third edition of *Origin* published in April 1861. See *Origin* 3d ed., pp. xiii–xix. CD had been deeply distressed by Owen's rejection of *Origin* and by his misrepresentation of several points in his unsigned review (see *Correspondence* vol. 8).

15 Daniel Oliver, a botanist in the herbarium and the librarian of the Royal Botanic Gardens at Kew, was one of the botanical editors of the new series of the *Natural History Review* (see Turrill 1959, p. 236). Oliver contributed a bibliographic review to the first number, listing and commenting on recent publications on phanerogamic plants (*Natural History Review* n.s. 1 (1861): 85–115). The bibliography became a regular feature of the journal.

To *Gardeners' Chronicle* [before 5 January 1861][1]

[Down]

Mr. James Drummond sent me a packet of seeds of this plant from Swan River,[2] with the following memorandum:—"The achenia of several small composite plants, more especially of that above named, are blown about by the wind till a shower of rain falls, when they attach themselves by a gummy matter to the soil by their lower ends, at the same time setting themselves perfectly upright. They ornament many a barren spot in this country throughout the dry season, and they are not easily removed even when the ground is flooded by thunder storms."

The achenia of the Pumilio are singularly-shaped bodies; the calyx (pappus) consists generally of nine scales or sepals, expanded like a flower, with each sepal beautifully ornamented by branching lines; the lower part, including the seed, is bent towards one side (see Fig. 1) at nearly right angles, and somewhat resembles a human foot in shape. The upper side or instep of this foot is smooth, but the toe and the sole, which is about 1-25th of an inch in length, is covered (see Fig. 2) with from 30 to 40 imbricated little bladders. Each bladder is oval and 1-200th of an inch in length, and is formed of thin structureless membrane, enclosing a hard ball of dry mucus or matter which becomes adhesive when moistened. The sole or the achenium is pitted where the bladders are attached[3] but these do not open into its interior. When the achenia are placed in water or on a damp surface, the bladders in a few minutes all burst longitudinally and discharge their contents, rendering a large drop of water very viscid. This viscid drop does not diffuse itself throughout the water, like gum, but remains surrounding the achenium. When dried it becomes stringy, and will again rapidly absorb moisture and swell. Spirits of wine does not cause the bladders to burst, and it renders the mucus slightly opaque.

If a pinch of these achenia be dropped from a little height on damp paper, the greater number fall like shuttlecocks, upright and rest on the sole; the bladders then quickly burst, and as the paper dries the seeds become firmly attached to it. Many achenia, however, drop so as to rest on one edge of the sole, and in this case the drying of the mucus pulls the upper edge of the sole down, and so places the flower-like calyx nearly upright. Any one looking at a piece of paper over which when damp a number of achenia had been scattered by chance would conclude that each one had been placed upright and carefully gummed.

Fig. 1. Side view.

Fig. 2. View of under side, seen a little obliquely.

If an achenium falls upside down, so as to rest on the tips of the calyx, the sole does not touch the damp surface, yet moisture is so rapidly absorbed by the sepals, that in seven minutes I have seen the bladders burst: in this case as the paper dries the exuded mucus dries on the surface of the sole and the seed is not fixed; but if subsequently it be blown the right way up on a damp surface, the mucus will soften and act and attach it firmly. In so dry a climate as Australia the existence of these little bladders of dried mucus, having a strong affinity for water and becoming highly viscid on that side alone of the achenium which alights on the ground, seems a pretty adaptation to ensure the attachment of the seed to the first damp spot on which it may be blown. Whether the tendency of the achenia to place themselves upright be of any service to the plant would be hard to determine; but it seems possible that the salver-shaped calyx, which, as we have seen, so rapidly absorbs moisture and carries it down to the lower surface of the achenia, might aid in utilising dew or showers of fine rain.[4]

Charles Darwin

Gardeners' Chronicle and Agricultural Gazette, 5 January 1861, pp. 4–5

[1] The date is given by the publication of the letter in the *Gardeners' Chronicle and Agricultural Gazette*, 5 January 1861. The letter was published under the heading 'Note on the Achenia of Pumilio Argyrolepis'.

[2] CD had begun corresponding with James Drummond, superintendent of the government botanic gardens in Perth, Western Australia, in May 1860 (see *Correspondence* vol. 8). At the suggestion of Joseph Dalton Hooker, CD had written to ask Drummond to observe the fertilisation mechanism of various species of Goodeniaceae that were native to western Australia. In his letter of 8 October 1860 (*Correspondence* vol. 8), Drummond included the information on the gummy seeds of *Distylis* quoted here by CD. Drummond also enclosed seeds of a composite identified by Hooker as *Pumilo argyrolepis*. See *ibid*., letters to James Drummond, 20 December [1860], and to J. D. Hooker, 26 December [1860], and letter from J. D. Hooker, 28 December 1860.

[3] This word was misprinted as 'attachedf' in the *Gardeners' Chronicle*.

[4] For Hooker's opinion that this case appeared to provide a good illustration of the operation of natural selection, see *Correspondence* vol. 8, letter from J. D. Hooker, 28 December 1860.

To Richard Kippist[1] 7 January [1861][2]

Down Bromley Kent
Jan 7.

Dear Sir

[Will] you be so good as to have ready for me tomorrow morning (Tuesday) Memoires presentes par Divers Savan a l'Acad. R. des Sciences: Paris.— Tom. IV.[3]

I hope to call for it myself; if I fail I will send Carrier for it on Thursday.—[4]

Yours very faithfully | C. Darwin

American Philosophical Society (B/D25.285)

[1] Richard Kippist was the librarian of the Linnean Society of London, from which CD often borrowed scientific works.

[2] The year is based on the 1859 watermark of the stationery and by 7 January falling on a Monday in 1861.

[3] It is not clear which paper CD wished to consult in the *Mémoires présentés par divers Savans à l'Académie Royale des Sciences de l'Institute de France. Sciences mathématiques et physiques* 4 (1833). This volume contains a paper on the natural history of insects by Léon Dufour (Dufour 1833), but it is not a work that CD ever seems to have cited. It is possible, that CD meant to write 'VI' instead of 'IV'. Volume 6 of this journal contains a paper by François Désiré Roulin (Roulin 1835) that CD cited a number of times in chapter 3 of *Variation*, which he prepared in the first three months of 1861 (see 'Journal'; Appendix II).

[4] There is no record that CD went to London on 8 January 1861. A weekly carrier service operated between Down and London on Thursdays (*Post Office directory of the six home counties* 1862).

From Jeffries Wyman 8 January [1861][1]

Cambridge [Massachusetts]
Jan 8th 1860

My Dear Sir

It gives me great pleasure to answer in so far as I am able your enquiries with regard to the rattlesnake.[2] Dumeril & Bibron have given in their Herpetol. Generale,[3] a somewhat confused description of the structure of the rattle & Czermack has published a memoir entitled "Uber den schallenzeugenden Apparat von Crotalus" Zeitsch. f. wissench. Zool. Vol VIII 1856.[4] This last I have not seen, & only know it through the short account taken from it in Van der Hovens Zoology—[5] As I have some good specimens of tails in alcohol, at the risk of sending you superfluous information, I shall give you some account of the examination of them, which I have been led to make by the questions contained in your last letter.

The first specimen is that of a foetus of Crot. durissus which was nearly mature, is nearly eight inches long, & was taken from the body of the parent. For this I am indebted to Prof Baird of the Smithsonn. Institution.[6] On the tail of the foetus *one rattle is already formed*, fits the end of the tail very closely & has a somewhat different shape from the subsequent ones in the same species. It is a true cuticular covering & is continuous with the cuticle on the rest of the body.

The other specimens are from adults of C. rhombifer, & are fortunately in a condition to show very well one stage in the formation of the rattle. I will state first of all, that the last three caudal vertebrae are very much compressed laterally, anchylosed, & covered with a somewhat thickened layer of integument. The body thus produced has three prominent ridges as at *a* in the figure, & is the matrix on which the rattle is formed.

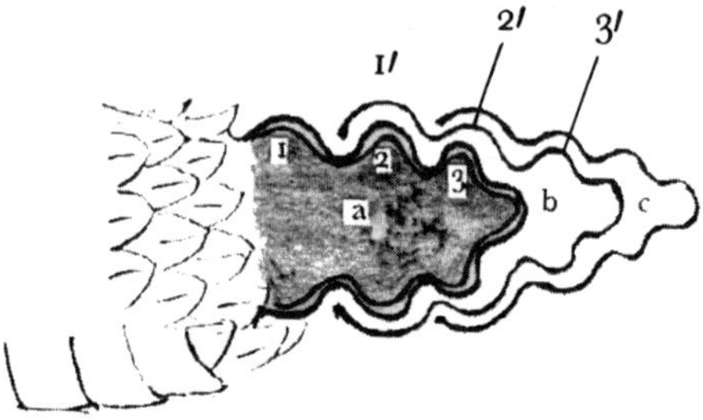

foetal

In two instances a new joint or shell of the rattle has just been completed, but still fits tightly to the matrix in every direction, to which it sustains the relation of common cuticle. When mature the covering thus secreted becomes detached from the matrix over its whole surface, & is forced back in such a manner that it now grasps only the second & third ridges instead of covering all of them. In this way the free space between it & its successor is gained, as at *b* & *c* I have never had an opportunity of witnessing the moulting of one of these snakes & cannot therefore describe the process by which the new formed joint is thus forced back— I *suppose* that in the first instance there is a shrinking of the matrix by which the rattle is detached & then, as each successive rattle is larger than its predecessor, the growth of the matrix which follows the separation pushes the rattle back. However this may be, one direct observation will be worth more than any or all suppositions. That the slipping of the rattle does take place I am sure, for numerous specimens which I have examined show it.

You mention the habit which the Trigonocephalus has, of vibrating its tail.[7] Perhaps it will be no news to you, but I will mention at a venture, that our common blacksnake (Coluber constrictor) has the same habit, & when its tail is made to vibrate against dried leaves, the sound which is produced is quite startling from its resemblance to that of the rattlesnake. The sound of the blacksnake I have heard— I am also informed by others, among them by Prof Baird that there are several species in our western country which have the habit as well.

Incomplete
Postmark: JA 22 61
DAR 89: 18–21

CD ANNOTATIONS
3.11 second & third ridges] 'ridges' *added pencil*
Top of first page: '(Prof? Wyman)' *pencil*
On cover: 'Lisotriton palmipes' *ink, del pencil*; 'Viperidæ | Penny Encyclop accounts of Fascination[8] | Wyman says Black Snake rattles its tail.' *ink*; 'Ch VIII'[9] *brown crayon*

[1] The year is provided by the postmark: Wyman erred in writing '1860'.
[2] See *Correspondence* vol. 8, letter to Jeffries Wyman, 3 December [1860]. Wyman, professor of anatomy at Harvard University, had begun a correspondence with CD at the suggestion of Asa Gray (see *ibid.*, letter from Asa Gray, 23 January 1860).
[3] Duméril and Bibron 1834–54.
[4] Czermak 1857.
[5] Hoeven 1856–8, 2: 259–60.
[6] Spencer Fullerton Baird was assistant to Joseph Henry at the Smithsonian Institution in Washington, DC.
[7] See *Correspondence* vol. 8, letter to Jeffries Wyman, 3 December [1860]. CD described the vibrating tail of this snake, noting its curious resemblance to that of the rattlesnake, in *Journal of researches*, pp. 96–7.
[8] CD refers to the article on American venom-snakes in the *Penny Cyclopædia* (1833–58), vol. 25–6, pp. 351–3, in which an account is given of the rattle as a device used to fascinate prey. See n. 9, below.
[9] CD refers to chapter 8, 'Difficulties on the theory of natural selection in relation to passages from form to form', in his 'big book' on species. In the discussion of habits that appear to have been acquired by one organism for the good of another, which thereby stood in opposition to the tenets of natural selection, CD mentioned the vibrating tail of *Trigonocephalus*. CD suggested that this habit, rather than having been acquired to warn other animals, was 'more likely to serve to paralyse by fear or fascinate its prey.' (*Natural selection*, p. 383 and n. 1).

To William Darwin Fox 9 January [1861]

Down Bromley Kent
Jan 9th

My dear Fox

I received yesterday from my Brother the two Inkstands for William & myself; & we are both very much obliged for your present.

I have filled mine this morning & it seems to work capitally, but I have not yet had much writing with it, as I have been merely correcting proofs—[1] The only fault is that I sometimes take too much ink, but with a little practice I shall get over this, & I have been eternally plagued with my Ink getting muddy, & if that is avoided I shall be for an equal eternity grateful to you,—as no doubt William will, who writes as detestably bad a hand as I do & my good Father did before me.—

We manage to keep Etty room of a good temperature & I do not think she suffers from the extreme & dreadful cold: she has been rather better these 4 or 5 days than usual. All my Boys are now at home & very jolly.[2] Farewell my dear Fox & I hope always to keep your Inkstand, as a memorial of you by me.—

Farewell | Your affect | C. Darwin

P.S. I believe you have been a pig-Breeder. I found a strange statement in a German Book that white Sows go with young a week longer or shorter (I forget which) than piebald or black Sows.—[3] I presume it is false: but many odd peculiarities are correlated with colour.—[4] Have you kept any exact record? Or do you know any careful Breeder that does?—

Do not think of answering this unless you have materials to judge by.—[5] It must be a cock & bull story, though given rather positively by goodish man.—

Postmark: JA 10 61
Christ's College Library, Cambridge (Fox 126)

1 CD was correcting proofs of the third edition of *Origin*, issued in April 1861. See 'Journal' (Appendix II).

2 According to Emma Darwin's diary, William Erasmus Darwin had come home from Cambridge University on 13 December 1860. George Howard and Francis Darwin, who were attending Clapham Grammar School in south London, stayed at Down until 4 February 1861.

3 CD was preparing chapter 3 of *Variation*, which discusses variations in pigs, cattle, and horses. The German work on pigs to which he refers may be Nathusius 1860. CD's abstract, bound in the annotated copy of Nathusius 1860 in the Darwin Pamphlet Collection–CUL, is dated 'Jan. 14/61'. However, the remark he cites has not been traced in this work.

4 CD discussed correlation of colour with other characteristics of organisms in *Natural selection*, pp. 299–300, and *Origin*, pp. 144–5.

5 Fox was apparently able to provide CD with information on this point. In the discussion in *Variation* of the gestation period in pigs, CD stated: 'The Rev. W. D. Fox has given me ten carefully recorded cases with well-bred pigs, in which the period varied from 101 to 116 days.' (*Variation* 1: 74).

To Joseph Dalton Hooker 15 January [1861][1]

Down Bromley Kent
Jan 15th

My dear Hooker

The sight of your hand-writing always rejoices the very cockles of my heart.—[2] I am glad to have seen H.C.W. letter; but it does not alter my disbelief.[3] I have put one sentence in my new Edit, in which I admit that the descendants of two genera might converge into one; but even then I shd. think they would form two sections of the new genus.[4] What he means by saying that it is a logical absurdity (I can understand that it might be thought improbable in any degree) that all forms shd. descend from a dozen or one form, I cannot conceive.—

I most fully agree to what you say about Huxley's article & power of writing.[5] What a smasher for Owen! The whole Review seems to me *excellent*. How capitally Oliver has done the resume of Bot. Books.[6] Good Heavens how he must have read! I am sorry he did not say a little on Sturr on Astrantia:[7] I have only skimmed it, but the manner in which he shows by map how a widely extended species breaks up into local forms seemed to me very pretty. What a silly & little feeling on H.C.W's part to suppose that Oliver would intentionally overlook the Cybele.—[8]

I quite agree that Phillips is unreadably dull.—[9] You need not attempt Bree,[10]— the man must be a conceited fool.— If you come across Dr Freke on "Origin of species by means of Organic Affinity",[11] read a page here & there just to see the maximum of ill-written unintelligible rubbish, which he tells the reader to observe has been arrived at by "induction", whereas all my results are arrived at only by "analogy". I see a Mr Neale has read a paper before Zoolog. Soc. on "Typical Selection", what it means I know not.—[12] I have not read H. Spence[r],[13] for I find that I must more & more husband the very little strength which I have. I sometimes suspect I shall soon entirely fail; my stomach now keeps bad nearly all day & night.— As soon as this dreadful weather gets a little milder, I must try a little water-cure. Have you read "Woman in White"—the plot is wonderfully interesting.[14] I can recommend a Book which has interested us greatly, viz "Olmsted, Journey in Back Country".[15] It is an admirably lively picture of man & Slavery in the S. States.

I see that your Sikkim Rajah has been again playing tricks.[16]

I am very glad to hear so good account of Mrs. H. & your children.[17] Etty has had a good fortnight & has got up the state she was in a month ago, & comes down almost every evening.—[18]

My dear Hooker | Yours affect | C. Darwin

Are you doing anything besides your gigantic Genera Plantarum (which I rather hate as I know it will absorb so much time & no doubt be most valuable)?[19] I always *much* like to know what you are about. In the Spring I hope you will find time & come here for a few days.— A sight of you is a real pleasure.—

God Bless you.—

DAR 115.2: 85

[1] Dated by the reference to the new edition of *Origin*, which was issued in April 1861, and by CD's discussion of the first number of the new series of the *Natural History Review*, published in January 1861.

[2] Hooker's letter has not been found.

[3] Hewett Cottrell Watson's letter to Hooker, dated 4 January 1861, is in the archive of the Royal Botanic Gardens, Kew. The letter offers further illustrations of what Watson called the 'convergence' of species. For earlier correspondence on this point, see *Correspondence* vol. 8, letter from H. C. Watson, [3? January 1860], and letter to H. C. Watson, [5–11 January 1860]. CD discussed Watson's views and his objection to natural selection and divergence in the revised American edition of *Origin* (*Origin* US ed., pp. 116–17*; see *Correspondence* vol. 8, Appendix IV, pp. 576–7); he also included the discussion in *Origin* 3d ed., p. 141:

> A distinguished botanist, Mr. H. C. Watson, believes that I have overrated the importance of the principle of divergence of character (in which, however, he apparently believes), and that convergence of character, as it may be called, has likewise played a part . . . Mr. Watson has also objected that the continued action of natural selection with divergence of character will tend to make an indefinite number of specific forms.

For CD's response to the first point, see n. 4, below. For his reasons for not believing in an unlimited proliferation of different forms, see *Origin* 3d ed., pp. 141–3.

[4] 'I will only say that if two species of two closely allied genera produced a number of new and divergent species, I can believe that these new forms might sometimes approach each other so closely that they would for convenience sake be classed in the same new genus, and thus two genera would converge into one; but from the strength of the principle of inheritance, it seems hardly credible that the two groups of new species would not at least form two sections of the supposed new single genus.' (*Origin* 3d ed., p. 141).

[5] T. H. Huxley 1861a. The article was published in the new series of the *Natural History Review* under Thomas Henry Huxley's general editorship.

[6] Daniel Oliver worked with Hooker at the Royal Botanic Gardens, Kew. Oliver contributed a bibliographic review of the botanical literature on phanerogamic plants published in the first nine months of 1860 to the *Natural History Review* n.s. 1 (1861): 85–115. See also letter to T. H. Huxley, 3 January [1861] and n. 15.

[7] Stur 1860. There is a copy of the work in the Darwin Library–CUL; it is a reprint of an article published in the *Sitzungsberichte der Mathematisch-Naturwissenschaftlichen Classe der Kaiserlichen Akademie der Wissenschaften (Wien)* 40: 469–524.

[8] Oliver's bibliographic review (see n. 6, above) did not include a notice of the third volume of Watson 1847–59, published in 1859.

[9] Phillips 1860. CD's annotated copy is in the Darwin Library–CUL.

[10] Bree 1860. CD's lightly annotated copy is in the Darwin Library–CUL.

[11] Freke 1861. There is a copy of this work in the Darwin Library–Down. Henry Freke had published a brief notice in October 1860 claiming to have preceded CD by publishing in 1851 the view that all plants and animals had descended from a single prototype. There is a copy of this notice (Freke 1860) in the Darwin Pamphlet Collection–CUL.

[12] Neale 1861. The Christian Socialist Edward Vansittart Neale believed that CD's theory relied too heavily on chance variations. Neale advocated the view that variations arose through an 'internal action of the living Power gradually modifying the constitution of the individual.' (Neale 1861, p. 5).

[13] The second number of Herbert Spencer's *First principles* (Spencer 1860–2) was published in January 1861. CD was a subscriber to this work. His copy, which is not annotated, is in the Darwin Library–CUL.

[14] Collins 1860.

[15] Olmsted 1860.

[16] Hooker undertook a botanical expedition in the Himalayas between 1848 and 1850, during which time he and another British subject were briefly imprisoned by the Sikkim Raja, Cho-phoe Namgyé. To secure their release, the British governor-general sent in an expeditionary force, an action that ended in territory belonging to the Raja being seized. See *Correspondence* vol. 4, letter to W. J. Hooker, [January 1850]; Hooker 1854, 2: 202–41; and L. Huxley ed. 1918, 1: 306–19. *The Times*, 14 January 1861, p. 6, carried a report on the recent dispatch of another British force following the kidnapping of workers from the hill station at Darjeeling. This incident eventually led to the imposition of a new treaty with Sikkim (*EB*).

[17] CD refers to Hooker's wife, Frances Harriet Hooker, and the Hookers' children, William Henslow, Harriet Anne, Charles Paget, Marie Elizabeth, and Brian Harvey Hodgson.

[18] Henrietta Emma Darwin was recuperating slowly after having contracted typhus fever in 1860.

[19] Hooker and George Bentham had begun work late in 1860 on an ambitious project to prepare a complete systematic catalogue of all the genera of flowering plants then known to botanists. The first part of volume one of the *Genera plantarum* (Bentham and Hooker 1862–83) appeared in 1862. See L. Huxley ed. 1918, 2: 18–23.

To John Murray 21 January [1861]

Down Bromley Kent
Jan 21.

My dear Mr Murray

Please return enclosed & tell me whether you approve.

As 4250 have been sold & 2000 will be printed, if number is to be retained, it will stand "Seventh Thousand".—[1]

I think it would be worth while to insert in any advertisement or notice "Third Edition with additions & Corrections", for the additions are important & I shall give list of them.—[2] I am sorry to say that they have already added 20 pages to Book & will I suppose add 30 altogether.[3] There will be moreover a *short* Historical Sketch.[4]

Please observe that I have corrected only to p. 216 & unless M[essrs] Clowes get on **much quicker** the Book (with Index &c) will not be out till end of March at soonest, which is a pity.—[5]

I hope you will be again generous & give me some copies, for I shall have anyhow to order a good many for my staunchest supporters.—[6]

Whenever you write please gratify my curiosity by telling me whether you have sold any fair number of my Journal.—[7] The scheme seemed to me rash.—

Yours very sincerely | C. Darwin

Endorsement: 'Jan 21— 1861'
John Murray (publishers)

[1] CD was working on the proof-sheets of the third edition of *Origin* (see *Correspondence* vol. 8, letter to John Murray, 22 [November 1860]. The new edition included 'Seventh Thousand' on the title page.

[2] Murray agreed to CD's suggestion: this information was carried on the title page of the third edition of *Origin*. The new edition also included a list of 'Additions and corrections, to the second and third editions', giving the page number and subject of the various changes that had been made (see *Origin* 3d ed., pp. xi–xii).

[3] The third edition of *Origin* ran to 525 pages, 35 more than the second edition.

[4] The third was the first English edition of *Origin* to include 'An historical sketch of the recent progress of opinion on the origin of species' (*Origin* 3d ed., pp. xiii–xix). This sketch was a modified version of the 'Preface' CD had prepared for the revised and augmented American edition and also included in the German translation, both of which were published in 1860 (see *Correspondence* vol. 8).

[5] The printing firm of William Clowes and Sons was Murray's preferred printer. The new edition was distributed early in April (see 'Journal'; Appendix II).

[6] For CD's presentation list for the third edition of *Origin*, see Appendix VII.

[7] Murray had issued a reprint of CD's *Journal of researches* in 1860, stating on the title page that it was the tenth thousand (see Freeman 1977, p. 36). In January 1860, CD recorded that Murray had sold 6,000 copies of this work (see *Correspondence* vol. 8, Appendix II).

To William Walmisley Baxter?[1] 22 January 1861

[Down]

One Pint of Tincture of Henbane.[2]

Down. Jan22[d]./61/ | C. Darwin

Auckland Public Library (RBR MS 29–30)

[1] Baxter was a dispensing chemist in Bromley, Kent.

[2] Henbane (*Hyoscyamus niger*) is a poisonous plant of the nightshade family. CD had begun a series of experiments in the autumn of 1860 in an attempt to determine the effects of various poisons on the insectivorous plant *Drosera rotundifolia* (see *Correspondence* vol. 8). He discussed the results of the trials with henbane in *Insectivorous plants*, pp. 84, 206.

To Daniel Oliver 22 January [1861][1]

Down Bromley Kent
Jan. 22.

My dear Sir

You mentioned to me Kurr on Nectaries[2] & in doing this you did me a great service,—that is if helping a man to ride his hobby-horse can be called a "great service"! To watch flowers & insects has been a very old hobby of mine.—[3] Some statements in Kurr lead me to ask you, to give me some information on one point, if in your power, & if it can be done without *much* trouble.— But I have no idea whether you have the plant alive at Kew, or whether dried plants would serve.— Half-opened flowers do not do, for pistil bends subsequently. The point is, that I as yet find it a rule that *if* the pistil is curved to one side, it curves to the side of the nectary, so that the stigma lies in the gangway to the secreting nectary or to the honey-receptacle.—[4] The converse proposition does not hold good; the nectary may be lie one side, without the pistil being curved, but in this case it nevertheless lies in the gangway.— One of the best cases is Saxifraga, in which in 2 or 3 species (names unknown to me) the nectary goes all round & the pistil is straight; but in S. sarmentosa the pistil is curved & the nectary is one sided & lies on side towards which pistil bends.— Now Kurr, as you will see in enclosed paper gives name of some species with one-sided & others with circular nectaries; can you find out at any time for me how the pistils are? Analogous facts occur in several *genera* of Proteaceæ & Caprifoliaceæ: can you aid me?—

The rule interests me in several ways, & any the most trifling generalisation, such as this, seems worth making out.— I have already got a good many facts.— But you must *not* waste much time on it.—

Ever yours very truly | C. Darwin

[Enclosure]

Kew

J. G. Kurr | Proteacea Will you keep this till flowering season if necessary

p. 32 | Grevillea & Hakea, Embothrium, Lomatia, Stenocarpus } nectary on one side. Is the pistil bent to that side?

I imagine the other genera have not lop-sided nectary.

p. 54 | Caprifoliaceæ in most (if I understand) the nectary on one side; but in some as Cornus & Viburnum not on one side— how are pistils?

p 57 In Saxifraga punctata, umbrosa & cuneifolia nectar secreted on *one side* whereas in S. decipiens, trifurcata, mutata cressifolia & granulata it surrounds the germen. How is pistil in the 3 first species. In S. sarmentosa the nectary is on one side & pistil is bent to the side.—

p. 71. In Melianthus major, nectary lateral; how is pistil?

Down House (MS 10: 5)

[1] Dated by the reference to Kurr 1833 (see n. 2, below).
[2] Kurr 1833. An annotated copy of this work is in the Darwin Library–CUL. Tipped into the volume are two pages of notes in CD's hand that are dated 14 and 19 January 1861, respectively.
[3] Some of CD's notes on this topic are in DAR 76. Several of the slips are dated from the 1840s.
[4] For CD's first expression of this general rule, see *Correspondence* vol. 8, letter to J. D. Hooker, 12 July [1860].

To Asa Gray 23 [January 1861][1]

Down
23.

My dear Gray

Excuse extreme brevity, as my poor girl has just begun severe attack.—[2] I was very glad to hear that you have decided to publish.[3] Please remember & *grant me favour* that I bear the whole small risk of loss[4] I am a rich man.—[5] You will not misunderstand me.—

The enclosed explains itself.[6] Price of 2^{s}—strikes me as too high.—[7] I copy list I made few days ago of *first* lot, which I would for myself distribute.—

I have just received letter from Prof. Wyman.—[8] I have not read it, but I see it is profoundly interesting to me.— Thank him.

Yours most cordially | C. Darwin

[Enclosure]

For Reviews	Hooker	Wollaston
Annals of N. Hist	Bentham	Sir H. Holland
Edin. New Phil. J.	Carpenter	Falconer
Athenæum	Huxley	Harvey
Sat. Review	Lubbock	M'Donnell
Gardeners Ch.	Self	L. Horner
Nat. Hist. R.	Henslow	Phillips
Zoologist.	Sedgwick	H. D. Rogers
Geolog. Soc.	Hopkins	Prestwich
Royal Soc.	B. of Oxford	Al. Decandolle
Linn. Soc	Herschel	Os. Heer
Athenæum Club	Whewell	Thomson of Calcutta.
Fleuriere (a Reviewer)[9]	Lyell	Blyth of do.
Private	C. Kingsley	Thwaites of Ceylon
S. P. Woodward	Bunbury	
H. C. Watson.	R. Chambers	

Gray Herbarium of Harvard University (12)

[1] The date is based on the reference to the publication of Gray's pamphlet on natural selection and natural theology (A. Gray 1861a). See nn. 3, 4, and 6, below.

[2] Henrietta Emma Darwin, who was recuperating from a long period of illness, was experiencing an attack of incessant vomiting (see following letter).

[3] In the autumn of 1860, CD asked Gray whether he would be interested in reprinting for distribution in England his series of three articles on *Origin* published in the *Atlantic Monthly* (A. Gray 1860a). See *Correspondence* vol. 8, letters to Asa Gray, 10 September [1860], 19 October [1860], 24 October [1860], and 11 December [1860], and letter to J. D. Hooker, 11 December [1860].

[4] CD and Gray had agreed to share the costs of printing the pamphlet (see *Correspondence* vol. 8, letter to Asa Gray, 11 December [1860]).

[5] In addition to his considerable private wealth, CD had received net profits of over £838 from the sale of the first two editions of *Origin* in 1860. See *Correspondence* vol. 8, 'Journal' (Appendix II).

[6] The enclosure was copied from a set of lists of people and establishments to which CD planned to send a presentation copy of the pamphlet (see Appendix III).

[7] The pamphlet, entitled 'Natural selection not inconsistent with natural theology. A free examination of Darwin's treatise "On the origin of species", and of its American reviewers' (A. Gray 1861a), was published in London by Trübner and Co. for sale at the price of 1*s*. 6*d*. CD's copy is in the Darwin Pamphlet Collection–CUL.

[8] Letter from Jeffries Wyman, 8 January [1861].

[9] According to CD's list of presentation copies for the third edition of *Origin* (see Appendix VII), 'A. de la Fleuriere' was the science writer for the weekly newspaper the *Leader and Saturday Analyst; a review and record of political, literary, artistic, and social events*. The publication was a continuation of the *Leader*, founded in 1850 by George Henry Lewes and Thornton Leigh Hunt. In the issue of 18 February 1860, p. 160, the *Leader and Saturday Analyst* carried a favourable notice of Thomas Henry Huxley's lecture on *Origin* at the Royal Institution, referring to CD's theory as 'a subject that will for some time to come be foremost in the minds of thinking men.'

To William Erasmus Darwin [24 January 1861]

Down
Thursday | 12° oclock

My dear William

The vomiting continued all yesterday till night, when it stopped & has not returned.[1] But she has taken gigantic dose of Calomel which has not acted, so we continue uneasy; but Mr Williams who has been just here, considers her better.[2] She eats absolutely nothing.—

I hope that your journey was prosperous & that you will, my dearest old fellow, enjoy yourself.—[3] We shall be glad of a note.

Your affect | C. Darwin

Postmark: JA 24 61
DAR 210.6

[1] After having experienced a period of good health (see letters to W. D. Fox, 9 January [1861], and to J. D. Hooker, 15 January [1861]), Henrietta Emma Darwin had again fallen seriously ill.

[2] Edward Augustus Williams, a surgeon in Bromley, Kent, was the Darwins' local physician. Emma Darwin's diary records that he visited Down House on 2 January and again on 19 January 1861.

[3] The cover of the letter is addressed to William in care of "Mrs Skipworth, Rothwell House Caistor, Lincolnshire". William was apparently staying with the family of a friend from Christ's College, Septimus Patrick Skipworth, whom he may have visited the previous summer (see *Correspondence* vol. 8, letter to W. E. Darwin, [30 July 1860]).

From William Duppa Crotch 25 January 1861

Uphill House | Weston Super Mare
Jany 25th/61

Dear Sir,

Suddenly after the first written date[1] I went to Berlin to work out some difficulties in the Hemiptera—[2] I had just returned from Shetland, & had gleaned two facts interesting to you, as I hope— The Shetland ponies during winter are much stinted in food & feed largely on sea-weed— as a result their stomach becomes so modified & reduced in size that few physiologists could guess the proprietor to have been a horse— Again it is well known that MacGillivray contended for physiological characters as invariable & affording specific characters when morphological results are useless—[3] Could he revive & see the Shetland Sea gulls or rather their membranous crop—that proventriculus on which he gloated—developing muscular fibre & becoming a complete gallinaceous gizzard under the influence of a change of diet from fish &c. to corn: What would he then say of physio- versus morphological char? one wd. change an obvious species, (so called—), the other, so far, wd. preserve it—tho' it is difficult to say what change some generations of grain diet wd. produce in the external organs—

In Berlin too I learned an important fact, I fancy new. M. Raymond took near Toulon several specimens of an eyeless beetle (Anophthalmus Raymondi) in a *cellar* belonging to an ancient monastery *long abandoned*)—[4] How did they get there & where from?

Lecturing to unbelievers is grand fun—especially as I find we hit pretty equally hard & when a man knows the thrust in *Hybrid* I find it difficult to parry & generally reply by a thrust in Selection— But I waste your time—

Yours sincerely | W. D. Crotch

DAR 47: 173–4

CD ANNOTATIONS
1.1 Suddenly … organs— 1.14] *crossed pencil*
3.1 Lecturing … Crotch 4.1] *crossed pencil*

[1] The date was originally written as '8 January 1860.' and then corrected by Crotch.
[2] Crotch, an avid entomologist, had a particular interest in Lepidoptera, Coleoptera, and Hemiptera.
[3] William Macgillivray, a noted systematist, was the author of a standard taxonomic work on British birds (Macgillivray 1837–52).
[4] The discovery of a new species of the genus *Anophthalmus* by E. Raymond was announced in the *Annales de la Société Entomologique de France* 3d ser. 8 (1860): li. CD was interested in *Anophthalmus* and other blind cave animals in connection with his belief that faculties no longer of use to an organism

would be likely to degenerate. See *Origin*, pp. 137–9. His explanation of how this could occur had been called into question by Andrew Murray in 1860, who cited the geographical distribution of species of *Anophthalmus* as an apparent obstacle to CD's explanation. See *Correspondence* vol. 8, letters to Andrew Murray, 28 April [1860] and 5 May [1860], and letter from Andrew Murray, 3 May 1860.

From Francis Walker 1 February 1861

Grove | Highgate
Febry. 1st. 61.

My dear Sir,

Mr. F. Smith has given me two Dipterous Species forwarded by you to him for the names.[1] One "aboundant on T. officinalis" is Dolichopus nigripennis, (Fallén) a common species The other "sucking Euphrasia" is *Empis* chioptera (Fallen) also common,[2] the male with milk-white wings, very different in appearance from the dark-winged female.—[3]

I was not able to attend to the Aphides last year, but I will endeavour to make use of your suggestions respecting them & to study as much as I can, the inquiry as to species.—[4]

Believe me, yours sincerely, | F. Walker

DAR 46.2 (ser. 3): 54–5

CD ANNOTATIONS
1.2 T. officinalis"] 'T.' *altered to* 'Tomentilla' *pencil*
1.3 The other . . . female.— 1.5] *crossed ink*
2.1 I was . . . F. Walker 3.1] *crossed pencil*
Top of letter: '*Ch 3*'[5] *brown crayon*

[1] CD often sent insect specimens he wished to have identified to Frederick Smith, entomologist at the British Museum. Francis Walker was the author of numerous catalogues of the insect collections of the British Museum and of a three-volume monograph on Diptera for *Insecta Britannica* (London, 1851–6). Smith identified for CD the bees and other Hymenoptera found to be agents in orchid pollination, and Walker named the Diptera (see *Collected papers* 2: 139).

[2] The Swedish entomologist Carl Frederik Fallén described *Dolichopus nigripennis* in Fallén 1814–27, 2: Dolichopodes, p. 15, and *Empis chioptera* in *ibid*., 1: Empidiae, p. 24.

[3] CD was investigating the role of insects in pollinating plants. *Tormentilla officinalis* (see CD's annotations, above) is another name for the herb *Potentilla tormentilla*, in which a ring-shaped nectary within the stamens is commonly visited by flies. *Euphrasia* has 'loose-pollen' flowers by which the visiting insect receives a shower of pollen from the anthers.

[4] No earlier letter from CD to Walker giving 'suggestions' concerning aphids has been found. For CD's expression of interest in the recent work on parthenogenesis in aphids, see *Correspondence* vol. 7, letters to John Lubbock, 30 [March? 1858] and [November 1858].

[5] The reference is to chapter 3 of CD's 'big book' on species, 'On the possibility of all organic beings occasionally crossing, & on the remarkable susceptibility of the reproductive system to external agencies' (see *Natural selection*).

To Charles Lyell 2 February [1861]

[Down]
Feb. 2[d]

My dear Lyell

I have thought you would like to read enclosed passage in letter from A. Gray (who is printing his Reviews as pamphlet & will send copies to England), as I think his account is really favourable in high degree to us.—[1]

"I wish I had time to write you an account of the very absurd lengths to which Bowen & Agassiz—each in their own way—are going.[2] The first denying all heredity (all transmission except specific) whatever.[3] The second coming near to deny that we are genetically descended from our gr[t]-gr[t]-grandfather; & insisting that evidently affiliated languages e.g. Latin Greek Sanscrit owe none of their similarities to a community of origin,—are all autochtonal.[4] Agassiz (foolish man) admits that the derivation of languages & that of Species or forms stand on the same foundation, & that he must allow the latter if he allows the former,—which I tell him is perfectly logical."

Is this not marvellous.?

Ever yours | C. Darwin

I sent Calcutta Review a couple of days ago.—[5]

Postmark: FE 3 61
American Philosophical Society (238)

[1] The letter from Asa Gray has not been found, but see the letter to Asa Gray, 23 [January 1861], which is probably a response to it. CD had asked Lyell whether he thought it advisable to publish Gray's series of articles on *Origin* ([A. Gray] 1860a) as a separate pamphlet for distribution in Britain (see *Correspondence* vol. 8, letters to Charles Lyell, 12 September [1860], 23 [September 1860], and 4 December [1860]). Lyell approved the proposal (*ibid.*, letter to Asa Gray, 24 October [1860]). See also Appendix III. As Lyell told one correspondent, 'Asa Gray's articles, all of which I have procured, appear to me to be the ablest, and on the whole grappling with the subject, both as a naturalist and metaphysician, better than anyone else on either side of the Atlantic.' (K. M. Lyell ed. 1881, 2: 341).

[2] Francis Bowen, professor of philosophy at Harvard University, and Louis Agassiz, professor of natural history at Harvard, had established themselves as vocal opponents of CD's theory of transmutation; both had written negative reviews of *Origin* ([Bowen] 1860a, Bowen 1860b, and Agassiz 1860) and had spoken out against the book at several meetings of the American Academy of Arts and Sciences in 1860 (see *Correspondence* vol. 8).

[3] The reference is to Bowen's remarks on the inheritance of certain traits made at a meeting of the American Academy of Arts and Sciences on 8 January 1861 and reported in the *Proceedings of the American Academy of Arts and Sciences* (Bowen 1861).

[4] At the same meeting at which Bowen made his remarks (see n. 3, above), Agassiz disputed the view of Cornelius Conway Felton, the president of Harvard University, that the Aryan family of languages derived from a common aboriginal language. As reported in the *Proceedings of the American Academy of Arts and Sciences* 5 (1860–2): 102, Agassiz expressed his 'general disbelief in the supposed derivation of later languages from earlier ones, he regarding each language and each race as substantially primordial, and ascribing the resemblances and coincidences of language to a similarity in the mental organization of the races.' See also Lurie 1960, p. 299. Lyell had recently read a thesis similar to Felton's in an essay by Friedrich Max Müller (Max Müller 1856) that had greatly impressed him (see K. M. Lyell ed. 1881, 2: 341–2).

[5] A favourable review of *Origin* appeared in the September 1860 issue of the *Calcutta Review*, a copy of which is in the Darwin Pamphlet Collection–CUL.

To William Bernhard Tegetmeier 3 February [1861]

Down Bromley Kent
Feb. 3^d.

My dear Sir

I am much obliged to you for telling me of your curious case.[1] I have never met with anything parallel; that is supposing the organs of generation are perfect. Everyone knows of the effect of castration on Deers Horns.—[2] And I have read somewhere of a Deer in a ship which without being emasculated failed to reproduce its Horns.[3]

I have also read of males of small birds, like Bull-finches (I forget what kind) which on the moult in menageries failed to reproduce the male plumage— This case comes near yours, but then it has been supposed that the male organs had failed in their action.— How are castrated Cocks in plumage? Their spurs are said to grow to enormous size.—

I am very glad you are going to breed from your Bird. If fertile it will add greatly to interest; & if the variety should be inherited it will be still more curious.—[4]

With many thanks to you for telling me of this curious case | Believe me | My dear Sir | Yours sincerely | C. Darwin

New York Botanical Garden Library (Charles Finney Cox collection)

[1] Tegetmeier's letter has not been found. He exhibited the 'curious case' at a meeting of the Zoological Society of London on 26 March 1861 (see *Proceedings of the Zoological Society of London* (1861), p. 102). CD described the case in *Variation* 1: 253:

> Mr. Tegetmeier has recorded the remarkable case of a brown-breasted red Game-cock which, after assuming its perfect masculine plumage, became hen-feathered in the autumn of the following year; but he did not lose voice, spurs, strength, nor productiveness.

[2] CD discussed the general phenomenon of the assumption of female secondary sexual characteristics by males as a result of castration in *Variation* 2: 51–2.

[3] See *Variation* 2: 158.

[4] Tegetmeier was successful in breeding from the cock. In *Variation* 2: 253, CD reported: 'This bird has now retained the same character during five seasons, and has begot both hen-feathered and male-feathered offspring.'

To Jeffries Wyman 3 February [1861][1]

Down Bromley Kent
Feb. 3^d.

My dear Sir

I have been extremely much interested by your letter on Rattle-snakes.—[2] The subject has always rather fascinated me. By the way can there be any truth in the many accounts of Rattle-snakes *fascinating* their prey?[3] Do they rattle when gazing at their prey & can the poor animals be paralised by fear?— I never could understand from drawing, the rattle, till seeing yours, which makes it quite clear. If not known, is it not a shame that you do not publish your diagram?[4] The whole contrivance & structure is much more complex than I expected; & I fear it will be impossible to show amongst other snakes any gradation except the vibrating the tail (I did not know of your Black snake having this habit, & am glad to hear it) & the little knob at the end of tail of Trigonocephalus. Nevertheless I will have a look some day at the tails of the Viperidæ in B. Museum for bare chance of any vestige of the concentric ridges which seem to play an important part in retaining the rattle. I presume each new rattle adheres closely to the "matrix" till the period of moulting.— It certainly is a very curious structure; & odiously wonderful to me if it serve merely to warn prey & so injure its possessor.—[5]

With respect to Electrical organs, have you seen M'Donnel's article in the new Nat. Hist. Review? the discovery (if proved) of these organs in head of Skates seems very interesting.—[6]

Thank you much for telling me your observations on the marks on the face of Vespidæ:[7] I will show your sketch & consult our great authority on these insects F. Smith of B. Museum.[8] I shall be curious to know whether anything analogous occurs with our wasps. I particularly value such cases of variation & have already got a large series of cases, & very important they seem to me. Those who believe in creation, will have to say this mark in this animal was thus created; in this other animal that very same mark is due to Variation.

With cordial thanks for your most interesting letter. Believe me, my dear Sir | Yours sincerely C. Darwin

Jeffries Wyman Jr

CD note:[9]

In accounting for origin of rattle snake no difficulty if it c^{d} be shewn, rattle of any use and intermediate forms. *Do they go in pairs?? [*interl in CD's hand*] Those that think that an animal admirably provided to find its own sustenance shall purposely have a structure rendering such provision partly futile must yet be startled that other snakes quite as venomous & I believe incomparably more active & again others equally inactive have no rattle. Now [*after del* 'Now I have shewn'] from habits of trigonocephalus we may suspect that the vibration of tail & consequent rattle is somehow useful, & in the minute terminal button we have first indication of intermediate form *as I long ago remarked [*interl*]. Has not rattle snake wide range— A small argument for utility of rattle or more properly that it is not injurious otherwise other venomous genera w^{d} have beaten this out of the field.

1 Dated by the relationship to the letter from Jeffries Wyman, 8 January 1861.
2 Letter from Jeffries Wyman, 8 January 1861.
3 See letter from Jeffries Wyman, 8 January 1861, n. 8.
4 Wyman subsequently published a short account on this subject (Wyman 1861). See also Dupree 1951.
5 See *Origin*, p. 201. See also CD note, above.
6 M'Donnell 1861. CD mentioned this discovery to Wyman in his letter of 3 December [1860] (*Correspondence* vol. 8).
7 The remarks were apparently made in the concluding section of Wyman's letter of 8 January 1861, which is now missing.

[8] Frederick Smith, a specialist on the Hymenoptera, was employed in the zoological department of the British Museum.

[9] The undated note is in DAR 205.5: 212. It is in Emma Darwin's hand except for one addition by CD.

To J. D. Hooker 4 February [1861]

Down Bromley Kent
Feb 4th

My dear Hooker

I was delighted to get your long, chatty letter & to hear that you are thawing towards science.[1] I almost wish you had remained frozen rather longer; but do not thaw too quickly & strongly. No one can work long like you used to do. Be idle; but I am a pretty man to preach, for I cannot be idle, much as I wish it, & am never comfortable except when at work. The word Holiday is written in a dead language for me, & much I grieve at it.—

We thank you sincerely for your kind sympathy about poor Etty, who about a fortnight ago had three terrible days of sickness & was given loads of calomel, which I always dread.[2] She has now come up to her old point & can sometimes get up for an hour or two twice a day.— Poor George has literally every tooth in his head, except a few lower incisors decayed: they have all gone suddenly together & been stopped & drawn by Mr Woodhouse & I fear this points to some deep flaw in his constitution, which was formerly indicated by his intermittent pulse.[3] Never to look to the future, or as little as possible, is becoming our rule of life.— What a different thing life was in youth, with no dread in the future;—all golden, if baseless, hopes.—

With respect to Athenæum I answered for increase in number of members; I knew nothing of crowding & did not like paying £3::s10·d.0 for each luncheons & about 3·3·0 has been my private tariff for benefits derived from Athenæum. I am sorry to have anything said against the splendid Library. As half the 100 would be elected by Committee; I looked at the result as *raising* the standard of Club.—[4]

I am glad to hear that you have been touching up your Arctic paper;[5] I shall be very curious to read that attentively. I am sorry for you, though you do not seem sorry for yourself, for your long job over the Cryptogamia collection.[6]

With respect to Nat. Hist. Review,[7] I can hardly think that Ladies would be so very sensitive about "Lizards guts"; but the publication is at present certainly a sort of Hybrid; & original illustrated papers ought hardly to appear in a Review. I doubt it's ever paying; but I shall much regret if it dies.[8] All that you say seems very sensible: but could a Review in strict sense of word be filled with readable matter?

I have been doing little, except finishing the new Edit. of Origin,[9] & crawling on most slowly with my volume on "Variation under Domestication,"[10]—& how much to give under each head puzzles me dreadfully.— I have, however, in preparation for Drosera in the summer, been trying a few little experiments with C. of Ammonia, on the secreting Hairs of Primula Sinensis;[11] & I find that the matter in the heads

is easily acted on, & segregates into spheres, with very small doses of the ammonia; & what surprises me is that a more prolonged dose causes the spheres to become re-diffused into opake granular matter.[12] I long to be at Drosera again: I cannot persuade myself that it is the weight of $\frac{1}{78,000}$ of a grain of solid substances which causes such plain movement; nor that it is in most of the cases the chemical nature; & what it is, stumps me quite.

Do you remember a tall Silver Fir in the field in front of this House. It was so ugly we have grubbed it up. The hole was 3 ft. 6 inches in diameter with touching roots extending much further. Rings of growth 110.— It seems to have been planted on little mound of made earth; & I have got several pounds of this earth from exactly under centre in a sort of cone, where it is hardly possible seeds could have got for last 60 or 80 years. I have saved this earth on purpose to spite you about seeds germinating. And I have a great advantage, for if any come up in my study it will be good case; if none do (as I rather fear) then I shall say there were no seeds in earth![13]

I had a long letter about a week ago from Asa Gray, but I did not send it, thinking you would not care for it, as it almost wholly is on Design & quasi theological. He tells me that two of my opponents are gone almost demented: Bowen denying that any deviation is ever inherited; & Agassiz maintaining that Greek Latin & Sanscrit are not affiliated but, like the races of men, are autochthonous![14] It is impossible to argue better for us.—

Thanks for offer of Richardson Polar R. but I sincerely hope it is not worth my reading.[15] I admire Framley as much as you can do.—[16] I presume you have read "Woman in White":[17] we read it to the children, & they were so interested. Did I recommend "Olmsteads Journey in Back Country":[18] it interested me extremely.—

I am glad to hear so good an account of your Willy.[19] A lull after a gang of children is very pleasant.

Ever yours affectionately | C. Darwin

Endorsement: '(1861)'
DAR 115.2: 87

[1] Hooker's letter has not been found. CD refers to Hooker's resolution, expressed in the letter from J. D. Hooker, [26 November – 4 December 1860] (*Correspondence* vol. 8), to restrict his activities with the aim of reducing the stress under which he worked. See also L. Huxley ed. 1918, 1: 534–7.

[2] See letters to Asa Gray, 23 [January 1861], and to W. E. Darwin, [24 January 1861].

[3] CD took George Howard Darwin to London on 28 January 1861, returning the next day. There is an entry in CD's Account book (Down House MS) on 24 February 1861 that reads: 'Woodhouse

George's Teeth £.17.*s*.17'. Alfred J. Woodhouse, a dentist at 1 Hanover Square, is listed in the 1861 *Post Office London directory*.

[4] The Athenaeum Club had circulated a questionnaire among its members asking whether, to meet its pressing financial needs, they would prefer to increase the annual subscription fees from 6 to 7 guineas or to expand the club's membership by 200. In the event of increasing the number of members, an election committee was to recommend 100 new members, with the remaining 100 to come from the existing waiting list. The membership voted for the latter proposal: additional members were elected and the entrance fee for new members was raised from 25 to 30 guineas ([Cowell] 1975, pp. 23, 127).

[5] CD refers to Hooker's paper on the distribution of Arctic plants, read at a meeting of the Linnean Society on 21 June 1860 and published in 1861 (Hooker 1861). See L. Huxley ed. 1918, 2: 26–31.

[6] This is probably a reference to Hooker's work on the cryptogamic collection at the Royal Botanic Gardens, Kew.

[7] See letter to T. H. Huxley, 3 January [1861].

[8] CD had learned about the proposed changes in the editorial programme of the *Natural History Review* from Thomas Henry Huxley, the new general editor of the periodical. See *Correspondence* vol. 8, letter from T. H. Huxley, [before 14 December 1860].

[9] The third edition of *Origin* was published in April 1861 (Freeman 1977, p. 78).

[10] According to his 'Journal' (Appendix II), CD was at work on chapter 3 of *Variation*, which covered 'Pigs—Cattle—Sheep—Goats'.

[11] CD intended to carry on with experiments on insectivorous plants, particularly species of *Drosera*, that he had begun the previous summer (see *Correspondence* vol. 8). He was especially interested in the phenomenon of inflection of the leaf hairs in response to mechanical and chemical stimuli, and in the molecular process of 'aggregation' that occurred in the fluid within the gland cells of the tentacles after stimulation. His investigations were continued over fifteen years and were eventually published in *Insectivorous plants* (1875).

[12] CD summarised the results of his experiments to determine the effect of carbonate of ammonia on the leaves of *Primula sinensis* in *Insectivorous plants*, pp. 348–50. His notes on this investigation, carried on in January and February 1861, are in DAR 54: 12–20.

[13] Hooker had long been sceptical of CD's claim that ancient seeds could retain their vitality and germinate. See especially *Correspondence* vol. 5, letters to *Gardeners' Chronicle*, 13 November [1855], and to J. D. Hooker, 14 November [1855].

[14] Asa Gray's letter has not been found, but see the letter to Charles Lyell, 2 February [1861], in which an extract from it is quoted.

[15] J. Richardson 1861.

[16] Trollope 1861.

[17] Collins 1860.

[18] Olmsted 1860.

[19] William Henslow Hooker, the Hookers' eldest son, was 8 years old.

From William A. Wooler[1] 4 February 1861

Darlington
4th Feby 61

Dear Sir

I have just recd. yours of the 2d. Inst and think it best to reply to it at once—[2] I may say you are under no obligation as it affords me much pleasure when time permits to write on these subjects— I am sorry to inform you that the Rabbits I presented the Revd G P Wilkinson of Harperley Park[3] are all dead—so my scheme there is frustrated— It is quite right as a general rule respecting all the young of the Himalaya being white & those of the Silver Greys Black—yet 5 years ago an odd young one of the Himalaya occasionally came slightly Grey but not so dark as the fur I sent you[4] I have also found this colour is more prevalent in the autumn than any other part of the year but my observations have been to confined to warrant any thing like an assertion on this head— The pair of old Himalayas (bred from the Himalaya Doe & sandy Grey buck) produces more & darker ones than I ever bread before— I have from time to times sent short a/cs on this subject to the Gardeners Chronicle but altho I have a short paper ready have not sent it to complete these accounts some years ago being very sceptical about these Rabbits being anything more than a variety I asked several times Questions in the Gardeners Chronicle &c to illicit information— one party in reply stated he understood they were Hungarians[5]

My observations of the curly feathered fowls being always Hens referred to those whose feathers were like those I sent you resembling somewhat a curl—that is neither the pure Frizzled or the cross but a few which comes either from the Frizzled hen & the cock (which is the cross between the Frizzled & silk) or from this cross entirely The cross produced Frizzled silk & it was not till the following year we ever got one with what I call curls—but as the experiment was never satisfactorily carried out by keeping each Hens eggs separate I am unable to distinctly say if they were produced from a pure Frizzled Hen or a Frizzled silk Hen— The Cock I know was a Frizzled silk— we generally only got one with the curls at a time & not even one out of every Hatch last year a Frizzled Hen & some Frizzled silk Hens were confined with the Frizzled silk cock & I should say at least 30 chickens hatched out of which 2 come with these curls—

Mr George Ferguson published a work (dedicated to the Earl of Derby) in 1854 styled

Ferguson on Rare Prize
& Domestic Poultry[6]

and I see his observation states that these Hybrids are sterile— Yet in experiment no 2 with Hybrid Hen (produce of Pheasant cock & Black red Game Hen) with Domestic cock of the same blood as the Black red Game Hen the mother produced a few unproductive eggs—

The mode he puts these experiments are not very clear his "no 21" is "Two male inter-hybrids" & he calls an inter hybrid the issue from a Hybrid Hen & Pheasant cock as in no 4—&c &c and from one of these "Inter hybrid males & a Pheasant Hen like no 26 *one male chick was reared*—and "no 31 Female inter hybrid the produce of no 11" (that is "Three males & one female inter hybrids moderate supply of non productive eggs") with Pheasant cock the issue was "one chick but did not survive the 6th. day" This last is incomprehensible as he states the issue of no 11 were *non productive* eggs—[7]

You will no doubt be aware that Horticulturists & Botanists affirm that the Coliflowers, Brussel sprouts &c &c & the Cabbage all have one origin the common wild plant— The same with the Peach & nectarine which I believe the same tree some times & at the same times produces both fruit— I forget where I have the authority for this but I think it is in the Gardeners Chronicle within the last 2 years[8]

apology for this scrawl | I am Dear Sir | Yours very truly | W. A. Wooler

C Darwin Esq

DAR 181.2

CD ANNOTATIONS
0.1 Darlington . . . curls— 2.12] *crossed ink*
5.1 Horticulturists . . . Esq 7.1] *crossed ink*
Top of first page: 'On Hybrids from Fowls & Pheasants being occasionally fertile' *ink*; 'Ch IX'[9] *brown crayon*

[1] The *Post Office directory of the county of Durham* for 1873 lists a William A. Wooler residing at Sadberge Hall near Darlington.

[2] CD's letter to Wooler has not been found. CD was collecting information on the various breeds of domestic rabbits in preparation for writing chapter 4 of *Variation*. He was particularly interested in the origin of the Himalayan breed because it seemed to throw 'some light on the complex laws of inheritance' (*Variation* 1: 108).

[3] George Pearson Wilkinson was curate of Thornley, Wolsingham, Durham (*Crockford's*).

[4] In *Variation* 1: 109, CD stated that Wooler had sent him a sample of the fur of a young Himalayan rabbit that was pale grey in colour rather than the normal white.

[5] CD gave a description of the Himalayan rabbit and a detailed account of its origin in *Variation* 1: 108–11. He stated that the breed was produced by a cross between silver-grey and Chinchilla rabbits.

[6] Ferguson 1854. There is an annotated copy in the Darwin Library–CUL.

[7] George Ferguson reported the results of his experiments in breeding common domestic fowl and pheasants in Ferguson 1854, pp. 200–5. CD's notes on this work are bound into the back of his copy. Referring to the discussion on page 201, he wrote in ink and underlined in pencil: 'Hybrids with Pheasant— *Lies*.—'. See also letter to W. B. Tegetmeier, 25 February [1861].

[8] CD discussed the origin of the nectarine from the peach and cited various accounts of one tree producing both fruits in *Variation* 1: 336–43.

[9] The reference is to chapter 9 of CD's 'big book' on species (*Natural selection*), which discussed the subject of hybridism.

To John Lubbock [before 5 February 1861][1]

[Down]
Saturday night

My dear Lubbock

Your paper seems to me very clear & interesting.[2] The subject is too deep for me to offer any remarks. Except indeed to raise my voice in highest tone against your quoting difficult German passage. Surely you ought to translate them; otherwise they will be skipped by 99 out of 100 readers.—[3]

I hardly know what to say or think about your concluding speculations. I should grieve to see them struck out; & such speculations always call vivid attention to any facts. Yet they seem rather bold & to rest on scanty basis.— I fear that they might lead the old sober coaches to call you rash.—[4]

Yet I daresay if I had written them myself, I sh^{d} have been enamoured with them & would not have given them up for any body. The passage on last long page seems somehow rather less bold than the actually concluding sentences. On last *long* page do you not leave impression that difference in eggs *does* go with agamogenesis?[5] All I can say is that I am loth to see so fine a speculation cancelled *at present*, yet I tremble for you.—

Yours affect^{ly} | C. Darwin

Down House

[1] The date is based on the reference to Lubbock 1861b, which was received by the Royal Society of London on 5 February and read on 21 February 1861.

[2] The reference is to Lubbock's detailed, histological study of the formation of the egg cell in various groups of invertebrates (Lubbock 1861b). The work resulted from a continuation of Lubbock's research into the processes of asexual reproduction in various insect groups. In two previous publications (Lubbock 1857 and 1859), he had described the two different kinds of eggs (agamic and true ova) found in *Daphnia* and other genera that reproduce parthenogenetically. In Lubbock 1861b, he showed that insect species that reproduce sexually also exhibit two different types of egg development, and hence that the phenomenon was not unique to organisms exhibiting parthenogenesis (Lubbock 1861b, p. 624).

[3] In the printed paper, Lubbock translated into English all passages cited from the works of German authors.

[4] CD probably refers to Lubbock's speculation, in the concluding remarks of Lubbock 1861b, that the existence of two different kinds of spermatozoa in several invertebrate species might provide a key to the problem of sex determination. He posed the question 'whether the two sorts of spermatozoa produce embryos of different sexes.' (Lubbock 1861b, p. 625). The process of fecundation (fertilisation), in particular the extent and manner of interaction between male and female elements, was a subject of great debate at the time. See Farley 1982.

[5] In both Lubbock 1857 and 1859, Lubbock had stated that there was no essential difference between the agamic eggs and the true ova of various insects (see *Correspondence* vol. 6, letter to John Lubbock, 27 October [1856]; and vol. 7, letter to John Lubbock, [November 1858]). In Lubbock 1861b, p. 624, Lubbock accentuated the histological differences that he found between the two kinds of eggs, stating that 'throughout the Annulosa there are two sorts of eggs, which are of an essentially different structure, and cannot, therefore, strictly speaking, be regarded as homologous with one another.' For contemporary naturalists, Annulosa comprised all the segmented animals, namely Crustacea, Mryriapoda, Arachnida, Insecta, and Vermes. Today these animals are classified within the phyla Arthropoda and Annelida. In modern usage, Annulosa is sometimes mentioned as a less preferred alternative name for Annelida (Rothschild 1965).

To J. D. Hooker 8 [February 1861][1]

Down
8th.

My dear Hooker

You mentioned in a former letter about rubbing in cod-liver oil.[2] Etty cannot take it internally. I asked our Doctor here & he had never heard of such a thing.[3] Emma & I would be much obliged for a little more information. How did you hear of it?[4] Is evidence pretty good of benefit derived from rubbing in. As it would be a rather nasty proceeding for Etty to be thus basted, we should begin with more confidence, if we knew more. Will you take trouble to send us a line.—[5]

Yours affect— | C. Darwin

How long ought the oil to be rubbed in can you tell at all?

Endorsement: 'Feby/61'
DAR 115.2: 86

[1] Dated on the basis of the endorsement and by the relationship to the letter to J. D. Hooker, 20 [February 1861]. See also n. 5, below.

[2] Hooker's letter has not been found.

[3] Probably Edward Augustus Williams, surgeon in Bromley, who was the Darwins' family physician. Williams had been treating Henrietta Emma Darwin throughout her illness of the past year. Etty, whose condition had been improving, had suffered a severe vomiting attack late in January (letter to W. E. Darwin, [24 January 1861]). Late in the summer of 1860, following a bout of typhus fever, she had experienced a hardening of her abdominal tissues that the doctors believed would only gradually subside. See *Correspondence* vol. 8, letter to J. D. Hooker, 7 August [1860]. Cod-liver oil was believed to improve the appetite and help the patient put on weight (see n. 4, below).

[4] Although the application of cod-liver oil as a treatment for various maladies was common on the Continent, it was not generally recommended by English physicians until the 1840s, after John Hughes Bennett published a book drawing the attention of English medical men to the beneficial effects of the oil (Bennett 1841). It was later adopted as a remedy for the loss of appetite associated with consumption. Although normally taken internally, cod-liver oil could be applied externally if the patient was unable to digest it (*EB*).

[5] Hooker's reply has not been found, but see the letter to J. D. Hooker, 20 [February 1861]. There is an entry in CD's Account book (Down House MS) on 15 February 1861 that reads: 'Ball. Cod Liver Oil 10s.6d'. There is a James Ball, oilman, of Duke Street, Grosvenor Square West, listed in the *Post Office London directory* for 1861.

To *Gardeners' Chronicle* [before 9 February 1861][1]

[Down]

I am much obliged to Mr. Marshall, of Ely, for his statement that the 15 plants of Fly Orchis (Ophrys muscifera) which does not grow in his neighbourhood, but which flourished in his garden, had not one of their pollen masses removed.[2] The Orchis maculata, on the other hand, which likewise does not grow in the neighbourhood, had all its pollen masses removed. Mr. Marshall is not perhaps aware that different insects haunt different Orchids, and are necessary for their fertilisation.[3] From the wide difference in shape of the flower of Orchis and Ophrys, I should have anticipated that they would be visited and fertilised by different insects. In Listera, for instance, it is chiefly Ichneumonidæ, and sometimes flies, which by day perform the marriage ceremony. In the case of most Orchids it is nocturnal moths. Orchis pyramidalis, however, is visited by Zygæna, and I have examined one of these day-sphinxes with three pair of pollen-masses firmly attached to its proboscis. There can hardly be a doubt that the Butterfly Orchis is visited by different moths from most of the smaller Orchids; and I have recognised its peculiar pollen masses attached to the sides of the face of certain moths. It is probable that the same kind of moths would visit all the species of true Orchis, which closely resemble each other in structure. Thus the Orchis conopsea, planted in a garden some miles from where any native plant grew, had its pollen masses removed; so this is a parallel case with that of O. maculata given by Mr. Marshall. I have also transplanted the rare Malaxis to a place about two miles from its native bog, and it was immediately visited by some insect, and its pollen-masses were removed. On the other hand, the Epipactis latifolia, growing in my garden and flowering well, had not its pollen-masses removed; though in its own home, several miles distant, the flowers are regularly visited and thus fertilised. We thus see that the seeds of an Orchid might be carried by the wind to some distant place, and there germinate, but that the species would not be perpetuated unless the proper insects inhabited the site. I have now Goodyera repens growing in my garden, and I shall be curious to see next summer whether our southern insects discover or appreciate the nectar of this Highland Orchid.[4]

C. Darwin

Gardeners' Chronicle and Agricultural Gazette, 9 February 1861, p. 122

[1] The date is given by the publication of the letter in the *Gardeners' Chronicle and Agricultural Gazette*, 9 February 1861. It was headed: '*Fertilisation of British Orchids by Insect Agency*'.

[2] CD refers to a letter by William Marshall of Ely published in the *Gardeners' Chronicle* on 20 January 1861 under the heading 'Fertilisation of British orchids by insect agency'. Marshall was responding to CD's notice, printed in the issue of 9 June 1860, in which CD requested readers to observe the bee or fly orchis (*Ophrys apifera* and *O. muscifera*) to see whether the pollen-masses of the flowers had been removed by insects, 'for it seems a strange fact that a plant should grow pretty well, as it does in this part of Kent, and yet during several years seldom be fertilised.' (see *Correspondence* vol. 8, letter to *Gardeners' Chronicle*, [4–5 June 1860]; *Collected papers* 2: 34).

[3] CD identified various insects that pollinated different species of orchids in *Orchids*, pp. 36–8. He mentioned Marshall's case on page 38.

[4] See *Orchids*, pp. 112–16.

To Richard Kippist[1] 13 February [1861?][2]

Down, Kent
Feb. 13th

... "I will send tonight by our Carrier[3] all the Books in Hand in an old Portmanteau (*which please return*) so ... they will come safely.[4]

With sincere thanks for your vy. kind assistance. | C. Darwin"

"I suppose De Blainville's Osteographie does not include Rabbits or Hares, as I much wish to see such if it were ever published."[5]

Incomplete
James Lowe Autographs Ltd (dealers)

[1] Kippist was librarian of the Linnean Society of London, from which CD often borrowed works that he needed to consult for his research.

[2] The year is suggested by the 1859 watermark of the stationery and by CD's reference to the carrier's leaving 'tonight'. George Snow ran a carrier service between Down and London that left Down at

2 A.M. every Thursday morning (*Post Office directory of the six home counties* 1862). In 1861, 13 February fell on a Wednesday. This date is further confirmed by the reference to 'Rabbits and Hares'. In 1861, CD was preparing the chapter on variation in rabbits for *Variation* (see 'Journal'; Appendix II).

[3] George Snow. See n. 2, above.

[4] See letter to Richard Kippist, 7 January [1861]. It is not known which books from the Linnean Society CD was returning.

[5] Blainville 1839–64.

To Leonard Horner 14 February [1861][1]

Down Bromley Kent
Feb. 14th

My dear Mr. Horner

I must just thank you for your note,[2] but I will take advantage of your kind & considerate offer of discussing the points referred to till we meet.[3] The latter point seems to me very intricate & I have often thought it over.—

Man does not cause any variation he only accumulates any which occur: I do not suppose that God intentionally gave the parent Rock-Pigeon a tendency to vary in size of Crop, so that man by selecting such variations should make a Pouter; so under nature, I believe variations arise, as we must call them, in our ignorance accidentally or spontaneously, & these are naturally selected or preserved from being beneficial to the successive individual animals in their struggles for Life.— I know not whether I make myself clear.—[4]

Believe me | My dear Mr Horner | Yours very sincerely | C. Darwin

American Philosophical Society (B/D 25.362)

[1] The date of the letter is based on the 1859 watermark on the stationery, and on the year provided in K. M. Lyell ed. 1890, 2: 300.

[2] Horner's letter has not been found. CD had received a letter from him soon after the publication of *Origin* expressing his general approval of the arguments presented in the book. See *Correspondence* vol. 7, letter to Leonard Horner, 23 December [1859].

[3] CD was in London on 21 and 22 February 1861. On 21 February, he gave an account to the Philosophical Club of the Royal Society of the results to date of his experiments on *Drosera* (Bonney 1919, p. 154). Horner was the chairman of the meeting (see Appendix IV).

[4] Horner was preparing to discuss *Origin* in his presidential address to the anniversary meeting of the Geological Society of London on 15 February 1861 (Horner 1861). See letter to Leonard Horner, 20 March [1861].

To William Erasmus Darwin 17 February [1861][1]

[Down]
Febr 17th

My dear Wm

I send off cheque without the delay of a Post, which is just starting

Yours affecty | C. Darwin

Duke University

[1] Dated by an entry in CD's Account book (Down House MS) written on 15 February 1861: 'William addition to allowance £25'. William was in his third year at Cambridge University.

To Asa Gray 17 February [1861][1]

Down Bromley Kent
Feb 17th.—

My dear Gray

I received your note of Feb 5th this morning.[2] I am most sincerely sorry to hear of your accident to your thumb, which I cannot but fear will be a very serious loss.[3] I will send your note soon to Hooker that he may hear about it.— I am glad to hear of the 250 copies now I presume at Trübners;[4] I wrote to him today about a few advertisements &c &c & to send me copies for distribution. I hope for my sake, as well as yours, that Murray & Lyell will not prove entirely true, that it is impossible to circulate a pamphlet in England.—[5]

I fear that the state of the U. States must stop all interest in everything not political.—[6] I will in few days enquire at my Bankers how I can repay you the £7.s10: I truly wish you would let me pay the whole £15. Will you not?[7] I have asked Trübner to forward C. Wright's article to proper quarter.[8] The Printers have been very slow with my new Edit. of Origin so that I have been able to insert notice of your Pamphlet with title in full, which I am especially glad of.—[9] The other day a very clever Lady was staying here & read your Reviews; & remarked "how extraordinary if such a writer has written *only* on science" I need not say that she was not a scientific Lady.—[10]

With respect to Pumilio, you are quite right[11] I knew nothing whatever on subject, but I looked at the seeds to amuse myself, & asked Hooker who told me a little about Sencio & he thought Lindley would like an account for G. Chronicle.[12] It was foolish of Lindley to put it so conspicuous: in my note to him I said that he might like to use it sometime just to fill up space.— It strikes me, however, as pretty case of adaptation for the ignorant, like myself.—

When I last wrote, after receiving your letter of Dec. 15th, I was in much distress about my daughter: the attack though very severe lasted only 4 days & she has now got up to state as before.—[13]

I have just reread your letter, which like all yours, interests me. You ask about Drosera: if you like to try anything, put the minutest atom (under a lens) on point of fine needle on any one single extreme marginal gland of a leaf, which has all the hairs equally expanded & watch it or look again in 10 minutes.— Or put fragment of Hair of your head & look in a hour's time. I intend trying many more experiments this summer & then publishing: I am doubtful on many points.[14] But the worst is that my health is failing much. I literally cannot listen to a novel for $\frac{1}{2}$ hour without fatigue. My good dear wife declares, I must go with our whole family (if my girl can be moved) for two months to Water Cure; & I fear I must, but it will be ruin to all my experiments.— I remember formerly having read your extremely curious observations on Tendrils, but I thank you for telling me of them.—[15]

With respect to Design &c you say that you suppose that I have "not brought forward my real objections against your views.—" I have no real objection, nor any real foundation, nor any clear view.— As I before said I flounder hopelessly in the mud.—

You have amused me much by your account of Agassiz's denying the community of descent of allied languages, & of Bowen denying heredity.[16] I cannot believe that Bowen is a strong man. What an odd & foolish fancy he must think it that all breeders of Race-Horses, Cattle & pigs &c should keep pedigrees, & would certainly prefer breeding from a poor animal of a good pedigree than from the finest of bad pedigree.— These men in fact work on **my** (I wish I dared say **our**) side.

Farewell. Do not forget to have look at rostellum of your Spiranthes.—[17]

Farewell.— I most truly hope that your Thumb will not cause you much inconvenience.

Most cordially & gratefully yours | Charles Darwin

If you want a book for light reading, the 1st. vol of Leslie's Life (the American Painter) is very nice.—[18]

Gray Herbarium of Harvard University (54)

1 Dated by the reference to plans for the distribution of A. Gray 1861a (see n. 7, below).

2 Gray's letter has not been located.

3 Gray lost part of his left thumb in a gardening accident (Dupree 1959, pp. 307–8).

4 CD refers to their arrangement to reprint as a pamphlet Asa Gray's three articles on *Origin* from the *Atlantic Monthly* ([A. Gray] 1860a). Gray had arranged for the Boston publishers Ticknor and Fields to print the pamphlet and send the copies to the English publisher Nicholas Trübner for distribution. See letter to Asa Gray, 23 [January 1861].

5 See *Correspondence* vol. 8, letter to Charles Lyell, 23 [September 1860], letter from Charles Lyell, 25 September 1860, and letter to Asa Gray, 24 October [1860].

6 *The Times* was carrying lengthy accounts of the worsening political situation in the United States. Abraham Lincoln, the Republican candidate who had been elected president on a platform of excluding slavery in the territories, had not yet been sworn in. Following the lead of South Carolina, six other states had seceded from the Union and withdrawn their representatives from Congress. Northern states were building up their militia and preparing for an expected attack on one of their Southern forts.

7 There is an entry in CD's Account book (Down House MS) on 25 February 1861 for £7 10*s*. under the heading: 'Asa Gray for Pamphlet'. CD and Gray had agreed to share the expense of publishing A. Gray 1861a (see *Correspondence* vol. 8, letter to Asa Gray, 11 December [1860]).

8 See following letter.

9 The advertisement for the pamphlet was inserted as a 'Postscript' at the end of the list of additions and corrections in the third edition of *Origin*, p. xii. The new edition was published in April 1861. See Freeman 1977, pp. 78–9, and Peckham ed. 1959, p. 57.

10 The 'very clever Lady' may have been 'Miss Latter', whose visit to the Darwins from 2 to 4 February 1861 was recorded in Emma Darwin's diary. Payments to Miss Latter entered into CD's Account book (Down House MS) suggest that she had served as a governess for the Darwin children from April 1859 until February 1860, when Camilla Ludwig assumed the position.

11 See letter to *Gardeners' Chronicle*, [before 5 January 1861].

12 John Lindley was editor of the *Gardeners' Chronicle*. See *Correspondence* vol. 8, letter to J. D. Hooker,

26 December [1860], and letter from J. D. Hooker, 28 December 1860. The genus *Senecio* includes ragworts and groundsel.

13 See letter to Asa Gray, 23 [January 1861].

14 In the event, CD did not publish the results of his study of *Drosera* until 1875 (*Insectivorous plants*).

15 The reference is to A. Gray 1858.

16 See letter to Charles Lyell, 2 February [1861].

17 See *Correspondence* vol. 8, letter to Asa Gray, 31 October [1860], in which CD asked Gray to examine the American species *Spiranthes autumnalis*. See also letter to Asa Gray, 17 September [1861].

18 The reference is to the edited volume of autobiographical recollections of Charles Robert Leslie (Taylor ed. 1860).

To T. H. Huxley 17 February [1861][1]

Down Bromley Kent
Feb. 17th

My dear Huxley

Wrights Review on my Origin will be sent from Trubner in a day or two to Williams & Norgate.[2] Asa Gray writes, he thinks that "if Huxley accepts it he sh^d prefix a list of Books on these subjects, viz Books which have *philosophised* on the origin of species. And Huxley is at liberty to add supplement, strike out ad libitum."—[3]

God knows what it will turn out, but Gray seemed much struck with it, from which I infer it will be metaphysical or theological.

A. Gray has republished his Reviews as a pamplet;[4] I have directed a copy & advertisement to be sent to Nat. Hist Review. For Asa Gray's sake (& my own) could you insert notice *literally* only of two or three length??[5] His Reviews have struck others besides myself as very able.—

I heard from Lubbock of your capital Lecture & splendid wipe at Anonymous attacks in Edinburgh.—[6] He will grow tired of attacking you.— I sincerely hope poor Mrs Huxley is rising in strength & spirits,[7] & I hope you are not killing yourself.—

My daughter is much the same & I have been lately in poor way.

I forget whether I expressed my admiration at the opening passages of your Review on Brain:[8] they were grand.

Adios.— | Ever yours | C. Darwin

This requires no sort of answer.—

Imperial College of Science, Technology and Medicine Archives (Huxley 5: 169)

1 Dated by the reference to A. Gray 1861a.

2 Asa Gray had asked CD whether Huxley might be interested in publishing in the *Natural History Review* a review of *Origin* by his friend Chauncey Wright. See *Correspondence* vol. 8, letters to Asa Gray, 11 December [1860] and 14 December [1860], and letter from T. H. Huxley, [before 14 December 1860]. The publishing firm of Nicholas Trübner served as the agent in England for many American works and was distributing Asa Gray's pamphlet (A. Gray 1861a). Williams and Norgate published the *Natural History Review*.

[3] The letter from Asa Gray has not been found.

[4] A. Gray 1861a.

[5] Huxley mentioned the pamphlet in his anonymous review of the third volume of Louis Agassiz's *Contributions to the natural history of the United States of America*, which included Agassiz's criticism of *Origin*. See [T. H. Huxley] 1861c, p. 433.

[6] Huxley delivered a lecture entitled 'On the nature of the earliest stages of the development of animals' at a Friday evening meeting of the Royal Institution of Great Britain on 8 February 1861. The abstract of the lecture, published in the *Notices of the Proceedings at the Meetings of the Members of the Royal Institution of Great Britain* 3 (1858–62): 315–17, gives no indication of any kind of polemical 'wipe' that may have been part of the oral presentation. The allusion is to the on-going enmity between Huxley and Richard Owen that intensified following the publication of Owen's anonymous review of *Origin* in the *Edinburgh Review* ([R. Owen] 1860b). Owen had taken the opportunity to criticise Huxley's February 1860 lecture on CD's theory at the Royal Institution (see *Correspondence* vol. 8). See also L. Huxley ed. 1900, 1: 190–2.

[7] Henrietta Anne Huxley had given birth to Leonard Huxley on 11 December 1860. The Huxleys' first-born child, Noel, had died in September 1860 (L. Huxley ed. 1900, 1: 151, 216).

[8] Huxley opened his paper with the following passage (T. H. Huxley 1861a, p. 67):

> As the biological sciences have grown in breadth and in depth, and as successive generations of naturalists have succeeded in penetrating further and further into the arcana of nature, the questions—In what relation does the thinker and investigator stand to the objects of his inquiries? What is the tie which connects man with other animated and sentient beings?—have more and more forcibly pressed for a reply.

To J. D. Hooker 20 [February 1861]

Down Bromley Kent
20th.

My dear Hooker

You will see by enclosed what an unfortunate accident Asa Gray has had.[1] You will see, also, that his pamphlet has arrived. When I get copy you shall, of course, have one.— I shall insert a few advertisements: one & copy to Gardeners Ch. Do you ever see Lindley could you persuade him to notice it by 2 or 3 lines?[2] Is there any Botanical Journal in which you could do same? I shd. much like for Asa Gray's sake (& indeed for my own) if I could get some sold.—

Many & cordial thanks for your admirable & clear letter about rubbing in oil.[3] We have begun.— Our Doctor was so struck with your letter that he begged for copy.— Etty continues about the same, but rather better; Now it is 4½ weeks since a vomiting attack.— My health has lately been very bad, & I am become so weak that I cannot listen for above ½ hour even to a novel without great fatigue. But today I am extra well & if I keep so tomorrow, I shall come up for Phil. Club.[4] I wish there was a chance of you; but probably I shall fail. Emma declares she will take the whole family to Malvern for 2 months this Spring, & I very much fear I must go.[5] It will be a cruel blow to me, as I must give up Drosera, crossing Primulas & many other experiments for this summer. & I had so much to do. But I cannot go on as I now am.—

Please remember the Apocynium.[6]

By Circular received today I see that your side will be victorious at Athenæum.[7] Please return sometime A. Gray's letter.

Farewell | My dear Hooker | Yours ever | C. Darwin

I received cargo today of 19 moths with pollen-masses of Orchids attached to them![8] Did you observe my little notice on transplanted orchids not being fertilised in G. Chronicle?[9] Nothing came up in earth from under Tree. Hang them![10]

Endorsement: '/Feb/61'
DAR 115.2: 88

[1] Asa Gray's letter has not been found. Gray had lost part of his left thumb in a gardening accident (Dupree 1959, pp. 307–8).

[2] A full-page notice of A. Gray 1861a, with an introduction and extracts, appeared in the *Gardeners' Chronicle and Agricultural Gazette*, 9 March 1861, p. 219. John Lindley, a close friend of Hooker's, was the editor of the gardening section of the journal.

[3] Hooker's letter has not been found, but see the letter to J. D. Hooker, 8 [February 1861].

[4] According to Emma Darwin's diary, CD went to London on 21 February 1861 and returned to Down the following day. He attended a meeting of the Philosophical Club of the Royal Society of London, at which he gave an account of his experiments on *Drosera* (Bonney 1919, p. 154). See also Appendix IV.

[5] CD refers to James Manby Gully's hydropathic establishment at Malvern, Worcestershire. Although he had derived benefit from Gully's treatment, CD had not returned since the death there of his daughter Anne Elizabeth Darwin in 1851 (see *Correspondence* vol. 5).

[6] Through correspondence with Daniel Oliver, CD had established the identity of an insectivorous plant that he remembered growing in his father's garden. Hooker had offered to send him samples of the plant, *Apocynum androsaemifolium*, a species of dogbane. See *Correspondence* vol. 8, letter to J. D. Hooker, 17 December [1860] and n. 4.

[7] See letter to J. D. Hooker, 4 February [1861].

[8] The gift may have been sent by Frederick Bond, who had offered to send CD moths that had pollinia attached to them (see *Correspondence* vol. 8, letter from Frederick Bond, [16? June 1860]). CD thanked Bond for sending him a large number of such specimens in *Orchids*, p. 35.

[9] See letter to *Gardeners' Chronicle*, [before 9 February 1861].

[10] See letter to J. D. Hooker, 4 February [1861].

To T. H. Huxley 22 February [1861][1]

Down Bromley Kent
Feb. 22[d]

My dear Huxley

On my journey home, a scheme occurred to me, which I think is worth your thinking about.—[2] Do you not think a little change would be the best thing for M[rs]. Huxley, if she could be induced to try it?[3] I have been talking with my wife & she joins heartily in asking whether M[rs]. Huxley would not come here for a fortnight & bring all the children & nurse. But I must make it clear that this House is dreadfully dull & melancholy. My wife lives upstairs with my girl & she would see little of M[rs]. Huxley, except at meal times, & my stomach is so habitually bad that I never spend the whole evening even with our nearest relations. If M[rs] Huxley could be induced to come, she must look at this house, just as if it were a country inn, to which she went for a change of air. There is this one advantage, that she would have no trouble, such as she would have if she were to go to an Inn. Also we are near; so that she c^{d} start from the Victoria station for Beckenham, whither we could send our carriage. And Lastly you could probably come down & spend the Sunday here. Now do think of this scheme. I know well that this house in its present state presents heavy disadvantages; but perhaps a little visit here might do Mrs Huxley good & give her strength.—[4]

My dear Huxley | Yours very sincerely | Ch. Darwin

P.S.[5] I have heard from Trübner that the Box from U. States did *not* contain any M.S. Review from Wright of Boston.[6] It seems to have been forgotten by some accident.

Imperial College of Science, Technology and Medicine Archives (Huxley 5: 157); DAR 145

1 Dated by the invitation for Henrietta Anne Huxley to visit Down House (see n. 4, below).

2 CD travelled to London on 21 February 1861, returning to Down on 22 February (Emma Darwin's diary). He attended a meeting of the Philosophical Club of the Royal Society on 21 February 1861 (Bonney 1919, p. 154).

3 Henrietta Huxley was recuperating from the birth of a child in December 1860 and suffering 'inconsolable grief' over the death in September 1860 of the Huxleys' first-born son, Noel (L. Huxley ed. 1900, 1: 216). Emma Darwin had experienced a similar trauma in 1851 (see *Correspondence* vol. 5).

4 Henrietta Huxley and the children visited the Darwins at Down House from 9 to 25 March 1861 (Emma Darwin's diary). In a letter sent to Henrietta Emma Litchfield and published in the biography of her mother, Henrietta Huxley recalled this visit (*Emma Darwin* 2: 186):

> in the early days of our acquaintance, just after we had lost our boy, she begged me to come to her and bring the three children and nurse, and I should have the old nurseries at Down. I first wrote that I was too ill to be out of my home, that I could not get downstairs till 1 o'clock. Her reply was, that that was the usual state of the family at Down, and I should just be following suit. What wonder that I had for her always the most grateful affection.

5 The postscript is no longer with the original letter but is included in the copy in DAR 145.

6 See letter to T. H. Huxley, 17 February [1861].

To Andrew Murray 23 February [1861][1]

Down Bromley Kent
Feb. 23^{d}

My dear Sir

I send by this Post, a pamphlet by Prof. Asa Gray on my "Origin," which several good judge think very well written.[2] Although the author praises me he does not fully concur, & I have thought that you might possibly like to look at it. Could you get it noticed, in 2 or 3 lines, in any Nat. Hist. periodical in Edinburgh?[3] I sh^{d} be very glad to have it a little known for Prof. A. Gray's sake, as well as my own.— I hope this request is not very indelicate; I am far from meaning that I want to see it praised, only just noticed. It is quite likely that you have not the power, even if you were willing to do so.—

I have just finished a corrected Edit. of the Origin, with some additions, I have told my publisher, when it is bound, to send you a copy.— I mention this because, if you looked at the altered parts, you will not find much alteration in the parts; which you attacked.[4] To the best of my judgment (however erroneous that may be) these parts did not seem to me to require correction. And if you were to notice these parts, you might think that I sent a copy to you out of Bravado, which I assure you is not the case, & would be vile taste on my part to you, who have attacked me with the candour & chivalry of a Knight of old.—

My "corrections" remind me that I heard too late from a M^r^ — Mr Crotch who has lately been in Berlin, that he saw the Anophthalmus Raymondi which had been taken in the cellars of a very ancient deserted monastery, near Toulon. I wish I knew whether the account has been published; it is very curious if true.[5]

Pray believe me | My dear Sir | Yours sincerely | C. Darwin

Nothing in this note need give you the trouble of answering it.—

R. D. Pyrah

[1] Dated by the reference to A. Gray 1861a.

[2] A. Gray 1861a.

[3] Murray, who practised law in an Edinburgh firm until 1859, was a member of the Royal Society of Edinburgh and formerly the president of the Botanical Society of Edinburgh. He was the author of a number of papers, many of which treated entomological topics. See letter from Andrew Murray, 3 March 1861.

[4] Murray had published a critical review of *Origin* (Murray 1860), founding several of his objections on entomological examples. He particularly questioned CD's explanation of the origin of blind cave insects. See *Correspondence* vol. 8, letters to Andrew Murray, 28 April [1860] and 28 [April 1860].

[5] See letter from W. D. Crotch, 25 January 1861.

To John Tyndall 23 February [1861][1]

Down Bromley Kent
Feb. 23^d^

Dear Tyndall

I send the Letters.[2] No 1 tells most.— It is only an *inference* that Mr Wedgwood saw congelation,[3] but it seems to me pretty clear that he did. Letters no 2. & 3 are hardly worth your looking at.— Letter 4 contains reference to protrusion of glaciers by the swelling of frozen unfiltrated water.— I doubt whether the letters are worth your looking at, but it has amused me to see the speculations of philosophers (my two Grandfathers) nearly 80 years ago on a subject which they so little understood. It seems that Dr D. communicated the fact to Dr Black by M^r^. Robert D. (who was my Father) & who lodged with Dr Black in Edinburgh.—[4]

Dear Tyndall | Yours very sincerely | Ch. Darwin

PS. | I send by this Post, a pamphlet which contains the best account, pleasantly written, of my Origin, if you chance to care to read it.[5]

Down House (MS 8: 3)

[1] The date is given by the reference to A. Gray 1861a.

[2] CD sent Tyndall four letters from his paternal grandfather Erasmus Darwin to his maternal grandfather Josiah Wedgwood I. The letters are in DAR 227, Boxes 1 and 6 (see Appendix V). CD may have mentioned these letters to Tyndall at the 21 February meeting of the Philosophical Club of the Royal Society, which they both attended (see Appendix IV).

[3] CD refers to Tyndall's interest in the phenomena of glaciation. Tyndall had discussed the results of his and Thomas Henry Huxley's study of glacier motion in a lecture at the Royal Society of London in January 1857 that CD attended. CD had been particularly interested by Tyndall's explanation of 'regelation', or the process by which moist fragments of ice become re-solidified. See *Correspondence* vol. 6, letters to T. H. Huxley, 17 January [1857] and 3 February [1857], and to John Tyndall, 4 February [1857]. Tyndall had recently published a book describing his observations and outlining his theoretical views on the subject (Tyndall 1860).

[4] CD's father, Robert Waring Darwin, had attended lectures at the University of Edinburgh before going to Leiden to study medicine. During his stay in Edinburgh, he apparently lodged with the professor of chemistry, Joseph Black, noted for his researches on the heat exchanged in the process of converting between the liquid and solid states of water.

[5] A. Gray 1861a. Tyndall's name appears on the presentation list CD drew up for the distribution of the pamphlet. See Appendix III.

To John Murray 24 February [1861]

Down Bromley Kent
Feb. 24.

My dear Sir

At last I am able to send last corrected sheet.—[1]

Please look at back of Title-page: if you object to my inserting my Geological works, strike them out;[2] I care little. Please look also, at Postscript p. XII; I most particularly hope you will not object to this insertion.[3]

Please forward the sheet without loss of time to M^{essrs} Clowes.[4] Please do not forget in any Advertisement to insert "*with Additions & corrections*". Is Third Edition with additions ever printed on Back of Cloth??

Will you oblige me by kindly distributing copies by enclosed list, which please do not lose.—[5]

I hope you will be generous & present me some copies. Has my Journal sold at all satisfactorily?[6]

My dear Sir | Yours very sincerely | C. Darwin

Endorsement: '/61'
John Murray (publishers)

[1] CD was correcting the proofs for a new edition of *Origin*. The third edition was published in April 1861.

[2] A listing of CD's previous publications was included on p. iv of the third edition of *Origin*. The list included *Journal of researches*, *Coral reefs*, *Volcanic islands*, *South America*, and *Living Cirripedia*.

[3] An advertisement for A. Gray 1861a was inserted on p. xii.

[4] The firm of William Clowes & Sons printed the first and subsequent editions of *Origin*.

[5] CD's list of presentation copies for the third edition of *Origin* is in DAR 210.18. It is transcribed in Appendix VII.

[6] Murray had brought out a new printing of CD's *Journal of researches* in 1860. It was advertised on the title page as being the 'Tenth thousand'. See *Correspondence* vol. 8.

To James Lamont 25 February [1861][1]

Down. | Bromley. Kent.
Feb. 25.

Dear Sir

I am extremely much obliged for your very kind present of your beautiful work "Seasons with the Sea-Horses";—and I have no doubt that I shall find much interesting from so careful and acute an observer as yourself.[2]

Dear Sir. | Yours sincerely obliged— | Charles Darwin.

P.S. I have just been cutting the leaves of your book & have been very much pleased & surprised at your note about what you wrote in Spitzbergen.— As you thought it out independently, it is no wonder that you so clearly understand Natural Selection, which so few of my Reviewers do or pretend not to do.[3]

I never expected to see anyone so heroically bold as to defend my bear illustration—[4] But a man who has done all that you have done must be bold!— It is laughable how often I have been attacked and misrepresented about this Bear— I am much pleased with your remarks and thank you cordially, for coming to the rescue.—

Copy
DAR 146

[1] The year is provided by the reference to Lamont 1861, which was published between 14 and 28 February 1861 (*Publisher's Circular* 1861, p. 102).

[2] CD's presentation copy of Lamont 1861 is in the Darwin Library–Down. For the earlier correspondence between CD and Lamont, see *Correspondence* vol. 8, letter from James Lamont, [23 February 1860], and letter to James Lamont, 5 March [1860].

[3] In describing the natural history of Spitsbergen, Lamont stated his opinion 'that an attentive study of the Arctic animals is capable of mightily strengthening the theory of ***progressive development***, first suggested by the illustrious Lamarck, and since so ably expounded and defended, under somewhat modified forms, by the author of the "Vestiges of Creation," and by Mr. Charles Darwin.' (Lamont 1861, pp. 271–2). Lamont illustrated this assertion by citing his view that the brown bear was the progenitor of the modern polar bear, speculating that brown bears had been transported by ice to Greenland and Spitsbergen. Assuming that those of the palest colour and with the greatest amount of external fat would have had the best chance of survival, he stated that the 'process of natural selection would do the rest, and *Ursus arctos* would, in the course of a few thousands, or a few millions of years, be transformed into the variety at present known as *Ursus maritimus*.' (Lamont 1861, p. 274). In a note, Lamont said that although these remarks were written in Spitsbergen before *Origin* was published, he 'did not claim any originality' for his views (*ibid.*, p. 275 n.).

[4] Citing CD's reference in *Origin*, p. 184, to the black bear seen swimming with its mouth open and catching insects in the water, Lamont stated that he saw no difficulty in such a bear developing through natural selection, as CD had said, into a creature more aquatic in structure and habits until it became as 'monstrous as a whale' (Lamont 1861, pp. 277–82). CD had altered the passage for the second edition of *Origin* and removed it from subsequent editions (Peckham ed. 1959).

To W. B. Tegetmeier 25 February [1861][1]

Down Bromley Kent
Feb. 25th

My dear Sir

Some time since you were so very kind as to say that you would lend me your collection of Fowl Skulls.[2] Can you now spare them for a few weeks? If you can, will you send them by enclosed address; as I hope to begin on Fowls in about a week's time. But I am *far* from decided what I shall do, & whether I shall enter in much detail or not. I have a mass of old notes; & till I read them over I do not know what I shall do; & therefore I do not know how far I shd. describe the Skulls.—[3] I hope to Heaven your Skulls (if you can spare them) are numbered or named on the Bone; for in comparing many it is most difficult to avoid losing loose tickets.— I have a good many of my own.—

Can you tell me what sort of man Ferguson the author of a Poultry Book is?[4] Has he had much experience? Is he honest? He gives wonderful table (which stretches my belief to cracking point) of Hybrids from Pheasant & Fowl.[5] Could he really have reared such a number of Hybrids?— Of course any man may blunder: but does he mean to speak the truth?[6]

By the way he asserts that the Hen of the Malay has only 10 tail-feathers.[7] Could you ascertain this for me, & allow me (if really true) to quote you in corroboration: perhaps he counted the feathers in some *one* abnormal specimen.—[8]

I fear I shall cause you a great deal of trouble in packing up the skulls, i.e. if you can spare them.

My dear Sir | Yours sincerely | C. Darwin

New York Botanical Garden Library (Charles Finney Cox collection)

[1] The year is provided by the reference to CD's writing his manuscript on fowls for *Variation*. According to his 'Journal' (Appendix II), he 'Finished Fowls' on 16 May 1861, having spent eight weeks on the subject.

[2] See *Correspondence* vol. 8, letters to W. B. Tegetmeier, 17 April [1860] and 20 July [1860].

[3] CD discussed the osteological differences of various breeds of fowls in *Variation* 1: 260–70.

[4] CD refers to George Ferguson, the author of *Ferguson's illustrated series of rare and prize poultry* (Ferguson 1854). The work had been recommended to CD by William A. Wooler (see letter from W. A. Wooler, 4 February 1861). There is an annotated copy of Ferguson 1854 in the Darwin Library–CUL.

[5] CD had long been interested in collecting authenticated instances of successful crosses between pheasants and common fowl (see *Natural selection*, pp. 434–5). Tegetmeier had previously put CD in touch with a breeder who had made such crosses. See *Correspondence* vol. 6, letter to W. B. Tegetmeier, 31 May [1856], and letters from Edward Hewitt, 18 December 1857 and 22 December 1857.

[6] A note written on the inside back cover of CD's copy of Ferguson 1854 gives an indication of Tegetmeier's response. It reads:

> I must be careful about trusting this man.— Mr Tegetmeier says not known as a Fancier. Mr Brent does not know, but says he offered to sell coops & Aviaries—so must at least have kept Birds.— Tegetmeier has convinced me whole Book a pack of lies & compilation

Bernard Philip Brent had frequently provided CD with information on the hybridisation of fowls and other domesticated birds. In *Variation* 1: 248 n. 36, CD stated, following his citation of Ferguson's book: 'This writer, I am informed, cannot generally be trusted. He gives, however, figures and much information on eggs.'

7 Ferguson 1854, p. 186.

8 Tegetmeier apparently did investigate this point further. See letter to W. B. Tegetmeier, 28 March [1861]. In *Variation* 1: 258–9, CD stated that the Malay hen had sixteen tail feathers.

To Asa Gray 26–7 February [1861][1]

Down Bromley Kent
Feb. 26th

My dear Gray,

My Bankers told me that this was the only plan by which they could transmit the £7.^{s}10.^{d}0: I hope it will not be troublesome to you.[2] M^{ess} Trübner have been most liberal & kind, & say that they shall make no charge for all their trouble. I have settled about a few advertisements; & they will gratuitously insert one in their own periodical.[3] If you have any opportunity, would it not be well to write & thank them? They say that 1^{s} would be sufficient price, but as they nevertheless recommended 1^{s}·6^{d}, I have, of course agreed. I have distributed 100 copies to men of science & sent some to Reviewers, Libraries &c. I am heartily rejoiced at this. I believe your pamphlet will do me & Natural Selection, right good service. I think the Title very good,[4] but I wish the type had been rather larger, yet this would have cost more. I have not reread it yet, but I see long new note on Heredity.[5] The share of freit of Box was only 12^{s}, & the advertisements will be a few pounds, for which I have told Trübner I am responsible; but I hope sale will cover this. Yet all tell me that pamphlets *will not* sell. By the way the 7·10·0 just covers cost of my 100 copies at 1^{s}·6^{d} each. If the pamphlet sh^{d} sell, I will inform you for chance of your being able to send over more. I wish heartily it would sell for my own sake & that some profit might be sent to you.—

Trübner informs me that in Box there was no M.S. bundle for me, (ie S. Wright's Review). Tr. has written to Liverpool to enquire whether it could have been lost when Box was opened at Custom House.— I hope it is not lost.— Perhaps you addressed it to Williams & Norgate or to Prof. Huxley.—[6]

My dear Gray | Yours most sincerely | Chas. Darwin

Feb. 27th.—

Gray Herbarium of Harvard University (57a)

1 Dated on the basis of an entry in CD's Account book (Down House MS) on 25 February 1861 recording the sum mentioned in the letter.

2 The payment was CD's share of the cost of printing 250 copies of Gray's articles on *Origin* (A. Gray 1861a). See letter to Asa Gray, 17 February [1861].

3 See letter to T. H. Huxley, 17 February [1861].

4 The title of Gray's pamphlet was 'Natural selection not inconsistent with natural theology. A free examination of Darwin's treatise "On the origin of species", and of its American reviewers'. CD's lightly annotated copy is in the Darwin Pamphlet Collection–CUL.

[5] A. Gray 1861a, p. 52 n. The footnote was a response to Francis Bowen's assertion, in Bowen 1861, that acquired habits were not heritable. Gray offered as support for CD's opposing views a long quotation from his own review of *Origin* in the *American Journal of Science and Arts* (A. Gray 1860b, pp. 165–8).

[6] CD refers to the manuscript of Chauncey Wright's review of *Origin*. Gray was sending it to CD to be forwarded to Thomas Henry Huxley for possible publication in the *Natural History Review*, published by Williams and Norgate. See letter to T. H. Huxley, 17 February [1861].

To Daniel Oliver 26 February [1861][1]

Down Bromley Kent
Feb. 26th

Dear Oliver

I am much obliged for your curious paper on the Hamamelideæ.[2] The points which you indicate of the complex affinities, wide distribution, small number of species seem all clearly to point to a ruined & bankrupt Family.[3] And how curious the structure of the wood![4] But you well indicate how difficult it is in such cases, especially when one does not know the use or function of a structure, to tell whether any point is due to inheritance or to subsequent acquirement, in which case the point would only be one of analogous resemblance. It has pleased me extremely to see you allude to such questions in so profound a paper; it is the best of all answers to those who pooh-pooh the whole subject.

I long to be at Drosera again,[5] but Heaven knows whether my health will last out, & whether I shall not have to retreat for two months to Malvern for Hydropathy.

But, if in your power, I shd. be extremely much obliged if you could procure for me a few plants of Dionæa & keep them at Kew for me & let me know what they cost that I may repay you for them.[6] I mean to attempt only a few comparative experiments with Drosera.—

There is another plant which I want much: would it give any extra trouble to order it at same time; if it would, I am not so silly as to expect you to do it for me. But I have thought it might be ordered & come with Dionæa & so give *no* more trouble. It is Cypripedium spectabile from N. America.— A neighbour says it seeded with him; from which I infer that probably some insect visited it; so that I am curious to make observations on it, having observed so many other Orchid Genera.

If you can easily help me, I believe you will, & pray believe me | Yours sincerely | Charles Darwin

Down House (MS 10: 39)

[1] Dated by the reference to Oliver's paper on *Sycopsis*, a new genus of Hamamelidaceae, a family of trees and shrubs (Oliver 1860). Oliver read it at a meeting of the Linnean Society of London on 15 March 1860.

[2] Oliver 1860.

[3] For CD's reasons for believing that such patterns indicate that a group is nearing extinction, see *Origin*, pp. 109–11. In his paper, Oliver mentioned CD's views on distribution, along with those of Alfred

Russel Wallace and Joseph Dalton Hooker, and suggested that such cases as the Hamamelidaceae might 'assist us towards a solution of some of the most important problems engaging the attention of botanists.' (Oliver 1860, p. 87).

[4] Oliver, after describing the histological structure of *Sycopsis*, pointed out that although the wood most nearly resembled that of the Conifera, the two groups shared almost no other characters. Referring to Hooker's view of the 'highness' of the specialised tissue of conifers and other gymnosperms, Oliver suggested that 'the histological character of the elementary tissue of the vegetative organs might afford a contrasting constancy' in comparison with the great differences in external characteristics, and hence be important for establishing affinity (Oliver 1860, p. 88).

[5] Oliver assisted CD with his studies of *Drosera* and other insectivorous plant species in the summer of 1860 (see *Correspondence* vol. 8).

[6] Oliver was able to obtain the plant specimens CD asked for. See letters to Daniel Oliver, 1 April [1861] and 1 May [1861].

To W. B. Tegetmeier 1 March [1861][1]

Down Bromley Kent
March 1

My dear Sir

Many thanks for your kindness about the skulls & for answering me about Ferguson.—[2] The account of the Hybrids is in his Poultry book & very obscurely written. I am always loth to believe that a man lies, yet I can hardly believe that he made so many crosses. No doubt hybrids which are generally sterile will sometimes breed, yet his account is very incredible. I began to suspect the whole Book to be a pure fiction, but having to write to Mr Brent about Rabbits, I asked him;[3] & he says he knows nothing of him, except that he offered a lot of coops & aviaries for sale, which shows at least that probably he did keep Birds.

I shall be curious to hear about tails of Malay Hens; & if 10 be true number, it would be well to look at Cock.— Will you ask another question for me, as I daresay you will see at Preston Rabbit fanciers;[4] viz whether E. S. Delamere (ie Dixon)[5] is correct in saying that in *Half*-lops, the ear which hangs down is longer & broader than that which sticks up?[6]

I find my rabbit skeletons, which I am now at work at so tough a job, that I shall not begin Fowls quite so soon as I expected.

I may mention a trifling little fact which has surprised me: I threw away somewhere about the premises 2 or 3 years ago some wax coloured with vermilion, & I find that a swarm which built its combs last summer must have somehow discovered a lump, for the combs are extensively & prettily coloured with vermilion.—[7]

With sincere thanks | Yours truly | C. Darwin

New York Botanical Garden Library (Charles Finney Cox collection)

[1] The year is provided by the reference to CD's work on rabbits and fowls for *Variation*, which he carried out early in 1861. See 'Journal' (Appendix II).

[2] The reference is to George Ferguson, author of a manual on poultry (Ferguson 1854). See letter to W. B. Tegetmeier, 25 February [1861].

[3] The letter to Bernard Philip Brent has not been found. See letter to W. B. Tegetmeier, 25 February [1861], n. 6.

[4] Tegetmeier was to be one of the judges of fowls and pigeons at the Preston poultry show, to be held on 6 and 7 March 1861 (*Cottage Gardener* 25 (1861): 357).

[5] Delamer 1854, p. 136. An annotated copy of this work is in the Darwin Library–CUL. Eugene Sebastian Delamer was the pen name of Edmund Saul Dixon.

[6] Tegetmeier apparently did verify this description. CD included the statement in *Variation*, citing Delamer 1854 as one of his sources and reproducing an illustration of the half-lop rabbit taken from this work (*Variation* 1: 107–8).

[7] When CD was studying the formation of bees' cells in 1858, Tegetmeier had recommended providing the bees with coloured wax to investigate their manner of cell construction. See *Correspondence* vol. 7, letter to W. B. Tegetmeier, 8 [June 1858]. CD acknowledged Tegetmeier's assistance in *Origin*, pp. 228–30. See also letter to W. B. Tegetmeier, 22 March [1861].

To George Rolleston 2 March [1861][1]

Down Bromley Kent
March 2d.

My dear Sir

I must have the pleasure of thanking you for your note.[2] I *quite* agree with your remarks on Prof. A. Gray's pamphlet:[3] it is in my opinion the ablest thing which has appeared.— It is very kind of you to send me the pamphlet, which I will shortly read.—[4] In a new Edit of Origin, which will immediately appear, I have enlarged a *little* on the relation of the most diverse types & on a few other subjects.[5]

Mr Paget of London told me of the singular fact of the discolouration of the skin in relation to the state of the supra-renal capsules. Also of the singular state of the teeth in inherited secondary Syphilis.—[6] But your fact of "clubbed finger nails in Cyanosis" is quite new to me. If you would not think me quite unreasonable, I shd. be very much obliged for a little more information. What is cyanosis? Is it due to disease of heart? Which way are nails modified. Does the modification appear at birth? Is it an inherited disease?[7]

Pray excuse me troubling you & believe me Dear Sir with many thanks | Yours sincerely | ⟨Charles Darwin⟩[8]

Wellcome Institute for the History of Medicine Library, Western MS 6119: 5

[1] The year is provided by the reference to A. Gray 1861a.

[2] Rolleston's letter has not been found.

[3] A. Gray 1861a.

[4] Rolleston had probably sent CD his essay 'On the affinities of the brain of the orang utang' (Rolleston 1861a), which appeared in the April issue of the *Natural History Review*.

[5] The third edition of *Origin* was published in April 1861.

[6] No letters from James Paget on these topics have been found, but CD thanked him for information on the 'case of teeth affected by syphilitic parents' in the letter to James Paget, [15 October – 19 November 1859] (*Correspondence* vol. 7). Paget was a surgeon and lecturer on physiology at St Bartholomew's Hospital, London. He was a member of the Philosophical Club of the Royal Society and frequently gave accounts of interesting medical cases at club meetings.

[7] CD included both Paget's case and that of Rolleston in his discussion of 'correlated variability' in *Variation* 2: 331–2. See also letter to George Rolleston, 7 March [1861]. Rolleston mentioned the correlation in Rolleston 1861b, pp. 487–8.

[8] The signature has been cut away.

From Andrew Murray 3 March 1861

Royal Horticultural Society, | South Kensington, W.
3 March *18*61

My dear Sir

Your letter of the 23[d] has followed me from Edinb. to London,[1] where I have come to reside—having been appointed Acting Secretary to this Society—D[r]. Lindley assuming the Honorary duties[2]

I shall have great pleasure in noticing the paper by D[r]. Gray & procuring my remarks to be inserted in the Edinb. Phil: Journal if my friend Balfour shall not have previously done it himself—[3] So far from demurring to it I feel myself highly honoured that you should ask me—and am greatly obliged by your promising me a copy of your New Edition—

I had not heard of the Anophthalmus Reymondi being found in the cellars of a deserted Monastery near Toulon—[4] You know its natural habitat is a grotto near Toulon—& it is only what one might expect that on a dark night it might wander out— Indeed that is one of the things that always puzzled me why these blind insects were never found out of the Caverns—

The *Anillus Cœcus* is perhaps even a more interesting case—first found under some large stones which lay at the bottom of a great heap of manure or decaying thatch w[h]. had lain untouched for years near Bourdeaux.— Unfortunately for the marvellous of the story it has since been found in other dark places—altho very rarely—

Yours sincerely | And[w]. Murray

Is the copy of D[r]. Grays paper for myself—or to be returned—[5]

DAR 47: 154–5

CD ANNOTATION

1.1 Your ... Edition— 2.5] *crossed ink*

[1] Letter to Andrew Murray, 23 February [1861].

[2] John Lindley had been an officer of the Royal Horticultural Society since 1822. He served as assistant secretary from 1827 to 1858, when he was appointed honorary secretary and member of council. He gave up these offices to take charge of the colonial department of the International Exhibition of 1862. (*DNB*.)

[3] John Hutton Balfour, professor of botany at Edinburgh University, was one of the editors of the *Edinburgh New Philosophical Journal*. CD had already sent him a copy of A. Gray 1861a (see letter to Asa Gray, 12 March [1861] and Appendix III).

[4] See letter to Andrew Murray, 23 February [1861], and letter from W. D. Crotch, 25 January 1861.

[5] See following letter.

To Andrew Murray 3 March [1861][1]

Down Bromley Kent
March 3[d]

My dear Sir

One single line to thank you for your note & information, & to say that of course I meant the pamphlet for you, if you think it worth accepting.

In Haste | My dear Sir | Yours very sincerely | C. Darwin

R. D. Pyrah

[1] Dated on the basis of the relationship to the preceding letter.

To John Murray 3 March [1861][1]

Down Bromley Kent
March 3[d].

My dear Sir

Many thanks for your satisfactory answers to all my queries.—[2] I hope to give no more trouble for a long spell.— I am making some progress, though slow with my Volume "on Variation under Domestication". But my object is to ask you to alter the address of Andrew Murray (in the list of my presentation copies) to "Royal Hort. Soc. S. Kensington".[3]

My dear Sir | Yours sincerely | C. Darwin

John Murray (publishers)

[1] Dated by the relationship to the letter from Andrew Murray, 3 March 1861.
[2] See letter to John Murray, 24 February [1861]. Murray's response has not been found.
[3] See letter from Andrew Murray, 3 March 1861. CD's list of presentation copies for the third edition of *Origin* is in DAR 210.18. For a transcription, see Appendix VII.

To Joseph Leidy 4 March [1861][1]

Down Bromley Kent
March 4th

Dear Sir

I received a few days ago your note of Decr 10th & your most generous present of a whole bundle of your publications, which I value most highly & am extremely glad to possess.—[2]

Your note has pleased me more than you would readily believe; for I have during a long time heard all good judges speak of your palæontological labours in terms of the highest respect.[3] Most palæontologists (with some few good exceptions) entirely despise my work; consequently approbation from you has gratified me much.— All the older geologists (with the one exception of Lyell, whom I look at as a host in himself) are even more vehement against the modification of species than are even the palæontologists. I have, however, been equally surprised & pleased at finding that several of the younger geologists who are now doing such good work in our geological survey, go with me & are as strong, as I can be, on the imperfection of the geological record.—[4]

Your sentence that you have some interesting facts "in support of the doctrine of selection, which I shall report at a favourable opportunity", has delighted me even more than the rest of your note.[5] I feel convinced that, though as long as I have strength I shall go on working on this subject, yet that the sole way of getting my views partially accepted will be by sound workers showing that they partially accept them. I say partially, for I have never for a moment doubted, that though I cannot see my errors, that much in my Book will be proved erroneous.—

Pray forgive this egotistical note & with cordial thanks for your letter & kind present, believe me Dear Sir, with sincere respect, Your obliged | Charles Darwin

Academy of Natural Sciences of Philadelphia

[1] Dated by the relationship to the letter to Asa Gray, 12 March [1861], in which CD mentioned having received Leidy's letter.

[2] Leidy's letter has not been found. The Darwin Pamphlet Collection–CUL contains seven articles by Leidy, all published before 1861. The pamphlets, which are lightly annotated, discuss various palaeontological specimens discovered in the United States.

[3] CD had become familiar with some of Leidy's zoological research when he was preparing his monograph on cirripedes. See *Correspondence* vol. 4, letters to Louis Agassiz, 22 October 1848, and to J. D. Dana, 8 October 1849.

[4] For a list of geologists CD believed to be supportive of his views, see *Correspondence* vol. 8, letter to J. D. Hooker, 3 March [1860]. See also letters to T. W. St C. Davidson, 30 April 1861, and to George Maw, 19 July [1861].

[5] Leidy was possibly referring to the discovery of a remarkable series of fossil horses and their near relatives in the Tertiary deposits of the Bad Lands of Nebraska. Although Leidy's descriptions of five new genera of the solid-hoofed ungulates and other vertebrate fauna were already prepared for press, publication was delayed until 1869 (see Leidy 1869, p. 24). In this work, however, he was only able to allude to how these forms related to 'the theory of the origin of species through successional development' (*ibid.*, p. 355; see also Ruschenberger 1892, p. 161, and A. Desmond 1982, p. 235 n. 2).

From Gideon Lincecum 4 March 1861

Long Point. Texas.
4th. March. 1861.

Charles Darwin. M. A.
Down Bromley Kent. England.
Dear Sir.

Your kind letter of 27th January, was just one month on the way.[1] I would not pester you again, but for the question contained in it.

You speak enviously of my "long career in wild countries". I might do the same, in regard to your oppertunities in the *tame* country of books and seminaries, pompous priests and legal superstition. But I don't, except your trip around the globe. If I can not have company whose minds are clearly free, I would prefer to go alone. And thus it has turned out with me through my long sojourn. I have had no associates, and my observations and conclusions, be them right or wrong, are not trammelled by the sway of other minds. Except five month's schooling, at a deserted log cabbin in the backwoods of Georgia, by an old drunkard, my mind has not been biased by training of any kind from designing man. In the canebrakes and unhacked forests on the borders of the above named state, with the muscogee Indian boys for my class mates, I learned my first lessons, in natures grand seminary.[2] Here arose my first thoughts on the subject that is now, by yourself, and by me subscribed to, denominated Natural Selection". Pardon this digression.

You ask, "Do you suppose that the ants plant seeds for the ensuing crop?"—[3] I have not the slightest doubt of it. And my conclusions have not been arrived at from hasty, or careless observation, or from seeing the ants do something that looked a little like it, and then guess at the results. I have, at all seasons, watched the same ant cities, during the last 12 years, and I know what I stated in my former letter is true. I visited the same cities yesterday, I find their crop of *ant Rice*, growing finely, exhibiting also, the signs of high cultivation; not a blade of any other kind of grass or weed, to be seen in 12 inches of the circular row of the Rice.

We have not only agricultural ants in Texas, but we also have a species, that are regular Horticulturalists. The mound, which is constituted of the sand that is thrown out from their cells below, and their extensive tunnels, is from one to two feet high, and sometimes spreding over an area of two or three square rods. It is upon this elevated ground that they plant their shade trees. They cannot stand our summer sun, and in those cities, of recent date, not having had time to grow the shade, no ant is ever seen on the mound, when the sun is 9 o'clock high in hot weather. They are compelled to perform their work during the night time, until the city is properly shaded. Neither can they travel out over the unshaded plains of a sunny day, to bring in the provisions for the indoor workers hence the necessity of tunnels, or underground passages to the trees, and patches of herbaceous plants that produces the leaves of which the food of this species, entirely consists. The sand which is taken out from their tunnels, is all thrown out upon the city mound. The excavation of these underground passages, are always commenced in the city, and extending outwards to some district—often four or five hundred yards—that produces plentiful crops of the kind of leaves upon which they feed; the amount of sand thrown out from them increases the elevation of the mound very considerably; the bore of the tunnel, for the purpose of allowing sufficient room for them to carry a piece of leaf through it that is as wide as a dime, sometimes larger, is generally an inch in diameter. Its outer terminus, most commonly, ends at two or three points, under the shade of a spreading tree, or in a garden or cornfield.

The holes where they come out, are always concealed, by being carefully, and ingeneously covered with dry leaves, bits of stick &c. When they enter a garden in this way, they seldom fail to ruin it, in spite of the efforts of the owner to prevent them. All kinds of fruit trees, many flowering shrubs and garden vegetables, are entirely trimmed of their leaves, and totally ruined. One gentleman on San Antonio River, had a very fine garden invaded by the cutting Ant, (the common name of the species in question) in great numbers, which were rapidly destroying his vegetables by night. As he had good irrigating facilities; he conceived the idea of surrounding his garden by a considerable ditch, and let a sluice of running water through it. All of which he accomplished in good stile, and for two or three days was boasting that he had outed the little pests. It was not long till he found that his garden was being cut up and damaged as bad as before he had surrounded it with water, but as they perpetrated their mischief at night, he could not discover how they had managed to cross his ditch. It was several days before he found the secret and then by accident found, at the root of a little flowering shrub in the garden the concealed terminus of one of their tunnels. On further search he discovered several other concealed holes. And now, forming a resolution that he would not be outdone by them, but that he would destroy them, cost what it might. So calling four or five negroes, with their impliments for ditching, set them to work, ditching and following the ant tunnels from inside the garden; he found that the tunnel passed at a dry depth beneath the water in the ditch, when it rose again within 18 inches to 2 feet of the surface, and at about that depth, he followed it to a large mound city of ants, distant four hundred yards. And now the only chance to destroy them was to dig out the city. This he was prosecuting at last account.

Their horticultural action is exhibitted in their nice judgement in the selection of the quick growing species of the heavy foilaged trees to plant for shade; of which they cultivate four or five kinds Celtis occidentalis, Viburnum dentatum, Ilex vomatoria, Zanthoxylum carolinanum, and the mustang grape vine; all beautifully ornamental. When they locate a city so far out on the bald prairie, that they cannot carry the seeds of the above named trees so great a distance through the grass, they collect and plant the seeds of the Argemone mexicana, a large, quick growing prairy weed, having spreading tops with large leaves, which under the ant culture, besides producing ample shade, is, with its large white flowers, quite ornamental.

It would require a considerable volume to describe this most interesting type of our ants. I will not bore you further on that topic. I have answered your single question; you must excuse me.

You may hower, say to your brother, that if he feels like it, and will put the proper questions to me, he will find, that these little Emmets can teach lessons to the genus homo even, that would be profitable to imitate.[4]

You may be too much engaged, but if you feel any interest in it, this correspondence need not cease.

I remain Dear Sir | Very truly thine | Gideon Lincecum.

Linnean Society (MS S.P. 604A)

CD ANNOTATIONS[5]
0.1 Long Point . . . 1861. 0.2] *underl and circled, pencil*
0.3 Charles . . . digression. 2.13] *crossed pencil*
7.1 It would . . . Lincecum. 10.1] *crossed pencil*

[1] CD's letter has not been found in the Lincecum collection in the University of Texas at Austin Archive. He wrote in response to the letter from Gideon Lincecum, 29 December 1860 (*Correspondence* vol. 8). According to his biographer, Lincecum made the following comment about CD's letter: 'To Sir Charles Darwin I am indebted for the most polite letter I ever received.' (Burkhalter 1965, p. 213 n. 5).
[2] The Muskogee Indians were the chief group in the Creek confederation, whose peoples were spread out over the south-eastern states of the United States until their resettlement in the Indian Territory (Oklahoma) in the 1830s. They were settled agriculturists with an elaborate social organisation (*EB*).
[3] In his letter of 29 December 1860 (*Correspondence* vol. 8), Lincecum had described the apparent cultivation of a grain-bearing grass by a local species of ant.
[4] Emmet is an archaic term for ant. CD may have told Lincecum the same anecdote about Erasmus Alvey Darwin that he related to Charles Lyell in his letter of 12 April [1861]. Referring to CD's interest in the social instincts and habits of ants, Erasmus predicted that someday they would find that ants even had 'bishops'.
[5] CD's annotations are editorial and relate to the reading of this and Lincecum's earlier letter at a meeting of the Linnean Society of London in April 1861 (see letter to George Busk, 5 April [1861]). The passages marked were those that CD suggested be deleted from the paper.

To Philip Lutley Sclater 4 March [1861]

Down Bromley Kent
March 4th

My dear Mr Sclater

You told me that I might have one of the Russian Rabbits to skeletonise (which Rabbits, both Bucks, I sent to Gardens).[1] Will you be so good as to give order to have one killed by throat being cut, *especially not knocked on Head*, & sent by address on next page.—[2]

I signed your certificate the other day at Royal Socy.[3]

I see there has been paper on skeleton of Hybrid Hare-Rabbit,[4] as I am *now* working at Skeletons of Rabbits, I shd. like to see this notice, if you *chance* to have spare proof.—

Yours very sincerely | Ch. Darwin

Endorsement: 'March 1861'
American Philosophical Society (239)

[1] Sclater was the secretary of the Zoological Society of London and involved in the management of the society's zoological gardens in Regent's Park. Presumably CD refers to the so-called Russian or Himalayan rabbit (*Variation* 1: 108). He described the osteological characters of this breed in *Variation* 1: 127–8.
[2] The page on which the address was written is missing.
[3] Sclater was one of fifty people proposed as candidates for membership of the Royal Society of London in March 1861. He was among the fifteen elected on 6 June (*Proceedings of the Royal Society of London* 11 (1860–2): 130–1, 233).
[4] Crisp 1861.

To George Rolleston 7 March [1861][1]

Down Bromley Kent
March 7th

My dear Sir

Though you so kindly tell me not to write, you must let me have the pleasure of thanking you, busy as you are, for taking so much trouble to explain to me, the curious point of cyanosis & the clubbed nails.[2] It may work in well, yet the case seems very complex.— I sincerely thank you.—

Yours very faithfully | ⟨Charles Darwin⟩[3]

How truly I agree with your remark on "borrowing organs" "prophetic types" &c &c![4]

I hope you will work out the point about the mucus tubes; & I wish you could at same time work out the relation of the mucus tubes in Annelids with tracheæ of Insects. This has, always, seemed to me a curious case.[5] | C. D.

Wellcome Institute for the History of Medicine Library, Western MS. 6119: 6

1 Dated by the relationship to the letter to George Rolleston, 2 March [1861].

2 Rolleston had written to CD informing him of a correlation between clubbed finger nails and cyanosis (morbid blueness of the skin), and CD had asked him for further information on the subject (see letter to George Rolleston, 2 March [1861]). Rolleston's reply has not been found.

3 The signature has been cut away.

4 Louis Agassiz discussed the concept of 'prophetic types' in the introductory volume of Agassiz 1857–62, pp. 116–18. There is an annotated copy of this work in the Darwin Library–CUL. Agassiz's view was that extinct, fossil groups exhibiting combinations of characters now found only in distinct animal groups were evidence of a creator's plan. For an earlier expression of CD's negative opinion of Agassiz's view, see *Correspondence* vol. 7, letter to T. H. Huxley, 13 [March 1859]. See also letter to Cuthbert Collingwood, 14 March [1861].

5 Rolleston was investigating the means by which certain lamellibranch molluscs distended their muscular feet, a process thought by some to involve the influx of water into a system of tubes homologous to the tracheae of insects. A paper on the subject by Rolleston and Charles Robertson was read before the Royal Society of London on 3 February 1859; CD may have read the abstract that was published (Rolleston and Robertson 1859). Further investigations led them to modify their views, which were published in a more detailed report (Rolleston and Robertson 1862).

To Alexander Goodman More 8 March 1861

Down, Bromley, Kent.
March 8, 1861.

My dear Sir

I am much obliged for your note and most kind offer of observing orchids.[1] Later on I will remind you on points connected with E. palustris; and I will further venture to trouble you for a few specimens of O. ustulata and Aceras, if they grow anywhere near you; for then I shall have examined large majority of British species.[2] I never understood whether E. palustris grew near you: I have always regretted that I did not ask you last Autumn to gather me two or three roots, which I would have attempted to grow in pot with lot of water. I have got many orchids in my garden including the rare Goodyera.[3]

I am not in the least surprised at your demurring to accept my notions on species. It took me long years before I converted myself; though daily thinking and observing on the subject. You ask why I should not draw a line and allow natural selection to do a little work and no more. I can give no direct answer to this. But I think you do not fully see how, as it seems to me, the subject may fairly be approached. Take the case of Light,—existence of Ether, and the existence of its undulations are both absolutely hypothetical or conjectural; but because this hypothesis explains and groups together a multitude of phenomena, it is now universally admitted as a true theory. So, as it seems to me, the descent of species with their modifications through natural selection groups together and fairly well explains many phenomena (as classification, morphology, rudimentary organs, embryology, partially Geograph. Distrib. and partially Geolog. succession), and *therefore* I believe in its truth. These phenomena are otherwise inexplicable, and my many hostile Reviewers have hardly attempted to improve my explanations, therefore I believe Natural Selection will after many years prevail.

My dear Sir | Yours very sincerely | C. Darwin.

Copy
DAR 146

[1] More's letter has not been found.

[2] CD had enlisted More's assistance in the summer of 1860 to help him procure specimens of orchids. More collected and sent to CD various species that grew near his home on the Isle of Wight. See *Correspondence* vol. 8. In *Orchids*, p. 95 n., CD thanked More for repeatedly sending him specimens of the 'beautiful Orchis' *Epipactis palustris*.

[3] CD referred to *Goodyera repens*, as 'this rare Highland Orchid' in *Orchids*, p. 112 n.

To J. D. Hooker 11 March [1861]

Down Bromley Kent
March 11th

My dear Hooker

Mrs. Huxley has come here with her children for a change of air, & Huxley has just gone back to London; & says he will come here next Sunday.[1] Can I persuade you to come on next Saturday for the Sunday.[2] A train leaves the new Victoria Station at 5° P.M. & gets into Bromley at 5° 30″ & my carriage shd. meet you & I would send you back early on Monday (if you wished to go) as Huxley must then go. If you can come it would be much pleasure to us all; but I hear from Huxley you are not very well.

Will you send me a line *by return of Post*; for this reason, I feel bound to ask Carpenter, who about a year ago expressed a wish to this effect, & I shd feel much safer if you were here, as I have been continually knocked up of late. *But you must not think of coming* if for any reason disinclined; & if you do not come, I shall not ask Carpenter. Perhaps in any case he would not come.—[3]

We have just made up our minds to go to Malvern about 1st of May for 6 weeks or two months, that I may have a dose of Hydropathy.—[4] We continue rubbing in the oil into Etty & hope & think it does her good.—[5]

Yours affecty. | C. Darwin

Endorsement: '61/'
DAR 115.2: 89

1 Henrietta Anne Huxley and her children, Jessie Oriana, Mariane, and Leonard Huxley, arrived at Down House on 9 March 1861. For the circumstances that prompted the visit, see the letter to T. H. Huxley, 22 February [1861]. In a letter Huxley wrote to Henrietta while she was at Down, he stated: 'It appears to me, that you are subjecting poor Darwin to a savage Tennysonian persecution. I shall see him looking like a martyr and have to talk double science next Sunday.' (L. Huxley ed. 1900, 1: 225).

2 According to Emma Darwin's diary, Thomas Henry Huxley visited Down on Sunday, 10 March 1861, returning to London the next day. He came back to Down the following Sunday, 17 March. Hooker had arrived on the previous day.

3 William Benjamin Carpenter came to Down on Sunday, 17 March 1861 (Emma Darwin's diary). For the deferral of Carpenter's proposed visit in 1860, see *Correspondence* vol. 8, letter to W. B. Carpenter, 17 June [1860].

4 CD refers to James Manby Gully's hydropathic establishment at Malvern, Worcestershire. The Darwins did not visit Malvern in 1861 but spent the summer in Torquay (see 'Journal'; Appendix II).

5 See letters to J. D. Hooker, 8 [February 1861] and 20 [February 1861].

To Asa Gray 12 March [1861][1]

Down Bromley Kent
March 12th.

My dear Gray

This note will be chiefly on business.— I received this morning Wrights Article:[2] I have hardly glanced at it, but rather fear it is too metaphysico-theological for me. Huxley will be here in few days, & I will hand it to him & he will decide.[3] If it does not suit him; what on earth shall I do with it? I have received, also, Procs. of American Academy in 4 vols.,—very many thanks for them.—[4]

I have received your note of Feb. 15th & am truly obliged to you for allowing me to bear very small risk of publication.[5] I am surprised & pleased at sale of new Edit. of Origin in America.— You will have received before this the Bill for £7.s10.d0.—[6] Your pamphlet will do the subject **very great** good, whether or no it sells.— I have had many letters about it; all full of praise—"truly admirable" says one, "& am lending my copy to one person after another". Another says, "has read nothing on the subject with anything like the satisfaction"— Another says he (ie you) "is a cunning fencer & believes in you entirely".— I wish this was half true; though I do *not* mean the cunningness, which is utterly mistaken & false.

I sent copies to Ward, Martineau & Mansell;[7] but had already sent copies to Hooker, Bentham, Harvey & Balfour;[8] so they will have duplicates. I was much pleased the other day by note from Leidy who professes adherence to natural Selection—[9]

Is any species of Cypripedium common with you? I wish you would just look whether pollen-masses are removed (by insects?).—[10]

Could you do me a great favour procure me a pamphlet published in 1850 in Charleston, entitled "Letter to J. Bachman on the Question of Hybridity in Animals by D. Morton".—[11] Or was it subsequently published in any Journal? I know his paper in Silliman (& poor it is.)—[12]

I am heartily glad to hear about your thumb.—[13]

I have had my stomach so extra bad of late, that we have resolved to go bag & baggage for 6 weeks or two months, before May 1st., to Malvern for Hydropathy, which is sole thing which revives me.—[14]

Forgive this extra dull note, & believe me your ever most truly & obliged | Charles Darwin

Gray Herbarium of Harvard University (52)

[1] The year is provided by the reference to A. Gray 1861a.

[2] See letter to T. H. Huxley, 17 February [1861] and n. 2.

[3] Chauncey Wright's review of *Origin* was not published in the *Natural History Review*, of which Thomas Henry Huxley was chief editor. See also letters to Asa Gray, 11 April [1861], and to T. H. Huxley, 22 May [1861].

[4] CD's copies of the *Proceedings of the American Academy of Arts and Sciences* are in the Darwin Library–CUL. In the preceding year, the journal had carried reports of a number of discussions about CD's theory.

[5] Gray's letter has not been found.

[6] See letter to Asa Gray, 17 February [1861] and n. 7.

[7] The botanist Nathaniel Bagshaw Ward was a fellow of the Linnean Society and of the Royal Society of London. Both James Martineau and Henry Longueville Mansel were noted for their theological and philosophical writings. See Appendix III for CD's presentation list for Gray's pamphlet (A. Gray 1861a). Ward appears on the list, but Martineau and Mansel do not.

[8] In a previous letter to Gray, CD included Joseph Dalton Hooker, George Bentham, and William Henry Harvey in a list of those to whom he planned to send a copy of A. Gray 1861a (see letter to Asa Gray, 23 [January 1861]). John Hutton Balfour may have received a personal copy or the one that CD sent to the *Edinburgh New Philosophical Journal*, of which Balfour was an editor. See Appendix III.

[9] Joseph Leidy's letter has not been found, but see the letter to Joseph Leidy, 4 March [1861].

[10] Gray's observations on the orchid *Cypripedium* were not sent to CD in time to be included in the first edition of *Orchids*. In the second edition of 1877, however, CD cited Gray's descriptions of several American species and quoted his statement that these orchids are 'excellently adapted to brush off the pollen from an insect's head or back.' (*Orchids* 2d ed., p. 229).

[11] S. G. Morton 1850a. CD had recorded this and other papers of Samuel George Morton published in the *Charleston Medical Journal* in the 'books to be read' section of his reading notebook (see *Correspondence* vol. 4, Appendix IV, *128: 165). A heavily annotated copy of the paper is in the Darwin Pamphlet Collection–CUL. Morton's essay endorsed the view that human races were separate species or 'types' and that interracial breeding produced 'hybrids'. It was heavily cited by pro-slavery adherents before the American Civil War (see Stanton 1960).

[12] S. G. Morton 1850a was a reply to a paper by John Bachman (Bachman 1850), which in turn commented on S. G. Morton 1847, a paper on hybridity published in the *American Journal of Science*

and Arts, also known as 'Silliman's journal'. The correspondence on this issue carried on throughout 1850 in the *Charleston Medical Journal*.

[13] Asa Gray had lost part of his left thumb in a gardening accident (Dupree 1959, pp. 307–8).

[14] CD refers to James Manby Gully's hydropathic establishment at Malvern, Worcestershire. The planned visit did not take place.

To P. L. Sclater 12 [March 1861]

Down Bromley Kent
12th.

My dear Mr Sclater

One line to thank you for the Rabbit received safely.[1]

I am very glad you like Asa Gray: I fully thought that **I** had sent you a copy, but have just looked at my list & find that by some accident your name omitted.[2]

You will receive in short time a copy of my new Edit. (corrected) of the Origin, which Murray tells me he will soon distribute.—[3] I am *most uncommonly* pleased, I can assure you, to hear that you are become "heretical" on species.— I perceived with no surprise that you were at first dead against me: in fact I cannot say that I respect anyone who has knowledge & can change his opinion suddenly on such a point.— If any fair & good opportunity occurs in any publication, I hope that you will say a word on **our** side: for I observe that those opposed write vehemently, & those on our side are silent; consequently the public have no means of knowing how many are on our side. But day before yesterday I had letter from a Professor, who *dares* not speak out.— Your note has pleased me very much: though I did not intend to have written all this.

Ever yours | C. Darwin

Endorsement: '*March 1861*'
American Philosophical Society (240)

[1] See letter to P. L. Sclater, 4 March [1861].

[2] A. Gray 1861a. For CD's lists of those to whom he intended to send a presentation copy, see the letter to Asa Gray, 23 [January 1861] and Appendix III.

[3] The third edition of *Origin* was published in April 1861 (see 'Journal'; Appendix II). For CD's list of presentation copies, see Appendix VII.

To Cuthbert Collingwood 14 March [1861][1]

Down Bromley Kent
March 14th.

Dear Sir

I am much obliged for your long letter, as I always like to know how naturalists view the subject. I feel not a shade of surprise at your entirely rejecting my views;[2] my surprise is that I have been successful in converting some few eminent Botanists, Zoologists, & Geologists. In several cases the conversion has been very slow & that is the only sort of conversion which I respect.— I *entirely* agree with you that there is no more direct proof of variation being *un*limited in amount than there is that it is strictly limited.— In a new & corrected Edit. of the Origin, which will appear in about a week or two, I have pointed this as emphatically as I could.—[3] I did not formerly explicitly say this (but indirectly in several places) because I thought it was obvious.

The manner in which I wish to approach the whole subject, & in which it seems to me it may fairly be approached, I can best illustrate by the case of Light.— The Ether is hypothetical, as are its undulations; but as the undulatory hypothesis groups together & explains a multitude of phenomena, it is universally now admitted as the true theory. The undulations in the ether are considered in some degree probable, because sound is produced by undulations in air. So natural selection, I look at as in some degree probable, or possible, because we know what artificial selection can do.— But I believe in Nat. Selection, not because, I can prove in any single case that it has changed one species into another, but because it groups & explains well (as it seems to me) a host of facts in classification, embryology, morphology, rudimentary organs, geological succession & Distribution.—

I have no space to discuss the many points alluded to in your letter.— I cannot see such perfection in structure as you do. In the new Edit. I have attempted to explain how it is that many low forms have not progressed to a higher grade of organisation.[4]

I did not allude to the very curious subject of "alternate generations", because I did not, & do not yet, see, how it has any *special* bearing on my views.—[5] I look at alternate generations, as not *essentially* differing from various stages in any one individual larva—a form of gemmation being merely added at some stage. Under this point of view I see no essential difference between alternative generation & metamorphosis: you, I presume, take some very different view.—

I forget what Agassiz says on subject.— I quite agree with you that Agassiz's Review is not in the least unfair.[6] He misunderstands me a good deal.— His "categories of thought", "*prophetic* types" & his views on classification are to me merely empty sounds.[7] To others they seem full of meaning.—

I received several months ago, & thank you for, a very curious pamphlet on representative forms, (or some such title) which interested me *very much*.—[8]

With my best thanks, I remain | Dear Sir | Yours very faithfully | Charles Darwin

I am much pleased at & grateful for the sentence which you kindly copy from a recent letter from Agassiz.— I once saw him, & was charmed with him.—[9]

British Library (Add. MS. 37725, ff. 6–9b)

[1] The identity of the correspondent and the year are based on the reference to CD's having received 'several months ago' a pamphlet on representative forms (Collingwood 1860a). See n. 8, below.

[2] Collingwood's letter has not been found. Presumably the letter raised many of the objections to CD's theory discussed in a paper Collingwood delivered before the Literary and Philosophical Society of Liverpool on 10 December 1860. In the paper, Collingwood contrasted Darwin's views on the origin of species with those of Louis Agassiz, as expressed in his critical review of *Origin* (Agassiz 1860). Collingwood concluded that Agassiz's criticism was 'one of the most conclusive and formidable

(against Darwin's theory) which had yet appeared.' (Collingwood 1860b, p. 81). Collingwood sent CD a copy of this paper, which is in the Darwin Pamphlet Collection–CUL.

3 *Origin* 3d ed., p. 89.

4 *Origin* 3d ed., pp. 133–43.

5 In his paper, Collingwood discussed Agassiz's view that the alternation of generations and other examples of polymorphism within a species were 'far more astonishing phenomena, strictly circumscribed between the natural limits of unvarying species, than the slight differences produced by the intervention of men, among domesticated animals' (Agassiz 1860, p. 153). Collingwood then stated that these phenomena had been ignored by CD. See Collingwood 1860b, pp. 92–5. The passages in question are marked in CD's copy of the paper.

6 Collingwood 1860b was written to defend Agassiz 1860 against the view expressed by Henry Hugh Higgins that Agassiz's review was 'quite unworthy of so distinguished a naturalist' (Higgins 1860, p. 42).

7 In Collingwood 1860b, pp. 83–5, Collingwood attempted to elucidate Agassiz's concept that species, while having no material existence, yet existed as 'categories of thought' (Agassiz 1860, p. 142). Agassiz discussed the concept of 'prophetic types' in the first volume of *Contributions to the natural history of the United States of America* (Agassiz 1857–62, 1: 116–18), a copy of which is in the Darwin Library–CUL (see also letter to George Rolleston, 7 March [1861] and n. 4).

8 A lightly annotated copy of Collingwood 1860a, inscribed 'Charles Darwin Esq with the Author's Compts.', is in the Darwin Pamphlet Collection–CUL.

9 CD may be referring to the 1846 meeting of the British Association for the Advancement of Science, at which Agassiz spoke to a meeting of the Ray Society that CD attended. See *Correspondence* vol. 4, letter to J. E. Gray, 18 December 1847.

From George Maw 15 March 1861

Benthall Hall. *Broseley.*
March 15th. 1861.

Dr Sir.

I hope you will not consider that I as a stranger to you am taking too great a liberty in sending you a copy of circular relating to the ⟨Tes⟩timonial that is being raised for Mr Newman.[1] I think naturalists are under a great debt of gratitude to him for organizing the collection of a vast amount of information & valuable facts connected with Natural-History. indeed he stands quite alone as the Naturalists Calenderer[2] & although I have no particular sympathy with Testimonials in general I think Mr Newman's peculiar services demand some marked expression of favor from his brother naturalists—

I think I know your "Origin of Species" almost by heart to me it is a most perplexing question on the one hand the history of man upon the Earth appe⟨ars to⟩ be a wonderful epitomy of what ⟨*two words*⟩ has gone before us.— Race struggling with race— Some getting the mastery— The stronger extinguishing the weaker then culminating & perhaps becoming unwieldly by their size & trying to include too great a diversity of components breaking up into different nations—isolated from each other by some peculiar character⟨s⟩ which though of advantage to them separately were incompatible in association—

Each of the young nations at first fill different spheres increase & multiply— all have not the same strength the more powerful outrun the others The weakest are wholly extinguished & those of middle power too strong to die & too weak to multiply go dwindling on from age to age as feeble communities bringing down to posterity antiquated characters which have been quite lost in the subdivision of the more vigorous races into modern nations. Then on the other hand the classification of mineral *Elements* seems to be a close repetition of that in organic nature— Group subordinate to group & unequal amounts of difference between different groups & the individuals of different groups. *this* cannot be the result of any active process. The grouping together of the Alkalis with very small amounts of difference is most striking not only by arbitrary measures of resemblance but by a most complete series of correlated affinities. Then again as opposed to this close resemblance there are isolated *individuals* as Carbon Nitrogen Silex &c. & isolated *groups* of closely related individuals as the Iodine group or the platina group amongst the metals— also sociable groups as the alkalis alkaline Earths. & True Earths. indeed just that *kind* of variety in the measure of affinity between individuals or groups that we find in the organic kingdoms. then again these are curious colateral relationships (as between the Heavy & light metals thro the resemblance of Lead to Baryta) which look most singularly like geneological affinity. These facts seem to throw a doubt on the cause you assign for the parallel phenomena in the organic kingdoms.

With great respect & many apologies for the liberty I have taken in writing to you | Believe me Dr Sir | very faithfully & obediently yrs | George Maw.

I forgot to state that it has been determined to purchase & present to Mr Newman a number of valuable works on Natural History which his own means will not enable him to become possessed of.

Charles Darwin Esqre

DAR 171.1(3): 95

CD ANNOTATIONS
0.3 D^r^ ... nations. 3.6] *crossed pencil*
Top of first page: '11'[3] *brown crayon, circled brown crayon*; '(good)' *brown crayon*; 'Diseases & Furniture can be classed naturally' *pencil*

1 Edward Newman was the editor of the *Zoologist* and the natural history editor of the *Field*.
2 As editor of the *Zoologist*, Newman saw to the publication and recording of large amounts of natural history data. He also personally contributed a great deal of zoological data: the *Royal Society catalogue of published scientific papers* lists 281 papers by Newman.
3 The number of CD's portfolio of notes on classification.

To George Maw 17 March [1861]

Down Bromley Kent
March 17th

Dear Sir

I am much obliged for your striking note. Your remarks on the races of man are interesting; but I have been particularly struck with your observations on the classification of mineral bodies. The idea had crossed my mind in a very vague & feeble manner: it is so difficult to be honest that I fear that I must have unconsciously banished the idea as disagreeable. I see now the full force of the difficulty & I will make a note not to forget this subject. I must own that classification may be closely like that due to descent & yet have no relation to it. I am grateful for your excellent criticism.—[1]

I am very sorry to say that at present I do not feel inclined to subscribe to the testimonial to Mr Newman. I have not a shade of feeling against Mr Newman; on the contrary I have much respect for his labours. But to t⟨he⟩ best of my judgment, it seems hardly a case for a testimonial. If one is given to him; I think there are others with at least equal claims— This is my only objection. I will, however, talk with some other naturalists, & I hope my objection may be got over.—[2]

Dear Sir | Yours sincerely obliged | Charles Darwin

P.S. | As you have been so great an observer, I ask two questions for *remote* chance of your having facts.— Have you sown seed of weeping trees & what were seedlings?

Have you observed any remarkable cases (excluding mere variegations of leaves) of branch or bud sports, (or as I shall call such cases "*bud*-variations") in leaves, flowers, or fruits.— e.g. nectarines on peach-trees—provence Roses on Moss-roses.—[3] Do not, pray, answer unless you have any facts.—[4]

Endorsement: '61'
Royal Horticultural Society

[1] CD included a reference to the analogy between mineralogical classification and classification in the organic realm in the fourth edition of *Origin* published in 1866 (*Origin* 4th ed., p. 488).

[2] The presentation of the testimonial to Newman was announced in the *Zoologist* 1861: 7457–62. CD was not listed among the subscribers.

[3] CD defined bud-variation as including 'all those sudden changes in structure or appearance which occasionally occur in full-grown plants in their flower-buds or leaf-buds.' (*Variation* 1: 373).

[4] CD included information provided by Maw about varieties of scarlet pelargonium showing unusual correlations in *Variation* 2: 330–1.

From Henry Walter Bates 18 March 1861

King St Leicester
18 March 1861

Dear Sir

At last the Ent. Society have printed my paper & I am enabled to send you ⟨a⟩ copy according to promise.[1]

I do not know whether the peru⟨sal⟩ will repay your trouble, & I cannot point out especial passages embodying the facts most useful & interesting to you; the⟨y⟩ are scattered over the whole,—perhaps the concluding observations & those under the head of P. Hierocles are most to the point.[2]

I think there are about 3 points of interest arising from the review of the species of Papilio—& the genus from the precision of the specific characters & the great amount of material existing in collections is well calculated to illustrate them.— These are

1 The derivation of the Amazonian fauna. I confess I was not prepared for the result to which I was obliged to arrive after a close examination of the species & their distribution—viz: that the Guiana region must ⟨ha⟩ve been the seat of an ancient & peculiar fauna transmitted through vast lapses of time; & that thence was derived the fauna of the Amazon valley.—[3] Also that it was still so rich in endemic species. Surely I am right in deriving the conclusion that there can have been no great extinction here ⟨d⟩uring the glacial epoch.[4]

2 The widely different variability of species when under different local conditions in localities widely apart— Let us suppose 4 species A B C D living at localities 1 & 2. A will be not in the least modified: B constant at 1 will be instable at 2: C will have become changed at 2 in all its individuals, but the change is so small that all will admit the difference to be that of a variety: whilst D will have become changed so much more considerably at 2 that every author will treat the form as a perfectly good species. Yet all the points of difference between D1 & D2 are similar to those between B1 & B2 and ⟨C⟩1 & C2, only they are greater in degree or more numerous

3 The permanency of local varieties after they have become established. It is still the favourite argument of our best naturalists that varieties will always return to their normal form & that they will inter breed & produce fertile offspring. This argument is derived from the observations of varieties produced by domesti⟨cation⟩—a false guide—such varieties are too rapidly made, to be compared with the slow alterations of the whole organism which takes place in nature & affects, I have no doubt, at length the reproductive elements.— In the genus Papilio there is a set of local varieties all connected by fine gradations of differences; & yet in one well established case two of these varieties exist in contact & do not show the slightest tendency to amalgamate.—[5] It is a case exactly parallel to what would be if we were to find in the wild state a series of graduated local varieties between the horse & ass;— I thought it likely I should find in the Natural History of the Horse & ass some data to prove the parallel & turned to a paper by Blyth lately published on the varieties of wild ass.[6] I was surprised to find how little was satisfactorily known on the subject & how uncertain & vacillating is the state of our knowledge of the species & varieties of these conspicuous animals.

With ⟨re⟩gard to the varieties of species involving modifications of their reproductive elements, a learned entomologist friend of mine, Mr Baly, has found in the chrysomelidæ, what he calls *specific differences* in the male organs of generation;[7] he finds they ⟨v⟩ary between very closely allied forms,—but then he reasons I think falsely,—he says that the fact of the difference in these organs proves distinctness of species & thus he is proving to be distinct species forms which all Entomologists had agreed to consider as varieties.— It is very amusing to read (In the proc. Ent^l^. Soc. I think September or October las) that he had received a large series of Donaciæ from N. America,— there were many doubtful forms amongst them &

he has applied the test of these organs to prove whether they are distinct species or no.[8]

I hope you will excuse this rather rambling letter & favour me with your opinion on my pap⟨er⟩ at your earliest convenience

Yours sincerely | H W Bates

DAR 160.1: 61

CD ANNOTATIONS
1.1 At last . . . epoch. 4.7] *crossed pencil*
6.1 3 The . . . animals. 6.16] *crossed pencil*
Top of first page: '2[d] Letter in Portfolio (4)— on Glacial action—Good.' *ink*

[1] Bates 1861a. Bates's earlier letter to CD has not been found, but see CD's response to it (*Correspondence* vol. 8, letter to H. W. Bates, 22 November [1860]). CD's copy of this paper has not been located.

[2] Bates 1861a, pp. 349–54 and pp. 341–2. In his concluding observations, Bates put forward the view that the *Papilio* species peculiar to the Amazon region were modifications of species from nearby regions such as Guiana. He cited Darwin's argument that during the glacial period the refrigeration extended into equatorial regions, thus enabling species of temperate zones to pass from one hemisphere to another. Bates claimed that if this were so, it would have resulted in the extinction of many forms peculiar to the equatorial regions, and that this was not borne out by the present distribution of *Papilio*. In the section on *P. hierocles*, he pointed out that the existence of forms that seemed to serve as connecting links between *P. hierocles* and the similar species *P. vertumnus* provided an argument for classifying one as a variety of the other; yet, the two forms, although found in close association, never hybridised (Bates 1861a, pp. 341–2 n.).

[3] This point is discussed in the concluding remarks of Bates 1861a, p. 352.

[4] This conclusion, discussed in Bates 1861a, p. 352 (see n. 2, above), contradicted CD's view that many tropical species became extinct during the worldwide cold period and were subsequently replaced by invading temperate flora and fauna (see *Origin*, pp. 376–9). For CD's response to Bates's observations, and his views on the problem of accounting for the distribution of tropical species, see the following letter and the letter to H. W. Bates, 26 March [1861]. CD mentioned the opposition of Bates's facts to his theory in the fourth edition of *Origin* (*Origin* 4th ed., p. 451).

[5] Bates refers to *P. hierocles* and *P. vertumnus*. See n. 2, above.

[6] Blyth 1859. An annotated copy of this paper, presented to CD by Edward Blyth, is in the Darwin Pamphlet Collection–CUL.

[7] Joseph Sugar Baly had published several papers describing species of Chrysomelidae, the leaf beetles. Baly was describing some of the insects collected by Bates.

[8] The discussion about using the generative organs of forms of *Donacia* (tree-dwelling beetles) as characters for taxonomic discrimination took place at a meeting of the Entomological Society of London on 3 September 1860 (*Transactions of the Entomological Society of London* n.s. 5 (1858–61), Proceedings, p. 128).

To J. D. Hooker 18 March [1861]

Down | Bromley, Kent.
March 18th.

My dear Hooker

I have been recalling my thoughts on the question whether the Glacial period affected the whole world contemporaneously, or only one longitudinal belt after another.[1] To my sorrow my old reasons for rejecting the latter alternative seem to me sufficient;[2] and I should very much like to know what you think. Let us suppose that the cold affected the two Americas either before or after the Old World; let it advance first either from north or south, till the Tropics became slightly cooled, and a few temperate forms reached the Silla of Caraccas and the mountains of Brazil. You would say I suppose that nearly all the tropical productions would be killed; and that subsequently, after the cold had moderated, tropical plants immigrated from the other non-chilled parts of the world. But this is impossible, unless you bridge over the tropical parts of the Atlantic,—a doctrine, which you know, I cannot admit; though in some respects wishing I could.[3] Oswald Heer would make nothing of such a bridge.[4]

When the glacial period affected the Old World; would it not be rather rash to suppose that the meridian of India, the Malay archipelago and Australia were refrigerated, and Africa not refrigerated? But let us grant that this was so; let us bridge over the Red Sea (though rather opposed to former almost certain communication between Red Sea and Mediterranean); let us grant that Arabia and Persia were damp and fit for the passage of tropical plants; nevertheless just look at the Globe and fancy the cold slowly coming on and the plants under the tropics travelling towards the equator; and it seems to me highly improbable that they could escape from India to the still hot regions of Africa, for they would have to go Westward with a little northing round the northern shores of the Indian Ocean. So if Africa were refrigerated first, there would be considerable difficulty in the tropical productions of Africa escaping into the still hot region of India.— Here again you would have to bridge over the Indian Ocean within so very a recent period; and not in the line of the Laccadive archipelago.— If you suppose the cold to travel from the southern pole northwards, it will not help us, unless we suppose that the countries immediately north of the northern tropic were at the same time warmer, so as to allow free passage from India to Africa which seems to me too complex and unsupported an hypothesis to admit.

Therefore I cannot see that the supposition of different longitudinal belts of the world being cooled at different periods helps us much. The supposition of the whole world being cooled contemporaneously (but perhaps not quite equally; S. America being less cooled than the Old World) seems to me the simplest hypothesis: and does not add to the great difficulty of all the Tropical productions not having been exterminated.

I still think that a few species of each still existing tropical genus must have survived in the hottest or most favourable spots, either dry or damp. The tropical productions though much distressed by the fall of temperature would still be under the same conditions of length of day, &c; and would be still exposed to nearly same enemies, as insects and other animals; whereas the invading temperate productions though finding a favouring temperature would have some of their conditions of life new, and would be exposed to many new enemies.— But I fully admit the difficulty to be very great—

I cannot see the full force of your difficulty of no known cause of a mundane change of temperature. We know no cause of continental elevations and depressions, yet we admit them. Can you believe, looking to Europe alone, that the intense cold which must have prevailed, when such gigantic glaciers extended on the plains of N. Italy was due merely to changed positions of land within so recent a period. I cannot. It would be far too long a story, but it could, I think, be clearly shown that all our continents existed approximately in their present positions long before the Glacial period, which seems opposed to such gigantic geographical changes necessary to cause such a vast fall of temperature. The Glacial period endured in Europe and N. America, whilst the level of the land oscillated in height fully 3000 feet; and this does not look as if changed level was the cause of the Glacial period. But I have written an unreasonably long discussion.— Do not answer me at length, but send me a few words sometime on subject.

My dear Hooker | Ever yours | Ch. Darwin

I have had this copied that it might not bore you too much to read it.—[5]

A few words more: when equatorial productions were dreadfully distressed by fall of temperature & probably by changed humidity & changed proportional numbers of other plants & enemies (though this might favour some of the species), I must admit that they *all* would be exterminated, if productions exactly fitted, not only for the climate, but for all the conditions of the equatorial regions during the glacial period existed & could everywhere have immigrated. But the productions of the temperate regions would have probably found under the Equator in their new homes & soils considerably different conditions of humidity & periodicity; & they would have encountered a new set of enemies (a most important consideration) for there seems good reason to believe that animals were not able to migrate nearly to the extent which plants did during the G. period. Hence I can persuade myself that the temperate productions would not entirely replace & exterminate the productions of the cooled Tropics; but would become partially mingled with them. I am far from satisfied with what I have scribbled. I conclude that there *must* have been a mundane glacial period; & that the difficulties are much the same whether we suppose it contemporaneous over world, or that longitudinal belts were affected one after the other.—[6] For Heaven's sake forgive me!

Endorsement: '1861'
LS(A)
DAR 115.2: 90

[1] CD and Hooker may have discussed this topic during Hooker's recent visit to Down. Emma Darwin recorded in her diary that Hooker arrived at Down on 16 March 1861. Although CD and Hooker had discussed the theory of the migration of plants during a worldwide glacial period on numerous occasions, Hooker was at this time particularly interested in working out details of the theory in connection with his study of the distribution of the Arctic flora. See L. Huxley ed. 1918, 2: 27.

[2] For CD's consideration of the possibility that the cold period affected various regions of the globe at different times, see *Natural selection*, p. 548. See also *Origin*, p. 374, where he stated that it seemed

probable that the cold 'was, during a part at least of the period, actually simultaneous throughout the world.'

3 For a clear statement of CD's reasons for not believing in the land-bridge doctrine of geographical distribution, see *Correspondence* vol. 6, letters to Charles Lyell, 16 [June 1856] and 25 June [1856].

4 The Swiss palaeobotanist Oswald Heer had invoked a hypothetical continuous tract of land connecting the Old and the New Worlds to explain the existence of plant and animal species common to both (Heer 1857, pp. 22–4).

5 The letter is in the hand of an amanuensis. CD made some corrections to the text and added the postscript in his own hand. CD's draft of the letter is in DAR 50 (ser. 5): 22–5.

6 CD expanded the discussion of the mundane glacial period in the fourth edition of *Origin* (*Origin* 4th ed., pp. 442–56).

To Leonard Horner 20 March [1861][1]

Down Bromley Kent
March 20th

My dear Mr. Horner

I am very much obliged for your Address, which has interested me much.[2] I have been particularly glad to see your excellent summary on metamorphism, for I was very ignorant of the recent researches.—[3] I thought that I had read up pretty well on antiquity of man; but you bring all the facts so well together in a condensed focus, that the case seems much clearer to me.—[4] How curious about the Bible![5] I declare I had fancied that the date was somehow in the Bible.— You are coming out in a new light as a Biblical critic!

I must thank you for your remarks on the Origin of Species (though I suppose it is almost as incorrect to do so, as to thank a Judge for a favourable verdict): what you have said has pleased me extremely.[6] I am the more pleased, as I would rather have been well attacked than have been handled in the namby-pamby, old-woman style of the cautious Oxford Professor.—[7]

I most sincerely hope that Mrs Horner is a little better;[8] & with my kindest remembrances to all your party, | pray believe me | My dear Mr Horner | Yours sincerely obliged | Charles Darwin

Emma sends her very kind remembrances.

Rt Hon. Lady Lyell

1 The year is provided by the reference to Horner's presidential address (see n. 2, below).

2 Horner, outgoing president of the Geological Society of London, delivered the presidential address at the society's anniversary meeting on 15 February 1861 (Horner 1861).

3 Horner devoted a large part of his address to a discussion of recent studies of the chemical basis of metamorphic phenomena in geology (Horner 1861, pp. xl–lx). He pointed out that the 'application of chemistry to the explanation of geological phenomena has hitherto received more attention on the continent than with us.' (Horner 1861, p. xl).

4 Horner examined in one section of his talk the 'Evidence of the early existence of the human race' (Horner 1861, pp. lx–lxxi).

5 CD refers to Horner's comment that the marginal notation in standard editions of the English Bible, which assigned the year 4004 B.C. as the beginning of the world, could be traced to the seventeenth

century biblical chronology of James Ussher, bishop of Armagh. Referring to recent discussions about the apparent conflict between Genesis and geology, Horner stated: 'The study of geology has become so general that those who are instructed in its mere elements cannot fail to see the discrepancy between this date and the truths which geology reveals.' On this basis, he advocated removing this erroneous date from future editions of the Bible in order to 'best express that entire ignorance of "The Beginning" which no human power will ever be able to dispel.' (Horner 1861, p. lxx).

6 Referring to *Origin* as a 'remarkable work', Horner praised CD's 'truly philosophic modesty and candour' in presenting objections to his theory and recommended 'to the careful study of geologists' the two chapters on the geological record (Horner 1861, pp. xxxix–xl).

7 The reference is probably to John Phillips, professor of geology at Oxford and Horner's predecessor as president of the Geological Society. The previous year, Phillips had briefly mentioned *Origin* in his presidential address and later criticised CD's views in his Rede lecture in Cambridge (Phillips 1860).

8 Ann Susan Horner was suffering from what was regarded as neuralgia. See K. M. Lyell ed. 1890, 2: 307.

To W. B. Tegetmeier 22 March [1861][1]

Down Bromley Kent
March 22[d]

My dear Sir

I ought to have answered your last note sooner; but I have been very busy.[2] How wonderfully successful you have been in breeding Pouters! you have good right to be proud of your accuracy of eye & judgment. I am in the thick of Poultry; having just commenced & shall be truly grateful for the skulls, if you can send them by any conveyance to the Nag's Head next Thursday.[3] If too busy, will you send them whenever you can, addressed, thus

C. Darwin E[sq]
care of Mr Mitchell
Per S. E Railway — Postman
Bromley, Kent

You ask about Vermilion wax: positively it was not in state of comb, but in solid bits & cakes, which were thrown with other rubbish not far from my Hives.—[4] You can make any use of fact, you like.—[5] Combs could be concentrically & variously coloured & dates recorded by giving for few days wax darkly coloured with vermilion & indigo, & I daresay other substances.

You ask about my crossed Fowls; & this leads me to make a proposition to you, which I hope cannot be offensive to you.— I trust you know me too well to think that I would propose anything objectionable to the best of my judgment. The case is this. For my object of treating Poultry, I must give sketch of several breeds with remarks on various points. I do not feel strong on subject. Now when my M.S. is fairly copied in an excellent hand-writing; would you read it over, which would take you at most an hour or two, & make comments in pencil on it; & accept, like a Barrister, a fee, we will say of a couple of guineas.—

This would be a great assistance to me; specially if you would allow me to put note, stating that you, a distinguished judge & Fancier, had read it over. I would state that you doubted or concurred as each case might be: of course striking out what you were sure was incorrect. There would be little new in my M.S. to you; but if by chance you used any of my facts or conclusions, before I published, I sh^d wish you to state that they were on my authority; otherwise I shall be accused of stealing from you.— There will be little new except that perhaps I have consulted some out of the way books, & have corresponded with some good authorities.

Tell me frankly what you think of this; but unless you will oblige me by accepting remuneration, I cannot & will not give you such trouble.— I have little that several points will arise which will require investigation, as I care for many points disregarded by Fanciers; & according to any time thus spent you will, I trust, allow me to make remuneration. I hope that you will grant me this favour.—[6]

There is one assistance which I will *now* venture to beg of you, viz to get me, if you can, another specimen of *old white* Angora Rabbit. I want it dead for Skeleton; & *not* knocked on Head.[7] Secondly I see in Cottage Gardener (March 19th. p. 375) there are (impure) Half-lops with one ear quite upright & shorter than the other lopped ear. I much want a dead one.— Baker cannot get one.—[8] Baily is looking out;[9] but I want two specimens. Can you assist me, if you meet any Rabbit fancier? I have had rabbit with one ear more lopped than the other; but I want one with one ear quite upright & shorter, & the other quite long & lopped.—[10]

My dear Sir | Yours sincerely | Ch. Darwin

New York Botanical Garden Library (Charles Finney Cox collection)

1 The year is provided by the reference to CD's study of fowl for his work on *Variation* (see 'Journal'; Appendix II).

2 Tegetmeier's letter has not been found.

3 Tegetmeier had agreed to lend his extensive collection of the skulls of various breeds of fowl for CD's study of the osteological differences between domesticated and wild poultry. The carrier that operated between Down and London used the Nag's Head Inn, Borough, as the point of collection.

4 See letter to W. B. Tegetmeier, 1 March [1861].

5 Tegetmeier had previously conducted experiments on the construction of bees' cells in which bees were provided with various colours of wax (see Tegetmeier 1858). His intention to carry out further experiments was mentioned in Tegetmeier 1859, p. 35, but he does not appear subsequently to have published the results of any systematic experiments on bee-cell construction.

6 There is an entry in CD's Account book (Down House MS) on 31 March 1861 that reads: 'Tegetmeier 10^s Science.' This probably pertains to specimens Tegetmeier sent to CD (see n. 7, below). A further entry on 13 June records that CD paid him the sum of £5 5*s*. 'for scientific assistance'. Tegetmeier's assistance in reading CD's manuscript is acknowledged in *Variation* 1: 225 n.

7 Tegetmeier sent CD the rabbit as requested. See letter to W. B. Tegetmeier, 28 March [1861].

8 Either Samuel C. or Charles N. Baker, poultry dealers in Chelsea. Over the past few years, CD had obtained different varieties of fowls from them.

9 John Baily was a poulterer and dealer in live birds with a shop in Berkeley Square, London.

10 In *Variation* 1: 119, CD stated that he had examined the skulls of three half-lop rabbits. He was interested in learning whether differences in the size and musculature of the ear had any influence

on the structure of the skull. He found this to be so, calling the case 'a remarkable case of correlation of growth'. He went on to state (*Variation* 1: 119):

> Who would have surmised that by keeping an animal during many generations under confinement, and so leading to the disuse of the muscles of the ears, and by continually selecting individuals with the longest and largest ears, he would thus indirectly have affected almost every suture in the skull and the form of the lower jaw!

To Daniel Oliver 23 March [1861][1]

Down Bromley Kent
March 23^{d}

My dear Sir

You told me formerly that you did not much care about my troubling you: I hope to Heaven you keep of same mind.— Will you ask M^{r}. Fitch to make two little diagrams for me for woodcuts; as by enclosed paper.—[2] And will you kindly give him the specimens, viz two common Primroses, one "pin-headed, as the Florists say with stigma at mouth of corolla: & the other with stamens at the mouth.— And cut them longitudinally with sharp scissors Mr Fitch could keep a little account against me.— I *think* I shall publish an account of my observations & experiments on Primula; i.e. on its dimorphous condition.[3] By the way you will find the pin-headed & *non*-pin-headed in any bank of Primroses.— I find that with P. Sinensis, the short-stamened flowers fertilise themselves, whereas the long-stamened flowers will not set without man's or insects' aid.—

I am now crossing largely Cowslips & polyanthuses. And this leads me to beg a second favour, viz to send me 2 or 3 flowers (not whole trusses) of pin-headed & non-pin headed, of any species of Primula, except P. vulgaris, Sinensis & Auricula.— If you have any other species pray fold me a few flowers in *double* green leaf, & pollen would come neither too dry or too wet for measurement.— If you can send please send names of species.—[4]

Yours very truly | Ch. Darwin

Down House (MS 10: 4)

[1] The year is provided by the reference to CD's paper on *Primula* (see n. 3, below).

[2] Walter Hood Fitch was the botanical artist at the Royal Botanic Gardens, Kew. The woodcuts he prepared were published in CD's paper on dimorphism in *Primula* (see n. 3, below).

[3] CD presented his paper on the dimorphic condition of *Primula* to the Linnean Society on 21 November 1861. It was published in the *Journal of the Procedeeings of the Linnean Society (Botany)* 6 (1862): 77–96 (see also *Collected papers* 2: 45–63). This paper was CD's first published study of dimorphism. CD stated that, although botanists recognised the existence of the long- and short-styled forms of *Primula*, they regarded the phenomenon as 'mere variability' (*ibid.*, 2: 45). He illustrated that this recognition was not specific to the botanical community by referring to florists' use of the terms 'pin-eyed' and 'thumb-eyed' to describe what CD believed to be two distinct forms of *Primula* (*ibid.*, 2: 45).

[4] Oliver sent CD the flowers of a number of different *Primula* species (*Collected papers* 2: 48).

To P. L. Sclater 23 March [1861]

Down Bromley Kent
March 23d

Dear Sclater

It is a shame to trouble so busy a man as you; but I shd. be very much obliged for a bit of information, which you can give me better than anyone.

Would it be too strong to say? that, from the known distribution of the Gallinaceæ it would be almost as improbable that a species of the genus Gallus should be endemic in S. America, as that a Humming Bird shd. be found in the Old World.— With respect to Africa, may I not say that no species is known there; and that it is not very probable that any species should have wandered so far from the Metropolis of the genus in India & the Northern Malay Isld.—[1]

Lastly can you tell me where in Proc. Zoolog. Soc. Gallus Temminckii of G R. Gray is described; for I cannot conceive what species is meant.[2]

Pray forgive & aid me.— If you were to say "yes" (supposing that I am right) it would suffice.—

Yours very sincerely | C. Darwin

Many thanks for your note received other day.— I had not heard of your paper at Oxford.—[3]

Endorsement: '*March 1861*'
American Philosophical Society (241)

[1] CD included these statements in the section of *Variation* on the parentage of fowls. In a note, he stated: 'I have consulted an eminent authority, Mr. Sclater, on this subject, and he thinks that I have not expressed myself too strongly.' (*Variation* 1: 237 and n. 27).

[2] The paper by George Robert Gray was printed in the *Proceedings of the Zoological Society of London* in 1849 (G. R. Gray 1849). In *Variation* 1: 235, CD stated that *Gallus temminckii* was thought by some to be a sterile hybrid.

[3] Sclater 1860. Sclater's note has not been found.

To John Crawfurd 25 March [1861][1]

Down Bromley Kent
March 25

My dear Mr Crawfurd

You sent me some time ago an essay "On the relation of the Domestic Animals to Civilization." You give some facts on Fowls to which I wish to refer.[2] Now will you be so kind as to tell me when this was published, as it has no name of Author nor title, & I do not know how to refer to it— What is its date? There is one other question which possibly you may be able & have kindness to answer viz is Sir W Jones date of 1200 BC esteemed correct for Institutes of Menu?[3]

It is a long time since I have troubled you, so forgive me I most sincerely hope you are well.— Pray believe me | yours sincerely | C Darwin

Copy
DAR 143

[1] Dated by the relationship to the letter to John Crawfurd, 7 April 1861.
[2] Crawfurd [1860a]. The paper, which surveys the history of the domestication of birds, was read at the 1860 meeting in Oxford of the British Association for the Advancement of Science; only an abstract of the paper appears in the published report (*Report of the 30th meeting of the British Association for the Advancement of Science held at Oxford*, Transactions of the sections, p. 155). The paper was subsequently privately printed. CD's annotated copy is in the Darwin Pamphlet Collection–CUL. CD cited Crawfurd [1860a] in *Variation* 1: 246 n.
[3] Jones trans. 1796. CD had obtained this information from Edward Blyth (see *Correspondence* vol. 6, letter from Edward Blyth, [*c.* 22 March 1856]). See also letter to John Crawfurd, 7 April 1861.

To H. W. Bates 26 March [1861]

Down Bromley Kent
March 26

Dear Sir

I have read your papers with extreme interest & I have carefully read every word of them.[1] They seem to me to be far richer in facts on variation, & especially on the distribution of varieties & subspecies, than anything which I have read. Hereafter I shall reread them, & hope in my future work to profit by them & make use of them. The amount of variation has much surprised me. The analogous variation of distinct species in the same regions strikes me as particularly curious. The greater variability of female sex is new to me. Your Guiana case seems in some degree analogous, as far as plants are concerned, with the modern plains of La Plata, which seem to have been colonised from the north, but the species have been hardly modified.—

I have been particularly struck with your remarks on the Glacial period.[2] You seem to me to have put the case with admirable clearness & with crushing force. I am quite staggered with the blow & do not know what to think. Of late several facts have turned up leading me to believe more firmly that the Glacial period did affect the Equatorial Regions; but I can make no answer to your argument; & am completely in a cleft stick. By an odd chance I had only a few days ago been discussing this subject, in relation to plants, with Dr. Hooker, who believes to a certain extent; but strongly urged the little apparent extinctions in the Equatorial regions. I stated in a letter some days ago to him, that the Tropics of S. America seem to have suffered less than the Old World.[3]

There are many perplexing points, temperate plants seem to have migrated far more than animals— Possibly species may have been formed more rapidly within Tropics than one would have expected. I freely confess that you have confounded me: but I cannot yet give up my belief that the Glacial period did to certain extent affect the Tropics.—[4]

Would you kindly answer me 2 or 3 questions if in your power.— When species (A) becomes modified in another region into a well marked form (C), but is connected with it by one (or more) gradational form (B,) inhabiting an intermediate region; does this form (B) generally exist in equal numbers with (A) & (C), *or inhabit an equally large area?*— The probability is that you cannot answer this question: though one of your cases seem to bear on it.—

Has Mr Baly published on Chrysomelidæ: if so, or when he does, would you give me reference to his paper?[5]

In Butterflies, in which the sexes are differently coloured, is the male or female most beautiful in *our* eyes? Do you know in Tropics any strictly nocturnal moths with gaudy colours? As with Birds, have you ever noticed that female butterflies make any selection of the male with which they copulate? Do several males pursue same female? Are butterflies attracted by gay colours, as it has been asserted Dragon-flies are. Any authentic facts on the courtship of Butterflies would be *most thankfully* received & quoted by me. But I can see how very improbable it is that anything sh$^{d.}$ have been observed.—[6]

You will, I think, be glad to hear that I now often hear of naturalists accepting my views more or less fully: but some are curiously cautious in running risk of any small odium in expressing their belief.

With cordial thanks & respect | Believe me my dear Sir | Yours sincerely | C. Darwin

(Have you received copy of new Edit. of Origin?)[7]

Endorsement: '1861'
Cleveland Health Sciences Library (Robert M. Stecher collection)

CD note:[8]

March 23$^{d.}$ 1861. Suppose world shortly before Glacial to have had present temperature; but we know before that it was hotter. Probably *the effect of this [*interl*] then would have been an ['N. Tropical' *del*] Equatorial Flora, which could live only in hottest regions; & a N. [*interl*] Tropical & S. Tropical Flora. [*illeg del*] fitted for less intense Heat, & distinct from each other, for they must have been separated from each other for long time by Equatorial Flora. During Glacial period, I suppose Equatorial Flora almost wholly annihilated. And the N. & S. Tropical Flora, which would not have suffered quite so much under Equator during Glacial period, to have occupied Equatorial regions. Their commingling would have made a rich Flora., & [*over* '—'] which now has become accustomed to rather hotter climate than that to which they were accustomed, ['during' *del*] previously to Glacial period.—

35° North

N. Tropical Flora

20 [*over* '25']° N.

Equatorial Flora nearly all exterminated by Glacial period.— Equator

20 [*over* '25']° S.

S. Tropical Flora distinct from North

35° S.

All this upset by Mr Bates paper.— Can species have been very largely developed since glacial period.— Certainly from Lat 25° to Equator productions now ought to have very uniform character. How is this? | In ['Brazil' *del*] S. America immigration must have been almost exclusively by lines of mountains.—

Crowning difficulty is why equatorial productions not all destroyed—as climate [*above del* 'some of temperate'] would have been best fitted for sub-tropical productions— Probably was & our present Tropical productions were ['the' *del*] sub-tropical before glacial— All Fauna & Flora much disturbed & a few temperate forms crossed owing to disturbance—taking advantage of high land—

With respect to S. America we have but little evidence of the ⟨probability⟩ of ⟨such forms⟩— Yet some seem to have crossed [*del illeg*] Region of Amazona ['seems' *del*] is greatest difficulty— Undoubtedly there is much to explain as yet

[1] Bates 1861a was printed in two separate issues of the *Transactions of the Entomological Society of London* for 1861.

[2] See Bates 1861a, pp. 352–3. See also letter from H. W. Bates, 18 March 1861. CD took Bates's work into account in the revisions he made to the discussion of the mundane glacial period in the fourth edition of *Origin* (*Origin* 4th ed., p. 451).

[3] Letter to J. D. Hooker, 18 March [1861].

[4] See CD's note transcribed following the letter.

[5] See letter from H. W. Bates, 18 March 1861 and n. 7.

[6] For Bates's replies to CD's questions, see the letter from H. W. Bates, 28 March 1861. CD cited Bates on many of these points in the fourth edition of *Origin* (*Origin* 4th ed., pp. 502–6). He also referred in *Descent* to Bates's published account of his studies of the natural history of the Amazon region (Bates 1863).

[7] The third edition of *Origin* was published in April 1861.

[8] The note is in DAR 50 (ser. 6): 26.

To J. D. Hooker 26 [March 1861]

Down Bromley Kent
26th

My dear Hooker

I must just write to thank you sincerely for your letter, & to say how rejoiced I am that it seems clear from your account that poor dear Henslow's sufferings are nearly over, whether or no his struggle is much prolonged.—[1] I would not have missed coming for anything if he had wished to see me, but in fact I am not equal to such an exertion. It is strange how immediately any mental excitement upsets & utterly prostrate; seeing George chloroformed for his teeth brought on my eternal sickness for 24 hours.—[2] Etty certainly improves & we have faith in your Oil:[3] next Monday will be 14 weeks since last regular attack. She gains a little in flesh, but sadly little in strength.— We have almost made up our minds to go by three stages to Torquay on 1st of June.—[4]

What a strange account you give of the chaos of articles in Henslow's House.— Give our kindest remembrances to Mrs. Hooker.— I am glad you work a little;— that alone drives cares away for a time. I shall be most anxious to read your Arctic Paper.—[5]

Farewell | C. Darwin

Endorsement: 'March 1861.'
DAR 115.2: 92

[1] Hooker's letter has not been found. He and his wife, Frances Harriet Hooker, the eldest daughter of John Stevens Henslow, were at Henslow's home in Hitcham, Suffolk. While on a visit to the south of

England, Henslow caught a cold that rapidly developed into bronchitis, aggravating a former ailment of his heart and lungs (*Gardeners' Chronicle and Agricultural Gazette*, 15 June 1861, p. 552).

[2] See letter to J. D. Hooker, 4 February [1861] and n. 3.

[3] See letter to J. D. Hooker, 8 [February 1861].

[4] The Darwins did not go to Torquay until July (see 'Journal'; Appendix II).

[5] Hooker 1861.

To William Samuel Symonds 26 March [1861][1]

Down Bromley Kent
March 26th

My dear Sir

I am much obliged for your kind present of your volume on Old Bones & for the very flattering manner in which you allude to my work.—[2]

Pray believe me | Yours sincerely & obliged | C. Darwin

American Philosophical Society (242)

[1] The identity of the recipient and the year are provided by the reference to Symonds 1861 (see n. 2, below).

[2] Symonds was the author of *Old bones; or, notes for young naturalists* (Symonds 1861). Symonds frequently cited CD's views in the book and referred to *Origin* as a 'remarkable work' (*ibid.*, p. 33). The book is not in the Darwin Library.

To J. D. Hooker 27 [March 1861]

Down Bromley Kent
27th

My dear Hooker

I had intended to have sent you Bates' article this very day.[1] I am so glad you like it. I have been extremely much struck with it. How well he argues & with what crushing force against the glacial doctrine. By the Lord I cannot wriggle out of it. I am dumb-founded; yet I do believe that some explanation some day will appear & I cannot give up equatorial cooling. It explains so much & harmonises with so much.— When you write (& much interested I shall be in your letter) please say how far Floras are generally uniform in generic character from 0° to 25° N & S.—[2]

Before reading Bates, I had become thoroughily dissatisfied with what I wrote to you.— I hope you may get Bates to write in Linnean.[3]

Here is a good joke: H. W. Watson (who I fancy & hope is going to review new Edit. of Origin) says that in first 4 paragraph of the Introduction, the words "I" "me" "my" occur 43 times![4] I was dimly conscious of this accursed fact.— He says it can be explained phrenologically which I suppose civilly means that I am the most egotistically self-sufficient man alive,— perhaps so.—[5]

I wonder whether he will print this pleasing fact; it beats hollow the parentheses in Wollastons writing[6]

I am **my** dear Hooker | Ever yours | C. Darwin
Do not spread this pleasing joke, it is rather too biting.—
(Many thanks about London Review: I will see how far I care for it—)[7]

Endorsement: '/March 1861.'
DAR 115.2: 93

[1] Bates 1861a.
[2] Hooker's response has not been found. The band of the tropics CD mentions would include the area of the north-eastern part of South America studied by Bates (see Bates 1861a, p. 223). See also letter to H. W. Bates, 26 March [1861] and CD note.
[3] See letter to J. D. Hooker, 18 March [1861]. Bates published his next paper in the *Transactions of the Linnean Society of London* (Bates 1861b).
[4] CD means 'H.C.' here: his reference just above to [H. W.] Bates probably prompted the error. No review of *Origin* by Hewett Cottrell Watson has been identified.
[5] Watson was a firm believer in phrenology; he had published on the subject in the 1830s (*DNB*).
[6] When the anonymous review of *Origin* appeared in *Annals and Magazine of Natural History* in February 1860, CD suspected, rightly, that Thomas Vernon Wollaston was the author because 'no one else in the world would have used so many parentheses.' (*Correspondence* vol. 8, letter to Charles Lyell, 15 and 16 [February 1860]).
[7] Hooker had apparently sent CD a copy of the most recent issue of the *London Review and Weekly Journal of Politics, Literature, Art and Society*. The periodical had begun publication in July 1860. For CD's opinion of this number, see the letter to J. D. Hooker, 23 [April 1861].

From H. W. Bates 28 March 1861

King St Leicester
28 March 1861

My Dear Sir

I received your kind letter today[1] & think it better to reply at once, as I am going to London on Monday for a few days study at the Insect room Brit. Mus. & shall not have another opportunity very soon. It gives me very great pleasure to find that my paper is likely to be useful to you. I am quite convinced that Insects offer better or clearer illustrations of the problems you occupy yourself with than any other class of animals or plants. It is so easy with them to obtain great series of examples & have them before you in a small compass, which is one advantage they have.—

I do not think the result I have come to regarding the persistence of the Guiano-Amazonian fauna can be shaken. I can see *no traces* of a migration of high-temperate-zone forms across the region. Now there are a number of genera of insects characteristic of the high-temperate zones, which are now common to S. America & N. America with Europe; but they have not a single representative in Amazonia, although some of them have representatives along the Andes of New Granada & Peru.[2] Your hypothesis only requires (as you have stated) a migration along belts of latitude & therefore I do not think that the conclusion I have come to affects your position much after all.— I am further confirmed in my view that no migration took place across Guiano-Amazonia by the fact that there is a migration going forward now, or has been within the historic period of species of a migratory genus of Butterflies, but although one or other of the forms has obtained a footing in most other regions of the world there is no trace of them in that country. I allude to the species of the genus Pyrameis of which there are two English species P. atalanta & P. cardui. The geographical distribution of the species of this genus is very peculiar & has often excited attention. The following is a sketch of the facts

P. atalanta is found throughout Europe & N. America as far South as Mexico. On the Southern slopes of the Himalaya, & in Madeira (!) it exists as P. Callirhoe considered by some as a local var. In *Java* it presents us with a further *var.* in P. Dejeanii: & in the Sandwich Islands another allied form, P. Tammeamea; It appears to go no further South than Java, & is *not* found in S. America. Pyrameis cardui is found throughout Europe, Africa to the cape of Good hope; Asia & the Asiatic islands to Australia & New Zealand, without varying.— In America it is found throughout the united states, in Venezuela it exists only as a var. P. Leachiana, throughout N. America it is found in conjunction with P. Huntera a very closely allied species; but the two have different areas, although existing together in some parts; To the west of the Andes from California to chili there exists another form of Cardui, viz P. Carye; and in the S. of Brazil another, P. Myrinna; in the Amazonian region not a trace of any of these forms is to be found. From these facts it would appear 1st. that Northern temperate forms more readily cross the equator in the old world than in the new; 2nd that they are *subjected* to far greater modifying or transforming influences in the New World than in the old, 3rd that the only passage made use of by them in the New world is the chain of the Andes. I should add that the species of Pyrameis are strongly migratory; many accounts at least of immense migrations of P. cardui having been published.—

When I read your book first, the chapter on Geogr. distr. of course struck me very much; but a doubt occurred to me whether the resemblance in the faunas & floras between the N. & S. temperate zones was of such a nature as to require the explanation you have given. I am quite ignorant of the nature of the curious resemblance in the plants of Tierra del fuego & am quite inclined to think my doubt is unfounded in this respect.

However, two considerations have presented themselves to me. 1st. there are in several classes of animals & plants groups which seem to have originated at a remote epoch of the worlds history, & which, in consequence of the *limited nature of their variability* have gone on repeating nearly similar specific forms in all countries where they exist up to the present time. In plants I am not aware it is any other class than the inferior Cryptogamia which are in this position; but in insects there are several groups. Two occur to me at the present time the Hesperiadæ in the Lepidoptera & the Libellulidæ (Dragonflies) in the Neuroptera. Similar species of these are found all over the world. I will not dwell too long on the subject or my letter will run to great length. I wish to suggest whether it is possible that the Tierra del fuego plants belong to such a group as I have described. I know that they are not found (as you have stated) in the tropical zones, but are the genera found there;

& are they of such a homomorphic nature (to coin a word) as I have described.—

The other consideration is that there are genera peculiar to the high temperate zones of both hemispheres which present in South America compact groups of species so very dissimilar to those of North America that it is out of the question that they could have been derived thence so lately as the glacial period. I will give you a strong & very intelligible case viz that of the Chilian species of *Carabus*, of which yourself discovered several of extreme beauty.[3] The genus Carabus at the present time is not found in the Northern hemisphere, further south than 28° nor in the Southern further North than 33°.— The Chilian species number 11 & form a group distinguished by peculiarities of form & sculpture;— Now there is not a species that at all approximates those of N. America; there is an utter dissimilarity between the two groups. The nearest approach to the Chilian forms is presented in species of the South of Europe (perhaps some may occur in E. Asia). How did the Chilian Carabi get there, & why are they so different from those of the rest of the world?— To me this is a most interesting fact; it seems to point to a far more ancient Glacial epoch, than the post-tertiary one.[4] The Carabi may be studied by any one in five minutes by consulting the "Bowringian collection" in the British Museum, where they are lucidly arranged.[5]

With regard to the question you ask me—whether an intermediate local form B is numerous & widely dispersed between the ranges of its extreme forms A & C: the facts that I have on the subject are numerous & rather ill digested at present. They are complicated in themselves & difficult. I know many instances of two local forms separated by a wide space without apparent natural barriers, untenanted by intermediate forms. There are many others (closely allied) which exist together on their mutual frontiers without blending. There are others which are very polymorphic in a central region, whilst in other localities E. W. N. & S. segregated into several well defined local varieties or admitted species. I will mention now one case only that meets your question. It is one, whose correctness I am quite sure of. On the dry soils, supporting for the most part a thinner forest growth, of the hilly regions of Guiana & Venezuela there exists a conspicuous Heliconia, H. Melpomene, it is abundant also in the central parts of the lower Amazons where the Hilly sandy country occurs & is found on the N. & S. Shores in extreme profusion. In the moist alluvial plains eastward to Pará & westward to Peru & Bolivia, not a single individual is to be seen. In its place, occupying exactly the same sphere as it were, is H. Thelxiope a form so strikingly peculiar in colours that no one has ever doubted it to be a perfectly distinct species; it swarms in individuals & is nearly constant in its specific characters. Now wherever this form comes into contact with Melpomene there exists a number of intermediate varieties many of which have been described as species: they *are rare*, & very *restricted in range*. They have puzzled Lepidopterists & it has been almost settled that they were hybrids. I am convinced they are not *hybrids*; I never saw Melpomene & Thelxiope in copula; and besides these pretended hybrids exist at Demerara & Cayenne, where Thelxiope *does not occur*. I am thoroughly convinced that Thelxiope is a local variety of Melpomene, having all the appearance of a species, but created by the influences to which it has been subjected, out of Melpomene; & that the intermediate forms are the gradations. I have found Melpomene in copula with these forms & perhaps it may sometimes be so with Thelxiope, but that would not affect the case much I think.

The group from which I have supplied this example is very interesting,—being restricted to S. America, which swarms with its species; & apparently a modern creation. I am now studying it with a view to writing upon it & I think I have got a glimpse into the laboratory where Nature manufactures her new species.[6]

—The other question you ask me relates to sexual selection.— I *have* seen Papilios attracted by bright colours in the forest, namely by the scarlet sepals of certain plants having inconspicuous or no flowers. I have also seen repeatedly many males following a female.— On this subject I will point out to you the following facts. In the Æneas section of the genus Papilio, the males are generally of extremely brilliant colours, velvety black vivid green, carmine, & red with opalescent reflections,—the females are plainer & so different from the males that they were generally held to be distinct species until I took them in copulâ. Now it is not *all* the species that present this disparity. In one P. Panthonus the two sexes are exactly alike in colours, the male only being a trifle brighter; from this species the divergence may be traced getting wider & wider from species to species up to P. Sesostris or P. Childrenæ, where the dissimilarity reaches its highest point. There is exactly the same phenomenon presented in the genus Epicalia & others. The females, however, vary in localities;—& although I thoroughly believe in your theory of sexual selection yet I think that local circumstances have some effect on colours. I do not however think that sunlight is the direct cause of vivid colours, although beautifully coloured male butterflies are almost always in the sun (genus Catagramma) whilst their drab partners are in the shade. I think the causes lie in the abundance of food, warmth & moisture of the atmosphere & even in the sluggish state of the atmosphere because the brightest coloured butterflies are not generally found within the influence of the Atlantic sea breezes, but in the sultry valleys of the Andes & the centre of the continent & these causes operate on the larva & so by correlation on the perfect insect.

I cannot say that I have noticed female butterflies pointedly selecting their male partners. The extremely rich male butterflies of the genus Catagramma live & sport together in the sun light all day, whilst their plainer coloured females are confined to the shades of the forest. The males too appear to be very much more numerous than the females,— I think *there is no* doubt now about this significant fact there must be many hundred males to one female of these butterflies, & in another genus Cybdelis where the males are also distinguished by rich colours the males are immensely more numerous than the females. In another beautiful genus, Megistanis no female has yet been found although the males are very numerous. In Callithea, however where the males are the most richly coloured of all butterflies, both sexes are found in equal numbers, but *here the females* are not *much* behind the *males* in *beauty*. The males which sport together all day in the sunlight disappear about 4–5 P.M. & I have watched them then flying off to the forest (towards the summits of the trees) where doubtless some of them find their mates.—

There are no strictly nocturnal Lepidoptera gaily coloured in the tropics: There is no scarcity of brilliantly col$^{d.}$ moths but they are all day fliers.

Perhaps the above facts will illustrate a little the problems of sexual disparity & beauty of colours. There is no phase of your theory I like better than its explanation of the subtle adaptations of organic beings. There is one topic which Entomology will help to illustrate viz. that of mimetic analogies. I have an immense number of facts on this subject. Some of these resemblances are perfectly staggering,—to me they are a source of constant wonder & thrilling delight. It seems to me as though I obtain a glimpse of an *intelligent motive* pervading nature, as well as of the mighty never-resting wonder working laws that regulate all things.—[7]

—I received the copy of 3^{d} edit. of the "Origin" a most valued present for which accept my thanks.— I received a letter from my friend Wallace last week he was going to Timor[8]

Believe me My Dear Sir Yours sincerely | H W Bates

There is a stupid error in my paper, the note referring to "Recent & *ancient* fossil shells". The secretary took upon himself to correct the proof, I saw it and forgot to erase it—[9]

DAR 160.1: 62

CD ANNOTATIONS
1.1 I received ... have.— 1.8] *crossed pencil*
9.1 I *have* seen ... female.— 9.4] *scored brown crayon*
9.14 & although ... colours. 9.15] *scored brown crayon*
14.1 H W Bates] 'H. W. Bates' *added ink*
Top of letter: '1' *pencil, circled pencil*
On cover: '*Bates*: Valuable letter Glacial: very good— *S. America⟨n⟩ [*interl*] Car⟨abus⟩ points to ancient glaci⟨al⟩ period.— Chalk glacial & Pa⟨ ⟩ do.— | Very good on intermediate vars. in intermediate zones.' *ink*; '(1)' *brown crayon*; 'Admirable on Sexual Selection. | Butterflies attracted by bright sepals.' *ink*; 'Bates' *brown crayon*

1 Letter to H. W. Bates, 26 March [1861].

2 New Granada was the name of the former Spanish colony that included present-day Colombia, Ecuador, Venezuela, and Panama.

3 See *Journal of researches*, p. 38. CD's South American beetles were described by George Robert Waterhouse and Frederick William Hope (see *Correspondence* vol. 2).

4 CD included Bates's information about South American *Carabus* species in the discussion of the mundane glacial period in the fourth edition of *Origin* (*Origin* 4th ed., p. 454).

5 The vast collection of Coleoptera amassed by John Charles Bowring was formally presented to the British Museum in 1863 (British Museum (Natural History) 1904–6, 2: 581).

6 Bates published these data in Bates 1861b, pp. 557–8. In describing the relationships between the various local species of Heliconidae of the Amazon valley, he referred to these forms as illustrating the 'process of the creation of a new species' and 'the manufacture, as it were, in process.' (Bates 1861b, p. 501).

7 Bates published some of his material on mimicry in butterflies and moths in Bates 1861b, pp. 502–15, stating that the explanation of mimicry 'seems to be quite clear on the theory of natural selection, as recently expounded by Mr. Darwin in the "Origin of Species."' (*ibid*., p. 511).

[8] Bates had first met Alfred Russel Wallace when Wallace was attending the Collegiate School in Leicester and the two went on entomological excursions together. In 1847, they decided to undertake a joint expedition to the Amazon, raising funds by offering specimen collections from little-explored areas. After four years in South America, Wallace returned to England in 1852, departing again in 1854 on an eight-year scientific expedition to the Malay Archipelago. See Marchant ed. 1916, 1: 24–36.

[9] The reference is to an incorrect citation in Bates 1861a, p. 226 n., of Samuel Pickworth Woodward's *A manual of the Mollusca; or, a rudimentary treatise of recent and fossil shells* (Woodward 1851–6).

To W. B. Tegetmeier 28 March [1861][1]

Down Bromley Kent
March 28

My dear Sir

I am in truth grateful to you for so frankly & kindly acquiesing in my proposal & request.[2] My M.S. will not be ready for a considerable time, as I work abominably slowly from many reasons.—

I can see that I shall have many points to investigate; & as you have opportunity you can investigate them; & I will have no scruple in asking, as you with real & true kindness will allow me to make up for your loss of time. You will have to endeavour to give me some idea of how much time I cause you to lose.—

Hearty thanks for what you have already done & written about Rabbits & Malays.— In rabbits my skeleton work has *much* interested me.— I am glad to hear that you are going to write on Hen-tailed cocks.[3] By an odd coincidence, yesterday I was thinking of an experiment which I sh^d. like to see tried, viz to cross a cock Hamburgh Hen-tail & Seabright Hen tailed-Hen,[4] & see whether the Hen-tail w^d be inherited under such trying circumstances; & likewise to see effect on colour of chickens.— I would gladly pay all direct & indirect expences, but I know it is very troublesome with poultry to make such crosses, & so hardly worth while. I would do it myself; but I have more experiments with plants in Hand, than I fear with my poor health I can attend to.—

With cordial thanks | Yours sincerely | C. Darwin

If you do **not** hear, you will understand that your skulls have come safe.—

The skulls have arrived & Live Rabbit—[5] Cordial thanks—

New York Botanical Garden Library (Charles Finney Cox collection)

[1] The year is given by the reference to CD's manuscript on fowls for *Variation*, which he finished in 1861. See 'Journal' (Appendix II).

[2] CD had asked Tegetmeier to read and make comments on the chapter on fowls that he was preparing (see letter to W. B. Tegetmeier, 22 March [1861]).

[3] Tegetmeier 1861.

[4] See letter to W. B. Tegetmeier, 2 April [1861]. Sebright fowl were named after their breeder, John Saunders Sebright (*DNB*).

[5] See letter to W. B. Tegetmeier, 22 March [1861].

To T. H. Huxley 1 April [1861][1]

Down Bromley Kent
Ap. 1st.

My dear Huxley

I return the 3 pamphlets by this post.— The Duke is very liberal & highly complimentary to me, but I do not think much of his argument—[2] Curious case of absence of organic remains is that of India, as described by Oldham.—[3]

I was extremely glad to see your notice in Athenæum.[4] It seemed to me very well done, but almost too civil. It is a good joke, that since Owen attacked me, I do not feel at all a good monitor, & feel more inclined to clap anyone on the back, then to cry hold hard!—[5] I wonder whether he will answer you.—[6] Oh Lord what a thorn you must be in the poor dear man's side.—

Ever yours | C. Darwin

We shd. be very glad sometime to hear how Mrs. Huxley goes on.—[7]

Imperial College of Science, Technology and Medicine Archives (Huxley 5: 162)

1 Dated by the reference to Huxley's 'notice' in the *Athenæum* (see n. 4, below).

2 George Douglas Campbell, duke of Argyll, discussed *Origin* in his presidential address delivered before the Royal Society of Edinburgh on 3 December 1860. The address was published in the *Proceedings of the Royal Society of Edinburgh* 4 (1857–62): 350–77. Campbell praised the 'rich collection' of facts CD marshalled in support of his theory, but stated that they 'bear a very small proportion to the purely speculative conclusions' that CD had drawn from them (*ibid.*, pp. 371–6).

3 The reference may be to a paper by Thomas Oldham on the fossil remains in the Cretaceous rocks of India published in the *Journal of the Asiatic Society of Bengal* (Oldham 1858). Oldham prefaced a list of recently identified fossil species by pointing out the absence in India of representatives of several genera that were common in the Cretaceous deposits of Europe (Oldham 1858, pp. 114–15).

4 The *Athenæum*, 30 March 1861, p. 433, included a critical notice by Huxley of a published abstract of Richard Owen's Royal Institution lecture of 19 March on 'The gorilla and the negro' (*ibid.*, 23 March 1861, pp. 395–6). Huxley pointed out that there were two 'important omissions' in the information given, which he believed obscured 'the real resemblances and differences which obtain between the brain of the highest ape and that of man.'

5 CD's reference is to Owen's hostile review of *Origin* ([R. Owen] 1860b).

6 Owen responded to Huxley's notice in the next issue of the *Athenæum*, 6 April 1861, p. 467. This in turn prompted a lengthy and pointed response by Huxley in the issue of 13 April 1861, p. 498. See letter to J. D. Hooker, 23 [April 1861].

7 Henrietta Anne Huxley was recuperating from the birth of her latest child and grieving over the death of her first-born son. She and the Huxleys' three children had visited the Darwins at Down from 9 to 25 March. See letter to T. H. Huxley, 22 February [1861].

To Daniel Oliver 1 April [1861][1]

Down Bromley Kent
Ap. 1.

My dear Sir

I never dreamed that primroses did not abound with you & am sorry to have given you extra trouble.—[2] I send by this Post, some flowers which I do not doubt will arrive safely & in fresh state.— Thanks for information about Oxalis.—

Please keep memorandum what I owe you for Cypripedium; & I can repay you when I know what the Dionæas cost, if any can be got—[3]

Yours most truly | C. Darwin

American Philosophical Society (243)

[1] Dated by the relationship to the letters to Daniel Oliver, 26 February [1861] and 23 March [1861]. The stationery carries an 1860 watermark.

[2] See letter to Daniel Oliver, 23 March [1861].

[3] CD had asked Oliver to obtain for him some *Dionaea* plants and a specimen of *Cypripedium spectabile* (see letter to Daniel Oliver, 26 February [1861]).

To W. B. Tegetmeier 2 April [1861]

Down Bromley Kent
April 2^d.

My dear Sir

Many thanks for all your interesting enclosures & note.[1] Your paper at Zoolog. Soc^y. seems to have been very successful.[2] I hope that you may be able to prove that your cock is still fertile & the peculiarity hereditary.—[3] It is most singular, the increased combativeness, & a good suggestion that of this cock beating others & so propagating its breed; but probability of this of course will depend on some of the offspring inheriting the peculiarity. I think you will find that the feathers on neck of the male G. varius though not hackles differ from those of hen: the other two known species have hackles.—

Until you know positively that your game-cock *always* produces "Hennies", the cross between it & a hen Sebright would hardly be fair.— I feared that a hen-tailed Hamburgh Cock would not easily be procured:[4] I would almost wager that such a cross would bring back the tail.—

The notice on Hares is done by some right good naturalist: the examination of their skulls would be out of my line; though I do want & must go to B. Museum for skulls of the largest & smallest species of Hares.—[5] You must have a power of clairvoyance; as shown by the offer of examination of skull & *skin* of the **wild** fawn-coloured rabbit; so much do I want **wild** vars. of the rabbit that this very morning I had letter telling me that I sh^d receive in a week a silver-grey from a Norfolk warren.—[6]

I presume I may clean the skull well.— But how can I get particulars about this rabbit, whether it is a breed or a *single* chance variety; & where it lives, & whether the case is true.— I must take in the Field, & will do so.—[7]

Many thanks for information about the young Pigeons; I shall quote this additional case to the others given by you.—[8]

By the way in my poultry M.S. I have been lately often "taking your name in vain": many little facts given by you, which I never thought would have come into play, have been very useful. Did you notice in my "Origin", I say a little on young Pigeons.—[9]

With respect to Herbert on Hybridity. He published an appendix to his Amaryllidaceæ 1837 (a large & expensive Book). Also Journal of Hort. Soc. Vol 2. (1847) p. 1 and p. 81.—[10]

I see you describe nostrils as peculiar in Polands; I think this *perhaps* ought to be figured— Could you procure & send me (by Carrier) a Polish Bird of any Breed. The whole bird might come dead.—[11]

I sent you order for 10^{s} for expences,; but I see that I cause you so much loss of time that we must have a settling before I send my M.S. & God knows when that will be done; so many interruptions occur.—[12]

With sincere thanks | Yours very sincerely | C. Darwin

Endorsement: '/61'
New York Botanical Garden Library (Charles Finney Cox collection)

[1] Tegetmeier's letter has not been found.

[2] Tegetmeier exhibited an adult male brown-breasted red game fowl that had developed female plumage at a meeting of the Zoological Society of London on 26 March 1861 (*Proceedings of the Zoological Society of London* (1861): 102).

[3] At the meeting (see n. 2, above), Tegetmeier stated that he was carrying out experiments to test whether the disposition of male fowl to adopt female plumage was hereditary. In *Variation* 1: 253, CD mentioned the case, stating that Tegetmeier's bird had now 'retained the same character during five seasons' and had produced 'both hen-feathered and male-feathered offspring.'

[4] CD had suggested that Tegetmeier might like to attempt this cross (letter to W. B. Tegetmeier, 28 March [1861]).

[5] Tegetmeier may have sent CD part of a paper by Edwards Crisp (Crisp 1861), which appeared in the same part (part 1) of the *Proceedings of the Zoological Society of London* for 1861 as the notice about Tegetmeier's unusual fowl. There is a copy of this part of the *Proceedings* in the Darwin Library–CUL with both papers lightly annotated. Crisp compared the anatomical structure of several specimens of hare, rabbit, and the so-called leporine or rabbit–hare hybrid. CD cited Crisp 1861 in *Variation* 1: 126 in his discussion of the relative skull capacities of domestic rabbits and hares.

[6] The letter has not been found. CD's specimen of a silver-grey rabbit came from the Thetford warren in Norfolk. That of a wild, 'fawn-coloured variety' (presumably the one provided by Tegetmeier) is said to have come from Scotland (*Variation* 1: 127).

[7] Tegetmeier was the editor of *The Field, the Farm, the Garden, the Country Gentleman's Newspaper*. This periodical is not listed among the journals in the Darwin Library–CUL or the Darwin Library–Down.

[8] See *Variation* 1: 170.

[9] *Origin* 2d ed., p. 445.

[10] Herbert 1837 and 1847.

[11] Tegetmeier was a specialist on Polish fowl (see *Correspondence* vol. 6, letter to W. B. Tegetmeier, 20 March [1856]). Skulls of Polish fowls are figured in *Variation* 1: 262, 263; the nasal bones are discussed in *ibid.*, p. 264.

[12] Tegetmeier had been asked to read, for a fee 'like a Barrister', CD's manuscript on fowls, intended for *Variation* (see letters to W. B. Tegetmeier, 22 March [1861] and 28 March [1861]). An entry on 30 March 1861 for 'Tegetmeier 10^{s} Science./' is recorded in CD's Account book (Down House MS). On 13 June, CD paid him £5 5*s*. 'for scientific assistance'. Tegetmeier's help is acknowledged in *Variation* 1: 225 n.

To H. W. Bates 4 April [1861][1]

Down Bromley Kent
Ap. 4th

My dear Sir

I have been unwell, so have delayed thanking you for your admirable letter.[2] I hope you will not think me presumptuous in saying how much I have been struck with your varied knowledge, & with the decisive manner in which you bring it to bear on each point,—a rare & most high quality, as far as my experience goes.— I earnestly hope you will find time to publish largely: before the Linn. Soc. you might bring boldly out your views on Species.— Have you ever thought of publishing your travels & working in them the less abstruse parts of your Nat. History? I believe it would sell, & be a very valuable contribution to Nat. History.—[3] You must, *also*, have seen a good deal of the natives.

I know well it would be quite unreasonable to ask for any further information from you; but I will just mention that I am now & shall be for a long time writing on Domestic variation of all animals. Any facts would be useful; especially any showing that *savages* take any care in breeding their animals; or in rejecting the bad & preserving the good—or any fancies which they may have that one coloured or marked dog &c is better than another.— I have already collected much on this head, but am greedy for facts.— You will at once see their bearing on Variation under Domestication.

Your observations on Carabus, with respect to Glacial period, seem very important.—[4] I daresay you know that some geologists have speculated on a Permian & even on a Challk Glacial period.— When considering plants of Australia, a suspicion crossed me that there must have been an ancient migration from N. temperate to S. temperate regions.— I feel sure that if you saw lists of plants in T. del Fuego—on *isolated* mountains of India, Java, Borneo, Abyssinia, S.E. Australia, Fernando Po, you would see that there *must* have been a very recent migration. This view is largely supported by plain geological facts.— I have a rather long M.S. discussion, well copied out, which Hooker has read, & which if you thought it worth your while, you might with welcome read;[5] but I doubt whether it would be worth your while.— Hereafter when I come to Geograph. Distrib. (& God knows when that will be) I will deeply consider all your most valuable remarks.—[6]

Thank you for facts on intermediate vars. in intermediate regions; I can see how complex the case is; & I hope before I come to that subject you will have largely published. But the case you give is excellent.—

Hardly anything in your letter has pleased me more than about sexual selection. In my large M.S (& indeed in Origin with respect to tuft of hairs—on breast of Cock-Turkey)[7] I have guarded myself against going too far; but I did not at all know that male & female butterflies haunted rather different sites. If I had to cut up myself in a Review, I would have worked & quizzed Sexual selection; therefore, though I am fully convinced that it is largely true, you may imagine how pleased I am at what you say on your belief.— This part of your letter to me is a quintessence of richness.— The fact about Butterflies attracted by coloured sepals is another good fact, worth its weight in Gold.[8] It would have delighted the heart of old Christian C. Sprengel,—now many years in his grave.—[9]

I am glad to hear that you have specially attended to "mimetic" analogies—a *most* curious subject.— I hope you will publish on it.[10] I have for a long time wished to know whether what Dr. Collingwood asserts, is true, that the most striking cases generally occur between insects *inhabiting the same country*.—[11]

Believe me | Yours most truly obliged | Ch. Darwin

Cleveland Health Sciences Library (Stecher collection)

1 Dated by the relationship to the letter from H. W. Bates, 28 March 1861.
2 Letter from H. W. Bates, 28 March 1861.
3 Bates published an account of his travels in the Amazon River basin in 1863 (Bates 1863).
4 See letter from H. W. Bates, 28 March 1861 and n. 4.
5 Joseph Dalton Hooker had read CD's manuscript on geographical distribution, originally intended for his 'big book' on species, in 1856 (see *Correspondence* vol. 6, letters to J. D. Hooker, 13 July [1856] and [16 October 1856], and letter from J. D. Hooker, 9 November 1856). Hooker also read the 'abstract' of this chapter that CD prepared for *Origin* (see *ibid.*, vol. 7, letters to J. D. Hooker, 2 March [1859] and 15 March [1859]). The original manuscript is in DAR 14 and has been published as chapter 11 of *Natural selection*.
6 CD planned, after finishing *Variation*, to devote a volume to the subject of natural selection in nature, a section of which was to treat geographical distribution. In the event, he included part of this material in subsequent editions of *Origin*.
7 *Origin*, p. 90.
8 See letter from H. W. Bates, 28 March 1861. A note in DAR 76: 18 refers to this point: 'In Bates letter in Portfolio (4) case of Butterfly attracted by bright sepals of flowers.—' CD cited Bates's point about male and female butterflies inhabiting different sites in *Descent* 1: 403.
9 CD had criticised Christian Konrad Sprengel's view that the colour of the corolla of nectar-producing flowers served as a guide to visiting insects in the *Gardeners' Chronicle* in 1841 (see *Correspondence* vol. 2, letter to *Gardeners' Chronicle*, [16 August 1841]; *Collected papers* 1: 142–5).
10 Bates's paper on the mimetic butterflies of the Amazon valley (Bates 1861b) was read at a meeting of the Linnean Society of London on 21 November 1861. CD repeated some of these observations in his unsigned review of Bates's paper (*Natural History Review* n.s. 3 (1863): 219–24; see *Collected papers* 2: 87–92).
11 The reference may be to Collingwood's recent paper on mimetic analogy, or what he called 'homomorphism' (Collingwood 1860c). CD had also read another work of Collingwood's on the same topic (Collingwood 1860a). See letter to Cuthbert Collingwood, 14 March [1861].

To J. D. Hooker 4 April [1861]

Down Bromley Kent
Ap. 4th

My dear Hooker

I have been confined to bed; otherwise I shd. have written yesterday to have thanked you for your note about poor dear Henslow.—[1] In two or three days do let me have literally one line. If he is well enough tell him how gratefully & affectionately I shall love & honour him to the day of my death.—

Ever yours | C. Darwin

Endorsement: '/1861'
DAR 115.2: 95

1 Hooker's letter has not been found. He was staying at the home of his father-in-law, John Stevens Henslow, who was seriously ill.

To Daniel Oliver 4 April [1861][1]

Down Bromley Kent
Ap. 4th

My dear Sir

Very many thanks for the Primulas (& about common Primrose) which I received this morning & examined.[2] Only one was dimorphic: but I have every reason to suppose the others would have been so, had you possessed a greater number of plants.— P. Siberica, however, seems to differ from all other species in not being dimorphic.—[3]

I have been deeply grieved to hear about Prof.— Henslow—

Yours very sincerely | C. Darwin

As P. Siberica is apparently so exceptional, would you ask Mr Croker, if you have other plants, which come into flower to send me 2 or 3 flowers (not whole trusses) for examination, in order that I may see whether they present any incipient trace of dimorphism.—

But if your plants have been increased all by offsets from same original plant, it will be no use to send them. As the character is permanent by offsets.

I do not know whether I shall succeed in making out the meaning of the dimorphism; but I have not been idle, for I have made much above 100 crosses with the pollen of the different sizes.

The lot sent has been most interesting to me.— Pray thank Mr Croker—[4]

Do not trouble yourself to write if you can at any time send me other specimens of P. Siberica—

Down House (MS 10: 29)

1 Dated by the relationship to the letters to Daniel Oliver, 23 March [1861] and 1 April [1861] and by the reference to John Stevens Henslow's illness.

2 CD had asked Oliver to send him specimens of various species of *Primula* (see letter to Daniel Oliver, 23 March [1861]).

3 In his paper 'On the two forms, or dimorphic condition, in the species of *Primula*, and on their remarkable sexual relations' (*Journal of the Proceedings of the Linnean Society (Botany)* 6 (1862): 77–96), CD mentioned that 'unless indeed this single specimen was anomalous', *P. sibirica* was not dimorphic (*ibid*., p. 81; *Collected papers* 2: 48–9).

4 Charles William Crocker was foreman of the propagating department at the Royal Botanic Gardens, Kew.

To George Busk 5 April [1861][1]

Down Bromley Kent
Ap. 5th

Dear Busk

In last nor of Journal of Linn. Soc. there is a marvellous account of Ants;[2] enclosed is its match, which if you think fit, might be read, with some such title as "Extracts from two Letters from G. Lincecum Esqre of Long Point Texas to Ch. Darwin Esqe, on the Habits of Ants".—[3] Please observe, I know nothing of writer.— But if you will take trouble to read the *whole* of these extraordinary epistles, I think you will be impressed with belief that the man does not *intentionally* tell lies. He paid the heavy postage on both.— I have struck out with pencil what ought *not* to be read.— If the facts are true, it is perhaps most marvellous instinct ever recorded.— Really I can almost believe the statements, after Kirby's account of the Ants bringing up the eggs of their imprisoned Aphides to the sun to be warmed & to be hatched early that they might be milked soon.—[4]

The only use of publishing such a paper in my estimation is that it might call some other observer's attention to the points.—

The whole letters are so odd that they are almost worth your reading,—such spelling,—such grammar! He evidently speaks the truth that he was never educated.—

You must use your own judgment whether to read,—I hardly know what to think.[5]

Pray believe me | Yours very sincerely | C. Darwin

P.S. In reading the 1st letter, attend to **his paging**; for the order goes very oddly.

Linnean Society

[1] Dated by the reference to the letters from Gideon Lincecum (see n. 3, below).
[2] The March issue of the *Journal of the Proceedings of the Linnean Society of London* 5 (1861): 217–18 carried an extract from a letter by Mrs Lewis Hutton giving an account of the habits of a species of Australian ant. Hutton described what she took to be an elaborate 'burial ritual' for soldier ants conducted by about 200 ants from the same nest.
[3] Letters from Gideon Lincecum, 29 December 1860 (*Correspondence* vol. 8), and 4 March 1861. Busk was secretary of the Linnean Society of London.
[4] Kirby and Spence 1815–26, 2: 89–90.
[5] An abstract of Lincecum's letter of 29 December 1860 was read at a meeting of the Linnean Society on 18 April 1861 and published in the *Journal of the Proceedings of the Linnean Society (Zoology)* 6 (1862): 29–31. Busk omitted Lincecum's description of the 'horticultural ant' given in the letter of 4 March 1861, stating that the account 'appears to be identical with the "Cutting Ant," *Œcidina nexucaba*, Sm., described by Mr. S. B. Buckley' (*ibid.*, p. 31). The reference is to Buckley 1860.

From John Innes[1] [before 6 April 1861][2]

with spiteful little savages. I think a bee never gets his sting withdrawn from human skin. I have often been stung and not touched them and the sting, if really inserted, has always remained behind. ⟨ ⟩ withdraw the sting with no difficulty—

We have only three weeks now to remain in these comfortable quarters. We shall go with much regret [3] We have made some plans. ⟨ ⟩ I do

Incomplete
DAR 48: 69

CD ANNOTATIONS
1.1 I think . . . skin. 1.2] *double scored pencil, brown crayon*; '(Rev J. Innes)' *written over text, pencil*
1.2 I have . . . behind. 1.3] *scored brown crayon, square bracket added at end, pencil*
2.1 We have . . . plans. 2.2] *crossed pencil*

[1] The identity of the correspondent is given by CD's annotation.
[2] Dated by the relationship to the letter from John D. Glennie Jr, 6 April 1861.
[3] Innes, having been unable to secure accommodation in Down village the previous year, was living elsewhere (see *Correspondence* vol. 8, letters to John Innes, 18 July [1860], 6 September [1860], 11 September [1860], and 28 December [1860]).

To Samuel Birch 6 April [1861]

Down Bromley Kent
April 6th

My dear Sir

You formerly gave me some valuable extracts about Poultry from Chinese Encyclops.—[1] I shd. be so much obliged if you could answer me 4 questions on enclosed paper. You might affix "yes" or "no" & "date" & return the paper, without troubling yourself to write. I enclose your old memorandums, as they will recall the Books in question.

Pray forgive me being so troublesome & believe me | My dear Sir | Yours sincerely obliged | Ch. Darwin

I enclose directed envelope so as to give as little trouble as possible.

[Enclosure]

What is probable age of the Japanese Encyclop? (marked with single red line)[2]

Am I right in believing that the Santsae or Chinese "Encyclop. of Three Principles", was compiled from other *ancient* works? I *think* you told me so.— Is this the case also with the Puntsaon?[3]

You give in note (marked with 2 red lines) an extract about Fowls with *black feathers & bones*. Is anything said about the feathers being hairy. I wish much to know, as I have extraordinary case of reversion in this breed.—[4]

Pickering in his Races of Man p. 374 speaks of a Head & neck of Fowl in a tribute-procession to Thouthmosis III. Can you tell me date of this King?[5]

Endorsement: '1861'
British Museum (Department of Western Asiatic Antiquities), Correspondence 1826–67: 1493

[1] In 1856, CD had asked Birch, then assistant keeper in the department of antiquities of the British Museum, whether he knew of any Chinese works that contained references to the varieties of domesticated animals and cultivated plants kept by the Chinese (see *Correspondence* vol. 6, letter to J. E. Gray, 14 January [1856]). Birch subsequently provided CD with translated extracts from such works (see *ibid.*, vol. 7, Supplement, letter to Samuel Birch, 8 April [1856]).

[2] Apparently Birch was unable to provide a firm date for the work. In *Variation*, CD mentioned references to dwarf fowl in an ancient Japanese encyclopaedia that Birch had located (*Variation* 1: 230 n. 5 and 247).

[3] In *Variation*, CD gave information on breeds of fowl taken from two ancient Chinese texts translated for him by Birch: 'a Chinese Encyclopædia published in 1609, but compiled from more ancient documents' and 'the Chinese Encyclopædia published in 1596, but compiled from various sources, some of high antiquity' (*ibid.*, 238, 247). The texts to which he refers are the *San ts'ai t'u hui* (1609), compiled by Wang Ch'i, which included sections on birds and beasts and plants and trees; and the famous pharmacopia *Pen ts'ao kang mu* (1593) of Li Shih-chen (*Cambridge history of China* 7: 774–7).

[4] CD probably refers to an 'extraordinary' case of reversion he had noticed in the progeny of a cross between a white silk hen and a black Spanish cock. The offspring were all black with 'blackish combs and bones'; as they grew, however, they developed characteristics resembling the wild *Gallus bankiva*, even though both breeds were thought to be ancient and were known to breed true. See *Variation* 1: 241–2.

[5] In *Variation* 1: 246 n. 33, CD wrote:

> Dr. Pickering, in his 'Races of Man,' p. 374, says that the head and neck of a fowl is carried in a Tribute-procession to Thoutmousis III. (1445 B.C.); but Mr. Birch of the British Museum doubts whether the figure can be identified as the head of a fowl.

The reference is to Pickering 1850.

From John D. Glennie Jr 6 April 1861

Stapleford Lodge | near Cambridge
Ap. 6. 1861.

Dear Sir

Your friend Innes has forwarded your letter of the 30th. Ulto: to me respecting "Bees."[1] If my memory serves me aright you & I have already met before & the same subject was then the magnet which drew us together. Were not you present at a lecture which I delivered on Bees at Southend Lewisham, & did not I lend you part of my MS. afterwards referring to the construction of cells?[2] But be this as it may I now gladly turn to the points of enquiry in your letter. They are, I hold, very fully answered in a letter I wrote to Notes & Queries in a reply to a correspondent who started the idea that "bees lose their life by stinging" is a "popular error" & asking "what foundation there was for it". I will therefore refer you to that (N. & Q. Vol XI. No. 295. page 489)[3] I had forgotten what gave rise to it or even the drift of my answer but on reading it now I am sure that it must be that to which Innes referred you.[4] I will now *add* a few remarks to my former statements which will touch the points of your letter, & you may make any use you please of what I state. Bees & wasps are not, *on this point* to be placed exactly in the same category. The bee never stings until previously brought to a state of intense excitement and when once the act has been committed she seems no longer mistress of her actions Were she able to *calm* herself immediately after planting her sting, and if the offended person would likewise be calm & allow the bee the time & opportunity to release itself (two contingencies *most* unlikely to concur) I believe it possible that the poor bee might escape with sting & life entire. But as soon as she has fixed herself by her sting to the flesh she begins to tug away & twist about till the bag of poison & sting together part from the rest of the body. But the moment *this* catastrophe has taken place, all the excitement gives way at once, the death-wound is felt & she does not even seem to care to escape the vengeance of her late enemy.

The wasp, on the contrary, goes to work with far more coolness & conscious strength. She is an idle adventurer, devoid of all the home ties of the busy bee, who lives by plunder whose hand is against every one & every one's hand against her. Her armour both offensive & defensive is proportionally stronger than that of the thrifty bee. The one is of the agricultural labourer, the other of the (Amazon warrior) type. If the wasp stings & has the *opportunity* of escape she will get away uninjured as is very often the case. Her stinging apparatus is much more firmly fixed than that of the bee & (I *have been told* but cannot *vouch* for it) the sting is either not barbed or much more slightly barbed than that of the bee.

To save my bees lives when operating upon them & unavoidably irritating them I use *thick* woollen gloves (ringwood gloves)[5] or two pair, & woollen trousers & coat (not fustian or corderoy or leather) thick enough to prevent the sting from reaching so deep as to the skin, because the barbs do not act to any injurious extent in clinging to the woollen texture & the bees sting & sting again to their hearts content without injury to me or to themselves.

I have been writing all this in the midst of a shower of conversation & repartee from some 6 different members of my family in the same room. Forgive therefore incoherence & lengthy verbiage. I believe I have answered your questions. If not, or I can help you on any other point, pray write again to me direct

Y^rs. truly | (Rev) John D Glennie Jun

DAR 48: 70–3

CD ANNOTATIONS
1.1 Your ... cells? 1.5] *crossed pencil*
1.9 & asking ... 489) 1.10] *scored brown crayon*
2.8 the sting ... bee. 2.9] *double scored brown crayon*
4.1 I have ... Jun 5.1] *crossed pencil*
Top of first page: 'On Bees stinging | (Try Pork or Chicken skin—)' *pencil*

[1] CD's letter has not been found. It may have been written at the suggestion of John Innes (see letter from John Innes, [before 6 April 1861]). Innes, perpetual curate of Down, was himself a beekeeper who had supplied CD with information on bees (see *Correspondence* vol. 6, letter to John Innes, [after 16 February 1857]).

[2] It is not known whether CD attended Glennie's lecture. CD was actively engaged in investigating the construction of bees' cells in 1858 (see *Correspondence* vol. 7).

[3] [Glennie] 1855. This was published in the first series of *Notes and Queries*. Glennie wrote the piece under the pseudonym J. D. Ottinge.

[4] This information may have been included in the missing portion of the letter from John Innes, [before 6 April 1861].

[5] Ringwood, Hampshire, had long been a centre for the manufacture of woollen gloves (Hewitt 1912).

To John Crawfurd 7 April 1861

Down Bromley Kent
April 7 61

My dear Mr. Crawfurd

Very many thanks for your note & kind present of the two pamplets[1] I felt sure the one I asked about, was by you: but I did not know whether you acknowledged it. I can now refer to it as read at Oxford in 186–[2] I will look to Book about Indian dates.—[3] I do not believe in metempsychosis nor in Genesis—& you are growing so orthodox, that you will end your days, I believe, in believing in the Tower of Babel—[4]

With sincere thanks | Yours very sincerely | C Darwin

PS I have just read & been much interested by your essay on the Horse[5] I thank you much for having sent it. What a store of knowledge you possess

Copy
DAR 143

[1] In addition to sending CD a copy of Crawfurd 1860b (see n. 5, below), Crawfurd apparently sent an abstract of Crawfurd [1860a], which was published in the *Report of the 30th meeting of the British Association for the Advancement of Science held at Oxford*, Transactions of the sections, p. 155.

[2] Crawfurd [1860a]. See letter to John Crawfurd, 25 March [1861], and n. 1, above. A dash for the year was left by the copyist.

[3] It is not known to what book CD refers, although it may be a work by the orientalist Horace Hayman Wilson. CD had asked Crawfurd about the dating of the 'Institutes of Menu' (see letter to John Crawfurd, 25 March [1861]). In *Variation* 1: 246, CD wrote concerning the domestication of fowls in India:

> In India it must have been domesticated when the Institutes of Manu were written, that is, according to Sir W. Jones 1200 B.C., but according to the later authority of Mr. H. Wilson, only 800 B.C., for the domestic fowl is forbidden. Whilst the wild is permitted to be eaten.

[4] Crawfurd's orthodox religious views were readily apparent in his review of *Origin* ([Crawfurd] 1859). See *Correspondence* vol. 7, letters to Charles Lyell, 2 December [1859], and to T. H. Huxley, [5 December 1859].

[5] Crawfurd 1860b.

To Daniel Oliver 9 April [1861][1]

Down Bromley Kent
Ap. 9th.

My dear Sir

Anytime when you happen to write to me, *but not write on purpose*, please tell me name of enclosed Fumaria or Corydalis— it grows in my garden in open air.— It exhibits a very little, pretty adaptation to Bees.—[2] The long nectary alone secretes honey; & when Bees are sucking, they push off the hood over the pistil & stamens; & then the pistil & stamens (*curved to side of nectary*) are rubbed against Bee's belly, & so cross would often be effected. As long as hood is kept pushed off the pistil & stamens, *it keeps open*, & when it springs back, it beautifully snaps together & again encloses the pistil & stamens.— Now if you push the opposite way, which Bees never do, as there is no nectar on that side, it is with difficulty that the hood can be pushed off the pistil & stamens, but when this is effected, the hood does *not keep open*, & so can never again enclose & protect the pistil & stamens.[3] It is worth looking at fresh flower, & trying with a pin.—

In Dielytra, where nectar is on *both* sides, pistil is straight, & the hood can be pushed off *both sides* with equal facility, as I have seen Bees do.— Here is much cry & little wool![4]

Yours very sincerely | C. Darwin

Down House (MS 10: 6)

[1] The year is given by a note dated 9 April 1861 relating to *Fumaria* (see n. 3, below).

[2] CD had begun experiments on the pollination mechanisms of Fumariaceae in 1858, though he did not publish on this topic until many years later. See *Nature* 9 (1874): 460 (*Collected papers* 2: 182–3) and *Cross and self fertilisation*.

[3] The observation referred to in the letter is described in a note dated 9 April 1861 (DAR 76 (ser. 2): 17):

> I have been looking at that Fumariæ dingy purple with one long nectary & one rudimentary— (the latter alone [*interl*] white). Plenty of sweet nectar in the long one; none in short.— Pistil curved to long-nectary side.— Here [*after del* 'I fi'] is beautiful contrivance, if hood be forced off, as Bee would in sucking, it keeps wide open as long as open & when it springs back it beautifully encloses, as at first, the anthers— It can be pushed the other way with difficulty, but then there is this marked difference the hood does *not* keep open; so that when loosed it does not enclose pistil & stamens, which are thus no longer protected. . . . Mr Oliver believes Corydalis solida (=C. bulbosa)

[4] This proverb from the Latin is equivalent to 'much ado about nothing'.

To Asa Gray 11 April [1861][1]

Down Bromley Kent

My dear Gray

I was very glad to get your Photograph:[2] I am expecting mine which I will send off as soon as it comes. It is an ugly affair, & I fear the fault does not lie with the Photographer.—[3]

I believe, but cannot swear, that I wrote & told you that Wrights Review had come through Sampson & Son.[4] I had time hardly to read it, before Huxley took it away. He much feared it was too general & not natural-Historical enough for him.[5] This was my impression, likewise; though I daresay it is very clever. What shall I do with it, if Huxley does not take it? I know no other Review to send it to.—

Since writing last I have had several letters full of highest commendations of your Essay:[6] all agree that it is by far the best thing written, & I do not doubt it has done the Origin much good. I have not yet heard how it has sold. You will have seen Review in G Chronicle.—[7] There is to be a Review by A. Murray in next Eding. New Phil. Journal.—[8] I received the Letter of Credit returned: I am pleased & surprised at Profit from the American Edit.[9] Remember that you are to be at no expence about your Essay. I presume nothing literary now sells in the troubled U. States.[10]

Poor dear Henslow, to whom I owe much, is dying; & Hooker is with him.—[11]

Many [thanks] for two sets of sheets of your Proceedings.[12] I cannot understand what Agassiz is driving at.—[13] You once spoke, I think, of Prof. Bowen, as a very clever man. I sh^{d} have thought him a singularly unobservant & weak man from his writings.[14] If ever he agrees with me on any one point, I shall conclude that I must be in error on that. He never can have seen much of animals or he would seen the difference of old & wise dogs & young ones.— His paper about hereditariness beats everything.[15] Tell a breeder that he might pick out his worst *individual* animals & breed from them & hope to win a prize; & he would think you not a fool, but insane.— I believe Bowen is a metaphysician & that I presume accounts for an entire want of common sense.

Please remember Spiranthes;[16] if you insert a culm of grass, remember before you withdraw it to bend or bow it towards rostellum. Please if you come across *wild* Apocynum, observe whether it catches flies as in England.—[17] I enclose my Photograph which has come rather crumpled, but I suppose can be ironed smooth.—[18]

My dear Gray | Yours most truly | C. Darwin—

P.S. I enclose a little Photograph made this morning by my eldest Son[19]

April 11th.—

Gray Herbarium of Harvard University (53)

[1] Dated by the reference to A. Gray 1861a.

[2] Gray's photograph has not been found in the Darwin Archive. For a contemporary photograph of Gray, see the frontispiece to *Correspondence* vol. 6.

[3] The photograph to which CD refers may be the one of him taken in the late 1850s (see the frontispiece to *Correspondence* vol. 8).

[4] See letter to Asa Gray, 12 March [1861]. CD refers to Sampson Low and his son Sampson Low Jr, who were partners in the publishing firm Sampson Low, Marston & Co. of Ludgate Hill, London (*Modern English biography*).

[5] Thomas Henry Huxley had agreed to consider publishing Chauncey Wright's review of *Origin* in the *Natural History Review*, of which Huxley was senior editor. The review did not appear in the journal.

[6] A. Gray 1861a.

[7] A review of A. Gray 1861a appeared in the *Gardeners' Chronicle and Agricultural Gazette*, 9 March 1861, p. 219. CD had helped to arrange the review (see letter to J. D. Hooker, 20 [February 1861]).

[8] No notice of A. Gray 1861a appeared in the *Edinburgh New Philosophical Journal*. Andrew Murray had told CD that he would notice Gray's pamphlet in the journal (see letter from Andrew Murray, 3 March 1861).

[9] According to CD's 'Journal' (Appendix II), he received £10 10*s*. from D. Appleton and Co. in May 1861 as his share of the profits on the American edition of *Origin*. However, this amount reflected that remaining after Gray had used part of the money to pay for the printing costs of A. Gray

1861a. In the letter from Asa Gray, 31 March [1862] (*Correspondence* vol. 10), Gray informed CD about this transaction: 'I took about £50 . . . which ⟨Ap⟩pleton & Co sent me for ⟨your⟩ book, to pay the printers with'. In May 1860, CD had received £21 17*s*. 6*d*. Gray had negotiated on CD's behalf with the publisher, who agreed to grant CD author's royalties on the edition (see *Correspondence* vol. 8).

10 The United States of America was on the brink of civil war. Hostilities broke out on 12 April 1861, when Southern troops fired on Fort Sumter, South Carolina (*EB*).

11 Joseph Dalton Hooker and his wife Frances Harriet Hooker were at John Stevens Henslow's bedside at the rectory in Hitcham, Suffolk. Henslow died on 16 May.

12 CD obviously forgot to write 'thanks' when he began a new page. Gray had sent CD copies of pages from volume 5 of the *Proceedings of the American Academy of Arts and Sciences*, which carried discussions relating to CD's theory of the origin of species. Both Francis Bowen, professor of natural religion, moral philosophy, and civil polity at Harvard, and Louis Agassiz, professor of natural history, had presented papers at a meeting of the Boston society on 8 January 1861 (see Bowen 1861). For CD's earlier mention of this discussion, see the letter to Charles Lyell, 2 February [1861].

13 See letter to Charles Lyell, 2 February [1861], n. 4.

14 Bowen had published two reviews of *Origin* in 1860. The first ([Bowen] 1860a) CD described as being 'clever & dead against me' (see *Correspondence* vol. 8, letters to J. D. Hooker, 18 [April 1860], and to Asa Gray, 25 April [1860]). The second (Bowen 1860b) CD repudiated as being 'absurd', 'monstrous', and 'rubbish' (see *ibid*., letter to Asa Gray, 26 November [1860]).

15 In Bowen 1861, Bowen denied the heritability of mental and physical traits (see letter to Charles Lyell, 2 February [1861]).

16 CD had asked Gray to make observations of the American species of *Spiranthes* (see *Correspondence* vol. 8, letter to Asa Gray, 31 October [1860]). Gray's observations on *Spiranthes* are cited in *Orchids*, p. 123 n.

17 CD did not discuss *Apocynum* in *Insectivorous plants*.

18 See n. 3, above.

19 The photograph of CD taken by William Erasmus Darwin, from the Gray papers in the Gray Herbarium of Harvard University, is reproduced as the frontispiece to this volume.

To J. D. Hooker 11 April [1861]

Down Bromley Kent
Ap. 11.

My dear Hooker

In a note received this morning, Huxley mentions poor dear Henslow's death.[1] Thank God his sufferings are over.— I need not say how truly I sympathise with Mrs. Hooker's & your own grief.— I think he was the best, the most noble character I ever knew. It will ever do one good to think of him.—

My dear friend | Yours affectionately | Ch. Darwin

Endorsement: '/61'
DAR 115.2: 96

1 The letter from Thomas Henry Huxley has not been found. CD had misunderstood Huxley's note: John Stevens Henslow was seriously ill but still alive. See letter to J. D. Hooker, 14 [April 1861].

To Charles Lyell 12 April [1861]

Down Bromley Kent
Ap. 12th

My dear Lyell

I have been most deeply interested by your letter. You seem to have done the grandest work & made the greatest step of anyone with respect to man.—[1] It is an especial relief to hear that you think the French superficial, deposits are deltoid & semi-marine.[2] But two days ago I was saying to a friend, that the unknown manner of accumulation of these deposits seemed the great blot in all the work done. I could not stomach debacles or lacustrine beds.— It is grand.— I remember Falconer told me that he thought some of the remains in the Devonshire Caverns were preglacial;[3] & this I presume is now your conclusion for the older celts with Cyrena & Hippopotamus.[4] It is grand. What a fine long pedigree you have given the Human Race!

I am sure I never thought of parallel Roads having been accumulated during Subsidence.—[5] I think I see some difficulties on this view; though at first reading your note I jumped at idea.— But I will think over all I saw there. I am (stomacho volente) coming up to London on Tuesday to work on Cocks & Hens & on Wednesday morning about $\frac{1}{4}$ before 10 I will call on you, (unless I hear to contrary); for I long to see you.—[6] I congratulate you on your grand work

Ever yours | C. Darwin

P.S. Tell Lady Lyell that I was unable to digest the funerial ceremonies of the ants;[7] notwithstanding that Erasmus[8] has often told me that I shall find some day that they have their Bishops. After a battle, I have always seen the ants carry away the dead for food. Ants display the utmost economy & always carry away a dead fellow creature as food.— But I have just forwarded two most extraordinary Letters to Busk, from a back-woodsman in Texas, who has evidently watched ants carefully & declares most positively that they plant & cultivate a kind of grass for store-food, & plant other bushes for shelter! I do not know what to think, except that the old Gentleman is not fibbing intentionally. I have left the responsibility with Busk whether or no to read the letters.[9]

Postmark: AP 13 61
American Philosophical Society (244)

[1] Lyell's letter has not been found. Since 1859 he had been actively investigating the geological evidence that pertained to the antiquity of man to include in a revised edition of his *Principles of geology*. In the event, Lyell published his findings in a separate volume (C. Lyell 1863). For CD's comments on some of Lyell's earlier findings, see *Correspondence* vol. 7, letter to Charles Lyell, 2 September [1859].

[2] Lyell had for several months been evaluating the notes he had made and specimens he had collected during visits to Amiens and Abbeville, France, in 1859 and to Abbeville again in 1860 (K. M. Lyell ed. 1881, 2: 341, 344). By determining more precisely the nature of the drift deposits of the valley of the Somme, in which large numbers of flint implements had been discovered, Lyell hoped to resolve some of the conflicting geological interpretations of these deposits, which provided critical evidence concerning the antiquity of man. For Lyell's determination that the Abbeville deposits were

'fluvio-marine' rather than lacustrine, see *ibid.*, p. 344. On the cover, Lyell wrote in part: 'C. Darwin Apr. 13 | 1861 Somme valley beds whether preglacial'.

[3] Hugh Falconer and other naturalists had been engaged since 1858 in a study of the fossil remains of a cave near Brixham in Devon (C. Murchison ed. 1868, 2: 486, 491–7). CD had been much interested in their finding flint implements beneath deposits containing the bones of extinct mammalian species. See *Correspondence* vol. 7, letters to J. D. Hooker, 6 October [1858] and 20 [October 1858].

[4] Lyell had found the extinct freshwater bivalve *Cyrena fluminalis* above a bed containing flint implements at Abbeville (K. M. Lyell ed. 1881, 2: 344), just as Falconer had discovered the bones of hippotomuses associated with worked flints in the caves of southern England. Lyell, however, was unclear about how these finds related chronologically to the glacial period. As he stated in a letter to Charles James Fox Bunbury of 26 April 1861 (*ibid.*, p. 344):

> The relation of man to the close of the glacial period is a point on which I have not yet made up my mind, but I suspect those beds in France may have begun before the northern drift of England was finished.

[5] The reference is to the so-called 'parallel roads' of Glen Roy, in Scotland. In 1839, CD published a paper in which he concluded that the roads were the remnants of former beaches formed successively as a result of the periodic *elevation* of the land (see *Collected papers* 1: 89–137). Lyell was reviewing the geological evidence to explain the phenomena of Glen Roy in connection with revisions he intended to make to his *Principles of geology*.

[6] According to Emma Darwin's diary, CD went to London on 16 April 1861 and returned to Down on 20 April.

[7] Mary Elizabeth Lyell had apparently drawn CD's attention to an account of 'funerial ceremonies' in Australian ants. See letter to George Busk, 5 April [1861] and n. 2.

[8] Erasmus Alvey Darwin.

[9] See letter to George Busk, 5 April [1861]. On the cover, Lyell wrote in part: 'Ants in Texas planting'.

To J. D. Hooker 14 [April 1861]

Down. Bromley Kent
14.

My dear Hooker

I have been pained at a mistake which I have made.[1] In a letter from Huxley, I misunderstood him that poor dear Henslow was dead, & I inferred that he had seen you & that you were at Kew. I believe & hope that this is all a blunder of mine.— But I wrote to Kew, thinking that you were there, & I fear that Mrs Hooker may have opened my note. This has pained. I much wish to hear how Henslow is going on. I cannot avoid hoping that he may be rallying, or is it really all over.—

But you must be worn out with nursing, so do not write. I know well how even writing a direction is intolerable sometimes. I go to London on Tuesday & daresay I shall hear news there.[2]

My dear old friend | Yours affect^y^. | C. Darwin

Endorsement: 'April/61'
DAR 115.2: 97

[1] See letter to J. D. Hooker, 11 April [1861] and n. 1.

[2] CD went to London on Tuesday, 16 April 1861 (Emma Darwin's diary).

Henry Walter Bates.
(Courtesy of the Royal Geographical Society, London.)

Erasmus Darwin. Portrait by Joseph Wright of Derby, *c.* 1770. (Courtesy of Darwin College, Cambridge.)

Josiah Wedgwood I. Portrait by Sir Joshua Reynolds, 1782. (Courtesy of the Trustees of the Wedgwood Museum, Barlaston, Staffordshire.)

To W. B. Tegetmeier 14 April [1861][1]

Down Bromley Kent
Ap. 14th

My dear Sir

Many thanks for Scotch Fawn Rabbit received yesterday & examined to day: I have been very glad to see it. How shall I return it? Can you find out *two* points for me,—was it *wild* & was it (as I suppose) a single odd individual?—[2]

The Angora has just been examined: before making skeleton, for several reasons (great size large ears &c) I suspected that it had been crossed with a large Fancy rabbit; & now I think this still more strongly.— Could you get me a pure *albino* rabbit? I shd be very glad of another specimen.

Also please do not forget a half-lop, with one ear *erect* & *smaller* than the other.—

I have got (but not yet read) the Field to day & see your curious Cock, which I have quoted in my M.S.[3] I have now done external characters & shall commence on skeletons.

I plainly see that I shall consume much of your valuable time.— I now enclose a string of questions: some few easy to answer but mostly troublesome & requiring special examination of several Fowls.—

You will see several apply to Shangais; also some to Polands these will be easy to you.— My M.S. will, also, be a much longer job for you than I had anticipated.[4] I have kept list of my Queries, which may shorten your answers.—[5] Could you visit any one who keeps several Shangais & Malays? All my queries are founded on what I have seen or heard; but it would take too much space to give my reasons for asking.—

Your assistance will be invaluable to me.—

My Dear Sir | Yours sincerely | Ch. Darwin

P.S. | I have just opened your Box with Skulls. What a magnificent collection![6] I suppose that I have a score more, so that I shall have good materials. I notice that some as Bantams, are not clean enough.— I will assume that I may boil them & clean them, though of course there is always some risk in this, in relation to the articulated bones at the base. But the loss would not be irremediable & I will take care. I must confess that some of my own became disarticulated, but this to me was no injury.—

How did you saw so neatly the skull of Poland?

[Enclosure][7]

(19) Though the cock which conquers, naturally gets first choice of wives, yet from analogy with other Birds, & from one account, I imagine the Hen has to be won or charmed to grant her favours.— Can you throw any light on this point, which interests me much.?— There can hardly be a doubt that the beauty of male Bird is to charm the female. Now it has occurred to me to ask you to observe, whether Hens yield to your Hen-tailed Game, as readily as to other cocks. He would be victor, & so have enormous advantage, but would his want of beauty at

all interfere with his amatory success, *supposing any other & gorgeous cock were present.*— I formerly thought of clipping and mutilating the feathers of a cock & observing, but I could not spare time to observe.

Do you know anyone who has kept many Hen-tailed Games & other Game-cocks, such a man might have observed on this point.—

New York Botanical Garden Library (Charles Finney Cox collection)

1 Dated by the relationship to the letter to W. B. Tegetmeier, 2 April [1861], and by the reference to CD's manuscript on fowls. CD was composing the chapter on variation in domestic and wild fowls for *Variation* (see 'Journal'; Appendix II).

2 CD included measurements taken from the wild, fawn-coloured rabbit from Scotland in the chart on osteological differences between various breeds published in *Variation* 1: 127.

3 *Variation* 1: 253. See letter to W. B. Tegetmeier, 2 April [1861].

4 Tegetmeier had agreed to read CD's manuscript on fowls (see letters to W. B. Tegetmeier, 22 March [1861] and 28 March [1861]).

5 The list of CD's queries has not been found; only part of the enclosure has been located.

6 CD had asked to examine Tegetmeier's extensive collection of the skulls and skeletons of various breeds of fowl (see *Correspondence* vol. 8, letters to W. B. Tegetmeier, 17 April [1860] and 20 July [1860]).

7 The enclosure is in DAR 84.1: 145. Tegetmeier's response to CD's queries is written on the back (see letter from W. B. Tegetmeier, [28 May 1861]). The annotations that CD made on the enclosure pertain to Tegetmeier's response: see the letter from W. B. Tegetmeier, [28 May 1861].

From George Rolleston 16 April 1861

Oxford.
April 16. 1861.

My dear Sir—

I write a single line to say that I do not see in your Index to the "Origin of Species" any reference to the point of the "Variability of Rudimentary organs"— This point you touch upon in Pages 149 & 168. And it may be illustrated in the concrete from Leuckart in the Cyclopædia of Anatomy Article "Vesicula Prostatica" p. 1424,[1] & from Leydig Zeitschrift Wiss. Zoologie ⟨ ⟩.1.38. à propos of the same structure—[2]

Yours very Truly | George Rolleston—

DAR 176

CD ANNOTATION

Top of first page: '10'[3] *red crayon*

1 Leuckart 1852. The article by the German comparative anatomist Rudolph Leuckart was published in volume 4 of Robert Bentley Todd's *The cyclopaedia of anatomy and physiology* (London, 1835–59). CD referred to this work in the discussion of rudimentary structures associated with the reproductive system of man in *Descent* 1: 31.

2 Leydig 1850.

3 The number of one of CD's portfolios of notes in which he kept information relating to rudimentary structures or abortive organs. Many of the notes in DAR 205.1 are so marked.

From P. L. Sclater 17 April 1861

Zoological Society of London, | 11, Hanover Square, | London, W.
April 17th *1861*

My dear Sir

In p. 422 of your new edition you say "Madeira does not possess one peculiar bird"—[1] This is not quite the fact. Out of the 99 birds given by Mr Vernon Harcourt in Ann. N.H. June 1855 as belonging to its Fauna *one—Regulus maderensis* is peculiar— it is of course an altered form of *R. auricapillus*.[2] Three others *Fringilla canariensis*, *Cypselus unicolor* and *Columba trocaz* if not peculiar are distinct from European species and only found elsewhere in the Canaries— I am not aware that Cyps. *unicolor* has been met with even there.

Very faithfully Your's | P. L. Sclater

DAR 205.3 (Letters)

CD ANNOTATION

Top of first page: '19'[3] *red crayon, circled red crayon*; '& Origin'[4] *red crayon*

[1] This statement in the third edition of *Origin* was unchanged from the first edition (*Origin*, p. 391).

[2] CD included this information in *Origin* 4th ed., p. 465:

> Almost every year, I am informed by Mr. E. V. Harcourt, many European and African birds are blown to Madeira; this island is inhabited by 99 kinds, of which one alone is peculiar, though very closely related to a European form;

The reference is to Edward William Vernon Harcourt, with whom CD had corresponded on this topic in 1856 (see *Correspondence* vol. 6, letter from E. W. V. Harcourt, 31 May 1856, and letters to E. W. V. Harcourt, 19 August [1856] and 23 August [1856]). An annotated copy of Harcourt's paper on the ornithology of Madeira (Harcourt 1855) is in the Darwin Pamphlet Collection–CUL.

[3] The number of CD's portfolio of notes on the geographical distribution of animals.

[4] See n. 2, above.

To Oswald Heer?[1] 20 April [1861?][2]

Down. | Bromley. | Kent. S.E.
April 20th

Respected Sir

I am extremely much obliged for the honour & kindness which you have done me by sending your "Untersuchungen &c".—[3] The subject has always seemed to me a most interesting & important one; & I hope soon to read your valuable Memoir.

With my most sincere thanks I have the honour to remain | Your obliged & obedt servant | Charles Darwin

Provenance unknown (listed in Sotheby catalogue, 24 June 1975)

[1] The recipient of the letter is conjectured on the basis of CD's having received a presentation copy of Heer's *Untersuchungen*, published in 1860 (see n. 3, below).

[2] The year is suggested by the stationery with the printed address 'Down Bromley Kent SE'; CD's first firmly dated use of this stationery occurs in letters of early May 1861 (see letter to T. C. Eyton, 3 May [1861]). CD used this stationery until 1869, although he sometimes wrote on plain paper giving his address as 'Down, Bromley, Kent' or simply 'Down'.

[3] CD's presentation copy of Heer 1860 is in the Darwin Library–CUL. It is annotated and has an abstract in CD's hand pinned on the inside of the back cover (see *Marginalia*).

To Frederick Wollaston Hutton 20 April 1861[1]

Down, Bromley, Kent.
April 20

Dear Sir

I hope that you will permit me to thank you for sending me a copy of your paper in the Geologist,[2] and at the same time to express my opinion that you have done the subject a real service by the highly original, striking and condensed manner with which you have put the case. I am actually weary of telling people that I do not pretend to adduce direct evidence of one species changing into another, but that I believe that this view in the main is correct, because so many phenomena can be thus grouped together and explained. But it is generally of no use, I cannot make persons see this. I generally throw in their teeth the universally admitted theory of the undulation of Light—neither the undulation, nor the very existence of Ether being proved,—yet admitted because the view explains so much.[3] You are one of the very few who have seen this and have now put it most forcibly and clearly.[4]

I am much pleased to see how carefully you have read my Book and what is far more important reflected on so many points with an independent spirit. As I am deeply interested in the subject (and I hope not exclusively under a personal point of view) I could not resist venturing to thank you for the right good service which you have done.

Pray believe me, Dear Sir | Yours faithfully & obliged | Charles Darwin

I need hardly say that this note requires no answer.

Copy[5]
DAR 145

[1] The date is established by the reference to Hutton 1861 (see n. 2, below).

[2] Hutton 1861. A lightly annotated copy of the paper is in the Darwin Pamphlet Collection–CUL.

[3] Although in letters written soon after the publication of *Origin* CD had compared his theory with that of gravitation (see *Correspondence* vol. 7), he subsequently suggested that the undulatory theory of light provided a better analogy (see *ibid*., vol. 8, letter to J. S. Henslow, 8 May [1860]). See also letters to Cuthbert Collingwood, 14 March [1861], and to George Maw, 17 March [1861]. Thomas Henry Huxley had mentioned this analogy in his comments on *Origin* at the Oxford meeting of the British Association for the Advancement of Science (*ibid*., vol. 8, letter to T. H. Huxley, 5 [December 1860]).

[4] For CD's further comments about Hutton's paper, see the letter to J. D. Hooker, 23 [April 1861].

[5] A transcription of this letter also appeared in Hutton 1887.

To Samuel Birch 21 April [1861][1]

Down Bromley Kent
April 21.

My dear Sir

I am very sorry that I should have troubled you when so busy & when only recovering from your illness, of which I had not heard.—[2] *Pray* give yourself no more trouble: it is by no means worth wasting your valuable time. You did give me date of the two Chinese Encyclopedias.—[3] And you have now answered some of my questions.—

With very sincere thanks.— Believe me | Yours very truly | C. Darwin

British Museum (Department of Western Asiatic Antiquities), Correspondence, 1826–67: 1493

[1] Dated by the relationship to the letter to Samuel Birch, 6 April [1861].
[2] See letter to Samuel Birch, 6 April [1861].
[3] See letter to Samuel Birch, 6 April [1861], n. 4.

To W. E. Darwin 21 and 22 April [1861?][1]

Down.
Ap 21^{d}

My dear W.

I return the enclosed. It seems a most wonderful cure, & if all is really as it appears, what thousands of lives this precious doctor has cost by not revealing his secret.[2]

Your affect.— | C. Darwin

It is now late in evening & Henrietta has had a much better day.—

22^{d} 12° | Henrietta goes on splendidly[3]

DAR 210.6

[1] The date is suggested by the reference to Henrietta Emma Darwin's illness (see n. 3, below).
[2] The reference has not been traced.
[3] This sentence was added in pencil. Emma Darwin's diary for 1861 is full of references early in the year to Henrietta's progress in recuperating from the illness she had been suffering from for many months (see letters to Asa Gray, 23 [January 1861] and to W. E. Darwin, [24 January 1861]). On both 21 and 22 April 1861, Emma recorded that Henrietta had a 'good day'.

To P. L. Sclater 21 [April 1861]

Down Bromley Kent
21.

Dear Sclater

As you are working at Birds of S. America,[1] & for my credit sake do oblige me & look at Birds in Zoolog. of Beagle p. 67 & see what I say on the 3 species of Opetiorhynchus & consider whether I am likely to have blundered when I observed difference of Habits of the species;[2] so at p. 74 on Scytalopus, when I specify difference of habits.[3] I see that I have made a horrid mistake in speaking of O. parvulus, which was a name temporarily attached, I believe, to one of the forms of O. Patagonicus.[4] I believe that Capt. Abbot must have confounded O. vulgaris & antarcticus,[5] which, as I remarked *on the spot,* are most closely similar except in habits.—

I *much* enjoyed my talk with you.—[6] This note obviously requires no answer & written chiefly to indicate my self-conceit—

Yours very truly | C. Darwin

The Opetiorhynchus from *Chiloe* **seems** case of intermediate variety[7]

Endorsement: '*April 1861*'
American Philosophical Society (245)

[1] Sclater may have informed CD that he was working on a catalogue of his collection of American birds (Sclater 1862). This volume served to bring together much of the information Sclater supplied in his many papers on birds of American origin, including that on the birds of the Falkland Islands (Sclater 1860–1). See n. 5, below.

[2] *Birds* constituted part three of *The zoology of the voyage of H.M.S. Beagle*. Species descriptions were provided by John Gould; CD gave information concerning habits and habitats. The reference is to *Opetiorhynchus patagonicus* (Tierra del Fuego), *O. antarcticus* (Falklands), and *O. nigrofumosus* (Northern Chile).

[3] In *Birds*, p. 74, CD described and compared *Scytalopus magellanicus* and *S. fuscus*.

[4] In *Birds*, p. 67, CD discussed the differences between *O. parvulus* and *O. vulgaris* as though *O. parvulus* was a species in its own right rather than a variety of *O. patagonicus*.

[5] In Abbot 1861, p. 154, Charles Compton Abbott claimed to have seen only one species of the genus *Cinclodes* (*Opetiorhynchus*), *C. antarcticus*, on the Falkland Islands. In Sclater 1860–1, p. 385, Sclater noted both *C. vulgaris* and *C. antarcticus* as being natives of the Falkland Islands.

[6] CD was in London from 16 to 20 April 1861 (Emma Darwin's diary). It is possible that CD met Sclater at a meeting of the Linnean Society of London that he attended.

[7] In *Birds*, p. 67, CD described a long-beaked variety of *O. patagonicus* that inhabited the island of Chiloe.

To J. D. Hooker 23 [April 1861]

Down
23d.

My dear Hooker

I am much pained to think of poor dear Henslow's state.[1] I will write again this evening on indifferent subjects.[2] I write now only to say that if Henslow, you thought, would really like to see me, I would of course start at once. The thought had once occurred to me to offer, & the sole reason why I did not was that the journey with the agitation would cause me probably to arrive utterly prostrated.

I shd. be certain to have severe vomiting afterwards, but that would not much signify, but I doubt whether I could stand the agitation at the time. I never felt my weakness a greater evil. I have just had specimen for I spoke a few minutes at Linn. Soc on Thursday & though extra well, it brought on 24 hours vomiting.[3] I suppose there is some Inn at which I could stay, for I sh[d] not like to be in the House (even if you could hold me) as my retching is apt to be extremely loud.—

I sh[d]. never forgive myself, if I did not instantly come, if Henslow's wish to see me was more than a passing thought.

My dear old friend | Your affect | C. Darwin

P.S. Judge for me: I have stated exact truth: but remember that I sh[d]. *never* forgive myself, if I disappointed the most fleeting wish of my master & friend to whom I owe so much.—

Endorsement: 'April/61'
DAR 115.2: 98

[1] Hooker was in Hitcham, Suffolk, at the bedside of his father-in-law John Stevens Henslow, who was seriously ill.
[2] See following letter.
[3] CD went to London on 16 April 1861 and attended a meeting of the Linnean Society of London on Thursday, 18 April. On 19 April, Emma Darwin recorded in her diary: 'Ch. poorly in London'.

To J. D. Hooker 23 [April 1861][1]

Down Bromley Kent
23[d].

My dear Hooker

I was very glad to get your letter this morning (to which I wrote brief answer), though it contained so melancholy an account of poor dear & honoured Henslow.[2] How strange & pathetic an account you give of his mental state; & how his kind feelings shine out.[3] He truly is a model to keep always before one's eyes. How I wish his sufferings were closed in the long & tranquil sleep of death. It must be very depressing to you, & I am glad you can read & think on other subjects.—

I quite agree with what you say on Lieut. Hutton's Review (who he is, I know not):[4] it struck me as very original: he is one of the very few who see that the change of species cannot be directly proved & that the doctrine must sink or swim according as it groups & explains phenomena. It is really curious how few judge it in this way, which is clearly the right way. I have been much interested by Bentham's paper in N.H.R;[5] but it would not of course from familiarity strike you, as it did me. I liked the whole,—all the facts on the nature of close & varying species. Good Heavens to think of the British Botanists turning up their noses & saying that he knows nothing of British plants![6] I was, also, pleased at his remarks on classification, because it showed me that I wrote truly on this subject in the Origin.—[7] I saw Bentham at Linn. Soc[y]. & had some talk with him & Lubbock & Edgeworth, Wallich several others.—[8] I asked Bentham to give us his ideas of species: whether partially with us or dead against us, he would write *excellent* matter. He made no answer, but his manner made me think he might do so, if urged; so do you attack him.[9] Everyone was speaking with affection & anxiety of Henslow.—

I dined with Bell at Linn. Club,[10] & liked my dinner: though it must be confessed they are rather a poor set of muffs, & I sat by the muffiest, viz Miers.—[11] But dining out is such a novelty to me that I enjoyed it. Bell has a real good heart.— I liked Rolleston's paper, but I never read anything so obscure & not self-evident as his "cannons".[12] I had a dim perception of the truth of your profound remark, that he wrote in fear & trembling "of God, man & monkeys", but I would alter it into God, man, Owen & monkeys.—[13] Huxley's letter was truculent & I see that everyone thinks it too truculent;[14] but in simple truth I am become quite demoniacal about Owen, worse than Huxley, & I told Huxley that I shd. put myself under his care to be rendered milder.[15] But I mean to try to get more angelic in my feelings; yet I never shall forget his cordial shake of the hand when he was writing as spitefully as he possibly could against me.[16] But I have always thought you have more cause than I to be demoniacally inclined towards him.—[17] Bell told me that Owen says that the Editor mutilated his article in Edinburgh R.[18] & Bell seemed to think it was rendered more spiteful by Editor; perhaps the opposite view is as probable: Oh dear this does not look like becoming more angelic in my temper.

I had splendid long talk with Lyell (you may guess how splendid, for he was many times on his knees with elbows on sofa & rump high in air) on his work in France: he seems to have done capital work in making out age of the celt-bearing beds; but the case gets more & more complicated. All, however, tends to greater & greater antiquity of man. The shingle beds seem to be estuary deposits.—[19]

I called on R. Chambers at his very nice house in St John's wood & had very pleasant half-hour's talk:[20] he is really a capital fellow. He made one good remark & chuckled over it, that the Laymen universally had treated the controversy on the Essays & Review, as a merely professional subject & had not joined in it, but had left it to the Clergy.—[21]

I shall be anxious for your next letter about Henslow.

Farewell with sincere sympathy | My old friend | C. Darwin

P.S. We are very much obliged for London R.[22] We like reading much of it, & the science is incomparably better than in Athenæum. You shall not go on very long sending it, as you will be ruined by pennies & trouble. But I am under a horrid spell to the Athenæum & Gardeners Chron, both of which are intolerably dull, but I have taken them in for so many years, that I *cannot* give them up. The Cottage Gardener for my purpose is now far better than G. Chronicle[23]

DAR 115.2: 91

[1] Dated by the relationship to the preceding letter and by an entry in the diary of Charles James Fox Bunbury (see n. 3, below).

[2] Hooker's letter has not been found. For CD's reply, see the preceding letter.

[3] In a diary entry of 24 April 1861, Charles James Fox Bunbury recorded an account of a recent visit to Henslow by Adam Sedgwick (Bunbury ed. 1891–3, *Middle life* 3: 264–5):

> Henslow is, it seems, inevitably dying: the doctors give no hope of his life, though he has lingered much longer than had been expected: but his intellect is perfectly clear, and he is in the most cheerful and delightful frame of mind, not merely resigned but happy to die.

4 Hutton 1861. See letter to F. W. Hutton, 20 April 1861. Frederick Wollaston Hutton was a young army officer with an interest in geology. He was a good friend of the geologist Andrew Crombie Ramsay, whom CD counted among the supporters of his theory (see *Correspondence* vol. 8, letter to J. D. Hooker, 3 March [1860]).

5 Bentham's paper, 'On the species and genera of plants, considered with reference to their practical application to systematic botany', was published in the April issue of the *Natural History Review* (Bentham 1861a). Although it was stated that the paper was extracted from one read before the Linnean Society of London on 15 November 1858, Bentham later related to Francis Darwin a slightly different account of its history (*LL* 2: 294):

> On the day that [CD's and Alfred Russel Wallace's] celebrated paper was read at the Linnean Society, July 1st, 1858, a long paper of mine had been set down for reading, in which, in commenting on the British Flora, I had collected a number of observations and facts illustrating what I then believed to be a fixity of species, however difficult it might be to assign their limits . . . Most fortunately my paper had to give way to Mr. Darwin's, and when once that was read, I felt bound to defer mine for reconsideration; I began to entertain doubts on the subject, and on the appearance of the 'Origin of Species,' I was forced, however reluctantly, to give up my long-cherished convictions, the results of much labour and study, and I cancelled all that part of my paper which urged original fixity, and published only portions of the remainder in another form, chiefly in the 'Natural History Review.'

In a note appended to his copy of this issue in the Darwin Library–CUL, CD wrote: 'Bentham on Classification Excellent'.

6 Charles Cardale Babington had made this comment about George Bentham. See *Correspondence* vol. 7, letter to J. D. Hooker, 21 July [1858].

7 Bentham held the view, as did CD, that classification was a matter of convention. In Bentham 1861a, p. 133, Bentham wrote:

> in nature there are no two individuals exactly alike in every respect. In all collections of individuals even when the immediate offspring of one parent, peculiarities will be found common to some and not to all. The species or collection of individuals thus defined, becomes, therefore, as arbitrary as the genus or collection of species, and reduces the rules of classification in the one case, as in the other, to little more than rules of convenience.

8 CD attended a meeting of the Linnean Society of London on Thursday, 18 April 1861. He refers to John Lubbock, Michael Pakenham Edgeworth, and George Charles Wallich.

9 Since the publication of *Origin*, CD and Hooker had been interested in learning Bentham's opinion of CD's views on the origin of species. See *Correspondence* vol. 7, letter to J. D. Hooker, 14 December [1859], and letter from J. D. Hooker, [20 December 1859]; and *ibid.*, vol. 8, letter to J. D. Hooker, 3 March [1860].

10 Thomas Bell was president of the Linnean Society.

11 John Miers was a specialist on the flora of South America.

12 In Rolleston 1861a, George Rolleston set out nine points of discrepancy between the human and the simian brain and suggested criteria to determine the value of these points as differentiating characteristics (Rolleston 1861a, p. 206).

13 Rolleston had dissected the brain of an orang-utan and identified what he took to be the hippocampus minor, which Richard Owen maintained was present only in the brains of humans. On this point, Rolleston's findings agreed with the view of Thomas Henry Huxley, who, in opposition to Owen, claimed that there were no essential differences between the brains of humans and the higher apes. Rolleston, although stating that the simian brain did not show 'those sharply differentiating characteristics which have been supposed to put it into a position of such marked inferiority to that of man', did not express clear support for Huxley's position (Rolleston 1861a, p. 215).

14 The reference is to an exchange of letters between Huxley and Owen printed in recent numbers of the *Athenæum*. See letter to T. H. Huxley, 1 April [1861]. Huxley's second letter, published in the 13 April issue (p. 498), referred to Owen's 'very serious errors respecting matters of anatomical fact'.

15 See letter to T. H. Huxley, 1 April [1861].

16 CD refers to his meeting with Owen soon after the publication of *Origin*. See *Correspondence* vol. 7, letters to Charles Lyell, [10 December 1859], and to Richard Owen, 10 December [1859] and 13 December [1859].

17 For CD's opinion that Owen had snubbed Hooker by focusing primarily on *Origin* and not discussing Hooker 1859 on its own merits in his review ([R. Owen] 1860b), see *Correspondence* vol. 8, letters to Asa Gray, 25 April [1860], and to J. D. Hooker, 26 April [1860].

18 The editor of the *Edinburgh Review*, in which [R. Owen] 1860b was published, was Henry Reeve. Thomas Bell had earlier told Hooker that Owen acknowledged authorship of the review (see *Correspondence* vol. 8, letter from J. D. Hooker, [20 April 1860]).

19 CD visited Charles Lyell during his recent stay in London (see letter to Charles Lyell, 12 April [1861] and n. 6).

20 CD had first met Robert Chambers in 1847 (see *Correspondence* vol. 4, letter to J. D. Hooker, [18 April 1847]).

21 *Essays and reviews* (1860), which comprised seven essays (six by liberal clergymen) addressing the implications of recent scholarship for traditional religious interpretations, had recently been severely attacked by the Church establishment. For a more detailed consideration of the controversy surrounding this work, see Appendix VI.

22 Hooker had sent CD the most recent issue of the *London Review and Weekly Journal of Politics, Literature, Art and Science* (see letter to J. D. Hooker, 27 [March 1861]).

23 Beginning with the number of 2 April 1861, the *Cottage Gardener and Country Gentleman* began a new series under the title *Journal of Horticulture, Cottage Gardener, and Country Gentleman. A journal of horticulture, rural and domestic economy, botany and natural history*. CD had subscribed intermittently to the journal since 1855 and regularly since January 1860. His copies are in the Darwin Library–CUL. There are numerous references in *Variation* to the journal as a source of information about various domesticated plants and animals.

To Jean Louis Armand de Quatrefages de Bréau 25 April [1861][1]

Down. | Bromley. Kent.
April 25.

My dear Sir.

I received this morning your "Unité de l'Espèce Humaine", & most sincerely do I thank you for this, your very kind present.[2] I had heard of and been recommended to read your articles, but not knowing that they were separately published did not know how to get them.— So your present is most acceptable, & I am very anxious to see your views on the whole subject of Species & variation; & I am certain to derive much benefit from your work. In cutting the pages I observe that you have most kindly mentioned my work several times.[3]

My views spread slowly in England & America; and I am much surprised to find them most commonly accepted by Geologists, next by Botanists and least by Zoologists.—[4] I am much pleased that the younger and middle-aged Geologists are coming round; for the arguments from Geology have always seemed strongest against me. Not one of the older geologists (except Lyell) has been even shaken in his views of the eternal immutability of species— But so many of the younger men are turning round with Zeal that I look to the future with some confidence.—[5]

I am now at work on "Variation under domestication" but make slow progress.—[6] it is such tedious work comparing skeletons—

With very sincere thanks for the kind sympathy which you have always shown me and with much respect | I remain, My dear Sir | Yours faithfully & obliged— | Charles Darwin.

I have lately read M. Naudin's paper; but it does not seem to me to anticipate me, as he does not shew how Selection could be applied under nature;[7] but an obscure writer on Forest Trees, in 1830, in Scotland, most expressly & clearly anticipated my views—though he put the case so briefly, that no single person ever noticed the scattered passages in his book—[8]

Copy
DAR 147

[1] Dated by the reference to Quatrefages de Bréau 1861 (see n. 2, below).

[2] An annotated copy of Quatrefages de Bréau 1861 is in the Darwin Library–CUL. See also *Marginalia* 1: 693–5. In his book, Quatrefages classified the human species as a separate kingdom (*règne humain*).

[3] In Quatrefages de Bréau 1861, pp. 37–9, the author discussed the views of CD and others on the question of species change. For Quatrefage's views of *Origin*, see *Correspondence* vol. 7, letter to J. L. A. de Quatrefages de Bréau, 5 December [1859].

[4] CD was pleased by the number of supporters he counted among geologists, including Charles Lyell, Andrew Crombie Ramsay, Joseph Beete Jukes, and Henry Darwin Rogers. See *Correspondence* vol. 8, letter to J. D. Hooker, 3 March [1860].

[5] CD believed that the general acceptance of his views depended on their being taken up by the younger generation of scientists (see *Correspondence* vol. 8, letters to T. H. Huxley, 2 December [1860] and 5 [December 1860]). He expressed his pleasure in the support of such men on several occasions (see letters to Joseph Leidy, 4 March [1861], and to J. F. W. Herschel, 23 May [1861]).

[6] CD was writing the early chapters discussing variation in different domestic animals that were eventually published in the first volume of *Variation* (see 'Journal'; Appendix II).

[7] For CD's earlier response to the suggestion that the French botanist Charles Victor Naudin had possibly anticipated some of his views in Naudin 1852, see *Correspondence* vol. 7, letters to J. D. Hooker, 21 [December 1859] and 23 [December 1859], and to Charles Lyell, 22 [December 1859]. CD included Naudin's name among those who had previously published on the modification of species in the historical sketch included in the preface to the revised and augmented American edition of *Origin* and in the German and third English editions. In his abstract of Quatrefages de Bréau 1861, CD noted the implicit reference to Naudin's so-called 'loi de retour' (law of return) as it applied to hybrid crosses (Quatrefages de Bréau 1861, p. 161).

[8] CD refers to Patrick Matthew. See *Correspondence* vol. 8, letter to *Gardeners' Chronicle*, [13 April 1860], and Appendix V.

To Thomas William St Clair Davidson 26 April 1861

Down Bromley | Kent
April 26th 1861.

My dear Sir

I hope that you will excuse me for venturing to make a suggestion to you which I am perfectly well aware it is a very remote chance that you would adopt— I do not know whether you have read my Origin of Species in that book I have made the remark, which I apprehend will be universally admitted that *as a whole* the fauna of any formation is intermediate in character between that of the formations above and below.[1] But several really good judges have remarked to me how desirable it would be that this should be exemplified and worked out in some detail, and with some single group of beings. Now every one will admit that no one in the world could do this better than you with Brachiopods.[2] The result might turn out very unfavourable to the views which I hold; if so, so much the better for those who are opposed to me—

But I am inclined to suspect that on the whole it would be favourable to the notion of descent with modification; for about a year ago Mr Salter in the Museum in Jermyn Street glued on a board some Spirifers &c. from 3 palæozoic stages and arranged them in single and branching lines with horizontal lines marking the formations (like the diagram in my Book if you know it) and the result seemed to me very striking though I was too ignorant fully to appreciate the lines of affinities.[3] I longed to have had these shells engraved, as arranged by Mr Salter, and connected by dotted lines, and would have gladly paid the expense, but I could not persuade Mr Salter to publish a little paper on the subject

I can hardly doubt that many curious points would occur to anyone thoroughly instructed in the subject, who would consider a group of beings under this point of view of descent with modification All those forms which have come down from an ancient period very slightly modified ought, I think to be omitted and those forms alone considered which have undergone considerable change at each successive epoch. My fear is whether Brachiopods have changed enough. The absolute amount of difference of the forms in such groups at the opposite extremes of time ought to be considered; and how far the early forms are intermediate in character between those which appeared much later in time— The Antiquity of a group is not really diminished, as some seem vaguely to think, because it has transmitted to the present day closely allied forms. Another point is how far the succession of each genus is unbroken from the first time it appeared to its extinction with due allowance made for formations poor in fossils.

I cannot but think that an important essay (far more important than a hundred literary Reviews) might be written by one like yourself and without very great labour—[4] I know it is highly probable that you may not have leisure, or not care for or dislike the subject, but I trust to your kindness to forgive me for making this suggestion. If by any extraordinary good fortune you were inclined to take up this notion—I would ask you to read my Chap X on Geological Succession— And I should like in this case to be permitted to send you a Copy of a new Edit just published, in which I have added & corrected somewhat in Ch IX and X.[5]

Pray excuse this long letter and believe me | My dear Sir, | Yours very faithfully | Charles Darwin.—

I write so bad a hand I have had this note copied.—

Copy
DAR 143

[1] This idea was discussed at length in chapters 9 and 10 of *Origin*. The 'remark' to which CD refers is probably his statement that 'the fauna of any great period in the earth's history will be intermediate in general character between that which preceded and that which succeeded it.' (*Origin*, pp. 333–4).

[2] Davidson was an expert on fossil brachiopods. He was the author of a multi-volume work on the British fossil species (Davidson 1851–86). Robert Chambers may have been among those who made this suggestion (see letter to Robert Chambers, 30 April [1861]).

[3] John William Salter was palaeontologist to the Geological Survey in London. For a discussion of Salter's palaeontological chart, displayed in the Museum of Practical Geology, see *Correspondence* vol. 8, letter to Charles Lyell, 4 May [1860] and n. 7, and letter from Charles Lyell, 7 May 1860.

[4] See letter to T. W. St C. Davidson, 30 April 1861 and n. 6.

[5] CD sent Davidson a copy of the third edition of *Origin* (see letter to T. W. St C. Davidson, 30 April 1861, and Appendix VII).

To the *Field* [before 27 April 1861][1]

[Down]

I should esteem it a great favour if some of your numerous readers would take the trouble to give me any facts on the colour of the two parents of true dun horses.[2] I mean by true duns, horses having a stripe or list along the the spine, and often transverse stripes on the legs, the general colouring being either a mouse-dun or a tint which may be described as a creamy bay or chesnut. I am aware from inquires made in Norway, where true dun ponies are extremely common, that one or both parents are there always duns; and so it is, as I am informed, with the dun ponies of Devonshire. But I have occasionally seen dun cart-horses and hacks, which did not seem to have the blood of any pony or cob in them. It is surprising how often I have vainly asked the parentage of such horses, and vainly made inquiries from breeders. I have myself seen one colt, bred from a black mare and bay horse, which might certainly be called a dun, and which had a narrow, but strongly defined, spinal stripe before it shed its first hair. I should be much obliged for any information on this subject; and likewise whether a dun horse or pony is always dun-coloured before it sheds its first hair.[3] Does the spinal stripe often disappear when the first coat is shed?—

Charles Darwin (Down, Bromley, Kent.)

The Field, the Farm, the Garden, the Country Gentleman's Newspaper 17 (1861): 358

[1] Dated by the publication of the letter in the 27 April 1861 issue of the *Field*. Appearing in the column 'Notes and questions on natural history', it was headed: 'Dun horses'.

[2] CD was seeking information on the parentage of horses for inclusion in *Variation*. He had briefly discussed the appearance of stripes in duns and other breeds as reflecting on reversion to an ancestral characteristic in *Origin*, pp. 163–7. For other such inquiries, see *Correspondence* vols. 6–8.

[3] See letter to the *Field*, [before 25 May 1861].

To Robert Chambers 30 April [1861][1]

Down, Bromley, Kent
Ap. 30.

My Dear Sir,

I must send you a line to thank you for your "Ice & Water" which I am sure will interest me much;[2] though I believe we split a little about solid glacier ice and icebergs.—[3] Thanks, also, for extract out of newspaper about Rooks & Crows;— I wish I dared trust it.[4]

I see in cutting pages half-an-hour ago, that you fulminate against the scepticism of scientific men.—[5] You would not fulminate quite so much, if you had had so many wild-goose chases after facts stated by men not trained to scientific accuracy. I often vow to myself that I will utterly disregard every statement made by any one who has not shown the world he can observe accurately. I wish I had space to tell you a curious history, which I was fool enough to investigate on almost universal testimony of Beans growing this year upside down.—[6] I firmly believe that accuracy is a most difficult quality to acquire.— I did not, however, intend to say all this.—

I very thoroughly enjoyed my half-hours talk at your pleasant house.—[7] I have been corresponding with Mr Davidson on the Genealogy of Brachiopods; & he will some day, I believe, discuss subject as we wish.[8] He has seen Salters table of species grouped like a tree.[9] Mr D. is not at all a full believer in great changes of species which will make his work all the more valuable.—

I have also written to Mr Jamieson urging him to take up Glen Roy.[10]

My dear Sir | Yours very Sincerely | C. Darwin

Copy
DAR 143

[1] Dated by the reference to Chambers's paper 'Ice and water' (see n. 2, below).

[2] Chambers 1861. The work is one of a series of five papers on various topics that were included in Chambers's *Edinburgh papers* (1861). CD's presentation copy of the paper is in the Darwin Pamphlet Collection–CUL.

[3] In an earlier work (Chambers 1853), Chambers upheld evidence for the effects of large-scale glaciation in Europe and dismissed indications of the action of icebergs, a view that CD favoured. For CD's caustic comments on this paper, see *Correspondence* vol. 5, letter to Charles Lyell, 7 June [1853].

[4] The extract has not been found.

[5] Questioning why the scientific community had dismissed earlier evidence for the antiquity of man, Chambers suggested that this was owing to 'the unsatisfiable scepticism about facts reported by others' motivated by scientists' fears about their reputations (Chambers 1861, pp. 40–1).

[6] See *Gardeners' Chronicle and Agricultural Gazette*, 27 April 1861, p. 390.

[7] During his recent stay in London, CD had visited Chambers at his home in St John's Wood (see letter to J. D. Hooker, 23 [April 1861]).

[8] See letter to T. W. St C. Davidson, 26 April 1861, and the following letter.

[9] John William Salter. See letter to T. W. St C. Davidson, 26 April 1861 and n. 3.

[10] CD's letter to Thomas Francis Jamieson has not been found, but see the letter from T. F. Jamieson, 3 September 1861, and letter to T. F. Jamieson, 6 September 1861. In his paper on the parallel roads

of Glen Roy (Jamieson 1863), Jamieson acknowledged CD's support: 'He furnished me with some useful maps and memoirs bearing upon the problem, and likewise indicated several points worthy of particular attention.' CD and Chambers had also published on Glen Roy (see *Correspondence* vol. 4). See Rudwick 1974 and Appendix IX.

To T. W. St C. Davidson 30 April 1861

Down. Bromley. | Kent
April 30th. 1861

My dear Sir

I thank you warmly for your letter.[1] I did not in the least know that you had attended to my work. I assure you that the attention which you have paid to it, considering your knowledge and the philosophical tone of your mind (for I well remember one remarkable letter you wrote to me, and have looked through your various publications) I consider one of the highest perhaps the very highest compliment which I have received—[2] I live so solitary a life that I often do not hear what goes on, and I should much like to know in what work you have published some remarks on my book.[3] I take a deep interest in the subject and I hope not simply an egotistical interest; so therefore you may believe how much your letter has gratified me; I am perfectly contented if anyone will fairly consider the subject whether or not he fully or only very slightly agrees with me.

Pray do not think that I feel the least surprise at your demurring to a ready acceptance; in fact I should not much respect anyone's judgment who did so that is if I may judge others from the long time which it has taken me to go round. Each stage of belief cost me years. The difficulties are as you say many & very great: but the more I reflect the more they seem to me to be due to our underestimating our ignorance. I belong so much to old times that I find that I weigh the difficulties from the imperfection of the geological record, heavier than some of the younger men.— I find to my astonishment and joy that such good men as Ramsay, Jukes Geikie and one old worker Lyell do not think that I have in the least exaggerated the imperfection of the record.[4] If my views ever are proved true our current Geological views will have to be considerably modified—

My greatest trouble is not being able to weigh the direct effects of the long continued action of changed conditions of life without any selection, with the action of selection on mere accidental (so to speak) variability— I oscillate much on this head but generally return to my belief that the direct action of the conditions of life has not been great. At least this direct action can have played an extremely small part in producing all the numberless and beautiful adaptations in every living creature.

With respect to persons belief, what does rather surprise me is that anyone (like Carpenter) should be willing *to go so very far* as to believe that all Birds may have descended from one parent, and not go a little farther and include all the members of the same great Division;[5] for on such a scale of Belief, all the facts in

Morphology and in Embryology (the most important in my opinion of all subjects) become mere Divine Mockeries. This leads me to remark how singularly few have judged the argument on right principles many complain that I have not proved that any one species changes into another and they ignore the fact that the view given, apparently groups together and explains many phenomena. No one urges as a fatal objection to the Theory of Light that the undulations in the ether cannot be proved—or the very existence of ether, yet because the undulatory theory explains much it is now universally admitted—

I cannot express how profoundly glad I am that some day you will publish your theoretical view on the modification & endurance of Brachiopodous species;[6] I am sure it will be a most valuable contribution to knowledge.— My weak health interferes much with my work, but I am never idle— I should never, however, be able to acquire sufficient knowledge to trace and appreciate the changes in any great group of beings, as you will be able to do with Brachiopods. I am at present at work on "Variation under Domestication" and slow and laborious work I find it, but I think it will throw some little light on the laws of variation— Have you read Hooker's Introduction to the Flora of Australia;[7] I think it would interest you?

Pray forgive this very egotistical letter, but you yourself are partly to blame, from having pleased me so much. I have told Murray to send a copy of new Edit to you and have written your name.—[8]

With cordial thanks. | Pray believe me | My dear Sir. | Yours very sincerely | Ch Darwin

As my Copyist writes so well, I do not apologise for length of this

There is only a little added & a little modified in Ch IX & X of new Edition of Origin.—

Copy
DAR 143

[1] Davidson's letter has not been found. It was a response to CD's letter of 26 April 1861.

[2] For the previous extant correspondence between CD and Davidson, see *Correspondence* vol. 6, letter to T. W. St C. Davidson, 23 December [1856], and letter from T. W. St C. Davidson, 29 December 1856.

[3] See letter from T. W. St C. Davidson, 3 May 1861. The remarks were published in Davidson 1861, an annotated copy of which is in the Darwin Pamphlet Collection–CUL.

[4] Andrew Crombie Ramsay, Joseph Beete Jukes, and the young Scottish geologist Archibald Geikie were all employed by the Geological Survey of Great Britain. CD mentioned several times that Charles Lyell was the only one of the 'older' geologists who was coming round to his views on the modification of species (see letters to Joseph Leidy, 4 March [1861], and to J. L. A. de Quatrefages de Bréau, 25 April [1861]).

[5] William Benjamin Carpenter upheld this position in one of his two reviews of *Origin* (Carpenter 1860, p. 404). See also CD's earlier criticisms of Carpenter's view (*Correspondence* vol. 8, letter to Charles Lyell, 27 and 28 April [1860], and letter from Charles Lyell, 18 September 1860).

[6] In Davidson 1851–86, 2: 212–13 (published in 1863), Davidson mentioned CD's views on species and varieties and stated his own belief that when the entire record of fossil brachiopods was known more

completely, 'a large proportion of them will be traced through their various modifications to a parent form in stages far more ancient than we are in many cases disposed to admit'. See also letter from T. W. St C. Davidson, 3 May [1861] and n. 6.

7 Hooker 1859.

8 CD had offered to send Davidson a copy of the third edition of *Origin*, published by John Murray (see letter to T. W. St C. Davidson, 26 April 1861). See also Appendix VII.

To John Murray 30 April [1861]

[Down]
Ap. 30th

My dear Sir

The enclosed was written before receiving yours this morning, for which I am much obliged & for your Bill for £480, for which I return you my thanks & send the Receipt.—[1]

The nine months' time seems to me perfectly reasonable. In the Receipt you speak of 2000 copies; in your note you say 2500; I shd. like to know which is correct; I presume 2000.—[2]

I am not surprised at the sale slackening, & I formerly thought the number which you printed rather bold, but I now find so many persons deeply interested in subject that I believe whole Edition will sell sooner or later.— *Pray remember to advertise with "additions & Corrections"*, for I find that *several* persons have read the Book twice & thrice, & it is not unlikely that some would buy the new Edition.— When next in London I will call on you to consult about illustrations on my Book on "Variation under Domestication" which slowly but surely progresses.—

With many thanks for your uniform kindness | Believe me | My dear Sir | Yours sincerely | C. Darwin

[Enclosure][3]

London. April 29/61 Received of John Murray Esqre the Sum of Four and eighty pounds[4] (by his promissery note) for the third Edition (2000 copies) of my Work on the *Origin of Species*

£480[5] Charles Darwin

Endorsement: '*1861. Ap. 30.*'
John Murray (publishers)

1 See the enclosure. The sum represents CD's share of the profits on the sale of the third edition of *Origin*. Murray, however, made an error in calculating the sum to be paid to CD. See letter to John Murray, 3 May [1861], and nn. 4 and 5, below.

2 The print-run of the third edition of *Origin* was 2000 copies (Peckham ed. 1959, p. 776).

3 The enclosure is in Box 37(3) in the John Murray archive.

4 CD altered the sum to read: 'Three' and 'seventy-two' pounds.

5 CD altered the figure to read: '372'.

To Daniel Oliver 1 May [1861][1]

Down Bromley Kent
May 1.

My dear Sir

Very many thanks for your two notes.—[2]

I sh^d. be very glad of three plants of Dionæa (for which I could pay & for Cypripedium)[3] & as I shall not go to sea for my daughter's sake till middle of June, the sooner I could have them the better, as I might try few experiments soon on them.[4] But how on earth can they be packed?

The best address will be

Ch. Darwin Esq^e
care of Mr Mitchell
per Railway Postman
Bromley
Kent

Pray thank Mr Crocker.—[5]

Also thank Mr Fitch (& ask him to keep memorandum.)[6] I will not have the outline which does capitally done on wood yet, for I have been very unfortunate in my experimental plants dying, & I may not come to any sure conclusion & in this case should not publish.—

Nevertheless if your P. farinosa sends forth long-stamened & short-stamened flowers, I sh^d be *infinitely* obliged for 2 or 3 flowers of each kind to examine folded up in 2 or 3 green leaves & sent by Post.[7]

Thanks for references about nectaries. Kurr's work you told me of before & I bought & have read it.—[8] You will now be very busy with your Lecture[9]

with many thanks | Yours truly | Ch. Darwin

Down House (MS 10: 7)

[1] Dated by the reference to CD's experiments on *Dionaea*, begun in the summer of 1860 (see nn. 3 and 6, below), and by the Darwins' planned seaside visit (see n. 4, below).

[2] Oliver's 'notes' have not been found.

[3] CD had asked Oliver to procure these plants for him (see letter to Daniel Oliver, 26 February [1861]).

[4] The Darwins were planning a visit to the seaside in the hope that it would improve Henrietta Emma Darwin's health. CD had been concerned about Henrietta's ill health since the spring of 1860. See *Correspondence* vol. 8; see also letter to J. D. Hooker, 8 [February 1861]. The visit had to be postponed until 1 July, when the family travelled to Torquay, Devon. See 'Journal' (Appendix II).

[5] Charles William Crocker was the foreman of the propagating department at the Royal Botanic Gardens, Kew. CD had solicited Crocker's assistance in obtaining some young *Dionaea* plants (see *Correspondence* vol. 8, letter to Daniel Oliver, 16 November [1860]).

[6] CD had asked Walter Hood Fitch to prepare drawings of *Primula* to illustrate the paper he was preparing on dimorphism in this genus (see letter to Daniel Oliver, 23 March [1861]).

[7] CD read a paper on the dimorphic condition of *Primula* at a meeting of the Linnean Society of London in November 1861. Although *P. farinosa* is not mentioned in the paper, it is included in the

revised and augmented version subsequently published in *Forms of flowers* (1877).

8 Kurr 1833. See letter to Daniel Oliver, 22 January [1861]. An annotated copy of this work is in the Darwin Library–CUL.

9 Oliver had been appointed professor of botany at University College London in November 1860. According to an advertisement in the *Athenæum*, 27 April 1861, p. 545, Oliver's course of lectures on botany ran from 1 May to the end of July 1861.

From T. W. St C. Davidson 3 May 1861

33 Park Crescent Brighton
3^d May 1861

My Dear Sir.

I have to express my warmest thanks for your very kind and instructive letter of April the 30th. as well as for the valuable present of the last edition of your admirable work "Origin of Species."[1] it arrived last night, but I have not yet had time to peruse the chapters IX & X which I will D.V. do in a day or two being extremely pressed for time with work for the Pal. Soc. & my Indian paper for the Geol. Soc.—[2]

I can assure you that I shall give *every attention* to the subject of *descent with modification*, as suggested in your to me most interesting letters. I have not time alas to day to enter into details concerning the Brachiopoda, but shall not lose sight of the object & must be content at this moment to send you 3 little rough tables of the known distribution of the Brachiopoda—and which I consider to be as near the truth as the present state of our information will permit[3] It is certain and no one can object to your statement relative to the great *imperfection of Geological record* (I was reading this morning some ob^s. upon the subject in the last number of the Geologist, for May 1861 just appeared).[4] The Geological record is therefore *very imperfect* and the Palæontological record still more so—& I admit that you have a strong hold there for your theory.

Having myself on account of my monographs been obliged to examine every layer (if I may be permitted the expression) in search of Brachiopoda and to trace as far as possible the first appearance, duration, modification & extinction of each form, have been constantly (not now & then *but almost every day*) at a loss or puzzled what to do with intermediate shapes. did they belong to one or to the next or most closely allied species was the constant question. & more the material increased greater was the difficulty. Such work is sufficient to convince even the most incredulous of the variability of species & how a same species may become considerably modified by time and peculiar conditions— I have not the slightest doubt that by far the greater number of species of Brachiopoda could be even now linked together The genera, perhaps also and the more our knowledge will extend the more closely will the whole become linked together 'Natura non facit saltum' was my belief in 1848 & was so stated by me when I described the British Silurian Species in the Bulletin de la Soc. geol de France'[5] and I believe that one might (tabulate the species as you propose) show a gradation & radiation, between many of the most extreme conditions.

We are very little acquainted with the rocks of the Cambrian period, but very few species of fossils have been found and that is our oldest record at present known— well in all we find *orthis* and *Lingula* only for the present, but any day might produce a larger number and especially so as you remark in the Lower Silurian the introduction of so many new genera or species. which we cannot believe to have been created for the first time then. Many of these may have been derived by modification from older strata (or from a period) still far more ancient than the upper Cambrian. When I conclude my work on the Brachiopoda I shall D.V. give a number of curious facts showing discent with modification, that many forms considered as distinct creation are only modifications.[6] Still there are many difficulties in the way of fully realising the theory of a single creation, and I can well understand the trouble & time you have required in acquiring your conviction. I believe that the facts in Morphology & in Embryology will when more fully worked out, will cast considerable light upon many points at present but little known with reference to species.—

I am *obliged* to conclude this letter written in *great haste*, but cannot do so without once more thanking most sincerely for your very valuable communications as well as for the present of the new edition of your remarkable work.

Believe me Dear Sir | Yours very sincerely | Tho Davidson—

P.S. I forgot to say that I have not read Hookers "Introduction to the flora of Australia",[7] but when I next go to London will endeavour to get the loan of it from the Royal Soc. lib^{y}.— I have not published yet any thing upon the subject of your theory as I am still continually thinking over the matter, intending to do so at the conclusion of my work for the Pal Soc. I merely said two or three words upon the subject in a little paper published in the Geologist of which I forward a copy.[8]

to C. Darwin Esq. FRS.

[Enclosure][9]

[Table I is entitled 'Geological distribution at present known of the Genera & Sub-genera of Brachiopoda.' A note at the top reads: 'about 1950 so-termed species of Brachiopoda are described and arranged as follows'. On the left-hand side of the table, Davidson listed genus and subgenus names of brachiopods, each followed by attribution details and the name of a species from the genus. On the right-hand side Davidson drew eleven columns, each headed with the name of a geological period, from the 'Upper Cambrian' to 'recent'. He indicated during which geological periods each genus or subgenus was extant by means of crosses entered in the appropriate columns. At the bottom of the table, Davidson wrote: 'N.B a number of So-termed new Genera recently proposed by Prof. J. Hall,[10] M^r^. Billings[11] and some others have been omited as their *value* is not yet determined, but in any case they would have to enter into one or another of these groups & would not alter the geological limits as here recorded.'

Table II is headed: 'Showing the first appearance at present known of each genus.' It also lists genus and subgenus names of brachiopods on the left; instead of crosses, Davidson used continuous bars spanning across the columns to indicate the duration of each form in the various geological periods.

Table III has the heading: 'Showing the at present known duration of each great group.' This table lists as the six 'great groups' of brachiopods: *Orthis*, *Spirifer*, *Atrypa*, *Rhynchonella*, *Terebratula*, and the 'Unarticulated group *Lingula Discina* Crania etc.'

At the bottom of Tables II and III, which are placed side by side on one page, Davidson summarised the salient points presented in the two tables:

> 1 From this table it would appear 1st that the large genera such as Orthis, Leptæna, Rhynchonella, Crania, discina, Lingula, Chonetes, Productus, Athyris, Spiriferina, Terebratula, Terebratella, Thecideum Waldheimia & Terebratella had the longest duration in time—
>
> 2. That a great many so termed genera have not been found represented beyond the Palæozoic range or did not depart the upper limit of the Permian formation.
>
> 3. That it was during the Palæozoic Period that the Brachiopodous forms were most numerous & greatest variation in generic character prevailed.]

DAR 99: 1–2

CD ANNOTATIONS

1.1 I have ... copy. 7.6] *scored pencil*
7.2 but when ... lib^y^.— 7.3] *cross in margin, pencil*
7.3 I have ... copy. 7.6] *cross in margin, pencil*

1 CD presented Davidson with a copy of the third edition of *Origin* (see letters to T. W. St C. Davidson, 26 April 1861 and 30 April 1861, and Appendix VII).

2 Davidson refers to his monograph on fossil Brachiopoda (Davidson 1851–86), which was published by the Palaeontographical Society of London; he also cites his study of fossil forms of India (Davidson 1862) 'D.V.' is an abbreviation for 'Deus vult', 'God wills it'.

3 See enclosure and n. 9, below.

4 The reference is to the second part of Hutton 1861, which puts forward explanations for the imperfection of the geological record.

5 Davidson 1848, p. 311.

6 In the next-published volume of his monograph, which appeared in 1863, Davidson mentioned the possibility that fossil brachiopods might provide evidence for CD's view of species and varieties (Davidson 1851–86, 2: 212–13). He did not return to the question, however, until the final volume of his monograph, published shortly before his death. There Davidson quoted from CD's letter of

26 April 1861 and stated (*ibid.*, 5: 387):

> In several subsequent letters Darwin reiterated his suggestions; and, although I have not neglected a request coming from so eminent a quarter, I am bound to state that I have found the subject beset with so many apparently inexplicable difficulties that year after year has passed away without my being able to trace, in a satisfactory manner, the descent with modification among the Brachiopoda which the Darwinian doctrine requires.

Davidson stated that he none the less believed in some form of the doctrine of species transmutation.

7 Hooker 1859. See letter to T. W. St C. Davidson, 30 April 1861.

8 Davidson 1861, pp. 58–9. A lightly marked copy of Davidson's paper is in the Darwin Pamphlet Collection–CUL. See letter to T. W. St C. Davidson, 26 April 1861.

9 Owing to the difficulty of typesetting Davidson's tables, only the table headings and a summary of their contents are provided.

10 Davidson may be referring to Hall 1860.

11 Elkanah Billings, a geologist employed by the Geological Survey of Canada, had recently described a number of new brachiopod genera (Billings 1859a and 1859b).

To Thomas Campbell Eyton 3 May [1861][1]

Down. | Bromley. | Kent. S.E.
May 3^{d}

Dear Eyton

I have received the skeletons this morning & am very much obliged to your kindness for sending them.—[2] I am truly sorry to give trouble; but one domestic cock in longest box with sliding lid has no sort of name. Can you tell me what it is, as it is useless without name.—

Also what is "Gungla" cock is G. Bankiva? or G. Sonneratii?[3]

The other Box contains the Hamburgh.— And the fourth Box contains Call Duck; so perhaps you will know what the fourth is.

I cannot say much for my own health, though I have been rather better lately— I write in a hurry to catch Post—

With sincere thanks | Yours very truly | C. Darwin.

American Philosophical Society (247)

1 Dated by the reference to skeletons of fowls. CD was composing his chapters on fowls for *Variation* (see 'Journal'; Appendix II).

2 CD acknowledged his indebtedness to Eyton for three of the fowl skeletons included in his discussion of the osteological differences between various breeds of fowl in *Variation* 1: 260.

3 See letter to T. C. Eyton, 6 May [1861].

To Luke Hindmarsh 3 May [1861][1]

Down, Bromley, Kent. S.E.
May 3rd

Sir

I hope that you will excuse the liberty which I take in writing to you. Several years ago I was very much interested by your excellent memoir on the wild

Chillingham cattle.[2] I am very anxious for information on one point; but whether you still retain interest on the subject, or could spare time to give or obtain for me this information, I know not. The point is the average number of animals which are annually killed. I presume that some account is kept, and it must be known how many have been killed during the last half-dozen or dozen years. When you wrote the herd was about 80; and I should wish to know how many during any period in which the slaughtered animals have been recorded, existed. My object is to ascertain the rate of increase of these cattle relatively to those on the Pampas in S. America.[3]

Hoping that you will forgive the liberty which I take and grant me this favour, I beg leave to remain | Sir | Your obliged and obed[t] serv[t]. | Charles Darwin

I saw a translation of the greater part of your paper lately in a French Periodical.[4]

Copy
DAR 145

[1] Dated by the relationship to the letter from William Hardy to Luke Hindmarsh, [8 May 1861], and to the letter to Luke Hindmarsh, 12 May [1861].

[2] Hindmarsh 1839. The citation was sent to CD by William Yarrell (see *Correspondence* vol. 2, letter from William Yarrell, [December 1838]).

[3] CD's interest in the Chillingham cattle arose both from their supposed similarity to the extinct *Bos primigenius* and from the example they provided of a breed long subject to interbreeding. See *Variation* 1: 81, 83–4,-4; see also letter from William Hardy to Luke Hindmarsh, [8 May 1861] and n. 4, and letter to Luke Hindmarsh, 12 May [1861] and nn. 3 and 4.

[4] The journal in question has not been traced.

To John Murray 3 May [1861]

Down. | Bromley. | Kent. S.E.
May 3[d]

My dear Sir

I am sorry for your sake, as well as my own, for the mistake. I return your Bill & sh[d] be obliged for a new one. I have altered the receipt, whether as altered it will suffice I know not.—[1] There were few corrections in the last Edition, though several pages added, yet I am surprised, at so large a reduction of profit, relatively to the last Edition, which since receiving your note, I have calculated.[2]

Kindly attend to the few points specified in my last note & believe me | My dear Sir | Yours very sincerely | C. Darwin

Endorsement: '*1861*'
John Murray (publishers)

[1] See the enclosure with the letter to John Murray, 30 April [1861]. Murray had originally calculated CD's share of the profits on the third edition of *Origin* to be £480; the revised figure was £372.

[2] CD received £636 13*s*. 4*d*. in royalties on the second edition of *Origin* (see 'Journal' (Appendix II) and Peckham ed. 1959, p. 775).

To P. L. Sclater 4 May [1861]

Down. | Bromley. | Kent. S.E.
May 4th

Dear Sclater

Immediately on the receipt of your note & kind offer of inserting in the Ibis a note about the habitats of the Falkland Birds,[1] I got out my catalogues & notes made on the spot.[2] But I am sorry to say that I can make nothing out definitely, for I always refer to numbers not knowing at the time the names of the Birds; so that if any specimens got wrongly named, it would only be possible to set matters right by looking to my original specimens.[3]

Accordingly I wrote to Mr Gray & to Gould to know if they had any of my specimens (which most unfortunately were given to Zoolog. Soc & afterwards all distributed), but they have not.—[4] Therefore I look at the case as hopeless. I cannot conceive how such a mistake could have occurred; but without fresh & distinct evidence, I do not see that I could do any good publishing a note. A false habitat is a positive mischief, worse than a species not appearing in a list; & therefore I shd. say after Capt. Abbotts careful work, it would be better for the two names to be considered as errors, than to be given without positive evidence.—[5]

I am sorry to have caused you this trouble, & am myself vexed that I cannot either prove myself right or confess to a great & heavy blunder.

Yours very sincerely | C. Darwin

P.S. | I had a letter the other day from Mr Swinhoe, & he tells me that he has sent you a new rock-pigeon & (as I understand) the *wild* Anser cygnoides;[6] but when next in London I must call on you & enquire about these two Birds,—which surprise me.— Perhaps the Pigeon may be the Himalayan Rock-pigeon.—

Endorsement: '*May 1861*'
American Philosophical Society (246)

1 Sclater's note has not been found, but see the letter to P. L. Sclater, 21 [April 1861]. Sclater was the editor of the *Ibis, a magazine of general ornithology*, published by the British Ornithologists' Union.

2 CD's catalogue of numbered ornithological specimens collected on the *Beagle* voyage is in DAR 29.2. A transcript of the catalogue, including annotations, was published by Nora Barlow (Barlow ed. 1963).

3 The numbers in the catalogue (see n. 2, above) correspond to CD's specimens, which are recorded and described in the *Beagle* zoological diaries (DAR 30 and 31).

4 George Robert Gray and John Gould both contributed species descriptions for *Birds*. CD presented the *Beagle* collection of mammals and birds to the Zoological Society of London in January 1837. After the society's museum was closed and its collections dispersed in 1855, most of the birds went to the British Museum, but some of CD's specimens were either destroyed or lost. See Porter 1985, pp. 1002–4.

5 See letter to P. L. Sclater, 21 [April 1861]. Charles Compton Abbott, in his account of the birds of the Falkland Islands, stated of *Cinclodes antarcticus* (Garn.) (a synonym for *Opetiorhynchus antarcticus*) (Abbott 1861, p. 154):

> I do not believe that there is any second species of this genus found in East Falkland, although Mr. Darwin states that *Cinclodes patachonicus* is "common" there. Nor have I ever seen or heard of any such bird as the *Scytalopus magellanicus*, also mentioned by Mr. Darwin.

[6] Robert Swinhoe, British consulate in Amoy, China, sent CD specimens of domesticated pigeons from China (see *Correspondence* vol. 5, CD memorandum, [December 1855], and *Variation* 1: 132 n. 1, 186). CD had sent Swinhoe a copy of the first edition of *Origin* (see *Correspondence* vol. 8, Appendix III). Swinhoe included the rock-pigeon, *Columba leucozonura*, and *Anser cygnoides* in the lists of birds he had collected in North China (Swinhoe 1861a, pp. 259–60, and 1861b, p. 344, respectively).

To the *Field* [before 4 May 1861][1]

Sir,—

From the recent investigations of Mr Tegetmeier, and from those of the older naturalists, most people who keep Polish fowls are aware that the tuft of feathers on their heads is supported by an extraordinary, almost hemispherical, protuberance of the front part of the skull.[2] This protuberance is accompanied by an equal change in the shape of the brain. Pallas and some of the older writers describe the Polish fowl as, in consequence, stupid;[3] but Mr Tegetmeier has shown that this is a mistake. The experience of many savage races of man proves that the external shape of the brain may be greatly modified by pressure, without the intellectual faculties being affected. Nevertheless, after having recently examined the skulls of Poland fowls, I am astonished that such profound changes should have produced no effect on the mental powers of these birds. Bechstein, writing in Germany in 1793, says, that although it is an error to suppose that fowls with such misformed skulls are stupid, yet that he had a white-crested black Poland which "was crazy, and which all day long wandered anxiously about."[4] I, also, formerly had a silver-spangled Poland which was curiously affected; she sometimes seemed lost in reverie, and allowed any one to approach so as even to touch her; she was solitary in her habits, and was extraordinarily deficient in the sense of direction. I have seen her stray hardly more than a hundred yards from her feeding place, and become completely lost; and she would then continue obstinately to try to get through a fence in a direction exactly opposite to her home. Now, will any of your correspondents who have long kept tufted fowls have the kindness to state whether they have observed any clear signs of deficiency in the mental powers of any of their birds? or, has any one ever seen a "crazy fowl," such as Bechstein describes, in any other breed?[5]

Charles Darwin. | Down, Bromley, Kent.

The Field, the Farm, the Garden, the Country Gentleman's Newspaper 17 (1861): 383

[1] The date is taken from the issue of the *Field* in which the letter appeared. It was headed: 'Influence of the form of the brain on the character of fowls'.

[2] William Bernhard Tegetmeier read a short paper on the 'remarkable peculiarities' of the skulls of feather-crested Polish fowls at the Zoological Society of London in November 1856 (Tegetmeier 1856). CD had provided him with a reference for this paper (see *Correspondence* vol. 6, letter to W. B. Tegetmeier, 20 March [1856], and n. 3, below).

[3] Pyotr Simon Pallas described the protuberance on the skulls of Polish fowls in Pallas 1767–80, pt 4, pp. 18–23.

[4] Bechstein [1789–95], 3: 400. CD's copy of this four-volume work in the Darwin Library–CUL is heavily annotated.

[5] A response to CD's query appeared in the 11 May 1861 issue of the *Field*, p. 404. Under the heading 'Polish fowls', the anonymous respondent stated: 'I have half a dozen black, with white toppings, and they certainly are tame or stupid. You may tread upon them—they don't seem to see well, and they seldom find the roosting-place, but crouch or perch anywhere.—'

From W. B. Tegetmeier 4 May [1861][1]

Muswell Hill | N
Saturday May 4.

My dear Sir.

I trust you will excuse my long delay in replying to your last but I have been very busy and your queries could not all be answered by myself.—[2] I enclose some replies.—[3] I am not quite sure as to what you call the first pruning will you let me know as it is desirable that I should be quite certain of your meaning.—

I was surprized at the Number of tail feathers in the Malays.[4]

I will send the answers to your other querys as I can observe them as I do not like sending you my impressions merely.

I am collecting facts respecting the Antwerps or flying birds of Belgium—. I have two volumes of a periodical devoted to the Pursuit—and have some of the very best birds. I have requested T T Parker Esq who has a number of these birds to send you an egg just chipped[5] would you be good enough to let me know what you think of the cerebral development in that stage.— I thought you would like to have the bird in that state as it would complete your series.—[6] These birds fly distances we have no idea of in this country their speed for short distances is a league a minute!!! The following is the account of a match from Bordeaux to Verviers.—

Birds liberated at Bordeaux Saturday July 22. at $4\frac{1}{2}$ AM
Temperature high. Wind NW
Distant direct 186 leagues.
1st bird arrived at Verviers on Sunday 23 July at 12.30
2nd on Monday 24 at 6 54 A.M

Mr Brent wrote some papers in the Field some time since on this subject, but I find his papers of little use to me.[7] I want facts and not theories.

Mr Huxley asked me to publish the paper in his journal,[8] but I am so busy that I do not think I can get it ready for the printer for some little time, so that if you wish to see it I can let you have the Memorandum in the first instance—

I am glad to see you using the Field—[9] With regard to your last Query.[10] I have seen ⟨moping⟩ semi idiotic polish but as a general rule those I kept had no particular alteration of mental characters some are as savage as a game cock.—[11] one I had would fight my hand just as a savage game cock will, if teased—

Believe me | Very truly Yours | W B Tegetmeier

C Darwin Esq.

I enclose two neck feathers from Duckwing and Black red Game

DAR 205.2 (Letters)

CD ANNOTATIONS
1.1 I trust . . . sending 3.2] *crossed pencil*
1.3 I am . . . meaning.— 1.4] *cross in margin, brown crayon*
3.2 you . . . series.— 4.6] *crossed ink*
4.4 would you . . . stage.— 4.5] *cross in margin, brown crayon*
6.1 Mr Brent . . . teased— 8.4] *crossed ink*
7.1 Mr Huxley . . . time, 7.2] *scored pencil*
7.2 so that . . . instance— 7.3] 'No' *added brown crayon, circled brown crayon*
8.1 With regard . . . characters 8.3] *scored brown crayon*
Top of first page: 'Pigeon flight' *under del* 'Polish [*illeg*]' *brown crayon*; '18'[12] *red crayon, circled red crayon*

1 The year is suggested by the discussion of domesticated fowls. CD was writing his chapter on fowls for *Variation*; he finished it on 16 May 1861 (see 'Journal'; Appendix II).
2 See letter to W. B. Tegetmeier, 14 April [1861] and the enclosure.
3 Tegetmeier's enclosure has not been found.
4 The number of tail feathers in various varieties of fowl is discussed in *Variation* 1: 258–9.
5 Neither Parker nor the fancier's periodical has been identified.
6 CD had formed a collection of newly hatched pigeons, preserved in alcohol, as a means of determining when the strongly marked characters of the adults first became manifest. See *Correspondence* vol. 6, letters to W. D. Fox, 15 March [1856], and to W. B. Tegetmeier, 24 June [1856], and letter from Charles Lyell, 1–2 May 1856, n. 10. See also *Origin*, pp. 445–6, and *Variation* 1: 178, 248–50.
7 Bernard Philip Brent was a frequent contributor to both the *Cottage Gardener* and the *Field* on various subjects pertaining to domesticated birds and animals.
8 Tegetmeier's paper on Antwerp carrier pigeons did not appear in the *Natural History Review*, of which Thomas Henry Huxley was chief editor.
9 See letters to W. B. Tegetmeier, 2 April [1861] and 14 April [1861]. CD had recently sent several letters to the *Field*, which were printed in 'The Naturalist' section in the column 'Notes and questions on natural history'. See letters to the *Field*, [before 27 April 1861] and [before 4 May 1861].
10 See letter to the *Field*, [before 4 May 1861].
11 CD cited Tegetmeier's information in *Variation* in the discussion of the effect of the remarkable bony protuberance of Polish fowls on their intelligence (see *Variation* 1: 257 and n. 58).
12 The number of CD's portfolio of notes on the means of dispersal of plants and animals.

To T. C. Eyton 6 May [1861][1]

Down. | *Bromley.* | *Kent. S.E.*
May 6th

Dear Eyton

Very many thanks for so promptly & kindly answering my queries.—[2] No doubt the skeleton was the Duckwing Game, & the wild Gallus G. Bankiva,—at least every bone agrees.—

As I have two good skeletons of Dorking it will not be worth while to trouble you to send yours (if found), or that of G. varius. On no account would I have the mounted Skeletons sent, though I cordially thank you for the offer.

I have now examined 25 skeletons & about 55 skulls.—[3] In the skulls I find some differences characteristic of the breeds; in the other Bones much *fluctuating* variability.

My impression, though rather vague, from what little I have seen & read is that the skeletons of species of the same restricted genus or sub-genus do not commonly differ but very slightly. I do not at all like to trouble you, but I certainly should much like to *quote* your opinion how far as a general rule, the skeletons (excepting the head) differ in allied species.— One fact has surprised me (& on this too I sh$^{d.}$ like to have your opinion), namely that though some parts of the skeleton besides the skull present plain differences, yet that all the bones of the wing & leg in all the Breeds are identically similar in configuration, though not in relative lengths & thickness.[4]

With best thanks Dear Eyton | Yours very sincerely | C. Darwin

P.S. | As I shall not come to the Ducks for a few weeks, I will venture to keep your specimens for that time.—

Heaven forgive me; I have just thought of another question, which I much wish answered. viz, whether the skulls of any Birds which have a *large* top-knot, like the Curassows have any protuberance, to support the top-knot?

Can you forgive me?

American Philosophical Society (248)

[1] Dated by the relationship to the letter to T. C. Eyton, 3 May [1861].

[2] Eyton's letter has not been found, but see the letter to T. C. Eyton, 3 May [1861].

[3] For CD's analysis of the osteological differences between various breeds of fowl, see *Variation* 1: 260–70. He states that he examined 27 skeletons and 53 skulls, three of which belonged to Eyton (*ibid.*, p. 260).

[4] CD reported this observation in *Variation* 1: 268–9, stating: 'it was truly surprising to see how absolutely every process, articulation, and pore agreed, though the bones differed greatly in size.' (*ibid.*, p. 269).

From John Murray 7 May 1861

May 7—1861

My Dear Sir

I enclose my amended note of hand for £372 for the 3rd edition of your work on Species[1]

This, as far as I can judge, is the amount of $\frac{2}{3}$^{d} profit arising from 2000 copies.[2] There are always contingencies arising—such as advertising expences, which make it difficult to fix beforehand the precise profits on any work.[3] Thus in the first Edition—the sum I gave you turned out far more than $\frac{2}{3}$^{d} or at least my share eventually fell below $\frac{1}{3}$ considerably.[4]

Moreover it must be borne in mind that it is not a mere rule of 3 sum to calculate profits of a book— Thus—the 3$^{rd.}$ Ed$^{n.}$ of 2000 Copies cost only £70. less than 3000 of the 3$^{rd.}$ edition. Moreover this last is increased by two sheets beyond the second edition 23 shts instead of 21.

I enter into these details to prevent any misunderstanding and hope I have been successful.

If we print any more editions I would prefer deferring the settlement untill the whole is disposed of when I could arrange the amount of $\frac{2}{3}$ds. profit to the uttermost farthing.[5]

I remain | Dear Sir | Y^rs. Very faithfully | (signed) John Murray

C. Darwin Esq

Copy
John Murray (publishers)

[1] See letters to John Murray, 30 April [1861] and 3 May [1861].

[2] According to the terms CD and Murray agreed upon for the publication of *Origin*, CD was to receive '$\frac{2}{3}$^ds of the net proceeds of the edition—payable by note of hand at six months from the day of publication' (*Correspondence* vol. 7, letter from John Murray, 1 April 1859).

[3] According to Murray's ledgers, Murray spent £45 on advertising the first edition of *Origin*, £60 4*s.* for the second, and £29 3*s.* 6*d.* for the third (see Peckham ed. 1959, pp. 775–6).

[4] CD received £180 as his share of the profits for the first edition; Murray recorded having made £57 4*s.* 2*d.* 'balance profit' (Peckham ed. 1959, p. 775).

[5] The fourth edition of *Origin* was published in December 1866; CD received his share of the profits in September 1867 after all copies of the edition were sold (Peckham ed. 1959, pp. 776–7).

From William Hardy to Luke Hindmarsh [8 May 1861][1]

Chillingham Castle | Alnwick.

Dear Sir

In reply to yours of the 6th. inst.[2] I cannot give you the particulars of what have been slaughtered of the Wild Cattle, since 1838, as they have only been in my charge for 10 years, during which time, I think the average number slaughtered, and killed by fighting, accidents &c, is about 10 a year, as Lord Tankervilli[3] wished the standing number to be 50.[4]

Fifteen months ago, Lord T— desired that the number should be increased to 70, since that time we have not slaughtered any; but three full grown bulls, have been killed by fighting, two others have been killed upon account of their old age, three calves have died from accidental deaths, getting into ditches &c, and at this time the Herd have increased to 59; this will shew you, free from accidents, and increase of 14 in fifteen months.

I do not perceive since I have known them, which is now 30 years, any indications of their diminishing in size, or in any of their peculiar characteristics, as I have kept an account of the weights of the slaughtered, for the last 10 years and they keep up to their average weight.—

Yours &c &c

Copy
DAR 46.1: 92

CD ANNOTATIONS
Bottom of letter: 'To L Hindmarsh Esqr May 8th 61/ | Hardy' *ink*
Verso of letter: 'Chillingham Cattle' *pencil, underl pencil*

[1] The date is given by CD's annotation.

[2] Hindmarsh had written to Hardy at CD's instigation. See letter to Luke Hindmarsh, 3 May [1861]. William Hardy was bailiff to Lord Ossulston, son of the earl of Tankerville (*Post Office directory of Northumberland* 1858).

[3] Chillingham Castle was the seat of Charles Augustus Bennet, earl of Tankerville. The copyist erred in transcribing the name.

[4] CD quoted this information in *Variation* 2: 119 as part of the discussion of the good effects of crossing and the detrimental effects of close interbreeding.

To John Murray 8 May [1861][1]

Down. | Bromley. | Kent. S.E.
May 8

My dear Sir

I write one line to say that I have received, with my thanks, your Bill of £372, & to say that your note is quite satisfactory.[2] I am sorry to have caused you trouble. I hope to be in London before very long & will call

In Haste | My dear Sir | Yours very sincerely | C. Darwin

John Murray (publishers)

[1] Dated by the relationship to the letter from John Murray, 7 May 1861.

[2] See letters to John Murray, 30 April [1861] and 3 May [1861], and letter from John Murray, 7 May 1861.

To W. E. Darwin 9 May [1861][1]

!!*Down. | Bromley. | Kent. S.E.*!!
May 9th.—

My dear William

I believe you have been much neglected about Letters, but I have been awfully overworked in that line, & so indeed has Mamma been.— Parslow sent off today by R.y a parcel for you with Boots &c.—[2] Mamma went up to Cumberland Place yesterday with Etty for her teeth.[3] Etty stood the journey well, but was much knocked up on her arrival.—

I am glad you are going on walking tour in Wales; but I cannot help fearing for your ancles—[4] for Heaven-sake give up, though it will be a grievous evil, if your ancles fail.— Do you think it would amuse you to look out for Glacial phenomena: if so I would send you my little paper on N. Wales, though I must have it back as I have no other copy.[5] If you shd. find any rare orchids send me some in bud & open, in any old tin Box, with damp moss round roots: they would be a treasure to me.—[6] Look out for little Dun Striped Ponys—[7] I suppose we shall certainly go to Torquay early in June; though I dread the journey for Etty & indeed a little for myself.—[8] All the Hensleighs & Eras. say that it is a charming place.—[9]

Tim has been staying here:[10] he is a helpless fellow, he let the Pony loose in the Westerham Rd. & the pony came home across country in gallant style, jumping everything & cleared the Hurdles & Hedge into the sand-walk. I have not had one game of Billiards since the Boys were here;[11] indeed the Table has been covered with skeletons of Cocks & Hens, & has been very useful for that purpose.

Mamma will not come back till next Wednesday, so I am rather dull; & I shall be all the duller when Miss Ludwig replaces Emily Thorley on Monday.—[12]

Good Bye | My dear old man | Yours affect^y | C. Darwin

Look at crest on Envelope.[13]

DAR 210.6

[1] Dated by the reference to the Darwins' stay in Torquay (see n. 8, below).

[2] Joseph Parslow was the Darwins' butler.

[3] Emma Darwin recorded in her diary that she went to London on 8 May 1861 and 'Went to Mr W.' on 9 May. Alfred J. Woodhouse, of Hanover Square, London, was the Darwin family dentist. CD's Account book (Down House MS) shows a number of payments to him in 1861, including one on 8 July 1861. Emma's brother Hensleigh Wedgwood and his family lived at 1 Cumberland Place, Regent's Park. Emma and Henrietta Emma Darwin returned to Down on 17 May.

[4] William had seriously injured one of his ankles while on a walking trip in the Lake District in the summer of 1859. See *Correspondence* vol. 7, letter to W. E. Darwin, 25 [August 1859].

[5] CD refers to 'Notes on the effects produced by the ancient glaciers of Caernarvonshire, and on the boulders transported by floating ice' (*London, Edinburgh, and Dublin Philosophical Magazine and Journal of Science* 21 (1842): 180–8; *Collected papers* 1: 163–71).

[6] CD was engaged in a study of the structural adaptations that facilitated the process of cross-fertilisation in orchids.

[7] CD added this note in the margin. CD was interested in collecting cases of stripes appearing in dun-coloured ponies as an illustration of reversion to ancestral characters (see *Variation* 1: 55–61). William had already provided CD with information about striping in a Belgian cart-horse (*ibid.*, p. 57).

[8] The Darwins left for a six-week stay in Torquay on 1 July 1861 ('Journal'; Appendix II).

[9] CD refers to Hensleigh and Frances Mackintosh Wedgwood and to his brother, Erasmus Alvey Darwin.

[10] Emma Darwin recorded in her diary that Alfred Allen (Tim) Wedgwood, the nineteen-year-old son of Hensleigh and Fanny Wedgwood, came to Down on 1 May 1861 and stayed until 6 May.

[11] According to Emma's diary, George Howard and Francis Darwin were home from school from 2 to 4 March 1861. They visited again from 20 to 22 May. Both were attending Clapham Grammar School in south London. CD had purchased the billiard table for Down House in 1859 (see *Correspondence* vol. 7).

[12] Camilla Ludwig was the Darwins' governess. Emily Thorley, the younger sister of a previous governess of the family, sometimes cared for the children on an interim basis. There is an entry in CD's Account book on 7 May 1861 for £10 10*s.* for 'Miss E. Thorly'.

[13] CD wrote this at the top of the letter above the address. He was drawing William's attention to the printed stationery he had recently begun to use.

To Luke Hindmarsh 12 May [1861][1]

Down, Bromley, Kent, S.E.
May 12th.

Dear Sir

I thank you sincerely for your prompt and great kindness, and return the letter, which I have been very glad to see and have had copied.[2] The increase is more rapid than I anticipated; but it seems rather conjectural; I had hoped that in so interesting a case some exact record had been kept. The number of births or calves reared till they followed their mothers would perhaps have been best datum. From Mr. Hardy's letter I infer then 10 must be annually born to make up the deaths from various causes.[3] In Paraguay, Azara states that in a herd of 4000 from 1000 to 1300 are reared, but then though they do not kill calves, but castrate the young bulls, no doubt the oxen would be killed earlier than the cows, so that the herd would contain probably more of the female sex, than the herd at Chillingham.[4]

There is not apparently any record, whether more young Bulls are killed than cows. If you should continue to keep notes on this interesting subject, which you have so well illustrated, it would be worth while to have the date of the last killed owing to the number exceeding 50, until one has to be killed from number exceeding 70. I am surprised that Lord Tankerville does not have exact record kept of deaths and sexes and Births, after a dozen years it would be an interesting statistical record to the naturalist and agriculturalist.

With my cordial thanks for your kindness, I beg leave to remain, Dear Sir, | Yours faithfully and obliged | Charles Darwin.

Copy
DAR 145

CD note:[5]

Azara. (Quadrupeds des Paraguay. Tom. 2. p. 368) states (very accurate now) that cattle on Estuaries *in Paraguay [*interl*] increase at 1/4 or 1/3 of total— So that where 4000 kept of *all ages* & all sexes, they produce annually from 1000 to 1300 calves. These are probably reared. They do not kill calves, but castrate— probably more cows than oxen & Bulls, as cows, though killed, are probably not killed so soon.— I do not know ['how' *del*] whether at Chillingham more Bulls killed than cows. Nor how many calves born.— At Chillingham the Bull, if Mr H. estimate be correct must be 1/5 to keep up number—

1 Dated by the relationship to the letter to Luke Hindmarsh, 3 May [1861], and to the letter from William Hardy to Luke Hindmarsh, [8 May 1861].

2 Letter from William Hardy to Luke Hindmarsh, [8 May 1861].

3 In *Variation* 2: 119, CD estimated the annual rate of increase of the Chillingham herd to be 'about one in five'.

4 The reference is to Félix d'Azara's descriptions of the fauna of Paraguay (Azara 1801). See CD's note, above. In *Variation* 2: 119, CD compared the rate of increase of the Chillingham herd with that of a half-wild South American breed, citing Azara's estimate of an annual rate of increase for the latter of 'one in between three and four'. CD used the comparison to support his argument that cross-breeding leads to increased fertility.

5 The note is in DAR 46.1: 91; it refers to Azara 1801, an annotated copy of which is in the Darwin Library–CUL.

To *Journal of Horticulture*[1] [before 14 May 1861][2]

Down, Bromley, Kent

Will Mr. Beaton, who has made such a multitude of most interesting observations on the propagation of plants, have the kindness to state whether varieties of the

John Stevens Henslow.
(Courtesy of the Royal Botanic Gardens, Kew.)

William Erasmus Darwin. Photograph by William Farren, *c*. 1861.
(Courtesy of the Darwin Museum, Down House.)

same species of Composite plants frequently cross each other by insect agency or other means?[3] For instance, will any of the Cinerarias, if kept apart from other varieties, breed true? but if standing near other varieties, will they generally, or almost certainly, produce a much greater diversity of coloured seedlings?

I saw an allusion by Mr. Beaton to this subject in *The Cottage Gardener* of last year with respect to Zinnias;[4] and from this allusion I infer that Zinnia sports much when kept separate.

As I am begging for information on the natural crossing of plants, I will likewise venture to inquire whether the great raisers of Hollyhocks find it necessary to keep each variety far separate from the others for raising seed. The late famous horticulturist, the Hon. and Rev. W. Herbert, when I visited him at Spofforth many years ago, remarked that he was much surprised (considering the structure of the flower and the relative periods of maturity of the pollen and stigma) how true some sorts of Hollyhocks bred, even when growing close to other varieties.[5] I have found this to be the case with some of the varieties, and cannot understand how it is possible. Mr. Beaton might, if he pleased, write an article, very valuable to physiological botanists and of some practical utility, on the natural crossing of varieties.[6] He might indicate in which genera crossing most commonly occurred, and in which it seldom or never occurred. For instance, I have observed Sweet Peas during several years and believe that they never cross;[7] and it is not easy to make an artificial cross, though I succeeded at last, but got no good in a horticultural point of view.—

Charles Darwin, *Down, Bromley, Kent.*

Journal of Horticulture, Cottage Gardener, and Country Gentleman, n.s. 1 (1861): 112

[1] Beginning with the issue of 2 April 1861, the *Cottage Gardener* changed its name to the *Journal of Horticulture, Cottage Gardener, and Country Gentleman. A journal of horticulture, rural and domestic economy, botany and natural history*. CD's copies of the journal,—issues from October 1855 through March 1856 and from January 1860 to February 1866,—are in the Darwin Library–CUL.

[2] CD's letter was published in the issue of 14 May 1861.

[3] The letter, printed under the heading: 'Phenomena in the cross-breeding of plants', was prefaced by the following editorial comment: 'Having received the following letter from Mr. Darwin we forwarded it to Mr. Beaton, and now publish it with his reply.' The reference is to Donald Beaton, a Scottish practical gardener with an interest in hybridisation who contributed regularly to the journal. The editors of the *Journal of Horticulture* were George W. Johnson and Robert Hogg.

[4] See *Cottage Gardener*, 10 April 1860, p. 19.

[5] CD had corresponded with William Herbert, a great authority on plant hybridisation, about the crossing of plants (see *Correspondence* vol. 2). CD visited Herbert, after his appointment as dean of Manchester, on 15 September 1845 (*ibid.*, vol. 3, 'Journal' (Appendix II)).

[6] CD's letter was followed by a long response from Beaton, which addressed several of the points raised by CD (*Journal of Horticulture, Cottage Gardener, and Country Gentleman* n.s. 1 (1861): 112–13). CD, however, did not find Beaton's information helpful (see letters to *Journal of Horticulture*, [17 May 1861], and to J. D. Hooker, 14 May [1861]).

[7] CD had for many years been investigating mechanisms of crossing in the Leguminosae in connection with his view that all organic beings must occasionally cross-fertilise. He found that *Lathyrus odoratus*,

the sweetpea, challenged this general rule. See *Natural selection*, chap. 3, especially pp. 68–71, and *Correspondence* vol. 6, letters to George Bentham, 30 November [1856], and to W. D. Fox, 22 February [1857].

To T. C. Eyton 14 May [1861][1]

Down. | Bromley. | Kent. S.E.
May 14th

Dear Eyton

I thank you sincerely for your wish to see me in London & for your kind invitation to Eyton.[2] I should most truly enjoy seeing you; but this very day I wished for two urgent reasons to go to London, but I have been so unwell lately that I was unable. If ever I am again at Shrewsbury, I will, if strong enough pay you a visit & see the place, at which I have spent many happy days.

On your return home, if you can spare time, (but I know the request is perhaps unreasonable & shall not be surprised at not hearing) I sh^{d}. be very glad of a brief answer, with permission to quote you, on the three points on which I asked.
(1) whether the skeletons of Birds, excepting skull, of the same restricted genus, do not generally very closely resemble each other. (2) whether the Wing & leg bones are generally very constant in form. (3) whether in largely crested Gallinaceæ the skull is protuberant to support the crest.—[3]

Dear Eyton | Yours very sincerely | C. Darwin

American Philosophical Society (249)

[1] Dated by the relationship to the letter to T. C. Eyton, 6 May [1861].
[2] Eyton's family seat was in Eyton, near Shrewsbury in Shropshire. CD and Eyton shared a common interest in natural history during their student days at Cambridge University. Eyton, in addition to keeping a large number of animals on his estate, had formed a fine museum collection that was particularly rich in bird species (see *Correspondence* vol. 6, letter to T. C. Eyton, 31 August [1856]).
[3] See letter to T. C. Eyton, 6 May [1861].

To J. D. Hooker 14 May [1861]

Down. | Bromley. | Kent. S.E.
May 14th

My dear Hooker

I have been putting off writing from day to day, as I did not wish to trouble you, till my wish for a little news will not let me rest. Unless by a prodigy poor dear Henslow is recovering, good God what a time his sufferings have been prolonged. But from your last note, I hope, I need not say sufferings, but only life.[1] What a miserable time you must have had.—

I am alone at present, as Emma has taken Etty to Dentist in London; & their stay in London, I hope, has done both good.[2] I sh^{d}. have gone up, but I have been rather extra ailing of late. I have no news to tell you, for I have had no interesting letters for some time & have not seen a soul.—

I have been going through Cottage Gardener of last year, on account chiefly of Beaton's articles:[3] he strikes me as a clever, but d—d cock-sure man (as L. Melbourne said)[4] & I have some doubt whether to be much trusted. I suspect he has never recorded his experiments at the time with care. He has made me indignant by the way he speaks of Gärtner, evidently knowing nothing of his work.—[5] I mean to try & pump him in Cot. Gard. & shall perhaps defend Gartner.—[6] He alludes to me occasionally & I cannot tell with what spirit. He speaks of "*this* M^r. Darwin", in one place, as if I were a very noxious animal.[7]

Let me have a line about poor Henslow pretty soon.

Farewell my dear Friend | Ever yours | C. Darwin

I fear there is something seriously amiss with Lyell.—[8]

Endorsement: '/1861.'
DAR 115.2: 99

[1] John Stevens Henslow, Hooker's father-in-law, died after a long illness on 16 May 1861.

[2] Emma and Henrietta Emma Darwin went to London on 8 May 1861 and returned to Down on 17 May (Emma Darwin's diary). See letter to W. E. Darwin, 9 May [1861] and n. 3.

[3] Donald Beaton contributed a weekly column to the *Cottage Gardener*. He was well known to both Hooker and his father, William Jackson Hooker, director of the Royal Botanic Gardens, Kew (see *Cottage Gardener*, 4 September 1860, p. 337).

[4] William Lamb, Lord Melbourne, had a reputation for having unconventional manners and speech 'interlarded with oaths' (*DNB*).

[5] CD had studied closely the hybridisation experiments carried out by the German botanist, Karl Friedrich von Gärtner. CD cited results from Gärtner's systematic crossing of various hybrids in *Natural selection* and in *Origin*.

[6] See letter to the *Journal of Horticulture*, [17 May 1861]. CD did not challenge Beaton's criticism of Gärtner's work until 1863, in a letter written to the *Journal of Horticulture*, [3 February 1863] (see *Calendar* no. 3966; *Correspondence* vol. 11).

[7] Beaton briefly mentioned CD's views on species in several of his regular articles in the *Cottage Gardener* (*Journal of Horticulture*). CD's reference is to Beaton's statement in 'Kew Arboretum and pleasure-grounds' (*Cottage Gardener*, 11 September 1860, pp. 352–3):

> For, you see, this Mr. Darwin has made a wonderful impression by his notions of how plants came first into being, and how such beings could hold on for so many ages . . . They have laws, reason, and logic for all this, and for ten times as much; and ten to one if some of us be not drawn from our own conclusions by that logic—and the logic of facts is the hardest of all to bear up against.

[8] See letter to J. D. Hooker, 18 [May 1861].

From Robert Edmond Grant[1] 16 May 1861

Dear Mr. Darwin,

On public grounds, and as an old fellow-labourer in the same rich field of philosophic inquiry, I avail myself of the opportunity afforded in publishing this brief outline of the primary divisions of the Animal Kingdom, to dedicate these pages to you in testimony of my admiration and approval of your late successful attempt to throw further light on that involved and obscure question, regarded by some inquiring minds as the mystery of mysteries of organic nature, the origin of species by natural law; or, as you have more happily and more definitely expressed the problem, "the origin of species by means of natural selection, or the preservation of favoured races in the struggle for life."

More than fifty years have now elapsed since the 'Zoonomia' of your illustrious ancestor, Dr. Erasmus Darwin, first opened my mind to some of "the laws of organic life,"[2] which he so clearly expounded, and so successfully applied to explain the abnormal phenomena of the human body; and nearly forty have already rapidly fled away since you and I were busied in exploring microscopically the delicate structures and the living phenomena of the lowest organisms abounding in the rich fauna of the Firth of Forth.[3]

But while I have been humbly occupied with the gleanings of our fellow-labourers in the cabinets and seminaries of Europe, and reporting the results to my youthful auditors,[4] you have been widely surveying the grand domain of nature with a learned spirit over the oceans, seas, islands, and continents of the globe, in every latitude and in every clime, nearly from the one pole to the other, and accumulating that rich store of thought and observation which entitles you of all men to aspire to the complete solution of those great, though obscure problems of organic nature, which have so long perplexed philosophers; a labour which you have so successfully commenced, and in which you have already surpassed all your predecessors, from the time of Plato.

Intellectual triumphs like yours, which have been hailed with the assent and applause of all competent unbiassed minds at home and abroad, while they charm away the clouds of mysticism which overhang some parts of our science and of philosophy, and obscure the greatest truths of nature, alone add permanent glories to the annals of our country in the great struggle for intellectual preeminence and ascendency among the nations of the earth. With one fell sweep of the wand of truth, you have now scattered to the winds the pestilential vapours accumulated by "species-mongers" over every step of this ever-varying, ever-charming part of nature's works; and your next movement will dispel the remaining clouds of "mystical supernatural typical intrusions" which still hang on the horizon of the sublime prospect, now opening to the view, of the natural animalization of the orbs of space by the same simple laws which govern the physical and chemical phenomena with such wondrous harmony throughout the rest of the material universe.[5]

I remain, my dear Sir, | With great respect and regard, | Yours most truly, | Robert E. Grant.

2 Euston Grove, Euston Square, London.
16th May, 1861.

Grant 1861, pp. v–vi

[1] It is not known whether Grant actually sent CD a copy of the letter or whether CD otherwise read it. The letter was published in Grant 1861, which was dedicated to CD. The book is not in the Darwin Library. The work is a brief outline of a course of lectures on zoology offered by Grant at University College London, where he was professor of zoology and comparative anatomy.

[2] E. Darwin 1794–6. CD's grandfather Erasmus Darwin expounded evolutionary views similar to those adopted by Grant (see A. Desmond 1984a). CD included Grant's name in the revised 'historical sketch' added to the third edition of *Origin* among those who had espoused evolutionary doctrines. CD referred to an 1826 paper in which he stated that Grant 'clearly declares his belief that species are descended from other species, and that they become improved in the course of modification.' (*Origin* 3d ed., p. xiv). According to the catalogue of CD's library prepared by H. W. Rutherford in 1908, CD had copies of a number of Grant's papers on marine invertebrates published in the 1820s, including Grant 1826; these may be in the Darwin Library–Down. A copy of Grant's book on comparative anatomy (Grant 1835–7) is in the Darwin Library–CUL.

[3] Grant had introduced CD to the study of marine invertebrates in 1826 and 1827 while he was a student at Edinburgh University. See *Autobiography*, p. 49, and *Correspondence* vol. 1, 'Journal' (Appendix I). For Grant's influence on CD's early zoological education, see Jespersen 1948 and Sloan 1985.

[4] Grant had studied at the Muséum d'histoire naturelle in Paris and was a follower of French materialism and philosophical anatomy. He maintained close contact with European naturalists through frequent visits to Paris and the Continent. For Grant's career after leaving Edinburgh in 1827, see A. Desmond 1984a and 1984b.

[5] For Grant's mature views on the development of life, see A. Desmond 1984b.

To *Journal of Horticulture* [17 May 1861][1]

Much obliged am I to Mr. Beaton for his very interesting answer to my question.[2] When Mr. Beaton says he does "not know of an instance of the natural crossing of varieties," I presume he intends to confine his remark to the plants of the flower garden;[3] for every one knows how largely the varieties of the Cabbage cross, as is likewise the case (as I know from careful trial) with Radishes and Onions. It was this fact which led me to suppose that varieties of flower-garden plants would naturally cross.[4]

I can quite understand, after reading Mr. Beaton's remarks, that it would be very difficult, perhaps impossible, to detect such natural crossing from the degree to which most of these varieties vary. I should, however, think that those who raise for sale seeds of distinct varieties of the Hollyhock, Stocks, &c., must know whether it is indispensable to keep the parent plants apart.[5]

I will not trouble Mr. Beaton again if he will have the kindness to procure for me answers on one or two points quoted in his paper (June 26, 1860) from the "king of British cross-breeders"—namely, whether I understand rightly that the white *Anemone apennina* seeding in a mass with the blue (*Anemone apennina?*) produced many pale shades?[6] For this seems to be a case of two varieties naturally crossing, though I want to know the fact for another reason—namely, because Anemone does not secrete nectar; and secondly, whether *Mathiola incana* and *glabra*, which the writer speaks of as "crossing freely," were artificially crossed.[7]

Mr. Beaton's statement (July 24, 1860) that if the pollen of five kinds of Geranium (I presume what botanists call varieties, and not what are called species, are here referred to) are placed on the stigma of a flower, one kind alone takes the lead and produces an effect, seems to me a most curious observation. It is, I fear, unreasonable to ask for a few precise cases on this head; for, as I gather from Mr. Beaton, it must be difficult to know whether one or more kinds have produced an effect, owing to the great variability of crossed varieties.[8]

I have been delighted to observe how strongly Mr. Beaton insists that "not a flower in a thousand is fertilised by its own immediate pollen."[9] This is a subject which I have attended to for the last twenty years.[10] From my experiments on a small scale I would not venture to put the case nearly as strongly as Mr. Beaton does; but on the other hand, some of the plants which Mr. Beaton advances as self-fertilisers seem, as far as I can trust my own observations, doubtful.[11] I will give one instance, as it might possibly induce some one to try the experiment. *Leschenaultia formosa* has apparently the most effectual contrivance to prevent the stigma of one flower ever receiving a grain of pollen from another flower; for the pollen is shed in the early bud, and is there shut up round the stigma within a cup or indusium. But some observation led me to suspect that nevertheless insect agency here comes into play; for I found by holding a camel-hair pencil parallel to the pistil, and moving it as if it were a bee going to suck the nectar, the straggling hairs of the brush opened the lip of the indusium, entered it, stirred up the pollen, and brought out some grains. I did this to five flowers and marked them. These five flowers all set pods; whereas only two other pods set on the whole plant, though covered with innumerable flowers. The seeds in these pods were bad, or else I had not skill to make them germinate.[12] I became so strongly convinced that insects would be found concerned in the fertilisation of these flowers, that I wrote to Mr. James Drummond, at Swan River in Australia, and asked him to watch the flowers of plants of this order; and he soon wrote to me that he had seen a bee cleverly opening the indusium and extracting pollen;[13] and a bee with its mandibles thus covered with pollen would very likely effect a cross between one individual and another of the same species. I have been told that this pretty plant, the *Leschenaultia formosa*, never sets seed in this country. I wish some skilful cultivator would rout up the pollen within the indusium in the manner described, and see whether he could not thus get seeds.—[14]

Charles Darwin, *Down, Bromley, Kent.*

Journal of Horticulture, Cottage Gardener, and Country Gentleman n.s. 1 (1861): 151

[1] The date is established by the reference in the following letter to CD's having sent this letter off to the *Journal of Horticulture* 'yesterday'. It was published in the issue of 28 May 1861 under the heading: 'Cross-breeding in plants. Fertilisation of Leschenaultia Formosa.' See also *Collected papers* 2: 42.

[2] See letter to *Journal of Horticulture*, [before 14 May 1861]. Donald Beaton's response to CD's queries was published directly following the letter.

[3] Beaton had written in response to CD's previous letter (see n. 2, above) (*Journal of Horticulture, Cottage Gardener, and Country Gentlemman* n.s. 1 (1861): 113):

> My own experience of variable plants was given last week, and I do not exactly comprehend what is meant by natural varieties, for all the so-called varieties in cultivation have been artificially obtained either by a change of cultivation, or by crossing with pollen such kinds or species as would sport from seeds under cultivation. These kinds I call variable plants, their own progeny being constantly variable in aspect, and just as variable when the pollen of another flower is applied to them. It is a difficult thing for a gardener to see or comprehend the meaning of what botanists call varieties, or natural varieties of plants, because there is no limit, or sign, or any other indication in their outward aspect to distinguish them from the oldest species on record, and there is nothing in the botanical structure of even a variable seedling to distinguish it from a genuine species.

[4] For CD's efforts to ascertain information on this point through his own experiments and through queries to various botanists and practical gardeners, see *Correspondence* vol. 5, letters to William and Julius Fairbeard, [October 1855 – May 1856], and to J. S. Henslow, 10 November [1855] and 12 November 1855; and *ibid.*, vol. 6, especially letter to J. D. Hooker, 19 July [1856].

[5] Beaton included hollyhocks among those plants that he called 'fixed varieties', 'as gardeners say when a variable plant comes true from seeds after sporting for some years' (*Journal of Horticulture, Cottage Gardener, and Country Gentleman* n.s. 1 (1861): 112).

[6] In his article on begonias in the *Cottage Gardener*, 26 June 1860, pp. 193–5, Beaton included the following statement on the authority of 'the king of all the British cross-breeders' (p. 194): '"I have the white *Anemone apennina*, which some folks do not believe in, and many pale shades from its seeding in a mass with the blue."'

[7] CD refers to Beaton's statement that '*Mathiola incana* and *glabra* cross freely notwithstanding Darwin's theory of species.' (*Cottage Gardener*, 26 June 1860, p. 195).

[8] The reference is to Beaton's article entitled 'Crossing flowers', published in the *Cottage Gardener*, 24 July 1860, pp. 253–5, in which he disputed the scientific view of 'superfoetation', or the fertilisation of a plant ovary by two or more kinds of pollen. Beaton gave as a counter-example the case of fertilisation in geraniums (*ibid.*, pp. 254–5):

> No flower on earth is more easy to prove by if more than one pollen can influence a cross than any of the common Geraniums. Their stigma is parted into five parts, and each part rolls back from the rest, or from the centre; and there are five seeds for every flower, corresponding to the five divisions of the stigma, or mother, as we say. Now, by applying five kinds of pollen, one kind to each division of the stigma, it is easy to conceive the possibility of each seed being influenced by that pollen only which dusted its corresponding division; and if the scientific explanation of the process by which the pollen reaches the ovum, or skeleton seed, were correct, superfœtation would be inevitable, and five kinds of progeny must be obtained from that flower so operated upon.

[9] Beaton made this statement in his article of 14 May 1861 (*Journal of Horticulture, Cottage Gardener, and Country Gentleman* n.s. 1 (1861): 113).

[10] See Notebook E, pp. 144, 183, and Questions & experiments notebook, pp. 2, 14 (*Notebooks*). See also *Correspondence* vol. 2, letter from William Herbert to J. S. Henslow, 5 April 1839.

[11] Beaton listed among those flowers 'every one of which must, of necessity, be fertilised by its own pollen in the great majority of instances' sweetpeas, bellworts (*Campanula*), and wheat (see *Journal of Horticulture, Cottage Gardener, and Country Gentleman* n.s. 1 (1861): 113).

[12] For descriptions of CD's experiments, see *Correspondence* vol. 8, letters to J. D. Hooker, 18 [April 1860] and 26 April [1860], and letters from J. D. Hooker, [20 April 1860] and [28 April 1860]. The results of some of CD's experiments on *Leschenaultia* are recorded in his Experimental book (DAR 157a).

[13] See *Correspondence* vol. 8, letters to James Drummond, 16 May 1860 and 20 December [1860], and letter from James Drummond, 17 September 1860. The letter in which Drummond related this information, however, has not been found.

[14] CD himself tried the experiment again in 1862 (see *Correspondence* vol. 10). He published a short note on the topic in 1871 (see *Gardeners' Chronicle and Agricultural Gazette*, 9 September 1871, p. 1166; *Collected papers* 2: 162–5).

To Charles William Crocker[1] 18 May [1861][2]

Down. | Bromley. | Kent. S.E.
May 18th

My dear Sir

I am extremely much obliged to you for so kindly writing to me & offering me information.[3] I have long been curious about Hollyhocks.—[4] The Dean of Manchester told me that some kinds came true, even when growing near others.[5] This surprised me greatly, as it can be shown to be almost certain that every flower must be fertilised by the pollen of a distinct flower. Hence I bought some good seed, & several kinds came up very true; the black-coloured kind came up absolutely true, & I saved its seed, & though a dozen different varieties were growing *close*, again all the seedlings were true. It can hardly be, judging from my *extremely* limited experience, that the flower always inherits its mother's colour, for I castrated & crossed a dark purple with pollen of yellow kind, & there was great variability in the seedlings, but one seedling alone showed plainly the yellow colour of its father. In case of most plants pollen of another variety seems to have a prepotent effect over a plant's own pollen, (as experiments made by myself have convinced me); & it has occurred to me whether in some very fixed vars. of Hollyhocks exactly the reverse was the case,—that is whether the pollen of a plant's own variety was not prepotent over that of other varieties.

The facts which I wish to know, are whether some or many of the variations of Hollyhock come true by seed. And secondly whether the great raisers of Hollyhock seed, who sell named kinds, whether they grow the varieties far apart to prevent crossing.[6] And thirdly, whether when vars. of Hollyhocks are artificially crossed & castrated, whether the colours of the seedlings are generally intermediate. I should be extremely much obliged for any information on these points or on the progress of your work of selection.—

I have not much time for reading, & have not yet seen your work, but must see it & your account of the vars. of Cassia.—[7]

I yesterday sent off another letter to Cot. Gardener:[8] but I begin greatly to doubt the great Mr. Beaton.— He makes, I am sure, plenty of mistakes in his late discussions, & is dogmatic in highest degree

My dear Sir, your sincerely obliged | Charles Darwin[9]

AL incomplete
Cleveland Health Sciences Library (Robert M. Stecher collection)

[1] The correspondent is identified by the relationship of the letter to the letter to C. W. Crocker, 1 June [1861].

[2] Dated by the reference to CD's letter to the *Journal of Horticulture* (see preceding letter).

[3] Crocker apparently wrote to CD in response to the letter to the *Journal of Horticulture*, [before 14 May 1861]. Crocker's letter has not been found.

[4] CD had experimented with crossing hollyhocks: details of his experiments are recorded in his Experimental book (DAR 157a). See also *Correspondence* vol. 5, letters to J. S. Henslow, 10 November [1855] and 12 November 1855.

[5] The noted hybridiser William Herbert was the dean of Manchester. See letter to *Journal of Horticulture*, [before 14 May 1861].

[6] CD did obtain information on this point, stating in *Variation* 2: 107: 'I have been informed that nurserymen who raise seed for sale do not separate their plants'.

[7] The work has not been identified.

[8] Letter to *Journal of Horticulture*, [17 May 1861]. CD questioned statements made in earlier articles by Donald Beaton.

[9] The last two lines of the letter are in the hand of another writer, apparently owing to someone having excised CD's signature.

To J. D. Hooker 18 [May 1861]

Down. | Bromley. | Kent. S.E.
18th

My dear Hooker

I was very glad to hear that poor dear Henslow is at rest.[1] I fully believe a better man never walked this earth. What a loss he will be to his parish! I can well believe how you will miss him. I well remember his saying before you married that if he could have picked out anyone for his son-in-law, it would have been you.— How kind he was to me as an undergraduate constantly asking me to his House & taking me long walks. I am thankful to think that at the time I fully enjoyed & appreciated his kindness. I suppose Babington will be Professor at Cambridge.[2] What a contrast!

Give our kindest remembrances to Mrs. Hooker.[3] My wife admired from her heart poor dear Henslow

Farewell my dear Friend | Your affect— | C. Darwin

By the way, thanks about Beaton: I have now read more of his writings & one answer to me in C. Gardener;[4] & I can plainly see that he is not to be trusted. He does not well know his own subject of crossing. He has learnt largely from Herbert, but exaggerates all Herberts opinions grossly.[5] Says not one plant in a thousand is ever fertilised by its own pollen! gives some exceptions & these are false & so on.— Coolly assumes that all Botanists have blundered & that it is an entire mistake that pollen-grains emit a tube which penetrates the stigma!!!—[6]

I like London R. so much better than Athenæum, that I will take it in. Will you send me only *one* more. But would it be worth while to take it in jointly?[7] I do not care about having it first— Perhaps not worth while.—

P.S. | I do not know what is the matter with Lyell,—[8] I suspect something surgical or unmentionable.— He has been confined to bed for some time & has required incessant nursing. Lady L. is in very good heart about him; but I heard that Doctor said he did not make so quick process as he liked, & this I did not much like.— A longish confinement & not rapid progress does not sound well; but

I hope there is no risk.— I wish I knew what was matter. Emma saw Lady Lyell a few days ago.—[9]

Endorsement: 'May 1861.'
DAR 115.2: 100

[1] John Stevens Henslow died on 16 May 1861.
[2] Charles Cardale Babington succeeded Henslow as professor of botany at Cambridge University. For CD's and Hooker's poor opinion of Babington's botanical work, see *Correspondence* vol. 7, letters to J. D. Hooker, 21 July [1858] and 30 [July 1858].
[3] Frances Harriet Hooker was Henslow's daughter.
[4] See letter to *Journal of Horticulture*, [before 14 May 1861]. Donald Beaton's response was published immediately following the letter.
[5] William Herbert, the dean of Manchester, was a great authority on hybridisation.
[6] CD refers to Beaton's article in the *Cottage Gardener*, 24 July 1860, pp. 254–5, in which Beaton described experiments on crossing geraniums and concluded that his results proved the 'impossibility of the pollen passing in grains in tubes of extreme tenuity to the embryo seed, which is the way it is explained by scientific men.'
[7] See second letter to J. D. Hooker, 23 [April 1861].
[8] Charles Lyell had fallen ill late in April following an excursion to Bedford to examine flint implements recently discovered in a gravel pit. In June, he travelled to Kissingen, Bavaria, to convalesce. See K. M. Lyell ed. 1881, 2: 343, 344.
[9] According to her diary, Emma Darwin visited Mary Elizabeth Lyell on 13 May 1861 during her stay in London.

To T. H. Huxley 22 May [1861][1]

Down. | Bromley. | Kent. S.E.
May 22^{d}

My dear Huxley

I had a note this morning from Asa Gray saying that Wright was very glad to hear by my first note that his Review had not arrived & was lost, & was very sorry to hear by my second note that it had arrived & was sent to Nat. Hist. Review.[2] So he does not estimate it highly.— I write now to ask you, when done with, (as I suppose it is) to send it me, as Wright asks me to keep it for him.—[3]

We have decided to move all our household on June 3^{d} or 10th to Torquay for 6 or 8 weeks for my daughters sake (who has been going on very well) for my own sake, who have been extra dyspeptic & fit for utter extermination.[4] If you feel inclined to be a very good man (but I do not urge it, as I know how hard you are worked) you will let us have a line to tell us how Mrs Huxley is, & how you yourself are.—

So poor dear Henslow is at last gone;[5] as good & true & noble a man as ever lived.—

Hooker tells me that the squib in Punch was by Sir P. Egerton, which astonishes me;— I did not think it very good.—[6] One of Lubbock's brothers, I fear, is dying, from being thrown out of carriage—, Head badly hurt & arm badly fractured[7]

Farewell | Yours most sincerely | C. Darwin

Imperial College of Science, Technology and Medicine Archives (Huxley 5: 164)

[1] The year is provided by the reference to the death of John Stevens Henslow (see n. 5, below).
[2] Asa Gray's letter has not been found, but see the letter to Asa Gray, 11 April [1861]. Gray had forwarded to CD a manuscript review of *Origin* by Gray's friend Chauncey Wright. CD had sent it to Huxley for possible inclusion in the *Natural History Review*.
[3] The copy of Wright's review has not been located. It was not published in the *Natural History Review*.
[4] The Darwin family had to postpone their departure for Torquay until 1 July 1861 ('Journal'; Appendix II).
[5] Henslow died on 16 May 1861.
[6] The reference is to a poem entitled 'Monkeyana' that appeared in *Punch, or the London charivari*, 18 May 1861, p. 206. A 'squib' on the relation of CD's species theory to the question of the origin of man, it was signed 'Gorilla, Zoological Gardens, May, 1861'. For a transcription, see Appendix VIII. Philip de Malpas Grey-Egerton, the Tory MP for South Cheshire, was a recognised authority on fossil fishes.
[7] John Lubbock's brother Montagu eventually recovered from the accident. See also letter to W. E. Darwin, [25 May 1861].

To John Frederick William Herschel 23 May [1861][1]

Down. | Bromley. | Kent. S.E.
May 23^d

Dear Sir John Herschel

You must permit me to have the pleasure to thank you for your kind present of your Physical Geography.[2] I feel honoured by your gift, & shall prize this Book with your autograph. I am pleased with your note on my book on species, though apparently you go but a little way with me.[3] The point which you raise on intelligent Design has perplexed me beyond measure; & has been ably discussed by Prof. Asa Gray, with whom I have had much correspondence on the subject.—[4] I am in a complete jumble on the point. One cannot look at this Universe with all living productions & man without believing that all has been intelligently designed; yet when I look to each individual organism, I can see no evidence of this. For, I am not prepared to admit that God designed the feathers in the tail of the rock-pigeon to vary in a highly peculiar manner in order that man might select such variations & make a Fan-tail; & if this be not admitted (I know it would be admitted by many persons), then I cannot see design in the variations of structure in animals in a state of nature,—those variations which were useful to the animal being preserved & those useless or injurious being destroyed. But I ought to apologise for thus troubling you.—

You will think me very conceited when I say I feel quite easy about the ultimate success of my views, (with much error, as yet unseen by me, to be no doubt eliminated); & I feel this confidence, because I find so many young & middle-aged truly good workers in different branches, either partially or wholly accepting my views, because they find that they can thus group & understand many scattered facts. This has occurred with those who have chiefly or almost exclusively studied morphology, geographical Distribution, systematic Botany, simple geology & palæontology. Forgive me boasting, if you can; I do so because I sh^d. value your partial acquiescence in my views, more than that of almost any other human being.—[5]

Believe me with much respect | Yours, sincerely & obliged | Charles Darwin

Royal Society (HS 6:17)

[1] The year is provided by the reference to Herschel 1861.

[2] Herschel 1861 was one of a series of articles that Herschel wrote for the eighth edition of the *Encyclopædia Britannica* (Edinburgh, 1853–60), slightly revised and published as a single volume. A copy is in the Darwin Library–Down.

[3] In Herschel 1861, p. 12, Herschel stated his belief that species change had not taken place either by gradual, progressive variation or by the sudden extinction of one race followed by the introduction of a new one; rather he believed that change had occurred 'by a series of overlappings, leaving the last portion of each in co-existence with the earlier members of the newer series'. To this discussion he appended the following note:

> This was written previous to the publication of Mr. Darwin's work on the Origin of Species, a work which, whatever its merit or ingenuity, we cannot, however, consider as having *disproved* the view taken in the text.

Having argued for the necessity for 'intelligent direction' behind species change, Herschel ended the note: 'with some demur as to the genesis of man, we are far from disposed to repudiate the view taken of this mysterious subject in Mr. Darwin's work'. Soon after the publication of *Origin*, a copy of which was presented to Herschel, CD heard that Herschel had referred to natural selection as the 'law of higgledy-piggelty'. See *Correspondence* vol. 7, letters to J. F. W. Herschel, 11 November [1859], and to Charles Lyell, 23 November [1859] and [10 December 1859].

[4] CD had greatly admired Asa Gray's attempt to reconcile natural selection with design in nature and arranged to have the series of articles in which Gray discussed his views reprinted and distributed in Britain. CD had sent a copy of A. Gray 1861a to Herschel (see Appendix III).

[5] In his *Autobiography*, CD stated that Herschel's *Introduction to the study of natural philosophy* (Herschel 1831) was one of the two books that had most influenced him, the other one being Alexander von Humboldt's *Personal narrative* (Humboldt 1814–29). He recalled that the two 'stirred up in me a burning zeal to add even the most humble contribution to the noble structure of Natural Science. No one or a dozen other books influenced me nearly so much as these two.' (*Autobiography*, p. 68). On the importance of Herschel's philosophical writings for CD's methodology, see Ruse 1975.

To J. D. Hooker 24–5 May [1861]

Down. | *Bromley.* | *Kent. S.E.*
May 24th

My dear Hooker

Thanks for your two notes.[1] I am glad that the Burial is over & sincerely sympathise & can most fully understand your feelings at your loss.[2] I grieve to think how little I saw of Henslow for many years. With respect to a biography of Henslow, I cannot help feeling rather doubtful, on the principle that a biography could not do him justice. His letters were generally written in a hurry & I fear he did not keep any Journal or Diary. If there were any vivid materials to describe his life as Parish-priest, & manner of managing the poor it would be very good.

I am never very sanguine on literary projects, I cannot help fearing his Life might turn out flat. There can hardly be marked incidents to describe.— I sincerely hope that I take a wrong & gloomy view; but I cannot help fearing. I would rather see no life than one that would interest very few. It will be a pleasure & duty in me to consider what I can recollect; but at present I can think of scarcely anything. The equability & perfection of Henslows whole character, I sh^{d}. think would make it very difficult for anyone to pourtray him.— I have been thinking about Henslow all day a good deal; but the more I think the less I can think of to write down. It is quite a new style for me to set about; but I will continue to think what I could say to give any, however imperfect, notion of him in the old Cambridge days.[3]

Pray give my kindest remembrances to L. Jenyns, who is often associated with my recollection of those old happy days.[4] It is really fearful to think of Babington in that Chair.[5]

With respect to your former note, thank you for telling me about Lyell: it makes me easy about him.[6] From what you say about London Review, it will be better to encourage Editor, & I have written to order it for myself.—[7] Two of your Himalayan Rhododendrons have flowered; R. formosum (I think) splendid dark crimson & R. Thompsoni(?) more like an Azalea—orange shading into Reddish orange.—[8]

I have just reread your note: perhaps I am wrong to give my first impression that it would be very difficult to give a good & vivid Biography; & that a flat Biography is a doubtful good. I hope I am wholly wrong; but I thought it best to give my first impression.— I will see what I can do; but it will be miserably poor.—

My dear Hooker | Ever yours | C. Darwin

25th— P.S. Please let me hear, as soon as you have decided anything about the Biography, as I will merely jot down (which I have begun to do) anything which occurs to me, but will not attempt to arrange till I hear. Two or three pages is the utmost which I could do; for I find that my recollections are not defined enough.

Endorsement: '1861'
DAR 115.2: 101

1 Neither of the letters from Hooker has been found.

2 The funeral of Hooker's father-in-law, John Stevens Henslow, was held on 22 May 1861. Henslow had died on 16 May. See L. Huxley ed. 1918, 2: 61.

3 CD did contribute a brief recollection of Henslow to the memoir edited by Henslow's brother-in-law Leonard Jenyns. See Jenyns ed. 1862, pp. 51–5; *Collected papers* 2: 72–4 (see also Appendix X).

4 CD and Jenyns had occasionally gone on entomological excursions together while CD was a student in Cambridge. After CD's return from the *Beagle* voyage, which Jenyns had himself considered undertaking and then declined, Jenyns agreed to describe the fresh-water fish CD had collected (*Fish*). See *Correspondence* vols. 1 and 2, and *Autobiography*, pp. 66–7.

5 Charles Cardale Babington succeeded Henslow as professor of botany at Cambridge University.

6 See letter to J. D. Hooker, 18 [May 1861] and n. 8.

[7] See letter to J. D. Hooker, 18 [May 1861]. The *London Review and Weekly Journal of Politics, Literature, Art and Science*, edited by Charles Mackay, began publication in July 1860.

[8] Hooker had collected many species of *Rhododendron* during his expedition to the Himalayas, 1848–50. His work entitled *The rhododendrons of Sikkim-Himalaya* (Hooker 1849) widened the interest of horticulturalists in using rhododendrons as garden plants. Hooker is credited with introducing the plant to Great Britain for widespread cultivation (Turrill 1963). Hooker named *R. thomsonii* after his friend Thomas Thomson, the co-author of the *Flora Indica* (Hooker and Thomson 1855).

To Bartholomew James Sulivan 24 May [1861][1]

Down. | Bromley. | Kent. S. E.
May 24th

My dear Sulivan

Many thanks for the Mendoza account which I have been glad to see.[2] It seems to have been one of the worst earthquakes on record.— It is very strange, as the Chili earthquakes, even that of Concepcion was not even felt at Mendoza.—[3] Many thanks, also, for your last note about the Geese's nests:[4] I suppose I was falsely informed

FitzRoy was so kind as to send me the last London Review (which I intend to take in) & I read the article on Genesis.—[5] I cannot say that it at all satisfied me. Almost any statement & fact can be reconciled by stretching the statements & facts; & this seemed to me the case throughout the article.— But I am weary of all these various attempts to reconcile, what I believe to be irreconcileable.

With many thanks | My dear Sulivan | Yours very sincerely | C. Darwin

J. A. Sulivan

[1] The year is provided by the reference to 'the Mendoza account' (see n. 2, below).

[2] The city of Mendoza, in western Argentina, was completely destroyed by an earthquake and the fire that resulted on 20 March 1861. The French naturalist Pierre Joseph Auguste Bravard was among the 10–12,000 victims (*EB*). The precise account mentioned in the letter has not been traced.

[3] CD had personally experienced the severe earthquake that occurred along the coast of Chile in February 1835 (see *Correspondence* vol. 1, letters to W. D. Fox, [7–11] March 1835, and to Caroline Darwin, 10–13 March 1835). He described its physical effects in *Journal of researches*, pp. 368–81.

[4] The letter has not been found.

[5] Robert FitzRoy had been captain on the *Beagle* voyage on which both CD and Sulivan served. The article referred to appeared in the 'Special supplement' of the *London Review and Weekly Journal of Politics, Literature, Art and Society* issued on 18 May 1861. The fifth part of the series 'Seven answers to the seven Essays and reviews', the article was entitled 'Answer to the fifth essay.— Mosaic cosmogony—genesis & geology—science vs. revelation' (*London Review* 2: 585–92). The reference was to the fifth essay of *Essays and reviews*, 'On the Mosaic cosmogony' (Goodwin 1860). Quoting a paragraph from chapter 9 of *Origin*, which described the fragmentary and imperfect condition of the geological record, the anonymous author stated that it was 'more reasonable and probable' to suppose that scientists have erred in their interpretation rather than that the revelation of Scripture is wrong. The article concluded with comments on the geology of Patagonia, in which both CD and FitzRoy's observations were mentioned.

To the *Field* [before 25 May 1861][1]

I am very much obliged to Mr Bennett for his information about Norwegian dun ponies;[2] but I received some years ago, through the Consul-General, Mr Crowe, the same account, which probably came from Mr Bennett.[3] The point on which I am anxious for information is, whether a cross between two coloured horses (neither of which are dun) ever produce duns. I believe that we could thus obtain some insight into the aboriginal colour of the horse. I have as yet only a single case of the parentage of a dun—namely, a bay horse and black mare. A German writer (Hofacker) on the breeding of horses gives the case of two chesnuts producing a "goldfalb", which, I believe, is a dun; and of a chesnut and brown producing a mouse-dun (mausrapp).[4] I hope "Eques" will fulfil his kind offer of giving more information on the subject.[5] I have collected a considerable body of evidence on the remarkable tendency of the offspring of a cross between differently-coloured breeds reverting to the colour of the aboriginal parent.[6] With pigeons, I made numerous crosses for this express purpose, and frequently got a near approach to the marks and colour of the wild rock-pigeon.[7] Again, I crossed a Spanish cock and white silk-hen; and one of the cockerels, although at first coal-black, in the autumn assumed the splendid red plumage of the wild jungle-cock (*Gallus bankiva*). Two young Cocks from the black Spanish and white game-hen assumed red neck and saddle hackles, so as partly to resemble a pile game cock. Mr Brent crossed two varieties of duck, and some of the ducklings assumed the plumage of the wild duck.[8] I could give other facts. For instance, it is asserted in work on poultry that hens raised from a cross between two breeds of fowls neither of which sit, are good sitters; and here we see a cross has brought back the proper aboriginal instinct of incubation. In my own experience, however, the crossed offspring from the Spanish cock and a Poland hen did not incubate. If anyone has any analogous facts to those above given, and would communicate them, I should be much obliged. The whole subject of the results of crossing distinct breeds is an interesting one under many points of view.—

Charles Darwin (Down, Bromley, Kent).

The Field, the Farm, the Garden, the Country Gentleman's Newspaper 17 (1861): 451

[1] Dated by the publication of the letter in the 25 May 1861 issue of the *Field*. It appeared under the heading: 'On dun horses, and on the effect of crossing differently coloured breeds'.

[2] CD refers to a notice headed 'Dun horses' and signed 'T. Bennett (Christiania)' in the 18 May 1861 issue of the *Field*, p. 431. It was a response to CD's letter to the *Field*, [before 27 April 1861]. Stating that 'I was some time ago requested to get similar information' on this subject (see n. 3, below), Bennett described the characteristics of the dun ponies of Norway.

[3] The letter from John Rice Crowe, British consul-general in Norway, has not been found, but see *Correspondence* vol. 6, letter to Laurence Edmondston, 2 August [1857] and n. 5. CD included information on striping in the dun horses of Norway in *Variation* 1: 58, citing Bennett's article in the *Field* as one of his sources (*ibid*., n. 31).

[4] Hofacker 1828. There is an annotated copy of the work in the Darwin Library–CUL.

[5] A response to CD's letter to the *Field*, [before 27 April 1861], signed 'Eques' appeared in the 11 May 1861 issue of the *Field*, p. 404; the author used the term 'mouse-dun' to describe the coloration of colts

bred from crossing a cream-coloured stallion and two yellow-dun mares. He promised to provide further information on the topic in a future number of the *Field*, which he did in the issue of 8 June 1861, pp. 493–4. There the writer corrected his earlier description of the colts, now calling them 'golden-dun, nearly what Hofacker (as quoted by Mr Darwin) calls gold-falb'.

6 Many of CD's notes on this topic are in DAR 205.7. For CD's use of experiments in breeding domestic animals as a means of gaining information on the laws of variation, see Bartley 1992.

7 CD's records of breeding experiments in which he crossed a number of different breeds of pigeon are in DAR 205.7: 166–89.

8 Bernard Philip Brent had provided CD with information about crossing a number of different domesticated birds. He is acknowledged on a number of occasions in *Variation*.

To W. E. Darwin [25 May 1861][1]

[Down]
Saturday Night

My dear William

John Lubbock called here today, & told me that lately a Gentleman had applied to him to recommend some young man, who had not been in business, as a *partner* in a Bank in some sea-port town in S. of England. He said that you had just occurred to him as perhaps liking the offer. It is, however very probable that the place may be filled up. I am astounded that such a place sh$^{d.}$ go begging for a week. The old Partner wants to retire this summer, & J. Lubbock's friend (about 35 years old) does not want to have whole business on his shoulders & therefore wants a Partner.—[2]

The share of Partner would be about (probably I sh$^{d.}$ think less) £500 per annum. Lubbock says the Bank he believes very safe & capable of extension, in which case profits of partners would of course increase.— As I understood **no** capital required; but a deposit, as guarantee of your probity, of 10,000£ required.—[3] Lubbock says he knows the Bank always keeps an almost uselessly large reserve stock. All this sounds (of course everything would require cautious investigation) very promising. On other hand there would be cruel drawback of your leaving Cambridge at once, & beginning life at once with regular work & very short holidays. But then contrast this with your waiting till 35 years old before you probably would get a brief.—[4] Banking is dull work & requires only sense, caution & judgment (all of which you have) & there is the enormous advantage of your being an independent man early in life. If you are a Barrister & do not succeed (& success is a great chance in that walk of life) you will hardly even have enough to marry & settle; for as far as I can tell after my & Emma's death each of you Boys will have about 600, or *at most*, £700 a year; & that is a very small income to marry & have children on.—

I do not wish at all to urge you to accept (if indeed the offer is still open); but think it over, & let me soon have your first impression.

I was so astonished at the communication, that I had not presence of mind to think over it. But I will try & see Lubbock tomorrow & write again.[5] Lubbock remarked that if you did think at all about it, you had better go down & see the man & place & hear particulars; & I sh$^{d.}$ certainly talk affair over with my solicitor.—[6] The offer strikes me as oddly good: how on earth is it that the Partners sh^d. not know some young man?— I wish to Heaven it had been one year later, after you had taken degree & then you might have tried the life for a year or so with hardly any detriment to the Law.— Indeed I suppose that you might continue eating your dinners at Lincoln's Inn.—[7] But, perhaps, the whole scheme will be repugnant to you & you had rather stick to the Law with all its disadvantages.

I will try & find out tomorrow where the place is.— The offer seems strangely advantageous contrasted with what Mackintosh Wedgwood works for, & not harder work, I should suppose, than Godfrey's place.—[8] Think deliberately & let me hear. The probability is immense that the whole will turn out moonshine; but I am sure it is far too good an offer not to be well considered.

Goodnight My dear William | Your affect. Father | C. Darwin

John L. brought a better account of Montagu tho' it sounds very bad as he is still almost insensible but his spine is not injured.[9] He said that he was considered out of danger

You need not write till you hear again tomorrow from me.

DAR 210.6

[1] This and the following letter are the first two in a series of letters that were exchanged between CD, William Darwin, and John Lubbock in May and June of 1861. All pertain to negotiations carried out in connection with William's appointment as partner in a Southampton bank. Postmarks on envelopes that contained three of the letters, preserved in DAR 210.6, confirm the dating.

[2] Lubbock's 'friend' was George Atherley, partner in the Southampton and Hampshire Bank, Southampton, for which Lubbock's bank, Robarts, Lubbock & Co., were the London agents (*Banking almanac* 1861).

[3] Prior to 1860, most of the banks of England and Wales were small, independent establishments, with the number of partners limited to six. When a partner died, his capital often went out of the business and a new partner had to be found, occasioning a period of uncertainty 'except in those instances when a son or a relation filled the vacancy' (*EB*).

[4] William was in his third year at Cambridge University, having entered at the beginning of the Michaelmas term in October 1858 (see *Correspondence* vol. 7, letter to W. E. Darwin, 15 [October 1858]). CD had mentioned the prospect of a legal profession for William while he was still a student at Rugby School planning on attending Cambridge University (see *ibid.*, vol. 6, letters to W. E. Darwin, [17 February 1857], and to Syms Covington, 22 February 1857). The two had agreed that William should take a degree from Cambridge University, sitting for the mathematical tripos, and then read for the law by entering one of the Inns of Court in London (see *ibid.*, vol. 8, letter to W. E. Darwin, [4 March 1860]).

[5] See following letter and letter to W. E. Darwin, [26 May 1861].

[6] CD's solicitors were David Rowland and William Mackmurdo Hacon of the firm Rowland and Hacon, 31 Fenchurch Street, London (CD's Address book, Down House MS).

[7] Standard procedure for studying for the legal profession in England involved the candidate being articled at one of the Inns of Court, such as Lincoln's Inn, to read for the law prior to sitting an examination. Thereafter the candidate would continue to read law for at least another year in the chambers of a practising barrister. There was also a residency stipulation, referred to as 'eating dinners' at the Inn, which for university men entailed dining there three times in each of four terms for a period of three years (*EB*).

[8] James Mackintosh Wedgwood, eldest son of Hensleigh and Frances Mackintosh Wedgwood, held a position in the Colonial Office from 1858. Godfrey Wedgwood, son of Francis and Frances Mosley Wedgwood, had joined his father in the family pottery works in Staffordshire (Wedgwood and Wedgwood 1980, pp. 263, 270).

[9] John Lubbock's nineteen-year-old brother Montagu had suffered a serious injury as a result of a carriage accident (H. G. Hutchinson 1914, 1: 178).

To John Lubbock [25 May 1861][1]

[Down]
Saturday night

My dear Lubbock

I have been thinking much of your communication about the Bank.[2] I was so astonished at it, that I did not collect my wits enough to ask all that I might.— I have just written a long letter to William, & that has made me feel that I have hardly given him data for even a first impression. I know that it is quite likely that the place may be already filled up; but it seems so good an offer as to be well worth deliberation.—

I sh$^{d.}$ very much like to ask you a *few* more questions. I do not like to ask you to call here, so if you will appoint any time whatever tomorrow I will drive down to High Elms, if you can see me.—[3] Just send verbal message at what hour.

It was *extremely* kind in you to think of William, but whether he will be inclined to give up the Law (as I should do under his circumstances), which he thinks he shall like, I know not.—

Yours very sincerely obliged | Ch. Darwin

I can with entire truth say of William that he has excellent sound practical sense & judgment. Though not very clever he was in VI form at Rugby & has got scholarship at Cambridge;[4] & has always borne good character at School. He has been steady & never extravagant. I never saw a shade of feeling bordering on untruthfulness or deceit in him. He is energetic in all his pursuits.—

Down House

[1] See preceding letter, n. 1.

[2] See preceding letter.

[3] High Elms, the home of the Lubbock family, was on the far side of Down village, approximately 1$\frac{1}{2}$ miles from Down House.

[4] William in 1857 was in the sixth form, the top year in Rugby School (see *Correspondence* vol. 6, letter to Syms Covington, 22 February 1857). William had won a scholarship to Christ's College, Cambridge in 1859 (see *ibid*., vol. 7, letter to W. E. Darwin, 7 July [1859]).

To Abraham Dee Bartlett[1] 26 May [1861][2]

Down. | Bromley. | Kent. S.E.
May 26

Dear Sir

The Bearer will deliver 3 Rabbits (if none dead on voyage) from Madeira.[3] Will you take charge of them for me & show this note to Mr Sclater?[4] They are zoologically very interesting for they have run wild on little Isl[d] of P. Santo, since year 1420;[5] & judging from 2 dead ones seen by me, they have become greatly reduced in size & modified in colour & in their skeletons.[6] I want much to see them alive, & to try whether they will cross freely with common Rabbits.— I am going immediately to leave home for two months.[7] Would there be any objection to your keeping them for some time & matching them with some other breed; or if you think fit, first try & get some purely bred.—[8]

I may perhaps be mistaken, but I was very much surprised at many of the characters of the two dead specimens which I saw.—

If anyone sh[d] die I sh[d] like its skeleton. Pray forgive me troubling you, but I know not what to do with them at present.—

If worth consideration, I would of course pay for their keep.—

In Haste— | Dear Sir | Yours very faithfully | Ch. Darwin

New York Botanical Garden Library (Charles Finney Cox collection)

1 Bartlett was the resident superintendent of the zoological gardens of the Zoological Society of London.
2 Dated by the reference to CD's leaving home 'for two months' (see n. 7, below) and by the year given in Bartlett ed. 1900, p. 331.
3 The live rabbits from the island of Porto Santo, in the Madeira Islands, were sent to CD by W. Haywood (*Variation* 1: 113).
4 Philip Lutley Sclater was secretary of the Zoological Society.
5 CD recounted the history of the breed in *Variation*. He stated that the rabbits had been turned out *circa* 1420 by Portuguese visitors to the recently discovered island (*Variation* 1: 112 and n. 23).
6 CD had previously examined the skeletons of two of the rabbits given to him by Richard Thomas Lowe via Thomas Vernon Wollaston. See *Correspondence* vol. 6, letter from T. V. Wollaston, [February 1856], and letter to W. D. Fox, 8 [June 1856].
7 The Darwins were planning to leave soon for Torquay (see letter to W. B. Tegetmeier, 28 May [1861]).
8 According to Bartlett's biographer, two rabbits were deposited at the zoological gardens on 21 May 1861 and entered in the catalogue as being females (see Bartlett ed. 1900, p. 332).

To W. E. Darwin [26 May 1861]

[Down]
Sunday

My dear William

Lubbock has lunched here & I have had long talk with him.[1] First for generalities. He thinks nowadays (& many have made same remark to me) that business is better line of life than Professions. no doubt there is always risk in business, but he thinks, if a person manages his own business, that the risk is much exaggerated. He remarked how much more leisure & less hard work a Banker has than a Barrister, if he gets any business; & how dreary the endless waiting for Business is. On other hand he acknowledges that Private Banks do not get on so well as formerly owing to prevalence of Joint-Stock Banks.[2] He would not tell me the Town, (as it w[d]. be breach of confidence), but it must be, I think, Brighton, Portsmouth or

Southampton. It is not a small & is a growing town. He says he likes *decidedly* the head Partner about 35 years old.

He says the usual thing after first start is for Partner to have one day holiday a week, & if that were sometime on Saturday you could leave on Friday evening & often come home or have little larks; & that there would be 2 or 3 weeks holidays in Autumn. The hours he supposes would be from 9 to 4: so you could have whole evening to yourself: & I fancy in the day there is time for letters & reading papers &c.— He spoke very strongly of its being safe & well conducted concern, capable of extension with energy.— Of course with extension profits would increase; & he *thought* that probably your share would increase, if you were energetic, even if the Bank did not increase.

He believes the place is not filled up, as he thinks he sh^{d}. certainly have heard: he had recommended a young man, a friend & connection of the Robarts (with which Bank, the Lubbocks are now joined)[3] & this friend would have joined, had not a better opportunity at the moment occurred. I sh^{d}. not have thought of this scheme for a moment, had not the source of information been unique, viz by Lubbock, who must know well such affairs & is certainly my sincere friend.— But here comes the worst aspect of the whole affair. The £10,000 would have to be made over absolutely to you: it would not be part of capital of Bank; but as being your absolute property, you as Partner, would lose it, if the Bank failed. As the interest of the £10,000 is more that I sh^{d}. give you as allowance you would have to pay me part back; but then besides the part you retained, viz £280, or £300, you would have about £500 from the Bank; so you would be at once an almost rich man with £800 a year![4] Nevertheless, I think this the worse part of affair as you thus risk more than half your fortune.— Of course, as Lubbock remarked, the Books of Bank, if you thought of the scheme, would have to be carefully examined by Lawyer & accountant. L. thought perhaps the Gentleman would allow you to join for year & see how you liked it, & how he liked you. Lubbock said he thought the demand quite reasonable that you sh^{d}. have 10,000 of your own before joining; as it would injure the Bank to have it said that a penniless young man had become partner: he said this when I asked what was the use of the £10,000, if not wanted as capital in the Bank.—[5]

L. further remarked when I demurred to risking the £10,000, that ultimately in the Bank, or in *any* Business (but not in a profession) there would be & must be some risk.— I can quite see that it would be a grievous blow to you to give up Cambridge at once, & the next year or two of free & pleasant life: but yet such a chance as this at least deserves deliberate consideration & will never recur. How would it be possible to keep your name on Christ's, supposing you were permitted to try the scheme for a year?[6] You must ponder well, & write **soon** & tell us what you think; & then I will instantly write to Lubbock, for him to find out whether the Place is still open.— This is a great affair, & most perplexing to decide. Mamma rather leans to your taking it, or trying it. There is always some risk for the future, & great present disappointment & a wrench in your premeditated plans of life; but it seems an unusually fair opening for a prosperous career. You would have plenty of time in evening to follow any pursuit & improve your mind— If you think at all about scheme; consult Mr. Wolstenholme (& tell him as much as you like)[7] how it would be to leave Cambridge for a year, if that be possible, & if Partner would permit the scheme of year's trial. It is a great crisis in your life. I am thankful that you are a sensible man— Your affect. Father | C. Darwin

Supposing that you were permitted to try for a year & did not like it, you would in that time thoroughily understand complicated accounts & see something of real business, & on my life that would not be education thrown away.— You might in evening continue reading a *little* law & some law would be useful, if you stuck to Banking.—[8]

Postmark: MY 27 61
DAR 210.6

1 See letter to John Lubbock, [25 May 1861].
2 The Banking Act of 1826 provided for the establishment of joint-stock banks other than the Bank of England, with the proviso that they be located outside of a 65-mile radius of London. This restriction was relaxed in 1833 (*EB*).
3 In 1860, the bank directed by John William Lubbock, of which John Lubbock was a partner, was amalgamated with that of Abraham John Robarts and partners to form Robarts, Lubbock & Co. (*DNB*). Following the Banking Acts of 1844 and 1845, no new banks were allowed to form, but existing ones were able to merge as a means of expanding their capital (*EB*).
4 William received, while a student at Cambridge, an annual allowance from CD of £260, in addition to £20 he received from the scholarship he had won (see *Calendar* no. 4639).
5 When CD came to draw up his will in September 1881, he included the details of this arrangement, directing that this advance be deducted from the amount William was to inherit: 'I some time since advanced and gave to my eldest Son the said William Erasmus Darwin the sum of *Four thousand five hundred pounds* to enable him to enter into partnership as a Banker with George Atherley of the County of the Town of Southampton Banker'.
6 William was a student at Christ's College, Cambridge.
7 The mathematician Joseph Wolstenholme, a fellow of Christ's College, was William's mathematical tutor. In 1862 he was moderator for the mathematical tripos examination (*Historical register of the University of Cambridge*).
8 The postscript was written on a separate slip of paper.

To Daniel Oliver 27 May [1861][1]

Down. | Bromley. | Kent. S.E.
May 27th

My dear Sir

Will you show this note to Mr Crocker & ask him if he is inclined to try a little experiment for me.—[2] Vinca major I believe never seeds in England & is said never to seed in Germany; & I find pollen so placed it could never get on stigma without insect aid.— I have pot of Vinca major & I have passed *fine* bristle between anthers (not cutting or touching the flowers) in same way as proboscis of moth would pass to nectary, near the sides of the corolla; pollen sticks to bristle &

a bristle thus covered from pollen of *one* flower is used for another flower.— I did this & I have 4 or 5 fine pods swelling, whereas every other seed-vessel shanked off soon after corolla dropped.—[3]

Now will Mr Crocker do this (& it will not take him 2 minutes) for any exotic Vincas, which naturally have never seeded; **mark** the $\frac{1}{2}$ dozen flowers thus treated, & hereafter let me know whether pods swelled & whether seed was obtained.—

My apocynum (or fly-catcher) is coming up; I thought it was as dead as mutton: I have been trying Vinca in preparation for this Plant.—

You once told me that Mr Crocker liked experiments; if so I wish he would insert fine brush, like a Bee into a few exotic Polygalas which do not naturally seed & mark flowers & see if pods swell; I *suspect* they require insect agency. By the way you have not sent me P. farinosa; I daresay you have been so busy that you forgot it.—[4]

I beg & pray you not merely to acknowledge this.

Yours most sincerely | C. Darwin

I am surprised to find Cowslips utterly sterile without aid.— bears on origin of Oxlips.—[5]

Down House (MS 10: 8)

[1] Dated by the relationship to the letter to Daniel Oliver, 1 May [1861].

[2] Charles William Crocker was the foreman of the propagating department at the Royal Botanic Gardens, Kew. Oliver, professor of botany at University College London, worked in the herbarium and served as the librarian at Kew.

[3] CD described this experiment in more detail in his letter to the *Gardeners' Chronicle*, [before 15 June 1861].

[4] CD asked Oliver to send him a specimen of *Primula farinosa* in the letter to Daniel Oliver, 1 May [1861]. A note in the margin of the letter, presumably in Oliver's hand, reads: 'These have not come out yet.'

[5] CD was experimenting with primroses and cowslips in connection with his investigation of the dimorphic condition of their pistils and stamens. He described the results of his study in a paper read before the Linnean Society of London in November 1861 (see *Collected papers* 2: 45–63). Having shown that cowslips required insect agency to effect pollination and set seed, CD advocated the view that oxlips were produced among the progeny of cowslips only as a result of cross-pollination with primroses (*ibid*., pp. 60–1).

To Edward Cresy 28 May [1861][1]

Down, Bromley, Kent
May 28th.

Dear Cresy

Many thanks for your pleasant note and enquiries about rats, and for your splendid present of the Railway Map, which I am very glad to possess.— I have never seen du Chaillu and am puzzled what to think.[2]

I am surprised to hear about Owen:[3] I thought his courage was as indomitable as his malignity.

I am very glad to hear that you like Asa Gray.[4] A correspondent (Lieut. Hutton who wrote good article on Nat. Selection in Geologist)[5] sent me enclosed Extract (need *not* be returned) and as you seem to care so much for subject I enclose it, as Sir J. Herschel sent me the book.[6]

Dear Cresy | Yours very sincerely | C. Darwin

We are off in 10 days or so.[7] In Autumn we hope we shall see you here.

Copy
DAR 143

[1] Dated by the reference to 'Asa Gray' (see n. 4, below).

[2] The reference is to the recently published *Explorations & adventures in equatorial Africa* by the French-born American explorer Paul Belloni Du Chaillu (Du Chaillu 1861). Upon his arrival in London, Du Chaillu, who had become a celebrity in the United States through his lectures and through newspaper accounts of his adventures, was invited by Roderick Impey Murchison to give a lecture at the annual meeting of the Royal Geographical Society on 25 February 1861. Du Chaillu spoke before a large and distinguished audience, attracted by the first public exhibition of the gorilla skins and skulls he had collected. Du Chaillu offered a first-hand account of the fierce nature of the beasts themselves and of the tribes of cannibals he had encountered. See Vaucaire 1930, pp. 125–30. However, the accuracy of Du Chaillu's account of his travels and the identification and description of the animals he collected was challenged by John Edward Gray, keeper of the zoological department of the British Museum, and several others in a series of letters published in the *Athenæum* late in May and in June 1861. This exchange sparked a public controversy that spilled over into the pages of *The Times* and other periodicals.

[3] Du Chaillu's lecture at the Geographical Society was preceded by remarks by Richard Owen, who gave a brief account of the anatomical structure of the gorilla and pointed out its similarity to human skeletons (see R. S. Owen ed. 1894, 2: 115–16). On 18 March, Du Chaillu spoke before the Royal Institution of Great Britain, followed the next day by a formal lecture by Owen, who gave a more detailed description of the anatomical structure of the skulls and brains of gorillas. Owen maintained that the skeleton of gorillas more nearly approximated that of humans than did those of orang-utans or chimpanzees. Nevertheless he stated that the structure of their brain was strikingly dissimilar to that of man. See *Athenæum*, 11 May 1861, pp. 621–3, and 23 March 1861, pp. 395–6; and Du Chaillu 1861, pp. 363–82. See also letter to T. H. Huxley, 1 April [1861].

[4] A. Gray 1861a. CD had recommended Asa Gray's articles on *Origin* to Cresy as by far 'the best Review' of the book (*Correspondence* vol. 8, letter to Edward Cresy, 12 December [1860]).

[5] The reference is to Hutton 1861. See letter to F. W. Hutton, 20 April 1861.

[6] John Frederick William Herschel had recently sent CD a copy of Herschel 1861 (see letter to J. F. W. Herschel, 23 May [1861]).

[7] The Darwins did not actually leave for Torquay until 1 July 1861 ('Journal'; Appendix II).

To James Hunt 28 May [1861]

Down. | Bromley. | Kent. S.E.
May 28th.

My dear Sir

I hope that you will be so good as to present my sincere thanks to the President & Council of the Ethnological Society for the distinguished honour which they have conferred on me in electing me one of their Honorary Fellows.[1] I am also very much obliged for the present of the Vol. I. of the Transactions.[2]

I beg leave to remain | Dear Sir | Your obliged & obed. servant | Charles Darwin

P.S. | Permit me to thank you personally for the very kind expression towards me in your letter.[3]

Postmark: MY 28 61
American Philosophical Society (250)

[1] Hunt was an honorary secretary of the Ethnological Society of London, whose stated purpose was 'the study of the human race in all its varieties, and in all the phases of its history and progress' (*Transactions of the Ethnological Society of London* n.s. 2 (1863): iii).

[2] The first volume of the new series of the *Transactions of the Ethnological Society of London* was published in 1861. CD's copy has not been located in the Darwin library.

[3] Hunt's letter has not been found. The official announcement of the award, dated 26 May 1861, is in DAR 230. It states that CD was elected an honorary fellow of the society 'in consideration of the services . . . rendered to the Science of Ethnology'.

From W. B. Tegetmeier [28 May 1861][1]

I believe that as a general rule a hen when with egg will yield to any cock I do not think she refuses (as with pigeons) when she requires impregnating

You can make tame hens squat to your hand by scratching the scapular region as with the feet of a cock

I had a tame cochin hen who used to squat to any thing that touched her back, much to the annoyance of my wife the contact of whose dress often produced the movement.

My hen cock was victor always and some of the hens constantly kept with him—a *very* handsome full feathered bird being in the same yard.— I have never heard that game cocks after being trimmed for fighting were not equal favorites with the hens

In fact strong cocks make the hens yield by stamping alternately with each foot and sometimes will tear open the side with the spur if the foot slips between the hens wing and her body.

The hens impregnated by the hen tailed game cock lay very fertile eggs.— I have just looked at two settings and found all the 21 eggs fertilized—so that impregnation must have been very effectually performed.[2]

Incomplete
DAR 84.1: 144

CD ANNOTATIONS
2.1 You . . . cock 2.2] *scored brown crayon*
3.2 much to . . . movement. 3.3] *scored brown crayon*; 'This shows so domesticated that all instinct lost.—' *added ink*
4.3 being . . . alternately 5.1] *scored brown crayon*
Verso of letter: 'Sexual Selection W. B. Tegetmeier. | May 28— 1861' *red crayon*

[1] The date is given by CD's annotation. The letter from Tegetmeier was written on the back of the enclosure sent with the letter to W. B. Tegetmeier, 14 April [1861]. See also following letter.
[2] See letter to W. B. Tegetmeier, 2 April [1861] and n. 3.

To W. B. Tegetmeier 28 May [1861][1]

Down. | *Bromley.* | *Kent. S.E.*
May 28th

My dear Sir

Very many thanks for all your information.[2] I do not care for any point in a hurry, if not thereby forgotten; for I have finished my M.S. & shall add nothing & correct nothing until you have read my M.S. & until I have scraped more information together.—[3]

I have two skulls of Sultans. Never mind about Ptarmigans from what you say.— Perhaps I shall get Creve-cœurs & Guelderlands from you or Mr. Baker,[4] from whom, however, I have had no answer.

I am particularly obliged for information about courtship of Fowls. It is clear that the instinct of our Hens, (from what you say about touching them) is so vitiated that one cannot judge how they would behave in a state of nature.—

I have not read Du Chaillu; you are severe on him; but I fear from all that I hear & read he deserves severity.—[5]

With many thanks | Yours very sincerely | C. Darwin

We hope to start, on account of my daughter's health, to Torquay on 10th. of June to stay 6 or 7 weeks there.[6] Whether my copyist will have copied my M.S. in time for me to look it over before that time I know not; if not, I must defer sending it till our return.

New York Botanical Garden Library (Charles Finney Cox collection)

[1] Dated by the relationship to the preceding letter and to the letter to W. B. Tegetmeier, 14 April [1861].
[2] See preceding letter.
[3] Tegetmeier had agreed to read CD's manuscript on variation in fowls intended for publication in *Variation*. See letters to W. B. Tegetmeier, 22 March [1861] and 28 March [1861].
[4] Either Samuel C. or Charles N. Baker, dealers in ornamental poultry.
[5] Paul Belloni Du Chaillu. See letter to Edward Cresy, 28 May [1861] and n. 2.
[6] In the event, the Darwins were unable to leave for Torquay until 1 July ('Journal'; Appendix II).

From Bernard Philip Brent 29 May 1861

Dallington | nr. Robertsbridge | Sussex
May 29th. 1861

Dear Sir,

In reply to your's of the 17th. inst respecting the courtship of Fowls etc. I will endeavour, to give you what information I can,[1] in the first place the sexual signs may be classed under three heads, as simple gallantries; general intercourse; and force; the simple gallantries offered by the cock to his hens, consists in bending himself sideways towards her, dropping one wing scraping it along the ground by a succession of quick jerks and striking it against his leg thus he draws up to her looking in her eye, uttering a self satisfied *tuck tuck a raw* in an assuring tone, then elevating himself raising his hackle, shaking his feathers or Crowing; he also scrapes the ground, picks up any substance calls the hens *chup chup* drops it to the first that comes, steps back a step or two, and looks up or round with a patronizing air. The General intercourse, is when both consent and understand each other, the cock elevates his head raises his hackle in a foile and offers the hen, if she accepts she stoops, spreads her legs, stiffens her whole frame raises the shoulders of the wings shuts up her tail and drops the feathers below her vent. the cock steps on holding her by the hackle.

When he uses force it is generally a strange hen or one that has been away for a time, dropping both wings and tail raising his hackle and saddle feathers he gives chase and catches her if he can

there are however many minor or less noticeable signs in use and well understood, much meaning is conveyed by the voice the eye and the actions—

for instance the cock looks up catches the eye of a hen at a distance she feels the effect of his movement and gives a slight totter or stagger her shoulders rise ever so slightly and he rushes over to her;

hens donot seem to have any great choice in the matter, their desire is the ruling passion, they will accept the gallantries of any cock who offers them, if they are laying, but there are many exceptions, an old hen will resist a very young chicken, or even a strange cock if her desire has been lately satisfied. there are some termagant hens, that will fight any stranger hen or cock, and if the cock is not man enough to knock the spirit of opposition out of her she will not submit to him till her desire compels her, or he takes her unawares but in such cases her spirit of opposition frequently, causes her to become a great bully, but she submits at once to the bird who can give her a good hiding—

where many fowls are kept and more cocks than one the head man takes the beat most to his liking and all the hens on his beat submit to him the others keep at a distance and entice what hens they can to follow them. any hen straying on to the others ground, will be accosted by the cock on to whose ground or into whose presence she may come if she is inclined she accepts his advances if not she runs away to the protection of the head cock where she knows the other dares not come. there are many cases in which it may appear that a preference is given by some hens. but it is I think generally, dependent on sexual desire—

I have had hens that were kept without a cock frequently stoop to me; laying hens if tame will almost always place themselves in position if gently handled, when kept from a cock that is provided they are so tame that their fears do not overbalance their sexual desires.— I have known where several hens were kept without a cock one hen, a laying one, took to treading the other hens out of pure wantonness,— I donot think a hen would care how ugly a cock was if he could only knock her spirit of opposition out of her if she showed it all fowls even the most timid pullet feel a desire to attack a stranger male or female, and if he was beaten by her she would leave him no peace. but if she is in a state of longing desire he has only to look her in the face and raise his hackle and down she squats quite rigid.

Ducks I know less of they always attack a stranger they meet face to face placing their heads low and beaks on their chests, they press against each other then seize hold of the feathers or strike with the wings till one gives in and utters the quack of fear. the drake speaks to his ladies with much the air of a consequential dandy he raises his head lifts up his bill utters his harsh *hush hush* sometimes he strikes the water with his bill making a sort of *boom* and at the same time cocking up his tail the ducks answer by rushing after him bobbing their heads quickly at their sides and uttering quick short quacks—

the sexual signals are bobbing the head in measured movements, the duck depressing herself before the drake. Geese also bob their heads the Gander constantly follows and watches the goose she does not seem to care anything for him he is continually when on the water dipping his head or asking her, if she consents she also dips her head this they repeat many times and they both become very excited at this head dipping till she sinks or flattens herself out on the water he seizes her by the back of the neck they tread among much noise all the old ladies have a word to say and it is generally finished up with a wash.

I know nothing of the Pea fowls but I believe the expanding of the train is a sexual ask, In turkeys the cock or gobbeler struts with expanded tail and drooping wings just as the Peacock does but it is no use unless the hen is in want then his show has some effect every gobble of the cock she answers by *puck* and at last lays herself down as he advances she squeaks and presses her beak to her chest, the cock mounts tramples her and she squeaks it is a long affair and when over the hen seems right glad to be off—

Guinea fowls are very expeditious and very shy that very few have ever seen them. the cock challenges everything walking on tiptoe, his back raised, wings slightly away from his sides head up or down or any where sometimes on his back, he will be mincing a bout as if he did not know what he was at the hen is standing quietly apparently very unconcerned crying *come back come back* she lays down I never detected her sign but he rushes over to her and it is accomplished in an instant if there are many the hens are all standing calling *come back come back* with heads raised that is their common call. the cock makes a kind of saw setting noise, and a kind of purring pipe;

In all cases of Poultry that I have noticed the male is continually on the look out and asking the females by voice and gesture when she feels inclined she gives the signal, a look or gesture it is at once understood

Males decidedly chose and show much preference, especially among those that pair in a wild state. the hen canary when she feels the desire stoops and calls, the cock flies to her and treads her, though perhaps at other times she fights and scolds him, and though he follows and is on the look out has to keep a respectfull distance, some small birds sing while they are chasing their mates. but that is more of a rape than matrimony. The hen rather dreads the song I never recollect seeing a hen call or stoop to any song but generally fly away, or if she was the master silence him by fighting him, when in separate cages the hen will answer the song by a call but that is not the sexual call.

Unmatched hen Pigeons if rank and tame will always acknowledge my cooing to them some I had when I could not afford to buy cocks, would come down from the top of the house acknowledge my cooing stoop for me and allow me to lay my hand on them, I should not think that the hen fowls or hen Pigeons selected me for my beauty, but using their signs or counterfeiting them they being in a state of longing could not resist the natural feeling.

thus the cock Pigeon coos he bows his head sweeps the ground with his tail the hen cares nothing about it, and if he comes in her place she fights, but if unmated or salacious she acknowledges by a twinkle of the eye a bow of the head and an action of the throat as if swallowing she will sweep forward with spread tail making a curtsey and slightly raising the shoulders of the wings. a wanton will often take the tread of a cock at such times with out pairing.

When paired they signal by pecking behind the wings, then bill, and tread, but Pigeons when paired are very constant, and occasionally do show a preference and an aversion but sexual desire will overcome the strongest aversion and when once thoroughly paired she will not forsake her mate unless an other beats him from her when she takes to the new comer thus if a Pigeon cannot keep his nests, and the conquerer has not a mate, he is sure to win the hen from the other, unless there is room for the pair to find other place, it is therefore not so much his being victorious, as his having her accustomed place—

To answer your enquiries as you put them, where two cocks are in one yard the victor would not get all the hens as the vanquished would take advantage of their desire when out of the conquerors sight. a few pugnacious hens, will only submit to a strange cock if he can thrash them, but the generality, are governed by desire not choice or the preferance if any must be very small.

In all birds that I have noticed, I fancy the males asks if the female is ripe she accepts or goes to the first male she sees or hears. I donot think the hen cares for the beauty, of the cock she may for his attentions, beauty may assist as a charm, perhaps may assist in exciting her feelings but I donot think they ever think about it or make any choice other than to satisfy their desire by the first offer. the only fact of choice I can call to mind is that an old hen will sometimes run away from a young officious bird and seek the protection of the old cock, just to escape the annoyances of the juvenile.

I donot think the conqueror has any advantage in the affections of the hens simply as being the conquerer but he has a better chance as being master of the walk, and where there are many officious young stags the hens will run to the old cock for peace and quietness A game hen will yield to a Bantam if she is in want.

I have no experience of Henney cocks but have never heard that their wives ever objected to them on account of dress.

As I have before said both hen fowls and hen Pigeons, when kept without cocks have acknowledged my imitations of their signs and have stooped to me which I think can hardly speak much for their ideas of beauty.

I have endeavoured to answer your enquiries, if I have misunderstood them I will endeavour to do better if you ask again, any information I can give I shall be pleased to forward.[2]

I am Dear Sir | Your's respectfully | B P Brent

C Darwin Esq[r]

DAR 84.1: 1–9

CD ANNOTATIONS
1.3 simple gallantries; general intercourse;] *underl blue crayon*
1.4 the simple gallantries] *square bracket added and underl blue crayon*
1.10 The General intercourse, 1.11] *underl blue crayon*
1.11 both . . . understand] *underl blue crayon*
2.1 When . . . generally] *underl blue crayon*
5.1 have . . . matter,] *underl blue crayon*
5.2 they will . . . gallantries] *underl blue crayon*
5.3 an old hen . . . or cock, 5.5] *scored blue crayon*
5.8 she submits . . . hiding— 5.9] *scored blue crayon*
12.1 In all . . . wild state. 13.2] *scored blue crayon*
13.2 the hen . . . matrimony. 13.6] *enclosed in square brackets blue crayon*
16.6 he is sure . . . place— 16.8] *scored blue crayon*
17.3 a few . . . their desire 18.5] *scored blue crayon*
18.6 an old hen . . . old cock, 18.7] *scored blue crayon*

[1] CD's letter has not been found. Brent, an authority on the breeding and care of fowls and other ornamental birds, had assisted CD's study of variation in domesticated animals since 1856 (see *Correspondence* vols. 6 and 7, and *Variation* 1: 132 n.).

[2] CD cited information from Brent's letter, which he described as 'almost an essay', in the discussion of female preference in sexual selection in *Descent* 2: 117–18. In spite of Brent's insistence on the lack of choice by domestic hens and the major importance of sexual readiness for breeding, CD explained this as due to 'the artificial states under which they have long been kept'.

To J. D. Hooker 30 May [1861]

Down. | Bromley. | Kent. S.E.
May 30th

My dear Hooker

I did misunderstand you about the Biography.—[1] This morning I wrote my recollections & impression of character of poor dear Henslow about the year 1830.[2]

I liked the job & so have written 4 or 5 pages, now being copied. I do not suppose you will use all,—of course you can chop & change as much as you like. If more than a sentence is used, I sh$^{d.}$ like to see a proof page, as I never can write decently till I see it in print. Very likely some of my remarks may appear too trifling; but I thought it best to give my thoughts as they arose for you or Jenyns to use as you think fit— You will see that I have exceeded your request; but, as I said, when I began I took pleasure in writing my impression of his admirable character.—

I send this to Kew, thinking that you will be there on Saturday. With your regular work, you will soon get, I hope, into your regular frame of mind & look at poor Henslow's death as one of inevitable events in this life.—[3]

Farewell | my dear friend | Your affect | C. Darwin

I believe we shall go to Torquay about 10th, though I may be prevented by some Business affairs.—[4]

Your Rhod. formosus looks quite gorgeous & illuminates our whole garden.—[5]

A month or two ago you said you would have some discussion about Bates paper & mundane Glacial action:[6] but you will have forgotten whole subject, unless you chance to have kept my letter; & for long will be too busy to discuss such matters.

We take in L. Review— Many thanks for having sent it.

P.S.[7] Have you many Saxifrages at Kew? When you are a little quiet, I wish you would make a little observation for me, viz whether any & what of the most hairy species catch minute flies, & *whether there* is *any bending of the hairs all round towards the fly*.— In S. umbrosa many flies are caught, but there is *no* bending & I do not believe the dead flies are any profit to the plant: but I find that the glands at ends of hairs which secrete viscid matter & catch flies are much more permeable to C. of Ammonia, than the *exterior* walls of the cells in the bark or skin; but far less permeable than the *internal* walls between cell & cell of bark. You will see at once that this concerns genesis of Drosera, for we have in Saxifrage the habit of catching flies to start from & the viscid glands, & we have a considerable amount of *potential* power of absorption of C. of Ammonia. I think it worth investigation whether some of the more hairy Saxifrages may not profit by caught flies.[8] Will you look for me?—[9]

Endorsement: '61/'
DAR 115.2: 102

[1] Hooker's letter has not been found. CD had thought that Hooker wanted him to write a biography of Hooker's father-in-law, John Stevens Henslow, who died on 16 May 1861. See letter to J. D. Hooker, 24–5 May [1861].

[2] CD's 'recollections' of Henslow, consisting chiefly of his memory of Henslow during his student days at Cambridge, were published in the *Memoir of the Rev. John Stevens Henslow* edited by Henslow's brother-in-law Leonard Jenyns (Jenyns ed. 1862). CD's sketch is reprinted in *Collected papers* 2: 72–4, and in Appendix X.

3 Hooker's grief at the loss of Henslow is recounted in letters written in April and May 1861. On 23 May Hooker wrote (L. Huxley ed. 1918, 2: 61):

> Henslow has left a blank in my existence never to be replaced. Quite apart from considerations matrimonial, H. had more influence over my life and conduct than any other man ... More than all this, I miss his knowledge of loads of matters bearing on Botany which I never knew or took up but through him, and of loads of kindred subjects in which I have keenly interested myself, ever since I knew him.

4 The 'Business affairs' that postponed the family's departure for Torquay related to William Erasmus Darwin's future career. See letter to W. E. Darwin, [25 May 1861].

5 See letter to J. D. Hooker, 24–5 May [1861], n. 8.

6 Bates 1861. See letter from H. W. Bates, 18 March 1861, and letters to H. W. Bates, 26 March [1861], and to J. D. Hooker, 27 [March 1861].

7 The postscript is written on a separate sheet of paper; although it is not clear whether it was sent with this letter, the reference in the letter to J. D. Hooker, 19 June [1861], to CD's query about saxifrages makes this likely.

8 In the summer of 1860, CD had begun investigating the ability of various plants to derive nutriment from insects captured in the hairs of their leaves. Insectivorous plants, such as the sundew *Drosera*, provided good illustrations of very specialised adaptations. CD discussed the absorptive powers of the hairy glands of *Saxifraga umbrosa* in *Insectivorous plants*, pp. 345–8.

9 Hooker does not seem to have carried out the observations that CD mentions. CD later asked him again for specimens of saxifrage (see letters to J. D. Hooker, 28 September [1861] and 18 October [1861]), which Hooker eventually supplied in the spring of 1862 (see *Correspondence* vol. 10, letter from J. D. Hooker, 17 March 1862 and letter to J. D. Hooker, [18 May 1862]).

From Emma Darwin [June 1861][1]

I cannot tell you the compassion I have felt for all your sufferings for these weeks past that you have had so many drawbacks. Nor the gratitude I have felt for the cheerful & affectionate looks you have given me when I know you have been miserably uncomfortable.

My heart has often been too full to speak or take any notice I am sure you know I love you well enough to believe that I mind your sufferings nearly as much as I should my own & I find the only relief to my own mind is to take it as from God's hand, & to try to believe that all suffering & illness is meant to help us to exalt our minds & to look forward with hope to a future state. When I see your patience, deep compassion for others self command & above all gratitude for the smallest thing done to help you I cannot help longing that these precious feelings should be offered to Heaven for the sake of your daily happiness. But I find it difficult enough in my own case. I often think of the words "Thou shalt keep him in perfect peace whose mind is stayed on thee".[2] It is feeling & not reasoning that drives one to prayer. I feel presumptuous in writing thus to you.

I feel in my inmost heart your admirable qualities & feelings & all I would hope is that you might direct them upwards, as well as to one who values them above every thing in the world. I shall keep this by me till I feel cheerful & comfortable again about you but it has passed through my mind often lately so I thought I would write it partly to relieve my own mind.

DAR 210.10

CD ANNOTATION
End of letter: 'God Bless you. C.D. | June 1861'

[1] Dated from CD's annotation. According to CD's daughter Henrietta Emma Litchfield, the letter was one of two that Emma wrote to CD on the subject of religion, the first composed shortly after their marriage (*Correspondence* vol. 2, letter from Emma Darwin, [*c.* February 1839]). See *Emma Darwin* 2: 173–5.
[2] Isaiah 26:3.

To C. W. Crocker 1 June [1861][1]

Down. | *Bromley.* | *Kent. S.E.*
June 1st.

My dear Sir

Pray do not in the least hurry yourself: any time will do for me, & I shall much value your information. I do not quite understand whether you are thinking of a general course of *scientific* experiments in crossing or only in relation to Hollyhocks.— I am pretty confident there is a fine open field for research in regard to crossing *varieties* which have been greatly neglected *under a scientific point of view*, though largely & loosely practised by gardeners. Species on the other have been largely experimented on. As you have lived so much abroad, German is probably quite familiar to you[2] (I wish it were to me) & I would most strongly advise you to get Gärtners *admirable* "Versuche ueber die Bastardzeugung. 1849" & study it.—[3] If ever you thought of taking up scientifically the subject of crossing & feel that you have indomitable patience, I would beg permission to make a few suggestions on experiments, which I would surely have tried, had time & health permitted.—

With hollyhocks, I think it would be very useful to ascertain the effects of crossing, especially of those varieties which commonly breed true. You probably know far better than I; but when I tried I castrated these flowers in bud, but I believe it would quite suffice to remove anthers before the stigmas expand; & then cover the flower with gauze or net.— It would be extremely interesting to test following point; impregnate a dozen flowers of, we will say, a yellow Hollyhock (if that kind breeds true) with a black or some widely different kind; & then fertilise 12 of the yellow flowers with pollen of yellow flowers from another plant, *performing every manipulation the same to the two sets*; & then observe whether any difference in number of seed, which ought to be Counted or weighed. This experiment wd. be very interesting on account of a wonderful statement on this head by Gärtner with respect to crossing white & yellow Verbascum.[4]

Are you a fancier of Pelargoniums? I could suggest some interesting experiments; but I must not run on.— Pray excuse me, & believe me— | My dear Sir | Yours very sincerely | C. Darwin

American Philosophical Society (251)

[1] Dated by the relationship to the letter to the same unidentified correspondent, 18 May [1861].
[2] Crocker spent some time as a gardener at the summer residence of Victoria, Princess Royal of Great Britain, near Berlin.
[3] Gärtner 1849.
[4] CD described Karl Friedrich von Gärtner's experiments in crossing white and yellow varieties of *Verbascum* in *Variation* 2: 105–6. Gärtner asserted that crosses between differently coloured varieties, either of the same species or different species, produced less seed than crosses between similarly coloured varieties.

To W. E. Darwin 1 [June 1861]

[Down]
Saturday 1st

My dear William

I have just heard from Lubbock.[1] What mysteries these men of business make!— I am not yet to tell you name of man or place.— But the place is in every respect *good* in business point of view & one of the nicest towns in England.— I send you the copy of Mr —— letter. (*please keep this copy*) & of my letter to John Lubbock.[2]

I have told Lubbock, that if affair progresses; but I need not tell you what I say for I send a copy of my note & you will now see how affair stands. When in London, we would go together to my solicitor's & have a talk there.—[3] You see it is not unlikely that nothing may come of whole affair—Or my solicitors may find something objectionable.—

I do so wish this affair had come on one year later; but time & tide waits for no man.— I wish that you had written on receipt of my second letter; but I infer from not hearing that you keep of same mind.—[4] Write on receipt of this.—

My dear old William | Your affect.— | C. Darwin

[Enclosure][5]

—May 30 1861

My dear Mr. Lubbock—

Many thanks for your note; I should like very much to know (suppose anything came from your proposition) when Mr Darwin cd arrange to join. I ask this as you say he is about to leave College, & I am very anxious that if any arrangement is come to, that it shd be completed without much delay.

At the same time I think it right to say that I have an offer from another Gentleman who will come down to see me in a day or two, but I doubt if he will be quite an eligible person. I conclude Mr. Darwin will not mind sticking well to business, as I wish to take it rather easier than I have done for the last 18 years.[6]

I leave it to your discretion to mention my name or not in the matter as you think best, or leave it untill you can answer my question as to time.

Believe me | Y^rs very sincerely

Postmark: JU 1 61
DAR 210.6

[1] The letter from John Lubbock has not been found.

[2] See enclosure. The copy of the letter to Lubbock (see following letter) is in Emma Darwin's hand. CD marked it at the top in blue crayon: 'From C. Darwin to J. Lubbock'.

[3] CD used the firm of Rowland and Hacon, solicitors with offices at 31 Fenchurch Street, London.

[4] No letter from William giving his initial reaction to the partnership offer has been found. CD refers to his letter to William of [26 May 1861].

[5] The enclosure is in DAR 210.6. It is a copy, in Emma Darwin's hand, of a letter written by George Atherley, partner in the Southampton and Hampshire Bank. At the top, CD wrote in blue crayon: '(Private) (Keep this)'. The second enclosure, also in DAR 210.6, was a copy of the following letter to John Lubbock, with blanks in place of 'Southampton' and 'Atherley'.

[6] CD altered the figure from '10' to '18' in blue crayon.

To John Lubbock 1 June [1861][1]

Down. | Bromley. | Kent. S.E.
June 1st

My dear Lubbock

Many thanks for your note.[2] I told William that if the proposition came to anything, he would have to join immediately, & I left the acceptance entirely to his good sense, & he agreed. I am sure he would work hard, for througout life whatever he has done, he has done energetically. He would go down to Southampton at any time, direct from Cambridge; though perhaps it would be best for him on the road to see you & come here. This term ends about 20th & no doubt he could get leave of absence for rest of term; & then he will have kept *all* his terms, & next year he probably could run up for three days & take his degree, for he has a natural wish (though in fact useless) just to take a degree after having completed his time & worked pretty hard.[3] You can, if you like, forward this note to Mr Atherley;[4] & I pledge my word not to mention Mr Atherley's name to anyone, not even to my son William, until the affair has progressed

Dear Lubbock | Ever yours truly | C. Darwin

P.S. | I think if the affair progresses the best plan will be to send to William per Telegraph to come to London & I would meet him there & have some talk with him & then he could go down to Southampton.

Down House

[1] See preceding letter, n. 1.

[2] Lubbock's letter has not been found.

[3] William left Cambridge without a degree in the summer of 1861. Having met the residency requirement of attending three terms for three years, he entered for the mathematical tripos examination of 1862 and gained a degree (*Historical register of the University of Cambridge*; F. Darwin 1914, p. 20).

[4] George Atherley was a partner in the Southampton and Hampshire Bank, Southampton.

To A. G. More 2 June 1861

Down, Bromley, Kent, S.E.
June 2, 1861.

My dear Sir

I think you told me you had Aceras;[1] I see in Babington it flowers in June.[2] Would you have the great kindness to send me a few specimens? About the 10th or shortly afterwards I go for my daughter's sake to Torquay; If you can send me Aceras, and if it be not now in flower, I would write from Torquay, and give my address. If you do actually send me Aceras and some day alight on O. latifolia, would you put in 2 or 3 specimens, *but not otherwise*. I have been wonderfully lucky in getting orchids, and this morning I examined O. aramifera and O. ustulata.[3] Does Habenaria viridis grow in I. of Wight?[4] When I get this latter and Aceras I shall have seen all that I can hope for. I presume Cephalanthera ensifolia is hopeless.[5] I have had Malaxis and Goodyera, and shall get, I believe O. hircina; so have I not been fortunate? I find in O. aranifera the pollen-masses do not fall out as in Bee-Orchis. If you pass any group of the latter pray glance at a few to see if pollen-masses removed; but I despair of ever making out this species. I hope you will be so kind as to observe manner of ingress and egress of insects, if you can see any visit E. palustris; and whether Labellum is irritable.[6] The examination of that species has been one of my greatest treats, which I owe to you. I fear I am very unreasonable; but this subject is a passion with me.

Believe me, my dear Sir, Yours sincerely | C. Darwin.

Copy
DAR 146

[1] None of More's letters to CD, with the exception of one written in 1881, has been found. See, however, the letter to A. G. More, 8 March 1861. *Aceras* is a genus of orchid whose separation from the genus *Orchis*, according to CD, 'is evidently artificial' (*Orchids*, p. 19). CD intended to continue his study of the flowering parts of orchids, begun in the summer of 1860, during his approaching holiday in Torquay.

[2] Babington 1851, p. 309. There is an annotated copy of the volume in the Darwin Library–CUL.

[3] These species are described in *Orchids*, pp. 60–3 and 31–3, respectively. CD thanked George Chichester Oxenden for providing him with fresh specimens of both (*ibid.*, p. 31 n. 61).

[4] CD described *Peristylus* (or *Habenaria*) *viridis* in *Orchids*, pp. 76–9. More, who lived at Bembridge on the Isle of Wight, was apparently unable to send him specimens; CD eventually obtained some from Bingham Sibthorpe Malden (see letter to B. S. Malden, 15–16 June [1861]).

[5] Only *Cephalanthera grandiflora* is described in *Orchids* (pp. 104–12).

[6] More did make observations of the visits of insects to *Epipactis palustris*, the marsh helleborine. In the description of the flower parts of this species, CD stated: 'So flexible and elastic is the hinge [of the labellum] that the weight of even a fly, as Mr. More informs me, depresses the distal portion' (*Orchids*, p. 99).

To P. L. Sclater 2 June [1861]

Down. | Bromley. | Kent. S.E.
June 2[d]

Dear Sclater

Perhaps you will have heard that I ventured to send to Gardens temporarily 2 Rabbits from P. Santo.[1] If they are like one brought by Wollaston for me in Spirits, they are curiosities.[2] They have been feral for 450 years! I have history of their origin: they all spring from one doe brought by Zarco.[3] The one examined by me differed in skull & shape of dorsal vertebræ; in size *greatly*; in general colouring above & below; in colour of upper side of tail & in ears not being edged with black.— In fact, if *these are the general characters*, it is a *new species*!![4]

I am in perplexity, for I am going (probably about 10th.) to Torquay for my daughters health with most of my servants, for two months; & I dare not leave these rabbits here. Will you speak about care being taken of them, & getting them to breed if buck & doe; if not to cross them with some odd coloured variety to see if quite fertile.[5] If they sh[d] die it would break my heart not to see the body fresh.—

Now would you be so very kind as to tell me whether they seem likely to live, & what Bartlett thinks of them, as appearing different from common English rabbit.

Pray forgive me & grant the favour.—

Dear Sclater | Yours very sincerely | C. Darwin

When I get to Torquay, I will send my address, in case one or both sh[d] die for Body to be sent me.—

Endorsement: 'June 1861'
American Philosophical Society (252)

[1] See letter to A. D. Bartlett, 26 May [1861]. Sclater was secretary of the Zoological Society of London and active in the direction of the society's gardens.

[2] Thomas Vernon Wollaston. See letter to A. D. Bartlett, 26 May [1861], n. 6.

[3] J. Gonzales Zarco was a Portuguese navigator who claimed Madeira for Portugal in 1419. In *Variation* 1: 112 n. 23, CD cited Kerr 1811–24, 2: 117 as the source of this information. See also letter to A. D. Bartlett, 26 May [1861], n. 5.

[4] In *Variation* 1: 115, CD stated: 'If the history of these Porto Santo rabbits had not been known, most naturalists . . . would have ranked them as a distinct species.'

[5] In *Variation* 1: 115, CD wrote that he found it remarkable that 'Mr. Bartlett could never succeed in getting these two rabbits, which were both males, to associate or breed with the females of several breeds which were repeatedly placed with them.' Abraham Dee Bartlett was the superintendent of the gardens of the Zoological Society.

To A. G. More 4 June 1861

Down, Bromley, Kent. S.E.
June 4, 1861.

My dear Sir

I fear that I am very troublesome, but in case you should find time to watch for a little E. palustris, I have thought it worth while to enclose a little memorandum.[1]

Yours sincerely | C. Darwin

[Enclosure]

Epipactis palustris

(1) When an insect visits this flower, does it alight on the labellum? and in entering flower does it depress the terminal portion and does this spring up again after the insect has entered?

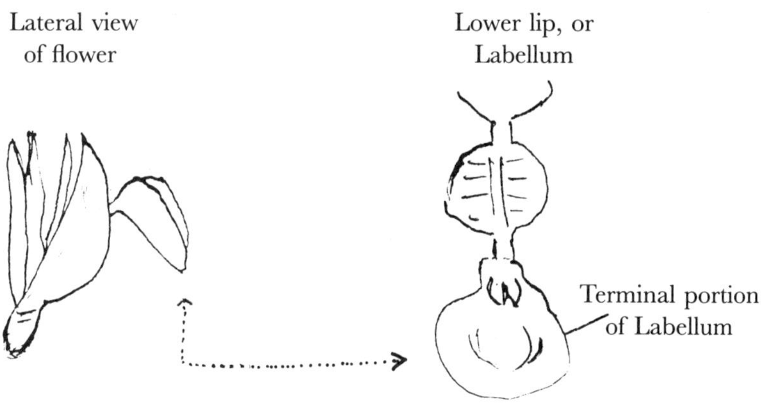

(2) *Especially*, when the insect departs does it crawl back by the labellum, or does it crawl out *at the top* of the flower?

(3) If you see one crawl out please observe whether the yellow pollen-masses are sticking to its Head or back: this would occur only in a lately opened flower. Observe whether it enters another flower and whether it does not there leave the pollen-masses sticking to the stigma, or at least some of the yellow pollen-grains, which would be easily visible.

(4) What kind of insect visits these flowers; if you should see several, please catch one by crushing flower in which it is in and send it.

(5) As the labellum in some foreign Orchids is irritable, tickle on calm day with bit of grass the middle part of Labellum near the joint of terminal part and see whether it excites any movement slow or quick. The best time to watch the Orchid would be a fine day after one or two cold gloomy days.

Mr. Oxenden of Broom Park is going to watch this flower for me, and I made these notes for his sake,[2] and then I thought I would send you a copy. If by any *extraordinary* chance the labellum should be irritable will you observe carefully what part must be touched.

Copy
DAR 146

[1] See letter to A. G. More, 2 June 1861 and n. 6.
[2] CD's correspondence with George Chichester Oxenden on this point has not been found.

To Asa Gray 5 June [1861][1]

Down. | Bromley. | Kent. S.E.
June 5th

My dear Gray

I have been rather extra busy, so have been slack in answering your note of May 6th—[2] I hope you have received long ago the 3d. Edit. of Origin.[3] Andrew Murray is an Entomologist & Horticulturist & now Secretary to Hort. Soc. of London.— He read a long & hostile & rather weak Review of Origin at the Royal Soc. of Edinburgh.—[4]

I have heard nothing from Trübner of sale of your Essay;[5] hence fear it has not been great: I wrote to say you could supply more.— I sent a copy to Sir J. Herschel; & in his new Edit. of his Physical Geography he has note on the origin of species, & agrees to certain limited extent; but puts in a caution on design, so much like yours that I suspect it is borrowed.—[6] I have been led to think more on this subject of late, & grieve to say that I come to differ more from you. It is not that designed variation makes, as it seems to me, my Deity "Natural Selection" superfluous; but rather from studying lately domestic variations & seeing what an enormous field of undesigned variability there is ready for natural selection to appropriate for any purpose useful to each creature.—

I thank you much for sending me your Review of Phillips.—[7] I remember once telling you a lot of trades which you ought to have followed;[8] but now I am convinced that you are a born Reviewer.— By Jove how well & often you hit the nail on the head. You rank Phillips' book higher than I do; or than Lyell does, who thinks it fearfully retrograde.— I amused myself by parodying Phillips arguments as applied to domestic varieties; & you might thus prove that the Duck or Pigeon has not varied because the Goose has not, though more anciently domesticated, & no good reason can be assigned why it has not produced many varieties.[9] With respect to F. Water, small area, compared with sea or land, I believe comes into play; rate of change & of extinction in F. Water having been much slower, hence Ganoid fishes are all fresh-water.—[10]

How true what you say about Matthew;[11] but I will not run on.—

I have been idling & working at Primula & think my experiments will explain their dimorphism:[12] now I much want one piece of information; I know that there are many cases of dimorphic plants; but are not the two forms always borne on same plant? Are there other cases of two forms *living mingled in nearly equal numbers*?— I have also been working on insect fertilisation of Orchids—beautiful facts—& I want information on Cypripedium. Have you Botanic Garden? Could you cover up a plant with net & leave one uncovered; *if it be one which sets seeds*, & see whether *protected* one sets seeds, & whether the pollen of the two after interval of time are in the same state.[13] Do not forget Spiranthes— look at flowers just opening: I am curious to know whether same curious structure as in our Spiranthes.—[14]

But I suppose you are all too overwhelmed with public affairs to care for science.—[15] I never knew the newspapers so profoundly interesting. N. America does not do England justice: I have not seen or heard of a soul who is not with the North. Some few, & I am one, even wish to God, though at the loss of millions of lives, that the North would proclaim a crusade against Slavery. In the long run, a million horrid deaths would be amply repaid in the cause of humanity.—[16] What wonderful times we live in.— Massachusetts seems to show noble enthusiasm.[17] Great God how I sh[d] like to see that greatest curse on Earth Slavery abolished.

Farewell. Hooker has been absorbed with poor dear revered Henslow's affairs—[18] | Farewell | Ever yours | C. Darwin

Gray Herbarium of Harvard University (60)

1 The year is given by the references to the publication of the third edition of *Origin* and to A. Gray 1861a and 1861b.

2 Gray's letter has not been found.

3 The third edition of *Origin* was published in April 1861. Gray's name appears on CD's list of presentation copies for this work (see Appendix VII).

4 Murray 1860. This paper was read before the Royal Society of Edinburgh on 20 February 1860. Before it was published, Murray sent CD a draft copy. For CD's comments on the paper, see *Correspondence* vol. 8, letters to Andrew Murray, 28 April [1860] and 5 May [1860], and to Charles Lyell, 1 [June 1860], and letter from Andrew Murray, 3 May 1860.

5 A. Gray 1861a (see Appendix III).

6 See letter to J. F. W. Herschel, 23 May [1861] and n. 3. In Herschel 1861, p. 12 n., Herschel stated that he could not accept 'the principle of arbitrary and casual variation and natural selection as a sufficient account, *per se*, of the past and present organic world'; instead he argued that 'an intelligence, guided by a purpose, must be continually in action to bias the direction of the steps of change'. All three of Gray's reviews of *Origin* ([A. Gray] 1860a, A. Gray 1860b, and 1860c) advocated the view that the acceptance of CD's theory of natural selection was compatible with a belief in design in nature. He stated, for example (A. Gray 1860b, p. 183):

> that if by successive origination of species and organs through natural agencies, the author means a series of events which succeed each other irrespective of a continued directing intelligence,— events which mind does not order and shape to destined ends,—then he ... has accumulated improbabilities beyond all belief.

7 Gray's review of Phillips 1860 (A. Gray 1861b) was published in the *American Journal of Science and Arts*. John Phillips had sent CD a copy of this work (see *Correspondence* vol. 8, letter to John Phillips, 14 November [1860]). An annotated copy of Phillips 1860, inscribed 'From the Author', is in the Darwin Library–CUL. The book, which was the substance of the Rede Lecture delivered by Phillips at the University of Cambridge in 1860, addressed the question of the origin of species and criticised the account provided by CD in *Origin*.

[8] CD on another occasion described Gray as 'a hybrid, a complex cross of Lawyer, Poet, Naturalist, & Theologian!' (*Correspondence* vol. 8, letter to Asa Gray, 10 September [1860]).

[9] Phillips argued that the fossil record, rather than indicating species change as CD's views required, provided evidence of species continuity. As an illustration, he pointed to the brachiopod *Lingula*, which 'recurs in all the systems of strata, and is still living' and yet 'gives no generic branches' (Phillips 1860, p. 212).

[10] Phillips 1860, pp. 108–14. Phillips argued that 'the comparative fewness of the freshwater races, and their great affinity over large tracts of the earth' counted against CD's views and concluded (*ibid*., pp. 110, 113–14):

> Circumstances have varied, ages have passed away, and yet every generic group exhibits at every step the same essential characters, and many of the little peculiarities . . . which cannot be consistent with accumulated tendencies to change.

Phillips cited ganoid fishes as an example of a group that had persisted unchanged since the Silurian era (*ibid*., p. 103).

[11] The reference may be to Patrick Matthew, who had claimed that he anticipated CD's concept of natural selection in his work on naval timber and arboriculture (Matthew 1831). CD conceded Matthew's priority (see *Correspondence* vol. 8, letters to J. D. Hooker, 13 [April 1860], and to *Gardeners' Chronicle*, [before 13 April 1860]).

[12] CD saw the dimorphic condition of *Primula* as inhibiting self-fertilisation and thus favouring the sexual union of distinct individuals of the same species. CD published the results of his study in 1862 (*Collected papers* 2: 45–63).

[13] Gray is cited in *Orchids* as having observed insects visiting *Cypripedium*, but no mention is made of the experiment suggested by CD.

[14] See *Correspondence* vol. 8, letter to Asa Gray, 31 October [1860]. See also letter to Asa Gray, 11 April [1861]. Gray's observations on *Spiranthes* are cited in *Orchids*, p. 123 n.

[15] The reference is to the outbreak in the United States of hostilities that marked the beginning of the Civil War. After the fall of Fort Sumter on 14 April 1861, President Abraham Lincoln signed a proclamation asking for volunteers for the Union Army, and the Confederate government also began mobilising its volunteer force (*EB*).

[16] CD had been a staunch opponent of slavery since his youth. This conviction had intensified as a result of his having observed first-hand the inhumanity of slavery in South America during the *Beagle* voyage. See, for example, *Correspondence* vol. 1, letter to Catherine Darwin, 22 May – 14 July 1833, and Barlow ed. 1933, p. 55.

[17] The first bloodshed of the war occurred on 19 April 1861, when troops from Massachusetts on their way to defend Washington, DC, were attacked by a mob in Baltimore (*EB*).

[18] Joseph Dalton Hooker's father-in-law, John Stevens Henslow, had died on 16 May 1861.

To Samuel Pickworth Woodward 5 June [1861][1]

Down. | *Bromley.* | *Kent. S.E.*
June 5th

My dear Sir

Thank you for your pleasant note.[2] Your nearest station is Bromley, or you would save ½ mile by going on to Bickley; but then by leaving London Bridge at 9°·10′ or 12°·30′ you find Buss at Bromley which would bring you to Keston some 3½ miles on road.— Bromley to Knockht Beeches 9½ miles.—[3]

Road through Keston, Down Cudham, much helped by footpaths.— But surely this wd be too great an exertion for you: you can beat me, I could not walk 4 miles hardly to save my life.—

We sh^d. be very happy to see you here for the day, but we have no attractions, & there are no Downs anywhere, & no O. fusca; we have some fly O. & some Bee O. but I have not yet seen latter.—[4]

But here is a puzzle, we had fixed positively all to start for Torquay on 10th.— Some Business in hand *may* interfere & at present I have no guess whether we shall be able to start on that day.—[5] If you like to chance a visit we sh^d be delighted to see you; but we may be all in confusion in packing up our whole household for move on Wednesday.—[6]

I agree about colonisation of Arctic regions,[7] but in some respects it seems to me a very perplexing question.

I fancied that the land-shells which I collected at St. Helena, from their position had quite recently become embedded & extinct.—[8] I believe my shells are at Geolog. Soc.—[9]

I want to beg a favour, if D^r. Günther is in London, to get him to give me a note, or reference to any paper, on the Silurus escaping from the Danube.[10] How was it ascertained?— Pray forgive haste, I have much to write this afternoon.—

I heartily wish your Boys & self a pleasant excursion whether they come into this part or not.— If you come cross Holwood Park by footpath— very pretty.—[11]

Yours very sincerely | C. Darwin

I just see by starting from Victoria St. by London Chatham & Dover R. you can go to St. Mary's Cray & then you would be $6\frac{1}{2}$ miles from Knockholt beeches— Below Knockholt lies Chevening Park—rather striking—[12]

British Library (Add. MS. 42579, ff. 230–32b)

1 The year is given by the reference to the planned visit to Torquay (see n. 5, below).

2 The letter from Woodward has not been found. He was an assistant in the geology department of the British Museum.

3 From the letter it appears that Woodward had proposed to visit CD in Down en route to Knockholt, Kent, near which is located the landmark known as the Knockholt Beeches, a 'remarkably fine' clump of old trees on the high ground of a chalk plateau (*Victoria history of the county of Kent* 1: 478). Knockholt is several miles south-east of Down.

4 According to the *Victoria history of the county of Kent* 1: 64, 'no British county excells Kent in the number of its orchids'. CD had examined the fly orchis, *Ophrys muscifera*, and the bee orchis, *O. apifera*, for his study of their pollinating mechanisms.

5 Negotiations were being conducted to secure William Erasmus Darwin a partnership in the Southampton and Hampshire Bank, Southampton (see following letter).

6 In the event, the Darwins postponed their departure for Torquay and remained at Down until the end of June (see 'Journal'; Appendix II).

7 For CD's views on 'colonisation' and the migration of species through the Arctic before and during the glacial period, see *Origin*, pp. 365–82.

8 CD believed that the land molluscs of St Helena had become extinct as a result of deforestation during the preceding century (see *Journal of researches* 2d ed., p. 488).

9 The location of the fossil shells collected by CD during the *Beagle* voyage is undetermined (see Porter 1985, pp. 996–7).

10 Albert Charles Lewis Gotthilf Günther, employed by the British Museum, was preparing a catalogue of the freshwater fish in the museum's collection (Günther 1859–70). *Silurus* is a genus of Eurasian

catfishes. Woodward complied with CD's request. See letter from A. C. L. G. Günther to S. P. Woodward, 14 June 1861.

[11] Holwood Park, near Bromley, Kent, was the seat of Robert Monsey Rolfe, Baron Cranworth.

[12] Chevening, the family seat of Philip Henry Stanhope, is a few miles north-west of Sevenoaks, Kent.

To W. E. Darwin 6 [June 1861]

Down.
6th.

My dear William.—

I enclose the £20.[1] No news about your affair.[2] I wish to God there was.— We clearly cannot go to Torquay on 10th.— I had long talk with J. Lubbock, who urged Mr. ——[3] to prompt action. I see, if affair progresses, there will be much negotiation & very great difficulties of all kinds. But we have a true friend & good adviser in J. Lubbock.— It seems that the great Houses of Robarts & Lubbock joined on mere verbal agreement![4] It seems that much *must* be left to honour & mutual good feeling; & there is this protection that partners are like man & wife & it hurts both their reputations to part.— I shall be so sorry if you lose your Welch tour:[5] but all is utterly dark.—

Good Bye my dear old sensible fellow, who will be the guardian of the Family. | C. D.

Second note

Since writing, I have received enclosed. Lubbock urges us to meet him at their Bank on Saturday.[6] Therefore I shall go up tomorrow (Friday) to 6 Queen Anne St.—[7] You come there either Friday night or in good time on Saturday morning; for I shall start at eleven oclock to go to City.

I cannot understand as —— is now negotiating with another party, why he wants to see us: but I suppose he has got some loophole to escape. Anyhow it is worth the trouble.— I was glad to hear that Mr —— is a University man—[8]

When we meet I will tell name & place; but I had better not break my word yet.—

I wonder whether it will come to anything!

Yours | C. D

[Enclosure]

My dear L.

Many thanks for yr note. The gentn I mentioned in my last letter has asked for a few days to think the matter over which I have of course conceded. Therefore perhaps Mr D. wd let the proposition as regards his son remain in abeyance for a few days I am not in such great haste as Mr D. in his note supposes, but if it proceeds I shd like to have it settled by or as soon as possible after the 1st July.

As regards the years trial it wd hardly do. as it wd not obviate my reasons for taking a partner at once, the motive for which I think you are aware of. As time however is every thing what say you to the plan (under any circumstances) of Mr D. & his son meeting me on Sat in London to talk matters over, & we c^{d} then mutually judge of each other. & he w^{d} have more time for consideration in the event of the affair progressing.

Postmark: JU 6 61
DAR 210.6

1 There is an entry for this sum in CD's Account book (Down House MS) on 6 June 1861 that reads 'Advanced to William (private)'.
2 The 'affair' concerned the possibility of William's joining the Southampton and Hampshire Bank. See especially the letter to W. E. Darwin, [25 May 1861].
3 George Atherley was a partner in the Southampton and Hampshire Bank. CD began to write the name but deleted it: he had promised John Lubbock to keep Atherley's name in confidence (see letter to W. E. Darwin, 1 [June 1861]).
4 In 1860, the bank directed by John William Lubbock, of which John Lubbock was a partner, was amalgamated with Robarts & Co. of Lombard Street, London to form Robarts, Lubbock & Co. (*Banking almanac* 1860, 1861).
5 William had planned to go to Wales on a walking tour (see letter to W. E. Darwin, 9 May [1861]).
6 The enclosure is a copy in Emma Darwin's hand of a letter from George Atherley to John Lubbock. For the text of Atherley's earlier letter to Lubbock, see the enclosure with the letter to W. E. Darwin, 1 June [1861]. The banking house of Robarts, Lubbock & Co. was located in the City of London.
7 The home of CD's brother, Erasmus Alvey Darwin, where CD usually stayed when in London.
8 Atherley was admitted pensioner at Trinity College, Cambridge, in 1837.

To John Lubbock 6 June [1861][1]

Down—
June 6th

My dear Lubbock.

Very many & sincere thanks. We will meet at your Bank on Saturday at 12 oclock.—[2]

To prevent mistakes I have written to M^{r}. A. to say so;[3] he will receive letter tomorrow morning.—

Yours most sincerely | C. Darwin

Down House (MS 7: 3)

1 The dating is established by the relationship to the other letters concerning the negotiations to secure William a partnership in a Southampton bank.
2 Lubbock was a partner in the private bank, Robarts, Lubbock & Co., located in Mansion House Street, London (*Banking almanac* 1861).
3 George Atherley was a partner in the Southampton and Hampshire Bank, Southampton.

To John Lubbock [8 June 1861][1]

Down. | Bromley. | Kent. S.E.
Saturday Night

My dear Lubbock

I fear that you will think me utterly unreasonable if I was to ask you to call here either before or after church & stay lunch if so inclined. I know that you cannot do more, or say more than you have done; but you cannot imagine what a relief it would be to me to have one more **final** talk with you over this business; for I know no one else in whose judgment on such matters I have confidence.—

I know I am utterly unreasonable to ask you to call; but I fear I shall be a good deal knocked up tomorrow.—

Good Heavens what a whirlwind of business your office appeared.[2] Pray forgive me & have charity, & a real charity it will be.—

Sir John Lubbock put the whole case with admirable clearness & helped me beyond measure.[3] If it were not for looking to future, I would close at once; but I cannot feel easy at future.—

Your most sincerely | C. Darwin

Down House (MS 7: 2)

[1] Dated on the assumption that the letter was written in the evening following the meeting on Saturday, 8 June 1861, at Robarts, Lubbock & Co. (see the two preceding letters). The eighth of June fell on a Saturday in 1861.

[2] See preceding letter and n. 2.

[3] Lubbock's father, John William Lubbock, was a partner in Robarts, Lubbock & Co. (*Banking almanac* 1861).

From Thomas Francis Jamieson 13 June 1861

Ellon. Aberdeensh.
13. June 1861.

My Dear Sir,

Although my botanical knowledge is of a very slender description I have no doubt I shall be able to get specimens for you of the *Listera cordata* & *Habenaria viridis*, both of which are to be found in this district or at least at no great distance.[1] I have gathered specimens of them myself but several years ago, & not just in this locality. The *Corallorhiza* has not hitherto been found, I believe, in this part of Scotland.

As you are interested in the change of habits of animals perhaps the following may be worth sending.

In 1845 when attending college in Aberdeen I was more addicted to the pursuit of birds than to classical learning & was always rambling about with a gun, & on one occasion shot a Kestrel (*Falco tinnunculus*) which had been previously wounded & had got one of its legs broken, and in skinning it I found the stomach unusually distended & on opening it found the contents to be remains of worms & also what appeared to be snails or slugs— It had apparently been obliged to betake itself to this unnatural *cuisine* from inability to catch its wonted prey. It had been observed haunting the outskirts of the town for some days previous by a fellow student who told me he had fired five shots at it. The following jotting from an old sort of ornithological journal which I used to keep, authenticates the fact. "1845, November

26th. Stuffed the Kestrel. Its stomach was much distended & filled with remains of worms, & as I supposed, snails &c. it having been compelled to feed upon these, one of its legs being broken."

A Gamekeeper here, (on whose veracity I can depend) told me what I thought a very curious physiological fact, viz that he has shot hares containing two broods of young of very different ages viz. one set nearly being littered & the others quite small.[2]

In the French scientific periodical the 'Cosmos' vol 14. p. 531 (being for 1859) there is a notice from an article in the Patrie by H. Berthoud for 10 May, regarding some curious modifications in the appetites of animals in a state of confinement or domestication.[3]

My Dear Sir | Yours very truly | Thos. F. Jamieson

Chas. Darwin Esq.

DAR 47: 171–2

CD ANNOTATIONS
0.1 Ellon . . . Scotland. 1.6] *crossed pencil*
4.1 A Gamekeeper . . . small. 4.4] *crossed pencil*
End of letter: 'See to McGillivray whether Kestrels get worms &c &c'[4]

1 Although few of the letters have been found, CD and Jamieson apparently corresponded frequently in 1861. In a letter to Charles Lyell dated 14 May 1861, Jamieson wrote: 'Mr. Darwin has in the kindest manner possible urged me to visit Glen Roy & has sent me copies of the principal papers on that celebrated locality' (Edinburgh University Library, MS Lyell 1, Gen. 112). See also letter to Robert Chambers, 30 April [1861].

2 Jamieson had become known as an agricultural expert while working as a factor on Scottish estates.

3 The report in *Cosmos: Revue Encyclopédique Hebdomadaire des Progrès des Sciences* 14 (1859): 531–2 summarised an article by Samuel Henri Berthoud, 'Fantaisies scientifiques', that had appeared in the French quarterly *La Patrie*. The article described cases in which animals held in captivity came to reject their natural foodstuffs and prefer altogether different ones; it also cited a case of maternal instinct being affected by domestication.

4 CD refers to Macgillivray 1837–52, an annotated copy of which is in the Darwin Library–CUL. The fact that earthworms form part of the kestrel's diet is mentioned in *ibid.*, 3: 331.

To Edward Walford 13 June [1861][1]

Down Bromley Kent
June 13th

Sir

I enclose the dates of the principal events of my Life, according to your request.—[2] I presume that I have given them much too fully, but you can use such as you think fit.—[3]

I have the honour to remain | Sir | Your obedt. servt. | Charles Darwin

E. Walford Esq

Ralph Colp Jr

[1] The year is provided by the reference to the biography of CD included in Walford ed. 1862.
[2] Walford was the editor of the 1862 edition of the biographical series *Men of the time* (*DNB*).
[3] The entry for CD in Walford ed. 1862 includes information on the educational institutions he attended, the *Beagle* voyage, his publications, and the medals awarded to him by the Royal Society and the Geological Society.

From Albert Charles Lewis Gotthilf Günther to S. P. Woodward 14 June 1861

Brit. Mus.
14. 6. 61.

My dear Sir

I have been so bothered with business of every kind that I could not answer your note before today.[1]

It appears that the "Wels" has been accidentally transferred by inundations into the Lake of Constance from the minor lakes north of it.[2] This, certainly, happened more than once, but after some time the individuals which had not been introduced in sufficient number, died out. Thus you may find, that already old Gesner mentions the Wels as a fish of the Lake of Constance.[3]

Nothing positive is heard of the fish up to the time of Hartmann, Helvetische Ichthyologie 1827, who states that the fish is transferred to the Lake by inundation.[4] After his time the fish appears to have died out again; for old fishermen stated to me at various parts of the eastern portion of the Lake, that they formerly never caught a Wels there, & that it was considered a great wonder when the first specimen was brought in to the town of Lindau. They asserted that this happened shortly after a "great water" & that this was the way of its introduction. From that time, however, the fishes multiplied, & they are not scarce now in the north-eastern part of the lake, whence I have procured young specimens.[5] See "Rapp, die Fische des Bodensee's. Stuttg. 1854."[6]

Yours very truly | A Günther

(Mr Woodward)

DAR 205.2 (Letters)

CD ANNOTATIONS

1.1 I have … today. 1.2] *crossed pencil*
5.1 (Mr Woodward)] 'To' *added above, pencil*
Top of first page: 'Keep' *pencil*; '18'[7] *blue crayon, circled blue crayon*
End of letter: 'Dr. Gunther | On Fish repeatedly [*interl*] introduced by inundation. | I believe a Silurus?' *pencil*

[1] Woodward had written to Günther at CD's request. See letter to S. P. Woodward, 5 June [1861]. The two naturalists worked in different departments of the British Museum.
[2] The sheat-fish (*Silurus glanis*) is also known as the wels, a word taken directly from the German. The fish is common in the Danube and in other rivers of eastern Europe (*OED*).

[3] Gesner 1558.
[4] Hartmann 1827.
[5] Günther had made an expedition to Lake Constance during his student days at the University of Tübingen (Gunther 1975, p. 233). While at university, he published two studies of the freshwater fish of Germany (Günther 1853 and 1855).
[6] Rapp 1854.
[7] The number of CD's portfolio of notes on the means of dispersal of plants and animals.

To W. B. Tegetmeier 14 June [1861][1]

Down. Bromley. Kent. S.E.
June 14th

My dear Sir

I send by this post my M.S. on Fowls:[2] you will find the reading a tough & dull job.— The notes are pinned on the back. You will have lots of time for I shall be away about 8 weeks.[3] Any very short corrections or doubts might be marked on margin; any longer remarks had better be on separate sheet, with reference to a pencil cross (x) & my page.— The Synopsis of characters will perhaps require closest attention.

Please observe that the style is not yet corrected. I have made several statements which I shd. not have ventured on, had I not known that you would kindly look the M.S. over. Moreover I hope to add & correct from further information which you will obtain for me.—

Please keep the M.S. in some shut & *safe* place.— It took me 8 weeks to write!![4]

Now forgive me for a bit of business: I enclose a cheque for five guineas, which perhaps you will think sufficient just to remunerate you for loss of time in looking over the M.S. & for the observations which you have already been so good to make for me.—[5]

Will you please acknowledge the safe receipt of the M.S. & this note.—

With sincere thanks for all your kindness | believe, my dear Sir | Yours sincerely | C. Darwin

The M.S. is splendidly copied.—[6]

New York Botanical Garden Library (Charles Finney Cox collection)

[1] Dated by the reference to CD's manuscript on fowls; according to his 'Journal' (Appendix II), CD finished writing it on 16 May 1861.
[2] Tegetmeier had agreed to read the chapter on domestic fowls that CD had prepared for inclusion in *Variation* (see letters to W. B. Tegetmeier, 22 March [1861] and 28 March [1861]).
[3] The Darwins spent eight weeks in Torquay, from 1 July to 27 August 1861 (see 'Journal'; Appendix II).
[4] This is also recorded in CD's 'Journal' (see Appendix II).
[5] An entry in CD's Account book (Down House MS) on 13 June 1861 records a payment to Tegetmeier of £5 5*s*.
[6] CD paid his copyist, Ebenezer Norman, £1 15*s*. 6*d*. for copying on 15 June 1861 (CD's Account book).

To the *Field* [before 15 June 1861][1]

I hope that you will permit me to return my sincere thanks to "Eques," of Argyllshire, for his information.[2] I have talked to many persons, but have never found any one who knew nearly so much about the inheritance of colours in horses. "Eques" asks me to define what is meant by dun—I wish I could. It seems that almost any colour that is not bay, chesnut, black, grey, or one of the roans, is sometimes called dun. Even in mouse-duns I have seen two very distinct tints. The commonest colour which, in my limited experience, I have heard called dun, is a bay or chesnut, more or less diluted with cream or clay tint. In India, the Kattywar breed, which is more striped than any other breed, is called dun; but, as I am informed by Col. Poole, the colour is generally between brown and black.[3] I now see that I should have put my question thus—Are horses, of any of the many indefinite tints commonly called duns, with a stripe along the spine, or with stripes across the legs, ever produced from parents neither of which are striped? for the first appearance or origin of the stripes is chiefly interesting for my purpose. I put the case—Are duns ever produced from parents neither of which are duns, because horses thus coloured are so frequently striped? and I thought I should more easily find out the parentage merely of the dun colour. What "Eques" says about the dirty mark on the withers, representing a single or double stripe, is exactly what I have observed. If "Eques" could find out, without much trouble, the colour of the dam and sire of his dun mare with the list and stripes, which was bred in Argyllshire, I should be very glad to hear it;[4] but I have already caused "Eques" very much trouble, and I beg permission again to thank him.—

Charles Darwin (Down, Bromley, Kent).

The Field, the Farm, the Garden, the Country Gentleman's Newspaper 17 (1861): 521

[1] Dated by the publication of the letter in the 15 June 1861 issue of the *Field*. It appeared under the heading 'Dun horses'.

[2] See letter to the *Field*, [before 25 May 1861] and n. 5.

[3] CD had written in 1858 to Skeffington Poole, a retired lieutenant-colonel from the first regiment light cavalry, Bengal, with queries on the colours and stripes of native Indian horses (see *Correspondence* vol. 7, letter to Skeffington Poole, 13 October [1858]). CD cited information on the subject provided by Poole in *Variation* 1: 58, 59.

[4] A response from 'Eques (Argylshire)' was published in the 27 July 1861 issue of the *Field*; the author stated that the dam of his dun mare had been dun, 'a bluer dun than herself', and the sire, brown (*Field*, 18 (1861): 91).

To *Gardeners' Chronicle* [before 15 June 1861][1]

I do not know whether any exotic Vincas seed, or whether gardeners would wish them to seed, and so raise new varieties. Having never observed the large

Periwinkle or Vinca major to produce seed, and having read that this never occurs in Germany, I was led to examine the flower. The pistil, as botanists know, is a curious object, consisting of a style, thickening upwards, with a horizontal wheel on the top; and this is surmounted by a beautiful brush of white filaments. The concave tire of the wheel is the stigmatic surface, as was very evident when pollen was placed on it, by the penetration of the pollen-tubes. The pollen is soon shed out of the anthers, and lies embedded in little alcoves in the white filamentous brush above the stigma. Hence it was clear that the pollen could not get on to the stigma without the aid of insects, which, as far as I have observed in England, never visit this flower. Accordingly, I took a fine bristle to represent the proboscis of a moth, and passed it down between the anthers, near the sides of the corolla; for I found that the pollen sticks to the bristle and is carried down to the viscid stigmatic surface. I took the additional precaution of passing it down first between the anthers of one flower and then of another, so as to give the flowers the advantage of a cross; and I passed it down between several of the anthers in each case. I thus acted on six flowers on two plants growing in pots; the germens of these swelled, and on four out of the six I have now got fine pods, above $1\frac{1}{2}$ inch in length, with the seeds externally visible; whereas the flower stalks of the many other flowers all shanked off. I wish any one who wishes to obtain seed of any other species that does not habitually seed try this simple little experiment and report the result.[2] I shall sow the seeds of my Vinca for the chance of a sport: for a plant which seeds so rarely might be expected to give way to some freak on so unusual and happy an occasion.

Charles Darwin,
Down, Bromley, Kent.

Gardeners' Chronicle and Agricultural Gazette, 15 June 1861, p. 552

[1] The letter was printed in the *Gardeners' Chronicle and Agricultural Gazette*, 15 June 1861, under the title 'Fertilisation of Vincas'. See also *Collected papers* 2: 40.

[2] A response to CD's letter appeared in the *Gardeners' Chronicle and Agricultural Gazette*, 27 July 1861, p. 699. The letter was signed 'C. W. C., *Royal Botanic Gardens, Kew*.' The author was Charles William Crocker, foreman of the propagating department at Kew. In fact, CD had already suggested that Crocker might like to undertake these experiments (letter to Daniel Oliver, 27 May [1861]). The text of Crocker's letter reads as follows:

> Following the suggestion made by Mr. Darwin at page 552, a week or two ago, I thought that I would try if the tropical kinds of Vinca could be induced to produce seed, which is never the case under cultivation if left to themselves. I impregnated eight flowers, and in the course of a few days had the satisfaction of seeing that the pistils in seven cases were swelling well. The erect double follicles are now in several instances more than an inch long; in one they are not yet ripe. The plant upon which I tried the experiment was the white-flowered variety of Vinca rosea. I used the pollen from the same plant as I wished also to see if this variety would reproduce itself by seed, or if it will revert to the normal colour of the species. I merely passed a hair down the tube of one flower after another as an insect might insert its proboscis in its search for nectar.

From B. P. Brent 15 June 1861

Dallington | n^{r}. Robertsbridge | Sussex.
June 15th. 1861

Dear Sir,

Being in conversation the other day with my Father respecting his crossing the Cacti, he said he found them quite fertile and averaged three seeds to a fruit, but the seeds varied very much in the period of germination, he had crossed some several times—[1] the following I took from his dictation as I thought it might interest you

⟨Respe⟩cting the fertility of hybrids:

1 | The Hybrid Cactus, Jenkensoni, reproduced with one parent viz Epiphyllum Speciosum;—

2 | The Hybrid Cereus Cactus, Mallisoni, reproduced, with Epiphyllum Ackermani;

3— | Cactus Jenkensoni, itself a hybrid of Epiphyllum Speciosum, and Cereus Speciosissimus—reproduced with the Hybrid Cereus Mallisoni, which was produced between Cereus Speciosissimus and Cereus Flagelliformis—

or it may be stated thus.

Epiphyllum Speciosum and Cereus Speciosissimus } produced Jenkensoni.
Cereus Speciosissimus and Cereus Flagelliformis } produced Mallisoni
} reproduced a compound hybrid not named—

Unfortunately my Father has no memoranda of his crosses nor did he have any of his productions named. but I think you may rely on the the truth of the abo⟨ve⟩

⟨A⟩ second Sow Guinea pig went ⟨to⟩ boar on the 13th. inst. I have ⟨p⟩ut her in a hutch by the fire and shall keep her as warm as I can, the one that went on May 25th. is out of doors in a north aspect and as cold as I can place her, so I hope to see if temperature has any effect.[2]

Miss Watts in a recent letter wished me to inform you she found the Crested fowls stupid but she puts in a word for her favourite Brahmas— I enclose that part of her letter.[3]

I am | Dear Sir | Your's sincerly | B P Brent

To C Darwin Esqr.

DAR 160.2: 300

CD ANNOTATIONS

0.3 Dear . . . times— 1.4] *crossed red crayon*
4.1 ⟨A⟩ second . . . effect. 4.4] *crossed red crayon*
5.1 Miss . . . Brent 6.1] *crossed ink*

[1] CD was particularly interested in cactus hybrids as they indicated that even species with marked external differences could produce fertile progeny (see *Natural selection*, pp. 412–13). For an earlier source of information on this point, see *Correspondence* vol. 6, letter to J. D. Hooker, 30 September [1857], and letter from Thomas Glover, 26 October 1857.

[2] CD wanted to know how changes in external conditions affected the fertility of organisms and their tendency to vary; aspects of the topic are discussed in several of the chapters of *Variation*. In *Variation*, 2: 161, CD stated that many domesticated animals 'can bear with undiminished fertility the most unnatural conditions'. No mention, however, was made of the results of Brent's breeding experiments.

[3] The enclosure from Elizabeth Watts has not been located. According to entries in the Catalogue of Down specimens (Down House MS), CD had received specimens of fowls from Watts via Brent. She is mentioned in *Variation* 1: 228 n. 2 as the author of the best account of sultans, a Turkish breed of crested fowls. The reference may be to the section on sultans in Watt's revised edition of William Charles Linnaeus Martin's *The poultry yard* (Martin 1855, pp. 78–80). CD was investigating whether the remarkable bony protuberances of various crested breeds of fowl in any way affected the individual's intelligence (see *Variation* 1: 262–6).

To Bingham Sibthorpe Malden 15–16 June [1861][1]

Down. | *Bromley.* | *Kent. S.E.*
June 15th

Dear Sir

I really hardly know how to thank you sufficiently for the Box of Orchids received this morning: they came beautifully fresh.—[2]

I felt an almost boyish delight at the H. viridis, for I thought that I shd. never see it.— By the way it ought not to be ranked with Habenaria.—[3]

Now I have seen & carefully every British orchid, which anyone can hope to see, except the Lizard, which I still hope either from your or Mr Oxenden's kindness to see.[4] And I will soon begin & write my paper.—[5]

I was, also, very glad to see the noble specimens of Aceras;[6] but these had not the little & curious monstrous flowers; & it must have been a strange chance that 5 out of six specimens sent me by Mr Oxenden were thus furnished. These little monstrous flowers illustrate structure of Habenaria. If in your wallks you meet Aceras look for little monstrous flowers[7] The old spec. of O. fusca were also very useful; as they confirmed what I observed in Mr. Oxenden's specimens that insects seldom visit this species & in consequence it seldom seeds.—

I know little of Botany, but it seemed to me that the fine purple orchids, with leaves not spotted & with hollow stems were O. latifolia (which I once saw) & the white ones O. maculata Would you have the kindness to tell me whether this is the case?

I was, also, glad to examine the state of pollen-masses of your Canterbury Fly ophrys.[8] In short, owing to your great kindness, I had a most interesting morning's work. If convenient to you, will you have a look at the pollen-masses of the Bee O. & especially of the var. arachnites to see whether any are removed in *oldish* flowers, or whether all (as here) have fallen on their own stigmas.[9]

With cordial thanks, pray believe me— yours sincerely | C. Darwin

If not thrown away will you kindly return me some time the slip from G. Chronicle.—[10]

June 16th P.S. I have just received & thank you for your note.[11] About the Lizard shall be private.— I thought O. militaris = O. fusca.—

American Philosophical Society (254)

[1] The year is given by the reference to CD's study of orchids (see n. 5, below).

[2] CD thanked Malden in *Orchids* for providing him with specimens of *Orchis fusca*, which is rare in Britain, and of *Peristylus* (or *Habenaria*) *viridis*, the so-called frog orchis (*Orchids*, pp. 43, 78). Malden also supplied CD with living *O. latifolia* and *O. maculata* (*ibid.*, p. 40).

[3] See *Orchids*, pp. 52–3, 76–9. See also letter to A. G. More, 2 June 1861.

[4] George Chichester Oxenden provided CD with a number of specimens of orchids (*Orchids*, p. 31). It was James Veitch of Chelsea, however, who first sent CD a specimen of the lizard orchis, *Catasetum saccatum* (*ibid.*, p. 214 n.).

[5] CD originally intended to publish the results of his study of orchids in the *Transactions of the Linnean Society of London*. It was not until September that he began to think of publishing the work as a book (see letter to John Murray, 21 September [1861]).

[6] Malden apparently sent CD four plants of *Aceras anthropophora* (*Orchids*, p. 40).

[7] This sentence and the preceding one were written in the margins of the paragraph discussing *Aceras* (see 'Manuscript alterations and comments'). CD briefly described the monstrous flowers of *Aceras* and mentioned their resemblance to the structure of *Habenaria chlorantha* in *Orchids*, p. 324 n.

[8] CD was interested in ascertaining whether insects ever visited the plants of *Ophrys muscifera* growing in Britain. He included the plants Malden sent in a tabulation of this study published in *Orchids*, p. 59.

[9] CD found *O. apifera* curious in exhibiting 'widely different means of fertilisation as compared with other species of the genus, and, indeed, as far as I know, with all other Orchids.' (*Orchids*, p. 63). Unlike most orchids, this species is well adapted to self-fertilisation. Of the flowers Malden sent, few had their pollinia removed (*ibid.*, p. 66).

[10] The reference may be to CD's letter printed in the *Gardeners' Chronicle and Agricultural Gazette*, 9 February 1861, p. 122, entitled 'Fertilisation of British orchids by insect agency'. See letter to the *Gardeners' Chronicle*, [before 9 February 1861].

[11] Malden's note has not been found.

To George Bentham 17 June [1861][1]

Down. | Bromley. | Kent. S.E.
June 17th

My dear Sir

You allude to spurless O. pyramidalis. Have you several specs.; and if so could you give me one spike with many well-opened flowers to examine.[2] I have for years been attending to insect fertilisation of Orchids, & I shd. **very much** from several curious reasons like to see what effect no nectary will have produced on the visits of insects. I once tried cutting off the nectaries; but nature's cutting off would be much better.—[3] If, as is very likely, you cannot spare me a spike, could you spare me one or two *hardly opened* flowers, that I may look & see whether gland is modified—[4]

Pray excuse me begging this favour & believe me yours very sincerely
C. Darwin

I have just been rereading your paper on Variation in Linn. Journal (where you allude to this Orchis)[5] & it does make me so wish you would write an essay on the origin of species.[6] It would be so valuable; You have such enlarged & matured views on all such points.—[7] If you send me a spike of O. pyramidalis, I shall have to spoil it, for I shall examine every single flower on it.—

Royal Botanic Gardens, Kew (Bentham letters: 697)

[1] The year is given by the reference to Bentham 1861a (see n. 5, below).

[2] Bentham's letter has not been found. He described an aberrant form of *Orchis pyramidalis* having no nectary in Bentham 1858, p. 501.

[3] CD gave an account of this experiment in *Orchids*, pp. 46–7. The object was to determine whether the flowers lacking a nectary would still be visited by insects, whose agency is necessary for the flower's pollination. His observations indicated that forms without a nectary were not visited as frequently by moths as those with a nectary (*ibid.*, pp. 47–8).

[4] Bentham sent CD some specimens of this species (see letter to George Bentham, 22 June [1861]).

[5] The reference is to Bentham's paper 'On the species and genera of plants', published in the *Natural History Review* (Bentham 1861a). CD's reference to the 'Linn. Journal' was perhaps prompted by a statement following the title of the article stating that the paper was extracted from one read before the Linnean Society of London on 15 November 1858. See also Jackson 1906, pp. 179, 180, and letter to J. D. Hooker, 23 [April 1861].

[6] As yet, Bentham had given no explicit indication of his opinion of *Origin* (see letter to J. D. Hooker, 23 [April 1861]). CD had been curious to know Bentham's views for some time (see *Correspondence* vol. 7, letter to J. D. Hooker, 14 December [1859], and letter from J. D. Hooker, [20 December 1859]).

[7] Bentham had prepared a long essay on species to serve as the introduction to his and Joseph Dalton Hooker's *Genera plantarum* (Bentham and Hooker 1862–83). The manuscript, which was never published, is in the library of the Royal Botanic Gardens, Kew.

To George Gordon 17 June [1861]

Down. | Bromley. | Kent. S.E.
June 17th

Dear Sir

You formerly did me the inestimable service of sending me Goodyera.[1] I have now examined almost every British Orchid, except Corallorhiza. I know it is most unlikely that you can aid me in this; but I have thought that you would forgive me writing for the bare chance. As it will save you trouble, I will assume if I do *not* hear that you cannot aid me.—[2] If you can, will you be so very kind as to let me have a line; as in one fortnight I must go somewhere to the sea-side: & as I shall take my microscope I could send you my address.

Pray forgive the liberty which I have taken in troubling you & with very sincere thanks for your former great kindness. Pray forgive me | Dear Sir | Yours sincerely obliged | Ch. Darwin

Goodyera is growing in a shady peat-bed in my garden & I have hopes it will flower with me!

Postmark: JU 17 | 61
Moray District Record Office (DDW 71/861/4)

[1] See *Correspondence* vol. 8, letter to George Gordon, 17 September [1860]. In *Orchids*, p. 112 n., CD stated: 'Specimens of this rare Highland Orchid were most kindly sent me by the Rev. G. Gordon of Elgin.'

[2] The orchid genus *Corallorhiza* is not described in *Orchids*.

To A. G. More 17 June 1861

Down, Bromley, Kent. S.E.
June 17, 1861.

Dear Sir

I suppose you are at home again, and I sincerely hope that you are nearly well again. I write chiefly to say that since writing to you I have been lucky enough to receive H. viridis.[1] You propose to look at Bee again and very glad I shall be to hear result. But I am convinced this flower will remain an everlasting puzzle to me.[2] I shall, also, be eminently grateful for observations on E. palustris.[3]

You mention Spiranthes: it is no use watching this, for I watched it last autumn at Eastbourne till I was sick:[4] it is, I have no doubt, visited by moths. Have you any entomologist near you? if you could persuade him to collect moths at night when near where the *Spiranthes* grows, he might find pollinia attached to proboscis and this would be a real treasure to me. By the way, I have just thought that I should be very glad if you would repeat one little experiment on Spiranthes and let me quote you as witness, if you succeed.

First open flower just to see position of parts.

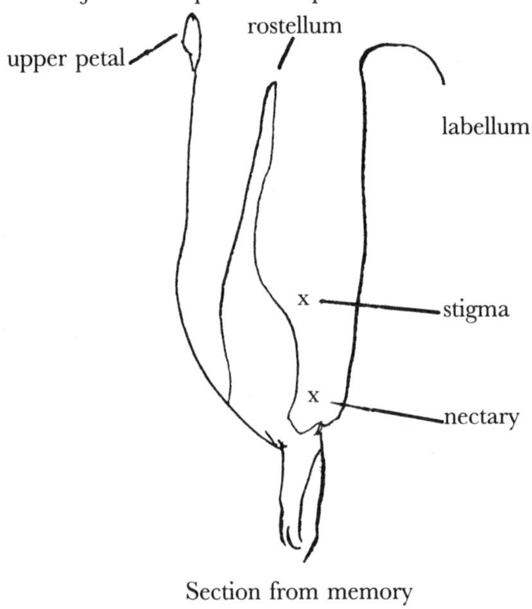

Section from memory

Take thin, *straight flexible* bristle or culm of grass, push it into nectary, to do which, on account of oblique projecting stigma, it must be pushed down near labellum.

Do this *often* and nothing will happen. Then before withdrawing the culm of grass, bow it by pushing down, so that convex surface touches the exact (but hidden) *median* line of rostellum and you will generally find that the pollinia (by *most* curious contrivance) stick to the culm. I was led to try this by reflecting that in act of withdrawing moths bow their probosces.

Lastly insert culm with the pollen-masses adherent, into the nectary of the same or another flower and you will find pollen-grain adherent on projecting stigma. The point on which I want a witness is the necessity of the bristle or culm being bowed so as to strike middle line of rostellum before it is withdrawn, in order that pollinia may be removed. You must be careful not to try flowers from which pollen-masses have been withdrawn by insects: it is best to choose the upper flowers which have been open only one or two days. Will you keep this note and try this little experiment?[5] I shall publish on British Orchids this late autumn.[6]

My dear Sir | Yours sincerely | C. Darwin

Copy
DAR 146

[1] CD had asked More whether he had access to *Habenaria* (or *Peristylus) viridis* (see letter to A. G. More, 2 June [1861]); shortly afterwards, CD received specimens from Bingham Sibthorpe Malden (see letter to B. S. Malden, 15–16 June [1861]).

[2] CD refers to the bee-orchis, *Ophrys apifera*. The 'puzzle' was that this species alone of all orchids appeared to be adapted to favour perpetual self-fertilisation. See *Orchids*, pp. 63–71. More's observations of bee-orchids growing on the Isle of Wight supported CD's own findings (*ibid.*, pp. 67–8).

[3] A description of *Epipactis palustris*, including More's observations on the adaptations of the flower for insect visits, is given in *Orchids*, pp. 95–102. See letters to A. G. More, 19 July 1861 and 13 August 1861.

[4] The Darwins spent nine weeks in Eastbourne in the autumn of 1860 (see *Correspondence* vol. 8, Appendix II).

[5] An account of this experiment is provided in *Orchids*, pp. 125–6, but More is not mentioned.

[6] At this point, CD was planning to write an article on orchids for publication by the Linnean Society of London. After his manuscript became too long to be printed in a journal, he decided to publish it in book form. *Orchids* was published in May 1862 (Freeman 1977, p. 112).

To *Journal of Horticulture* [before 18 June 1861][1]

D of Deal,[2] states, and, apparently, he is corroborated by Mr. Lightbody,[3] that when Auriculas throw up side blooms these keep pretty true to their character;[4] but that when they throw up a heart bloom— that is, from the axis of the plant, the flower, no matter what may be the colour of its edging, "is just as likely to come in any other class as in the one it belongs to." This seems to be an extremely curious observation. It shows that some little light could be thrown on the laws of variation, if the many acute observers who read *The Journal of Horticulture* would contribute their knowledge on such points. I wish "D." would have the kindness to give a few more details, such as out of so many heart blooms so many lost their character, and so many kept true; giving also the proportion in the side blooms which kept true.

As I am appealed to, I will make a few observations on this subject; but I have no doubt others could throw more light on the question. Professor Moquin-Tandon asserts, that with irregular flowers, as Snapdragons, the terminal flower in the axis of the plant is more apt to become regular, or peloric as botanists say, than the other flowers.[5] I once found a Laburnum tree with the terminal flower on each raceme nearly regular, having lost its pea-blossom structure.[6] With many Pelargoniums (I have one at present in my greenhouse, but I know not its name), the central flower in each truss every year comes regular, loses the two dark patches of colour on the two upper petals, and, what is very curious, loses the nectary, which may be seen in all the other flowers cohering to the flower-stalk. In the common Carrot the central floret in the umbel is dark purple and very different from the others; and I find that this central little flower is extremely variable. Are there not other cases of species which habitually have the central flower different from the others? It must, however, be confessed, that Mr. Masters, a high authority on such subjects, disputes that peloric flowers are apt to be central;[7] but it seems to me extremely improbable that the several recorded cases should be due to chance, and all these facts seem to hang together and to indicate that in the flower nearest the axis there is a tendency to differ from the others, or to be variable, or to revert to a hypothetical regular form—that is, as I should look at it, to revert to the structure of a remote ancestor. The curious case of the Auricula apparently falls into this same group of facts.

I hope that some of your correspondents will state whether in the case of single buds sporting, as has so often occurred with Pelargoniums, it has been observed that such sports occur more frequently on one part of the plant than on another. I suppose it is not so, or it would have been noted. Having alluded to the central flower in certain Pelargoniums which have lost the two dark patches of colour and the nectary, I would venture to ask some skilful observer to try whether this flower could be made by artificial fertilisation and by pulling off *some* of the adjoining flowers to yield seed. The stigma should be fertilised with pollen from, if possible, a peloric flower on another plant; and access of other pollen should, of course, be prevented. Peloric flowers have generally been found quite sterile; but Willdenow got seeds from a peloric Snapdragon, and the peculiarity was inherited:[8] hence it is possible, though not probable, that a new strain of quite symmetrically-flowered Pelargoniums might be thus raised. Experiments are tedious and very often fail; but it would be well worth while for any man endowed with plenty of patience to collect seed from the central floret and from the other florets on the same truss of any ordinary Pelargonium, and sow them separately and see which varied most. Of course, all the flowers should be fertilised by the same pollen and subsequently protected from insects.

The same law which causes the heart bloom on an individual Auricula not to keep so true as a side bloom, might cause the seedlings of the central flower of a

Pelargonium or other plant to vary more than the seedlings from the other and exterior flowers. This would be a secret worth discovering and revealing.—
Charles Darwin, Down, Bromley, Kent.

Journal of Horticulture, Cottage Gardener, and Country Gentleman n.s. 1 (1861): 211

1 CD's letter appeared in the 18 June 1861 issue of the *Journal of Horticulture* under the heading: 'Cause of the variation of flowers'.

2 'D., Deal' was a regular contributor to the *Journal of Horticulture*. CD was responding to a query that appeared in 'D's' article entitled 'The Auricula bloom of 1861', *Journal of Horticulture*, 4 June 1861, pp. 174–6, in which the following query appeared:

> How comes it to pass, that if an Auricula throws up a side bloom it is pretty sure to be in character; but that if it be from the heart of the flower ... it is just as likely to come in any other class as the one it belongs to? ... Can Mr. Darwin, Mr. Beaton, or anybody enlighten me on the first of these points?

Donald Beaton wrote a regular column for the *Journal of Horticulture*. CD and Beaton had been carrying on a public correspondence in the pages of the journal that touched upon a range of topics, including crosses, sports, and varieties. See letters to the *Journal of Horticulture*, [before 14 May 1861], and to J. D. Hooker, 14 May [1861].

3 In his contribution to the *Journal of Horticulture* (see n. 2, above) 'D., Deal' referred to a communication he had had with a Mr Lightbody of Falkirk. George Lightbody was a grower of auriculas who raised numerous varieties (R. Desmond 1977).

4 *Primula auricula*, commonly known as bear's ear, was a great favourite with flower fanciers owing to the large number of trusses it produced. The summer of 1861 was particularly noted for the fine bloom of auriculas.

5 Moquin-Tandon 1841, p. 192. CD's annotated copy of the volume is in the Darwin Library–CUL.

6 CD had mentioned this case of pelorism to Joseph Dalton Hooker many years earlier (see *Correspondence* vol. 4, letter to J. D. Hooker, [12 June 1847]; and vol. 6, letter to J. D. Hooker, 14 July [1857]).

7 The letter from Maxwell Tylden Masters discussing this point has not been found, but see the letter to M. T. Masters, 13 April [1860] (*Correspondence* vol. 8).

8 Karl Ludwig Willdenow. The reference has not been traced.

To J. D. Hooker 19 June [1861][1]

Down. | Bromley. | Kent. S.E.
June 19th.

My dear Hooker

I send by this post Journal of Hort. on account of strange little fact about position of flower causing variability, with my comment on it, which Editor has placed foolishly conspicuous.—[2] Please return the Paper.—

I suppose you have hardly yet got up your arrears of work; but in a week or two screw out time & let me hear a little news about yourself & your doing & about poor dear Henslow's family.—[3] I hope & suppose that Jenyns keeps his intention of Biograph. notice.—[4] The notices in G. Chronicle on Henslow have been, I think, very well done & bring out capitally his noble character & doings.—[5] By the way how frightfully dull the old Chronicle has become: I really think I must give it up.— Your suggestion of London Review has answered *admirably*: several of the scientific articles have been capital: & I gloat over not having that extra stupid Athenæum to read.—[6] I have been indexing for self Linn. Journal & I foresee great trouble in the complex arrangement of Bot. & Zoology & in those Supplements. Why on earth cannot they be worked into regular series?— What confounded wrong references will arise![7] I am glad to see that Bentham is new President.—[8]

Etty has been going on very well, though just at present having a drawback.— Some gain of strength has *certainly* been coincident with rubbing in the oil, which my wife has steadily continued since February.[9]

In about a fortnight we go to sea, to Devonshire, & when there I will tell you my address. We sh^d. have started more than a week ago; but have been delayed owing to troublesome business, viz offer of partnership in a small country Bank for William, which I now think will be arranged & will be a small, but pretty safe provision for him.— It was started by John Lubbock.[10] I have leaned for him to accept partly for leisure for other occupations.— He is getting very fond of Botany & of dissecting plants;[11] but I am dreadfully puzzled what to recommend him to do or what Books to get.— I hope that your & Bentham's book will give characters of Families & **Synopsis of Families**.—[12] Can you give me any suggestions for Willy? I have told him to get Lindleys Vegetable Kingdom & dissect plants of as many orders as he can get.[13]

I have, owing to many interruptions, not been going on much with my regular work (though I have done the very heavy jobs of variation of Pigeons, Fowls, Ducks, Rabbits Dogs &c) but have been amusing myself with miscellaneous work.—[14] I have been very lucky & have now examined almost every British Orchid fresh, & when at sea-side shall draw up rather long paper on the means of their fertilisation for Linn. Soc^y & I cannot fancy anything more perfect than the many curious contrivances.—[15] I shall never rest till I see a Catasetum eject pollen-masses,[16] & a Mormodes twist its column.[17] Can you tell me what would be best nursery to apply to for a plant just coming into flower; & secondly in summer do you think they would travel in safe state for me to examine?— Please answer this, if you can. For I am got intensely interested on subject & think I understand pretty well all the British species.—[18]

Farewell my dear old friend | Your affect | C. Darwin

I daresay you never had time to glance at flies caught by Saxifrages:[19] I must some time get a set of species to examine; for I find that the hairs of S. rotundifolia have much prompter power of absorption of C. of Ammonia, than those of S. umbrosa.—

Endorsement: '/1861'
DAR 115.2: 103

[1] The endorsement is confirmed by the reference to the death of John Stevens Henslow on 16 May 1861.

[2] See letter to *Journal of Horticulture*, [before 18 June 1861]. CD's letter appeared as the first item on the front page of the issue of 18 June 1861. The editors of the journal were George William Johnson and Robert Hogg.

[3] Hooker and his wife, Frances Harriet Hooker, stayed at the home of Frances's father, J. S. Henslow, during his final illness.

[4] Leonard Jenyns published a memoir of Henslow, his brother-in-law, in 1862, to which CD contributed a short piece (Jenyns ed. 1862, pp. 51–5; *Collected papers* 2: 72–4). See letters to J. D. Hooker, 24–5 May [1861] and 30 May [1861], and Appendix X.

[5] A three-part obituary of Henslow appeared in the *Gardeners' Chronicle and Agricultural Gazette*, 1 June, 8 June, and 15 June 1861.

[6] Hooker had suggested that CD might like to subscribe to the *London Review and Weekly Journal of Politics, Art and Society* (see letter to J. D. Hooker, 27 [March 1861] and n. 6).

[7] The *Journal of the Proceedings of the Linnean Society of London* began publication in 1856. Each year separate numbers for botanical and zoological papers were issued, the two sections being independently paginated; occasionally supplementary numbers were also published that were intended to form supplementary volumes. See Gage 1938, pp. 120–1. CD's copies of most of the numbers from volumes 1 to 8 (1856–64) are in the Darwin Library–CUL. His index to the journal is in DAR 75: 38–46.

[8] George Bentham was elected president of the Linnean Society at the anniversary meeting of 24 May 1861 (Jackson 1906, p. 188).

[9] Henrietta Emma Darwin had been ill for over a year. To help her recuperate, Hooker had recommended that she be rubbed with cod-liver oil (see letter to J. D. Hooker, 8 [February 1861]).

[10] John Lubbock was acting as intermediary in negotiations between CD and a Southampton banker, George Atherley, to arrange a partnership for William Erasmus Darwin in the Southampton and Hampshire Bank. See letters to W. E. Darwin, [25 May 1861] and 6 [June 1861].

[11] William Darwin had developed an interest in botany while he was being tutored before entering Cambridge University (see *Correspondence* vol. 7, letter to W. E. Darwin, [3 May 1858]).

[12] CD refers to the *Genera plantarum* that Hooker and Bentham were preparing (Bentham and Hooker 1862–83). The work contains, in addition to a description of the diagnostic characters distinguishing each genus, a conspectus of the striking characters of each natural order (see Green 1914, pp. 500–2).

[13] The third edition of John Lindley's *Vegetable kingdom* was published in 1853.

[14] CD refers to his preparation of chapters for inclusion in *Variation*.

[15] CD recorded in his journal for 1861: 'During stay at Torquay did paper on Orchids.' The Darwins visited Torquay from 1 July to 27 August 1861. See 'Journal' (Appendix II).

[16] CD mentioned his not yet having examined the lizard orchis, *Catasetum saccatum*, in the letter to B. S. Malden, 15–16 June [1861]. Not until October was he able to obtain specimens of *Catasetum* and *Mormodes* and observe their peculiar adaptations for promoting cross-pollination (see letters to J. D. Hooker, 11 October [1861] and 13 October [1861]).

[17] Like *Catasetum*, the species of the orchid genus *Mormodes* possess a mechanism for ejecting the pollinia. As CD described in *Orchids*, the flower parts of *Mormodes* act in a coordinated way such that when the column or pedicel to which the pollinium is attached is touched, the piece bends entirely backwards releasing the pollinium like a pendulum, and 'the whole body is projected perpendicularly up in the air' (*Orchids*, p. 257).

[18] Hooker wrote in the margin: 'Parker & Williams | Holloway'. The reference is to the nursery of Robert Parker and Benjamin Samuel Williams on Seven Sisters Road in Holloway. Williams was a specialist on orchids (R. Desmond 1977).

[19] See letter to J. D. Hooker, 30 May [1861].

To Frederick Smith 19 June [1861][1]

Down. | Bromley. | Kent. S.E.
June 19th

My dear Sir

The pollen-masses attached to the Bombus hortorum positively belong to no

British orchid & I believe I may say as positively that they belong to an exotic orchid of the group of Epidendreæ.—[2] It would be worth knowing whether this Bee was caught very far from a hot-house.— I fear you have not one of the sand-wasps with pollen-masses attached alluded to; else I sh[d]. so much like to see it. It would be very easy to ascertain whether the leaf on the S. American Wasp was pollen.— I suppose there is no chance of any of the insects in your own collection having pollen-masses attached.—

Will you kindly remember & ask M[r] Walker.[3]

With many thanks for your interesting note, | pray believe me | Yours very sincerely | C. Darwin

American Philosophical Society (255)
Endorsement: 'Wednesday'

[1] Dated by the reference to CD's study of orchids, which was published in May 1862. The year is confirmed by Smith's endorsement 'Wednesday': in 1861, 19 June fell on a Wednesday.

[2] A passage in *Orchids*, p. 164, mentions the case discussed in the letter:

> Sir W. C. Trevelyan sent to Mr. Smith of the British Museum (who forwarded to me) a Bombus hortorum—caught in his hot-house, where a Cattleya was in flower—with its whole back, between the wings, smeared with dry viscid matter, and with the four pollinia attached to it by their caudicles, and ready to be caught by the stigma of another flower if the bee had entered one.

CD refers to Walter Calverley Trevelyan.

[3] Smith and the entomologist Francis Walker were both members of the Entomological Society of London (*Transactions of the Entomological Society of London* n.s. 5 (1858–61): xx).

To George Bentham 22 June [1861][1]

Down. | Bromley. | Kent. S.E.
June 22[d]

My dear Bentham

Very many thanks for the specimen, which sufficed for one purpose & would hardly do, if I had a whole spike, for my other purpose.— I will write to Tenby.[2] I daresay the "Supplements" may be unavoidable, but I cannot help thinking they will be a fertile source of wrong references, from which I often suffer in an outrageous manner.[3]

By the way I may mention one great difficulty which I have on my subject of Orchids; viz that several species, which are habitually visited by insects (moths almost certainly) never secrete within their nectaries or on the Labellum an atom of honey. I have now daily watched O. morio for 23 days and tried all sorts of devices. C. C. Sprengel and Kurr make similar statements about this and other species.[4] Orchis Conopsea & some other species secrete abundantly; and this makes me so much wish to see spec. of O. pyramidalis without nectarys. C. C. Sprengel & Kurr believe that these orchids exist by an organised system of deception & that moths go on visiting them in hopes of finding nectar;[5] I cannot swallow the belief of such roguery.[6]

Do you know whether nectar is ever *secreted* & *reabsorbed* promptly?— I am utterly puzzled: I have watched the nectaries till $11\frac{1}{2}$ at night & all as dry as a bone.—

Here is a piece of presumption! I must think that you are mistaken in ranking Hab. chlorantha as a var. of H. bifolia:[7] the pollen-masses & stigma differ more than in most of the best species of Orchis. When I first examined them, I remember telling Hooker that moths would, I felt sure, fertilise them in a different manner; & I have just had proof of this in moth sent me with the pollinia (which can be easily recognised) of H. chlorantha attached to its proboscis, instead of to the sides of its face, as in Hab. bifolia.—[8]

Forgive me scribbling this way; but when a man gets on his hobby-horse, he always is run away with.— Anyhow nothing here requires any answer; unless, indeed, you chance to have any light about the nectar.

Yours very sincerely | Chas. Darwin

P.S. | I have just received owing to your or Hooker's kindness a copy of your Elementary Principles of Botany, which pleases him much & for which I thank either you or Hooker.—[9]

Royal Botanic Gardens, Kew (Bentham letters: 692, 690)

[1] Dated by the relationship to the letter to George Bentham, 17 June [1861].

[2] CD had asked Bentham to send him specimens of *Orchis pyramidalis* (see letter to George Bentham, 17 June [1861]). Bentham apparently told CD how to obtain further specimens from Tenby, South Wales (see letter to Charles Lyell, 6 August [1861]).

[3] CD may have made the same complaint about the organisation of the *Journal of the Proceedings of the Linnean Society* to Bentham as he had to Joseph Dalton Hooker, (see letter to J. D. Hooker, 19 June [1861]). Bentham had recently been elected president of the Linnean Society.

[4] Sprengel 1793, pp. 403, 404, and Kurr 1833, p. 28. There are annotated copies of both works in the Darwin Library–CUL.

[5] Christian Konrad Sprengel discussed this point in Sprengel 1793, pp. 403–4, using the term 'Scheinsaftblumen' to describe these flowers. Johann Gottlob Kurr also used Sprengel's term (Kurr 1833, p. 28). In *Orchids*, p. 45, CD translated the word as 'sham-nectar-producers'.

[6] According to CD's views, it would be an anomaly if a few species among the orchids, which generally relied upon insect visitation to facilitate fertilisation, did not secrete any nectar to attract potential insect visitors. He believed that insects would soon learn that their efforts would not be repaid. See *Orchids*, pp. 44–53.

[7] Bentham 1858, p. 512.

[8] Frederick Bond had sent CD these specimens (see *Orchids*, p. 35). See the description of *Habenaria bifolia* in *Orchids*, pp. 88–91, where this point is discussed in detail.

[9] Bentham 1861b.

To George Gordon 22 June [1861][1]

[Down][2]
June 22[d]

Dear Sir

I write one line to thank you most sincerely for your kind note & efforts to assist me.[3] Corallorhiza would certainly be most valuable to me. I sh[d]. especially wish for a few flowers in large bud.—[4]

I shall remain here for another week or 10 days. Immediately that I reach Devonshire, I will send my address.— Unfortunately on account of business I cannot fix my day of departure;[5] nor my address at sea-side till we get there; but it will be Torquay.—

With very sincere thanks | Pray believe me | Dear Sir | Yours truly obliged | Ch. Darwin

I think my observations on the fertilisation of Orchids are new & interesting.

Whenever I publish I will do myself the pleasure of sending you a copy of my paper.—

P.S We have just been able to fix to start on July 1st. but our journey to Torquay on account of my poor daughter will take three days[6]

Moray District Record Office (DDW 71/861/5)

1 Dated by the reference to the Darwins' planned holiday in Torquay. They stayed there from 1 July to 27 August 1861 (see 'Journal'; Appendix II).

2 The address has been excised from the letter.

3 Gordon's response to CD's letter of 17 June [1861] has not been found. CD acknowledged his assistance in *Orchids*, p. 112.

4 The orchid genus *Corallorhiza* is not described in *Orchids*.

5 CD was engaged in negotiations to secure William Erasmus Darwin a partnership in the Southampton and Hampshire Bank, Southampton.

6 According to Emma Darwin's diary, the Darwins spent the night of 1 July 1861 in Reading en route to Torquay.

To J. D. Hooker 22 June [1861]

Down. | Bromley. | Kent. S.E.
June 22d

My dear Hooker

Your note told me a wonderful lot of news, & gives excellent advice about William;[1] but upon my life I am sorry I wrote to you, for how overworked & out of spirits you must be. I never in my life I heard so many calamaties in one letter.— I do not know what "eczema" is; but I hope it is nothing very serious from which your Father is suffering.—[2] Poor Harvey & so recently married![3] And poor Sinclair, whom I saw at your House.—[4]

If you think of it, pray tell Oliver how sorry I am he is unwell. He must work too hard: I thought so before from his mass of knowledge.—[5] And then Mrs. Huxley, of whom I had not heard for a long time.—[6] You give a poor account of Mrs. Hooker.[7] My God what a world it is.

Farewell, Your affect. | C. Darwin

Give my kind remembrances to Thomson: I did not know that he was coming home.—[8] There are lots of things I should like to write about, but won't bother you.—

Pray do not write a word more about Willy: you have said **plenty** to keep him going.—

P.S. I have just received your *second* note. I thank you heartily, but it is too bad that you sh$^{d.}$ waste your time.— If you remember it return sometime Journ of Horticulture[9]

Endorsement: '/61.'
DAR 115.2: 84

[1] Hooker's letter has not been found. The advice probably pertained to William Erasmus Darwin's interest in botany. See letter to J. D. Hooker, 19 June [1861].

[2] William Jackson Hooker was severely troubled by eczema in the last years of his life and was particularly ill in the summer of 1861 (Allan 1967, pp. 207, 208).

[3] William Henry Harvey, professor of botany at Trinity College, Dublin, had married in April 1861. According to his biographer, he soon afterwards suffered a misfortune: 'While lecturing at Glasnevin, the first symptoms of hæmorrhage from the lungs appeared, and were succeeded by a severe attack, from the effects of which he never entirely recovered.' (Anon. 1869, p. 340).

[4] Andrew Sinclair, formerly colonial secretary in New Zealand, had drowned in a river in New Zealand (*DNB*). CD had met him in 1858 when Sinclair was visiting England (see *Correspondence* vol. 7, letter from J. D. Hooker, 15 January 1858).

[5] Daniel Oliver was Hooker's assistant in the herbarium of the Royal Botanic Gardens, Kew. CD had solicited Oliver's advice and assistance many times over the past year, first through Hooker and then directly (see *Correspondence* vol. 8, and this volume, letters to Daniel Oliver, 26 February [1861] and 1 May [1861]). Oliver had recently been appointed professor of botany at University College London.

[6] Henrietta Anne Huxley was recuperating slowly following the birth of Leonard Huxley in December 1860 and the death of the Huxleys' first-born child, Noel, in September 1860. Henrietta had visited Down earlier in the year (see letter to T. H. Huxley, 22 February [1861]).

[7] Frances Harriet Hooker was grieving over the death of her father, John Stevens Henslow, who died in May 1861 (see Allan 1967, p. 208).

[8] Thomas Thomson, a close friend of Hooker's, was professor of botany and superintendent of the Calcutta botanic garden. He had travelled with Hooker during his visit to India in the late 1840s and was co-author with Hooker of a volume on the flora of India (Hooker and Thomson 1855). Because of ill health, Thomson returned to Britain in 1861 (*DNB*).

[9] CD sent Hooker a copy of the 18 June 1861 issue of the *Journal of Horticulture*, which contained CD's letter describing a number of observations concerning laws of variation. See letters to *Journal of Horticulture*, [before 18 June 1861], and to J. D. Hooker, 19 June [1861].

From Hugh Falconer 23 June 1861

31 Sackville S$^{t.}$ W
23 June. 1861

My Dear Darwin

I have been to Adelsberg Cave[1]—and brought back with me a live *Proteus anguinus*, designed for you from the moment I got it—i.e. if you have got an aquarium, and would care to have it.[2] I only returned last night from the Continent and hearing from your brother, that you are about to go to Torquay, I lose no time in making you the offer.[3] The poor dear animal is still alive—although it has had no appreciable means of sustenance for a month—and I am most anxious to get rid of the responsibility of starving it longer. In your hands it will thrive—and have a fair chance of being developed without delay, into some type of the Columbidæ—say a Pouter or a Tumbler.

My Dear Darwin, I have been rambling through the north of Italy, and Germany lately. Every where have I heard your views, and your admirable essay canvassed—the views of course often dissented from, according to the special bias of the speaker—but the work—its honesty of purpose—grandeur of conception—felicity of illustration—and courageous exposition—always referred to in terms of the highest admiration. And among your warmest friends, no one rejoiced more heartily in the just appreciation of Charles Darwin than did,

Yours very truly | H Falconer

DAR 99: 3–4

[1] This famous stalactite cave, the largest in Europe, is located near the town now known as Postojna, Slovenia.

[2] The blind, colourless, perennibranchiate amphibian *Proteus anguinus* was first discovered in the Adelsberg cave (*EB*). CD had mentioned *Proteus* in *Origin* as one of several cave animals anomalous in not being related to non-cave-dwelling forms of the same continent and described such animals as 'wrecks of ancient life' (*Origin*, p. 139). He had been challenged in this interpretation by Andrew Murray (see *Correspondence* vol. 8, letters to Charles Lyell, 10 January [1860], and to Andrew Murray, 5 May [1860]).

[3] The Darwins left for Torquay on 1 July 1861 (see 'Journal'; Appendix II). CD's brother, Erasmus Alvey Darwin, lived in London.

To Henry Bence Jones 23 June [1861][1]

Down. | Bromley. | Kent. S.E.
June 23^d^

My dear D^r^. Bence Jones

I am not an M.D. Should the University of Breslau think fit to honour me, I shall feel much gratified—[2]

With my thanks for your note, pray believe me | Yours very sincerely | Ch. Darwin

Walter R. Benjamin Autographs (dealers)

[1] The year is provided by the reference to the University of Breslau's honouring CD (see n. 2, below).

[2] The degree of Honorary Doctor of Medicine and Surgery was awarded to CD by the University of Breslau in August 1861 (see letter from the University of Breslau, 4 August 1861). He received the diploma, transmitted by the secretary of the Prussian legation, in 1862 (see *Correspondence* vol. 10, letters from H. Albertof, 10 February 1862 and 13 February 1862). The diploma is in DAR 229.

To Hugh Falconer [24 June 1861][1]

Down

My dear Falconer.—

I have just received your note, and by good luck a day earlier than properly, and I lose not a moment in answering you, and thanking you heartily for your offer of the valuable specimen; but I have no aquarium and shall soon start for Torquay, so that it would be a thousand pities that I should have it.[2] Yet I should certainly much like to see it, but I fear it is impossible. Would not the Zoological Society be the best place? and then the interest which many would take in this extraordinary animal would repay you for your trouble.

Kind as you have been in taking this trouble and offering me this specimen, to tell the truth I value your note more than the specimen. I shall keep your note amongst a very few precious letters. Your kindness has quite touched me.

Yours affectionately and gratefully, | Ch. Darwin.

LL 2: 375

[1] The date as given in *LL* 2: 375.

[2] Falconer had offered CD a live *Proteus anguinus*, a blind cave salamander (see letter from Hugh Falconer, 23 June 1861).

To John Lubbock 24 June [1861][1]

Down. | *Bromley.* | *Kent. S.E.*
June 24th.

My dear Lubbock

I should have written before had I had any definite news to tell you about the Southampton affair.[2] There have been delays of all sorts; but I hope & think all is settled. William is gone down today to Southampton with an accountant & as I think we understand each other at last, I fully believe & hope all will be in a few days definitely settled.[3] Mr. A. demurs about guaranteeing the £500 per annum, & has certainly rather changed his mind;[4] but as my solicitor thought that this was too much to expect beyond the three or four first years, of course I shall give up.[5] This is the one point alone not settled. William has become quite anxious for success; & I fully expect will work hard & make a good man of business & *do you credit*.— We shall always feel grateful to you for your great kindness to us in every possible way about this affair.— I think there is every prospect of its turning out a capital affair for William.—

We hope to start next Monday for Torquay.[6] I wish I had any news to tell you, but I have literally not seen a soul; nor had any interesting letters. I heard indeed from Hooker chiefly about Botanists & Bot. matter.— He gives a very poor account of Mrs. Huxley,[7] as he saw H. at Club, where 25 dined.—[8] I heard also from Falconer who has brought (for me, but I have refused) a live blind Proteus from caves of Adelsburgh.[9] He says he met many interested about Natural Selection & origin of species.— I have been very desultory in my work of late & have been chiefly dissecting orchids & shall draw up my paper whilst at sea-side.—[10]

I sincerely hope that you are enjoying yourself; anyhow you have splendid weather.[11] If you see Forchammer give him my kind remembrances, & pray present my respects to Steenstrup & say that I have not forgotten his extreme kindness in sending me cirripeds.[12]

Farewell give my kind remembrances to Busk.[13] This is dreadfully dull note. Farewell | My dear Lubbock | Yours most truly | C. Darwin

I hear very good accounts of your Brother.—[14] Grand doings, I heard, at the wedding.—[15]

Down House

1 Dated by the reference to the Darwins' holiday in Torquay (see n. 5, below).

2 The 'Southampton affair' concerned the arrangements to secure a partnership in a Southampton bank for William Erasmus Darwin. See letters to W. E. Darwin, [25 May 1861], [26 May 1861], and 1 [June 1861], and to John Lubbock, [25 May 1861] and 1 [June 1861].

3 At an early stage in the negotiations, Lubbock suggested that before William signed the articles of partnership, the bank's account books 'would have to be carefully examined by Lawyer & accountant' (see letter to W. E. Darwin, [26 May 1861]).

4 George Atherley was a partner in the Southampton and Hampshire Bank.

5 CD's solicitor was William Mackmurdo Hacon of the London firm Rowland & Hacon.

6 The Darwins left for an eight-week holiday in Torquay on 1 July 1861 (see 'Journal'; Appendix II).

7 See letter to J. D. Hooker, 22 June [1861], n. 6.

8 The Philosophical Club of the Royal Society had met on 20 June 1861. CD, Joseph Dalton Hooker, and Thomas Henry Huxley were all members; Lubbock had been elected to the club in November 1860. The June meeting was the last time that members dined at the Thatched House Tavern on St James' Street, which closed later in 1861. See Bonney 1919, pp. 54, 55.

9 See letter from Hugh Falconer, 23 June 1861.

10 CD began writing his paper on orchids while in Torquay (see 'Journal'; Appendix II). Although he originally intended to publish the material in the journal of the Linnean Society, CD in fact issued it as a book in 1862 (*Orchids*).

11 Lubbock was in Denmark, examining some of the sites of ancient human habitation known as 'Kjökkenmöddings' (kitchen-middens). He contributed an account of his findings to the August issue of the *Natural History Review* (Lubbock 1861c).

12 Lubbock visited both Johan Georg Forchhammer, professor of geology at the University of Copenhagen, and Johannes Japetus Smith Steenstrup, professor of zoology, during his visit to Denmark. The two were members, along with the archaeologist Jens Jacob Worsaae, of a committee formed to study the geology, biology, and archaeology of the Danish shell-heaps and grave mounds. Forchhammer and Steenstrup had assisted CD with his study of fossil and living cirripedes (see *Correspondence* vols. 4, 5, and 7, Supplement).

13 George Busk accompanied Lubbock to Denmark to study the ancient crania found in the various mounds (see Lubbock 1861c, p. 494).

14 Montagu Lubbock, John Lubbock's nineteen-year-old brother, was convalescing after suffering a serious injury in a carriage accident (Hutchinson 1914, 1: 178).

15 Nevile Lubbock, aged twenty-two, married Harriet Charlotte Wood on 18 June 1861 (*Gentleman's Magazine*, July 1861, p. 82).

From Robert Colgate[1] 25 June 1861

Hempsted House | Eastbourne
June 25th. 1861

My dear Sir

I wrote as I promised you to my friend Mr. Boys in New Zealand and have just received a reply dated April 4th.[2] I fear the little he has said will not be sufficient to be useful to you.[3] He says:—"I cannot answer your numerous queries with sufficient accuracy for a naturalist—that is to say with the positive assertions from actual observation that would be required—but I can tell you that bees have only reached as far up the country as Rangiora[4] within the last twelve months, and that *white* clover did seed before bees were introduced, but *red* clover to the best of my belief did not— At all events the latter has become much more abundant during this last year. There are but very few insects indiginous to New Zealand that appear to supply the place of bees or perform their functions. I have never been able to succeed with French beans and scarlet runners during the many years I have been out here until this year, when I have had them in abundance which is I know attributable to the bees, having watched them. I have also noticed that bees prefer white clover to red; In the paddock in front of my house white and red clover being pretty evenly mixed, you would see a bee upon most of the white clover flowers and not one upon the red."[5]

The rest of his letter is private gossip. If this scanty information is of any interest to you I shall be very glad.

I have been very much interested in your book on natural selection which you were so kind as to send me.[6] I think I ought to have immediately thanked you on the receipt of it but I was expecting this New Zealand letter long before this. Will you think it presumption in me if I say that I think you have a slight error at p. 77. about field-mice—being destroyed by cats—[7] I believe that cats never seek *field*-mice and if they come across one and kill they will not eat it. Owls and hawks are the great enemies of field-mice.

I wonder if we shall have the pleasure of seeing you at Eastbourne again this summer—[8] I hope Miss Darwin is better and that M^rs. Darwin is well

I remain dear Sir | Yours faithfully | Robt. Colgate

DAR 76 (ser. 2): 171–2

CD ANNOTATIONS
1.1 Boys] *underl blue crayon*
1.7 but *red* . . . not— 1.8] *double scored blue crayon*
1.16 and not one upon the red."] *double scored blue crayon*
2.1 The rest . . . this. 3.3] *crossed blue crayon*

1 Colgate was a surgeon in Eastbourne. CD apparently made his acquaintance during the seven weeks the Darwins spent there in the autumn of 1860.

2 The letter was from John Cowell Boys.

3 CD apparently mentioned to Colgate his interest in learning whether the introduction of European hive-bees had affected the pollination of clover and other leguminous flowers in New Zealand. CD had already received information from a nurseryman in Christchurch, William Swale, on this subject. See *Correspondence* vol. 7, letter from William Swale, 13 July 1858, and letter to the *Gardeners' Chronicle*, [before 13 November 1858].

4 A borough near Christchurch, South Island, New Zealand.

5 In *Origin*, pp. 94–5, CD stated that only humble-bees visited red clover, for hive-bees could not reach the nectar.

[6] Colgate's name is included in the presentation list CD drew up for the third edition of *Origin* (see Appendix VII).

[7] Colgate refers to CD's statement in *Origin* that the number of humble-bees in a district was determined by the population of fieldmice, which destroy their comb and nests, and that:

> the number of mice is largely dependent, as everyone knows, on the number of cats ... Hence it is quite credible that the presence of a feline animal in large numbers in a district might determine, through the intervention first of mice and then of bees, the frequency of flowers in that district!

(*Origin*, p. 74; *ibid.*, 3d ed., pp. 76–7). CD did not modify this statement in subsequent editions of *Origin* (see Peckham ed. 1959, p. 157).

[8] The Darwins had taken a house at 15 Marine Parade, Eastbourne, from 22 September to 10 November 1860. See *Correspondence* vol. 8, Appendix II.

To George Gordon 4 July [1861][1]

2. Hesketh Crescent | Torquay
July 4th

My dear Sir

I must thank you for your extraordinary kindness to me about Corallorhiza.[2] Mr Watson forwarded to me your note to him[3] & he suggested Prof. Balfour to whom I have written.[4] I fear that there is no chance of my succeeding; but I am sure I am not the less obliged to you for your great kindness.—

My dear Sir | Yours truly obliged | Charles Darwin

Moray District Record Office (DDW 71/861/8)

[1] The year is given by reference to the Darwins' stay in Torquay.

[2] Gordon had offered to assist CD in acquiring specimens of the orchid *Corallorhiza*. See letter to George Gordon, 22 June [1861].

[3] The letter from Hewett Cottrell Watson has not been found. Gordon may have written to him after receiving CD's letter of 17 June [1861].

[4] John Hutton Balfour was professor of botany at Edinburgh University and director of the Royal Botanic Garden.

To George Dickie [5 July 1861][1]

2. Hesketh Crescent | Torquay
Saturday

Dear Sir

I thank you most sincerely for the Listera, which I have been *very glad* to examine.[2] But most unfortunately from being packed in wood Box, instead of tin, the moss was almost dry, & even the buds partly withered, so that the rostellum would not show any signs of the curious vital action peculiar to this genus.— This is the more unfortunate

Incomplete
Liverpool Public Library

[1] The dating is based upon the reference in the following letter to CD's having received Dickie's *Listera* specimens 'yesterday'. Although 6 July 1861 was a Saturday, CD clearly states in the following letter that he also wrote to Dickie 'yesterday'.

[2] Dickie, professor of botany at Aberdeen University, sent CD specimens of *Listera cordata* (see *Orchids*, p. 152). He had previously supplied CD with information about another plant that grew near Aberdeen (see *Correspondence* vol. 6, letter from George Dickie, 1 December 1856).

To George Gordon 6 July [1861][1]

2. Hesketh Crescent Torquay
July 6th

My dear Sir

I thought I should never have trespassed on your kindness again.— But pray listen to my case, pity me & help me if you can.— Mr Jamieson promised to send me Listera cordata, but writes that he cannot this year find it.[2] I then wrote to Prof. Dickie[3] & I received yesterday some specimens sent in a *wood* little Box, consequently dry & crushed from being packed in compact moss & almost useless to me, but just sufficient to show me a certainly new point of structure. I wrote again to Prof. Dickie yesterday to beg him to send me more,[4] but it is hopeless as he said after long search he could find only one little specimen *with large buds* (which are necessary for me).—

In the dead of the night it occurred to me that it might not be quite out of flower with you (if it grows near you) & I remembered your inexhaustible good nature, & the result is this note

Incomplete
Moray District Record Office (DDW/71/861/26)

[1] The year is given by reference to the Darwins' stay in Torquay.

[2] See letter from T. F. Jamieson, 13 June 1861.

[3] The letter has not been found.

[4] See preceding letter.

To J. D. Hooker [6 July 1861][1]

2. Hesketh Crescent | Torquay
Saturday

My dear Hooker.

Your note offering Catasetum followed me here:[2] hearty thanks, but I have written to Veitch,[3] (whose name you mentioned with Parker & Williams[4]) for 4 kinds; if he fails, I will let you know as I wish intensely to see Catasetum. I also much wish to see one of the Arethuseæ.[5] I hope all your sick friends & Father are going on well.[6] We slept at Reading & got here on Tuesday night & are settled in capital house with pretty view. Etty stood the Journey well; but about Exeter we were all in a frightful & laughable state of prostration & Port Wine alone saved our lives.

I hope your work will soon be lighter: but till it is I will not bother you with writing.— Thanks for Journal of Hort. returned.[7] I find this journal has **far** more contributors who will observe & report than Gardeners Chronicle. I get capital information in almost every number.—[8]

Yours affect | C. Darwin

Endorsement: 'July 61.'
DAR 115.2: 104

[1] The date is established by the endorsement and by the reference to the family's having just arrived in Torquay. The Darwins left for Torquay on Monday, 1 July 1861 (see 'Journal'; Appendix II).
[2] Hooker's letter has not been found, but see the letter to J. D. Hooker, 19 June [1861].
[3] CD received specimens of several varieties of orchid from Veitch & Son, Royal Exotic Nurseries, Chelsea. CD especially acknowledged the assistance and generosity of James Veitch Jr in *Orchids*, pp. 158 n., 214 n. Veitch & Son were among the leading growers of orchids in England (see Stewart ed. 1992).
[4] Robert Parker and Benjamin Samuel Williams were proprietors of a nursery in Seven Sisters Road, Holloway, London. CD thanked Parker for sending him an 'extremely valuable series of forms' in *Orchids*, p. 158 n. Along with Veitch & Son, this nursery was noted for developing successful techniques for the cultivation of orchids (see Stewart ed. 1992).
[5] CD subsequently wrote to the *Gardeners' Chronicle* requesting orchid specimens from the Arethuseae tribe (see letter to the *Gardeners' Chronicle*, [before 14 September 1861]). Hooker eventually sent CD some specimens early in 1862 (see *Correspondence* vol. 10, letter from J. D. Hooker, [18 January 1862]).
[6] See letter to J. D. Hooker, 22 June [1861].
[7] See letter to J. D. Hooker, 22 June [1861] and n. 9.
[8] See the letters to the *Cottage Gardener* and, under its new title, the *Journal of Horticulture* in this volume. CD's copies of this periodical are in the Darwin Library–CUL.

To A. G. More 7 July 1861

2 Hesketh Crescent, | Torquay.
July 7, 1861.

My dear Sir

I have been looking over my notes on E. palustris and should much like to see a few more flowers just on point of opening.[1] Should you think me very unreasonable if I were to ask you to send me a few: I have enquired and am told it does not grow here— I have never heard whether it grows near you: if it does and you could visit the spot twice, I would ask you to try a little experiment, viz. to cut off the terminal and movable division of the Labellum in 6 or 8 flowers which had not *quite* opened and which therefore could not have been visited by insects; and mark these flowers with little bits of thread and then see if these set pods as well as the other flowers. But this little experiment would be useless if this is a species which does not freely in ordinary cases set pods. I would not on any account have you take much trouble to try this; but if I had the opportunity I should try it for bare chance of its showing that this terminal portion of Labellum was of use, as

I suspect, in guiding and aiding some unknown insect in its proper function of fertilising the flowers.[2]

My dear Sir | Yours sincerely | Charles Darwin

Copy
DAR 146

[1] *Epipactis palustris* is described in *Orchids*, pp. 95–102. More is thanked for having 'repeatedly' sent CD 'fresh specimens of this beautiful Orchid' (*ibid.*, p. 95 n.).

[2] More was able to comply with CD's request. For the results of this experiment, see the letter to A. G. More, 19 July 1861 and n. 5.

From Tiberius Cornelius Winkler 7 July 1861[1]

Monsieur!

Mon mémoire sur les Poissons fossiles d'Oeningen, couronné par la société hollandaise des sciences à Harlem, ayant paru, je désire vivement vous témoigner, par l'envoi d'un exemplaire de cet ouvrage, le respect que m'inspirent vos travaux scientifiques.[2] De l'un de ces illustres travaux, de votre "Origin of species", j'ai eu l'honneur de faire la traduction hollandaise.[3] Je profite de l'occasion qui m'est fournie par l'obligeance de Sir Charles Lyell, de vous offrir en même temps un exemplaire de cette traduction.[4] Je prends la liberté de vous prier de vouloir bien me faire savoir si ces deux ouvrages sont parvenus dans vos mains.

Agréez, Monsieur, l'assurance de ma considération très distinguée.

T. C. Winkler

Harlem 7 Juillet 1861
A Monsieur Charles Darwin.

Edinburgh University Library (Lyell 1: 6084)

[1] For a translation of this letter, see Appendix I.

[2] Winkler 1861. There is a copy of this work in the Darwin Library–Down.

[3] Winkler's translation of *Origin* into Dutch (Winkler trans. 1860) is in the Darwin Library–CUL.

[4] Winkler's letter to CD was enclosed in a letter to Charles Lyell, dated 8 July 1861 (Edinburgh University Library (Lyell 1: 6085)), in which Winkler asked Lyell to forward the works mentioned in the letter to CD.

To W. D. Fox 8 July [1861]

2. Hesketh Crescent | Torquay
July 8th

My dear Fox

Your note has followed us here, where we arrived 4 or 5 days ago.—[1] As we wished to try 6 or 8 weeks of sea-side for my eldest girl we thought we would make this awfully long (to us) stretch & come here. We are charmed with the view from this crescent & with the walks all around & we have got a very good House. But is no joke bringing 16 souls & $\frac{3}{4}$ tun of luggage so far.— My eldest girl is still a sad invalid; but certainly improves: she gets up twice every day now & can walk one or two hundred yards.[2]

I am glad for myself to have this outing & change for I have been a poor wretch for many months. You say not one word about yourself. Are you not a pretty sort of man? It seems indeed strange to hear of your having two daughters married.—[3]

Poor dear Henslow's death has been a sad loss to many. He wrote me by dictation a most kind note from his death-bed.—[4] L. Jenyns is going to write a biographical notice of him.—[5]

I shall not go to Manchester;[6] but it would be a great temptation to get a sight of you.— How long it is since we met! Farewell my dear old friend | Yours affect^y | C. Darwin

My eldest son is going to join as Partner in a Bank at Southampton: I had so good an offer it seemed a pity to reject it.—[7]

We hear that the Darwins of—(Elston) are coming here.—[8]

Postmark: JY 8 61
Christ's College Library, Cambridge (Fox 131)

1 Fox's letter has not been found. The Darwins left for Torquay on 1 July 1861, spending the night en route in Reading. See 'Journal' (Appendix II).

2 Henrietta Emma Darwin was recuperating slowly from typhus fever suffered in the spring of 1860.

3 Two of Fox's daughters by his first marriage had recently married. Eliza Ann Fox married Henry Martyn Sanders, vicar of Skidby, Hull, and Harriet Emma Fox married Samuel Charlesworth Overton (*Darwin pedigree*).

4 The letter from John Stevens Henslow, apparently written shortly before his death on 16 May 1861, has not been found.

5 Jenyns ed. 1862 (see Appendix X).

6 The British Association for the Advancement of Science was to hold its 1861 meeting in Manchester.

7 CD was arranging for his son William Erasmus Darwin to become a partner with the Southampton and Hampshire Bank. See letters to John Lubbock, 10 July [1861], 1 August [1861], and [2 August 1861].

8 Elston Hall, near Newark, Nottinghamshire, was the seat of the senior branch of the Darwin family. CD's grandfather Erasmus Darwin was born at Elston Hall. The head of the family was Francis Rhodes Darwin, husband of Charlotte Maria Cooper Darwin, the granddaughter of Erasmus Darwin's older brother, William Alvey Darwin. Francis Rhodes took the Darwin family name in 1850 under the provisions of the will of his brother-in-law Robert Alvey Darwin. See *Darwin pedigree* and Freeman 1978.

To *Journal of Horticulture* [before 9 July 1861][1]

I hope that you will grant me a little space to thank some of your correspondents and Mr. Beaton for his interesting information how to test the effects of different kinds of pollen on the divisions of the same stigma of a Pelargonium, for my special purpose of ascertaining whether one *variety* is prepotent over another.[2] I fear that the Scarlet Pelargoniums include at least two wild forms, which botanists would rank as distinct *species*. If Mr. Beaton is at any time writing on these plants, perhaps he would tell us what he knows about the wild parent of the Horseshoe and other Scarlets.

I am very glad that "P." sent a list of his Perlargoniums with the central flower regular; for I was not aware how common the case was.[3] Will "P." be so obliging as to observe and report whether any of the regular central flowers set seed—that is, if the kinds specified are such as ever produce seed?

With respect to the fertilisation of Wheat: several years ago I examined the flowers day by day, and came to the same conclusion as that which "H.C.K." expresses so forcibly.[4] Mr. Beaton apparently does not much venerate botanical authorities,[5] but he might easily quote a long list of great names to show that Wheat is always fertilised in the bud;[6] what has misled so many botanists I cannot imagine. But stranger assertions of the same kind may be met with: for instance, that cruciferous plants are generally fertilised before the flower opens![7] As I am away from home I write without my notes;[8] but I remember that the Chinese have the singular belief that certain varieties of Wheat are always fertilised in the night-time. Col. Le Couteur, who attended so carefully to the varieties of Wheat, entertains no doubt that the different varieties, when growing near each other, cross.[9] On the other hand, a full account has been published of a large number of varieties, I think 150, which were cultivated close together in some continental garden during several years, and never crossed each other.[10] This account has much perplexed me; and I have sometimes been tempted to doubt whether any eye, however accurate, could have distinguished so many varieties, and that, perhaps, after all the varieties did cross. Mr. Beaton might advance this case in support of his belief that Wheat is fertilised in the bud.

As Mr. Beaton alludes to some mistake which he had made, might I venture to suggest to him to punish himself by telling sooner than he intended by what means he can produce from pollen of the same flower placed on the stigmas of the same variety two different sets of seedlings?[11] That is a mystery which it is tantalising to wait for.—

Charles Darwin, *Down, Bromley, Kent.*

Journal of Horticulture, Cottage Gardener, and Country Gentleman n.s. 1 (1861): 280–1

[1] Dated by the publication of the letter in the 9 July 1861 issue of the *Journal of Horticulture*. It was printed under the title 'Effects of different kinds of pollen'.

[2] The reference is to responses to CD's letter to the *Journal of Horticulture*, [before 18 June 1861]. Donald Beaton, who wrote a regular column for the journal, had discussed some of the points CD raised in an article entitled 'Cross-breeding plants' printed in the 25 June 1861 issue (*Journal of Horticulture* n.s. 1 (1861): 232–4).

[3] The list of *Pelargonium* varieties sent by 'P.' appeared under the heading 'Regular flowers in Pelargoniums' in the 2 July 1861 issue of the *Journal of Horticulture*, p. 253.

[4] CD refers to the letter from 'H.C.K.,— *Rectory, Hereford*.' printed under the title 'When is wheat fertilised?' in the 2 July 1861 issue of the *Journal of Horticulture*, p. 257. The writer was Henry Cooper Key,

rector of Stretton-Sugwas, Herefordshire. He referred the reader to Beaton's response to CD's letter of [before 14 May 1861], which was printed immediately following CD's letter (*Journal of Horticulture*, n.s. 1 (1861): 112–13). Beaton claimed, 'No kind of Wheat has ever been naturally crossed and never can be' and stated his belief that the seed was fertilised while the ear was still within its sheath, before the wheat blossomed (*ibid.*, p. 113). According to 'H.C.K.', Beaton's view was 'entirely in error', and he gave his reasons for believing that wheat was fertilised after the ear had risen out of the sheath and the blossoming process had begun.

5 In a number of his articles over the past year or so, Beaton had made several disparaging references to the incursion of botanical science on practical gardening, warning gardeners in one instance against 'building up systems of practice on a baseless, so-called-scientific foundation' (*Cottage Gardener*, 21 February 1860, pp. 311–13).

6 CD studied the literature treating the fertilisation of wheat and other plants and carried on his own investigations in 1856 and 1857. One of the 'great names' to which he refers was Jean Louis Auguste Loiseleur Deslongschamps. See *Correspondence* vol. 6, letters to J. D. Hooker, 13 July [1856], and to Asa Gray, 18 June [1857].

7 The reference has not been traced.

8 CD and his family were staying in Torquay (see 'Journal'; Appendix II).

9 John Le Couteur was the author of a well-known work on wheat (Le Couteur 1836).

10 CD referred to this experiment in *Variation* 1: 314–15, where he cited the account given in Loiseleur Deslongchamps 1842–3, pp. 45, 70.

11 Beaton had first referred to producing seeds of different varieties of geranium by placing the pollen of five different geraniums on the stigma of one flower in an article published in the *Cottage Gardener*, 24 July 1860, pp. 253–5. He repeated this claim in his article on 'Cross-breeding plants' in the 25 June 1861 issue of the *Journal of Horticulture*, pp. 232–4, stating that before giving this information, he challenged 'young gardeners to exercise their wits in trying to account for such a thing or how to do it.' (p. 234), repeating the challenge the next week (p. 257) and discussing the problem of cross-breeding in the 9 July 1861 issue (pp. 272–3). Beaton responded to CD's query more fully in an article entitled 'Fertilisation of wheat' published in the 23 July 1861 issue of the *Journal of Horticulture*, pp. 311–13.

To Charles Lovegrove 9 July [1861?][1]

Down
July 9th

My dear Sir

I thank you sincerely for the trouble which you have so kindly taken for me.—[2] I am sorry not to have the pedigree of your pretty pony: but the information sent will be very useful to me, more especially as it confirms what I heard from Norway, and did not know whether fully to believe.—[3]

With my best thanks, Believe me My dear Sir | Yours very faithfully | Charles Darwin

Barton L. Smith

1 The year is suggested by the 1860 watermark of the stationery and by the relationship of the letter to others written in 1861 (see n. 3, below). However, CD was not at Down House in July 1861: the family was visiting Torquay (see 'Journal'; Appendix II).

2 It is possible that Charles Lovegrove was the individual who provided CD information mentioned in *Variation* 1: 58: 'I have heard of a cob imported from Norway which had the shoulder as well as the other stripes well developed.' Lovegrove was the owner of Petley House in Down village (*Post Office directory of the six home counties* 1862).

[3] CD had written to several people in 1857 to inquire about the striping of native ponies of Norway (see *Correspondence* vol. 6, especially letters to the *Gardeners' Chronicle*, [before 13 June 1857], and to J. D. Hooker, 22 August [1857]). He had recently initiated a further search for information on this point in the *Field* (see letters to the *Field*, [before 27 April 1861], [before 25 May 1861], and [before 15 June 1861]). For the discussion of striping on the legs and shoulders of dun-coloured ponies as an indication of the colouring of the aboriginal horse, see *Variation* 1: 55–61.

To John Lubbock 10 July [1861][1]

2. Hesketh Crescent | Torquay
July 10th.

My dear Lubbock

I am heartily sorry to trouble you, but I am in a complete fix about the Southampton affair. Mr. Atherley[2] has sent the articles of Partnership to my solicitors (Messrs. Rowland & Hacon 31 Fenchurch St)[3] who say that Mr. A "has stated terms much favouring himself": they urge me to consult Sir John Lubbock "on the *fairness* of the terms", but I do not like to trouble him,[4] but perhaps you will talk the case over with him. You will see pencil notes by my solicitor on the margin. I am very ignorant, but the articles strike me as drawn up as if to get my son as a mere locum tenens.[5]

My solicitor says "But what does appear to me to be the most important difficulty in the negociation is Mr Atherleys requirement of a reserved right to introduce his son into business. For your son to assent to this is tantamount to his agreeing to become the partner of an unknown person, who *may* be either a knave or an idiot (of course this is only a possible case). And the objection to this seems to be so great that I should myself on no account,—on no consideration,—enter into any such engagement".— I should have thought that it was quite natural that Mr. A. should have had power of introducing his son; but I understood it was somehow to have been with the consent of my son.—[6]

To my utter ignorance it seems an absolutely fatal difficulty that my son, in case of Mr A.' death, say in course of 2 or 3 years, before my son can have made any connection, should be *compelled* to purchase Mr. A.' share; for he would have in such case to close the Bank. I am perfectly conscious that I have no claim to ask you to consider this troublesome business; if you cannot spare time, I must put the affair entirely in my solicitor's hands & unless he can get terms which seem to him fair, I must, I grieve to say, reject the offer, much to my son's disappointment. I would come up & discuss the business with you, but it is an awful two days journey for me: or William would run up.[7] To learn your general impression would be an *immense* advantage to us but pray remember that this affair, though to Sir John Lubbock or you a small concern, is of great importance to us.

Forgive me & believe me— Yours most truly | C. Darwin

Down House

[1] The year is provided by the Darwins' stay in Torquay (see n. 7, below) and by the reference to the negotiations concerning William Erasmus Darwin's future position.

[2] George Atherley was seeking a new partner to join him in the Southampton and Hampshire Bank, Southampton. See letters to W. E. Darwin, [25 May 1861], [June 1861], and 6 [June 1861].

[3] David Rowland and William Mackmurdo Hacon.

[4] John William Lubbock, John Lubbock's father, was a partner in the London bank Robarts, Lubbock & Co.

[5] Locum tenens: 'One who holds office temporarily in place of the person to whom the office belongs' (*OED*).

[6] Atherley's son was only eight years old (see letter to John Lubbock, 14 July [1861]).

[7] According to Emma Darwin's diary, William Darwin joined the family in Torquay on 9 July 1861. The Darwins had arrived a week earlier, on 2 July (see 'Journal'; Appendix II).

To Frances Julia Wedgwood 11 July [1861][1]

[Torquay]
July 11

Some one has sent us 'Macmillan'; and I must tell you how much I admire your Article;[2] though at the same time I must confess that I could not clearly follow you in some parts, which probably is in main part due to my not being at all accustomed to metaphysical trains of thought.[3] I think that you understand my book perfectly, and that I find a very rare event with my critics. The ideas in the last page have several times vaguely crossed my mind.[4] Owing to several correspondents I have been led lately to think, or rather to try to think over some of the chief points discussed by you. But the result has been with me a maze—something like thinking on the origin of evil, to which you allude.[5] The mind refuses to look at this universe, being what it is, without having been designed; yet, where one would most expect design, viz. in the structure of a sentient being, the more I think on the subject, the less I can see proof of design.[6] Asa Gray and some others look at each variation, or at least at each beneficial variation (which A. Gray would compare with the rain drops which do not fall on the sea, but on to the land to fertilize it) as having been providentially designed.[7] Yet when I ask him whether he looks at each variation in the rock-pigeon, by which man has made by accumulation a pouter or fantail pigeon, as providentially designed for man's amusement, he does not know what to answer;[8] and if he, or any one, admits [that] these variations are accidental, as far as purpose is concerned (of course not accidental as to their cause or origin); then I can see no reason why he should rank the accumulated variations by which the beautifully adapted woodpecker has been formed, as providentially designed. For it would be easy to imagine the enlarged crop of the pouter, or tail of the fantail, as of some use to birds, in a state of nature, having peculiar habits of life. These are the considerations which perplex me about design; but whether you will care to hear them, I know not.

Incomplete
LL 1: 313–14

[1] The year is that given by Francis Darwin in *LL* 1: 313.

[2] Frances Julia Wedgwood, the eldest daughter of Hensleigh and Frances Mackintosh Wedgwood, was the author of the unsigned article entitled 'The boundaries of science. A second dialogue', published

in *Macmillan's Magazine* in July 1861 (*Wellesley index*). The first part of the dialogue appeared in the June 1860 issue ([Wedgwood] 1860–1).

3 The articles take the form of a dialogue between the religiously orthodox Philocalos and the defender of Darwin's views, Philalethes. Topics discussed include the implications of the theory of natural selection for such metaphysical questions as the connection between the material and spiritual realms, the division between reason and faith, and the relationship of the Creator to his creation. Frances Wedgwood's 'metaphysical trains of thought' had been strongly influenced by the religious teachings of Thomas Erskine and his friend John Frederick Denison Maurice, one of the founders of Christian Socialism. She had spent several months with followers gathered at Erskine's estate at Linlathen, in Forfarshire, Scotland, in 1859. See Wedgwood and Wedgwood 1980, pp. 254, 267.

4 The closing pages of the second part of the dialogue address the question of whether there is design in nature and, if so, how this informs religious belief and scientific inquiry ([Wedgwood] 1860–1, 4: 244–7).

5 The problem of the existence of evil in the world forms part of the discussion mentioned in n. 4, above.

6 Philalethes argues that while there are 'unmistakeable evidences of *design*' in nature, any attempt to make the apparent harmony in nature's operations a complete one by bringing man into consideration is bound to fail ([Wedgwood] 1860–1, 4: 244–5). The discussion continues:

> *Philoc.* You mean to say, then, that somewhere or other there is a misfit between man and his dwelling-place; that, if the machinery was ever perfect, some evil power has laid its hand on the mainspring, and deranged the working ever since? *Philal.* Yes; and that influence lies wholly without the boundaries of science, which is exclusively occupied with the machinery itself, and can take no account of any influence from without. Go back as far as we will, therefore, science only shows us the working of the present order of things.

7 Asa Gray presented this view in the third part of his three-part series of articles on CD's theory published in 1860 and reprinted in A. Gray 1861a.

8 See *Correspondence* vol. 8, letter to Asa Gray, 26 November [1860].

To J. D. Hooker 13 July [1861]

2. Hesketh Crescent | Torquay
July 13th

My dear Hooker

I am glad to hear of Worthing & I hope it will do Mrs. Hooker good.—[1] I hope Harvey is better:[2] I got his Review of me a day or two ago,[3] from which I infer he must be convalescent: it very good & fair: but it is funny to see a man argue on the succession of animals from Noahs Deluge; as God did not then wholly destroy man, probably he did not wholly destroy the races of other animals at each geological period![4] I never expected to have a helping hand from the Old Testament.

I have no suggestions for Mr. Salvin;[5] but I heartily concur about St. Martha & once before urged some collectors to visit it.[6] The alpine forms would be very interesting. Just say a word to him to attend to domestic animals & plants.— By the way if he talks the language, ask him as personal favour to me to enquire a little in the wilder regions whether any of the farmers take any pains in breeding from the better animals or such as strike their fancy. But I fear such information (& I have much) would come too late for me.—

I am got profoundly interested in Orchids & think I have made out homologies of the sticky so-called gland of the pollinia, & of the rostellum to each other clearly.

Veitch & Co have never even answered my letter & I much fear I shall get nothing.[7] If they would but say "no", I would write elsewhere. I much want a Cattleyea or some one of the Epidendreæ, as I have examined a Humble-Bee with the pollinia of Cattlyea attached to its back.—[8] Really the contrivances in Orchids beat, I think, any animal.

I am ashamed to say that I have not read Du Chaillu;[9] for I have lost the art of reading & am either idle or writing

Farewell my dear Hooker. How I wish you were at this charming place | Adios.— C. Darwin

William is dissecting & drawing like mad.—[10]

Endorsement: '/61'
DAR 115.2: 105

[1] The Hookers were apparently intending to visit Worthing, on the Sussex coast, in the hope that the seaside would benefit Frances Harriet Hooker, who was grieving over the recent death of her father John Stevens Henslow.

[2] William Henry Harvey was suffering from a 'hæmorrhage from the lungs' (see letter to J. D. Hooker, 22 June [1861] and n. 3).

[3] Harvey 1861.

[4] In his essay on the 'natural evolution of organic species', Harvey argued that it is 'more probable that the *living horse* is a lineal descendant of some *fossil horse*, than that God first *wholly destroyed* one horse and then *re-created* from the dust a slightly improved horse.' (Harvey 1861, p. 150). Harvey used the same argument for man, assuming the preservation of 'the line of Adam's posterity through Noah.'

[5] Osbert Salvin was preparing to leave in the autumn of 1861 on a natural history expedition to Guatemala (*DNB*).

[6] Santa Marta is a port city in northern Colombia, bordering on the Sierra Nevada de la Santa Marta mountain range.

[7] See letter to J. D. Hooker, [6 July 1861]. The nursery of Veitch & Son of Chelsea specialised in orchids.

[8] See letter to Frederick Smith, [19 June 1861]. Having examined almost all the genera of British orchids, CD was beginning to study various genera of foreign orchids to compare and contrast their various pollination mechanisms. For the results of CD's study of the Epidendreae, see *Orchids*, pp. 159–66.

[9] Du Chaillu 1861.

[10] Having been interested in botany for a number of years, William Erasmus Darwin was currently studying the anatomy of plants by means of dissection. He had joined the Darwins in Torquay on 9 July 1861 (Emma Darwin's diary). See letters to J. D. Hooker, 19 June [1861] and 22 June [1861]. Two of William's botanical notebooks are in the Darwin Archive–CUL (DAR 117 and 234).

To George Maw 13 July [1861][1]

2. Hesketh Crescent | Torquay
July 13th

Dear Sir

Your kind note & Review[2] & Uriconium[3] have been forwarded from my home & have reached me here today.— I have cut the pages & have glanced over the pages (which shall hereafter be carefully studied) & can see that all your criticisms are written in a perfectly fair & kind tone. I thank you sincerely. I will not trouble you with any discussion, but will consider all your points.

I think it is a pity to mingle science & religion;[4] but that you say you have done from conscience; & I entirely believe you.— I did not enter on case of man, from its extraordinary difficulty & from believing that the case of man will follow that of other animals, whenever my views are finally rejected or admitted.— This rejection or admission will require, I have always seen & I now see still more plainly, many years.—

My opponents would have lost nothing if they had all treated me as fairly as you seem to have done.

With my sincere thanks | pray believe me | Dear Sir | Yours very faithfully | Charles Darwin

Endorsement: '61'
Royal Horticultural Society

[1] The endorsement is confirmed by the reference to Maw's papers (see nn. 2 and 3, below).

[2] Maw's letter to CD has not been found. His review of *Origin* appeared in the July 1861 issue of the *Zoologist* (Maw 1861a). There is a heavily annotated copy of the review in the Darwin Pamphlet Collection–CUL. See letter to George Maw, 19 July [1861] and CD note.

[3] Uriconium was the name of a Roman settlement, near Wroxeter in Shropshire, which was currently undergoing excavation. Maw, who lived nearby, was greatly interested in antiquarian subjects. He probably sent CD a copy of his article on the mosaic pavements found at this site (Maw 1861b).

[4] Maw accused CD of escaping responsibility for having produced a theory opposed to revelation by his use of generalisations and a kind of 'sentimental theology' (Maw 1861a, pp. 7609–10).

To John Lubbock 14 July [1861][1]

2. Hesketh Crescent | Torquay
July 14th.

My dear Lubbock

Pray give my sincere thanks to Sir John Lubbock for his note, & yourself accept my cordial thanks.[2] Your notes have been of greatest *possible* assistance to me. I felt so entirely ignorant that I did not know whether my solicitor was quite unreasonable.— I have now written to him, telling him that we require merely fair terms, to be as civil as possible, to push no point to an unreasonable extent, & finally that we are anxious to join. I hope Mr. Atherley will modify some of the points. What you seem to think struck me that there was too much provision for the distant future; as Mr Atherleys son is only 8 years old; I shd. have thought (& have so told my solicitor) that all such points might have been left open.—

I hope most deeply that the negotiation will not fall through: I must be guided to a large extent by my solicitor but I have urged him not to be too particular.

Anyhow I am sure you have done everything, far more than I could have expected, to help us.— I should hope that we shall have no need to trouble Sir

John or you again; but if perplexed, I will accept your kind offer.—
Yours very gratefully | Charles Darwin

Endorsement: '1861'
Down House

[1] The endorsement is confirmed by CD's stay in Torquay (see 'Journal'; Appendix II).
[2] The letter from John William Lubbock has not been found. Negotiations connected with the arrangements for William Erasmus Darwin's banking partnership, had run into difficulty over the terms to be included in the contract. See letter to John Lubbock, 10 July [1861].

From Henry Fawcett 16 July [1861][1]

Bodenham | Salisbury
July 16th.

My Dear Sir,

I feel that I ought not to have so long delayed writing to thank you for your very kind letter to me about my article on your Book, in Macmillan's Magazine[2]

I was particularly anxious to point out that the Method of Investigation pursued was in every respect, philosophically correct; I was spending an evening last week with my friend Mr. John Stuart Mill and I am sure you will be pleased to hear from such an authority that he considers that your reasoning throughout is in the most exact accordance with the strict principles of Logic. He also says, the Method of investigation you have followed is the only one proper to such a subject.[3]

It is easy for an antagonistic reviewer when he finds it difficult to answer your arguments to attempt to dispose of the whole matter by uttering some such commonplace, as "This is not a Baconian induction".[4]

I expect shortly to be spending a few days in your neighbourhood and if I should not be intruding upon you I should esteem it a great favour if you will allow me to call on you, and have half an hours conversation with you.

As far as I am personally concerned I am sure I ought to be grateful to you, for since my accident nothing has given me as much pleasure as the perusal of your Book. Such studies are now a great resource to me.[5]

Believe me to be, | My Dear Sir | Yours very truly | Henry Fawcett.

C. Darwin Esq.

DAR 98 (ser. 2): 29–30

CD ANNOTATION
Top of first page: 'Keep.' *blue crayon, circled blue crayon*

[1] The year is given by the reference to Fawcett 1860 (see n. 2, below).
[2] CD's letter to Fawcett, commenting on his article 'A popular exposition of Mr. Darwin on the Origin of species', published in the December 1860 issue of *Macmillan's Magazine* (Fawcett 1860), has not been found. There is an annotated copy of the review in the Darwin Pamphlet Collection–CUL.

CD mentioned Fawcett's review, however, in a letter to Thomas Henry Huxley, 5 [December 1860] (*Correspondence* vol. 8).

3 In the fifth edition of his *System of logic, ratiocinative and inductive, being a connected view of the principles of evidence, and the methods of scientific investigation* (Mill 1862), the philosopher John Stuart Mill referred to CD's 'remarkable speculation' on the origin of species as an 'unimpeachable example of a legitimate hypothesis.' For Mill, natural selection was 'not only a *vera causa*, but one proved to be capable of producing effects of the same *kind* with those which the hypothesis ascribes to it', and as for Adam Sedgwick's charge (in [A. Sedgwick] 1860) that CD had violated the rules of induction, Mill asserted that this was 'unreasonable' (Mill 1862, p. 18 n.):

> The rules of Induction are concerned with the conditions of Proof. Mr. Darwin has never pretended that his doctrine was proved. He was not bound by the rules of Induction, but by those of Hypothesis. And these last have seldom been more completely filled.

Mill's immediate response to *Origin* is given in a letter to Alexander Bain, 11 April 1860 (Mineka and Lindley eds. 1972, p. 695).

> [Darwin's book] far surpasses my expectation. Though he cannot be said to have proved the truth of his doctrine, he does seem to have proved that it *may* be true which I take to be as great a triumph as knowledge & ingenuity could possibly achieve on such a question.

For an interpretation of Mill's position concerning CD's 'hypothesis', see Hull 1973, pp. 27–8.

4 This statement appeared in Fawcett's article (Fawcett 1860, p. 83), where he attributed it to the author of the review of *Origin* in the *Quarterly Review*. The author of the review, Samuel Wilberforce, criticised CD throughout for having violated 'true philosophy', which he equated with Baconian induction. For CD's response to this charge, see *Autobiography*, p. 119. See also Hull 1973.

5 Fawcett, a fellow of Trinity Hall, Cambridge, had lost his sight in a shooting accident in 1858 (*DNB*).

To J. D. Hooker 17 [July 1861]

2. Hesketh Crescent | Torquay
17th

My dear Hooker

Hearty thanks for your invitation for Willy, who would have deeply enjoyed it & longs for the wonderful good it would have done him;[1] but his time draws near to go to Southampton; if that ever takes place, for there are 2 or 3 sudden hitches in our negotiations.—[2]

I was really ashamed to bother you about Catasetum; I have now written to Parker & Williams for chance.[3]

I sh^d be very glad of a Catleya **or** Epidendron,[4] & specially (from what A. Gray writes) for an Arethusa or one of that section.—[5] Have you *Mormodes*; ☞ with some buds that would complete my desiderata for comparison.[6] If the racemes were cut off & packed in **tin**-cannister with little slightly damp moss, they might be sent by Post, & I would pay postage: possibly Arethusa might come in pot, as that seems most important for me.— We shall be nearly 5 weeks more I really think that my paper will be a curious one, though not of much importance.—

I am sorry to hear that Henslow did not leave more property: in old days he spoke to me, as if he expected in after days to be well off.—[7]

What a bad job this loss of your Examination fees, now that your family increasing in expence, & expences will increase.—[8]

I hope you will not work too hard, for you have very much on your shoulders.— Farewell. Did you read Asa Gray's letter to me.—[9] The P.S. (need not be returned) is worth reading—

Yours affect | C. Darwin

Etty is going on well & continues to absorb Cod-liver oil— we think this best advice of yours of any we have had.[10]

Endorsement: 'July | 61'
DAR 115.2: 106

[1] Hooker may have invited William Erasmus Darwin to join him on a botanical excursion. CD had sought Hooker's advice on how best to foster William's current interest in botany (see letters to J. D. Hooker, 19 June [1861], 22 June [1861], and 13 July [1861]).

[2] The contractual arrangements for William's partnership in a Southampton bank had encountered a number of problems. See letters to John Lubbock, 10 July [1861] and 14 July [1861].

[3] See letters to J. D. Hooker, 19 June [1861] and [6 July 1861].

[4] See letter to J. D. Hooker, 13 July [1861].

[5] CD had received a letter from Asa Gray in which he discussed the pollination mechanisms of various orchids. Gray's letter has not been found, but see the letter to Asa Gray, 21 July [1861].

[6] CD was interested in examining several different species of the sub-tribe Catasetidae to compare their pollination mechanisms. Called by CD 'the most remarkable of all Orchids' (*Orchids*, p. 211), the Catasetidae eject their pollinia, which often become attached to the bodies of visiting insects (*ibid.*, p. 212).

[7] Hooker's father-in-law John Stevens Henslow had died on 16 May 1861. Henslow's benefice at Hitcham, Suffolk, had been worth £1179 per annum (*Journal of Horticulture* n.s. 1 (1861): 138). His natural history collection was sold at auction shortly after his death (see *Field* 18 (1861): 159).

[8] The reference to examination fees has not been identified. At the time, Hooker was an examiner in botany for the East India Company medical board, a twelve-year appointment that had begun in 1854 (L. Huxley ed. 1918, 1: 385). He was an examiner for the Apothecaries' Company Medal (*ibid.*, 385) and also for the London University, a position he held until 1864 (*ibid.*, 537).

[9] See n. 5, above.

[10] Hooker had recommended that Henrietta Emma Darwin be rubbed several times a day with cod-liver oil in an attempt to improve her appetite and promote her recovery from the after-effects of typhus fever (see letter to J. D. Hooker, 8 [February 1861]).

To A. G. More 17 July 1861

2 Hesketh Crescent, Torquay.
July 17, 1861

My dear Sir

If you are quite well and take walks, and if you are not utterly sick and weary of me and my requests, I should be very glad of 3 or 4 spikes of O. latifolia and O. maculata, *with a few of the lower flowers* withered. The point is this, that I find in the Kentish specimens that the O. latifolia is far less visited by insects than O. maculata; and as some Botanists doubt whether they are distinct species it would be rather curious to show them insects distinguish them.[1] The specimens might be sent in brown paper, without trouble of Box, by putting them in water I could easily count the flowers with pollen-masses in and out. You will see that it is not worth much trouble. I am now writing my paper and I fancy that the points are sufficiently curious to make you not regret having with so much patience and kindness helped me.

I found the other day a lot of Bee Ophrys with flowers nearly all withdrawn, and with the glands of the pollinia *all* in their pouches. All facts point clearly to eternal self-fertilisation in this species; yet I cannot swallow the bitter pill.[2] Have you looked at any this year?

Yours very sincerely | C. Darwin

Copy
DAR 146

[1] CD believed that the method of pollination of an orchid was a good indicator of specific identity (see letter to George Bentham, 22 June [1861] and n. 8). Although CD was unable to find conclusive evidence distinguishing *Orchis latifolia* from *O. maculata* on the basis of the number of pollinia removed by insects, he continued to consider them distinct species (see *Orchids*, p. 42).

[2] In appearing to be an 'eternal hermaphrodite', the bee orchis, *Ophrys apifera*, challenged CD's 'law' that all organic beings are occasionally fertilised by other individuals. CD conceded that this species was 'specially adapted for self-fertilisation'. Nevertheless, he did not 'swallow the bitter pill', arguing that although he had never observed an instance in which a plant's pollen masses had failed to reach its own stigmas, the morphology of the flower parts suggested to him that the bee-orchis might occasionally be cross-fertilised (see *Orchids*, pp. 65, 70, and *Orchids* 2d ed., p. 57).

To Hewett Cottrell Watson [17 July 1861][1]

[Torquay]

Begin by saying that I have but little information on variation— Bronn & Mr Maw state often or generally mingled.—[2]

The difficulty to know what to call vars & what species,—hopeless—[3] But your suggested plan to take list of forms which some call vars & some species ? print list?—[4]

Now most valuable information to know how far down to be called incipient species, not only for my object but generally to know how to settle this work, if habits & distribution of these were well attended to.—

Point. does each pair or two in any district familiar to observer grow mingled or separate & in what stations.[5] Or does one grow in one country & one in another? Where they mingle are they equally distinct Cause effects of condition or crossing. About the frequency.— Forms such as albinoes or monstrosities not apparently propagated to be excluded.—

Then about polymorphic genera, about the vars of these make **perhaps** a distinct case.— but it would be interesting to know range & station of these extremely varying forms.—

Any facts about rare var. being known in same locality for period, ie if propagated by seedlings—

Draft
DAR 185: 49

[1] The date is given by a note written by Watson on a copy he made of part of the letter received from Darwin, of which the draft is printed here. Watson wrote on a slip tipped into his presentation copy of the first edition of *Origin*: 'C Darwin in lre of July 17. 1861' (see n. 5, below). The book is in the collection of Texas Tech University Library, Lubbock, Texas. Watson also refers to 'your letter of 17th.' in the letter from H. C. Watson, 24 July 1861.

[2] In his review of *Origin*, George Maw stated that varieties were usually found mingled with their parent forms, just as different genera are closely associated 'without engaging in that mutually extirpating struggle upon which Mr. Darwin's theory of progression so much depends.' (Maw 1861a, p. 7600). In his criticism of CD's theory included as the final chapter of his translation of *Origin*, the German palaeontologist Heinrich Georg Bronn pointed out that variations used to distinguish different varieties of a species are not generally clearly marked but consist of several different deviating characters (Bronn trans. 1860, p. 503).

[3] Since the publication of *Origin*, Watson and CD had been corresponding concerning various aspects of the theory. Although Watson accepted natural selection as the mechanism for species change, he argued that in addition to CD's principle of the divergence of species from the original type, there also appeared to have been a 'convergence' of species. According to Watson, this principle was necessary in order to limit the vast numbers of different specific types that would otherwise proliferate. See *Correspondence* vol. 8, letters from H. C. Watson, [3? January 1860] and 10 May 1860. CD responded to Watson's points in the new material added to the third edition of *Origin*, pp. 141–3.

[4] See the earlier correspondence between CD and Watson on this topic, especially the letter from H. C. Watson, [after 23 March 1858] (*Correspondence* vol. 7). See also letter from H. C. Watson, 24 July 1861.

[5] The part of the original letter discussing this point has survived in a copy Watson made under the heading '*Distribution of closely allied species or subspecies*' (see n. 1, above). The text reads:

> The point I should so much like to see discussed, is how often do such allied forms (held species by some, vars. by other botanists) live mingled in same spot,—& how often in same district, but in different stations,—& how often in different geographical districts. I presume you could discuss these points only in a *few* cases. It has surprized me how little information I have been able to find on such subjects with respect to very closely allied forms.— The habitats & range of the forms in the protean or polymorphous genera, such as Rubus &c, might, I should think, be discussed in comparison with the somewhat more clearly defined forms—

To Peter Martin Duncan? 18 July [1861][1]

2. Hesketh Crescent | Torquay
July 18th.

Dear Sir

It would be a real pleasure to me to give you any information in aid of your most difficult & laborious undertaking.[2] But I am sorry to say that I am quite unable to answer any of your questions. It is now nearly 30 years since I began examining living corals,[3] but other pursuits interfered & I have utterly forgotten what little I knew. I remember attending a little to the effects of tranquil & disturbed water on their growth; but I cannot remember to what conclusion I came; but as a general rule, I think nearly all the species were distinct. I remember being impressed with the conviction that the classification of the stony corals would be very difficult.[4]

From all that I know of Dana, I shd. feel much inclined to place much trust in him. What a misfortune it is that his health has so much failed![5]

I am very sorry that I am so entirely incompetent to render any assistance, & remain, Dear Sir, with my best wishes | Yours very faithfully | Charles Darwin

American Philosophical Society (257)

[1] The year is provided by CD's stay in Torquay (see 'Journal'; Appendix II).
[2] Duncan was preparing to undertake a study of the British fossil corals. His monograph was eventually published by the Palaeontographical Society of London (Duncan 1866–72).
[3] CD had collected and made notes on a number of different corals and coralline algae during the *Beagle* voyage. His descriptions of corals are included in his Zoological diary (DAR 30 and 31), and several of his drawings of these specimens are in DAR 29.3. A few of his observations on corals were published in *Coral reefs*. See Porter 1985. For a discussion of CD's interest in studying coral structure and generation processes, see Sloan 1985, pp. 104–9.
[4] The stony corals or madrepores (Scleractinia) are common forms in coral reefs. CD had been urged by his friend Charles Stokes to study shallow-water corals such as *Fungia* during his voyage. See *Correspondence* vol. 2, letter to Charles Stokes, [1839–September 1842]; and vol. 7, Supplement, letter to Charles Stokes, [January – March 1842]. See also Sloan 1985, p. 104.
[5] The American zoologist and geologist James Dwight Dana had published a description of the corals and actinia collected during the voyage of the North American Exploring Expedition (Dana 1848). Of the 483 corals described, 229 were new, and Dana's classification of this difficult group has remained of value to the present day (*DSB*). CD had corresponded with Dana during the preparation of his monograph on the Cirripedia (see *Correspondence* vols. 4 and 5). In 1859, Dana's health had forced him to give up his scientific pursuits for an extended period, and he had not yet fully recuperated from his illness.

To George Maw 19 July [1861][1]

2. Hesketh Crescent | Torquay
July 19th

Dear Sir

I have now read your Article carefully & I must thank you again for the perfectly fair & liberal (or rather kind) spirit you show throughout.[2] I shd. much like to discuss some points; but I am not well & have come here for entire rest.— I must, however, just remark on the uncommon skill & fairness with which you have indicated many of strongest points in my favour. You have misunderstood me in only very few & not very important points, which is more than I can say of almost any of my Reviewers. You put capitally & very originally many points,—such as the opposed view on classification & on Homologies.—[3] I hope, but do not yet see my way, to lessen the force of your objections on these heads.—[4]

I do not think so much of some of the difficulties which you rate very highly: I find many of the very best of the younger geologists, viz Ramsay, Geikie, Jukes think that I have not spoken a bit too strongly on imperfection of the geological Record.—[5] I think you will find on reflexion & enquiry that you expect much more than can possibly be told under present state of geological knowledge (p. 7583) on period of coming in of Marsupials in Australia.—[6] The evidence you require can be adduced to limited extent in N. America, still less in S. America & strongly in Europe.— You put capitally case of mammary glands: oddly I lately determined to advance that case, owing to light thrown on *possibility* of such a transition by Prof. Wymans researches on Batrachians.—[7]

In your remarks & criticisms on every form not everywhere varying & producing a chaos of forms, I think you underrate the fact (if any part of my view is correct) that natural selection will only occasionally act when there is a place vacant which can be better filled up by selected & accumulated variations. Unless this process goes on, we should have mere individual variations, or at most such cases as Rubus, Salix &c.[8]

But I must not, for my sake & your sake go on scribbling: & I daresay I have not made myself intelligible.— Anyhow I can thank you & that was my sole object in beginning this note, so pray believe me Dear Sir | Yours very faithfully | Charles Darwin

Your review must have cost you much thought & labour.—

Royal Horticultural Society

CD note:[9]

I say in individual because leg lengthens jaw lengthens; & I believe such extends to most or all vertebrata— Mr Maw will say because leg formed by same law jaw will be formed. *Perhaps so. I argue only from known correlations & to be discovered!! [*interl*] But it seems to me, supposing each separately created this is a different consideration, from accounting for legs & jaws in all ['mammals' *del*] vertebrata by correlation— The correlation of growth applies only to individuals. But Mr Maw says there is correlation of colour in distinct genera. Laws of variation & some degree of genetic relation may account for this, as well as adaptation to ['sam' *del*] similar purposes.— [*blue crayon*]

All these remarks are a mere fallacy—it is mere assumption of Creation as a plan [*pencil*]

1 Dated by the reference to Maw 1861a and by CD's stay in Torquay in July 1861.

2 See the letter to George Maw, 13 July [1861], in which CD acknowledges having received a presentation copy of Maw's review of *Origin* (Maw 1861a). CD's annotated copy of the review is in the Darwin Pamphlet Collection–CUL. On the back cover, he wrote: '(Must be consulted before new Edit. of Origin)'.

3 Maw pointed out that CD laid great stress on how the facts of classification (in the degrees of resemblance between species of allied genera, families, and classes) and those of morphology (in the unity of type between members of the same class) lent support to his view of true genealogical relations between species. He attempted to show, however, that CD's principle of 'correlation of growth' could be explained just as well by the belief in special creation as it could by genealogy. He cited the similarity in colour exhibited by the flowers of plants in unrelated genera and the correlation of the form of the head to the form of the legs in animals (e.g., animals with long heads generally have long legs) whether related or no (Maw 1861a, pp. 7585–7). In the margin of the concluding page of this discussion, CD wrote, alternately: 'good' and 'no'. In response to Maw's interpretation of his principle of correlation of growth, CD wrote in pencil: 'But I introduce correlation only as far as known.— I apply it to parts of same individual & not to distinct individuals—'. At the bottom of the page, CD wrote: 'It may be truly said that what I ['ha' *del*] admit "correlation" would account for homologies; but I suspect that these correlations are applicable'.

4 See CD's note following the letter.

5 CD refers to the support for his views expressed by Andrew Crombie Ramsay, Archibald Geikie, and Joseph Beete Jukes, all of whom were attached to the Geological Survey of Great Britain. See *Correspondence* vol. 8, letters from A. C. Ramsay, 21 February 1860, and from J. B. Jukes, 27 February 1860.

6 Maw argued against CD's law of the succession of types, stating that if marsupials did not exist in Australia during the period they inhabited Europe, then:

> we have a strong plea for assuming the subsequent special creation of the modern Australian kangaroo and its allies; and if the far isolated continents of Europe and Australia were contemporaneously peopled with marsupials, it affords a striking exception to the evidence in support of the limitations of special types to special geographical areas.

(Maw 1861a, pp. 7583–4). CD marked this passage in his copy and wrote in blue crayon: 'In Europe several stages | S. America probably distinct stage'.

7 Maw used the mammary gland as an example of an organ that could not have developed gradually, in stages, stating (Maw 1861a, p. 7596):

> it is impossible to fancy a process of generation intermediate between the oviparous and viviparous, to account for the gradual adaptation of the young to suit the gradual perfection of the milk-producing organs.

At the top of this page, CD wrote: 'food in young stored up—, as we see in fish— Mem. Wyman on young Batrachian'. The reference is to a paper by Jeffries Wyman (Wyman 1859).

8 These are plant genera noted for the great number of species and varieties they include. In his review, Maw asked why, after a variety became a new species, variation should cease (Maw 1861a, p. 7597). CD wrote in the margin: 'Does not cease but not selected.'

9 The note is tipped into CD's copy of Maw's review.

To A. G. More 19 July 1861

2 Hesketh Crescent, | Torquay
July 19, 1861.

My dear Sir

Very sincere thanks for the E. palustris which will be very useful, as I have 2 or 3 doubtful points to examine.[1] I am greatly obliged for your patience in watching the flowers;[2] I know from experience it is tedious work. From analogy I should doubt moths being the visitants.[3] The irritability must be from what you say a false scent, yet a drop of Chloroform certainly caused first a slight movement forwards and then backwards; but this I now infer was mechanical from endosmosis or exosmosis.[4] Were your observations made in sunshine? I can see it would be very difficult to detect any movement owing to great flexibility and elasticity of the part.

Hearty thanks for the experiment of cutting off the distal portion of the Labellum;[5] your friend could see whether the pods of these flowers swelled in course of few weeks as well as those of the unmutilated flowers; but it would be well to open them to look for seeds.[6] My notion is that the springing back of the elastic labellum leads the insects to crawl out of the *upper* part of the flower and not backwards over the labellum. I am sure for removing the pollinia the rostellum has to be brushed upwards and outwards.[7]

Again believe me My dear Sir | Yours sincerely obliged | Charles Darwin

Copy
DAR 146

1 See letter to A. G. More, 7 July 1861.

2 CD thanked More for repeatedly sending him specimens of the marsh orchis, *Epipactis palustris*, in *Orchids*, p. 95 n. More had apparently observed that when flies visited this orchid, their weight caused part of the labellum to become greatly depressed (see *Orchids*, p. 99).

[3] CD discovered after the publication of *Orchids* that hive-bees were 'the chief agents in fertilisation' of this orchid (see *Orchids* 2d ed., p. 99).
[4] These terms denote the passage of fluid inwards and outwards, respectively, through a porous membrane (*OED*).
[5] CD had asked More to carry out this experiment (letter to A. G. More, 7 July 1861).
[6] CD reported the results in *Orchids*, pp. 101–2 n.:

> As it is quite possible that I may have overrated the importance of the peculiar structure of the labellum, I asked Mr. A. G. More to remove the distal half of the labellum from some flowers before they had expanded, but I was too late in my application. He was able to try only two flowers, which were situated near the summit of the spike. These flowers formed seed-capsules, which were certainly small; but this may have been owing to their position. Unfortunately also these capsules shed most of their seed in being sent to me; so that I could not ascertain whether the seeds were well formed. Of the few seeds which did remain within these two capsules many were shrivelled and bad.

[7] In *Orchids*, pp. 99–100, CD provided a full account of how he arrived at this understanding of the pollinating mechanism of *Epipactis palustris*.

To Henry Fawcett 20 July 1861

You could not possibly have told me anything which would have given me more satisfaction than what you say about Mr. Mill's opinion.[1] Until your review appeared I began to think that perhaps I did not understand at all how to reason scientifically.[2]

Incomplete
Stephen 1885, p. 100

[1] Fawcett had told CD about a conversation he had had with John Stuart Mill in which Mill endorsed the logic of CD's reasoning in *Origin*. See letter from Henry Fawcett, 16 July [1861] and n. 3.
[2] Both Adam Sedgwick and Samuel Wilberforce in their reviews of *Origin* had criticised CD's method of reasoning ([A. Sedgwick] 1860 and [Wilberforce] 1860). Fawcett's review published in *Macmillan's Magazine* (Fawcett 1860) countered these criticisms.

To Charles Lyell 20 July [1861][1]

2. Hesketh Crescent | Torquay
July 20th

My dear Lyell

I sent you two or three days ago a duplicate of a good Review of the Origin by a Mr. Maw,[2] evidently a thoughtful man; as I thought you might like to have it, as you have so many.— By the way I was pleased, considering how many have attacked me on "Induction" &c, to hear the other day from the blind H. Fawcett, that he had spent an evening with J. Stuart Mill "who considers that your reasoning througout is in the most exact accordance with the strict principles of logic. He also says the method of investigation followed is the only one proper to such a subject."

Considering how high an authority he is, this pleases me much, & I think you will be pleased.—[3]

Lady Lyell[4] & you will be glad to hear that Etty improves a little.[5] This is a quite charming place & I have actually walked I believe good two miles out & back, which is a grand feat.— I saw Mr Pengelly the other day & was pleased at his enthusiasm.[6]

I do not in the least know whether you are in London. Your illness must have lost you much time, but I hope you have nearly got your great job of the new Edition finished.[7] You must be very busy, if in London, so I will be generous & on honour bright do not expect any answer to this dull little note.— I am doing nothing, except drawing up a long paper on the fertilisation of Orchids, which will, I think, be curious;[8] but I almost wish I could have been completely idle here. When I shall ever complete my book on "Variation under Domestication" Heaven only knows, for I do not.—[9]

Farewell my dear Lyell; I heartily hope that all things go well with Lady Lyell & you.— | Yours most truly | C. Darwin

Endorsement: '1861'
American Philosophical Society (258)

1 The endorsement is confirmed by CD's stay in Torquay (see 'Journal'; Appendix II).
2 Maw 1861a.
3 Letter from Henry Fawcett, 16 July [1861]. After the publication in 1843 of his *System of logic*, John Stuart Mill became recognised as the leader of the empirical school of philosophy (*DNB*).
4 Mary Elizabeth Lyell.
5 The Darwins were visiting the seaside resort of Torquay in the hope that it would improve the health of their daughter, Henrietta Emma Darwin. Emma Darwin's diary is filled with references to Henrietta's recuperation while at Torquay in July and August 1861.
6 The geologist William Pengelly, a regular correspondent of Lyell's, was directing excavation work being carried out in the Brixham cave, near Torquay. Over the past three years the bones of extinct mammals and other remains had been discovered in deposits in the cave. Pengelly was keeping Lyell informed about the coal-plants found in deposits near Bovey Tracey, Devon (see K. M. Lyell ed. 1881, 2: 346).
7 The sixth edition of Lyell's *Elements of geology* did not in fact appear until 1865. The delay was due to his decision to publish the results of his study of the antiquity of man, originally to be included in the *Elements*, as a separate volume (C. Lyell 1863).
8 CD was intending to summarise the results of his study of the pollination mechanisms in orchids in a paper to be contributed to the Linnean Society of London (see letter to J. D. Hooker, 19 June [1861]). In the event, he published his results in a small volume in 1862 (*Orchids*).
9 *Variation* appeared in two volumes in 1868.

To Asa Gray 21 July [1861][1]

2. Hesketch Cresent | Torquay | Devon.
July 21.

My dear Gray

I have been indolent, otherwise I sh^d.^ have thanked you heartily for taking trouble about getting me the 3 Hybrid pamphlets,[2] which I received just before starting here for the sea for my daughter's sake who improves a little.— And now I have to thank you for your pleasant & useful note of July 2d.—[3]

I am now writing my Orchid paper & am glad to hear what you say about Cypripedium[4] & that you will have a look at Spiranthes:[5] remember that you must look at flower just opened & not visited by moth. I hope Hooker will send me an Arethusa.[6]

I am astonished at Drosera filiformis misbehaving: my experiments were all tried late in summer.[7] Did you expect a rapid or visible movement? that very rarely takes place: the *minutest* atom of raw meat placed on a *single* gland, *which is covered with viscid secretion*, shows motion best.— If gland is dry no movement takes place. I hope to finish my observations on Drosera this Autumn.[8] From what I have seen of Cypripedium insigne my difficulty is to know what induces an insect to place its head so far back in flower:[9] if you find many plants in flower watch for a little time for *bare* chance of an insect's visit.—

This diagram gives, as you know, the structure of the 2 forms of Primula;

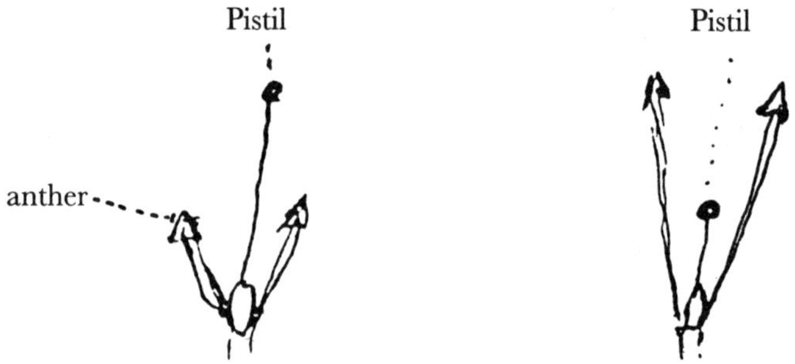

now if you know of analogous cases, I shd. be *particularly* obliged if you would name them & allow me to quote you:[10] also, if they are hardy plants & good seeders I would experiment on them.— I could not read one name given.—

How I shd. like half-an-hours conversation on Design: nothing else would make us quite understand each other.[11] As no one has aided the subject of natural selection & the knowledge of my Book so much as you, I must tell you what has pleased me much after the many attacks on me for neglecting "Induction" "Baconian philosophy" &c. We in England think John Stuart Mill the highest authority on such subjects, & he said lately to a friend, who wrote to me, as follows. "he considers that your reasoning throughout is in the most exact accordance with the strict principles of logic. He also says the method of investigation you have followed is the only one proper to such a subject."—[12] My wife's remark on reading this, was "why you know nothing about Logic".— Is it not (I mean Mill's & not my wife's sayings) very satisfactory to me?—

I was very glad of your P.S. on the state of your country; one values a private note far more than a dozen public letters. After carefully reading Olmstead's last Book I never doubted the North would conquer the South.[13] But then what is to follow? From Olmstead & Russell's letters in Times,[14] I cannot believe that the

South would ever have fellow-feeling enough with the North to allow of government in common. Could the North endure a Southern President? The whole affair is a great misfortune in the progress of the World; but I shd not regret it so much, if I could persuade myself that Slavery would be annihilated. But your president does not even mention the word in his Address.—[15] I sometimes wish the contest to grow so desperate that the north would be led to declare freedom as a diversion against the Enemy. In 50 or 100 years your posterity would bless the act.— But Heaven knows why I trouble you with my speculations; I ought to stick to Orchids.

Hooker, I fear, is as usual overworking himself: I seldom hear from him. This place is charming; but I am not as idle, as I ought to be.

Farewell | My good & kind friend | Yours most truly | Ch. Darwin

I am trying to make out the good of two forms of Thyme.[16]

We shall be here about 4 weeks more

Gray Herbarium of Harvard University (61)

[1] The year is provided by the Darwins' stay in Torquay in 1861.

[2] CD had asked Gray to send him a copy of S. G. Morton 1850a (see letter to Asa Gray, 12 March [1861]). There is an annotated copy of this paper and a copy of S. G. Morton 1850b in the Darwin Pamphlet Collection–CUL. The third pamphlet has not been identified.

[3] The letter from Gray has not been found.

[4] For the information CD sought from Gray concerning American species of *Cypripedium*, see the letter to Asa Gray, 5 June [1861]. For Gray's response, see *Correspondence* vol. 10, letter from Asa Gray, 10 November 1862.

[5] Gray's observations of *Spiranthes gracilis* and *S. cernua* are mentioned in *Orchids*, p. 123 n. When he came to review *Orchids* in 1862, Gray mentioned his examination of *Spiranthes* and demurred somewhat to CD's view of the implications of the flower's structure (see A. Gray 1862, p. 427).

[6] See letter to J. D. Hooker, 17 [July 1861]. CD was unable to examine any 'living flowers' of the Arethuseae before the publication of his study (see *Orchids*, pp. 269–70).

[7] CD carried out many experiments on various species of *Drosera* and other insectivorous plants during the summer of 1860 (see *Correspondence* vol. 8); he continued to study these plants well into the next decade. His notes on the sensitivity of the leaf hairs to nitrogenous substances are in DAR 54, DAR 60.1, and DAR 60.2.

[8] For a description of the experiments CD had suggested Gray might carry out on *Drosera*, see the letter to Asa Gray, 17 February [1861]. Having given a preliminary report on his study of insectivorous plants to the Philosophical Club of the Royal Society in February 1861, CD hoped to complete his researches in the autumn and write a complete account. However, *Insectivorous plants* was not published until 1875. CD's description of *D. filiformis* is given on p. 281.

[9] In *Orchids*, pp. 270–6, CD described the morphology of the flower parts of *Cypripedium*. This orchid is commonly known as the lady's-slipper orchid owing to the slipper-like appearance of the lip or lower-most petal of the flower (the labellum). CD proposed, on the basis of inserting a bristle into the labellum and noting where the pollen adhered, the probable means whereby insects could effect self- or cross-fertilisation (*ibid.*, pp. 274–6).

[10] CD intended to draw up an account of his study of heterostyly in primroses and other plant genera. His paper, 'On the two forms, or dimorphic condition, in the species of *Primula*, and on their remarkable sexual relations', was read to the Linnean Society in November 1861 and published the following year in the *Journal of the Proceedings of the Linnean Society (Botany)* 6 (1862): 77–96. See also *Collected papers* 2: 62. Gray supplied CD with a number of analogous cases included in this paper (see letters from Asa Gray, [27 and 29 August] and 2 September [1861], and 11 October 1861).

[11] CD and Gray had been corresponding over the past year about their differing views on natural selection and, in particular, whether such a mechanism could be reconciled with a belief in design in nature. In his pamphlet on this subject (A. Gray 1861a), Gray suggested that natural selection could be merged with natural theology if one assumed that variations were themselves 'guided' or 'designed'. CD, however, told Gray that he himself could not 'look at each separate thing as the result of Design' (*Correspondence* vol. 8, letter to Asa Gray, 26 November [1860]).

[12] Letter from Henry Fawcett, 16 July [1861].

[13] Olmsted 1860. CD had read Frederick Law Olmsted's *A journey in the back country* late in 1860 and found it to be 'a remarkably interesting Book.—' (*Correspondence* vol. 8, letter to Asa Gray, 26 November [1860]). Olmsted was highly critical of Southern society, arguing that the system of slavery on which it was based led to an inefficient economic order as well as a morally corrupt social order. Further, he denigrated the ability of the South to defeat the North, basing his argument on the problems faced by the South with regard to its military strength and economic resources.

[14] William Howard Russell served as the special war correspondent for *The Times*, contributing, beginning in April 1861, a regular column entitled 'The Civil War in America'. Although *The Times* was sympathetic towards the Southern cause, Russell's own sympathies were strongly with the North owing to his extreme opposition to the institution of slavery (*DNB*). See also Brogan ed. 1975.

[15] In his address before a special session of the United States Congress on 4 July 1861, President Abraham Lincoln stated that the administration had no intention, either 'directly or indirectly, to interfere with Slavery in the States where it exists.' This position was consistent with the Republican platform in the election of 1860, which opposed on constitutional grounds the right of southern states to secede from the union, but did not seek to interfere with existing institutions in those states. See McPherson 1988, p. 312.

[16] CD's notes on observations made on thyme, dated '1861. June 5', are in his Experimental book (DAR 157a, p. 72). CD summarised some of the results of his study of *Thymus* in *Forms of flowers*, pp. 298–305.

To John Tyndall 24 July [1861][1]

2. Hesketh Crescent | Torquay
July 24.

My dear Tyndall

I write a single line to say that it was not of the least consequence your retaining the letter for as long as you liked.— I shall be most happy at any time to send you the letters again, if you like to publish any passages.—[2]

We are all here idling for health-sake & enjoying the beautiful scenery & sea air—

By the way the other day at dinner we were all sticking bits of ice together by their points, marvelling at the phenomenon & talking of you—[3]

Yours most sincerely | Ch. Darwin

Down House (MS 8: 4)

[1] The year is given by reference to the Darwins' stay in Torquay during the summer of 1861 (see 'Journal'; Appendix II).

[2] The reference is to letters from Erasmus Darwin to Josiah Wedgwood I that CD had sent to Tyndall (see letter to John Tyndall, 23 February [1861], and Appendix V). The letters discussed physical aspects of the congelation of ice.

[3] Tyndall used the principle of 'regelation' to account for the way in which a glacier could be moulded under its own pressure. According to this principle, when ice is broken into fragments under pressure, the fragments, becoming moist through melting, reunite when they again come into contact. See Tyndall 1860, pp. 351–2.

From H. C. Watson 24 July 1861

Thames Ditton S.W.
24 July 61

My dear Mr Darwin

Tho' you desire me not to answer your letter of 17th. I will do so after a sort.[1] Assume that the quasi-popular monthly botanical journal "The British Botanist" tries its luck next year,—which is likely, not quite certain.[2] In such case, could you address a letter to it, much like the one to me, pointing out the observations you deem desirable to be made on the associated or dissevered distribution of allied subspecies. This would not only justify me in citing instances, but probably elicit facts from other botanists, bearing on the same question. In such a matter the knowledge or experience of any one individual should be confirmed or corrected by the eyes & understandings of others, who know what to distinguish, & what names to use.

When you are again within reach of Cybele Britannica Supplt., if you will refer to the middle paragh. of page 45 & examples in sequence,—you will see an allusion to one of the difficulties which prevent an accurate tracing out of the localities & distribution of subspecies; viz the dubiety of the records, arising from the diverse applications of the same name,—to A, to B, to C, or inclusively to ABC.[3]

Your letter was found here on my return from a day in town, bringing Maw's review of the 'Origin' with me.[4] Though aware of his intention to review, & even having recommended him to wait the 3d. edition before printing, I did not know whether he would be in support or opposition. He is more opposed than I had expected to find him. Surely Maw himself misconceives your views where he corrects Bree at the bottom of page 7578?[5] As you think that existing species can change into other species, by a very gradual accumulation of slight variations, you do really believe in & contend for transmutation of highly developed forms into some other (more or less different) highly developed forms,—which Maw seems to suppose outside your doctrine.

Is not Mr. D. himself slightly contradictory of himself, in his remarks on Habenaria bifolia & chlorantha?[6] If they "deserve to be called distinct species" because fertilized in a different manner by Moths,—a test to separate species & varieties is found or supposed.— Why cannot two varieties be fertilized differently, as well as two species? De facto, however, you are probably right; because by species you really mean only wider varieties or races, and each additional difference helps to widen.

Believe me, very faithfully Yours | Hewett C. Watson

P.S. You must just now be a close neighbour of my cousin Hercules Scott,[7] who is likely to look on you as a dreadful person; being something more opposed to your views than Maw is in the parag[h]. divided between pages 7609 & 7610,[8] that is, if he knows them at all.

DAR 181.1

CD ANNOTATIONS
3.1 Your . . . likely to 6.2] *crossed ink*
Top of first page: 'Keep' *blue crayon, circled blue crayon*; '(Varieties)' *pencil*

1 Letter to H. C. Watson, [17 July 1861].

2 Watson had planned since 1857 to launch a monthly journal called the *British Botanist*. Although two circulars were printed announcing the journal as 'intended chiefly for records in British Botany, descriptive, critical and topographical', the plan never came to fruition. See Allen 1986, pp. 71–3.

3 Watson had sent CD a presentation copy of the privately printed supplement to his four-volume *Cybele Britannica* (1847–59). There is an annotated copy of Watson 1860 in the Darwin Library–CUL. The reference is to a section giving new data concerning the distribution of British plants throughout 38 sub-provinces defined by Watson. Watson noted that botanists should be aware of whether the species name they were using in their local floras was 'the less restricted old (aggregate) species' or 'the more restricted recent (segregate) species'. One of the examples he gave was the following (Watson 1860, pp. 45–6):

> *Orchis bifolia* was long held to be one single species, and by some botanists it is still so regarded. It is treated as a single species in various Floras, local lists, etc. But latterly it has been more usually subdivided into two reputed species, *Orchis* (Habenaria, or Gymnadenia) *biflora* and *chlorantha*, two quasi-species slightly different in technical character. When the name *bifolia* is found in an old list, it may now be quite impossible to say with confidence which of the two modern semi-species was intended thereby.

There is a list in DAR 206 compiled by CD from Watson 1860 that indicates the distribution and range of related species.

4 Maw 1861a.

5 The reference is to Charles Robert Bree's negative critique of *Origin* (Bree 1860). Bree's book was one of four books reviewed in Maw 1861a, including the third edition of *Origin*.

6 CD had apparently referred to the differences between *Habenaria chlorantha* and *H. bifolia* in his letter to Watson of [17 July 1861], of which only a draft and a fragment survive; for a similar reference, see the letter to George Bentham, 22 June [1861]. In *Orchids*, p. 89, CD stated that the 'large butterfly orchis' *H. chlorantha* and the 'lesser butterfly orchis' *H. bifolia* 'differ in a great number of characters, not to mention the difference in general aspects and in the stations inhabited'.

7 The reference is to the botanist Hercules R. Scott.

8 In this passage of George Maw's review of *Origin*, Maw confessed that he was distressed by the apparent conflict between CD's theory and revelation, and particularly by CD's inference that the human species was 'genealogically allied to brute life' (Maw 1861a).

From H. C. Watson [after 24 July 1861][1]

P.S. As to Veronica humifusa & Serpyllifolia.— Some botanists hold the former a good species. And the facts read either way.

V. humifusa perishes under given circ̃es where V. Serpyllifolia flourishes, & naturally occurs under diff$^{t.}$ circ̃es.

Therefore it is a species of itself.— Or, therefore a var$^{y.}$ has a different climatal adaptation— Or, therefore a var$^{y.}$ is caused by difference of climate & soil acting on a succession of generations.[2]

Incomplete
DAR 47: 162

CD ANNOTATION
Bottom of page: 'Ch 7.'[3] *brown crayon*

[1] The date is suggested by the relationship between the contents of the postscript and topics discussed by George Maw in Maw 1861a (see n. 2, below). CD and Watson had been corresponding about the commingling of varieties and their possible divergence into new species. See the preceding letter and letter to H. C. Watson, [17 July 1861].

[2] The genus *Veronica* was cited by Maw as an instance in which species and varieties co-exist in the same geographic locality, seemingly without the 'mutually extirpating struggle upon which Mr. Darwin's theory of progression so much depends.' (Maw 1861a, p. 7600). In CD's copy of Maw's review (Darwin Pamphlet Collection–CUL), this passage is marked, and CD wrote next to it in pencil: 'This shows that no advantage'.

[3] The reference is to chapter seven, entitled 'Laws of variation: varieties and species compared', of CD's 'big book' on species (*Natural selection*).

To John Lubbock 26 July [1861][1]

2. Hesketh Crescent | Torquay
July 26th

My dear Lubbock

I am much obliged to you for so kindly seeing M$^{r.}$ Hacon from whom I heard this morning.—[2] It was *entirely* Mr Hacon's wish & suggestion that William should from the first have the power of giving one year's notice.— Having heard from him what you say I have written to him to agree to the seven years.— So little do I understand the advantages & disadvantages of the several points, that I had got to wish for a longer term than seven years.— So I felt forced to put myself into Mr Hacon's hands. I consider now that in fact everything is settled; & I am heartily glad to think that you will have no more trouble.[3] William has never for an instant fluctuated in his strong wish to join. Pray repeat my sincere thanks to Sir John.—[4]

I am enjoying myself & can walk a couple of miles out; & am doing a very little work in drawing up my Orchis paper.— Whether we shall stop here long I am very doubtful, for it is excessively damp & does not seem to suit Etty or my wife, which is a terrible evil & what to do & where to go Heaven only knows.

My dear Lubbock | Yours gratefully | Charles Darwin

Down House

[1] The year is given by the Darwins' stay in Torquay (see 'Journal'; Appendix II).

[2] CD's solicitor, William Mackmurdo Hacon of Rowland and Hacon, 31 Fenchurch Street, London, was negotiating the contractual details connected with William Erasmus Darwin's joining a Southampton bank. See letter to John Lubbock, 14 July [1861].

[3] The arrangements for making William Darwin a partner in the Southampton and Hampshire Bank, which began in May, were concluded in August (see letter to John Lubbock, 10 August [1861]).

[4] John Lubbock's father, John William Lubbock, had given CD advice about the legal provisions of the contract (see letter to John Lubbock, 14 July [1861]). Father and son were both partners in the London bank Robarts, Lubbock & Co.

To J. D. Hooker 27 July [1861][1]

2. Hesketh Crescent | Torquay
July 27th

My dear Hooker

You cannot conceive how the Orchids have delighted me.[2] They came safe, but box rather smashed: cylindrical old cocoa or snuff cannister much safer.— I enclose Postage.— As on account of movement I shall allude to what I suppose is Oncidium, to make *certain* is enclosed flower with crumpled petals this genus.—[3] Also I most specially want to know what enclosed little globular brown Orchid is. I have only seen pollen of a Cattleya on a Bee,[4] but surely have you not unintentionally sent me what I wanted most (after Catasetum or Mormodes) viz one of the Epidendreæ?!—[5] I *particularly* want (& will presently tell you why) another spike of this little Orchid, with *older* flowers, *some* even *almost* withered, as the rostellum evidently does not come into action soon: I think only way would be to tie spike to thin elastic stick & jam stick in cannister the stick also serving to keep piece of *damp* blotting paper or anything else from *touching* the flowers. Why I want to study rostellum is that the pollen is intermediate between the two great types, as I daresay may be well known.

(A) is waxy with no threads— (B) consists of a mass of pollen-grains united by elastic threads, & has exactly the structure of pollinia of Epipactis. So that (B) is in part functionally a caudicle, in structure a mass of pollen. I believe this explains the genesis of the caudicle of Orchis. By abortions of parts from this pollinium, almost any other could be made. Hence I am inclined to look at this as the prototype, & I am intensely curious to understand well the rostellum.[6]

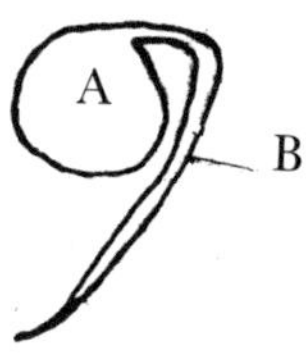

Did you not think that Stelis was all in bud; I did, for the sepals were so exactly closed; but *as yet*, I have only *one* flower, that on apex, which had not been visited by some insect & had its pollinia removed. Can you tell me *whence* (what country) this Stelis comes, as it is so attractive to British Insects?—[7]

Can you *easily* tell me name of larger *brown* orchis in bud with curved peduncles? I have not yet examined it though I have worked all day.—

I grieve to hear that I cannot see an Arethusa (Vanilla belongs to same type I see in Lindley's Veg. K.) for then I sh$^{d.}$ have seen all *great* leading types.—[8] You have given me intense enjoyment.—

Yours affect^{ly} & tired | C. Darwin

I am sorry to hear about death of Baby.—[9] William soon joins Bank. William has been examining seeds of Groundsell,—wonderful case[10]

DAR 115.2: 107

[1] The year is given by the Darwins' stay in Torquay from 2 July to 26 August 1861 (see 'Journal'; Appendix II).

[2] CD had asked Hooker to supply him with several orchid species that he had not yet examined (see letter to J. D. Hooker, 17 [July 1861]).

[3] See *Orchids*, pp. 192–3, where the 'depression' or movement of the pollinium of *Oncidium* is described. According to CD's view, this mechanism had been selected for in various orchids because it enabled them better to position the pollinia on insects and thus optimise the chances of the pollinia subsequently striking the stigmatic surface (*ibid.*, p. 357).

[4] See letter to Frederick Smith, 19 June [1861] and n. 2.

[5] *Cattleya* and other genera of the tribe of exotic orchids Epidendreae are described in *Orchids*, pp. 159–66.

[6] The final chapter of *Orchids* discusses the homologies of the flowers of orchids. In addition to tracing the gradation in the structure of the various organs such as the rostellum and the pollen-masses, CD also described what he believed to be the process by which the caudicle (the stalk on which the pollen-masses rest) was formed. The particular case mentioned in the letter is described in *Orchids*, p. 325.

[7] *Stelis* is a genus of orchids found in tropical America and the West Indies. In *Orchids*, p. 167, CD described how the three sepals of *Stelis racemiflora*, after expanding, gradually close 'with perfect exactness and shut up the flower, so that it is scarcely possible to distinguish an old flower from a bud'.

[8] CD was unable to examine any living flowers of the tribe Arethuseae, to which the genus *Vanilla* belongs, before *Orchids* was published in May 1862 (*Orchids*, p. 269). The reference is to Lindley 1846, which went through several editions. It is not known which edition CD used, although it is likely to have been the third edition (Lindley 1853), in which the section on orchids was expanded. John Lindley was the first botanist to prepare a detailed classification of the orchid family and was an authority on orchid cultivation. Lindley divided the family into seven tribes: Malaxeae, Epidendreae, Vandeae, Ophreae, Arethuseae, Neotteae, and Cypripedeae. With the exception of one genus, *Malaxis*, all the British orchids are confined to the Ophreae and Neotteae.

[9] The reference has not been traced.

[10] Legal arrangements were being made for William Erasmus Darwin to join the Southampton and Hampshire Bank. CD had previously asked Hooker to recommend a programme to direct William's current enthusiasm for botanical dissection (see letter to J. D. Hooker, 19 June [1861]). The 'seeds' (achenes) of the common groundsel, *Senecio vulgaris*, characterised by being covered with numerous soft, feathery tufts of hairs, are responsible for the plant's great success in propagating rapidly and throughout the year (*EB*). Hooker had earlier described for CD how in this and other Compositae, the characteristics of the achenes facilitate propagation. See *Correspondence* vol. 8, letter from J. D. Hooker, 28 December 1860.

To J. D. Hooker [28 July – 10 August 1861][1]

2. Hesketh Crescent | Torquay

Read this if ever you have time & inclination—

My dear Hooker

Many thanks for names & all information & for your enquiring about Epithecia.[2] Veitch sent me a grand lot this morning.[3] What wonderful structures! I have now seen enough, & you must not send me more, for though I enjoy looking at them **much**, & it has been very useful to me seeing so many different forms, it is idleness. For my object each species requires studying for days.— I wish you had time to take up the group. I would give a good deal to know what the rostellum is, of which I have traced so many curious modifications.[4] I suppose it cannot be one of the stigmas: there seems a great tendency for two lateral stigmas to appear?

My paper, though touching on only subordinate points will run, *I fear*, to 100 M.S. folio pages!!![5] The beauty of the adaptations of parts seems to me unparalleled. I shd. think or guess waxy pollen was most differentiated. In Cypripedium which seems least modified & a much exterminated group, the grains are single.[6] In *all others* as far as I have seen they are in packets of 4; & these packets cohere into many wedge-formed masses in Orchis; into 8, 4 & finally two. It seems curious that a flower shd. exist which could *at most* fertilise only 2 other flowers, seeing how abundant pollen generally is: this fact I look at as explaining the perfection of the contrivances by which the pollen, so important from its fewness, is carried from flower to flower. By the way Cephalanthera has single pollen-grains; but this seems to be a case of degradation for rostellum is utterly aborted.[7] Oddly the columns of pollen are here kept in place by very early penetration of pollen-tubes into the *edge* of the stigma: nevertheless it receives more pollen by insect agency.—

Epithecia has done me one good little turn: I often speculated how the caudicle of Orchis had been formed. I had noticed slight clouds in its substance half way down: I have now dissected them out, & I find they are pollen-grains fairly embedded & useless. If you suppose the pollen-grain to abort in the lower half of the pollinia of Epipactis; but the parallel elastic threads to remain & cohere; you have the caudicle of Orchis & can understand the few embedded & functionless pollen-grains.—[8]

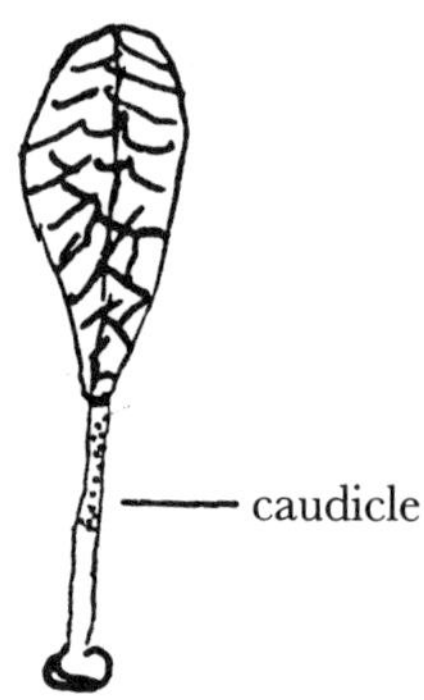

I must not look at any more exotic orchids; hearty thanks for your offer. But if you would make one single observation for me on Cypripedium, I should be glad. Asa Gray writes to me that the *outside* of pollen-masses is sticky in this genus:[9] I find that the whole mass consists of pollen-grain immersed in a sticky brownish thick fluid. You could tell by a mere lens & pen-knife.— If it is, as I found it, pollen could not get on stigma without insect-aid.—[10] Cypripedium confounds me much: I conjecture that drops of nectar are secreted by surface of Labellum beneath the anthers & in front of stigma, & that the shield over the anthers & form of Labellum is to compel insects to insert their probosces all round *both* organs.[11] It would be troublesome for you to look to this; as it is always bothersome to catch the nectar secreting; & the *cup* of Labellum gets filled with water by gardeners watering.—

I have examined Listera, ovata, cordata & nidus avis; the pollen is uniform: I suspect you must have seen some observation founded on mistake from the penetration & hardening of sticky fluid from rostellum, which does penetrate the pollen a little.—[12]

It is mere virtue which makes me *not* wish to examine more orchids; for I like it far better than writing about varieties of cocks & Hens & Ducks.—[13] Nevertheless I have just been looking at Lindleys list in Veg. K. & I cannot resist one or two of his great Division of Arethuseæ, which includes *Vanilla*.[14] And as I know so well the Ophreæ, I sh^{d} like (God forgive me) any *one* of the Satyriadæ, Disidæ, & Corycidæ.—

I fear my long lucubration will have wearied you; but it has amused me to write, so forgive me.— Ever your affect | C. Darwin

I am very sorry that I have failed to see Catasetum & Mormodes; I think I could have madout all & means of ejection of pollinia. Let me know affinities of Epithecia. I marvel often as I think over the diversity & perfection of the contrivances.—

There is a *Dutch*(!) Translation of "Origin" come out.—[15]

DAR 115.2: 109

[1] Dated by the relationship to the preceding letter and by the fact that CD had received a specimen of *Epithecia* from Hooker by 11 August 1861 (see n. 2, below, and letter to J. D. Hooker, [11 August 1861]).

[2] *Epithecia* is another name for *Epidendrum*. CD had asked Hooker to send him orchids from the tribe Epidendreae in the letter to J. D. Hooker, 27 July [1861]. In the same letter CD asked Hooker to confirm the names of some of the orchids he had sent.

[3] James Veitch Jr, of Veitch & Sons Royal Exotic Nurseries, Chelsea, is acknowledged in *Orchids* for having provided CD with specimens of exotic orchids (*Orchids*, p. 158 n.)

[4] At the end of the Introduction to *Orchids*, CD included a section giving definitions of terms, in which he described how the upper stigma is modified into 'an extraordinary organ, called the *Rostellum*, which in many Orchids presents no resemblance to a true stigma. The rostellum either includes or is formed of viscid matter' (*Orchids*, p. 6).

[5] Owing to the length of his manuscript, CD published his material on orchids in book form (*Orchids*) rather than in a scientific journal.

[6] CD described the orchid genus *Cypripedium* in *Orchids*, pp. 270–6, noting their single pollen grains on p. 271.

[7] In *Orchids*, pp. 155–6, CD discussed how the transition from orchid flowers with well-developed rostellums to those of *Cephalanthera*, with their aborted rostellums, might have arisen 'with each gradation useful to the plant'.

[8] CD reiterated this view of the development of the caudicle in *Orchids*, pp. 327–8.

[9] Asa Gray's letter has not been found, but see the letter to Asa Gray, 21 July [1861] and n. 4.

[10] CD discussed the necessity for insect aid in the pollination of *Cypripedium* in *Orchids*, pp. 273–6.

[11] See *Orchids*, pp. 279–80.

[12] Hooker had studied the rostellumn of *Listera ovata* and published a paper on the subject (Hooker 1854b). The point to which CD refers, however, is not discussed in this work. In *Orchids*, p. 139, CD

stated that Hooker had described the structure of this species 'minutely, and of course correctly', though he had not attended to the role of insects in its fertilisation, whereas Christian Konrad Sprengel had 'misunderstood both the structure and the action of the rostellum'.

13 CD had been writing up his materials on variation in fowl and ducks for inclusion in *Variation*. He completed the manuscript in May (see 'Journal'; Appendix II).

14 See letter to J. D. Hooker, 27 July [1861], n. 8.

15 Winkler trans. 1860.

From William Branwhite Clarke [August 1861][1]

Whilst I am writing, you will, I hope, pardon my allusion to another remark in your work, which I think must apply to myself. In p 373 you say "If one account which has been published can be trusted, we have direct evidence of glacial action in the South-eastern corner of Australia"—[2]

I think here you refer to a passage in one of my Reports to the Government of N.S.W. dated Eden 6 Mar 1852. and published in the Parl^y. Bl. Books. If so, I would now refer to p 167. in my little book, lately published, a copy of wh. is at the Geol. Soc^ys library, entitled "Researches in the Southern Gold Fields of N.S.W"—[3] On that page is a note respecting rocks polished by wallabies.

Till I saw those rocks, I had always firmly believed, that the *polish* at Jincabillee, Rock Flat and Aston Ck, was the result of *snow*: and even now I am of opinion such is the case. But I have no doubt whatever about the Bolivia granite. I enclose you a minute slice of the surface, that you may examine it, and a stereo scopic view of the locality.

In the same little book is a chapter on Climate & Vegetation in the Australian Alps (p 221–234) in which you will find mention of distinct *rocas perchés*.[4] I saw several of these which I am quite sure of. And in a coom on the Snowy Plain (not far from the Gangarlin River) I came upon an assemblage of rounded blocks of Granite which so much resembled a moraine, that I have no doubt of that either. Furthermore, in Norfolk Island, the surface is partly covered by transported blocks of greenstone. This evidence, therefore, of ice in *two ways* may I think be admitted.

I have also speculated on the possibility of auriferous deposits being in some instances *glacial*—i.e. deposited after the rocks had been subjected to the action of ice and snow. At the head of the Encumbene River (where the recent Kiandra gold workings are established p 117 &c) there are heaps of what the diggers call "made ground" i.e accumulations of detritus collected either by water of the rivers or of floods or in some other way since the gold was disintegrated from the rocks. And in these heaps are patches of shale or clay, containing impressions of plants which I believe to be of Miocene age. Now if you see what I say elsewhere, on p 86, p 94, p 105, you will think with me, that there were in ancient times more ⟨evi⟩dences of the abode and action ⟨of⟩ ice than we now perceive.[5]

I make no apology for thus addressing you—because, as you will observe, my object is on your own behalf.—

We have living here, an old ship-mate of yours *Conrad Martens*, who has sometimes named you to me. He is now looking old. He was my first church-warden here in 1846.—[6]

You will probably recollect a paper of mine in the Tasmanian Journal on Showers of Dust—you commented on it I think before the Geol. Socy.[7] I now know that the pumice which is also mentioned forms, in places along the coast, beds beyond and above high water mark: and must, therefore, be considered as belonging to Raised Beaches. So that the Pumice period is [*illeg*].

I have filled my paper with an almost unreadable scrawl—but I write against time. Pray believe me to be, | My dear Sir, | Yours very truly | W. B Clarke

Incomplete
DAR 161.2: 171

CD ANNOTATIONS
1.1 Whilst ... Australia"— 1.4] *crossed blue crayon*
2.1 I think ... wallabies. 2.5] *scored blue crayon*
4.6 Norfolk Island,] *underl blue crayon*
6.1 I make ... Clarke 9.2] *crossed blue crayon*
Bottom of first page: 'Glacial—' *blue crayon, circled blue crayon*; 'W. B. Clarke—Aug 1861' *blue crayon*

1 The date is provided by CD's annotation.

2 Clarke, who had emigrated to New South Wales in 1839, carried out an extensive survey of the geological and mineralogical deposits of Southern Australia. He refers to *Origin*, p. 373.

3 Clarke 1860. Clarke had discovered gold in Australia in 1841, and in July 1860 the governors of the Australian colonies officially recognised him as the first to have discovered gold on the continent (*DNB*).

4 CD had long been interested in erratic boulders as evidence of past glacial action and of a former worldwide cold period. He had already heard of such boulders occurring in New Zealand (see *Correspondence* vol. 6, letter to W. B. D. Mantell, 3 April [1856]).

5 In the next edition of *Origin*, CD slightly altered his statement regarding evidence of glaciation in south-eastern Australia to credit Clarke as the authority for the information (*Origin* 4th ed., p. 443).

6 Conrad Martens joined the *Beagle* as draughtsman in Montevideo in 1833, replacing Augustus Earle. Martens prepared a number of pencil sketches and watercolours from the voyage. He settled in Australia in 1835, becoming a distinguished landscape painter. Clarke was the minister of the parish church in Willoughby, New South Wales.

7 In his paper 'On the fine dust which often falls on vessels in the Atlantic', read to the Geological Society of London on 4 June 1845 (*Collected papers* 1: 199–203), CD cited one of Clarke's papers on the same topic published in the *Tasmanian Journal* 1 (1842): 321–41.

To Charles Lyell [1 August 1861][1]

2. Hesketh Crescent. Torquay
Thursday

My dear Lyell

Emma has gone a little tour with Etty,[2] but I have forwarded Lady Lyell's letter with the sad account of Mrs. Longfellow.—[3] I am sorry to have troubled you, but sh^{d} be much obliged for Dutch Translat. to be left at Q. Anne St.—[4] I am surprised at a *Dutch* Translation—[5]

I declare that you read the Reviews on the Origin more carefully than I do.— I agree with all your remarks.— The point of correlation struck me as well put, & on varieties growing together; but I have already begun to put things in train for information on this latter head, on which Bronn also enlarges.—[6]

With respect to Sexuality, I have often speculated on it, & have always concluded that we are too ignorant to speculate— no physiologist can conjecture why the two elements go to form the new being; & more than that why nature strives at uniting the two elements from *two individuals*; what I am now working at, viz Orchids, is admirable illustration of the law.— I sh^{d} certainly conclude that all Sexuality had descended from one prototype.[7] Do you not underrate the degree of lowness of organization at which sexuality occurs, viz in Hydra: & still lower in some of the *one-celled* free Confervæ which "conjugate", which good judges (Thwaites) believe is simplest form of true sexual generation.[8] But the whole case is a mystery.—

There is another point on which I have occasionally wished to say a few words.— I believe you think with Asa Gray that I have not allowed enough for the stream of variation having been guided by a Higher power.—[9] I have had lately a good deal of correspondence on this head.[10] Herschel in his Phy. Geograph. has sentence with respect to the Origin something to the effect that the higher law of providential arrangement sh^{d} always be stated.[11] But astronomers do not state that God directs the course of each comet & planet.— The view that each variation has been providentially arranged seems to me to make natural selection entirely superfluous, & indeed takes whole case of appearance of new species out of the range of science. But what makes me most object to Asa Gray's view, is the study of the extreme variability of domestic animals.— He who does not suppose that each variation in the Pigeon was providentially caused, by accumulating which variations man made a Fantail, cannot, I think, logically argue that the tail of the Woodpecker was formed by variations providentially ordained.— It seems to me that variations in the domestic & wild conditions are due to unknown causes & are without purpose & in so far accidental; & that they become purposeful only when they are selected by man for his pleasure, or by what we call natural selection in the struggle for life & under changing conditions. I do not wish to say that God did not foresee everything which would ensue; but here comes very nearly the same sort of wretched embroglio as between free-will & preordained necessity.

I doubt whether I have made what I think clear; but certainly A. Gray's notion of the course of variation having been led, like a stream of water by Gravity, seems to me to smash the whole affair. It reminds me of a Spaniard whom I told I was trying to make out how the Cordillera were formed; & he answered me that it was useless for "God made them". It may be said that God foresaw how they would be made. I wonder whether Herschel would say that you ought always to give the higher providential Law, & declare that God had ordered all certain changes of level that certain mountains should arise.— I must think that such views of

Asa Gray & Herschel merely show that the subject in their minds is in Comte's theological stage of science.—[12]

I have one other very distinct subject. William will, I apprehend, now certainly join M^r. Atherley's Bank, & it would be of real importance to him to get any good introduction in or near Southampton:[13] can you aid me? it would be a real service.—

Of course I do not want any answer to my quasi theological discussion: but only for you to think of my notions, if you understand them.

I hope to Heaven your long & great labours on your new Edit. are drawing to a close.[14]

Farewell | My dear Lyell | Yours most truly | C. Darwin

Very kind remembrances to all your party.—[15]

Endorsement: '2 Augt 1861'
American Philosophical Society (259)

[1] Dated by the endorsement. The first of August fell on a Thursday in 1861.

[2] Emma Darwin's diary records that she and Henrietta Emma Darwin, accompanied by Hope Elizabeth Wedgwood (the youngest daughter of Hensleigh and Fanny Mackintosh Wedgwood), went on a tour of the Dartmoor area of Devon from 30 July until 5 August. Their tour included visits to Ashburton, Holne Chase, Lushleigh, Whyddon Park, Chagford, Fingle Bridge, and Drewsteignton. According to Henrietta Litchfield, this was the only tour Emma Darwin 'ever took without the family in all her married life.' (*Emma Darwin* 2: 178).

[3] Frances Elizabeth Longfellow, wife of the American poet Henry Wadsworth Longfellow, was burned to death in her home in Cambridge, Massachusetts, after her dress caught fire. She was the sister of Mary Mackintosh, the wife of Robert Mackintosh, who was the brother of Fanny Mackintosh Wedgwood.

[4] The address is that of CD's brother, Erasmus Alvey Darwin, where CD stayed when in London.

[5] Winkler trans. 1860. The Dutch palaeontologist Tiberius Cornelius Winkler had sent a copy of the translation to Lyell to be forwarded to CD. See letter from T. C. Winkler, 7 July 1861.

[6] CD probably refers to George Maw's review of the third edition of *Origin* (Maw 1861a). The points he refers to are addressed in Maw's review and in Heinrich Georg Bronn's criticism of *Origin* in chapter 15 of Bronn trans. 1860. See letters to George Maw, 13 July [1861], and to H. C. Watson, [17 July 1861].

[7] On CD's views on the origin and functional importance of sexual dimorphism in evolution, see Ghiselin 1969 and Hodge 1985.

[8] Thwaites 1847.

[9] Asa Gray argued in A. Gray 1861a that natural selection and natural theology were consistent if one conceived of the 'stream' of variations, from which nature selects, as having been 'guided' or 'designed'. For the correspondence between Gray and CD on this point, see *Correspondence* vol. 8, letters to Asa Gray, 3 July [1860], and 26 November [1860]. See also letters to Asa Gray, 5 June [1861] and 21 July [1861].

[10] In addition to the letters to Gray (see n. 9, above), see the letters to J. F. W. Herschel, 23 May [1861], and to F. J. Wedgwood, 11 July [1861].

[11] Herschel 1861, p. 12 n.

[12] Auguste Comte viewed the development of knowledge as having progressed through three stages: theological, metaphysical, and positive. CD had read an extensive review of the first two volumes of Comte's *Cours de philosophie positive* (Comte 1830–42) in 1838 (see *Correspondence* vol. 2, letter to Charles Lyell, [14] September [1838]).

[13] William Erasmus Darwin was preparing to become a partner in the Southampton and Hampshire Bank.

[14] The sixth edition of Lyell's *Elements of geology* did not in fact appear until 1865. The delay was due to Lyell's eventual decision to publish the results of his study of the antiquity of man as a separate volume (C. Lyell 1863) rather than as a section within *Elements*.

[15] The Lyells were staying in Folkestone, Kent. Among their party was George Bentham (K. M. Lyell ed. 1881, 2: 347–8).

To John Lubbock 1 [and 2] August [1861][1]

2. Hesketh Cres[t] | Torquay
Aug 1st

My dear Lubbock

I sent my Boys hunting for Lepisma, but they only brought an Isopod with long tail.[2] I will myself go soon & hunt, but I have had a bad day or two. By the way the cause was a too long visit to Wollaston, which at the time I much enjoyed, but paid for.[3] He has a very nice little house & is working hard; but his life is sadly too solitary. He has not a naturalist to associate with. He seems pretty strong, but looks fearfully delicate.—

I pity you about S. & W. but one always feels those sorts of blunders most oneself.[4]

I will send 3 guineas soon to Essay Fund:[5] I ought to give 5, but I feel too stingy & with my party of 18 money goes like water. My Brother, who is here, will subscribe & I daresay will give five guineas.—[6]

I have had another awfully long letter from my Lawyer, who says the other Lawyer does not understand him: confound them both;[7] I suppose they will agree soon. The points seem to me of very little importance.

Pray do not say that if the life does not suit William you shall reproach yourself: nobody could possibly have been kinder than you have been or given more cautious advice.—

My wife & Etty are gone for a little tour of a week to the higher lands of Dartmoor[8]

Goodnight | My dear Lubbock | Ever yours | C. Darwin

P.S. | I grieve to say that since above was written I have had a long letter from M[r] Hacon. He wants very much to see you again, & I have venture to say that he might call on you tomorrow (*Saturday*) or Monday & get your opinion.— My judgment being almost worthless, on these points makes the negotiations wonderfully difficult.— God knows whether M[r] Hacon is too particular; but without your aid it is clear to me the whole affair will be smashed; & I cannot but fear that M[r]. Atherley's patience (*& yours*) will be utterly exhausted. Yet it seems that it would be madness on my part to tell my solicitor he is giving trouble for no good, as I cannot understand the force of the difficulties.—

Endorsement: '61'
Down House

[1] The extended date is suggested by the postscript, which CD obviously wrote on the morning of Friday, 2 August 1861.

[2] *Lepisma* is a genus of bristletails (Thysanura) commonly known as the silver-fish; it is a primitive insect frequently found under logs in woods. Lubbock was engaged in a study of the Thysanura, which he published in 1862 (Lubbock 1862).

[3] Thomas Vernon Wollaston, a friend of both CD and Lubbock, had recently moved from London. He lived at Kingskerswell, Newton Abbot, near Torquay (*DNB*).

[4] The reference may be to an error in Lubbock 1860, a paper describing fifteen new species of Entomostraca. Each of the descriptions gives the locality where the specimen was taken, often listing 'S. lat.' and 'W. long.'. However, no erratum for this paper was printed.

[5] CD's Account book (Down House MS) has an entry dated 1 August 1861 for 'Essays and Reviews' to the amount of £3 3*s*. (three guineas). A subscription fund was being raised to defend the authors of *Essays and reviews*. In December 1861, two of the clergymen who had contributed essays, Henry Bristow Wilson and Rowland Williams, were tried for heresy by the Court of Arches. They were found guilty and suspended from their livings. For further information on the controversy surrounding *Essays and reviews*, see Crowther 1970 and Ellis 1980. See also Appendix VI.

[6] Erasmus Alvey Darwin remained with the Darwins in Torquay until 22 August 1861 (Emma Darwin's diary).

[7] The reference is to the ongoing negotiations to secure William Erasmus Darwin a partnership with George Atherley in the Southampton and Hampshire Bank. CD's solicitor, William Mackmurdo Hacon, had questioned some of the contractual provisions drawn up by Atherley's solicitor. See also following letter.

[8] Emma and Henrietta Emma Darwin spent a week touring the Dartmoor area of Devon (Emma Darwin's diary).

To John Lubbock [2 August 1861][1]

[Torquay]
Friday

P.S.

My dear Lubbock

This is a P.S. to my letter posted a few hours ago.

Mr Hacon seems to think that Mr Atherley wishes for William for only some 10 or 12 years until his son is grown up, & then that William is to be turned off without permission to join any Bank in Hampshire.— Now the question is, if you will be so very kind as to advise Mr Hacon, whether, (as it seems to me) this is being too suspicious, or whether anything can be done reasonably to save William from this which no doubt would be as bad an ending to his life as could well be.— If he had power of joining any other Bank in Southampton *if dismissed by Mr. A.* there would be some very slight check to his being dismissed without grounds.

But I do not know whether it is usual or wise to look so far ahead. It seems to me there must be risk; both sides must be free after a certain period.— If we *knew* that Mr A. intended to keep William as a partner only till his son grew up, it would not be worth joining.

Oh good Heavens how much more difficult Business is than Natural History!

Ever yours | C. Darwin

Pray, I beg you, not trouble yourself to write to me, I shall hear from Mr Hacon what you say, & that is enough & far more than enough trouble for you.—[2]

Down House

[1] Dated by the relationship to the preceding letter.
[2] See letter to John Lubbock, 9 August [1861].

From the University of Breslau 4 August 1861[1]

Euer Hochwohlgeboren beehrt sich die unterzeichnete Facultät die ganz ergebenste Anzeige zu machen, daß sie bei der am 2ten. 3ten. und 4ten. August d. J. vollgezogenen Feier des 50 jährigen Bestehens und Wirkens der Universität Ihnen honoris causa die Würde eines Doctors der Medicin und der Chirurgie verliehen hat.[2]

Die Facultät hat geglaubt der Jubelfeier einen erhöhten Glanz dadurch zu verleihen, daß sie Ihren in der Wissenschaft hochgeehrten Namen in die Reihe ihrer Ehren-Doctoren aufnahm und beehrt sich Ihnen das über den Promotions-Lob ausgefertigte Diplom, so wie das bei der Jubelfeier der Universität im Auftrage der Facultät veröffentlichte Programm ganz ergebenst zu überreichen.[3]

Breslau den 4ten. August 1861.
Die medicinische Facultät der Königlichen Universität.
H. Barkow, Z.Z.D. Betschler Lebert Middeldorpf[4]

An den Magister der freien Künste, | Mitglied geologischen und Koniglichen Gesellschaft Herren Dr. Carl Darwin | Hochwohlgeboren in London.

DAR 230

[1] For a translation of this letter, see Appendix I.
[2] The University of Breslau (present-day Wroclaw, Poland) was incorporated in 1811 out of what was formerly a Jesuit College. Breslau was the capital of the Prussian province of Silesia.
[3] The diploma was sent to CD in 1862 (see *Correspondence* vol. 10, letters from the Prussian Legation, 10 February 1862, and from H. Albertof, 13 February 1862). It is in DAR 229.
[4] Hans Karl Leopold Barkow was dean of the medical faculty at Breslau and Julius Wilhelm Betschler, Hermann Lebert, and Albrecht Theodor Middeldorpf were professors in the medical faculty (see Kaufmann ed. 1911).

To Charles Lyell 6 August [1861][1]

2. Hesketh Crescent | Torquay
Aug. 6

My dear Lyell

Mr Bentham has sent me a spurless O. pyramidalis, which you & he picked up by a strange coincidence.[2] As Bentham says he is going to leave his lodgings, I know not where to write & thank him. If you see him, please thank him.—[3]

But I write, also, to beg you, if you should stumble on another, send it in a little tin cannister; for the specimen was so utterly smashed as to be useless.[4]

The specimen to be of use to me sh^d. have lower flowers & old & half-withered; in this condition it would be highly useful.—

I fear, however, from what I could make out of the state of the flowers that they are irregular monsters.—[5] I have written to Tenby in vain for this variety, which would be so useful to me.—[6]

Ever yours | C. Darwin

Royal Botanic Gardens, Kew (Bentham letters: 698)

[1] Dated by the Darwins' stay in Torquay, Devon, from 2 July until 26 August 1861 ('Journal'; Appendix II).

[2] CD had written to George Bentham to ask him to look out for specimens of *Orchis pyramidalis* that lacked nectaries (letters to George Bentham, 17 June [1861] and 22 June [1861]).

[3] The Benthams visited Folkestone in Kent, where the Lyells were also staying, during the first week in August (Jackson 1906, pp. 187–90).

[4] There is a reference to CD's request in a letter Lyell wrote to his nephew, Leonard Lyell, on 7 August 1861 (K. M. Lyell ed. 1881, 2: 348):

> I believe Aunt Mary told Mama about a letter of Mr. Darwin's, wanting Mr. Bentham to tell him how to get from Tenby in Wales a variety of *Orchis pyramidalis*, which you remember gathering here, 'without spurs,' which mamma one day pointed out to me as so characteristic of the flowers of that plant. The first specimen which Mr. Bentham gathered in the Warren when talking with me of C. Darwin's letter, was this very variety and monstrosity. But though we have picked two or three dozen since, not one of them departed from the usual type, and I fear they are all gone off now, which is a pity, as Darwin has written to me for another without spurs.

The references are to Lyell's wife, Mary Elizabeth Lyell, and Leonard's mother, Katherine Murray Lyell.

[5] CD discussed the 'monstrous flowers' produced by *O. pyramidalis* in *Orchids*, pp. 47–8.

[6] Bentham seems to have suggested that CD try to obtain specimens of the orchid from someone living near Tenby, South Wales (see letter to George Bentham, 22 June [1861]).

To John Lubbock 9 August [1861][1]

2. Hesketh Crescent | Torquay
Aug 9th.

My dear Lubbock

Your kindness has been so unbounded & your assistance so valuable that it is no use trying to thank you. You have laid William & myself under an enduring obligation.—[2] We have not heard this morning from M^r. Hacon (as I had hoped) his final opinion;[3] & we are trying to keep open till tomorrow morning, till we see Mr Hacon's deliberate opinion; but in fact we shall be far more guided by your general impression, & I think (unless Mr Hacon's argument should seem very strong) that we shall agree.— The clause certainly seems rather harsh, but a hundred, or rather a thousand, to one, it would never signify.—

When I have written to Mr Atherley tomorrow finally, I will tell you the result.—

I have been twice hunting Lepismas with very little success,—but William went yesterday & this morning & with much difficulty caught 7 or 8; only one is large. It really is almost impossible to get them without injuring them.—[4]

I send them by this post in tin-Bottle— Please send the Bottle sometime to my House, as it belongs to one of the Boys. If you want more, we will go again to the one spot where we can find them.—

Pray give our thanks to Sir John & Mr Ellice[5] & believe me.— Yours cordially obliged | Charles Darwin

I have just had long letter on Species from Kingsley[6]

Down House

[1] The year is given by the Darwins' stay in Torquay (see 'Journal'; Appendix II).

[2] Lubbock, a partner with his father John William Lubbock in the London bank Robarts, Lubbock & Co., had informed CD in June that a Southampton banker, George Atherley, was seeking a new partner. He wondered whether William Darwin, who was in his final year of studies at Cambridge University, might be interested in the offer. CD had subsequently relied on the Lubbocks' advice and mediation concerning the negotiations conducted by his solicitor, William Mackmurdo Hacon. See the letters to John Lubbock beginning with that of [25 May 1861].

[3] See letter to John Lubbock, 1 [and 2] August [1861].

[4] Lubbock asked CD and his sons, who were keen entomologists, to collect specimens of *Lepisma*, the silver-fish (see letter to John Lubbock, 1 [and 2] August [1861]).

[5] Russell Ellice was a partner in Robarts, Lubbock & Co. (*Banking almanac* 1861).

[6] The letter from Charles Kingsley has not been found. An avid naturalist, Kingsley had written to CD shortly after the publication of *Origin* to express his initial approval of the work and his intention to read it carefully at a later date (see *Correspondence* vol. 7, letter from Charles Kingsley, 18 November 1859).

To John Lubbock 10 August [1861]

2. Hesketh Crescent | Torquay
Aug 10th

My dear Lubbock

I heard from Mr Hacon this morning; & he with some hesitation advised acceptance. In fact we had made up our minds, from what you said, that unless Mr H. had been vehement against acceptance, to accept; so I telegraphed to him to write by this day's post to Mr. Atherley our submission & acceptance. So I hope from the bottom of my heart that all is settled.[1]

Ever yours truly obliged | Ch. Darwin

Endorsement: '1861'
Down House

[1] CD was attempting to finalise the arrangements for William Erasmus Darwin to enter into partnership with George Atherley in the Southampton and Hampshire Bank (see letter to John Lubbock, [2 August 1861]).

To J. D. Hooker [11 August 1861]

2. Hesketh Crescent | Torquay
Sunday

My dear Hooker

Could you spare me another spike of Apothecia & send it in tin box, or in oilsilk in wooden box[1] I sh$^{d.}$ wish one in very early bud, but one in full flower would be better than none?— Veitch has sent me a lot of magnificent Cattleyas all in full bloom;[2] & as in some respects the pollinia are little modified, so the rostellum & means of fertilisation are simpler than in any other Orchid. In fact these flowers lead me to look at the rostellum as (in this the simplest case) part of the stigma very little modified & not distinct from the true stigma.—[3] I want to see how the tissues are in the bud.—

The stigma secretes in Cattlya an enormous amount of viscid matter; I have said secrete, but it seems to me rather that the whole surface of the stigma deeply resolves itself into viscid matter; & this is all that the rostellum or upper part of the stigma does; & this viscid matter is forced by insects into the mouth of the anther & so withdraws the pollinia.—[4] I certainly sh$^{d.}$ very much like to see rostellum of Apothecia in early bud.

When I get home on Augst 28$^{th.}$ could you lend me Lindleys great work on Orchids to which he refers in Veg. Kingdom, as explaining nature of Parts;[5] to my sorrow I fear I ought not to publish, without seeing what is known: I much fear that it is a very precious Book.—

Ever yours | C. Darwin

Endorsement: 'Aug 11/61'
DAR 115.2 : 108

[1] CD intended to write 'Epithecia', a synonym for the orchid genus *Epidendrum*. CD thanks Hooker for telling him about 'Epithecia' in the following letter.

[2] James Veitch Jr provided CD with species of exotic orchids (*Orchids*, pp. 158 n. and 214 n). He was a partner in James Veitch & Sons Exotic Nursery in Chelsea, which specialised in growing orchids and other rare plants.

[3] CD described in detail the homology of the organs of orchids in the final chapter of *Orchids*, in which he argued that the rostellum had evolved from the third stigma of the flower. See *Orchids*, pp. 307–23.

[4] CD illustrated the structure of this orchid in *Orchids*, p. 161.

[5] The reference is to John Lindley's statement in Lindley 1853, p. 174: 'The general structure of Orchids, briefly embodied in the foregoing description, has been treated of at such length in the prefatory matter of the *Illustrations of orchidaceous plants*, that it is unnecessary to do more than refer the reader to that work.' Lindley had provided an introduction and notes to the detailed drawings of orchids prepared by Franz Bauer, artist to Joseph Banks at the Royal Botanic Gardens, Kew (Bauer 1830–8).

To J. D. Hooker 13 [August 1861]

2. Hesketh Crescent | Torquay
13$^{th.}$

My dear Hooker

I write merely to thank you for your note.[1] I really will give no more trouble for you must be sadly overworked. I am glad you will have a trip in Scotland.—

You give a very poor account of M^rs. Hooker & I grieve to hear it. I had hoped the sea would have invigorated her.[2] The sea has certainly done Etty great good; but I still pin my faith to your oil, which is never neglected[3]

Many thanks for telling me about Epithecia.[4] What a curious account you give of Owen.[5]

Farewell my dear old friend. I hope your Health will not fail. | Yours affect | C. Darwin

I really think William will make a Botanist & it pleases me much.—[6] If you were to ask *Crocker*(?) would he not send me a Vanilla & so you have no more thought or trouble on subject[7]

Endorsement: 'Aug/61'
DAR 115.2: 110

[1] Hooker's letter has not been found. It was a response to the preceding letter.

[2] Frances Harriet Hooker had been unwell since the death in May of her father, John Stevens Henslow (see Allan 1967, p. 208). The Hookers had visited the seaside at Worthing, Sussex, in the hope of facilitating Frances Hooker's recovery (see letter to J. D. Hooker, 13 July [1861]).

[3] CD refers to cod-liver oil, which Hooker had suggested might help Henrietta Emma Darwin recuperate from the after-effects of typhus fever. See letter to J. D. Hooker, 8 [February 1861].

[4] See letter to J. D. Hooker, [11 August 1861].

[5] The reference to Richard Owen has not been traced. Owen had recently published a three-part series in the *Athenæum* 'On the scope and appliances of a national museum of natural history' (Owen 1861b), which outlined his views on the separation of the natural history collections from the British Museum. CD, Hooker, and others opposed Owen's plans. See *Correspondence* vol. 7, Appendix VI, and Rupke 1988.

[6] William Erasmus Darwin was keenly interested in botanical dissection. See letters to J. D. Hooker, 19 June [1861], 22 June [1861], and 13 July [1861].

[7] Charles William Crocker was foreman of the propagating department at the Royal Botanic Gardens, Kew. Since first telling Hooker that he much wished 'to see one of the Arethuseæ', which included the genus *Vanilla*, CD had repeatedly been told that this was not possible (see letters to J. D. Hooker, [6 July 1861], 17 [July 1861], and 27 July [1861]). This tribe of orchids was the only one out of seven that CD had not yet studied.

To Charles Lyell 13 [August 1861]

2. Hesketh Crescent | Torquay
13th.

My dear Lyell

Very many thanks for the Orchids, which have proved extremely useful to me in two ways I did not anticipate, but were too monstrous (yet of some use) for my special purpose.[1]

When you come to "Deification" ask yourself honestly whether what you are thinking applies to the endless variations of domestic production which man accumulates for his mere fancy or use. No doubt these are all caused by some unknown law, but I cannot believe they were ordained for any purpose; & if not so ordained under domesticity, I can see no reason to believe that they were ordained in a state of nature. Of course it may be said that when you kick a stone, or when a leaf falls from a tree, that it was ordained before the foundations of the world were laid, exactly where that stone or leaf should lie. In this sense the subject has no interest for me.—[2]

Once again many thanks for the orchids; you must let me repay you, what you paid the Collector.—

Ever Yours | C. Darwin

Endorsement: 'August | 1861.'
American Philosophical Society (260)

[1] See letter to Charles Lyell, 6 August [1861]. CD mentioned Lyell's assistance in sending him specimens of 'spurless' *Orchis pyramidalis* in *Orchids*, p. 41.
[2] See letter to Charles Lyell, [1 August 1861].

To A. G. More 13 August 1861

2 Hesketh Cresent, | Torquay.
Aug. 13, 1861.

My dear Sir

As I asked you some time ago to be so kind as to have a look at Spiranthes[1] I write one line to say that I have found here hundreds of specimens, and have looked at so many I am sure of point referred to.[2] I was partly in error and partly correct. I shall be extremely curious to hear whether the experiment, which you so very kindly tried, on E. palustris has any influence on its setting seed.[3]

With many thanks for all your valuable assistance. | Yours sincerely | Ch. Darwin.

Copy
DAR 146

[1] See letter to A. G. More, 17 June 1861.
[2] The point about *Spiranthes* that CD wanted More to confirm was the necessity of the 'bristle or culm being bowed so as to strike middle line of the rostellum before it is withdrawn, in order that pollinia may be removed' (see letter to A. G. More, 17 June 1861). See also *Orchids*, pp. 126–7.
[3] See letter to A. G. More, 19 July 1861.

To John Lubbock 14 August [1861]

2. Hesketh Crescent | Torquay
Aug. 14th.

My dear Lubbock

I thank you sincerely for your very kind expressions towards me.[1] Your talking of living at Brighton made me realise how much I have enjoyed your friendship & what a loss your absence would be to me. By the way I have not heard where your new House is.—[2] William & my two little Boys have had this morning a grand hunt after Petrobi & have caught about a dozen, which I send by this same Post in a large old cannister.—[3]

With respect to the Bank, I have been thinking over what you say & will not forget it, but the Union Bank has on two or three occasions been so obliging to me, that I should feel it intolerably shabby to leave them without sufficient motive. It would make me feel almost guilty.—[4]

We shall return home, I am glad to say in about a fortnight, & I suppose by that time William will be a Banker![5]

Dear Lubbock | Ever yours very truly | Charles Darwin

Endorsement: '1861'
Down House

1 Lubbock's letter has not been found.

2 Lubbock and his wife Ellen Frances Lubbock had lived with his parents at High Elms, near Down, Kent since their marriage in 1856. Expecting the birth of their third child in December 1861, the family planned to move, eventually settling in Chislehurst, Kent (Hutchinson 1914, 1: 52).

3 Lubbock had asked CD and his sons to help him collect specimens of bristletails, members of the primitive insect group Thysanura. The genus *Petrobius* is not described in the paper Lubbock was preparing (Lubbock 1862).

4 CD banked for many years at the Union Bank of London, Prince Street, Mansion House, London.

5 William Erasmus Darwin joined the Southampton and Hampshire Bank as a partner in the autumn of 1861, moving to Southampton in October (Emma Darwin's diary).

To John Obadiah Westwood 15 August [1861][1]

2. Hesketh Crescent | Torquay
Aug. 15th

My dear Sir

Your note of the 12th has been received here today.[2] Although as a general rule it seems to me best to deposit specimens in the Brit. Museum & I bitterly regret I did not send there all my specimens,[3] yet from what you say it gives me real pleasure to concur with Prof. Bell in sending my Crustaceans to the Oxford museum,[4] which under your guardianship I do not doubt will become a grand repository.[5]

You may remember sending me two Bees with pollen-masses attached.[6] I am now drawing up a long paper on the fertilisation of Orchids. As the Bees had to be returned I could not ascertain to what species the pollinia belonged. You said in your note you thought you had moths with pollen-masses attached: if you have any specimen with such attached *to the head or base of the proboscis* (especially if only a single pollen-mass) & would allow me to steam the moth & so remove the pollen-mass, (which I could gum again on head), you would confer a great service on me & which should be acknowledged in my paper.— Of certain Orchids I have examined plenty of attached pollen-masses, but it is of much consequence to me to see some other kinds.— If in your power perhaps you will kindly assist me.[7]

I shall be here 10 days more & then return to Down.—
My dear Sir | Yours sincerely | C. Darwin
Can you give me reference to Morren's paper, to which you alluded?[8]

Oxford University Museum (Hope Entomological collections)

[1] The year is given by the Darwins' stay in Torquay (see 'Journal'; Appendix II).
[2] Westwood's letter has not been found.
[3] Following CD's return to England in 1836 from the *Beagle* voyage, he arranged for various parts of his natural history collection to be given to specialists for scientific description. As a result, the *Beagle* specimens were dispersed and eventually housed in a number of different institutions. See Porter 1985. At the time of his return, CD did not have a good opinion of the state of the zoological collections of the British Museum under the direction of the keeper John Edward Gray (see *Correspondence* vol. 1, letter to J. S. Henslow, [30–1 October 1836]; vol. 2, letter to Leonard Jenyns, 3 December [1837]; and vol. 5, letter to J. D. Dana, 8 May [1852]).
[4] CD had entrusted the Crustacea and Reptilia collected on the *Beagle* voyage to Thomas Bell, professor of zoology at King's College, London. Bell, however, was only able to describe the reptiles (*Reptiles*).
[5] Westwood had been appointed Hope professor of zoology at the University of Oxford in December 1860 (A. Z. Smith 1986). Before then he had been curator of the extensive entomological collection of Frederick William Hope, donated to the university in 1849. Westwood was arranging to purchase Bell's zoological collections to enrich the Hope Entomological collection housed in the newly constructed Oxford Museum (see Chancellor *et al.* 1988).
[6] See *Correspondence* vol. 8, letter to J. O. Westwood, 9 July [1860].
[7] Westwood sent CD some wasps with pollen-masses attached, but the pollen-masses were not those of orchid species (see letter to J. O. Westwood, 4 September [1861]).
[8] Morren 1836. See letter from J. O. Westwood, 26 September 1861.

To Charles Lyell 21 August [1861]

2. Hesketh Crescent | Torquay
Aug 21.

My dear Lyell

I know the page well & like many others I remember somewhere quoting it.—[1] You are very honest to append a note; for when I skimmed it first over, I did not see any was necessary for those who believe in the modification of species.— But you show it is perhaps necessary.—[2] The most trifling modification (& leaving out "Hence") of present sentence would make it strictly correct; for the adaptation of species to travel widely over existing continents, will necessarily adapt them for occasional still wider transportation to new lands. I have used in Origin this argument to account for very wide range of F. Water productions.[3] Your note strikes me as very good & very true: I have really no criticism, except a trifling one near end in pencil, which I have inserted on account of dominant & important species generally varying most.— You speak of "their views" rather as if you were a 1000 miles from such wretches; but your concluding paragraph shows that you are one of the wretches.—

I am pleased that you approve of Hutton's Review.[4] It seemed to me to take a more philosophical view of the manner of judging the question than any other

Review.— The sentence you quote from it seems very true: but I do not agree with the theological conclusion.—[5] I think he quotes from Asa Gray, certainly not from me;[6] but I have neither A. Gray or Origin with me.— Indeed I have over & over again said in Origin that n. selection does nothing without the variability: I have given whole chapter on laws & used strongest language how ignorant we are on these laws.— But I agree that I have somehow (Hooker says it is owing to my title)[7] not made the great & manifest importance of previous variability plain enough.— Breeders constantly speak of Selection as the one great means of improvement; but of course they imply individual differences; & this I sh$^{d.}$ have thought would have been obvious to all in Nat. Selection; but it has not been so.—

I have just said that I cannot agree with "which variations are the effects of an unknown law, ordained & guided without doubt by an intelligent cause on a preconceived & definite plan". Will you honestly tell me (& I should really be much obliged) whether you believe that the shape of my nose (eheu) was "ordained & guided by an intelligent cause"[8] By the selection of analogous & *less* differences, fanciers make almost generic differences in their pigeons, & can you see any good reason why the natural selection of analogous individual differences should not make new species? If you say that God *ordained* that at some time & place a dozen slight variations should arise, & that one of them alone should be preserved in the struggle for life, & that the other eleven should perish in the first, or few first, generations; then the saying seems to me mere verbiage.— It comes to merely saying that everything that is, is ordained.

Let me add another sentence.— Why should you or I speak of variation as having been ordained & guided more than does an astronomer in discussing the fall of a meteoric stone. He would simply say that it was drawn to our earth by the attraction of gravity, having been displaced in its course by the action of some quite unknown laws.— Would you have him say that its fall at some particular place & time was "ordained & guided without doubt by an intelligent cause on a preconceived & definite plan"? Would you not call this theological pedantry or display? I believe it is not pedantry in the case of species, simply because their formation has hitherto been viewed as beyond law,— in fact this branch of science is still with most people under its theological phase of development.—[9] The conclusion which I always come to after thinking of such questions is that they are beyond the human intellect; & the less one thinks on them the better. You may say, then why trouble me? But I sh$^{d.}$ very much like to know clearly what you think.

Ever yours | C. Darwin

The sentence at your page 3. marked X by me, reads rather roughly; but it is highly presumptuous in to criticise the style of one who unintentionally always writes Miltonic verses!

(We return Home on Monday 26th

Postmark: AU 21 61
American Philosophical Society (261)

1 The page referred to can be identified from the annotation Lyell made on the cover of CD's letter: 'C. Darwin | returning M.S. on Sicily newer than species inhabiting it— & on deification & Natural Selection—'. The fact of the flora and fauna of Sicily being of greater antiquity than the country itself is discussed in the fifth edition of Lyell's *Manual of elementary geology* (C. Lyell 1855a), p. 160. In the sixth edition, which Lyell was then preparing, he added a note to a subsequent sentence on the same page, indicating that his views on the creation and immutability of species had been modified since CD's theory of natural selection had been made known to him (see n. 2, below).

2 The note referred to was appended to the following passage (C. Lyell 1855a, p. 160 and C. Lyell 1865, p. 194):

> It is obviously expedient that the terrestrial and fluviatile species should not only be fitted for the rivers, valleys, plains, and mountains which exist at the era of their creation, but for others that are destined to be formed before the species shall become extinct; and, in like manner, the marine species are not only made for the deep and shallow regions of the ocean existing at the time when they are called into being, but for tracts that may be submerged or variously altered in depth during the time that is allotted for their continuance on the globe.

The note reads (C. Lyell 1865, pp. 194–5 n.):

> The three last pages, on "The Newer Pliocene Strata of Sicily," are given verbatim as they appeared thirty years ago in the first edition of the Principles of Geology (vol. iii. p. 115., 1833). The last sentence, marked with inverted commas, was couched in language implying my adherence to the theory that each species was originally created such as it now exists, and was incapable of varying so as to pass into a new and distinct species. In my recent work on the Geological Evidences of the Antiquity of Man, I have shown (chaps. xxi. to xxiv.) that Mr. Darwin's theory of natural selection removes many of the principal difficulties which stood in the way of Lamarck's doctrine of transmutation; and had I inclined as much in 1833 towards embracing Mr. Darwin's views as I do now, I should have expressed myself somewhat differently. But I have thought it best not to re-cast a passage which has been so often cited, both by writers who opposed and approved of it. The main proposition which seemed so startling in 1833, namely, that species in general may be older than the lands and seas they inhabit, is now the creed of almost every geologist, whether he adopts or rejects the theory that species may be indefinitely modified in their organization under the influence of new conditions in the animate and inanimate world.

3 See *Origin*, pp. 383–8.

4 Hutton 1861. A lightly annotated presentation copy of Frederick Wollaston Hutton's review is in the Darwin Pamphlet Collection–CUL.

5 Hutton's review of *Origin* (Hutton 1861) was published in the April and May 1861 issues of the *Geologist*. In the June issue a letter criticising Hutton's article was published, to which Hutton replied in the July issue (*Geologist* 4 (1861): 247–50, 286–8). Hutton stated that his review had concentrated only on 'the scientific arguments both for and against' CD's theory and had not touched on 'points connected with theology', but that to respond to his critic he felt compelled to discuss the theological implications of the theory (*ibid.*, p. 286).

6 In his letter, Hutton stated (*Geologist* 4 (1861): 286):

> I cannot understand why natural selection has been so often mistaken for a cause, when it is evidently the effect of the "struggle for life" acting on variations in species, which variations are the effects of an unknown law ordained and guided, without doubt, by an Intelligent Cause "on a preconceived and definite plan."

He referred any reader interested in 'proofs' for this argument to Asa Gray's series of articles on *Origin* (A. Gray 1861a), but his quotation is not taken from this article.

7 Joseph Dalton Hooker's comment does not appear to have been made in a letter. For an earlier expression of CD's doubt as to whether he had selected the right term for his theory, see *Correspondence* vol. 8, especially the letter to Asa Gray, 26 September [1860].

[8] See n. 6, above.
[9] See letter to Charles Lyell, [1 August 1861] and n. 12.

From Asa Gray [27 and 29 August] and 2 September [1861][1]

to carry down the pollinia, taken from one flower on the point of a fine needle, to the stigma of another flower, have failed,—the pollinia lodging above the stigma or over on the dorsal face.[2] I enclose a spike of *S. cernua*.

Sept. 2^{d}. That *Drosera filiformis* should show no sensitiveness is perhaps not remarkable, for it is not well arranged for fly-catching.[3] But I have got some D. rotundifolia, & tried it with atoms of fresh fish laid on the viscid tips, but I could detect *no movement* either immediate or later. Yet I could see that some of the leaves had their bristles turned inwards and had been catching flies. I have been to-day into the country and have brought in some fres⟨h⟩ *D. rotundifolia*, to try again to-morrow.

The most marked *diœcio-dimorphous* flowers I know are in Rubiaceæ, and the most familiar cases are in *Houstonia* (Oldenlandia, but the old name & genus will be restored) specially our *H. purpurea* and the everywhere common *H. cærulea*, and *Mitchella repens*. *Spermacoce virginiana* & *Diodia teres*, also show it. I have no data for a list; but the cases are numerous. I will make any observations you will indicate next spring.[4]

What you say of *Mill* well accords with what Prof. Henry of Smithsonian Inst^{n}. says of your investigations., your method, &c,—and he is the best judge of the logic of investigation I know of.[5]

Incomplete
DAR 110 (ser. 2): 76

CD ANNOTATIONS
1.1 to carry . . . to-morrow. 2.6] *crossed ink*
3.2 Oldenlandia] *underl red crayon*
4.1 What . . . know of. 4.3] *crossed ink*

[1] The date is established by CD's statement in the letter to Asa Gray, 17 September [1861], that he had just received Gray's letter 'of Augt 27th & 29th & Sept 2^{d}'.

[2] CD had asked Gray, in his letter of 21 July [1861], to 'have a look at Spiranthes' in order to confirm his description of the orchid's flower structure. Gray had apparently found differences in the sexual parts of American species of *Spiranthes*. See also letter to Asa Gray, 17 September [1861].

[3] CD had suggested that Gray might like to observe the sensitivity of *Drosera* leaves to substances placed on them (see letter to Asa Gray, 17 February [1861]). The thread-leaved sundew, *D. filiformis*, is common in wet, sandy, coastal areas from Massachusetts southward. See also letter to Asa Gray, 21 July [1861].

[4] CD had asked Gray whether he knew of any dimorphic plants in which the two flower forms occur on separate plants living commingled in the same area (see letter to Asa Gray, 5 June [1861]). In his *Manual of the botany of the northern United States*, Gray defined the term 'diœciously dimorphous' in the section describing the Rubiaceae, stating that in 'several genera, such as Mitchella, Oldenlandia, &c., the flowers, although perfect, are of two sorts in different individuals', one with 'exserted' (protruded)

stamens and short styles and the other 'included' stamens and long styles (A. Gray 1856, p. 171 n.). The term had first been defined in Torrey and Gray 1838–43, 1: 38. See also letter from Asa Gray, 11 October 1861.

[5] See the letter to Asa Gray, 21 July [1861], in which CD mentioned that John Stuart Mill considered CD's reasoning in *Origin* to be 'in the most exact accordance with the strict principles of logic.' The physicist, Joseph Henry, was the secretary and director of the Smithsonian Institution in Washington, DC.

From George Maw 27 August [1861][1]

Benthall Hall, | nr Broseley.
27th. Augt. 62

Dear Sir

I just write a line to thank you for yr. letter of the 19th. July & the consideration you have given my review.[2] I should have written sooner but only got back from an absence of 5 weeks on the continent last Thursday.

Several fresh points have occurred to me since I last wrote & as soon as I can find time to note them down I shall take the liberty of troubling you with a further communication[3] Pray do not however feel any necessity to reply as you have doubtless something better to occupy yr time than to answer all the communications that are made to you on the subject of yr book

Believe me Dear Sir | very truly yrs. | Geo Maw.

Charles Darwin Esqr

an immense number of cases of correlation of organization have occurred to me.—[4] all the fancy varieties of scarlet Pelargonium that have contracted or imperfect leaves have contracted flowers[5] castration of many male animals—as the ox tribe where the contour of the 2 sexes is notably distinct producing the contour of the female is another remarkable instance of correlated organization— The number of isolated functions that become co-modified at Parturition—as Production of Milk in the mother—Respiration of Fœtus—Independent circulation in fœtus—completed organization of heart of fœtus all taking place at the time of independent existence looks wonderfully like preordination—or when we remember that the imperfection of any one of these processes would be death to the fœtus, how difficult is it to believe that they can have been accumulated by gradual modifications.

DAR 99: 11–12

CD ANNOTATIONS

0.3 Dear Sir … yrs. 3.1] *crossed brown crayon*
5.1 an immense … flowers 5.3] *crossed pencil*; *double scored brown crayon*
5.3 castration … organization— 5.5] *double scored brown crayon*
5.5 The number … fœtus, 5.10] *double scored brown crayon*
Top of first page: 'Ch 7.'[6] *brown crayon*
Bottom of last page: 'For *Selection* of [*over* '&'] ⟨Cotemporous⟩ variation' *blue crayon*

CD note:[7]

Suppose period of gestation to become shorter from any cause (& it varies in cattle for instance) then if milk did not come sooner, the young would die & conversely: & so bond formed.— But when leaves of Geranium are contracted, flower becomes contracted from homology—petals & leaves being similar & probably being affected by similar causes— (what affects one would affect all) & this would take place almost inevitably, without the contraction of mouth of corolla was in some way highly injurious.—

1 Although the letter is dated '1862', Maw refers to CD's letter of '19th July', and from Maw's mention of his review of *Origin* (see n. 2, below), the subject matter seems to relate to the letter to George Maw, 19 July [1861], which was written from Torquay.

2 Letter to George Maw, 19 July [1861]. The reference is to Maw's review of the third edition of *Origin*, published in the *Zoologist* (Maw 1861a).

3 Maw's earlier letter to CD, which accompanied the presentation copy of Maw 1861a, has not been found, but see the letter to George Maw, 13 July [1861].

4 In his review of *Origin*, Maw challenged CD's assertion that evidence showing 'correlation of growth' was left unexplained by the view of the independent creation of each species (*Origin* 3d ed., p. 466). Maw argued that if the term 'correlation of growth' were exchanged for 'correlation of organisation', then such phenomena could be accounted for on the assumption 'that the uniformity of model, upon which the more important parts are formed, is the result of special creation working out a consistent proportion of the integral part' (Maw 1861a, pp. 7586–7).

5 See CD note, above. CD later used this example of correlation of growth in *Variation* 2: 330.

6 CD refers to chapter 7 of his 'big book' on species; the chapter, entitled 'Laws of variation', includes a section on the correlation of growth (*Natural selection*, pp. 297–304).

7 The note is written on the verso of the last page of the letter (DAR 99: 12). It seems to have been written in response to the letter from George Maw, 30 June 1862, and is thus also transcribed with that letter (see *Correspondence* vol. 10).

To W. B. Tegetmeier[1] 29 August [1861][2]

Down. | Bromley. | Kent. S.E.
Aug 29th

My dear Sir

One line to say that I have returned home, with some benefit to my health & much good to my daughter's.—[3] I am in no hurry, but when ready you can return my M.S.—[4] Kindly take the trouble to register it, & mark it "a M.S for printing"— The string had better pass over the ends of parcel.

My dear Sir | In Haste yours sincerely | C. Darwin

Victoria and Albert Museum, Baillie collection, L3736–1959, Pressmark 86.QQ.1

1 The correspondent is identified by the reference to returning CD's manuscript. CD had asked Tegetmeier to read the discussion of variation in fowls intended for *Variation*. CD sent the manuscript to Tegetmeier before leaving for Torquay (see letter to W. B. Tegetmeier, 14 June [1861]).

2 The year is established by the reference to CD's having just returned home. The Darwins arrived back at Down House, after spending eight weeks in Torquay, on 27 August 1861 (see 'Journal'; Appendix II).

3 The Darwins had visited the seaside in the hope that it would help Henrietta Emma Darwin recuperate from a prolonged illness.

4 See n. 1, above.

To J. D. Hooker 30 August [1861]

Down. | Bromley. | Kent. S.E.
Aug. 30th

My dear Hooker

I was thinking of writing to you today, when your note with the Orchids came.[1] What frightful trouble you have taken about Vanilla:[2] you really must not take an atom more; for the orchids are more play than real work. I have been much interested by Epidendrum & have worked all morning at them (for Heaven sake do not corrupt me by any more); & as far as I can make out, in the bud the rostellum has througout nearly the same structure as the exterior & next to exterior surface of the stigma.— Can you tell me the names of 3 of the 4 sent; or at least of the *first* in enclosed list & specimens:[3] The first alone much signifies; but even that is not worth much trouble. When at Torquay I made out well, what I never could understand, how the pollinia in early bud become permanently attached to the rostellum or so-called sticky gland.[4] But I won't run on.—

I wish to God you could give a better account of Mrs. Hooker; I heartily sympathise with you; it must be most depressing to you, my dear old friend.[5]

Etty has returned much stronger: I declare she did the first day's journey to Salisbury better than I did:[6] I was so knocked up that I gave up Manchester:[7] I cannot stand such fatigue & am in fact a man of seventy years old.— How I should enjoy to have one more Oxford time with you.—[8] If ever you can find time to run down here for a few days do, I beg you, propose it. Perhaps later in Autumn Mrs. Hooker would come with you. This is just the house for invalids

My dear Hooker | Ever yours | C. Darwin

P.S. | I suppose I ought to read what Lindley says on structure of orchids: so if you will send me the part with Introduction of his work;[9] I shd be *greatly* obliged: I enclose address for safest channel.

I should like sometime to hear how Oliver is?[10] What is the matter with him?

Endorsement: '1861.'
DAR 115.2: 111

[1] Hooker's letter has not been found, but see the letter to J. D. Hooker, [11 August 1861].

[2] See letter to J. D. Hooker, 13 [August 1861]. CD had repeatedly asked Hooker to help him procure a specimen of *Vanilla*, a genus belonging to a tribe of orchids CD had not yet examined.

[3] The enclosure has not been found.

[4] CD examined the young flower-buds of *Epidendrum cochleatum* and *E. floribundum*. In *Orchids*, pp. 310–11, he described the structure of the external cells of the rostellum and stigma of *Epidendrum*.

[5] Frances Harriet Hooker was recuperating slowly from the loss of her father, John Stevens Henslow, in May 1861.

[6] The Darwins left Torquay on 26 August 1861 and spent one night in transit in Salisbury, before returning to Down on 27 August. They had spent eight weeks at the seaside in the hope of improving Henrietta Emma Darwin's poor health.

[7] The British Association for the Advancement of Science was to meet in Manchester from 4 to 11 September 1861.

[8] CD and Hooker had attended the British Association meeting in Oxford together in 1847. See *Correspondence* vol. 4.

[9] CD had asked to borrow Bauer 1830–8 from Hooker. The work included notes and prefatory remarks on the structure of orchids by John Lindley.

[10] Daniel Oliver, Hooker's assistant in the herbarium of the Royal Botanic Gardens, Kew, had been ill for some time. Before his illness, Oliver had been assisting CD with his botanical researches.

To George Maw 31 August [1861][1]

Down. | *Bromley.* | *Kent. S.E.*
Aug. 31st

Dear Sir

I write merely to say how grateful I shall feel for any facts on correlation or criticisms.— Will you sometime please explain a little more fully about the contracted corolla of the Pelargonium with the narrow leaves.—[2]

You put the difficult case of correlated changes in structure at Birth in a very striking manner. Such cases of correlation, however, interest me less than that of the Pelargonium: for according to my notions, (whether false or true) if the period of gestation became from any cause (& it does vary) shorter, if the milk did not flow, the inheritance of the new variation would be stopped by the death of the progeny; & thus a bond of apparent correlation *might* arise: but, I presume that no one will doubt that the case of the pelargonium is due to petals & leaves being homologous & similar at a very early period & being similarly affected by some unknown cause.—[3]

Cases of old or injured females assuming male attributes is perhaps even more curious than the effects of castration on males.—

I shall be truly glad to hear any facts, when you have leisure & inclination to write to me & I beg leave to remain | Dear Sir | Yours sincerely | Charles Darwin

Endorsement: '1861'
Royal Horticultural Society

[1] Dated by the endorsement and by the relationship to the letter to George Maw, 19 July [1861].

[2] See letter from George Maw, 27 August [1861].

[3] See letter from George Maw, 27 August [1861], CD note.

From George Rolleston 1 September 1861

Oxford.
Sept 1. 1861.

My dear Sir.

I ought to have written to thank you for the copy of the Third Edition of the Origin of Species which I received some months, I am ashamed to think, ago.[1] It was my Intention to read it all through before writing in acknowledgment of your kindness but as I have not quite succeeded in doing this I have determined to put off writing no longer.

I suppose you will have seen a couple of Papers on "Sexual Limitation in Hereditary Disease" in the Medico Chirurgical Review for April & July.[2] In case you should not, I will make an extract of the fact brought forward in that paper by Mr Sedgwick. Medico Chir. Review. April. 1861. p. 484. "M. Geschreift[3] informs us that among the patients in the Eye Infirmary at Brussels were two Brothers Microphthalmic on the *left* side whose Father had lost his eye 15 years before his marriage in consequence of purulent ophthalmia whilst serving in the army of Holland."

It is not as bearing upon such facts as those of Brown Séquards production of Epilepsy in the way of transmission[4] or those of the moles suffering as you describe Page 144 from inflammation[5] & transmitting to their offspring the condition into which they have themselves been brought that such a fact, if it be one, is of chief importance as it seems to me but as bearing upon another question, which is, Does not the nervous system and its larger appendages such as eye and ear furnish (like the shortfaced tumblers) an exception to the law of inheritance at corresponding periods of the life of parent & offspring?[6]

Sœmmering held that the human Brain attained its full size at three years of age,[7] & this view even if not quite the truth, must shew as it was not put forth by a mere speculative man, that it is to be expected that the organ when about to be modified at all must be modified early— Again it is certain that the parts of the Brain in the Classes, Man & Ape, which are distinctively characteristic (so far as there are such parts) of the two classes are the earliest or amongst the earliest to shew themselves in the two Classes generally. Gratiolet, who has made the Convolutions his speciality, lays down in his Mémoire Sur les Plis Cerebraux, the following Rule in Italics: Les Parties qui doivent une jour dominer absolument apparaissent les premières bien que leur perfection ne s'acheve qu'après le developpement complet de tous les autres systêmes—[8]

Psychical manifestations in the offspring of both educated animals & educated men may be adduced either to shew the conformity or to prove the non conformity of the Nerve system to the usual Law—

With reference to your remarks on the rarely observed and recorded variations of internal and important organs (at pages 47 & 180) I have been watching in a great number of Insectivora & Chiroptera for variations in the number of their liver lobes. The common Plecotus Auritus has every now & then a liver just like man's, & the hedgehog shews also occasionally a very considerable tendency to concentrate the multitudinous tongues into which its liver is divided. The Bat is a very remarkable instance as I find its multifid liver is exactly reproduced in the Vampire Phyllostoma Hastatum, and as its usual condition contrasts as strongly with that of the Vespertilio noctula as it is possible for it to do, that condition being nevertheless that of great similarity to the organ in man which every now & then the Plecotus itself also mimics—[9] This I think bears upon what you say at Page 180.[10]

I must beg your pardon for writing at such length but as I have several times said on similar occasions I trust you will not take the trouble to reply to what requires no answer nor acknowledgment. Thanking you again for your sending me your Book

I am | Yours very Truly | George Rolleston—

DAR 176: 207

CD ANNOTATIONS

1.1 I ought . . . longer. 1.5] *crossed blue crayon*

2.1 "Sexual . . . Disease" 2.2] *underl blue crayon*

3.5 Does . . . offspring? 3.8] *scored blue crayon*

4.4 modified early—] *underl blue crayon*; '?' *added blue crayon*

4.5 characteristic . . . generally. 4.7] *double scored blue crayon*

4.10 perfection] *split over lines in MS*: *hyphen added blue crayon*

5.3 Nerve] 'nerve' *added ink*

Top of first page: 'Prof G. Rolleston' *blue crayon*; 'Laws | **Heredity** & Embryology | Variation of internal organs in wild animals | Inheritance of *eye*. Keep | Analogous variation'[11]

[1] Rolleston's name is included in CD's list of presentation copies of the third edition of *Origin* (see Appendix VII).

[2] W. Sedgwick 1861. William Sedgwick stated that little was understood about the hereditary transmission of disease, quoting CD's statement that the 'laws governing inheritance are quite unknown' (*Origin* 3d ed., p. 13). Sedgwick provided a number of cases of 'sexual limitation in hereditary disease' as a preliminary contribution to this subject (W. Sedgwick 1861, p. 477).

[3] Rolleston mistranscribed the name given in Sedgwick's paper: the reference is to Ludovic Anton Gescheidt.

[4] CD had been greatly interested by the successful attempt of Charles Édouard Brown-Séquard to transmit experimentally induced epilepsy through several generations of guinea-pigs (Brown-Séquard 1859–60). See *Correspondence* vol. 8, letter to Jeffries Wyman, 3 October [1860].

[5] Rolleston refers to the South American burrowing rodent *Ctenomys*, whose blindness CD believed to have been caused by the 'inflammation of the nictitating membrane' (*Origin* 3d ed., p. 154).

[6] The reference is to CD's rule that at whatever period of life a peculiarity first appears, it tends to appear in the offspring at a corresponding age (*Origin* 3d ed., pp. 14, 476).

[7] Samuel Thomas Soemmering.

[8] Gratiolet 1854, p. 83. 'The parts which will in time absolutely dominate are the first to appear, although they only reach perfection after the complete development of all other systems.' In a recent article (Rolleston 1861a), Rolleston had quoted extensively from Gratiolet's study of the brain of man and the primates.

[9] Rolleston reported on his study of variation in the lobes of the liver at the meeting of the British Association for the Advancement of Science later in September 1861 (Rolleston 1861c).

[10] *Origin* 3d ed., p. 180:

> A considerable part of the difficulty in recognising a variable species in our systematic works, is due to its varieties mocking, as it were, some of the other species of the same genus. A considerable catalogue, also, could be given of forms intermediate between two other forms, which themselves must be doubtfully ranked as either varieties or species; and this shows, unless all these forms be considered as independently created species, that the one in varying has assumed some of the characters of the other, so as to produce the intermediate form. But the best evidence is afforded by parts or organs of an important and uniform nature occasionally varying so as to acquire, in some degree, the characters of the same part or organ in an allied species.

[11] CD's annotations were obviously made at different times, for they are written in several different mediums: blue crayon, red crayon, and ink. Precise description obscures the meaning of the text.

From T. F. Jamieson 3 September 1861[1]

Dear Sir

I returned a few days ago from a trip to Lochaber where I spent a fortnight and now hasten to present you with some of the results of my visit,[2] and I may at once state that all I saw tended to impress upon me the conviction that these parallel roads have been formed along the margin of freshwater lakes and finding the marks of ice action so plain over the whole district I cannot help thinking that Agassiz hit upon the true solution of the problem when he pronounced these marks to be the effect of glacier-lakes.[3] I attentively examined the entrance to Loch Treig, and found both sides of the gorge to present the clearest evidence of most intense glacial action, and that to heights of many hundred feet above the like-rounded rocks, scores, flutings & perched blocks abound and all these phenomena are most conclusively seen to have been effected by a great volume of ice flowing down the valley now occupied by the Lake, and issuing out by this gorge into Glen Spean. Mr Milne states that the glacial markings here point S.W.[4] this is quite true as regards the east side of the gorge, but on the opposite or W. side they point S.E.

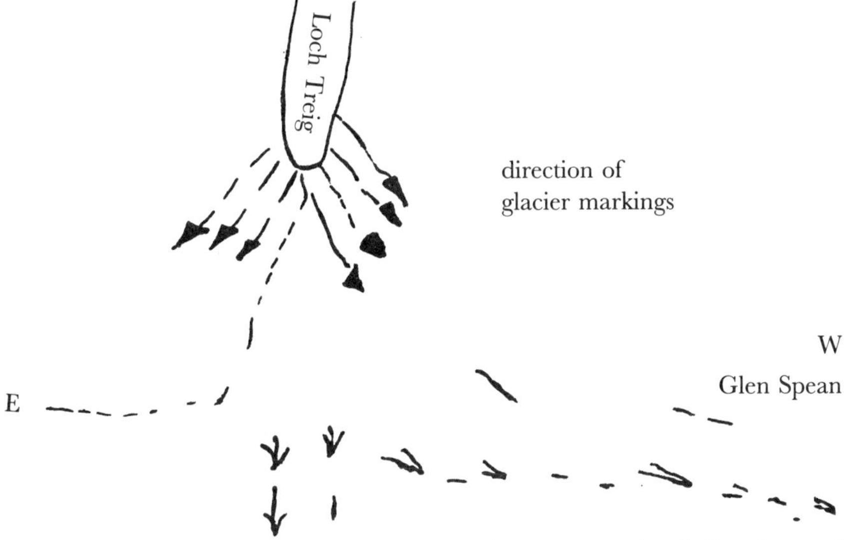

Mr Milne says he could find no transverse scratches on the rocks of Glen Spean. I was however more fortunate and found them on the north side of the Spean opposite the entrance to Loch Treig, as Agassiz had correctly described. The profusion of blocks on this side of Glen Spean over an immense expanse is very striking—in fact the whole ground about the mouth of Loch Treig is a perfect study of glacial action. I could see no line of erosion due to water action on either side of the entrance and have no doubt that the rounding you noticed was simply the glacier action which is there very conspicuous & striking.[5] I indeed noticed some scores & furrows on this level of the shelf. I should also mention that the protrusion of

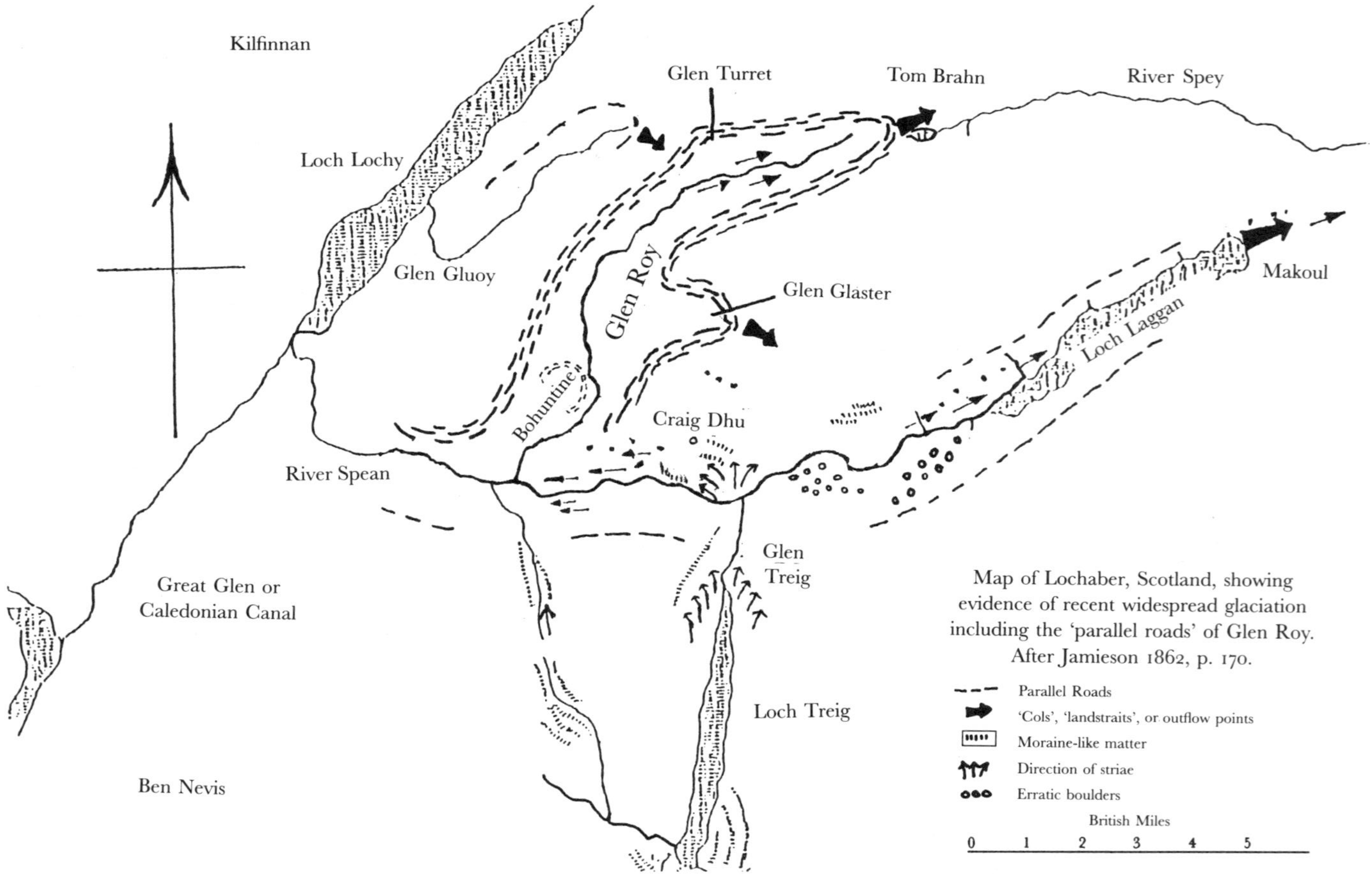

Map of Lochaber, Scotland, showing evidence of recent widespread glaciation including the 'parallel roads' of Glen Roy. After Jamieson 1862, p. 170.

glacier from Loch Treig across Glen Spean would, by resting on the north side of that valley, block up the outlet at the head of Glen Glaster so that this Glen Glaster outlet does not (as Mr. Chambers supposed) present any great difficulty on the Glacier theory.[6]

I went up to the head of Loch Treig and explored a wild glen which runs from the S.W. corner of the Lake to Spean-bridge & called the Larig or Larig Leachach.

There is no trace of any line round the head of Loch Treig, nor in fact round any part of it, altho' Sir Lauder Dick says it is visible on the W. side.[7] There are some indications of alluvial detritus here & there towards the lower end of the Col.—but a marked absence of such at the upper end & instead I found hillocks of what seems to be Moraine matter abounding in blocks. There is an absence of all deltas at the mouths of the streams that come into the head of the lake at levels above the present upper limit of its waters which leads me to think that the upper end of the lake must have been occupied by the glacier during the time when the waters in Glen Spean stood at the level of the lowermost line of Glen Roy. In traversing the Larig glen I found much evidence of glacier action but not lines like those of Glen Roy.

The only place where I observed any wearing away of solid rock due to the action of the waters that had formed these Glen Roy lines was at the head of Glen Roy or rather what has been called the head of Lower Glen Roy, where there is a sort of pass caused by a set of rocky eminences in the hollow of the glen.

Here also there had been previous glacial action so that it is not easy to determine how much is properly due to the effect of water. This pass is a notable place for wind (as I can testify, having been nearly blown off my feet there) which coming up the long hollow of the glen rushes over this high narrow pass with immense force, and when the water stood at the levels of the two uppermost lines of Glen Roy it would lash against these rocks with considerable force during gales from the S. W.

On the face of the hill marked Tom Brahn in your map,[8] the shelves are more broadly & rudely marked than any where else that I saw. This broad marking however consists of a protrusion of loose matter, and I could find no noticeable erosion of rock along any of the lines. The middle line crosses a mass of solid rock in situ here which I examined with the view of detecting traces of such erosion but the angles were quite similar to such as are usually to be found on weathered rocks not exposed to the flow of water.

At the West end of Loch Laggan there is clear evidence of the water having formerly stood at a height of probably 40–50 feet above their present level, there being a great bank of alluvial matter to that height which seems to have been the delta of the Gulban river Along the shores of Loch Laggan however there is no clear trace of any line that I could see, most of it indeed is so encumbered with block & rocks that the mark of a line could scarcely be expected. At the E. end there is also evidence of the lake having formerly stood at a higher level than at present and that to such a height as would, I have no doubt, cause its waters to flow eastward by Makoul towards Spey at the pass of Makoul (or Muckul as it has been written by some) or rather a little to the Eastward of it there is clear evidence of a large stream of water having flowed out towards the basin of the Spey, although I should describe it somewhat differently from Mr Milne. There has no wall-sided gorge been cut there is a ravine due to a series of rocky eminences occupying the middle of the valley. Here also there had been previous glacial action, but the subsequent effects of water flowing eastward are quite unmistakeable. The glacial scoring has been washed out & smooth winding curves with a tendency to form *potholes* are seen, whilst quantities of well rounded pebbles like cocoa nuts or cannon balls lie on the water-worn surfaces, and heaps of shingle in the recesses and sheltered spots to the East of the rock masses. This shingle is *intensely* water-worn & has been used in some places for making the road. I satisfied myself that this could not have been caused by tidal action but that it was due to a current flowing downward and Eastward—[9] In the rocky narrows a little above the height of the summit level where tidal action would also infallibly have left its trace I found stony detritus but presenting none of the water-rolled complexion of that which I have described. As there is no clear line along the E. end (or indeed almost any where else) of L. Laggan I doubt much whether this Makoul outlet has ever been properly connected by levelling with the lower line of Glen Roy, although I think it is highly probable that it does coincide with the level of that line. The outlet at the head of Glen Glaster does coincide with the middle line of Glen Roy & here also there appears to have been an outflow of water. The outfalls at the head of Glen Gluoy & Loch Spey I was unable to visit. In looking up Glen Roy from the Gap, I was much struck by the extreme neatness & precision of the lines which seemed to me very unlike what might be expected from the shore of a lake subject to tidal action. Very different from the appearance of true old coastlines which I had seen on the West coast of Argyleshire last summer, & also from those depicted by M. Bravais.[10] But what seemed to me even more important evidence in this respect was the wonderfully fine preservation of the deltas at the mouth of some of the streams near the head of Glen Roy. These deltas have the appearance of being lodged in the waters of a placid lake, even in a stagnant pool, so undisturbed is the outline of some of them. This seems inexplicable to me had the lake been an arm of the sea, subject to the flux & reflux of the tides.

Again the delta at the mouth of the Turret is out of all proportion too large for the size of the stream. I estimated the waters of the Roy as 3 or 4 times greater than those of the Turret at their point of junction whereas the Turret delta is 2 or 3 times bigger than the Roy delta. This I think can be explained only by supposing the Turret delta to be partly due to the outflow from Loch Gluoy. The glacier streams that had blocked up the mouth of Glen Spean have probably been those issuing from Glen Arkarg & the other glens at the mouth of the Caledonian Canal. At Fort Augustus the traces of glacial action are very noticeable.

In Glen Spean to N.E. of Bridge of Roy the scores run E. & W. but further up they sweep beautifully round into the mouth of Glen Treig opposite which they run N. & S. The glacial markings on the syenitic granite on the N. side of Glen Spean opposite the entrance to Loch Treig are amongst the finest specimens of ice-work I have seen, this with the heaps of moraine matter & the perfect wilderness of boulders made me stare with astonishment how any one, after Agassiz had drawn attention to all this, could go on the ground & yet deny that there had been any glacier here![11] I do not suppose there is any place in Britain where the traces of a great ice stream are more complete.

There are however some facts connected with the former glaciation of Lochaber very remarkable but they refer to a period anterior to the formation of the parallel lines, and although very extraordinary they are consistent & harmonise with what I have seen in other parts of Scotland. They indicate however a climate more like that of the *poles* than any thing else we have at present on the face of the globe, & the state of things which they disclose seems to me one of the most curious & inexplicable in the whole range of the geological record[12]

I have written the above notes very hastily but shall be happy to answer any queries you may wish to put. I may mention that I had most villainous weather, which prevented me making so long excursions as I otherwise might have done; and it was only by going doggedly to work with a waterproof & umbrella that I could get any thing done at all.

Yours ever | Thos. F. Jamieson

Ellon, Aberdeens[h].
3[d]. Sept. 1861.

Copy
Kinnordy MS, Charles Lyell's journal VII, pp. 75–92

[1] The original letter has not been found. The text has been taken from a copy entered into Charles Lyell's journal and headed: 'Letter from Mr. Jamieson of Ellon to Cha[s]. Darwin'. It is also printed in Wilson ed. 1970, pp. 518–23.

[2] Lochaber is a district in southern Inverness-shire, near Fort William, Scotland; see map, p. 248. CD had visited the region in June 1838 in order to investigate the famous parallel terraces or 'roads' of Glen Roy. His paper, 'Observations on the parallel roads of Glen Roy, and of other parts of Lochaber, with an attempt to prove that they are of marine origin', was published in the *Philosophical Transactions of the Royal Society of London* (1839): pt 1, 39–81 (*Collected papers* 1: 87–137). CD had encouraged Jamieson to 'take up Glen Roy' earlier in the year (see letter to Robert Chambers, 30 April [1861], and Appendix IX, letter from T. F. Jamieson to Charles Lyell, 14 May 1861). See also Barrett 1973 and Rudwick 1974.

[3] Louis Agassiz had visited Lochaber in the company of the geologist William Buckland in the summer of 1840, after attending the Glasgow meeting of the British Association for the Advancement of Science. Agassiz subsequently published a brief notice stating his belief that the parallel roads of Glen Roy had been formed by glacial lakes (Agassiz 1840b). Two years later, he published a more complete report on the glacial phenomena of Scotland in which several pages were devoted to a discussion of the parallel roads of Lochaber (Agassiz 1842).

[4] Milne 1849. The Scottish advocate and geologist David Milne (later Milne-Home) visited Glen Roy in 1845. His paper on the parallel roads, read to the Royal Society of Edinburgh, was privately

printed in 1847 and published in the *Transactions of the Royal Society of Edinburgh* in 1849. There is an annotated copy of the earlier version in DAR 139: 5. Although Milne had originally subscribed to CD's view that the terraces were the remnants of former sea-beaches successively formed as the land was elevated, he became convinced, after visiting the region, that the roads were lacustrine in origin. See Rudwick 1974, pp. 140–5.

5 CD stated in his paper (see n. 2, above) that the water-worn rocks at the entrance of Loch Treig and at the head of Lower Glen Roy had helped to convince him that marine action had been involved in the formation of the parallel roads. These rocks appeared 'as if they had lain for centuries on a sea coast ... washed by the ancient waves' (*Collected papers* 1: 92).

6 In Chambers 1848, pp. 109–10, Robert Chambers outlined Louis Agassiz's glacial theory of the formation of the parallel roads (see n. 3, above). Chambers then claimed that a fatal objection to this theory was that it did not take into account 'an opening at the head of Glen Glaster, by which all water above the No. 3 shelf must have been discharged' (*ibid.*, p. 110).

7 The Scottish writer Thomas Dick Lauder published one of the earliest scientific accounts of the Lochaber terraces, suggesting that the roads were the remnants of former freshwater lakes (Lauder 1823).

8 In the map of Lochaber that accompanied CD's paper (taken from the map published in Lauder 1823, the hill labelled 'Tombhran' is shown at the juncture of Glen Roy with Upper Glen Roy, near the intersection with Loch Spey.

9 CD did not visit the Pass of Muckul, which Milne identified as a major overflow channel for the supposed former lake in Glen Spean (see Rudwick 1974, p. 140). After reading Milne 1847a, however, CD still believed the site represented a channel of the sea (*Correspondence* vol. 4, letter to Charles Lyell, 8 [September 1847]):

> I look at the pass of Muckul ... as a channel just kept open, & the head of glen Roy (where there is a great Bay silted up) & of Kilfinnen, (at both which places, there are level-topped mounds of detritus *above* the level of the terraces) as instances of channels filled up at the stationary levels.

For a comparison of the differing interpretations given to the same phenomena by proponents of the lacustrine hypothesis versus the marine hypothesis, see Rudwick 1974, pp. 138–9, Table 1.

10 In Bravais [1843?], the French geologist Auguste Bravais provided several drawings showing the lateral profiles of sections of the Norway coastline that exhibited lines marking the position of ancient sea levels. A shortened, English version of the work, published in the *Quarterly Journal of the Geological Society* (Bravais 1845), did not include the diagrams.

11 CD had made a similar statement about Roderick Impey Murchison after both had visited North Wales in 1842. As CD told Charles Lyell, whereas he had seen obvious signs of glaciation in the region, Murchison saw 'no traces of glaciers, but only the trickling of water & of the roots of the Heath! It is enough to make an extraneous man think geology from beginning to end a work of imagination & not founded on observation ... But I confess I am astonished, so glaringly clear after two or three days did the evidence appear to me.—' (*Correspondence* vol. 2, letter to Charles Lyell, [5 and 7 October 1842]). See also *Autobiography*, p. 70.

12 Martin Rudwick has noted that 'as Jamieson clearly suspected, the relevant actualistic analogues [of the Lochaber roads] were not the glaciers of the Alps but those of the Polar regions.' (Rudwick 1974, p. 153 n. 98).

To Charles Hardy?[1] 4 September [1861][2]

Down. Bromley. Kent. S.E.
Sept. 4th

[To an unnamed correspondent[3] about the visiting by bees of red clover for honey. He tells his correspondent, who has written to him offering fresh and interesting material, that in the third edition of the *Origin*, he has modified certain statements

Thomas Francis Jamieson.
(Courtesy of the Scottish Agricultural College, Aberdeen.)

Louis Agassiz, *c.* 1861.
(Courtesy of the Museum of Comparative Anatomy, Harvard University.)

made in the first editions[4] and he asks him if he can find out if the clover in the Lowestoft district was a second crop.]

"If it were in full bloom at the end of August, it would be nearly certain, as Clover frequently flowers end of June— will you oblige me? . . ."[5]

Dawsons of Pall Mall (dealers)

[1] The recipient is identified from earlier correspondence about hive-bees visiting red clover. CD had asked Hardy to 'communicate any further observations on subject to me—' (see *Correspondence* vol. 8, letter from Charles Hardy, 23 July 1860, and letter to Charles Hardy, 27 July [1860]).

[2] The year is suggested by the reference to CD's having modified statements for the third edition of *Origin*. The third edition was published in April 1861.

[3] The summary of the letter text is taken from Dawsons's catalogue of 2 August 1974.

[4] In the third edition of *Origin*, CD changed a passage given in *Origin*, pp. 94–5, which states that hive-bees 'can easily suck the nectar out of the incarnate clover, but not out of the common red clover, which is visited by humble-bees alone; so that whole fields of the red clover offer in vain an abundant supply of precious nectar to the hive-bee.' On the basis of information supplied by Hardy, CD added: 'I have been informed, that when the red clover has been mown, the flowers of the second crop are somewhat smaller, and that these are abundantly visited by hive-bees.' (*Origin* 3d ed., p. 100). See *Correspondence* vol. 8, letter from Charles Hardy, 23 July 1860.

[5] In the fourth edition of *Origin*, published in 1866, CD followed the sentence reporting Hardy's information (see n. 4, above) with the disclaimer: 'I do not know whether this statement is accurate' (*Origin* 4th ed., p. 107).

To J. O. Westwood 4 September [1861][1]

Down. | Bromley. | Kent. S.E.
Sept. 4th

My dear Sir

I thank you cordially for all the great trouble, which you have so kindly taken for me.— Positively I have never seen your paper by Morren, or heard of its existence, except in your previous note.—[2] I will write to Gardeners' Chron. & see whether Lindley knows.—[3]

I do not possess Todds Cyclop. of Phy. &c:[4] is there an article on the fertilisation of Orchids?

I received last night the two Sphexes, & I thank you much for so kindly sending them; but unfortunately the pollen-masses do not belong to Orchids (I have removed only one) but to Asclepias, which abounds in N. America, & requires, equally with Orchids, insect-aid.—[5]

I am really sorry to trouble you, but how can I return the insects if not by Post? Shall I leave them at the British Museum in Dr Gray's hands, addressed to you, when I next go to London?[6] But I shall not be in London for a month or two.— When you write to tell me what to do, will you tell me about Todd.—

With many thanks | Yours sincerely | Charles Darwin

Oxford University Museum (Hope Entomological collection)

[1] Dated by the relationship to the letter from J. O. Westwood, 26 September 1861.
[2] See letter to J. O. Westwood, 15 August [1861].
[3] See the first letter to the *Gardeners' Chronicle*, [before 14 September 1861]. The botanist John Lindley was the editor of the *Gardeners' Chronicle*.
[4] Robert Bentley Todd edited *The cyclopedia of anatomy and physiology* (1835–59).
[5] CD had asked Westwood to send him specimens of insects with orchid pollen-masses attached to their probosces (see letter to J. O. Westwood, 15 August [1861]). *Sphex* is a genus of solitary wasp.
[6] John Edward Gray was keeper of the zoological collection of the British Museum.

To George Rolleston 5 September [1861][1]

Down. | Bromley. | Kent. S.E.
Sept. 5th

My dear Sir

You are very kind in telling me not to write; & though I am too hard worked, it is a pleasure to me to write, otherwise I would not do so.— Your note is a real gold-mine of facts & suggestions, *all* new to me.[2] I am very glad to hear of Gratiolets remark,[3] for I remember being very anxious to know how the case was; & I remember wishing to know how early an organ modified in a very unusual manner in its class appeared, relatively to its homologue in other members of the same class: & I thought that a Bat's wing would be good case, but I could not find any description.—[4]

Your cases of analogous variations are very valuable to me, & indeed all your note.— I do not know whether Craniologists can be at all believed, who assert that the Head alters in shape after peculiar studies in the adult.— I may mention a little personal anecdote. I worked hard during voyage of the Beagle & was a very idle sportsman before; & when I returned after 5 years, my Father almost immediately exclaimed, "Why the shape of your head is altered."—

I was much interested by your papers in Nat. Hist. R.[5] & I was pleased to see in Silliman's Journal that your article was specially noticed.—[6]

I have had a hard morning's dissection at some plants & am tired & will not amuse myself, by scribbling any more.

With hearty thanks for favours past & *to come*! | Yours sincerely | Charles Darwin

Royal College of Physicians

[1] Dated by the relationship to the letter from George Rolleston, 1 September 1861.
[2] Letter from George Rolleston, 1 September 1861.
[3] Gratiolet 1854. See letter from George Rolleston, 1 September 1861 and n. 8.
[4] See *Correspondence* vol. 6, letters to T. H. Huxley, 5 July [1857] and 9 July [1857], and letter from T. H. Huxley, 7 July 1857.
[5] Rolleston published two papers in the *Natural History Review* in 1861. The first, 'On the affinities of the brain of the orang utang' (Rolleston 1861a), was published in April. For CD's opinion of the work, see the letter to J. D. Hooker, 23 [April 1861]. The second, 'On correlations of growth, with a special example from the anatomy of a porpoise' (Rolleston 1861b), did not appear until October 1861.

[6] In the notice of the new series of the *Natural History Review* published in the July 1861 issue of the *American Journal of Science and Arts* 2d ser. 32 (1861): 130–2, Rolleston 1861a was singled out, being described as 'a well-considered and in every way most admirable communication.'

To J. D. Hooker 6 September [1861]

Down
Sept 6th

My dear Hooker

One line to say that the magnificent Book arrived quite safe.[1] I fear it will take me some little time to go through it.— Many thanks for names of Orchids. I am heartily glad to hear that Mrs Hooker is better & that you will sometime come to Down.—

If Oliver is at Kew remember me very kindly to him. Tell him I have got the Dionæas through his suggestion from Chichester & shall begin to observe tomorrow.[2] By the way if Oliver is at Kew, will you ask him to be so kind as to tell me anything he can about origin or habitat of Primula ciliata & *P. ciliata var. purpurata* Is it a wild var? They differ oddly in one important respect & I am very curious to know what this var. is.[3]

I think that you will think that I have made out the meaning of Dimorphism in Primula, satisfactorily & a very odd case it is & has caused me much labour in artificial crossing.[4]

Ever yours | my dear Hooker | C. Darwin

Orchids are inexhaustible in contrivances: Dendrobium chrysanthemum was given me yesterday & it has such a splendid contrivance, so that if an insect fail in removing the pollinia by the sticky matter, the pollinia are pitched right over the rostellum & always alight on the stigma, so nicely adapted is the force of the spring.[5]

Endorsement: '1861.'
DAR 115.2: 112

[1] Bauer 1830–8. See letter to J. D. Hooker, [11 August 1861] and n. 5.

[2] Daniel Oliver, assistant in the herbarium of the Royal Botanic Gardens, Kew, had been helping CD with his study of the sensitivity of the leaves of *Dionaea* and other insectivorous plants to nitrogenous substances. CD's observations and experiments on *Dionaea* are described in *Insectivorous plants*, pp. 286–320.

[3] See letter to Daniel Oliver, 11 September [1861].

[4] CD read his paper 'On the two forms, or dimorphic condition, in the species of *Primula*, and on their remarkable sexual relations' to the Linnean Society on 21 November 1861. See *Collected papers* 2: 45–63.

[5] The pollination mechanism of *Dendrobium chrysanthum* is described in *Orchids*, pp. 172–8.

To T. F. Jamieson 6 September [1861]

Down. | Bromley. | Kent. S.E.
Sept 6th

Dear Sir

I thank you sincerely for your long & very interesting letter. Your arguments seem to me conclusive.[1] I give up the ghost. My paper is one long gigantic blunder.[2]

I suppose & hope that you will publish an account of what you have observed.[3] The case seems very interesting. What a wonderful record of the old icy lakes do these shores present! It really is a grand phenomenon. I have been for years anxious to know what was the truth, & now I shall rest contented, though ashamed of myself.— How rash it is in science to argue because any case is not one thing, it must be some second thing which happens to be known to the writer.—[4]

I will take the liberty to forward your letter to Sir C. Lyell, as I am sure he would like to read it.—[5]

With very sincere thanks. Pray believe me, my dear Sir | Yours sincerely | Ch. Darwin

Did I not say that you would be able to settle the question?—[6]

Endorsement: '6. Sept. 1861'
National Library of Scotland (MS. 5406, ff. 167–8)

[1] See letter from T. F. Jamieson, 3 September 1861.

[2] CD refers to his paper 'Observations on the parallel roads of Glen Roy, and of other parts of Lochaber in Scotland, with an attempt to prove that they are of marine origin', published in the *Philosophical Transactions of the Royal Society of London* (1839): pt 1, 39–81. The paper presented evidence supporting a marine rather than a lacustrine theory to explain the phenomena. For CD's reluctance to accept that his explanation was a 'gigantic blunder', see Rudwick 1974.

[3] Jamieson published the results of his study of the parallel roads of Glen Roy in 1863 (Jamieson 1863).

[4] In his *Autobiography*, p. 84, CD called his paper on Glen Roy 'a great failure' and stated that he was 'ashamed of it', partly owing to his faulty scientific methodology:

> Having been deeply impressed with what I had seen of the elevation of the land in S. America, I attributed the parallel lines to the action of the sea; but I had to give up this view when Agassiz propounded his glacier-lake theory. Because no other explanation was possible under our then state of knowledge, I argued in favour of sea-action; and my error has been a good lesson to me never to trust in science to the principle of exclusion.

[5] In his paper, Jamieson referred to the encouragement he had received from CD and from Charles Lyell (Jamieson 1863, p. 240):

> Sir Charles Lyell suggested to me, when in London in 1861, to visit the district in order to get more evidence on the subject; and this was further urged by Mr. Darwin, who candidly admitted that, not having glacier-action in view when he was there, he had since, to some extent, doubted his own observations. He furnished me with some useful maps and memoirs bearing upon the problem, and likewise indicated several points worthy of particular attention.

[6] With the exception of the letters preserved in the Lyell correspondence, CD's letters to Jamieson have not been found. See, however, Jamieson's correspondence with Lyell, in which he mentions CD (Appendix IX).

To Charles Lyell 6 September [1861]

Down Bromley Kent
Sept 6th.

My dear Lyell

I think the enclosed is worth your reading.[1] I am smashed to atoms about Glen

Roy. My paper was one long gigantic blunder from beginning to end.[2]
Eheu. Eheu. | Ever yours | C. Darwin

Endorsement: '1861'
American Philosophical Society (262)

[1] Letter from T. F. Jameson, 3 September 1861.
[2] See preceding letter and Rudwick 1974.

To *Journal of Horticulture* [before 10 September 1861][1]

[Down]

Really obliged should I be if you could tell me the names of the parents of Gladiolus gandavensis. Also, whether the six following varieties—Eldorado, Canasi, Ophir, Linné, Brenchleyensis, and Vulcain, are the progeny of G. gandavensis by itself, or of G. gandavensis crossed by some other species? If the history of these six varieties be not known, their appearance may, perhaps, to instructed eyes tell their probable origin.—[2]
Charles Darwin.

Journal of Horticulture, Cottage Gardener, and Country Gentleman n.s. 1 (1861): 453

[1] Dated by the publication of the letter in the 10 September 1861 issue of the *Journal of Horticulture*.
[2] The answer to CD's query, an account of the parentage of *Gladiolus gandavensis*, was printed immediately following the letter; both query and answer appeared under the heading 'Parents of some gladioli'. The response was signed 'D.B.', possibly Donald Beaton, who wrote a regular column for the journal. However, the author did make the following statement: 'Mr. Beaton once or twice misstated the nativity of Gladiolus gandavensis, saying it was an Australian seedling, while it merely flowered there, with Mr. Bidwell, for the first time.'

To Charles Lyell 10 September [1861]

Down. | *Bromley.* | *Kent. S.E.*
Sept. 10th

My dear Lyell

I have been very much interested by your long letter (to be forwarded to-morrow);[1] but these subject have so much gone out of my head that I have nothing to suggest.[2]

I cannot think much of the absence of organic remains.[3] There were none in the vast gravel beds of Patagonia with sea-shells on surface; & none in the till of T. del Fuego, also with shells on surface;[4] though it may have been that these till beds were formed when T. del F. was like Greenland & afterwards subsided.— Yet I confess from what little I have of late read of Greenland I have suspected what you seem now inclined to admit & what Chambers so vehemently urges.[5] Is it certain that abundance of sea-living & swimming animals is any guide to shells &c living at the bottom? I cannot think that animals could live where icebergs are habitually grounded. I have remarked on this in Geolog. Transact Vol. VI p. 421,[6] but I refer to this only because a statement by Wrangell seems interesting on the tranquil space of shallow sea within that where the icebergs ground.[7]

With respect to former state of England, the Boulder which I describe at height of 803 ft in middle of Staffordshire has always seemed to me a striking case: it is in my paper on Glaciers of N. Wales in the Phil. Mag. 3^d series Vol. XXI. p. 186.—[8]

I pity you having to alter & modify on this great subject: I always marvel at & admire your industry & care.—[9]

I wish this note was better worth sending. I grieve over poor dear Glen Roy.—

Ever yours | C. Darwin

Endorsement: '1861'
American Philosophical Society (263)

[1] Lyell's letter has not been found. CD forwarded it to Thomas Francis Jamieson, who referred to it in his letter to Lyell of 11 September 1861 (see Appendix IX).

[2] See letter from T. F. Jamieson, 3 September 1861, and letters to T. F. Jamieson, 6 September [1861], and to Charles Lyell, 6 September [1861].

[3] One of the difficulties for those who viewed the 'parallel roads' of Glen Roy as having been formed by the sea was the absence of marine fossils in the area. In his 1839 paper on Glen Roy, CD stated that he 'attentively examined, with the expectation of finding fragments of sea shells, the matter accumulated on the shelves, and more especially the thicker beds of gravel and sand which occur at lower levels; but I could not discover a particle' of such evidence (*Collected papers* 1: 113). CD argued, however, that the preservation of marine fossils was an exceptional rather than a common phenomenon. In his letter to Lyell of 11 September 1861 (see Appendix IX) Jamieson stated: 'After what I saw in Glen Roy nothing short of the finding of marine remains will readily convince me that the sea has occupied Lochaber since the period of these shelves to a height exceeding some 600 or at *most* 800 feet.'

[4] CD had published these observations in his paper 'On the distribution of erratic boulders and on the contemporaneous unstratified deposits of South America', *Transactions of the Geological Society of London* 2d ser. 6 (1842): 415–31 (*Collected papers* 1: 145–63). In this paper he accounted for the absence of marine remains in these deposits by postulating the disturbance of the accumulating till by 'icebergs being lifted up and down by the tides' (*ibid.*, p. 152).

[5] The reference is to Chambers 1853, in which Robert Chambers presented evidence of glaciation in Scotland and parts of England. Chambers argued against the prevailing geological opinion, put forward most prominently by Lyell and Roderick Impey Murchison, that the various indications of the action of ice in these areas, such as striation, resulted not from the action of glaciers but from floating icebergs that were subject to tidal motions. CD had also supported the iceberg interpretation. See especially *Correspondence* vol. 5, letter to Charles Lyell, 7 June [1853].

[6] See n. 4, above.

[7] Wrangel 1840, p. 257.

[8] 'Notes on the effects produced by the ancient glaciers of Caernarvonshire, and on the boulders transported by floating ice', *London, Edinburgh, and Dublin Philosophical Magazine and Journal of Science* 21 (1842): 180–8 (*Collected papers* 1: 163–71). The erratic boulder at Ashley Heath in Staffordshire is cited as proof of the submergence of western England during the period in which floating ice abounded (*ibid.*, p. 170 n. 11).

[9] Lyell discussed new evidence concerning glacial phenomena in Britain in C. Lyell 1863.

To John Murray 10 September [1861]

Down. | Bromley. | Kent. S.E.
Sept. 10th

My dear Sir

Will you be so good as to send off immediately by Post, if possible, or by quickest channel, a copy of last Edit. of Origin to

Mad.elle Clemence Auguste Royer
2. Place de la Madeline
Lausanne
Switzerland

As she has agreed with a Publisher for a French Translation.—[1]

My Book seems exciting much attention in Germany, judging from the number of discussions sent me.[2] To my surprise a Dutch Translation was sent me lately.—[3]

I have never seen an advertisement of the new Edition, which seems to me a great pity as it was certainly considerably improved. I hope that the B. of Oxford & Co. have not made you sorry that you published it. I have sometimes thought that this might be the case.

My dear Sir | Yours sincerely | Charles Darwin

Endorsement: '*Sept. 10th 1861*'
John Murray (publishers)

[1] Clémence Auguste Royer's French translation of *Origin*, published by Guillaumin et cie and V. Masson et fils, Paris, appeared in 1862 (Freeman 1977). Royer, who had taught in a girls' school in Wales for a year, was living in Lausanne, where she offered lectures to women on philosophy and science. Prior to reading *Origin*, Royer had become a convert to Jean Baptiste Pierre Antoine de Monet de Lamarck's evolution theory, and in the preface to her translation she advocated an evolutionary view of human origins and development (*DSB*). See also Harvey 1987.

[2] The Darwin Pamphlet Collection–CUL contains several 'discussions' of CD's theory sent to him before September 1861 by German authors. These include: Köstlin 1860, Meding 1861, Pelzeln 1861, Schleicher [1861], and Wagner 1860. For discussions of the reception of *Origin* in Germany, see Montgomery 1974, Corsi and Weindling 1984, and Pancaldi 1984.

[3] Tiberius Cornelius Winkler's Dutch translation of *Origin* appeared in 1860 (Winkler trans. 1860). See letter from T. C. Winkler, 7 July 1861. A copy of the translation is in the Darwin Library–CUL.

To Daniel Oliver 11 September [1861][1]

Down. | Bromley. | Kent. S.E.
Sept. 11th

My dear Sir

I was pleased to see your handwriting again, & sincerely hope that you are quite recovered from your long illness. Take warning from me & do not work too hard.— Very many thanks for information about P. ciliata & *var.*—[2] I was sure it could not be an ordinary variety; it differed too much & I was utterly perplexed what to do about it.—[3] I sh.d have much liked to have seen P. farinosa, but I **well** knew that you could not send it, else you would have done so.—[4]

I hate changing work & I have so many irons in the fire that I have stopped looking after Dionæa.— I return your nice sketches which make me quite envious. I do not of course mean to attempt any minute work at structure; merely a few functional observations relatively to Drosera (which alas I have resolved to put off for another year, for I have amused myself too long over Primulas & Orchids).—[5]

I have made out very little on Dionæa: merely that the leaf behaves very differently when a fly, or only bit of cork & nothing is caught.— When a fly is caught much acid mucous is poured out as in Drosera. And the leaf must absorb so as to perceive when it has caught a fly— The "tortoises" *seem* to be the secretors & absorbers & appear to me to be strictly homologous with the glands in Drosera.— It is odd that meat, flies, sol. of C. of ammonia, will not excite movement or secretion; the sensitive filament must be touched to set all in action. The sensitiveness of the filament is prettily adapted: drop of water falling on them, or strong wind through pipe produces no effect; but a touch by a woman's Hair held with one inch length free will suffice. There is a pretty difference with Drosera: the latter does not care for a *single* rough touch with even a needle, but a weight left on the gland of $\frac{1}{70,000}$ of a grain will excite movement; & this is good for plant, for it has to clasp an insect when resting on & sticking to the gland: it would be lost labour to Drosera to close when merely touched by large insect.— In Dionæa, a far lighter single touch by mere hair causes movement; but a *much greater* weight, if put on delicately, may be left on the filament, without exciting movement than with Drosera.—

But I must not run on. | Pray believe me | Yours very sincerely | Charles Darwin

Down House (MS 10: 30, 66)

1 Dated by the relationship to the letter to J. D. Hooker, 6 September [1861].

2 Oliver's letter has not been found. CD had asked Joseph Dalton Hooker to pass on to Oliver his query about *Primula ciliata* and *P. ciliata* var. *purpurata* (see letter to J. D. Hooker, 6 September [1861]).

3 In his paper 'On the two forms, or dimorphic condition, in the species of *Primula*, and on their remarkable sexual relations', *Journal of the Proceedings of the Linnean Society (Botany)* 6 (1862): 77–96 (*Collected papers* 2: 45–63), CD acknowledged having received specimens of *Primula ciliata* and *P. ciliata* var. *purpurata* from Oliver. Of the latter, CD stated that the case was 'hardly worth giving, as the variety *purpurata* is said to be a hybrid between this species and *P. auricula*' (*Collected papers* 2: 48).

4 *P. farinosa*, the bird's-eye primrose is common in damp woods in northern England and southern Scotland. It is not described in CD's paper on the dimorphic condition of *Primula*. By the time CD published a fuller account of this study in 1877, however, he had been able to examine specimens of this species. See *Forms of flowers*, pp. 45, 224, and 273.

5 Oliver had greatly assisted CD's study of the sensitivity of the leaves of *Dionaea* and other insectivorous plants to various substances (see *Correspondence* vol. 8). CD continued to investigate this topic for a number of years, not publishing his results until 1875 (see *Insectivorous plants*). CD's notes on *Primula* are in DAR 108 and 111, and those on orchids are in DAR 70.

To *Gardeners' Chronicle* [before 14 September 1861][1]

I have been endeavouring during several years to make out the many contrivances by which British Orchids are fertilised through insect agency. I am very anxious to examine a few exotic forms. Several gentlemen have kindly sent me specimens; but I have not seen one of Lindley's grand division of Arethuseæ, which includes the Limodorideæ, Vanillideæ, &c.[2] If any one would have the kindness to send me a few flowers and buds of any member of the group, packed in a small tin canister, by post, addressed as below, he would confer a very great favour on me. Would you have the kindness to inform me, if in your power, whether the late Professor Morren has published anything (and where) on the fertilisation of Orchids by insect agency?[3]

Charles Darwin, *Down, Bromley, Kent.*

Gardeners' Chronicle and Agricultural Gazette, 14 September 1861, p. 831

1 The letter was published in the *Gardeners' Chronicle and Agricultural Gazette*, 14 September 1861, p. 831. See also *Collected papers* 2: 41.

2 The reference is to the section on orchids in Lindley 1853. John Lindley's 'fifth Tribe' of orchids, Arethuseae, is only briefly mentioned in *Orchids*, pp. 269–70. CD was not able to examine 'any living flowers'; reasoning from statements about allied forms, he suggested that 'mechanical aid' was necessary for their fertilisation.

3 An editorial note followed CD's letter, probably written by the editor John Lindley:

> [We are unable to answer this question, and must refer it to others. After searching through Morren's multitudes of pamphlets, we find nothing on Orchids except an academical dissertation on Orchis latifolia, and some remarks on the causes of the movements in the lips of Megaclinium.]

The reference is to Belgian botanist Charles François Antoine Morren. In *Orchids*, p. 270 n., CD cited Morren 1839 on the necessity of artificial fertilisation of *Vanilla* in the East Indies owing to the absence there of the insect that is the agent of fertilisation in its country of origin, Mexico. See also letters to J. O. Westwood, 15 August [1861] and 4 September [1861].

To *Gardeners' Chronicle* [before 14 September 1861][1]

A writer in your columns (p. 699) states that he caused Vinca rosea to seed at the Royal Gardens, Kew, by imitating the action of an insect in inserting its proboscis, as I had succeeded with the common Periwinkle.[2] By implication it may be presumed that V. rosea had not previously seeded at Kew. But another writer, "F. A. P." (p. 736) states that his Vincas seed profusely.[3] Mr. Horwood, gardener to G. H. Turnbull, Esq., of this place,[4] has just been so kind as to bring me a small plant of Vinca rosea with nine flowers fertilised by the insertion of a horse-hair, and it now bears nine fine pods. Mr. Horwood says he has grown many plants for the last eight or nine years, and never before saw a pod. What can be the cause of the difference in the results obtained on the one hand by "F. A. P.", and on the other by the writer from Kew and Mr. Horwood? Will "F. A. P." have the kindness to state, if he sees this notice, whether his plants were in a greenhouse with the windows left open, so that the moths could get access at night?[5]

Charles Darwin, *Down, Bromley, Kent.*

Gardeners' Chronicle and Agricultural Gazette, 14 September 1861, pp. 831–2

[1] The letter was printed in the *Gardeners' Chronicle and Agricultural Gazette*, 14 September 1861, pp. 831–2. See also *Collected papers* 2: 41.

[2] See letter to the *Gardeners' Chronicle*, [before 15 June 1861] and n. 2. The author of the notice was Charles William Crocker, foreman of the propagating department at the Royal Botanic Gardens, Kew.

[3] The note, published in the *Gardeners' Chronicle and Agricultural Gazette*, 10 August 1861, p. 736, reads:

> I am surprised at "C. W. C.'s" assertion in your number for July 27, that "Tropical Vincas never produce seed under cultivation if left to themselves," for I find both the white and pink kinds seed most profusely; in proof of this I enclose a small spray with seed pods on it. They sow themselves in the neighbouring pots, but the produce has never been different from the parent plants. *F. A. P.*

[4] George Henry Turnbull resided at the Rookery in Down. CD thanked Turnbull in *Orchids*, p. 158 n., for allowing him 'the free use of his hot-houses' and for giving him 'some interesting Orchids'; he also thanked Turnbull's gardener, 'Mr. Horwood', for help with some of his observations.

[5] No response to CD's letter has been found.

To Charles Lyell [15 September 1861][1]

Down. | *Bromley.* | *Kent. S.E.*
Sunday Evening

My dear Lyell

Many thanks for sending me your most interesting correspondence.[2] What a capital man, Jamieson is! The letters came this morning & shall be forwarded tomorrow, which is the same as if posted by London Monday general post. I *could* not read them this morning for I was in middle of experiments on Dionæa, which could not be stopped.[3]

It is a grand subject you are discussing, & light must soon appear.— I wish I could help in the least, but I cannot. The absence of deltas on upper level of the Lochaber shelves seems to me quite sufficiently accounted for by small amount of land above the shelves on the lake or sea theory. Your argument of if submergence 1200 ft. in Perthshire since chief glacialisation, how about Glen Roy, is very striking.[4] I have just been glancing at my Glen Roy paper;[5] & there yet seem to me several points of very difficult explanation. (1) The levelled land-straits with terrace-like fringes *above* the level of the highest shelf. (2) the beds of pebbles on very gently inclined surface between Glen Turret & Glen Gluoy; & beds of pebbles on former islet (Meal Roy or Mealderry) at level of Lower shelf at the mouth of Glen Roy. When one sees what little power Glacial sea has to form pebbles, the case seems very odd in Glacial *lake*. (3) The fringe of detritus on sides of Spean & Roy valleys, about 60 ft thick all down the valley.; and buttresses of gravel on sides of Hills.— Is it possible that the river when it began to flow over the embankment of ice would cut its channel so slow & so drain the lake so slowly as to allow of all this great sloping accumulation? I remember perfectly well my surprise at **vast contrast** of state of valleys in N. Wales & Lochaber.[6] But I suppose ice-lakes must be true cause.

What Jamieson says about climate not being icy if Scotland &c &c were all submerged 2000 or 3000 ft goes for nothing, when one thinks of Georgia in S. watery

Hemisphere.—[7] In former letter you speak of Scotland having been perhaps much loftier so as to account for great glaciers; this rubs against my notions, owing to the great extension & prevalence of Glacial phenomena. It always seems to me, until the contrary can be shown, to be safest to look at the great Glacial period as simultaneous.—

My dear Lyell | Yours most truly | C. Darwin

You quote Heer;[8] I suppose that you know that Wollaston has overly bad opinion (all told me when at Torquay)[9] of his Entomology & declares he is not to be at all trusted; & that Lowe says he has made *great* mistakes about Madeiran fossil plants.—[10]

American Philosophical Society (264)

[1] The date is given by the relationship to Lyell's letter to Thomas Francis Jamieson, 14 September 1861 (see Wilson ed. 1970, p. 523). The following Sunday was 15 September.

[2] Lyell had apparently sent CD Jamieson's letter of 11 September 1861 (see Appendix IX) as well as Lyell's response, of which only the draft is extant (see n. 1, above). Lyell, Jamieson, and CD had been conducting a three-way correspondence for several months concerning the geological interpretation of the 'parallel roads' of Glen Roy in Scotland.

[3] See letter to Daniel Oliver, 11 September [1861].

[4] This point is discussed in the draft of Lyell's letter to Jamieson, 14 September 1861 (see Appendix IX). Lyell was in the process of revising the discussion of Glen Roy for a new edition of his *Elements of geology*. In the event, he introduced this discussion in his study of the geological evidences for the antiquity of man (C. Lyell 1863). By arguing for an initial submergence of the highlands of Perthshire, he was attempting to synthesise Jamieson's theory of glaciation with CD's view of submergence and gradual elevation to account for the phenomena of the parallel roads of Glen Roy. Jamieson, however, had pointed out that a major problem with this hypothesis was the absence of marine remains in Glen Roy (see Jamieson's letter to Lyell, 11 September 1861; Appendix IX).

[5] 'Observations on the parallel roads of Glen Roy, and of other parts of Lochaber in Scotland, with an attempt to prove that they are of marine origin', *Philosophical Transactions of the Royal Society of London* (1839): pt 1, 39–81 (*Collected papers* 1: 89–137).

[6] CD studied the glacial phenomena in North Wales in June 1842; he had visited Lochaber in 1838 (see *Correspondence* vol. 2). For a fuller account of CD's views of the evidence supporting the glacier-lake theory, and his reasons for believing that it failed to explain a number of important points, see *Correspondence* vol. 4, letters to Charles Lyell, 8 [September 1847] and [11 October 1847], and to the *Scotsman*, [after 20 September 1847]. For a map of Glen Roy, see p. 248.

[7] See Jamieson's 11 September 1861 letter to Lyell (Appendix IX). In his paper on Glen Roy, CD accounted for the erratic boulders of Lochaber by assuming that they had been deposited on beaches or shelves by floating icebergs and subsequently exposed when the sea retreated as a result of the elevation of the land (see *Collected papers* 1: 121–2). CD described the snow-clad island of South Georgia in the South Atlantic in *Journal of researches*, p. 273.

[8] Oswald Heer, a specialist on the Tertiary remains of Europe, was currently visiting England at Lyell's request to study the fossil plant remains recently found near Bovey Tracy in Devon (see K. M. Lyell ed. 1881, 2: 346–7, 349–50).

[9] CD had visited the entomologist Thomas Vernon Wollaston in July 1861 during the family's stay in Torquay (see letter to John Lubbock, 1 August [1861]).

[10] Richard Thomas Lowe was an expert on the natural history of Madeira, where he had served as chaplain from 1832 until 1852. Heer, who had made an extended visit to Madeira for health reasons, published a well-known paper on the island's fossil flora (Heer 1857).

To Asa Gray 16 September [1861][1]

Down Bromley Kent
Sept. 16th

My dear Gray

In the whirl of your public affairs, science may be forgotten, or if not forgotten you may have no inclination to write. But if so inclined I sh^{d}. be very glad to have a little information on any cases of dimorphism, like that of Primula, & to be allowed to quote you.—[2] To make sure that you may understand what I want to know: I give this beautiful diagram of the two forms of Primula.

I think I have made out their good or meaning clearly. The pollen of A is fitted for stigma of B & conversely. The individuals are divided into two sets or bodies, like the males & females of Quadruped; but both in Primula are hermaphrodites: but I will not enter on details, as I will soon send a paper to Linn. Soc.^{y}.—[3] I should be *eminently* glad to know of other analogous cases. Are the two forms ever borne on same plant? Thyme is a different case: as the one form is simply female, stamens having aborted; these females, however, I find to be the most productive of seed.[4] Some of the species of Linum offer this case:

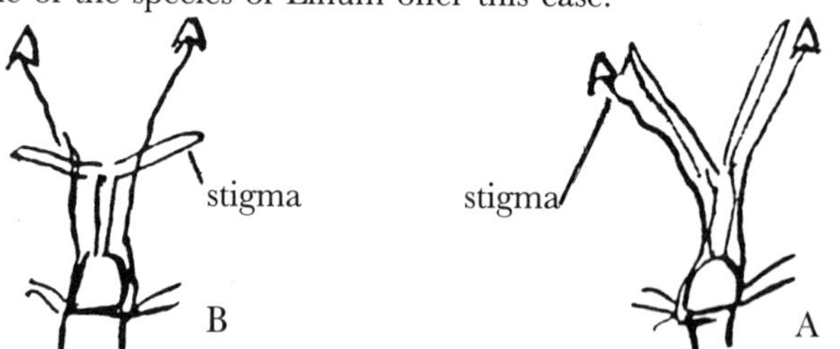

I am rather too late to experiment fairly; but I am almost sure that pollen of A is *absolutely* sterile on stigma of A: but good for stigma of B: whereas pollen of B is good for its own stigma B & for stigma of A.—[5]

This subject interests me much, so do help me if you can; for I have some *very faint* hopes that it may throw some light on Hybridisation;[6] I have sown seeds of Primulas raised in very scanty numbers from stigmas fertilised by what I call a homomorphous union; ie by pollen from plant of same form.—

If you have looked at Spiranthes let me hear:[7] but if your Spiranthes are like S. speciosa figured by Bauer:[8] they are very different from our S. autumnalis;

which by the way at Torquay I had the good luck to see Bees fertilise: I have almost finished my long Orchis paper,[9] & then I must go back to my true work on Cocks & Hens, fowls & rabbits.[10] Eheu Eheu, what much better fun observing is than writing.

By the way I have just been amusing myself in looking at Dionæa in aid of my Drosera work. How curious it is to see a fly caught & how beautiful are the adaptations compared with Drosera. But I must not run on about my hobby Horses.

Farewell | My dear Gray | Yours very sincerely | Charles Darwin

Gray Herbarium of Harvard University (73)

[1] The year is given by the reference to CD's statement that he will 'soon send a paper' on *Primula* to the Linnean Society of London (see n. 3, below).

[2] CD had been actively studying what he called the 'dimorphic condition' of *Primula* since the spring of 1860 (see *Correspondence* vol. 8). Some of his notes and experimental records on *Primula* and other dimorphic plants are in DAR 108 and 111.

[3] CD read his paper 'On the two forms, or dimorphic condition, in the species of *Primula*, and on their remarkable sexual relations' to the Linnean Society on 21 November 1861. See *Collected papers* 2: 45–63.

[4] CD had studied thyme plants earlier in the summer and made notes on wild thyme growing near Torquay when visiting the area in July and August 1861 (see DAR 109: 18–20 and DAR 157a, p. 72).

[5] CD referred only briefly to heterostyly in two *Linum* species in the *Primula* paper, stating that he intended to carry out further hybridisation experiments on the two forms (*Collected papers* 2: 62–3). In January and February 1862, he summarised the results of these experiments in his notes (DAR 108: 26–8). His paper on this topic, 'On the existence of two forms, and on their reciprocal sexual relation, in several species of the genus *Linum*', was read to the Linnean Society in February 1863 (see *Collected papers* 2: 93–105). In this paper, CD stated that he had put some pollen from the short-styled plants on the stigmas of long-styled plants and, although 'it was late in the season, namely September 15th', ultimately obtained 'six fine capsules', which he planted the following spring (*Collected papers* 2: 94).

[6] Sterility and hybridisation are discussed in the conclusion of the *Primula* paper (*Collected papers* 2: 61):

> The simple fact of two individuals of the same undoubted species, when homomorphically united, being as sterile as are many distinct species when crossed, will surprise those who look at sterility as a special endowment to keep created species distinct. Hybridizers have shown that individual plants of the same species vary in their sexual powers, so far that one individual will unite more readily than another individual of the same species with a distinct species. Seeing that we thus have a groundwork of variability in sexual power, and seeing that sterility of a peculiar kind has been acquired by the species of *Primula* to favour intercrossing, those who believe in the slow modification of specific forms will naturally ask themselves whether sterility may not have been slowly acquired for a distinct object, namely, to prevent two forms, whilst being fitted for distinct lines of life, becoming blended by marriage, and thus less well adapted for their new habits of life.

[7] CD had asked Gray to examine the flower structure of American species of the orchid *Spiranthes*. See *Correspondence* vol. 8, letter to Asa Gray, 31 October [1860]; see also letters to Asa Gray, 11 April [1861] and 5 June [1861].

[8] Bauer 1830–8.

[9] CD was intending to publish his work on orchids in the journal of the Linnean Society of London. After the manuscript was finished, however, he decided to publish it as a small book (see letter to John Murray, 21 September [1861]).

[10] CD refers to the manuscript he was preparing to support his view that there was extensive variation present in domesticated plants and animals on which both artificial and natural selection could operate. *Variation* appeared in two volumes in 1868.

To Asa Gray 17 September [1861][1]

Down
Sept 17th

Read First [2]

My dear Gray.—

I thank you sincerely for your very long & interesting letter, political & scientific, of Augt 27th & 29th & Sept 2d, received this morning.—[3] I agree with much of what you say & I hope to God we English are utterly wrong in doubting (1) whether the N. can conquer the S. (2) whether the N. has many friends in the South & (3) whether your noble men of Massachussetts are right in transferring your own good feelings to the men of Washington. Again I say I hope to God we are wrong in doubting on these points. It is number (3) which alone causes England not to be enthusiastic with you. What it may be in Lancashire I know not, but in S. England cotton has nothing whatever to do with our doubts.— If abolition does follow with your victory, the whole world will look brighter in my eyes & in many eyes. It would be a great gain even to stop the spread of Slavery into the Territories:—if that be possible without abolition, which I shd. have doubted. You ought not to wonder so much at England's coldness, when you recollect at the commencement of the war how many propositions were made to get things back to the old state with the old line of Latitude.—[4] But enough of this: all I can say is that Massachussetts & the adjoining States have the full sympathy of every good man whom I see; & this sympathy would be extended to whole Federal states, if we could be persuaded that your feelings were at all common to them.— But enough of this. It is out of my line, though I read every word of news & formerly well studied Olmsted.[5]

My other letter enclosed was unfortunately written last night:[6] I send it because I am yet not sure that you understand what I want. (I have just looked at your "Manual" & now see that the case of the Rubiaceæ is exactly the same as in Primula & Linum).[7] Did you know that in any case the pollen of the one form was not fitted to fully fertilise its own stigma?,—that sterility ensued of about the same degree as when allied species are crossed? This fact seems to me to make the case very interesting. I shd. be very grateful for any other cases in other Orders: I have found an old note of yours in which you say cases abound in "Rubiaceæ, Borragineæ Labiatæ &c." *Could you send me seed of any* such *Rubiaceæ*? though these would be bad plants to experiment on; yet I could to certain extent. I was looking at Galium criniatum this summer & found many flowers exclusively male: but I did not think of looking at pollen of the hermaphrodite flowers.

I fear I shall weary you; but I must write a little about Spiranthes.[8] At Torquay I was able to examine (& cover up) growing plants; before I had examined only cut flowers. In my note to you I blundered somewhat.—[9] I probably spoke of the "canoe" as being embedded within the rostellum; so it is in early bud, in so far as that the back of "canoe" is covered with large cells with viscid matter, which burst & thus attach to pollinia,—a process which I could never before understand.— Just as you say, when flower first opens only a hair or needle can be passed down & this, though *straight*, surely removes the pollinia; in this early condition of the flower you will never find pollen on stigma; but after 24–48 hours, the Labellum moves a little away & then position of nectar & oblique projecting stigma, allow the tips of pollen-masses beautifully to strike stigma.— If pollen-masses have not been removed in early period, the bowing of the proboscis is necessary for their removal.— The Bees which I saw at work (one had 5 canoes stuck to its proboscis, one over the other) alight at base of spike & go spirally up it; when they get to *upper* flowers the pollinia are attached to proboscis; they then fly to another plant & always alighting at base, they insert the pollinia in the lower & *more expanded* flowers & leave *masses* of pollen on stigma as I saw. You do not seem to have noticed the rupturing of the front of rostellum, with the most delicate touch, which seemed to me a vital action for the requisite touch was too delicate for mechanical action.—

I assure you that I tried D. rotundifolia & longifolia so often & showed the leaves to so many persons, that there can be no mistake in my observations.—[10] For some time (I cannot say how long without looking to my notes) after catching a fly the leaf cannot act. This is incidentally of use to plant for whilst the greater number of glands are dry, any useless object like bit of moss or cinder which has been clasped gets easily jerked or blown off.— I long to complete my work on Drosera; but must out of virtue defer it till next year; otherwise I shall never, with my small power of work, get even one volume of my larger work finished.—[11] I wish you would observe D. filiformis; I can hardly believe that any Drosera does not digest animal food.— I have found best plan is to try leaves which have opened after plants have been planted in Soup-Plate. Heaven forgive me, if you cannot, for scribbling at such length.—

Your question what would convince me of Design is a poser. If I saw an angel come down to teach us good, & I was convinced, from others seeing him, that I was not mad, I sh^d. believe in design.— If I could be convinced thoroughily that life & mind was in an unknown way a function of other imponderable forces, I sh^d. be convinced.— If man was made of brass or iron & no way connected with any other organism which had ever lived, I sh^d perhaps be convinced. But this is childish writing.—

I have lately been corresponding with Lyell, who, I think, adopts your idea of the stream of variation having been led or designed. I have asked him (& he says he will herafter reflect & answer me) whether he believes that the shape of my nose was designed.[12] If he does, I have nothing more to say. If not, seeing what Fanciers have done by selecting individual differences in the nasal bones of Pigeons, I must think that it is illogical to suppose that the variations, which Nat. Selection, preserves for the good of any being, have been designed. But I know that I am in the same sort of muddle (as I have said before) as all the world seems to be in with respect to free will, yet with every supposed to have been foreseen or preordained.

Farewell my dear Gray, with many thanks for your interesting letter— Your unmerciful correspondent | C. Darwin

Our English Holly (& all the cultivated vars.) are absolutely dioicous: I have just been looking at your Manual: Could I anywhere find out a fuller account of the state of some of the species of your Ilex?? for instance whether the female flowers have any pollen; (ours has anther but *no* pollen) & whether the male flowers ever set seeds. Our male trees never bear a berry. This would make a good case of gradation.— Having observed our Holly years ago I have always felt curious about the steps by which it became dioicous.[13]

Gray Herbarium of Harvard University (72)

[1] Dated by the reference to the letter from Asa Gray, [27 and 29 August] and 2 September [1861], only part of which is extant. The year is further confirmed by CD's reference to his letter 'written last night' (see preceding letter).

[2] CD sent this letter in the same envelope as the preceding letter.

[3] See letter from Asa Gray, [27 and 29 August] and 2 September [1861].

[4] Throughout 1861, both parties of belligerents in the American Civil War refrained from avowing that the war had anything to do with slavery and rather focused on the issue of states' rights versus the preservation of the union. See McPherson 1988, pp. 311–12.

[5] CD was an avid reader of the 'Foreign intelligence' columns of *The Times*, which carried regular reports on the political and military developments in the United States. The special correspondent covering the American Civil War for *The Times* was William Howard Russell. CD also refers to Frederick Law Olmsted, whose books describing conditions in the southern slave states (Olmsted 1856, 1857, and 1860) CD had read. See *Correspondence* vol. 4, Appendix IV, 128: 23, 25; and vol. 8, letter to Asa Gray, 26 November [1860]. In his reminiscences of his father, read at the banquet of the 1909 Darwin Centenary celebrations in Cambridge, William Erasmus Darwin recalled CD's 'hatred of slavery' (*Emma Darwin* 2: 169):

> I remember his talking with horror of his sleepless nights when he could not keep out of his mind some incidents from Olmsted's *Journeys in the Slave States* [Olmsted 1856], a book he had lately been reading; and in his many letters to Professor Asa Gray he alludes to slavery with the utmost detestation.

[6] See preceding letter.

[7] A. Gray 1856, p. 80. The passage is marked in CD's copy in the Darwin Library–CUL. Gray had told CD in his letter of [27 and 29 August] and 2 September [1861] that the 'most marked *diæcio-dimorphous* flowers I know are in Rubiaceæ'.

[8] CD had asked Gray to 'have a look at Spiranthes' in the letter to Asa Gray, 21 July [1861]. Gray had apparently done so and failed to confirm CD's observations. The portion of Gray's letter that discusses the details of his observations is missing (letter from Asa Gray, [27 and 29 August] and 2 September [1861]).

[9] See *Correspondence* vol. 8, enclosure to the letter to Asa Gray, 31 October [1860].

[10] CD had begun studying *Drosera* and other insectivorous plants during the summer of 1860. See *Correspondence* vol. 8.

[11] The reference is to CD's work on the variation of plants and animals under domestication, which was published in two volumes in 1868.

[12] See letter to Charles Lyell, 21 August [1861].

[13] A note in DAR 47: 75 reads: 'Asa Gray Manual p. 263 Ilex. it seems that the N. American species have not their sexes nearly so much separated.—' CD only briefly mentioned this case when he published a full account of his study of dimorphism (*Forms of flowers*, pp. 297–8). For Gray's response to CD's query about American species of holly, see the letter from Asa Gray, 11 October 1861.

To Henry Fawcett 18 September [1861][1]

[Down]
September 18

My dear Mr. Fawcett,—

I wondered who had so kindly sent me the newspapers, which I was very glad to see;[2] and now I have to thank you sincerely for allowing me to see your MS.[3] It seems to me very good and sound; though I am certainly not an impartial judge. You will have done good service in calling the attention of scientific men to means and laws of philosophising. As far as I could judge by the papers, your opponents were unworthy of you. How miserably A. talked of my reputation, as if that had anything to do with it.[4] . . . How profoundly ignorant B. must be of the very soul of observation![5] About thirty years ago there was much talk that geologists ought only to observe and not theorise;[6] and I well remember some one saying that at this rate a man might as well go into a gravel-pit and count the pebbles and describe the colours. How odd it is that anyone should not see that all observation must be for or against some view if it is to be of any service!

I have returned only lately from a two months' visit to Torquay, which did my health at the time good; but I am one of those miserable creatures who are never comfortable for twenty-four hours; and it is clear to me that I ought to be exterminated. I have been rather idle of late, or, speaking more strictly, working at some miscellaneous papers, which, however, have some direct bearing on the subject of species; yet I feel guilty at having neglected my larger book. But, to me, observing is much better sport than writing. I fear that I shall have wearied you with this long note.

Pray believe that I feel sincerely grateful that you have taken up the cudgels in defence of the line of argument in the "Origin;" you will have benefited the subject.

Many are so fearful of speaking out. A German naturalist came here the other day, and he tells me that there are many in Germany on our side; but that all seem fearful of speaking out, and waiting for some one to speak, and then many will follow.[7] The naturalists seem as timid as young ladies should be, about their scientific reputation. There is much discussion on the subject on the Continent, even in quiet Holland,[8] and I had a pamphlet from Moscow the other day by a man who sticks up famously for the imperfection of the "Geological Record," but complains that I have sadly *understated* the variability of the old fossilised animals![9] But I must not run on.

With sincere thanks and respects, | Pray believe me, | Yours very sincerely, | Charles Darwin.

Stephen 1885, pp. 100–2

[1] Dated by the reference to the reports printed in the *Manchester Examiner* and *Manchester Guardian* (see n. 2, below).

[2] The *Manchester Examiner* and the *Manchester Guardian* of 9 September 1861 both carried reports of the discussion that followed Fawcett's address to the British Association for the Advancement of Science. The British Association met in Manchester from 4 to 11 September 1861. There are copies of both reports in DAR 226.1: 92–3.

[3] The title of Fawcett's paper, as it appeared in the abstract printed in the *Report of the ... British Association for the Advancement of Science held at Manchester*, was 'On the method of Mr. Darwin in his treatise on the origin of species' (Fawcett 1861). In the abstract, however, it was reported that Fawcett's original title was: 'That the method of investigation pursued by Mr. Darwin, in his Treatise on the Origin of Species, is in strict accordance with the principles of logic.'

[4] The report of the discussion given in the *Manchester Guardian*, 9 September 1861, p. 6, cites William Crawford Williamson as saying that 'while he would not say that Mr. Darwin's book had caused him a loss of reputation, he was sure that it had not caused a gain.' Williamson was professor of natural history, anatomy, and physiology at Owens College, Manchester. See also *ML* 1: 195 n.

[5] An editorial insertion that follows the mention of 'B.' in the copy of the letter published in Stephen 1885 reads: '[who had said that Darwin should have published facts alone]'. In *ML*, where this letter is reprinted, Francis Darwin identified 'B.' as Edwin Lankester (*ML* 1: 195 n.). The *Manchester Guardian*, 9 September 1861, p. 6, cites Lankester as having stated: 'The facts brought forward in support of the hypothesis had a very different value indeed from that of the hypothesis.' Lankester was professor of natural history at New College, London.

[6] For a discussion of the strictures concerning the use of theory in early Victorian science, see Morrell and Thackray 1981. Fawcett defended CD's use of hypothesis and induction (see letter from Henry Fawcett, 16 July [1861] and nn. 3 and 4).

[7] The visitor from Germany has not been identified.

[8] A Dutch translation of *Origin* by Tiberius Cornelius Winkler was published in 1860 (see letter from T. C. Winkler, 7 July 1861). For an account of the reception of *Origin* in Holland, see Bulhof 1974.

[9] The pamphlet, *Übergänge und Zwischenvarietäten*, was by Hermann Adolfovich Trautschold of the Petrovsky Academy in Moscow. It was a privately printed reprint of a paper originally published in the Moscow University Bulletin (1860). The annotated copy is in the Darwin Pamphlet Collection–CUL. Trautschold stated that for several years he had been aware of the existence of many varieties and transitional forms in the fossil species in Jurassic formations near Moscow but laid no value on the specimens until he read the German translation of *Origin* (Bronn trans. 1860). His paper describes a series of what he believed to be transitional varieties and species of ammonites. Beside a list of various transitional varieties (p. 5), CD wrote in pencil: 'All these vars. seem to be found in same middle bed'. But next to a subsequent description of a series of forms transitional between *Ammonites biplex* and *A. humphriesianus*, CD wrote: 'But here we have cases of vars. leading to a species in a lower bed.—'

From H. C. Watson 20 September 1861

Thames Ditton S.W.
Sept 20th 61

Dear Mr Darwin

I have some recollection of intimating a distrust of Mr. Sidebotham's recorded experiments on Primroses, &c. though I cannot now hit upon the intimation.—[1] However, I find his two papers in Phytolst. The first (Vol. 2, p. 887)[2] is of no moment, except as calling forth a P.S. mem. by Newman or Luxford, to the effect that the Primula Jacquinii—(Bardfield oxlip) comes true from seed in second

& third generations.[3] I have again looked over his second paper, & am still unwilling to rely on his accuracy;[4] & likely enough the objections which I should now make to the record, are such as were made before, & fell under your notice.

First, I do not believe that he had botanical skill or experience adequate to name his var^s. & species correctly; especially, as to distinguishing P. Jacquinii from Cowslip & umbellate Primrose.

Second— He states a result which no one else has hitherto shown; namely, that P. Jacquinii produced Primroses, & a Cowslip produced P. Jacquinii. That these (unusual, at least) results should *both* have come out at *same time* from *single* experiments, seems in the highest degree unlikely. I have tried P. Jacquinii for (I think) 16 descents and never got a primrose from it; nor has *it* ever sprung from seed of Cowslip or Primrose during 20 years of observation & experiment. I prefer to think either that he misnamed the plants, or mingled seed by some chance.

Third— He writes that in order "to prevent all errors as much as possible, I collected the seed myself, and again sowed it on beds, where no Primulæ had been previously grown". He does not tell us whether these beds were near or wide apart, so that dew-worms &c. could not carry a seed from one to another. Nor does he show that any precaution was taken to prevent the seeds of Oxlips, primroses, cowslips (already in the garden) from being chance-mingled in the soil of those beds. The experimenter who thinks that "all errors as much as possible" are guarded against by the course he describes, must have very imperfect notion of the precautions which *are* necessary in such experiments. Had he been operating with exotic plants, never before grown in the garden, he might have assumed absence of seeds other than those purposely sown on his "beds".

Fourthly— There are other instances, which show that M^r. S. is unreliable in his botanical statements; i.e. he hazards statements in print, which are not true; although I think from carelessness & imperfect knowledge, rather than mala fides (Phytol. III, p. 144— Cybele Brit. II p. 168.)[5]

It is for yourself to decide how far I am warranted in distrust on such grounds.

In the paper to which you refer,—Phytol. III, p. 703–5,—Mr Sidebotham says nothing about "artificial fertilization". He simply writes, "to prevent any hybridization from the visits of insects, I protected them" (the oxlips, cowslips, primroses) "with glass during the early part of their flowering" (p. 704).

Cowslips, primroses, & polyanthuses are so usually grown, & have been so long grown in gardens, that we should never feel sure of their seeds not being in the ground, unless under some special circ̃es. My own course has been, for particular experiments, to sow the seeds in flower pots, filled with mould from asparagus beds, or peat-earth dug where no Primula can be found. I dare not trust to the ordinary earth of the garden; for I see Primulæ spring up among potatoes, & other unlikely crops.

The asparagus beds, here very carefully kept weeded, & invariably forked over in spring, are extremely unlikely to have Primula seeds in them. Still, a chance seed *might* be brought to them on a tool, blown thither by wind, perhaps even come in the old horse-litter used as a manure dressing. Thus if I sowed primrose-seed in a flower-pot filled from an asparagus bed, & got 49 primroses & one cowslip, I should distrust my own experiment. But when I sow the seeds of an intermediate form in such soil, & find *several* primroses, *several* cowslips, & *several* intermediates, & look in vain for a primula on the Asparagus bed,—then I rely on the results. This result, I believe, is the only one that we can yet rely upon; namely, that an intermediate form can produce cowslips, primroses, & intermediates; vice versa cowslips & primroses can each produce the intermediates. As yet, distrust all else.[6]

I hope that some benefit was found from your stay at Torquay. I have just returned from a few weeks of indolence in the Isle of Wight, & the better for it.—

Believe me, always very truly | Hewett C. Watson

C. Darwin | Esqre

DAR 181.1

[1] CD had apparently written to Watson inquiring about the reliability of Joseph Sidebotham's experiments on *Primula*. Watson's 'intimation' has not been traced, although he had criticised Sidebotham on a different topic, accusing him of making 'mis-statements' that he said were attributable to an 'intellectual rather than moral deficiency' (see *Phytologist* 3 (1848): 145).

[2] Sidebotham 1847.

[3] The editorial comment reads: 'We have lately had an opportunity of observing seedlings produced in the second and third generation from Primula Jacquinii (*Newm.*) or Jacquin's Primula, and find that they are in all respects identical with the parent plant. We again express a hope that the names of *elatior* and *oxlip* will not be applied to this species.' Edward Newman was the general editor of the *Phytologist*, while George Luxford was the managing editor and a frequent contributor of articles (see *Phytologist* 1 (1844): v–vi).

[4] Sidebotham 1849.

[5] See n. 4, above, and Watson 1847–59, 2: 168.

[6] CD referred to Sidebotham's work in the notes to his paper 'On the two forms, or dimorphic condition, in the species of *Primula*, and on their remarkable sexual relations', read before the Linnean Society of London on 21 November 1861 (*Journal of the Proceedings of the Linnean Society (Botany)* 6 (1862): 77–96; *Collected papers* 2: 45–63). Discussing the question whether the primrose (*Primula vulgaris*), the cowslip (*P. veris*), and the oxlip (*P. elatior*) were distinct species or only varieties of one species, CD stated his belief that oxlips were produced by cross-fertilisation between primroses and cowslips. In a note, he made reference to Sidebotham 1849 (*Collected papers* 2: 63 n. 6):

> Mr. Sidebotham (Phytologist, vol. iii. pp. 703–5) states that he protected his plants from crossing; but as he gives in detail all the precautions which he took, and says nothing about artificial fertilization, we may conclude that he did not fertilize his plants. As he raised very numerous seedlings, he would have had to fertilize many flowers, if they had been really well guarded against the visits of insects. Hence I conclude that his results are not worthy of trust.

To John Murray 21 September [1861]

Down Bromley Kent
Sept. 21st

My dear Sir

Will you have the kindness to give me your opinion, which I shall implicitly follow.— I have just finished a very long paper intended for Linn. Socy. (the title is enclosed)[1] & yesterday for the first time it occurred to me that **possibly** it might be worth publishing separately, which would save me trouble & delay.— The facts are new & have been collected during 20 years & strike me as curious.[2] Like a Bridgewater Treatise the chief object is to show the perfection of the many contrivances in Orchids.[3] The subject of propagation is interesting to most people, & is treated in my paper so that any woman could read it. Parts are dry & purely scientific: but I think my paper would interest a good many of such persons who care for Nat. History, **but no others**. In a few days an artist is coming here to make from 20–30 small woodcuts.—[4]

As far as I can calculate a paper contains about 29,000 words; in small page *with rather open type*, about 205 words to page, I calculate the matter would make 131 page, but with division into chapter say *at most* 135 pages. So it would be a very little Book, & I believe you think very little books objectionable. I have myself **great** doubts on subject. I am very apt to think that my Geese are Swans; but the subject seems to me curious & interesting.

I **beg** you not to be guided in the least in order to oblige me, but as far as you can judge, please give me your opinion.— If I were to publish separately, I would agree to any terms, such as half risk & half profit, or what you liked; but I would not publish on my sole risk, for to be frank, I have been told that no Publisher whatever, under such circumstances cares for success of Book.— I would pay myself for all drawing on the wood, but not for cutting.— I sh[d]. send *rough M.S* to be printed on Slips & would pay for extra corrections.— I sh[d] require & pay for 30 or 40 copies.— But if this little Book were to fail, it occurs to me that it might injure sale of my future larger Books.— In fact I am utterly in doubt.— Please give me your impression.—[5]

My dear Sir | Yours sincerely | C. Darwin

Endorsement: '*1861—Sep 21*'
John Murray (publishers)

[1] CD began preparing his manuscript on the pollinating mechanisms of various species of orchids during his stay in Torquay in July and August (see 'Journal'; Appendix II). As he states, he had always envisioned publishing the work in one of the journals of the Linnean Society of London.

[2] CD described the history of his study of orchids in his *Autobiography*, p. 127:

> During the summer of 1839, and, I believe, during the previous summer, I was led to attend to the cross-fertilisation of flowers by the aid of insects, from having come to the conclusion in my speculations on the origin of species, that crossing played an important part in keeping specific forms constant. I attended to the subject more or less during every subsequent summer; and my interest in it was greatly enhanced by having procured and read in November 1841, through the advice of Robert Brown, a copy of C. K. Sprengel's wonderful book, *Das entdeckte Geheimniss der Natur*.

CD's annotated copy of Sprengel 1793, dated August 1841, is in the Darwin Library–CUL. Michael Ghiselin points out that the account given above overlooks CD's earlier interest in pollination biology that was perhaps stimulated by his reading of Erasmus Darwin's *Botanic garden* (E. Darwin 1789–91). See Ghiselin 1984, pp. xvii–xviii. See also the reference to studying orchid fecundation in CD's Questions & experiments notebook, p. 15 (*Notebooks*).

[3] The 'Bridgewater Treatises' is the collective name given to eight works on natural theology published between 1833 and 1836 under the terms of a bequest made to the Royal Society of London by Francis Henry Egerton, the 8th earl of Bridgewater. According to Egerton's stipulations, each author was to write and publish a treatise 'On the power, wisdom and goodness of God, as manifested in the creation'. Several of the volumes became classic works in the apologetic writings of natural theology (*EB*; see also Topham 1993). For a discussion of *Orchids* as a treatise showing how organic structure could be explained on the basis of natural selection rather than divine purpose, see Ghiselin 1984.

[4] George Brettingham Sowerby Jr prepared the illustrations used in *Orchids* (see *Orchids*, p. vi).

[5] See letter from John Murray, 23 September 1861.

To Charles Lyell 22 September [1861]

Down Bromley Kent
Sept. 22[d]

My dear Lyell

I have read M[r] Jamieson's last letter, like the former ones, with *very great* interest.[1] What a problem you have in hand! It beats manufacturing new species all to bits.— It would be a great *personal* consolation to me, if M[r] J. can admit the sloping Spean Terraces to be marine; & would remove one of my greatest difficulties, viz. vast contrast of Welch & Lochaber valleys.[2] But then,, as far as I dare trust my observations, the *sloping* terraces run far up the Roy valley so as to reach not far below the Lower shelf.[3] If the sloping fringes are marine & the shelves lacustrine, all I can say is that nature has laid a shameful trap to catch an unwary wretch.— I suppose that I have underrated power of lakes in producing pebbles: this, I think, ought to be well looked to. I was much struck in Wales on carefully comparing the glacial scratches under a lake formed by moraine, (& which must have existed since glacial epoch) & above water, & I could perceive *no* difference.[4] I believe I saw many such beds of good pebbles on level of Lower shelf, which at the time I could not believe could have been found on shores of Lake. The Land-straits & little cliffs above them to which I referred were quite above the highest shelf—: they may be of much more ancient date than the shelves. Some terrace-like fringes at head of the Spey strike me as very suspicious.

M[r]. J. refers to absence of pebbles at considerable heights: he must remember that every storm, every deer, every Hare which runs tends to roll pebbles down hill, & not one ever goes up again.— I may mention that I particularly attended to this on S. Ventana in N. Patagonia,—a great, isolated, rugged Quartz mountain, 3000 ft high, & I could find not one pebble, except on one very small spot, where a ferruginous spring had firmly cemented a few to face of mountain.—[5]

If the Lochaber Lakes had been formed by a ice-period posterior to the (marine?) *sloping* terraces in the Spean; would not M[r]. J. have noticed gigantic moraines across the valley, opposite opening of L. Treig? I go so far as not to like making the elevation of the land in Wales & Scotland considerably different with respect to ice-period; & still more do I dislike it with respect to E. & W. Scotland. But I may be prejudiced by having been so long accustomed to plains of Patagonia. But the equability of level (barring denudation) of even the Secondary formations in Britain, after so many ups & downs, always impresses my mind, that, except when crust cracks & mountains are formed, movements of elevation & subsidence are generally very equable.—[6]

But it is folly my scribbling thus.— You have a grand problem, & Heaven help you & M[r] Jamieson through it. It is out of my line now a-days, & above & beyond me.—

Ever my dear Lyell | Yours truly | C. Darwin

(This is dreadfully untidy & useless, but I am too tired to rewrite it.—)

Endorsement: 'Sept[r] 1861'
American Philosophical Society (265)

[1] The reference is to Thomas Francis Jamieson's letter to Lyell, 19 September 1861 (see Appendix IX).

[2] Jamieson had written: 'I see no difficulty at present in allowing that the sea might have occupied Glen Spean &c after the era of these lakes up to 500 or 600 feet above the present coast line, in which case these accumulations might be partly of marine origin.' (see Appendix IX). CD discussed this point in his paper, 'Notes on the effects produced by the ancient glaciers of Caernarvonshire, and on the boulders transported by floating ice', published in the *Edinburgh New Philosophical Journal* 33 (1842): 352–63 (*Collected papers* 1: 163–71). Andrew Crombie Ramsay espoused the view that there had been a period of glacial activity in Wales after the deposition of marine drift (Ramsay 1852, p. 376).

[3] See CD's Glen Roy notebook, pp. 31–2 (*Notebooks*), in which he noted that the sloping terraces of Glen Spean were 'difficult to explain on ⟨formation⟩ deposition in lake'. In his paper on the 'parallel roads' of Glen Roy, CD described the accumulation of 'alluvial' gravels and other debris that formed flat-topped 'buttresses' below the level of the lowest 'road' in Glen Spean. He argued that if the valley had been filled with an arm of the sea, into which the river transported its debris, and if the land was rising in a succession of short steps, the river would have deposited its material in deltaic form. Therefore, as the sea retreated from the valley, the river would erode a gorge through material previously deposited and, at the same time, deposit more material further downstream at the point where it entered the sea at its new, lower level. With this account, CD believed he had demonstrated the case for a gradual fall of the level of water in Glen Spean from the lowermost 'road' towards the present sea level. See 'Observations on the parallel roads of Glen Roy, and of other parts of Lochaber in Scotland, with an attempt to prove that they are of marine origin' (*Collected papers* 1: 89–137); see also Rudwick 1974, pp. 121–4.

[4] CD mentioned this observation in his paper on the ancient glaciers of Caernarvonshire, North Wales, p. 354 (see n. 3, above).

[5] See *South America*, p. 79.

[6] In his work on Glen Roy and other geological phenomena, CD based his ideas on Lyell's theory of crustal elevation and subsidence as the *vera causa* of the phenomena. For a discussion of CD's work on Glen Roy within the context of his wider conception of natural history, see Rudwick 1974, pp. 153–74.

To A. G. More 23 September 1861

Down, Bromley, Kent. S.E.
Sept. 23, 1861.

My dear Sir

As I have drawn up my paper on the fertilisation of Orchids, I should be very glad to hear about the capsules of Ep. palustris. Will you, therefore, apply to your friend?[1] If ripe, I should like to see any capsules produced by the mutilated flowers and *others for comparison*. In some other experiments on Orchids I have found that the external size of capsule is a very poor criterion of production of *good* seeds. Will you ask your friend to observe how far the untouched flowers have produced capsules. And please inform me how many flowers you mutilated.[2]

Pray excuse my asking you to take all this trouble about my hobby-horse. I am very curious on the point, though I shall not be surprised to hear that I have much exaggerated the importance of the Labellum.[3]

I hope that you have quite recovered your health. And pray believe me | My dear Sir | Yours sincerely and obliged | Charles Darwin.

Copy
DAR 146

[1] See letter to A. G. More, 13 August 1861.
[2] See letter to A. G. More, 1 October 1861.
[3] In the description of *Epipactis palustris* in *Orchids*, CD noted that the 'curiously shaped' labellum was divided into two parts joined together by a hinge and suggested that the distal half formed 'an excellent landing-place for insects' (*Orchids*, p. 98).

From John Murray 23 September 1861

50A. Albemarle Street, London, W.
Septr 23 *1861*

My Dear Sir

I have considered attentively your very obliging & considerate note touching the publication of a little Vol. originally intended for Linnæan-Transactions—on the Fertilization of Brit. Orchids by Insects.[1] I have no hesitation in offering to print & publish the work including Illustrations at my own sole cost & risque giving you one half of the profits of every Edition. On hearing from you I will insert the announcement on my next Quarterly List—[2]

I am | My Dear Sir | Yours very sincerely | John Murray

Charles Darwin Esq

DAR 171.3(1): 331

CD ANNOTATION
Verso of letter: 'who will cut Title (if altered)' *ink*

[1] See letter to John Murray, 21 September [1861].
[2] *Orchids* was published by John Murray in May 1862.

To Joseph Reay Greene?[1] 24 September [1861][2]

Down
24 September

'I was but lately wishing to read some treatise up to the present mark, on these animals. I have cut the pages & can clearly see that your work will much interest & instruct me . . .'

Sotheby (21 December 1965)

[1] The identity of the correspondent is suggested by the catalogue description, which states that CD was 'returning thanks for the gift of his correspondent's treatise on Coelenterata'. There is a copy of Greene's *Manual of the sub-kingdom Cœlenterata* (the second part of Greene 1859–61) in the Darwin Library–Down.

[2] The year is established by the year of publication of the volume mentioned in the letter (see n. 1, above).

To J. D. Hooker 24 September [1861]

Down. | *Bromley.* | *Kent. S.E.*
Sept. 24th

My dear Hooker

I have kept Lindleys Orchid rather long,[1] but I hope I may keep it for about a fortnight longer, as Sowerby is coming to draw & until he has finished with my specimens, as the great Plates will be useful.—[2]

I have been acting, I fear that you will think, like a goose & perhaps in truth I have.— When I finished a few days ago my Orchis paper, which turns out 140 folio pages!! & thought of the expence of woodcuts, I said to myself I will offer the Linn. Socy to withdraw it & publish it as pamphlet. It then flashed on me that perhaps Murray would publish it, so I gave him a cautious description & offered to share risks & profits. This morning he writes that he will publish & take all risks & share profit & pay for all illustrations.[3] It is a risk & Heaven knows whether it will not be a dead failure; but I have not deceived Murray & told him that it would interest those alone who cared much for Natural History.

I hope I do not exaggerate the curiosity of the many special contrivances. It will save me a deal of trouble, as I shall send rough M.S & correct on slips. It will make a *very* little Book. Whenever you write please tell me Mr Fitch's Christian name; as he is to do on wood a couple of Diagrams for my Primula paper which, thank Heavens is finished for Linn. Soc.[4] It has run out pretty long; I fear I am getting terribly lengthy.

Whenever you write be sure tell me how Mrs. Hooker is, & when you start for Scotland—[5] I hope it may be soon, for I am sure you must want rest & a change.— I saw your letter about Schlagenweits: what a row the affair has made.[6]

Farewell.— Etty goes on splendidly, & still sucks in the Oil.—[7] Farewell | My dear Hooker | Ever yours | C. Darwin

Endorsement: '/61'
DAR 115.2: 113

[1] CD had borrowed Bauer 1830–8, to which John Lindley contributed notes and prefatory remarks on the structure of orchids. See letters to J. D. Hooker, 30 August [1861] and 6 September [1861].

[2] George Brettingham Sowerby Jr prepared the drawings included in *Orchids*, many of which were copied from Bauer 1830–8.

[3] See letter to John Murray, 21 September [1861], and letter from John Murray, 23 September 1861.

[4] Walter Hood Fitch was the botanical artist at the Royal Botanic Gardens, Kew. CD's paper 'On the two forms, or dimorphic condition, in the species of *Primula*, and on their remarkable sexual relations' was read to the Linnean Society on 21 November 1861 and published in the *Journal of the Proceedings of the Linnean Society (Botany)* 6 (1862): 77–96. See also *Collected papers* 2: 45–63. The two figures included in the paper are unattributed. They were subsequently included in CD's fuller account of dimorphism, *Forms of flowers*.

[5] Frances Harriet Hooker had not been well since the death in May 1861 of her father, John Stevens Henslow.

[6] The reference is to a controversy in the pages of the *Athenæum* concerning a hostile review of *Results of a scientific mission to India and High Asia, undertaken between the years 1854 and 1858, by order of the court of directors of the Honourable East India Company*, by the Schlagintweit brothers, Hermann Rudolph Alfred, Adolph, and Robert. The reviewer was highly critical of the calibre of the scientific results of the Schlagintweits, whose appointment he said was 'one of the most gigantic jobs that ever disgraced the annals of science', and he accused the Royal Society of London of complicity in agreeing to the appointment of the Schlagintweits over more qualified English officers in India (*Athenæum*, 17 August 1861, p. 215). Responding to the criticism of their part in the affair, Roderick Impey Murchison and Edward Sabine of the Royal Society sent letters to the *Athenæum* that were printed in the 7 September 1861 issue (p. 320). Sabine stated that the Royal Society had not approved the appointment of the Schlagintweits but only formed a sub-committee (consisting of himself, Hooker, and CD) to offer suggestions concerning the proposed mission. Hooker gave his own account of the proceedings, defending the action of the Royal Society committee, in a letter printed in the 21 September 1861 issue, p. 374. Before the brothers had left on their mission in 1854, CD had asked them to make some inquiries for him in India (see *Correspondence* vol. 6, letter to J. D. Hooker, 22 August [1857], and letter from Robert Schlagintweit, 25 September 1857).

[7] On Hooker's recommendation, the Darwins were rubbing cod-liver oil into Henrietta Emma Darwin's skin in the hope that it would accelerate her convalescence from typhus fever. See letter to J. D. Hooker, 8 [February 1861].

To John Murray 24 September [1861]

Down. | Bromley. | Kent. S.E.
Sept. 24th

My dear Sir

I am very much obliged for your note & very liberal offer.—[1] I have had some qualms & fears. All that I can feel sure of is that the M.S. contains many new & curious facts & I am sure the Essay would have interested me, & will interest those who feel lively interest in the wonders of nature: but how far the Public will care for such minute details, I cannot at all tell. It is a bold experiment, & at worst, cannot entail much loss: as a certain amount of sale will, I think, be pretty certain.—[2] A large sale is out of the question. As far as I can judge generally the points which interest me, I find interest others; but I make the experiment with fear & trembling,—not for my own sake, but for yours.—

I do not see how I can improve enclosed title: if you alter it, *please let me see alteration:*—[3]

When the drawings are made on wood, I will communicate with you about their cutting, as I shall be then ready to print.—

With hearty thanks | Yours very sincerely | Ch. Darwin

I think this little volume will do good to the Origin, as it will show that I have worked hard at details, & it will, *perhaps*, serve [to] illustrate how natural History may be worked under the belief of the modification of Species.—

P.S. | Do you think of little Book with Cloth Back or Pamphlet in paper? I ask because if former, sh^{d}. you like an Orchis in Gold; for in that case I could get Mr Sowerby to draw one—for the woodcut, which I shall use will hardly do—[4] Please let me have one line, if you wish for an ornament.—

[Enclosure]

On the various contrivances by which British Orchids are fertilised by Insects; & on the good effects of Intercrossing. by Charles Darwin M.A. FR.S. &c.[5]

Endorsement: '*1861—Sep 24*'
John Murray (publishers)

[1] See letter from John Murray, 23 September 1861.

[2] Half of the copies printed (under 2000) were sold by August 1862, but thereafter the edition sold slowly (Freeman 1977, p. 112). See also *Correspondence* vol. 10, letter to John Murray, 24 August [1862].

[3] See the enclosure and n. 4, below.

[4] *Orchids* was published in hardback and bound in plum-coloured cloth with a gilt orchid on the cover (Freeman 1977, pp. 112–13).

[5] The title was altered slightly: instead of 'British Orchids', it reads 'British and Foreign Orchids'.

From H. W. Bates [before 25 September 1861][1]

[Reports on some of the results of his study of various Amazonian insects.][2]

Incomplete
DAR 160.3: 63 (fragile)

[1] Dated by the relationship to the following letter.

[2] The letter is so badly damaged that only occasional words or parts of sentences are legible.

To H. W. Bates 25 September [1861]

Down. | Bromley. | Kent. S.E.
Sept. 25th

My dear Sir

Your letter, like every one that I have received from you, has been a mine of wealth & has interested me greatly.[1] But first for the most important point, viz your Book of Travels, & I hearily rejoice that you intend publishing.[2] I sh^{d}. think that you could not have a more respectable or pleasanter publisher than Mr Van Voorst.—[3]

At the same time, I apprehend, there can be no doubt that Murray has **much** greater power of getting large distribution. Murray has the character of being a very liberal paymaster & I am sure I have found him most liberal & pleasant to deal with. Of course every publisher will be cautious with a new author. If your Book should turn out popular (& there is some little mere luck in this) it would be a thousand pities that it sh^d. not be in hands that could press its sale. I cannot remember that Mr V. Voorst has ever had very large sale for any of his Books.— On other hand Mr V. V. would better appreciate your scientific character than Mr Murray.—[4] This is all the advice I can give; except that I rejoice I went to Murray with the Origin.—[5]

By the way here is case in point! Colburn did not value my Journal of Researches & would never, I am sure, have published a second Edit. I took it from Colburn & sold it to Murray & it has long & *great* sale up to present day.—[6] — If you sh^d. decide on Mr Murray, & if you would so like I shall be most happy to write to M^r. Murray & can most truly tell him how much I appreciate the force of intellect & knowledge & style of your letters to me.—[7] What terms he would offer you I cannot conjecture: he would, no doubt, wish to see some M.S.— As an old hackneyed author let me give you a bit of advice, viz to strike out every word, which is not quite necessary to connect subjects & which would not interest a stranger. I constantly asked myself, would a stranger care for this? & struck out or left in accordingly.— I think too much pains cannot be taken in making style transparently clear & throwing eloquence to the dogs. I hope that you will not think these few words impertinent.— (I would sell only 1st. Edition.)—[8]

Now for a few words on Science.— Many thanks for facts on Neuters. You cannot tell how I rejoice that you do not think what I have said on subject absurd.—[9] Only 2 persons have ever noticed it to me!— Viz the bitter sneers of Owen in Eding. Review, & my good friend & supporter Sir C. Lyell, who could only screw up courage to say "well you have manfully faced the difficulty".—[10]

What a wonderful case of Volucella, of which I had never heard; I had no idea such a case occurred in nature: I must get & see specimens in Brit. Museum.—[11] I hope & suppose you will give a good deal of Nat. History in your Travels; everyone cares about Ants— more notice has been taken of Slave Ants in the Origin that of any other passage.—

I fully expect to delight in your Travels. Keep to simple style as in your excellent letters,—but I beg pardon I am again advising.

What a capital paper yours will be on mimetic resemblances.—[12] You will make quite a new subject of it.— I had thought of such cases as a difficulty. & once when corresponding with D^r. Collingwood, I thought of your explanation; but I drove it from my mind, for I felt that I had not knowledge to judge one way or the other. D^r. C., I think, states that the mimetic forms inhabit same country; but I did not know whether to believe him.—[13] What wonderful cases yours seem to be.— Could you not give a few woodcuts in your Travels to illustrate this?— I am tired with hard's day's work—so no more, except to give my sincere thanks & hearty wishes for the success of your Travels.—

My dear Sir | Yours sincerely | C. Darwin

Endorsement: '1861'
Cleveland Health Sciences Library (Robert M. Stecher collection)

1 See preceding letter.
2 CD had encouraged Bates to write an account of his eleven-year-long travels in the Amazon basin in South America (see letter to H. W. Bates, 4 April [1861]).
3 John Van Voorst, the owner of Van Voorst publishing house in Paternoster Row, London, was a major publisher of works on natural history.
4 Bates's work, *The naturalist on the river Amazons*, was published by John Murray in 1863.
5 Charles Lyell had served as an intermediary in CD's negotiations with John Murray over the publication of *Origin*. See *Correspondence* vol. 7, especially the letters to Charles Lyell, 28 March [1859] and 30 March [1859], and to John Murray, 31 March [1859].
6 The publishing firm of Henry Colburn, of Great Marlborough Street, London, had published CD's account of the *Beagle* voyage (*Journal and remarks*) in 1839 as volume 3 of Robert FitzRoy's *Narrative of the surveying voyages of His Majesty's Ships Adventure and Beagle*. Later in the same year, Colburn reissued the work separately under a new title, *Journal of researches* (see *Correspondence* vol. 2). For CD's displeasure with Colburn over the handling of the publishing of *Journal of researches*, see *Correspondence* vol. 3, letters to John Murray, 17 March [1845] and 12 April [1845]. Murray had issued a new printing of the work, the tenth thousand, in 1860 (see Freeman 1977, pp. 31–40).
7 See *Correspondence* vol. 10, letter to John Murray, 28 January [1862].
8 Bates was unaware of the practice of selling the publication rights for a specific edition (see letter from H. W. Bates, 30 September 1861).
9 The reference is to the discussion in *Origin* of how natural selection could account for the existence of neuter insects, a problem that CD admitted had at first seemed 'insuperable, and actually fatal to my whole theory' (*Origin*, p. 236). Using the sterile worker ants as his model, CD suggested that even though they were unable to propagate their own kind, 'selection may be applied to the family, as well as to the individual, and may thus gain the desired end.' (*ibid.*, p. 237).
10 The reference is to Richard Owen's review of *Origin* ([R. Owen] 1860b, pp. 524–6). For a reference by Charles Lyell to the discussion in *Origin* of 'working ants', see *Correspondence* vol. 7, letter from Charles Lyell, 3 October 1859.
11 The information on *Volucella*, commonly known as the hover-fly, was included in the letter from H. W. Bates, [before 25 September 1861]. See also letter from H. W. Bates, 30 September 1861. In November 1861, CD visited the British Museum and examined the exhibition of South American insects collected by Bates (see letter to J. D. Hooker, 25 November [1861]).
12 A paper by Bates on mimetic butterflies was read before the Linnean Society of London on 21 November 1861 (Bates 1861b).
13 The reference may be to Cuthbert Collingwood's paper on what he called 'homomorphism' or 'representative form' (Collingwood 1860a). Collingwood had sent CD a copy of the work, which is in the Darwin Pamphlet Collection–CUL. See letter to Cuthbert Collingwood, 14 March [1861]. CD may also have read another of Collingwood's papers on this topic (Collingwood 1860c).

From J. O. Westwood 26 September 1861

Oxford
26 September | 1861

My dear Mr Darwin

I have succeeded in finding the reference to Morren's paper[1] which appeared in the Horticulteur Belge about 1835 & of which an abstract was given in the Trans.

Ent. Soc. 1. Proc. p xliv & xlix[2] It is entitled "On the Agency of insects in causing Sterility in flowers by the removal of the masculine organs observed among the Asclepiadeæ." The insects observed were our common white butterflies the tarsi of which were loaded with the pollen masses of the plants in question—not Orchideæ.[3]

If you could conveniently let my little box be left either at Van Voorst's Paternoster Row—or the British Museum I should get it in time[4]

Yours very truly | J. O. Westwood

DAR 181.1

CD ANNOTATIONS

2.1 If . . . time 2.2] *crossed ink*

Top of first page: 'Dichogamy'[5] *brown crayon*

Verso of last page: 'Mr Westwood sent me 2 large Hymenopt. insects *from N. America [*interl*] with numerous pollinia on their feet; & Mr Gosse told me he had seen probosces of Sphingidæ with them in N. America.'[6] *ink*

[1] Westwood had first mentioned a paper by Charles François Antoine Morren on orchids and insects in a letter (now missing) of 1860 (see *Correspondence* vol. 8, letter to J. O. Westwood, 9 July [1860]). CD had asked Westwood to give him a precise reference to the paper (see letters to J. O. Westwood, 15 August [1861] and 4 September [1861]). CD also asked for this information in the first of two letters to the *Gardeners' Chronicle*, [before 14 September 1861].

[2] See Morren 1836.

[3] Westwood published this information in the *Gardeners' Chronicle and Agricultural Gazette*, 5 October 1861, p. 889. CD did not refer to this paper of Morren's in *Orchids*, but he did cite another of Morren's papers (Morren 1839). See *Orchids*, p. 270 n.

[4] CD had asked Westwood to send him any specimens of moths that had orchid pollen-masses attached to them (letter to J. O. Westwood, 15 August [1861]). Westwood had sent CD two insects, but CD discovered that the pollen-masses did not actually belong to orchids (see letter to J. O. Westwood, 4 September [1861]). Westwood refers to the publishing house of John Van Voorst.

[5] Dichogamy is the condition in which the stamens and pistils of a hermaphrodite plant mature at different times, thereby preventing self-fertilisation. Most of CD's notes on dichogamy are in DAR 49, 76, and 205.8.

[6] Philip Henry Gosse spent twelve years in North America. His work on the entomology of Newfoundland remained unpublished, but his observations on the natural history of Canada were published in Gosse 1840 (*DNB*).

From J. D. Hooker [28 September 1861][1]

Royal Gardens Kew

Australian Plants naturalized on the Nilghieries[2] & turning out the native trees

x Eucalyptus amygdalinus, Lab.
x —— pulverulentus Sims
—— sp?
Acacia pulchella,
x —— dealbata Link

x —— melanoxilon (the common firewood of hillstation)[3]
—— longifolia Willd.
—— marginata Br.
x —— verticillata Willd. var ruscifolia (makes hedges commonly)
x —— stricta, Willd.
—— decipiens, Br.
Albizzia lophantha

x Those marked thus seed most abundantly & are propagated thus— the others may or may not do so— it is *believed* that all do— more enquiries to be made
J. Hooker

DAR 205.4 (Letters)

CD ANNOTATIONS
1.13 'Albizzia lophantha'] 'They are spreading over miles of Hills' *added ink*
Margin: '20'[4] *blue crayon, circled blue crayon*
Bottom of page: 'Sept 28—1861' *ink*

1 The date is provided by CD's annotation.
2 The Nilgiri Hills are located in south-eastern India, due east of Calcutta. Hooker was particularly familiar with the botany of India, having spent two years there, 1848–50.
3 Probably a reference to the British hill-station and sanatorium at Darjeeling, which Hooker visited (see *Correspondence* vol. 4, letter from J. D. Hooker, 3 February 1849).
4 The annotation refers to CD's portfolio of notes on the geographical distribution of plants.

To J. D. Hooker 28 September [1861]

Down
Sept. 28th

My dear Hooker

What a good soul you are not to sneer at me, but to pat me on the back. I have greatest doubt whether I am not going to do in publishing my paper, a most ridiculous thing. It would annoy me much, but only for Murray's sake, if the publication were a dead failure.[1] I hate to hear you so falsely abuse yourself & say that you are only fit for descriptive work.[2] For years & years you have been my Public & my judge, & I care more for your opinion in Nat. History, than for all the rest of the world. So do not vilify yourself;—though by Jove you cannot alter my opinion of you.—

Many thanks for list of Australian trees;[3] I am so glad that you are stirring up people in this *most important* subject. It makes the case wonderfully odd of the S. species not having naturally spread northward.—[4]

It is, I believe, true that Glen Roy shelves (I remember your Indian Letters) were formed by Glacial lakes.[5] I persuaded Mr Jamieson an excellent observer to go & observe them; & this is his result.—[6] There are some great difficulties to be explained; but I presume this will ultimately be proved the truth. Speaking of the

Australian trees reminds me that Mr Bates (who wrote so well against spreading of Lepidoptera during Glacial period)[7] is going to publish his Travels with Natural Hist. notes. I heard from him a few days ago, & was so much pleased to find that he who has specially attended to the castes of neuter ants, is quite satisfied with my discussion & explanation of this most difficult case.—[8] Owen sneered bitterly at what I wrote on this subject.—[9]

I am going to beg for help, & I will explain why I want it.—

You offer Cypripedium; I sh^{d}. be very glad of a specimen; & of any good sized Vandeæ, or indeed any Orchid; for this reason; I never thought of publishing separately & therefore did not keep specimens in spirits, & *now* I sh^{d}. be very glad of a few woodcuts to illustrate my few remarks on Exotic orchids.— If you can send me any, send them by Post in tin-cannister on middle of day of Saturday Oct 5th; for Sowerby will be here.—[10]

Secondly: have you any *white* & yellow vars. of Verbascum, which you could give me or propagate for me or *lend* me for a year. I have resolved to try Gärtners wonderfull & repeated statement, that pollen of white & yellow vars, whether used on the varieties or on *distinct* species has different potency.[11] I do not think any experiment can be more important on Origin of species; for if he is correct, we certainly have what Huxley calls new physiological species arising.[12] I sh^{d} require several species of Verbascum besides the white & yellow varieties of the same species.— It will be tiresome work, but if I can anyhow get the plants it shall be tried.

Thirdly, can you give me seeds of any Rubiaceæ of the suborder Cinchoneæ, as Spermacoce, Diodia, Mitchella, Oldenlandia.— Asa Gray says they present two forms like Primula.—[13] I am sure that this subject is well worth working out. I have just almost proved a very curious case in Linum grandiflorum which presents 2 forms A & B.— Pollen of (A) is perfectly fertile on stigma of A.—

But pollen of (B) is absolutely barren on its own stigma; you might as well put, so much flower on it; it astounded me to see the stigma of (B) purple with its own pollen; & then put a few grains of similarly looking pollen of (A) on them; & the germens immediately & always swelled; those not thus treated *never* swelling.[14]

Thirdly; can you give me any *very* hairy Saxifraga.—(for their function)

I send a resumè of my requests to save you trouble.— Nor would I ask for so much aid, if I did not think all these points well worth trying to investigate—

My dear old friend, a letter from you always does me a world of good.— And the Lord have mercy on me what I return I make—

Yours affect^y. | C. Darwin

Endorsement: '/61'
DAR 115.2: 114

[1] The reference is to CD's intention to publish the results of his study of orchid pollination as a book rather than an article. See letter to J. D. Hooker, 24 September [1861]; see also letter to John Murray, 21 September [1861], and letter from John Murray, 23 September 1861.

[2] Hooker and George Bentham were preparing the first part of the first volume of *Genera plantarum*, which was intended to be a systematic compilation of all known plant genera (Bentham and Hooker 1862–83). CD had been encouraging Hooker to write a general work on philosophical botany (see *Correspondence* vol. 8, especially letter to J. D. Hooker, 8 February [1860]).

[3] See preceding letter.

[4] CD referred to this difficulty in *Origin* in the chapter on geographical distribution of plants and animals during the glacial period (*Origin*, pp. 379–81).

[5] CD refers to the letters Hooker sent him during his expedition to the Himalayas, 1848–50. In a letter of 1849, Hooker described the geology of Sikkim, particularly noting the phenomenon of terraced hillsides, which he explained as resulting from former glacier lakes (see *Correspondence* vol. 4, letter from J. D. Hooker, 3 February 1849).

[6] See letters from T. F. Jamieson, 13 June 1861 and 3 September 1861. See also letter to Robert Chambers, 30 April [1861]. Thomas Francis Jamieson published the results of his examination of the geology of Glen Roy in 1863 (Jamieson 1863). See also Appendix IX.

[7] In *Origin*, pp. 365–79, CD put forward the view that a refrigeration of equatorial regions during the glacial period had allowed migration of temperate species from one hemisphere to the other, with tropical species suffering a certain amount of extinction. Bates disputed this view in Bates 1861, pp. 352–3: having found that among the butterflies of the Amazon region there were many endemic species with restricted ranges, he concluded that no significant degree of extinction could have occurred. See also letters from H. W. Bates, 18 March 1861 and 28 March 1861.

[8] Letter from H. W. Bates, [before 25 September 1861]. This letter is damaged and largely illegible, but see the letter to H. W. Bates, 25 September [1861]. Bates had studied Brazilian ants; some of his observations were published in Bates and Smith 1855.

[9] [R. Owen] 1860b, pp. 524–6. For CD's response to Richard Owen's 'malignant' review of *Origin*, see *Correspondence* vol. 8, letters to T. H. Huxley, 9 April [1860], to Charles Lyell, 10 April [1860], and to Asa Gray, 25 April [1860].

[10] According to Emma Darwin's diary, George Brettingham Sowerby Jr arrived at Down on 7 October 1861. An entry in CD's Account book (Down House MS) records a payment to Sowerby for ten days of work in preparing the woodcuts for *Orchids*.

[11] Many of Karl Friedrich von Gärtner's hybridisation experiments on *Verbascum*, the results of which are tabulated in Gärtner 1849, pp. 724–8, consisted of crossing differently coloured species with each other or crossing white- and yellow-flowered varieties of both *V. lychnitis* and *V. blattaria* with other *Verbascum* species. Gärtner also referred to his results in *ibid.*, pp. 92 and 180–1, though his use of the terms species and varieties is confusing, a point noted by CD (see *Marginalia*, p. 260). Gärtner had also mentioned his *Verbascum* experiments in Gärtner 1844, pp. 137–8. There are annotated copies of both volumes in the Darwin Library–CUL. CD discussed Gärtner's work on this aspect of *Verbascum* hybridisation in *Variation* 2: 105–6, but made no mention of his own experiments on the genus. Instead, he cited experiments conducted by John Scott to show that 'in the genus Verbascum, the similarly and dissimilarly-coloured varieties of the same species behave, when crossed, like closely allied but distinct species' (*Variation* 2: 107).

12 In his review of *Origin*, Huxley distinguished between what he termed 'morphological' species, 'distinctly definable from all others, by certain constant ... morphological peculiarities' ([T. H. Huxley] 1860, p. 543), and 'physiological' species. He stated that the best way to distinguish between two true physiological species and two varieties was to attempt to hybridise the two: true species would either be infertile *inter se* or produce infertile offspring, whereas two varieties would give rise to fertile progeny (*ibid.*, pp. 552–3). He went on to state that 'as the evidence now stands, it is not absolutely proven that a group of animals, having all the characters exhibited by species in nature, has ever been originated by selection, whether artificial or natural.' (*ibid.*, p. 567).

13 See letter from Asa Gray, [27 and 29 August] and 2 September [1861].

14 CD gave an account of the experiments he carried out in 1861 on *Linum grandiflorum* in a paper on dimorphism in *Primula* read to the Linnean Society of London on 21 November 1861 (see *Collected papers* 2: 62–3) and in a paper he eventually published on dimorphism in species of *Linum* ('On the existence of two forms, and on their reciprocal sexual relation, in several species of the genus *Linum*', *Journal of the Proceedings of the Linnean Society (Botany)* 7 (1864): 69–83; *Collected papers* 2: 93–105). See also *Forms of flowers*.

From H. W. Bates 30 September 1861

King St Leicester
30 Sept. 1861

My Dear Sir

Before you go to the B.M. to inspect the Volucellæ flies I wish to give you more exact information about them.[1] The species which resembles Humble bees is the Volucella bombylans (authors) and the constant variety is V. plumata (Meigen) (= V. mystacea Fabr.)— One of these resembles Bombus lapidarius, the other B. terrestris.— These are the only two varieties found in this neighbourhood and until recently the only two described. They were always considered as distinct species until Lepelletier de St Fargeau observed them to copulate,—[2] they are not the sexes, however;—as there are males & females in each form. The female lays its eggs & the larvæ are bred on the nests of Humble bees.—

It has been lately shown that these two forms are united by a series of varieties or connecting links to the number of five; one of which V. hæmorrhoidalis (Zetterstedt) has been described as a distinct species. They are found on the continent but I believe they are not found in all localities together with the two chief forms. I have never observed them.

Another species of Volucella, V. inanis resembles a wasp; its ova & larvæ are found in wasps' nests.—

A third common species is V. pellucens this in its dress is entirely unlike either of the other two species. I cannot find any account of its larva. It would be interesting to know whether it is parasitic also. The perfect insect has not the slightest resemblance to a Hymenopterous insect, but is of the same general appearance as the bulk of the Volucellæ species of which I took many in S. America.

It would appear then that V. pellucens retains the facies of its genus, but that the other two have been widely diverted in this respect from their normal type, in strict connection with their habit of depositing their ova in the nests of wasps and Bombi. It is not necessary to explain to you that I mean—the adult insects being exposed to the attacks of the Bombi near their nests,—their resemblance is explicable on the theory of Nat. Selection.

Precisely the same phenomenon occurs in genera of butterflies in S. America where some species wear the dress of species belonging to widely different families. But here the object of the analogical resemblance is different. Here it is because the one set of species is exposed in the adult state to a persecution by insect enemies from which the other set is free.[3]

This entire class of facts is extremely numerous in insects. But they are very complex. It requires the most subtle & flexible nature of mind to understand & explain them. But whilst they are unexplained on the rational theory, they remain one of the strongest arguments to the believers in the miraculous in creation.

Dr Collingwood's paper "on recurrent animal forms" does not touch this class of facts, either in description or explanation.[4] All that he can say is that similar habits &c will produce analogical resemblance between species belonging to widely different families. But ours are not general resemblances they are close specific resemblances as far as external appearance is concerned. But, as I have shown in the paper I mentioned to you,—the causes which produce close specific resemblances do not & cannot operate except in those forms which are already made generally similar by the operation of other causes.[5]

I am extremely obliged to you for your advice regarding my proposed book. It will be *very* useful to me. It was quite new to me that I could agree with a publisher to sell him only the 1st edition. I shall find it very difficult, I am afraid, to write in a simple style. I have begun to try to write a chapter or two—but I find it very perplexing, in the mass of details contained in my journals, to select what would be likely to be most interesting to the public. It will never do, I think to narrate even the chief events consecutively of 11 years travel & residence. I went backwards & forwards. In your voyage there was something similar;[6] & I shall try to work mine into a compact whole but can never imitate the wonderful art you have shown in doing so. When I have finished a chapter or two I will apply to Murray & then shall feel much obliged to you if you would write him a word or two by way of introduction.[7]

Yours sincerely | H W Bates

DAR 205.10 (Letters)

CD ANNOTATIONS
9.1 I am ... introduction. 9.11] *crossed blue crayon*
Top of first page: 'All about Volucella' *pencil*

[1] CD had expressed a desire to see the exhibition at the British Museum of Amazonian insects collected by Bates in South America (see letter to H. W. Bates, 25 September [1861]). Bates had written to CD about the fly *Volucella* in the letter from H. W. Bates, [before 25 September 1861].

[2] Amédée Louis Michel Lepeletier de Saint-Fargeau was an eminent French entomologist. His observation of *Volucella* was cited in Macquart 1834–5, p. 479 (see letter from H. W. Bates, [1 December] 1861).

[3] Bates refers to mimetic butterflies. See Bates 1861a and 1861b. See also letter to H. W. Bates, 25 September [1861].

[4] Collingwood 1860c. See also letter to H. W. Bates, 25 September [1861].

[5] Bates 1861b (see letter to H. W. Bates, 25 September [1861] and n. 12).

[6] In *Journal of researches*, CD departed from strict chronological order by conflating natural history observations that were made over a period of years (see *Journal of researches* 2d ed., p. 40).

[7] See *Correspondence* vol. 10, letter to John Murray, 28 January [1862].

From Charles Lyell 30 September 1861

London
30 Sept. | 1861

My dear Darwin

I am afraid you will not have by you a spare copy of your welsh glacier paper alluded to in Jamieson's last—vol. 33 Edinb. Phil. Journ. 1842[1] if you have I shall be very glad of it— if not I will try & buy but can at any rate borrow it.

I gather from Trimmer[2] & E: Forbes[3] 1st. that the marine drift of Moel Tryfane with its 12 recent shells is not Crag, not even Norwich crag, but more modern & of the glacial period— 2dly. that the rocks on which this elevated post pliocene, marine drift rests are polished & were glaciated before the marine shells strata were superimposed— whether this glaciation was by land-ice or icebergs &c is a question I cannot answer—

3dly that the submergence implied by the shells which occur at a height above 1300 ft. is according to Ramsay above 2000 ft. because the same continuous, stratified drift reaches to more than 2000 ft.[4]

Now I begin to have little doubt that the Glen Roy roads are later than the great submergence, if Scotland was ever submerged like Wales as very likely it was.

But how to make these three periods, 1st. Continental ice or Greenlandic period, 2dly chief submergence, 3d. Glen Roy or glacier-lake period coincide with the period of Elephas antiquus Rhinoceros leptorhinus & hippopotamus major as contrasted with the (later?) period of the mammoth, tichorhinus (or wooly) rhinoceros rein-deer & musk buffalo?[5]

I now believe that at Bedford both these groups of mammalia were after the marine boulder-clay of that middle-county district.[6]

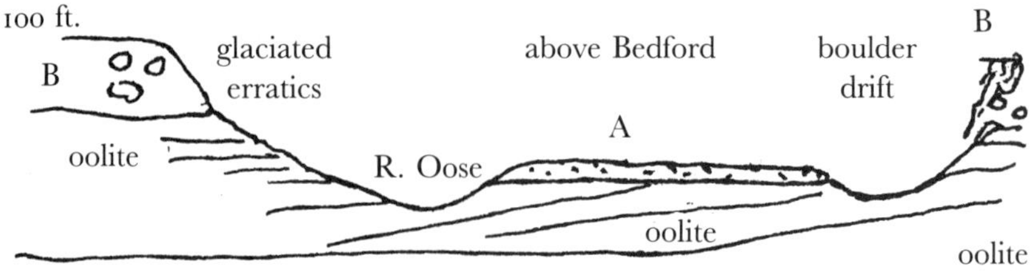

A is gravel containing freshw^r. shells & both sets of quadrupeds in different parts of the valley, probably not contemporaneous, but both posterior to B the northern drift & containing the wreck of said drift which was therefore anterior to the valley B. B. whereas the gravel A was posterior to excavation of valley.

At the bottom of A are human flint implements of the Amiens type[7] used by men who lived when the valley was not so deep by 30 feet as now where the Oose flows.

These men lived, perhaps with Hippopot. major—after the elevation of the sea-bottom of the boulder clay of Bedford.— Was this last coeval with the marine drift of Wales? If so what were the relations in time between your last Lanberris glaciers & the excavat^n. of the valley at Bedford & the mammoth of that ilk.[8]

Falconer & Prestwich place El. antiquus as well as mammoth of the caves of S. Wales as post glacial.[9]

I hope you will not merely wish Jamieson & me joy of having to solve these problems but give us an hypothesis at least about some of them.[10]

You may use the sea for your fringes above the pass of Mackul now that we give you a marine submergence *before* the shelves.

To save my copying these queries for Mr Jamieson be so kind as to return this letter that I may forward it to Ellon—

In the Norfolk cliffs Elephas antiquus was anterior to the erratics & glacial till—so it was near Zurich in Switzerland anterior to the largest Alpine erratics. But it seems to me to have been at Bedford a post-glacial beast.

Some of our Scotch peat bogs are surrounded by till & yet they & subjacent marl contain like the Danish peat bogs no extinct mammalia— Why is this.— The Danish peat lies also in depressions in drift & unstratified moraine-like boulder clay—

Were these glacial formations so modern that the fauna which immediately succeeded, a rich one, contained no extinct mammalia?

believe me | my dear Darwin | ever sincerely yours | Cha Lyell

Edinburgh University Library (Lyell 1: 2813–16)

1 'Notes on the effects produced by the ancient glaciers of Caernarvonshire, and on the boulders transported by floating ice', *Edinburgh New Philosophical Journal* 33 (1842): 352–63. The paper originally appeared in the *London, Edinburgh, and Dublin Philosophical Magazine and Journal of Science* 21 (1842): 180–8 (*Collected papers* 1: 163–71). Thomas Francis Jamieson referred to the paper in his letter to Lyell, 27 September 1861 (see Appendix IX).

2 Trimmer 1831. In his paper on glaciation in North Wales, CD stated that Joshua Trimmer 'was one of the earliest observers of the scores and other marks on the rocks of North Wales.' (see *Collected papers* 1: 170 n. 5). Moel Tryfan is a hill in North Wales that is 1192 feet above sea level.

3 In Forbes 1846, Edward Forbes discussed the distribution of fossil marine shells and some extant British forms collected by Trimmer on Moel Tryfan. The confused mixture of species from different sea-depths suggested to him that some disturbing force had been in operation (Forbes 1846, p. 384). Lyell cited Trimmer and Forbes on this point in C. Lyell 1863, p. 267.

4 Ramsay 1852, p. 372.

[5] Lyell discussed these points in chapters 13 and 14 of C. Lyell 1863, both of which discussed the 'Chronological relations of the glacial period and the earliest signs of man's appearance in Europe'. All of the species mentioned are post-Pliocene forms whose bones had been discovered in association with human artefacts (see *ibid*., pp. 156–9, 283). The topics Lyell mentions relate to his attempt, in writing up the results of his study of the antiquity of man, to synthesise the current knowledge of geological phenomena in the periods immediately antecedent and posterior to the first appearance of human remains.

[6] Lyell had visited the Biddenham gravel-pit near Bedford in April 1861, where remains of *Elephas antiquus*, *E. primigenius*, *Rhinoceros tichorhinus*, and *Hippopotamus major* were found in deposits that also contained rough flint implements used by ancient peoples (see K. M. Lyell ed. 1881, 2: 344 and C. Lyell 1863, pp. 163–6); these deposits extended back into the post-Pliocene period (see C. Lyell 1863, p. 207).

[7] Large numbers of ancient flint implements had been discovered in the gravel-pits near Amiens in France in the early 1850s. These, taken in conjunction with those previously uncovered near Abbeville in France and the findings of similar flint implements in association with the bones of extinct mammals in caves near Torquay, Devonshire, had convinced many naturalists that the antiquity of humankind was greater than had previously been thought (see C. Lyell 1863, pp. 94–105).

[8] CD described the clear signs of glaciation in the vale of Llanberis, North Wales, in his paper on the ancient glaciers of Caernarvonshire (see n. 1, above; see also *Collected papers* 1: 169–70).

[9] Falconer 1860. Both Hugh Falconer and Joseph Prestwich were experts in identifying the remains of extinct mammals in ossiferous caves.

[10] See letter to Charles Lyell, 22 September [1861].

To Charles Lyell 1 October [1861]

Down. | Bromley. | Kent. S.E.
Oct 1st

My dear Lyell

Thank you for the most interesting correspondence.[1] What a wonderful case that of Bedford. I thought the problem sufficiently perplexing before, but now it beats anything I ever heard of. Far from being able to give any hypothesis for any part, I cannot get the facts into my mind.— What a capital observer & reasoner Mr Jamieson is.[2] The only way that I can reconcile my memory of Lochaber with the state of the Welch valleys, is by *imagining* a great barrier, formed by a terminal moraine, at the mouth of the Spean, which the river had to cut slowly through, as it drained the lowest Lake, after Glacial period.— This would, I can suppose, account for the sloping terraces along the Spean. I further presume that sharp transverse moraines would not be formed *under* the waters of the Lake, where the glacier came out of L. Treig and abutted against opposite side of valley.— A nice mess I made of Glen Roy![3] I have no spare copy of my Welch paper;[4] it would do you no good to lend it, I suppose.— I thought that there must have been floating ice on Moel Tryfane.— I think it cannot be disputed that the last event in N. Wales was Land Glaciers.— I could not decide where action of Land Glaciers ceased & marine glacial action commenced at the mouths of the valleys.—[5]

What a wonderful case the Bedford case.—[6] Does not the N. American view of warmer or more equable period *after* great Glacial period become much more probable in Europe?—[7]

But I am very poorly today & very stupid & hate everybody & everything. One lives only to make blunders.— I am going to write a little Book for Murray on orchids[8] & today I hate them worse than everything so farewell & in a sweet frame of mind, I am | Ever yours | C. Darwin

Endorsement: '1861'
American Philosophical Society (266)

[1] Letter from Charles Lyell, 30 September 1861, and, presumably, some correspondence from Thomas Francis Jamieson (see n. 2, below).

[2] Jamieson and Lyell had been corresponding about the glacial phenomena of Scotland, with Lyell forwarding Jamieson's letters on to CD. A major point of the correspondence was Jamieson's argument that the so-called parallel roads of Glen Roy had been formed by glacial lakes rather than by arms of the sea, as CD supposed. See letters to Charles Lyell, [15 September 1861] and 22 September [1861], letter from Charles Lyell, 30 September [1861], and Appendix IX.

[3] 'Observations on the parallel roads of Glen Roy, and of other parts of Lochaber in Scotland, with an attempt to prove that they are of marine origin' *Philosophical Transactions of the Royal Society of London*, pt 1 (1839): 39–81; *Collected papers* 1: 89–137. See also Rudwick 1974.

[4] See letter from Charles Lyell, 30 September [1861] and n. 1.

[5] See letter from Charles Lyell, 30 September [1861]. CD believed that the erratic boulders near the summit of Moel Tryfan in North Wales had been transported on floating ice at a time when Moel Tryfan was submerged (see *Collected papers* 1: 167–9). In his letter to Lyell of 27 September 1861, Jamieson pointed out that CD's explanation of the physical features of Caernarvonshire was inconsistent with his explanation of the formation of the parallel roads of Glen Roy (see Appendix IX). For Jamieson's views on the sequence of glaciation, marine submergence, and re-elevation of land in Scotland, see Jamieson 1860a and 1862.

[6] See letter from Charles Lyell, 30 September 1861 and n. 6.

[7] James Dwight Dana had proposed a second, post-glacial warm period in North America on the basis of certain fossil remains uncovered in post-Pliocene deposits. Assuming a migration of northern plant species both before and after the worldwide cold period, Asa Gray had adopted this view to help account for the general similarity of European and American floras (A. Gray 1859). For the discussion of this point in CD's correspondence, see *Correspondence* vol. 7, letters to J. D. Hooker, 11 May [1859], and to Asa Gray, 11 November [1859] and 24 December [1859].

[8] John Murray published *Orchids* in May 1862.

To A. G. More 1 October 1861

Down, Bromley, Kent. S.E.
Oct. 1, 1861.

My dear Sir

I am sure that your kindness to me has been very great. I am sincerely sorry to hear that you still suffer in your health. I fear to trust in any way to two flowers alone. The capsules were certainly small, but then they were near the summit of the spike. I ought to have asked to have had the capsules gathered separately and put in paper; for they all opened on the journey and shed their seeds. The few seeds remaining in the two capsules had a larger proportion of bad seeds than those that were shed; but here again I dare not trust, for this may be due to the bad seeds not being in any case so readily shed.[1]

With my respected thanks[2] for all your kindness, I remain | My dear Sir, | Yours sincerely | C. Darwin

P.S. I mean to print a very little book on the fertilisation of Orchids by insects; and I will do myself the pleasure of sending you a copy whenever it is printed.[3]

Copy
DAR 146

[1] CD had asked for More's assistance in his study of the orchid *Epipactis palustris* (see letters to A. G. More, 4 June 1861, 17 June 1861, and 23 September 1861). Believing that the unusual hinged structure of the labellum in this species served to guide visiting insects to brush against the rostellum, thereby promoting cross-pollination, CD asked More to remove part of the labellum from several flowers to see whether these would still set seed that was fertile. For CD's description of More's results, see *Orchids*, p. 101.

[2] The copyist seems to have made an error in transcribing CD's handwriting.

[3] More's name appears on CD's presentation list for copies of *Orchids* (DAR 210.18). See also *Correspondence* vol. 10, Appendix III.

To H. W. Bates 2 October 1861

Down,
Oct. 2, 1861

My dear Sir

I write one line to thank you for your curious additional information on the Volucella; It will probably be years before I use the facts, but I shall be very curious to see the specimens and have so curious a case in my mind.[1]

I am very glad to hear that you have begun your Book;[2] it will no doubt be a difficult Book to write and will require much rewriting; but I fully believe that you will succeed and make an excellent Book— I am heartily glad that you have undertaken the work. At any time I will write to Mr. Murray.

My dear Sir | Yours very sincerely | C. Darwin

I have been so far rather idle of late, that I have been indulging in some extraneous work, and am going to publish a little work on the Fertilization of Orchids by Insects.

Copy
DAR 143

[1] See letter from H. W. Bates, 30 September 1861. CD does not appear ever to have made use of the information on *Volucella*. In the fourth edition of *Origin* he included a lengthy discussion of Bates's cases of mimicry in certain butterflies of the Amazon valley, referring to these as the 'most remarkable case of analogical resemblance ever recorded' and an 'excellent illustration of the principle of natural selection' (*Origin* 4th ed., pp. 502–6).

[2] See letter to H. W. Bates, 25 September [1861]. Bates's description of his travels, *The naturalist on the river Amazons*, was published by John Murray in 1863.

To John Murray 3 October [1861][1]

Down. | Bromley. | Kent. S.E.
Oct 3[d]

My dear Sir

I write in haste to know whether you object to my little Orchid Book being of size of Hutcheson on Dog Breaking, which I think is good size & which I mention because you published it.—[2]

Do you want ornament for cover???[3] I care not myself.—

In Haste | My dear Sir | Yours sincerely | C. Darwin

P.S | I was forced to order Blocks, & fixed this size, as Artist comes on Monday;[4] but the page might be large; but I sh[d] prefer little Book not to look pretentious.—

John Murray (publishers)

1 Dated by the reference to the publication of *Orchids* (see letter from John Murray, 23 September 1861).

2 Murray had published William Nelson Hutchinson's *Dog breaking* in 1848. There is an annotated copy of the second edition (W. N. Hutchinson 1850) in the Darwin Library–CUL. See also letter to John Murray, 5 October [1861].

3 See letter to John Murray, 24 September [1861].

4 George Brettingham Sowerby Jr arrived at Down on 7 October 1861 to prepare the illustrations for *Orchids* (Emma Darwin's diary).

To J. D. Hooker 4 October [1861]

Down Bromley Kent
Oct 4th.

My dear Hooker

Will you have kindness to read enclosed & look at diagram.— Six words will answer my question.— It is not an important point; but there is to me an *irresistible* charm in trying to make out homologies.— You know the membranous cup or clinandrum, in many orchids, behind the stigma & rostellum: it is formed of membrane which unites the filament of the normal dorsal anther, with the edges of the pistil. The Clinandrum is largely developed in Malaxis, & is of considerable importance in retaining the Pollinia which, as soon as flower opens, are quite loose.—

The appearance & similarity of tissues &c at once give suspicion that these lateral membranes of the clinandrum are the two other & rudimentary anthers, which in Orchis & Cephalanthera &c &c exist as mere papillæ. Here developed & utilised.—

Now for my question; exactly in middle of filament of *normal* anther, & exactly in middle of the lateral membranes of clinandrum, & running up to same height, are quite similar bundle of spiral vessels; ending upwards almost suddenly.— *Now is not this structure good argument that I interpret the homologies of sides of Clinandrum rightly*?[1]

Yours affect. | C. Darwin

I find that the great Bauer does not draw very correctly!! and good Heavens what a jumble he makes on functions.[2]

[Enclosure]

True Dorsal anther in bud with pollinia seen within cells.—

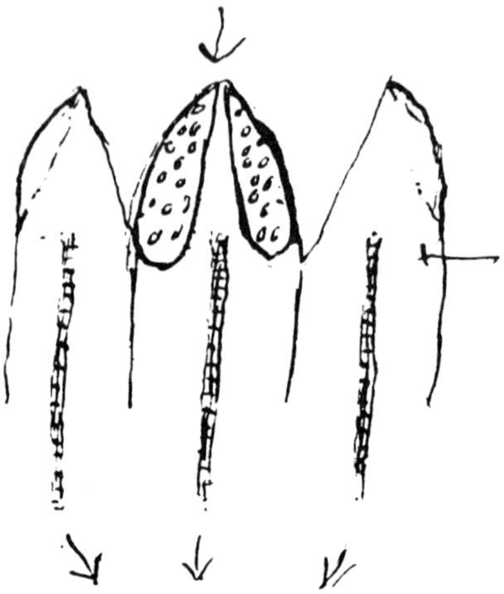

Membranous sides of Clinandrum folding round & united to edges of Pistil

Bundles of Spiral vessels

(Back of Clinandrum of Malaxis, *laid out flat*.)

Endorsement: '/61'
DAR 115.2: 115, DAR 115.1: 76

[1] Hooker's response has not been found, but see the letter to J. D. Hooker, 6–7 October [1861]. CD discussed the homologies of orchid flowers in the concluding chapter of *Orchids* and included a sentence describing the results of his dissection of *Malaxis* (*Orchids*, pp. 298–9):

> In the young flower-bud of Malaxis paludosa, the close resemblance between the membranes of the clinandrum and the fertile anther, in shape, texture, and in the height to which the spiral vessels extend, is most striking: it is impossible to doubt that in these two membranes we see two rudimentary anthers.

[2] Franz Andreas Bauer, botanical illustrator at the Royal Botanic Gardens, Kew during the period of Joseph Banks's superintendence, had prepared a major work illustrating orchids and describing the functions of the flower parts (Bauer 1830–8). CD had recently borrowed the book from Hooker. See letters to J. D. Hooker, [11 August 1861], 6 September [1861], and 24 September [1861]. Several of the illustrations in *Orchids* were copied from Bauer's drawings.

To John Murray 5 October [1861][1]

Down. | *Bromley.* | *Kent. S.E.*
Oct. 5.

My dear Sir

I think it very good idea to match the Origin; but then I must beg you to have large type & lines wider apart, otherwise the Book will look ridiculously small; but of that hereafter.

The advertisement reads well; I always spell Fertil**is**ation & not Fertili**z**ation—[2] I will attend to outside ornament.[3]

Yours sincerely | C. Darwin

Private collection

[1] The year is given by reference to the details concerning the publication of *Orchids*. See letters to John Murray, 21 September [1861] and 3 October [1861].

[2] An advertisement for *Orchids* appeared in the column 'Our weekly gossip' in the *Athenæum*, 5 October 1861, p. 448: 'Mr. Charles Darwin has prepared for publication a small work, containing his experience "On the Fertilization of British Orchids by means of Insects." It will form a sort of sequel to his work, the "Origin of Species".'

[3] See letter to John Murray, 24 September [1861]. An embossed illustration of a *Cycnoches* flower appeared in gilt on the cover of *Orchids* (see *Orchids*, p. 266).

To John Gould 6 October [1861][1]

Down. | Bromley. | Kent. S.E.
Oct. 6th.

Dear Gould

I am extremely much obliged to you for your present of the Trochilidæ, of which I have read every word (except the Synonyms) from your pleasant Introduction to the end.—[2] It certainly is a grand case of local distribution; & likewise of diversified adaptation. I was particularly glad to see one sentence, which I shall some day use, on the close alliance of the Species in the large genera.—[3] I see that you allude to the crossing of Birds in a state of nature; I, for one, repudiate this notion.—[4] One of the points which has interested me most (which you will not approve of) is the number of "races" or doubtful species. I think I shall extract all these cases; as it will show those persons who are quite ignorant of Nat History, that the determination of Species is not a simple affair.—[5]

I congratulate you on the completion of your magnificent work; & fully believe that your release from this labour will merely "enable you to go on with something else not less interesting".

With many thanks for your valuable gift, which has interested me much, pray believe me Dear Gould | Yours sincerely | Charles Darwin

Cambridge University Library (Add 4251/330)

[1] Dated by the reference to the presentation copy of Gould 1861 (see n. 2, below).

[2] Gould sent CD a presentation copy of *An introduction to the Trochilidæ, or family of humming-birds* (Gould 1861); an annotated copy, dated '4th. October 1861.', is in the Darwin Library–CUL. CD is mentioned in the preface as one of the naturalists who had 'paid especial attention to the Humming-Birds' (Gould 1861, p. ii).

[3] Gould 1861, p. 19: 'Apart from development, I observe that in the Humming-Birds, as in some other groups to which I have paid particular attention, the species of one genus are much more numerous than those of others, and that, whenever this is the case, the genus usually comprises many closely allied species.' This related to CD's conclusions about variation in large and small plant

genera, discussed in *Origin*, pp. 55–9. CD does not appear to have quoted Gould's sentence in any subsequent work. For a discussion of the importance in CD's work of variation in large genera, see Kohn 1985.

4 Gould referred to the three humming-bird species found on the island of Juan Fernandez, stating that 'nothing like a cross or intermixture' had ever been observed; describing the humming-birds of Jamaica, he stated: 'Nothing like interbreeding between two species appears to occur in this island' (Gould 1861, pp. 8, 9).

5 Gould held that all the humming-bird species he described had 'really distinctive and constant characters' and claimed: 'I have never observed an instance of any variation which would lead me to suppose that it was the result of a union of two species' (Gould 1861, p. 5). Inside the back cover of his copy of Gould 1861, CD wrote in pencil: 'M[r] G. *says he has [*above del* 'seems'] never noticed vars; but then he admits some slight individual variation, & if he find 2 forms from 2 districts ever so slightly different, they are called species.— In the same district *at same time*, sure to be similar variation hardly can occur'. An arrow points to the facing page, where CD gave a listing of page numbers under the heading: 'All references seen here abstracted. Those not struck out, all refer to slight variations & doubtful species.—'

To J. D. Hooker 6–7 October [1861]

Down Bromley Kent
Oct 6th

My dear Hooker

It is impossible to thank you enough. I was almost mad at the wealth of orchids.—[1] Several *invaluable* to me.— I cannot read enclosed ticket. Do not care about ticket for flower too much withered.[2] I send postage stamps.— What trouble it must have taken you!! I sh[d]. rather like another spec[n]. of *Masdevallia*; but I do not care much.—

The column of Cynoches was broken; everything else **perfect**.— I fear you would have no time for even simplest observation on Cynoches.— But I will for chance tell case. Loddiges told me "that Mormodes, if touched, would not fail to fertilise itself once in a thousand times".—[3] James Veitch tells me of some strange movement.—[4]

I sh[d]. much like to quote you on subject.— Is it that the pollinia slew round, when touched, & enter stigmatic cavity? Or does the lower part of column bend, when a large insect alights on the labellum or touches base of column;— I can conceive by this movement that pollinia would be left on insect's back & ultimately fertilise another flower.[5] But former supposition seems more probable, & would be of highest importance to me as it would be case of self-fertilisation like that of Bee Ophrys:— it is, however, very clear that there is sticky gland so that insects would sometimes remove pollinia.— The pollinia of your specimen were glued to the labellum.— Veitch sent me several spec[s] of Mormodes, but column of all broken & pollinia gone!! Eheu Eheu!!

Yours most gratefully | C. D.—

I must set to work again.—

Monday morning.— I have been at work all morning— you cannot tell how interesting & useful your specimens.— The buds of Pleurostalis have saved me from such a confounded blunder.[6] I am very anxious to hear, whether the spiral vessels, corresponding to each division of the Clinandrum is more of an argument, concerning their nature, than in being merely one point more of general resemblance. I find they come in in important manner in many Orchids.—[7]

They form the wings to the column in Cattleya & Oncidium.—[8] The worst of my work is that I have no conception what is new & what quite old in my work.— I fear I have been a precious fool to undertake it; but anyhow it has amused me.—

Endorsement: '/61'
DAR 115.2: 116

[1] Hooker's letter has not been found. CD had asked Hooker to send him specimens of several different plants, including a number of foreign orchids (see letter to J. D. Hooker, 28 September [1861]).
[2] This sentence was added later in blue crayon.
[3] Conrad Loddiges was a nurseryman in Hackney. In his 'big book' on species, CD cited Loddiges as saying that 'not one in a hundred' of the pollen-masses of *Mormodes* 'would miss hitting the stigmatic surface' (see *Natural selection*, p. 66).
[4] CD had obtained a number of exotic orchids from James Veitch, partner with his father and brothers in J. Veitch and Sons Nursery, Chelsea.
[5] CD subsequently described in similar terms the adherence of a viscid disc to the head rather than to the back of an insect in the process of pollinating *Mormodes* (see *Orchids*, pp. 258–60).
[6] CD refers to *Pleurothallis*, two species of which are briefly discussed in *Orchids*, pp. 166–7.
[7] See letter to J. D. Hooker, 4 October [1861].
[8] See *Orchids*, p. 299.

To Daniel Oliver 8 October [1861][1]

Down. | *Bromley.* | *Kent. S.E.*
Oct 8th

My dear Sir

I am perfectly ashamed to trouble Hooker again.— Would you have kindness to look at well-opened flower of Stanhopea saccata & see whether any nectar is in hollowed out base or cup of Labellum; or in other species of Stanhopea, if they have hollowed out base or cup to Labellum.—

I have particular reason to enquire; but I daresay the plant will have none, though it *ought* to have some.—[2] But Nature, as Agassiz says, does not lie,[3] & therefore it must have nectar,—else a theory of mine is wrong which is clearly impossible!—

In Haste | Ever yours | C. Darwin

Down House (MS 10: 31)

[1] The year is given by the relationship to the preceding letter and to the letter to J. D. Hooker, 13 October [1861].
[2] The idea that *Stanhopea* flowers '*ought*' to have nectar in the base of the labellum is explained in *Orchids*, pp. 282–3: 'As in Epipactis the cup at the base of the labellum serves as a nectar-receptacle, I expected to find that the analogous cup in Stanhopea, Acropera, &c., would serve for the same purpose; but I never could find a drop of nectar in it.'

[3] CD was fond of quoting this aphorism attributed to Louis Agassiz. See, for example, *Correspondence* vol. 6, letter to Asa Gray, 1 January [1857].

From Daniel Oliver [after 8 October 1861][1]

[Kew]

I find no trace of nectar or other secretion in the hollow base of labell of St. Saccata[2]

The papillae wh. clothe it are more or less moist, but no accumulation as a drop—

Draft
Down House (MS 10: 31v.)

[1] The draft of Oliver's letter is written on the back of the preceding letter.
[2] See preceding letter.

From Asa Gray 11 October 1861

Cambridge, Mass.
Oct. 11th. 1861

My Dear Darwin

It was through pure forgetfulness at the moment of writing my letter last August, that I did not endeavor to supply to you such instances as I could recollect (for I have no memoranda on the subject) of what Dr Torrey and I long ago (in Fl. N. Amer. 2, p. 38, 39) called *diœcio-dimorphism*.[1]

In Rubiaceæ I think this dimorphism was first somewhere mentioned by D. Don, but I have no reference.[2] We are most familiar with it in *Houstonia* (a genus which I have proposed to restore, in which it occurs in all our N. American species, and, as you will see in Wight & Arnott, Prodr. Fl. Pen. Ind. 1. p. 408, in some Indian ones.[3] In the living plants I have formerly noticed it in *H. purpurea*, and it is every year noticed familiarly in our most abundant *H. cærulea*. Of this I shall try to send you some tufts—seedlings of this year—which you can blossom in early spring, or in your house in winte⟨r.⟩ We must run the chance of your getting ⟨both⟩ subsexes. Wight & Arnott (see p. 439) ob⟨served⟩ it in *Knoxia*. Another familiar case is furnished by our pretty little *Mitchella repens*; this is recorded by Dr. Torrey in Flora of State of New York, 1. p. 315.[4] This would be a good plant for you to experiment upon, if I had only thought of marking plants the two sorts.

Diodia Virginica (and I suppose *D. teres*) also shows it.

The dimorphism may perhaps occur in many *Rubiaceæ*, but I have reason to think it that the far greater part show no such tendency. And *included* or *exsert* stamens are pretty good and constant generic characters in much of the order[5] *Mitchella* is an interesting instance for you from its relationship (with *Nertera*) to *Coprosma*, one of the few diœcious genera of Rubiaceæ, and in which the stamens are long-exserted in the male flowers, the styles or stigmas, in the female.

In these and in all genuine cases of the kind, the two sorts of flowers are always borne on different roots.

Upon consideration, I can say nothing of *Labiatæ* or *Borragineæ* from my own observation. But I believe that Dr. Torrey can tell us about the latter. ⟨I⟩ have just written to ask him.[6]

⟨In⟩ *Labiatæ*, what dimorphism I have observed—and without particular examination—is merely such as in *Nepeta Glechoma*, where the stamens of the *earlier flowers* appear to abort. I know not whether they do, or whether it is a case of precocious fertilization, such as I will presently call your attention to: but as the flowers are otherwise normal, I suppose it is not a case of this.

A case in point, which will interest you, occurs in *Rhamnus*—a genus which has both truly hermaphrodite and polygamo-diœcious species.

Rhamnus lanceolatus, of our Southern & Middle & Western States, bears, on different trees, two kinds of flowers which differ *in the pistil only*;—i.e. as far as appeared to ordinary investigation the *stamens* are the same in both; but in one the style is short and included; in the other long and exserted. In the latter the flowers are subsolitary in the axils and fruitful; in the former the flowers are more numerous and clustered, and not so fruitful. Yet they do mature some fruit. Here you have an initial state of dimorphism, *apparently* affecting only the female organs. (See Gray, Genera Am. Bor. Illustr 2, p. 180, f. 168.)[7] Your close observation might likely enough find a difference in the pollen.

Next take certain portions of the genus *Plantago*,—upon which, instead of writing details, I refer you to my *Manual* p. 269,[8] and to my observations in *Pacific Rail-Road Surveys*, vol. 4. p. 117,[9] or, in the Extra copies of Botanical Report on Dr. Bigelow's collection, p. (117) 61.)[10] Hooker possesses both the volume and the separate paper. I only add that, in *Plantago Virginica* I have since seen in herbaria, the staminate sort maturing seed, but generally it has gone to *dioicism*, while in *P. Patagonica* the two sorts are fruitful, perhaps almost equally so. (If you have difficulty in referring to the memoir above-mentioned, I will have the foot-note copied for you.)

Ilex opaca, the analogue of your Holly, I find in dried specimen, that the female flowers have stamens with filaments as long and anthers as large as those of male flowers, but in a flower-bud examined the anthers have *no pollen*. The male flowers have no pistil.[11]

Our Hollies (*Ilex* & *Prinos*) must be more particularly examined, to see whether anthers of any fertile flowers bear pollen. I should say they did, from general impression, and from my description in *Manual*, which was condensed from MSS, prepared long ago for Flora of N. America. But we descriptive botanists have not been careful nor exact enough for your purposes.

In our loose observations we never should have noticed, in *Primula, Houstonia*, &c—any difference in the *pollen* of the two sorts of flowers. I should much like to have you tell me *what the difference is*, in the pollen, and I will make observations next spring, upon *Houstonia*. The discovery that the *pollen* of *one is good for the pistil of the other, but not for the pistil of its own flower, is most important.*— I should rather expect this, but I want to know more about the *fact*—how you made it out, &c &c

You should next turn your attention to a very different sort of dimorphism, which is almost equally common, perhaps,—one which looks to *close-* instead of *cross* fertilization. I allude to such cases as that occuring in the European *Impatiens nolitangere*, where it was discovered by Weddell, & published by Ad^r. Jussieu (Malphigiaceæ, p. 85),[12] and in our American two species, by D^r. Torrey. See about this, Gray Genera Illustr. 2, p. 132, f. 152, 153.[13]

Here all the fruit in the early part of the season comes from pistils fertilized by their own pollen precociously, in the young bud. But at or after midsummer, many/some} of the conspicuous full-grown flowers are fertile,—their ovary lengthening enough to push out the stigma beyond the connivent scales, so that pollen of the same flower, or of other flowers brought by insects, can get access to it.

Malpighiaceæ, Violaceæ, Leguminosæ, &c—furnish very numerous instances of such precociously-fertilized flowers.— these flowers always far more fertile than others.— And a similar case in *Specularia perfoliata* has very long been known.

I did not know of diœcio-dimorphism in *Linum*. I have suspected it in *Oxalis*, from the differences in relative length of the stamens & styles of some species.

Incomplete
DAR 109: 82–3, DAR 110 (ser. 2): 117, DAR 111: 83

CD ANNOTATIONS
0.3 My Dear ... tendency. 4.2] *crossed red and blue crayon*
4.6 in the female.] *closing square bracket blue crayon*
5.1 In these ... of this. 7.5] *crossed red and blue crayon*; 'Holly—' *added in margin red crayon*
9.1 *Rhamnus lanceolatus*,] *opening square bracket red crayon*
12.1 Our Hollies] 'Holly' *red crayon, circled red crayon*
12.2 I should ... purposes. 12.5] *crossed red crayon*
13.1 In our ... &c 13.6] *crossed blue crayon*
14.3 I allude ... 153. 14.6] 'Impatiens' *red crayon*; 'Cleistogamy' *red crayon, circled red crayon*
17.1 I did ... species. 17.2] *crossed red crayon*
Top of first page: 'Other Genera | Houstonia—Diodia Knoxia' *red crayon*

[1] See *Correspondence* vol. 8, letter from Asa Gray, [10 July 1860]. CD had subsequently queried Gray again about his knowledge of dimorphic plants (see especially letters to Asa Gray, 5 June [1861], 21 July [1861], and 16 September [1861]). The reference is to Torrey and Gray 1838–43.

[2] The botanist David Don had served as librarian to the Linnean Society of London from 1822 to 1841 and as professor of botany at King's College, London from 1836 until his death in 1841. Gray may be referring to a paper by Don that describes Rubiaceae as providing the best example among plants of extensive variation in the sexual organs (Don 1837).

[3] Wight and Arnott 1834.

[4] Torrey 1843.

[5] This sentence was added at the bottom of the page and keyed to the text with an asterisk. CD cut off this portion of the letter, which was filed separately from the other two portions. It is now in DAR 110 (ser. 2): 117.

[6] The botanist John Torrey was the United States assayer in New York. He had recently moved to the campus of Columbia College, New York.

[7] Gray 1848–9.
[8] There is an annotated copy of Gray 1856 in the Darwin Library–CUL.
[9] Torrey 1857; this work includes many plant descriptions provided by Gray.
[10] John Milton Bigelow made the botanical collections for the Pacific railroad survey of 1853–4. The greater part of his collection was submitted to Torrey for examination and description (Torrey 1857, p. 62). The different sections of the survey report were also printed separately (Johnston 1943).
[11] The section from 'Ilex opaca' to 'no pistil.' was written in the margin and keyed to the text with an asterisk.
[12] Hugh Algernon Weddell was a former pupil of Adrien Henri Laurent de Jussieu. The reference is to Jussieu 1843.
[13] Gray 1848–9.

To J. D. Hooker 11 October [1861]

Down.
Oct 11th.

My dear Hooker

Mr Veitch most generously has sent me 2 splendid buds of Mormodes, which will be capital for dissection but I fear will never be irritable;[1] so for the sake of charity & love of Heaven do, I beseech you, observe what movement takes place in Cychnodes & what part must be touched.[2] Mr. V. has also sent me one splendid flower of Catasetum, the most wonderful orchid I have seen: Sowerby has made capital sketches, but I have not yet ventured to touch it.—

Did you or Oliver not tell me that your Mr Crocker[3] had made some observations on movement either of Catasetum or Mormodes? They wd. be very useful to me.—

Ever yours | C. Darwin

Endorsement: '/61'
DAR 115.2: 117

[1] CD mentioned in *Orchids* p. 265, that the nurseryman James Veitch sent him specimens of *Mormodes*.
[2] See letter to J. D. Hooker, 6–7 October [1861]. On the back of the letter there are three diagrams and some notes in Hooker's hand. These describe the slow movement of the rostellum in the orchid *Miltonia* and insect removal of the pollen 'gland' in *Odontoglossum*. CD included these species in his discussion of the Vandeae tribe of orchids, illustrating the response of parts of the flower to insect behaviour (*Orchids*, pp. 178–210). Hooker's letter is missing, but see the letter to J. D. Hooker, 15 October [1861].
[3] Charles William Crocker was the foreman of the propagating department at the Royal Botanic Gardens, Kew.

To Asa Gray [after 11 October 1861][1]

[Down]

run on to you in a most unreasonable way.—

I am so glad that you will look to some of your Rubiaceæ, & I hope that may find time to make a few experiments.— Thanks for notes about your Hollies, & I hope you will look a little to them.[2] There is to me incomparably more interest in observing than in writing; but I feel quite guilty in trespassing on these subjects, & not sticking to varieties of the confounded cocks, Hens & Ducks.—[3] I hear that Lyell is savage at me.— I shall never resist Linum next summer.—[4]

What you say about our keeping in our intrenchments & firing long shots about Design has made me laugh.—[5] I suspect I am more cowardly than you, as I ought to be, as I do not feel sure of my ground.— Here is my answering long shot about the cream-jug-nose:[6] I should believe it to have been designed (as I did formerly each part of each animal) until I saw a way of its being formed without design, & *at the same time* saw in its whole structure (as in homologies, embryology, rudimentary organs, distribution &c) evidence, of its having been produced in a quite distinct manner, ie by descent from another cream-jug whose nose subserved, perhaps, some quite distinct use. When I think of my beloved orchids, with rudiments of five anthers, with one pistil converted into a rostellum, with all the cohesion of parts, it really seems to me incredibly monstrous to look at an orchid as created as we now see it. Every part reveals modification on modification.[7] But enough & more than enough.—

Farewell, my dear Gray, with cordial thanks for your never ending kindness.— Yours most sincerely | C. Darwin

Of course I will send you my Orchid opusculus.—

Incomplete
Gray Herbarium of Harvard University (51a)

1 Dated by the relationship to the letter from Asa Gray, 11 October 1861. See also letter from Asa Gray, [27 and 29 August] and 2 September 1861, and letter to Asa Gray, 17 September [1861].

2 See letter from Asa Gray, 11 October 1861, and letter to Asa Gray, 17 September [1861]. CD made use of this and many of the other cases cited in Gray's letter when he came to write up a full account of his study of dimorphic plant species (see *Forms of flowers*).

3 CD had been working intermittently since January 1860 on preparing a treatise discussing variation in domesticated animals and plants (*Variation*). Since July 1861, however, he had halted this work in order to write up the results of his studies of orchid pollination and dimorphism in *Primula* (see 'Journal'; Appendix II).

4 Having already examined *Linum* plants grown in the summer of 1861, CD carried out further experiments in 1862 and his paper on dimorphism in *Linum* was read before the Linnean Society of London on 5 February 1863 (see *Collected papers* 2: 93–105). See also letter to J. D. Hooker, 28 September [1861].

5 This discussion was apparently included in the part of the letter from Asa Gray, 11 October 1861, that is now missing.

6 See letter to Asa Gray, 17 September [1861]. See also letter to Charles Lyell, 21 August [1861].

7 CD discussed in the final chapter of *Orchids* the range of modification of particular species from the archetypal plan of orchids.

To W. E. Darwin 12 October [1861][1]

Down
Oct 12th

My dear William

I am not going to write you a decent note, for I am half-dead with working with Mr Sowerby at the Orchid drawings.[2] It has been the devil's own job. But I am anxious to hear whether you have signed the Articles of Partnership.[3] I rather fear that we were not very civil to Mr Hacon in not writing to him & asking him whether it was all right your joining before he announced that all was ready.[4] In fact, I suppose, it could make no difference; in the case of having to do with a gentleman like Mr Atherley. No doubt a rogue could have got articles altered. Do not leave this note about.— You must at once buy some good lock-up thing.— Shall you not want the certificate that you have £5000 consols?[5] will this have to be produced on both sides.— Let me hear about the signing pretty soon.—

You cannot tell how much we are interested & amused by your letters. Everything sounds prosperous as yet: of course there will be drawbacks, as in every thing else in this world. I wish the winter was not coming on, as that will make it duller for you; but I am so glad to hear about your walk. What a nice considerate man Mr Atherley must be; pray give him my kind remembrances.— Enquire whether Sir H. James is in Southampton & whether you would like me to write to him. He is a pleasant & distinguished man.[6] I think, if I write, the best & civillest will be, for me to ask permission for you to call on him.

Mr Sowerby left this afternoon & comes back on Tuesday night, when I shall have 5 or 6 more very hard days' work.[7] The drawings will be pretty successful; but I feel sure that my little book will be too difficult & too uniform for the Public & I almost wish I had never thought of separate publication. So I must chance it now.—

Miss Pugh arrived this evening for a few days.[8] Crocket flourishes;[9] how I wish you were here my dear old fellow to give a few more crockets in a mournful spirit. We delight in your letters—

Farewell | your affect. Father | C. Darwin

DAR 210.6

[1] The year is given by the reference to the preparation of drawings for *Orchids* (see nn. 2 and 7, below).

[2] George Brettingham Sowerby Jr arrived at Down House on 7 October 1861 to prepare the illustrations for *Orchids* (Emma Darwin's diary).

[3] William Darwin was in Southampton making the final arrangements to join the Southampton and Hampshire Bank as a partner (Emma Darwin's diary). The conditions of the partnership had been first established in May 1861 and were finalised in August.

[4] William Mackmurdo Hacon, CD's solicitor, had been hesitant to accept all the provisions requested by the bank's partner, George Atherley. See letters to John Lubbock, [2 August 1861] and 10 August [1861].

[5] 'Consols' is an abbreviation for consolidated annuities, a form of government stocks. Consols offered a fixed rate of interest, usually low, giving the lender the advantage in the principal funded. CD had advanced William £4500 from his inheritance to serve as a guaranteed deposit toward the partnership in the Southampton and Hampshire Bank (Down House MS 11:16). See also letter to W. E. Darwin, [25 May 1861].

[6] Henry James, director-general of the Ordnance Survey, lived in Southampton, where the survey was based.

[7] See n. 2, above. According to Emma Darwin's diary, Sowerby stayed until 12 October 1861. He returned to complete the work on 15 October 1861. An entry in CD's Account book (Down House MS) for 21 October 1861 records a payment of £10 16*s*. to 'Mr Sowerby for 10 days'.

[8] Miss Pugh had been governess to the Darwin children in 1856 and 1857 and afterwards occasionally visited the family.

[9] This is probably a misspelling of 'croquet'. An entry for 4 September 1861 in Emma Darwin's diary reads 'Etty played Croqet', with the 'q' written over 'ck'.

To J. D. Hooker 13 October [1861]

Down Bromley Kent
Oct 13th

My dear Hooker.

Thanks for note.[1] It seems that I cannot exhaust your goodnature. I have had the hardest day's work at Catasetum & buds of Mormodes, & believe I understand at last mechanism of movements & function.— Catasetum is beautiful case of slight modification of structure leading to new function. I never was more interested in any subject in my life, than in this of Orchids.— I owe very much to you.—

Now for points to attend to.—

Does viscid surface of rostellum (or gland) in Cycnoches or Mormodes project beyond stigmatic cavity; or does it not lie (as in bud & as in mature Catasetum) within roof of stigmatic cavity. But you must not touch, you must judge by looks.—

2d. First touch base of Column; then touch basal edge of a curtain which depends (I suppose) in front of stigmatic cavity; then touch round smooth surface in middle over stigmatic cavity; & lastly touch either of two slight prominent knobs on side of stigmatic cavity: This latter point, I expect, will be sensitive & will cause pollinia to be ejected with force, gland foremost.— I judge from Catasetum: Which is the sensitive point I want most to know. Touch lightly with needle or bristle; & secondly whether gland-end is not projected foremost?— I cannot believe that pollinia would hit own stigma??? I can imagine that it may fertilise itself & this would be most important point for me.

I have written to Mr Veitch begging him to let me buy flowers, if any more open & I will send Parslow up to carry them down in hand.[2] Would it not save you trouble if I were to do same i.e. send servant to Kew? that is if you can spare me a flower or two.— Perhaps this would be best for me.—

If any flower of great tribe of Arethuseæ shd. open, please remember how much I wish to see one.—[3] Hearty thanks for information about vascular vessels;[4] I shall do a bit more work at them.

If it does not trouble Oliver will you ask him to look once again within sack of Stanhopea; for it is not very likely that there shd. be nectar till after flower has opened for some days & been well shone on.[5]

Your affect. | C. Darwin

See how pretty it is; an insect could not touch sticky gland of Catasetum, from its sticky surface lying close to roof of stigmatic chamber; consequently it is projected, sticky end first, with wonderful force, right on head or shoulders of an insect crawling on the Labellum.—

Endorsement: '/61'
DAR 115.2: 118

[1] Hooker's letter has not been found. It was a response to queries in the letters to J. D. Hooker, 4 October [1861] and 6–7 October [1861].
[2] James Veitch, a nurseryman in Chelsea, supplied CD with several foreign orchids. Joseph Parslow was the butler at Down House.
[3] CD was anxious to examine species belonging to the Arethuseae tribe to compare their flower structure with that of the other major tribes of orchids. However, he was unable to obtain any specimens before the publication of *Orchids* (see *Orchids*, p. 269).
[4] CD had asked Hooker whether the existence of 'spiral' vessels running up the middle of the lateral membranes forming the clinandrium in *Malaxis* could indicate that the membranes were modified anthers (see letter to J. D. Hooker, 4 October [1861]). Hooker's reply has not been found. However, he apparently encouraged CD to trace the longitudinal course of the vascular tissue in this and other orchids (see letter to J. D. Hooker, 27 October [1861], and *Orchids*, pp. 290–2).
[5] See letter to Daniel Oliver 8 October [1861], and letter from Daniel Oliver, [after 8 October 1861].

To J. D. Hooker 15 [October 1861]

Down Bromley Kent
15th

My dear Hooker

If your Acropera case shd. really turn out as you suspect, it would be very curious & would greatly interest me,—especially, as I chanced in another case to speculate on the possibility of such action. Unless wind by agitating slightly the flower would effect this, I shd. not believe in it. Wind is, necessary to self fertilisation of Bee Ophrys.—[1] I hope you will look to this. Try, if by any chance by pricking inside of Labellum, whether it be irritable.— How I shd. like to see one of the Bulbophyllum or other orchids with irritable Labellum!

I shall be particularly glad to see Acropera, as you kindly propose to send it. From analogy with Epipactis palustris, I shd. *suspect* its use was to direct insects in & out of flower in a different direction.—[2]

I have seen Odontoglossum & Miltonia: & as you say they have nothing remarkable; except indeed that in Miltonia the caudicle has not usual movements of depression. By the way the mechanism of this movement is rather curious & all lies in the bit of exterior membrane of rostellum, which is removed with gland. It is rather odd that I have not found one accurate or intelligible description of the so-called gland in Ophreæ, & still less in the Vandeæ

Catasetum is too long a story, & would weary you, though it is a pleasure to me to write. You may rely that the ejection of pollinia is *special* contrivance; & the plant would never set a seed without it; for an insect could never remove pollinia without this action.[3] Mr Smith is very partially right about the kind of Elasticity.[4] The irritation which excites the rupture of rostellum (this rupturing is common to many Orchids) is transmitted along the curled horns, or as I think I shall call them the antennæ. The pollinia are ejected with sticky gland foremost & will surely stick on the head of insect, when it touches tips of the antennæ, which stand over, what seems to be the nectar-receptacle.—

Ever yours | C. Darwin

☞ You must let me send Parslow for a Mormodes when open.— I am very curious to see it[5]

P.S. I have again changed my mind & sh^{d}. be *much obliged* for 2 or 3 **very young** buds of any of the Vandeæ Oncidium w^{d} be best: for I want to see whether the attachment of caudicle is effected in them as in Spiranthes: if you can send me buds, please do not send them for a week, when I shall have finished with Mr Sowerby & drawings.—[6]

Could you lend me Reichenbach;[7] I sh^{d} be very glad to see it; if so please send it by enclosed address.—

You speak of pollen-masses of Acropera not sticking to stigma; this is often or generally the case if you pull straight out of stigma; but if you pull at right angles, as an insect could do in flying away, the elastic part of caudicle is stretched & drawn over the edge of stigmatic cavity, & (though so *wonderfully* strong in some cases) ruptures.

Most cordial thanks for all your kindness & writing so long a letter. I am glad to hear of your Arctic paper.[8]

Your affect | C. D.

Endorsement: 'Oct 61'
DAR 115.2: 119

1 Hooker had apparently examined the parts of an *Acropera* flower at Kew and noted the unusual structure of the labellum (see letter to J. D. Hooker, 18 October [1861]). CD discussed *Acropera* in *Orchids*, stating: 'this genus for a long time remained the opprobrium of my work. All the parts seemed determinately contrived that the plant should never be fertilised.' (*Orchids*, p. 203).

2 CD refers to the labellum of *Acropera* (see letter to J. D. Hooker, 18 October [1861]). In *Orchids*, he called this part 'an extraordinary organ, baffling description' and stated that the two upper petals served as 'lateral guides leading into the hood-like upper sepal' (*Orchids*, pp. 204–5).

3 CD devoted an entire chapter in *Orchids* to a description of the mechanism by which the pollinia were ejected in *Catasetum*, which he called 'the most remarkable of all Orchids' (*Orchids*, pp. 211–48).

4 John Smith was foreman and curator at the Royal Botanic Gardens, Kew (Allan 1967).

5 These two sentences are written along the left margin of the page. Joseph Parslow was the butler at Down House.

6 George Brettingham Sowerby Jr was preparing the illustrations for *Orchids* (see letter to W. E. Darwin, 12 October [1861]).

7 Reichenbach 1852. See letter to J. D. Hooker, 18 October [1861].

8 Hooker 1861.

To W. E. Darwin 17 [October 1861][1]

Down
17th

My dear William.

This is mere note of business.— I have written to Mr Hacon to urge him to send the Deed to M^{r}. Atherley.—[2]

I have written to Sir H. James, asking him whether you may call on him; but do not call till you hear again from me.—[3] But I daresay he will call on you.— Mr Bonham Carter has written to Embley & you must wait till you hear from the Nightingales, before you go there.[4] It is evident that you will have plenty of acquaintances; but I am not sorry that you should know two or three respectable persons on your own account.—

What a magnificent seigneur you are in your house.—[5] Take care you are not ruined. But I daresay it is good policy starting rather extra superb.— I cannot help wishing you were in the Bank. It is impossible that Mr Hacon could have really wished to stop the affair.— Lord I wonder what his Bill will be.—

It is a good job that you get on smoothly with old Fall—[6]

Good Night.— I am nothing but an Orchis.—

Your affect Father | C. D.

DAR 210.6

[1] Dated by the relationship to the letter to W. E. Darwin, 12 October [1861].

[2] CD refers to his solicitor, William Mackmurdo Hacon, and to George Atherley, William's partner in the Southampton and Hampshire Bank (see letter to W. E. Darwin, 12 October [1861]).

[3] Henry James, director-general of the Ordnance Survey, resided in Southampton (see letter to W. E. Darwin, 12 October [1861]).

[4] Embley Park, about five miles north-west of Southampton, was one of the homes of William Edward Nightingale, the father of Florence Nightingale. CD probably refers to Henry Bonham-Carter, who was a relative of the Nightingales.

[5] William had taken rooms in the lodging house of Mrs William Pratt at 1 Carlton Place, Southampton (*Post Office directory of Dorsetshire, Wiltshire, and Hampshire* 1867).

[6] Phillip Carteret Fall had been a partner in the Southampton and Hampshire Bank until his retirement prior to William's appointment (*Banking almanac* 1861).

To J. D. Hooker 18 October [1861]

Down Bromley Kent
Oct 18th

My dear Hooker.

The Book of Reichenbach is that which *you* told *me* of "de pollinis Orchid. genesi et structura: Leipsic. 4to."—[1]

I am so glad that you can let me have a flower (two if possible) of Catasetum; & for buds of Oncidium.

Please thank Oliver for looking to nectary of Stanhopea.—[2]

The Acropera came *quite* fresh. The use of strange Labellum has, as might have been expected, utterly beaten me; the viscid disc forms a cap at end of filamentary rostellum & this I have not seen before; & to remove pollinia well, this must be upturned & brushed upwards; & I fancy Labellum may be concerned in leading insect to depart by this course. But there is one point which considerably interests me, namely that the opening in all the flowers of the stigmatic cavity was too small for end of pollen-mass to enter. Hence I suspect it subsequently enlarges; this I sh^{d}. much like to ascertain & therefore sh^{d}. be *very glad* of a few **old** flowers.—[3]

Why I wish this, is, because I have found other structures compelling old flowers to be fertilised always by younger flowers.— Spiranthes is a very good case; & Asa Gray who has been looking at N. American species, & did not know of movement of Labellum, was astonished at facility with which pollen-masses could be removed, & at impossibility, as it seemed to him, of the flower being ever fertilised.—[4] It is funny on what trifles the fertilisation of Orchids depends; Humble-bees alight at bottom of spike & crawl *up* Spiranthes; if they were to crawl *down*, very few flowers would be fertilised.— I long for Catasetum to test points of sensitiveness; Sowerby has made very nice drawings of outside & section.[5]

I have just been collecting seed of 2 wild species of Verbascum: some time tell me whether you can give me white & yellow **var^{s}**. of any species: & perhaps I could buy some other species; though I can find none in my few Nursery-Garden Catalogues. I am sure you will agree in importance of testing Gärtner's statement about the potency of their pollen.[6] Also can you give me seed of any Cinchoneæ? & a *hairy* Saxifrage.[7] You see I take you at your word & *fully* trust your goodnature.

Yours affec^{y} | C. Darwin

Endorsement: '/61'
DAR 115.2: 120

1 Reichenbach 1852. See letter to J. D. Hooker, 15 October [1861].

2 See letter to Daniel Oliver, 8 October [1861], and letter from Daniel Oliver, [after 8 October 1861].

3 See letter to J. D. Hooker, 15 October [1861]. CD discussed this point in *Orchids*, stating that the stigmatic chamber in *Acropera* was so narrow that the pollen-masses could 'hardly be forced in', making pollination appear a great difficulty (*Orchids*, p. 205).

4 Asa Gray's comments on *Spiranthes* were included in the missing section of the letter from Asa Gray, [27 and 29 August] and 2 September [1861]. See CD's response to Gray's remarks in his letter to Asa Gray, 17 September [1861]. For CD's discussion of the means of pollination in *Spiranthes*, see *Orchids*, p. 123.

5 George Brettingham Sowerby Jr's drawings of *Catasetum saccatum* appear in *Orchids*, pp. 216 and 217.

6 Gärtner 1844, pp. 137–8, and Gärtner 1849, pp. 92, 180–1, 724–8. Karl Friedrich von Gärtner had shown that the pollen of white and yellow varieties of particular *Verbascum* species had different potencies when crossed with other *Verbascum* species (see letter to J. D. Hooker, 28 September [1861] and n. 11). CD mentioned Gärtner's findings in his paper 'On the two forms, or dimorphic condition, in the species of *Primula*, and on their remarkable sexual relations' (*Collected papers* 2: 59) and in *Variation* 2: 93, 104–7. For CD's extensive notes on Gärtner 1844 and 1849, see *Marginalia*.

7 See letters to J. D. Hooker, 19 June [1861] and 28 September [1861].

To John Lindley 18 October [1861][1]

Down. | Bromley. | Kent. S.E.
Oct. 18th

My dear D^{r}. Lindley

I thank you sincerely for your information.[2] Your tracing shows that the species is saccatum beyond a doubt.—[3] I wrote to Gardener's Chron. as I thought it would trouble you less to answer in that way.[4] I am very much obliged.— I have been extremely much interested with Catasetum, and indeed with many exotic Orchids, which I have been looking at in aid of an opusculus, on the fertilisation of British Orchids. I very much fear that in publishing I am doing a rash act; but Orchids have interested me more than almost anything in my life. Your work shows that you carefully understand this feeling.[5]

Yours very sincerely & obliged | Charles Darwin

Royal Botanic Gardens, Kew (Lindley letters, A–K: 193)

[1] The year is provided by the reference to the publication of *Orchids*.
[2] Lindley's letter has not been found.
[3] *Catasetum saccatum*, the lizard orchid, is discussed in *Orchids*, pp. 214–31.
[4] Lindley was the editor of the *Gardeners' Chronicle*.
[5] Lindley had a special interest in orchids and was the author or editor of a number of works in which the morphology and taxonomy of orchids was described. CD often cited Lindley's works in *Orchids*.

To Charles Lyell 20 October [1861]

Down. | *Bromley.* | *Kent. S.E.*
Oct. 20th

My dear Lyell

Notwithstanding the Orchids, I have been very glad to see Jamieson's letter: no doubt, as he says, certainty will soon be reached.[1]

With respect to the minor point of Glen Roy, I cannot feel easy with a mere barrier of ice; there is so much sloping stratified detritus in the valleys: I remember that you somewhere have stated that a running stream *soon* cuts deeply into a glacier.— I have been hunting up all old references & pamphlets &c on shelves in Scotland & will send them off to Mr. J., as they may possibly be of use to him; if he continues the subject. The Eildon Hills ought to be especially examined. Amongst M.S. I came across a very old letter from me to you, in which I say "If a glacialist admitted that the sea, before the formation of the shelves, covered the country (which would account for the land-straits above the level of the shelves) & if he admitted that the land gradually emerged; & if he supposed that his lakes were banked up by ice alone, he would make out, in my opinion, the best case against the marine origin of the Shelves".—[2] This seems very much what Mr J. & you have come to. The whole Glacial Theory is really a magnificent subject.—

I have been working terribly hard lately (for me) at Orchids. The subject is, I fear, too complex for the Public & I fear I have made a great mistake in not keeping to my first intention of sending it to Linnean Socy.; but it is now too late, & I must make the best of a bad job.—

Yours affecy | C. Darwin

Endorsement: 'Oct. 24. 1861—'
American Philosophical Society (268)

[1] The reference is to a letter to Lyell from Thomas Francis Jamieson, 14 October 1861 (see Appendix IX). Jamieson was investigating the 'parallel roads' of Glen Roy. He had found evidence to support the view that the 'roads' were the beaches of former glacial lakes, not, as CD had maintained in a paper of 1839, beaches made by the sea at different periods during the elevation of the land. See letter from T. F. Jamieson, 3 September 1861.

[2] See *Correspondence* vol. 4, letter to Charles Lyell, 8 [September 1847]. CD's version is a paraphrase of the original letter.

To John Murray 21 October [1861]

Down. | *Bromley.* | *Kent. S.E.*
Monday | Oct 21

My dear Sir

Mr G. B. Sowerby has been here for 10 days making drawings of Orchids. We found it impossible to draw on wood; & it was absolutely necessary first to make perfectly shaded drawings on paper, which are now completed & of course I have paid for them.[1] But I have not paid for the drawings being transferred to wood but am quite ready to do so, when the cost can be proportioned. I should like Mr Sowerby to do this, as he will do it best, & further I think it would be a great advantage if he were to certain extent to superintend the cutting, as he says he will.

But I particularly want to call your attention to the fact that a **first-rate** cutter **must** be employed. I fear that the expence will be considerable.— I have asked Mr Sowerby to call on you tomorrow (Tuesday) with the drawings for you to settle how the job is to effected. If not done excellently well, the whole thing will be a failure.— The best woodcuts I have ever seen are in Prof. Owen's British Fossils, (Van Voorst) done by Mr W. Bagg of 63 Gower St;[2] but Mr Sowerby has never heard of him.—

I am very anxious that the cutting shd. be done quickly at once; for immediately that I have proofs of the cuts, I am ready for press.—

My dear Sir | Yours very sincerely | Charles Darwin

Endorsement: '—1861'
John Murray (publishers)

[1] An entry in CD's Account book (Down House MS) for 21 October 1861 records a payment of £10 16*s*. to 'Mr Sowerby for 10 days'.

[2] This information appears in the preface to Owen 1846. William Bagg, artist, is listed in the *Post Office London directory* for 1846.

To the *Journal of Horticulture* [before 22 October 1861][1]

[Down]

Would Col. Newman, who has so carefully attended to the habits of humble bees, have the kindness to state whether the queen humble bees are fertilised in the air or on the ground? I have a special reason for wishing to know this little fact, and whether the fertilisation does not often take place as late as in September?—[2]

C. Darwin.

Journal of Horticulture, Cottage Gardener, and Country Gentleman n.s. 2 (1861–2): 76

[1] Dated by the publication of the letter in the 22 October 1861 issue of the *Journal of Horticulture*.
[2] The letter was published under the heading 'Is the female Bombus fertilised in the air?'. CD's query was probably prompted by a note by Henry Wenman Newman published in the 15 October 1861 issue of the *Journal of Horticulture* (pp. 40–1), in which Newman questioned whether parthenogenesis occurred in bees and suggested that the eggs of the queen bee were fertilised by the 'aura' of the drone. For Newman's response to CD's queries, see the following letter.

From Henry Wenman Newman [before 22 October 1861][1]

The queens or females of the humble bees are not fertilised in the air, and the act of fertilisation takes place either in the nest or on some flower, or on the ground.[2]

I have made observations, more or less, since the year 1798, and give the result—scarcely half a dozen instances in sixty years.

1st. I have seen one couple come out of the nest of the *Bombus terrestris*; these flew away instantly, and were united until I lost sight of them.

2nd. I observed one male and a female come out of the nest of the *Bombus lucorum*, they separated near the entrance of the nest in the earth. The male was unable to fly, I examined him and found evident signs of recent connection.

3rd. The best view I ever had was at Upton Hall, in Northamptonshire, where I saw a male come to a young queen which was busy on a flower; they clung together, and went about 10 yards to another flower, and I watched these for about ten minutes until they separated; the male continued united during their flight, and both used their wings apparently with great ease.

4th. I saw a couple of *B. hortorum* on a flower, and watched them until they separated; they took one flight and separated in about seven minutes from the time I saw them first. All these cases were in the month of August.

5th. The next case was of *B. muscorum*. I watched a male of the *B. muscorum* among some long grass—at first I thought it was coming out or going into its nest, but on going very near I observed the bee alight near a young female of the same species, and after showing signs of kindness they flew off in union, but so rapidly that I could not follow to observe them again; this last was about the first week in September. The *B. muscorum* is the latest in appearing and the latest in nidification. I have seen nests of the *B. muscorum* as late as the last week in August, when the combs contained unhatched drones; and the latter almost always hatched before the young queens.

Wasps.— I have never except once observed the fertilisation of the wasp near Thornbury.[3] On the 7th of September, 1847, I was walking with my bailiff, Mr. Cossham, of Thornbury, when we saw a curious long-looking insect about 20 feet in the air; on its coming very near (it was gradually descending) it proved to be a male and female wasp united. Each of the wasps was trying to fly in a different direction. They fell to the ground, and I immediately crushed them both so as to kill them, but not to mutilate them at all. I had them carefully packed and sent to the

Committee of the Entomological Society, in London, directed to their Secretary; but as ill luck would have it, the box was intercepted by some letter-carrier or other person, and never reached its destination. The queen wasp was more than double the size of the male.—

H. W. Newman, *late Lieut.-Col.* (*Commandant*) *the South Gloucester Militia, Hillside, Cheltenham.*

N.B.— In late seasons such as 1816 and 1860, and in a few springs such as May, 1837–1838, I have noticed nearly one month difference in the hatching of the drones or male Bombi, and I have no hesitation in saying that the fertilising of the young queens, particularly of the *B. muscorum* may be retarded until the second week in September.

Journal of Horticulture, Cottage Gardener, and Country Gentleman n.s. 2 (1861–2): 76–7.

[1] Dated by the publication of the letter in the 22 October 1861 issue of the *Journal of Horticulture*.
[2] See the preceding letter, which was printed immediately before Newman's response.
[3] Thornbury Park, Gloucestershire, had formerly been Newman's home.

To W. E. Darwin 22 October [1861][1]

Down Bromley Kent
Oct 22d.

My dear William

The enclosed letter about the lease concerns you.[2] Of course I do not charge you in my will, interest on the £4500.—[3]

Today Mr John Smith called with message from Mr Fox, that you had promised subscription for prize for Rifle shooting:[4] I did not know what to do, but said I thought as Captain you ought to give rather more than others & asked Mr Fox to settle & I would pay him.—[5] It was dreadful bad day today for the match; as it rained torrents & blew; but our little Boys stood it out & said the men did not care for it. Berens was there.—[6] Down came off triumphant.— Town got first prize scoring eleven: John Fox second prize; & Parslow third of a sovereign.—[7] We are all proud & pleased, & James looked triumphant when he told me.— Mr Smith spoke very strongly, & sincerely I believe, what a loss you would be, as you made such a good Captain.—[8]

I have been looking over my Banking Book & the cheque for 1.1.0 to Mr Fletcher Lutwege has never been drawn at the Bank to this day; & as it was dated on June 10th.; it must either have been lost or Mr Lutwege never wanted the money.—[9] Did he ever acknowledge it? If not, perhaps it would be worth while your writing to him, as he would think you shabby never noticing his letter; but if you do not care, I do not.— I think it would have been presented before this, if not lost or stolen by some one who feared to use it.—

I am alone: Etty went up yesterday & Mamma today to London, for the Concert.—[10] We had jolly Letters from the Boys, & George has suddenly become a first rate correspondent. He is mad to see all your letters; & they want to know whether Southampton is a very dangerous place for garrotting!!—[11]

I hope I shall hear from you tomorrow. I am getting rested after 10 days tremendous work with Mr Sowerby.[12] I have just had a splendid Orchis, which shoots out its pollen-masses like an arrow on to insect.—

We are in despair about Horses & cannot get a pair that will do.—

Good Night | My dear old fellow | C. Darwin

DAR 210.6

[1] Dated by the reference to the London concert and to the illustrations for *Orchids* (see nn. 10 and 12, below).

[2] The enclosure has not been found. The letter was probably from CD's solicitors, Rowland and Hacon, of Fenchurch Street, London (see letter to W. E. Darwin, 12 October [1861]).

[3] Upon joining the Southampton and Hampshire Bank as partner, William was required to deposit £4500 as a guarantee of probity. CD advanced William this amount out of the total sum he was to receive as an inheritance (Down House MS 11: 16 and CD's will).

[4] CD refers to John Smith of Court Farm, Down (*Post Office directory of the six home counties* 1862). Thomas Samuel Fox was ensign for the Rifle Volunteer Corps for Farnborough, Kent, of which William Darwin was captain. The corps had been formed in 1859 (see *Correspondence* vol. 7, letter to John Lubbock, 17 December [1859]). In June 1861, CD had paid £1. 1*s*. 6*d*. as the 'Commission for William Captaincy'. For information about the volunteer movement, see Beckett 1982.

[5] There is an entry in CD's Account book (Down House MS) recording a payment to the 'Rifle Corps' for £1. 10*s*. on 9 November 1861.

[6] Edward Osmond Berens was lieutenant of the Farnborough Rifle Volunteer Corps.

[7] Daniel Town, a carpenter in Down, is listed in the *Post Office directory of the six home counties* for 1862. John Fox may have belonged to the firm Fox & Sons, brewers and maltsters, of Farnborough, Kent. The Darwin's butler, Joseph Parslow, had joined the Rifle Club at its foundation (*Correspondence* vol. 7, letter to John Lubbock, 17 December [1859]).

[8] William Darwin had recently moved to Southampton and was planning to resign his commission in the Farnborough Rifle Volunteer Corps.

[9] There is an entry in CD's Account book on 6 June 1861 that reads: 'Prize for Rifle shooting F. Lutwidge 1/1/'. Charles R. Fletcher Lutwidge was a lieutenant in the Tunbridge Wells corps of the Kent Rifle Volunteers. See also letter to W. E. Darwin, [27 October 1861].

[10] Emma and Henrietta Emma Darwin attended a performance of Jacob Ludwig Felix Mendelssohn-Bartholdy's *Elijah* oratorio at Exeter Hall, London, on Tuesday evening, 22 October 1861 (Emma Darwin's diary). The major attraction of this performance was the return to the London stage of Johanna Maria Lind, also known as Jenny Lind (*Athenæum*, 26 October 1861, p. 548).

[11] Francis and George Howard Darwin, aged thirteen and sixteen, respectively, attended Clapham Grammar School in Clapham, south-west London. According to the *Annual register* (1852), Chronicle, pp. 78–9: 'The crime of robbery by means of suffocation, and known as "garotte", from the Spanish mode of execution, have become exceedingly common'.

[12] George Brettingham Sowerby Jr had been at Down House preparing the illustrations for *Orchids* (see letters to W. E. Darwin, 12 October [1861], and to John Murray, [21 October 1861]).

To T. H. Huxley 22 October [1861][1]

Down. | Bromley. | Kent. S.E.
Oct. 22d

My dear & good Agent.

If Mr Campbell will really work, he would be a treasure beyond price.[2] But I have no faith in anyone having patience enough to experiment on another's man's suggestion. I have written to him to enquire.— When you see H. Spencer, pray thank him.—[3]

I have got my drawings done & can now breathe freely.[4] By the way you audacious agent to doubt about Catasetum: I will turn you off without a character, for do not think I am such a fool as to be grateful for a thousand good acts

Ever yours | C. Darwin

Imperial College of Science, Technology and Medicine Archives (Huxley 5: 177)

[1] The year is provided by CD's reference to having completed the illustrations for *Orchids* (see n. 4, below).

[2] Mr Campbell has not been identified.

[3] Herbert Spencer and Huxley were close friends.

[4] George Brettingham Sowerby Jr had been at Down House preparing the illustrations for *Orchids* (see letters to W. E. Darwin, 12 October [1861], and to John Murray, [21 October 1861]).

To J. D. Hooker 23 October [1861]

Down Bromley Kent
Oct. 23.

My dear Hooker.

I am heartily glad to hear for my own sake, & not for yours, that there is one human being more troublesome than myself.— It is really good of you to tell me that you like making the Garden useful.—

It is *plants* of Hedyotis or allies which must flower & seed with me to be of any use. It is plant of hairy Saxifrage. And plants of vars. & species of Verbascum:[1] why I asked again was because if you could not help me I would begin & enquire at nurseries.—

I enclose list, as that may save you time as a Memorandum.— You will see that *now* I sh^{d} be most glad of any genus of Orchid, which I have not seen.—

I can well believe that the way you are treating the Genera Plantarum, must be *most* interesting.[2] I have looked in Lindley about Gnetum;[3] what a curious form your new one must be,—what a fine living fossil, preserved from past times.[4] I like to hear of embryology revealing its affinities. You will understand why I want so much some Cinchoneæ or perhaps Rubiaceæ when you hear my paper on Primula & Linum.[5]

Would you ask Mr Smith whether he has ever seen Mormodes or Cycnoches eject its pollinia as Catasetum does?[6]

The Box has just arrived & plant quite safe & splendidly packed! I was astounded at the sight of the Box. My dear Hooker, you cannot tell how you have delighted me. I can now try many little experiments.— I shall send plant tomorrow to a neighbour's Hot-house to secure the flowers opening & being vigorous.—[7] I should have hated publishing without examining more than one flower.— Most cordially do I thank you. I look at Catasetum as at the top of the tree; beating even Listera. How on earth we shall ever succeed in packing the plant up to return it, I hardly know.—

Good Night. Ever yours | C. Darwin
Reichenbach on Poll. Orch. not in Linn. or Royal Soc.^{y}.[8]

[Enclosure 1][9]

Rather **Young** buds of Oncidium; not *very* young.

Old flowers of Acropera; & I sh^{d} rather like a bud

Any orchid genus which I have not seen, I give list of genera seen by me

---

Masdevallia: I sh^{d} also like a Dendrobium

Seeds or **plant** which would be best) of Hedyotis & any allied forms, to experiment on when in flower.

Verbascum *vars* (yellow & white) & some hardy species

Plant of Hairy Saxifrage

Seeds of Adlumia (one of the Fumariaceæ)

[Enclosure 2]

Exotic Genera of Orchids, which I have seen

(*Malaxeæ*)
Pleurothallis
Stelis
Microstylis
Dendrobium
(*Epidendreæ*)
Epidendrum
Barkeria
Cattleya
(Vandeæ)
Vanda
Phalænopsis
Sarcanthus
Aerides
Angræcum
Oncidium
Odontoglossum
Miltonia
Stanhopea
(Vandeæ continued)
Lycaste
Catasetum
Mormodes
Cycnoches
Rodriguezia
Calanthe
(Arethuseæ)
OOO
(Cypripedium.)

Endorsement: '/61.'
DAR 115.2: 121, 126a, 124a

1 See the first enclosure. CD was intending to initiate a series of breeding experiments with these plants in the coming growing season (see letter to J. D. Hooker, 18 October [1861]). *Hedyotis* is an Asian genus belonging to the Rubiaceae. CD had heard from Asa Gray that several genera of Rubiaceae exhibited dimorphism, a phenomenon CD had studied in *Primula* and other plants (see letter from Asa Gray, 11 October 1861). In *Saxifraga*, CD was interested in learning 'whether & what of the most hairy species catch flies' (see letter to J. D. Hooker, 19 June [1861]).

2 Hooker and George Bentham were preparing a systematic compilation of all known vascular plant genera, the *Genera plantarum*. The first part of the first volume of Bentham and Hooker 1862–83 appeared early in 1862.

3 Lindley 1853, pp. 232–4.

4 Hooker had recently examined the cones of an unusual plant, found growing in south-west Africa, that had been sent to Kew together with a sketch of the plant. Hooker identified the plant as belonging to the Gnetaceae and named it *Welwitschia* after its discoverer, Friedrich Welwitsch. In Hooker 1863, pp. 3–4, Hooker described this new genus, stating that its discovery was:

> the most wonderful, in a botanical point of view, that has been brought to light during the present century; for an attentive study of the structure of its vascular system, as well as of its reproductive organs, and of the evidences we have of its functional peculiarities and mode of development, will disclose in all these points very singular anomalies, which even appear in some instances subversive of theoretical axioms hitherto considered as fundamental in Botany.

5 CD read a paper on dimorphism in *Primula* to the Linnean Society of London on 21 November 1861 (see *Collected papers* 2: 45–63). In this paper, he briefly mentioned the results of his study of the two forms of *Linum*. Two years later, on 5 February 1863, he gave a fuller account of dimorphism in *Linum* (see *ibid*., pp. 95–105).

6 John Smith was curator at the Royal Botanic Gardens, Kew. CD described the genus *Catasetum* as 'the most remarkable of all Orchids' (*Orchids*, p. 211). Exclusively a male form, *Catasetum* requires mechanical aid for its pollen-masses to be transported to female plants. According to CD's account, nature 'has endowed these plants with, what must be called for want of a better term, sensitiveness, and with the remarkable power of forcibly ejecting their pollinia to a distance. Hence, when certain definite points of the flower are touched by an insect, the pollinia are shot out like an arrow' and attach to the insect, which effects pollination when it subsequently visits a female plant (*Orchids*, pp. 212–13).

7 In *Orchids*, p. 158 n., CD thanked his neighbour, George Henry Turnbull of The Rookery, Down, for allowing him to use his glasshouse.

8 CD had asked to borrow Reichenbach 1852 from Hooker (see letters to J. D. Hooker, 15 [October 1861] and 18 October [1861], and to John Lindley, 25 October [1861]).

9 On the back of this enclosure, Hooker made a brief list of CD's requests.

To Charles Lyell 23 [October 1861]

Down
23d.

My dear Lyell

I ought to have returned your note by this morning's post, but I was too busy dissecting.—[1] I suppose everyone takes your view that the water flowed out at head of valley when the lake existed.[2] The "intermediate shelf" in Glen Roy, seemed to me on careful examination as plain as any shelf whatever; but I think I remember Mr. J. did not think so; it has been noticed by everyone who has visited the valley.[3] There is no outlet known corresponding with this shelf; but as Mr Milne says there may be;[4] & the whole valley ought to be searched for lateral outlets between the two upper shelves.— A man might spend his life there. I hope M[r]. J. will go there again; for it is an opprobrium to British Geologists, that it sh[d]. not be settled beyond dispute.[5]

My difficulty is the sloping mass of matter, stratified & thick, at bottom of valley below lowest horizontal shelf: I think the river must have delivered detritus at infinitely many levels, by opening on a lake or arm of sea.— If there was terminal moraine at mouth of Spean to be *slowly* cut through all would be explained: I can hardly think ice would suffice. But if it were the sea, I cannot help a sneaking hope that the sea might have formed the horizontal shelves.—

Ever yours | C. Darwin

Endorsement: 'Oct[r] 1861'
American Philosophical Society (269)

[1] Lyell's letter has not been found.

[2] One set of phenomena critical for those who speculated about the formation of the so-called parallel roads of Glen Roy, in Lochaber, Scotland, was the existence of 'cols' at the head of glens, coincident in level with the various shelves. For those who viewed the shelves as the shores of former glacial lakes, these cols were watersheds or over-flow points, where the rain-water drained from the lakes. For CD, however, who held that the shelves were the remnants of marine beaches formed when an arm of the sea was present in the Lochaber glens, these areas had been 'land-straits' between former islands (see 'Observations on the parallel roads of Glen Roy, and of other parts of Lochaber in Scotland, with an attempt to prove that they are of marine origin', *Philosophical Transactions of the Royal Society of London* (1839): pt 1, 39–81; *Collected papers* 1: 89–137). When Lyell came to write up his account of the parallel roads of Glen Roy, he retracted his previous support for CD's theory, given in C. Lyell 1841, and adopted the glacier-lake explanation offered by Thomas Francis Jamieson. According to Lyell, these cols did indeed appear to be watersheds 'such as would have been formed if the waters of the rivers had been made to flow east, or out of the upper ends of the supposed glacier-lakes, instead of escaping at the lower ends, in a westerly direction, where the great blockages of ice are assumed to have occurred.' (C. Lyell 1863, pp. 259–60).

[3] The existence of a shelf immediately between the two upper shelves in Glen Roy was an observation that favoured CD's theory, for, as he states in the letter, no 'col' had ever been found coincident in level with this intermediate shelf. What CD says is not, in fact, accurate. As he himself pointed out in his 1839 paper, Thomas Dick Lauder did not note the existence of this intermediate shelf (see Lauder 1823 and *Collected papers* 1: 112).

[4] Milne 1847a, 1847b, and 1849.

[5] Jamieson made two geological excursions to Lochaber, first in August 1861 and again in July 1862 (see letter from T. F. Jamieson, 3 September 1861, and Jamieson 1863, p. 240).

From T. F. Jamieson 24 October 1861

Ellon Aberdeens[h].
24 Oct 1861

My Dear Sir,

It would be unpardonable in me not to thank you for the valuable papers &c on Glen Roy which you have been so good as send me—[1]

I found the locality so interesting & instructive that I hope to be able to pay another visit to it.[2] I saw Loch Spey from a distance standing on the top of Tom Brahn,[3] or rather endeavouring to stand, for it was a gale like to blow the teeth out of one's head— Possibly the high lying terraces or buttresses you refer to, may be explained by glacial action, for it seems, from Dr. Hooker's account, that similar terraces abound in the higher Himalayan valleys, wh. he refers to that agency— Lakes, he says, form between the *sides* of glaciers & the flanking hill occasionally—[4]

—The great sloping terraces at the mouth of Glen Spean I did not get examined, & am therefore unable to say much about, but if the ice-dam in that direction consisted of a glacier stream occupying the line of the Caledonian canal, great accumulations might be pushed into the dammed lake by short glaciers descending from the little hollows of Ben Nevis &c. & the hill streams of ditto

One of the greatest difficulties in the way of the glacier theory seems to me to arise from the supposition of the pressure of a column of water say 600 feet or so deep or more, being retained by an ice barrier. the glacier lakes of Switzerland seem to be retained chiefly by moraine matter, but in one case by ice alone— in the Himalaya near Karakorum pass also a large glacier lake—the largest I have heard of—seems to have been dammed by ice alone[5]

I am | My Dear Sir | Yours very faithfully | Thos. F. Jamieson

C. Darwin Esq

Edinburgh University Library (Lyell 1: 2828)

1 See letter to Charles Lyell, 20 October [1861]. Jamieson acknowledged CD's assistance in his published account of the parallel 'roads' of Glen Roy (Jamieson 1863, p. 240). See also letter from T. F. Jamieson, 13 June 1861, letter to Robert Chambers, 30 April [1861], and Appendix IX.

2 Jamieson made two visits to Lochaber, the first in August 1861 and the second in July 1862 (see Jamieson 1863, p. 240).

3 The hill Tom Brahn (or Tombrahn) overlooks the head of upper Glen Roy (see *Collected papers* 1: 87 and the map in this volume, p. 248).

4 Hooker 1854a, 1: 242–4; 2: 116–21. See also letter to J. D. Hooker, 28 September [1861].

5 Jamieson discussed the objections to postulating a former ice-dam in Glen Roy, and mentioned the glacier lakes of the Himalayas in his paper (Jamieson 1863, pp. 257–8).

To W. B. Clarke 25 October [1861][1]

Down. | *Bromley.* | *Kent. S.E.*
Oct. 25

My dear Sir

I thank you cordially for your very kind expressions towards me & for your letter which has deeply interested me.[2] I had not heard of your account of Glacial action in Australia:[3] had I heard of it, I should have alluded to the case in different terms.[4] But I saw an account in a short paper by a man, whose name I had never heard.[5] I have the reference somewhere, but I could not find it without a long search. Your name has of course been familiar to me for years.[6] When I come to the subject again, I will refer to your account, for the subject interests me extremely.[7] Indeed but a few months ago I wrote to Mr Gould in Tasmania to enquire.[8] There are great difficulties in believing in a mundane cooler period; but it would throw a flood of light on Geographical Distribution. I have only just sketched the subject in the Origin & have a long M.S. discussion drawn out, but Heaven knows whether I shall ever live to publish half my M.S.—[9] I should very much like to see the Granite striated. Also would you be so very kind as to take the trouble to tell me whether you refer to the penal settlement, Norfolk Is[d], where there were so many greenstone boulders, & are you sure that there was no greenstone close by?— Were the boulders angular?[10] This would be a very extraordinary case.— No subject interests me more than the Glacial period.

Many thanks for the Photograph which has pleased me much; for it has vividly recalled to my mind the views from the Blue Mountains—[11]

I had heard a little about New Zealand; & from what I saw there I always expected old rocks to be there discovered: in my own mind I excluded N. Zealand from the term oceanic Isl[d], as being connected by a Bank; but I ought to have been more guarded: I looked at it as at Madagascar & the Falkland Islands. Bear Is[d] is a real exception.— I am very glad to hear of your new discoveries of Secondary fossils in N. S. Wales.[12] I have for some time thought, that the geology of distant countries would help on the progress of the Science more than anything else; & in this, you have been an earnest worker.— Most cordially do I wish you all success.—

When you next see my old ship-mate M[r] Martens pray remember me very kindly to him;[13] I have two of his sketches now hanging up in one of my rooms.—[14] You have attended to so many branches of Nat. History, that I daresay you are a Botanist. There is one little experiment, which would be of great use to me, if you would be so very kind as to try it; namely to cover up any species of the Goodeniaceæ under a net, just before the flowers open, so as to prevent any Bee or other insect visiting it, & observe whether it sets seeds as well as an unprotected plant. I have made experiments on plants of this Order in my greenhouse, & I am very curious on this head.— Mr Drummond writes to me from Swan River that he has seen Bees open the indusium & extract pollen.—[15] It is unreasonable to even hint it but I am also very curious to know whether Eucalyptus & Mimosa seed well when insects are excluded. It is a subject which I have worked at much.— I wrote to D[r]. Müller, but he seems to be too busy to help me.—[16] I know that you are very busy, but then I fancy, that you have indomitable energy.—

You will see a short discussion on dioicous Trees in Origin, which will show *one* reason, why I want to know whether they are not crossed naturally by insects, & thus rendered more fertile.—[17] I am now preparing for press a little Book on this subject on Orchids. But I must apologise for scribbling on about my hobby-horses, & remain with much respect—

Yours very sincerely | C. Darwin

P.S. | Can you tell me whether the introduction of the Hive-bee has lessened the numbers of a small native-bee? Do you know of any papers on the effect of introduced animals or plants on the native productions.— This seems to me a most interesting subject generally neglected.

Mitchell Library (MSS 139) 36X, pp. 263–72

[1] Dated by reference to the publication of *Orchids*.
[2] Letter from W. B. Clarke, [August 1861].
[3] Clarke 1860.
[4] CD refers to his discussion of geographical distribution during the glacial period in *Origin*, pp. 365–82.
[5] The reference is to CD's statement about the evidence of glacial action in Australia (*Origin*, p. 373): 'If one account which has been published can be trusted, we have direct evidence of glacial action in the south-eastern corner of Australia.'
[6] Clarke had been a student at Jesus College, Cambridge, and had studied geology with Adam Sedgwick. He became a fellow of the Geological Society in 1826 (*List of the Geological Society of London* 1864). Clarke frequently published memoirs in geological journals and corresponded regularly with many of CD's acquaintances (see Jervis 1945).
[7] In *Origin* 4th ed., p. 443, CD cited Clarke as the authority for his statement that there appeared to be 'clear traces of former glacial action on the mountains of the south-eastern corner of Australia.'
[8] Charles Gould, son of the ornithologist John Gould, had emigrated to Tasmania in 1859 and was a surveyor to the Tasmanian Geological Survey (*Aust. dict. biog.* and Tree 1991, p. 209).
[9] The reference is to CD's 'big book' on species (see *Natural selection*, pp. 528–66).
[10] See letter from W. B. Clarke, [August 1861]. CD believed that the occurrence of angular boulders, at some distance from their original location provided evidence for transportal by icebergs rather than by glaciers (see *Collected papers* 1: 163–71).
[11] CD visited the Blue Mountains of New South Wales in 1836 during the voyage of the *Beagle* (*Journal of researches*, pp. 436–42).
[12] On 20 November 1861, Clarke read a paper before the Philosophical Society of New South Wales on recent geological discoveries in Australasia. He cited CD's statement in *Origin*, p. 308, that 'not one oceanic island is as yet known to afford even a remnant of any palæozoic or secondary formation', stating that CD was in error in applying this to New Zealand (Clarke 1861, p. 6). In the fourth edition of *Origin*, CD included a caveat concerning New Zealand (*Origin* 4th ed., pp. 372–3). Clarke sent CD a copy of his paper in 1862 (see *Correspondence* vol. 10, letter from W. B. Clarke, 16 January 1862), which is in the Darwin Pamphlet Collection–CUL.
[13] Conrad Martens, draughtsman on the *Beagle* from December 1833 until August 1834, settled in Australia in 1835 (see Keynes 1979, pp. 2–3). Martens had been Clarke's first church warden after his appointment in 1846 as rector of St Thomas's Church, St Leonards, New South Wales (see letter from W. B. Clarke, [August 1861], and Jervis 1945).
[14] CD purchased two of Martens's water-colours in January 1836 when the *Beagle* visited Sydney (see *Correspondence* vol. 1, letter to Susan Darwin, 28 January 1836). The water-colours are reproduced in Keynes 1979: No. 150, 'The *Beagle* in Murray Narrow, Beagle Channel' (pp. 116, 395), and No. 193, 'River Santa Cruz' (pp. 201, 397).
[15] See *Correspondence* vol. 8, letter from James Drummond, 8 October 1860. See also letter to *Journal of Horticulture*, [17 May 1861].
[16] CD's correspondence on this point with Ferdinand Jakob Heinrich von Mueller, director of the Botanic Gardens in Melbourne, Australia, has not been found.
[17] See *Origin*, p. 100.

To John Lindley 25 October [1861][1]

Down. | *Bromley.* | *Kent. S.E.*
Oct. 25th

My dear Lindley

I am infinitely obliged for your letter, answering everything so clearly & so fully.—[2]

Do **not** read enclosed without you like, I send it on mere *chance* of being new to you.—

I sh^d. be most grateful for *any* orchids (especially Dendrobium) in flower & **bud**: if it be a small parcel *tin cannister* best way by *Post*, & you must let me repay postage: if too large, address.

C. Darwin
care of Down Postman
Bromley
per Railway **Kent**

I terribly want any one of your "Arethuseæ."[3]

Mr James Veitch has been *most* generous;[4] I did not know that you had spoken to him: if you see him pray say I am truly grateful. I dare not write to a live Bishop or a Lady;[5] but if I knew the address of "Rucker"?? & might use your name as introduction I would write.[6] I am half mad on subject.— Hooker has sent me many exotics; but I stopped him, for I thought I sh^d. make a fool of myself; but since I have determined to publish, I much regret it.

I will not trouble you for the Books, for systematic works are out of my line.—

Hooker is trying to buy Reichenbach on the pollen & if he gets it will lend me, for it is in no public Library & I do not suppose would be worth my buying.[7]

With cordial thanks for your great kindness.—

Yours very sincerely | Ch. Darwin

Royal Botanic Gardens, Kew (Lindley letters, A–K: 194)

1 The year is given by the reference to Reichenbach 1852 (see n. 7, below).
2 The letter from Lindley has not been found. It was presumably a response to the letter to John Lindley, 18 October [1861].
3 Of the seven tribes of orchids described in Lindley 1853, the Arethuseae was the only tribe members of which CD was unable to examine.
4 James Veitch, a Chelsea nurseryman, supplied CD with many species of exotic orchids.
5 See letter to Dorothy Fanny Nevill, 12 November 1861.
6 Sigismund Rucker, West Hill, Wandsworth, is acknowledged several times in *Orchids* for having 'repeatedly' sent CD valuable orchids (see especially *Orchids*, p. 158 n.).
7 CD refers to Reichenbach 1852 (see letters to J. D. Hooker, 15 [October 1861], 18 October [1861], and 23 October [1861]).

To Lionel Smith Beale[1] 26 October [1861][2]

Down. | Bromley. | Kent. S.E.
Oct. 26th

Dear Sir

Should the Subscribers determine to make a present to Mrs Queckett,[3] I should have great pleasure in putting my name down for five Guineas:[4] but I am sorry to say that I have, I suppose peculiar opinions in regard to Testimonials, & if the Subscribers determine on one, I must decline subscribing, as although I have a sincere respect for Prof. Queckett, in my judgment his scientific claims are not of that *eminent* merit deserving a public Testimonial. I am heartily sorry to differ from so many better judges than myself, but each man must act on his own judgment.— Under these circumstances it will not of course do to put my name down on the committee.[5]

Dear Sir | Yours faithfully | Charles Darwin

Scripps College Library

[1] The recipient is identified from an entry in CD's Account book (see n. 4, below).

[2] The year is provided by reference to the death of John Thomas Quekett (see n. 3, below).

[3] Isabella Mary Anne Quekett was the widow of John Thomas Quekett, professor of histology and conservator of the Hunterian Museum of the Royal College of Surgeons. Quekett had died, aged forty-six, on 20 August 1861 (*DNB*).

[4] There is an entry in CD's Account book (Down House MS) on 13 January 1862 that reads: 'Dr Beale Queckett Testimonial 5/5'. Beale, professor of physiology and general and morbid anatomy at King's College, London, was, like Quekett, a noted microscopist.

[5] Beale was the honorary secretary of a working committee of the Royal College of Surgeons formed to enlist subscriptions to provide an educational endowment for Quekett's four sons (see *Quarterly Journal of Microscopical Science* n.s. 2 (1862): 63).

To W. E. Darwin [27 October 1861][1]

Down Bromley Kent
Sunday

My dear William

Your jolly letter has delighted us.—[2] The irons went some time since to Mr Wood.[3] I gave your message to Parslow about the Prize. Town did not win any prize at Sevenoaks, but was high.—[4] Tell me whether you mean to write to Mr Lutwege Fletcher.—[5] Lord Hardinge is trying to unite Down & the Crays corps; & in this case you will be able to resign captaincy.[6] I enclose Sir H. James: he is worth knowing; but he ought to have offered to call on you; as it is, you must call on him.—[7]

Mamma came home last night with bad headach but she is brisk today: she left Etty very brisk in London.—[8] Aunt Catherine is here.—[9] Mamma thinks 11^d^ cheap for meat— Your House is an awful sum.—

I do not quite understand about Mr A. Mr H. you & rent of Bank.—[10] I have asked M^r^. Hacon about your making a will.—

I sh^d^. think the Deed had of course be better kept in your iron-chest.— M^r^. A. seems rather bitter against Mr Hacon.

In Haste | Your affect | C. Darwin

DAR 210.6

[1] Dated by the references to Emma and Henrietta Emma Darwin's trip to London and to Emily Catherine Darwin's visit to Down House (see nn. 8 and 9, below).

[2] William's letter has not been found.

[3] George Wood was a farmer in Down (*Post Office directory of the six home counties* 1862).

[4] Joseph Parslow, butler at Down House, had won third prize at a Rifle Volunteer Corps musketry event held at Down on 22 October 1861 (see letter to W. E. Darwin, 22 October [1861]). Daniel Town was a carpenter in Down, (*Post Office directory of the six home counties* 1862). Sevenoaks is a town eight miles south-east of Down.

[5] Charles R. Fletcher Lutwidge. See letter to W. E. Darwin, 22 October [1861] and n. 9.

[6] See letter to W. E. Darwin, 22 October [1861]. The 'Crays' are three small villages to the north of Down. Charles Stewart Hardinge was an officer in the Westerham corps of the Kent Rifle Volunteers. The small volunteer units in this area of Kent were eventually amalgamated to form the 1st Volunteer Battalion of the Royal West Kent regiment. See Beckett 1982.

[7] CD had written to the surveyor Henry James of Southampton introducing William (see letter to W. E. Darwin, 12 October [1861]).

[8] Emma and Henrietta Emma Darwin had gone to London on 21 October 1861 and Emma returned on 26 October (Emma Darwin's diary).

[9] Emily Catherine Darwin, CD's younger sister, stayed at Down House from 25 to 28 October 1861 (Emma Darwin's diary).

[10] William Darwin had taken lodgings at 1 Carlton Place, Southampton. CD refers to George Atherley, William's partner in the Southampton and Hampshire Bank, 25 High Street, Southampton, and to William Mackmurdo Hacon, CD's solicitor. Hacon had represented the Darwins in the legal negotiations preceding the agreement on the terms of William's contract.

To J. D. Hooker 27 October [1861]

Down Bromley Kent
Oct 27.

My dear Hooker

Acropera is a beast—stigma does not open; everything seems contrived that it shall *not* be anyhow fertilised. There is something very odd about it, which could only be made out by incessant watching on several individual plants.[1]

I never saw the very curious flower of Canna; I sh[d] say pollen was deposited where it is, to prevent inevitable self-fertilisation: you have no time to try smallest experiment, else it w[d]. be worth while to put pollen on some stigmas, (supposing that it does not seed freely with you).— Anyhow insects would probably carry pollen from flower to flower, for Kurr states that tube formed by pistil stamen & "nektar-blatt" secretes (I presume internally) much nectar;[2] thanks for sending me this curious flower.

Now I want much some wisdom; though I must write at considerable length, your answer may be *very* brief. (You did me a wonderful good turn in telling me to trace down the vessels in Malaxis.—)[3] In R. Brown's admirable paper in Linn. Transacts,[4] he suggests (& Lindley cautiously agrees) that flower of Orchids consists of 5 whorls, the inner whorl of the two whorls of anthers being all rudimentary, & when the Labellum presents ridges, 2 or 3 of the anthers of both whorls being combined with it.[5] In the ovarium there are six bundles of vessels. R. Brown judged by transverse sections. It occurred to me after what you said to trace the vessels longitudinally & I have succeeded well.[6]

Look at my diagram (which please return for I am transported with admiration at it), which shows the vessels which I have traced, one bundle to each of 15 theoretical organs & no more.[7] You will see the result is nothing new, but it seems to confirm strongly R. Brown, for I have succeeded (perhaps he did, but he does not say so) in tracing the vessels belonging to each organ in front of each other to the same bundle in the ovarium: thus, the vessels going to lower sepal, to side of Labellum, & to one stigma (when there are two) all distinctly branch from one ovarian bundle. So in other cases, but I have not completely *traced* (only seen) that going to Rostellum.— But here comes my only point of novelty; in all orchids, as yet looked at, (even one with so simple a labellum as in Gymnadenia & Malaxis) the vessels on the two sides of the labellum are derived from the bundle which goes to lower sepal, as in Diagram. This leads me to conclude that Labellum is *always* a compound organ. Now I want to know, whether it is conceivable that the vessels coming from one main bundle should penetrate an organ, (the labellum) which receives its vessels from another main bundle? does it not imply that all that part of the labellum which is supplied by vessels coming from a lateral bundle must be part of a primordially distinct organ, however closely the two may have become united? It is curious in Gymnadenia to trace the *middle* anterior bundle in the ovarium, when it comes to orifice of nectary it turns & runs right down it, then comes up the opposite side & runs to apex of Labellum, whereas each side of nectary is supplied by vessels coming from the bundles coming from the lower sepals: hence even the thin nectary is essentially, I infer, tripartite; hence its tendency to bifurcation at its tip. This view of the Labellum *always* consisting of 3 organs (I believe 4, when thick as in Mormodes at base) seems to me to explain its great size; & tripartite form, compared with the other petals. Certainly if I may trust vessels, the simple Labellum of Gymnadenia consists of three organs soldered together.—

Forgive me for writing at such length; a very brief answer will suffice.

I am desperately interested in subject; the destiny of whole human race is as nothing to the course of vessels of Orchids.—

Your insane friend | C. Darwin

What plant has most complex single stigma & pistil: the most complex, I, in my ignorance, can think of is in Iris: I want to know whether anything beats in modification the rostellum of Catasetum.[8] Tomorrow I mean to be at Catasetum— Hurrah. What species is it? it is wonderfully different from that which Veitch sent me—which was C. saccatum: a gardener here thinks it is Cat. *globigerum*?[9] According to the vessels an Orchid-flower consists of 3 sepals & 2 petals free; & of a compound organ (the labellum) consisting of one petal & of 2 (or 3) modified anthers; & of a second compound body consisting of 3 pistils, one normal anther & 2 modified anthers often forming the sides of the clinandrum.—[10]

Endorsement: '/61.'
DAR 115.2: 122

[1] See letters to J. D. Hooker, 15 [October 1861] and 18 October [1861].

[2] Kurr 1833, p. 26. *Canna* is briefly mentioned in *Orchids* in the discussion of the pollen-masses of orchids. A member of a family allied to orchids, *Canna* is fertilised, according to CD, by its pollen being 'deposited on the pistil, close beneath the stigma' (*Orchids*, p. 324).

3 Hooker's letter has not been found, but see the letters to J. D. Hooker, 4 October [1861] and 13 October [1861].

4 Brown 1831, pp. 696–701. There is a copy of the paper in the Darwin Pamphlet Collection–CUL. Brown published a preliminary description of the structure of orchid flowers in Brown 1810, 1: 310.

5 Lindley 1853, pp. 183–3a. John Lindley illustrated his description with a diagram that CD modified for inclusion in *Orchids* (see *Orchids*, p. 292).

6 See letters to J. D. Hooker, 4 October [1861] and 6–7 October [1861]. In *Orchids*, pp. 290–1, CD explained the importance of the spiral vessels in making out the homologies of orchids:

> As spiral vessels are developed at a very early period of growth, which always gives much value to an organ in making out homologies; and as they are apparently of high functional importance, ... it appeared to me, guided also by the advice of Dr. Hooker, to be worth while to trace upwards all the spiral vessels from the six groups surrounding the ovarium.

7 The diagram has not been found. The one associated with this letter (dated 22 October [1861] by Francis Darwin) in *ML* 2: 274 was probably enclosed with the letter to J. D. Hooker, 8 November [1861]. The lost diagram probably resembled that included in the letter to J. D. Hooker, 10 November [1861].

8 CD awarded *Catasetum*, with its 'wonderfully modified' sensitive rostellum, 'the palm of victory' for perhaps being the 'highest' form of orchid (see *Orchids*, pp. 333–5).

9 In *Orchids*, p. 214 n., CD thanked James Veitch, the Chelsea nurseryman, for supplying him with his first specimen of *Catasetum saccatum*. The gardener CD refers to may be Mr Horwood, the gardener of CD's neighbour George Turnbull. CD thanked Horwood in *Orchids*, p. 158 n., for helping him with some of his observations. *C. globiflorum* is not discussed in *Orchids*.

10 See letter to J. D. Hooker, 10 November [1861], and *Orchids*, pp. 291–4.

To T. H. Huxley 31 October [1861][1]

Down. | *Bromley.* | *Kent. S.E.*
Oct. 31st

My dear Huxley

In a little Book, just published, called the Three Barriers (a theological hash of old abuse of me) Owen gives to the Author a new resume of his Brain doctrine;[2] & I thought you would like to hear of this. He ends with a delightful sentence. "No science affords more scope or easier ground for the caviller & controversialist; & these do good by preventing scholars from giving more force to generalisations than the master propounding them does, or meant his readers or hearers to give".—[3]

You will blush with pleasure to hear that you are of some use to the Master.

Ever yours | C. Darwin

I have had more Catasetums, all right you audacious "caviller".—[4]

Mr. Campbell vows he will work; if he will, he shall have lots—[5]

Imperial College of Science, Technology and Medicine Archives (Huxley 5: 194)

1 The year is given by CD's reference to the publication of *The three barriers* (see n. 2, below).

2 The anonymous work *The three barriers: notes on Mr. Darwin's "Origin of Species"* was published in August 1861 (*Publishers' Circular* 1861). The author was the Scottish Presbyterian clergyman Gilbert Rorison (*NUC*). A review of this work by Samuel Pickworth Woodward, which appeared in the *Critic*, 30 November 1861, is in CD's 'Scrapbook of Reviews' (DAR 226.1); Woodward incorrectly identified the author as 'George' Rorison. Huxley and Richard Owen were engaged in a public dispute over

the question of the zoological classification of man. In particular, they disagreed about whether or not there was an anatomical difference between the brain of humans and that of the higher apes. Owen argued that the human brain was uniquely distinguished by virtue of the presence of the hippocampus minor; Huxley maintained that this structure was also present in other members of the Quadrumana. See also letters to T. H. Huxley, 1 April [1861], and to J. D. Hooker, 23 [April 1861].

[3] [Rorison] 1861, p. 163. The discussion to which CD refers is in an appendix entitled 'On the human and brute brains', which presents, 'with Professor Owen's kind permission', Owen's criticism of Huxley's article 'On the zoological relations of man with the lower animals' (T. H. Huxley 1861a).

[4] See letter to T. H. Huxley, 22 October [1861].

[5] See letter to T. H. Huxley, 22 October [1861].

To J. D. Hooker 1 November [1861]

Down. | Bromley. | Kent. S.E.
Nov 1st

My dear Hooker

One line to **beg** you to look at Dendrobium macrophyllum(?) (this was sent me as *D. chrysanthum*) & tell me whether the Labellum is at all on *upper* or *lower* side of column *with respect to the ground*.—[1] Labellum embraces column, of course I mean place of attachment.[2] This point is of highest importance with respect to mechanism of flower.— I have been writing to Lindley on subject;[3] if you can do pray observe this—

If Bolbophyllum in fine weather has Labellum a little apart from column & when touched closes up against Column, I can well understand use of movement; could you ascertain this.—[4]

It is no use attempting to thank you for all your extraordinary kindness.

Post going | C. D.

I have looked at the Dendrobium & the contrivance is most remarkable & well deserves proving.—

I have no Cypripedium in spirits for Ducts & it would be I fear very difficult.—

Endorsement: '/61.'
DAR 115.2: 123

[1] *Dendrobium macrophyllum* is not mentioned in *Orchids*, but *D. chrysanthum* is discussed (*Orchids*, pp. 172–8). CD thanked Sigismund Rucker for sending him specimens of *Dendrobium* in *ibid*., p. 158 n.

[2] This sentence was written on the page opposite the letter text, with a line pointing to its proper location in the text.

[3] See letter to John Lindley, 1 November [1861].

[4] CD examined four species of *Bolbophyllum*, which are briefly described in *Orchids*, pp. 169–72. He was particularly interested in whether or not the labellum of *Bolbophyllum* was irritable (*ibid*., p. 172).

To J. D. Hooker 1 November [1861]

Down.
Nov. 1.

My dear Hooker

You may believe how deeply your letter has interested me. How very very kind you are, busy as you are, to give up so much time to me.—[1]

Since I wrote I have traced bundles to superior (so called) sepal, anther & rostellum all to one bundle most clearly; & in a confluent stigma two bundles to 2 lower sepals; so homologies of this part now clear & certain. I am uneasy at not understanding why R. Brown did not trace the vessels longitudinally; he clearly expresses this view of Rostellum & stigma, but gives no evidence.—[2]

I have examined several other very simple Labellums, & the vessels *always* run as stated: Anthers with pollen have been seen on Labellum of Bee Ophrys; I will try at a thick Labellum for the missing bundle, but doubt success; it is so difficult when object opake I must take thick longitudinal slice & soak in nitric acid. Many of your hints will be most useful: I will study your letter over & over again.—

If you really can spare another Catasetum, when nearly ready, I shall be most grateful: had I not better send for it? The case is truly marvellous; the (so-called) sensation or stimulus from a light touch is certainly transmitted through the antennæ for more than 1 inch, through at least 73 cells, *instantaneously*; & a membrane is ruptured (as in Listera) & double elasticity comes into play & pollinia are shot out with sticky gland always foremost.[3] A cursed insect or something let my last flower off last night!— Another very young spike is coming forward. But there are many trials which I should wish to make. It is singular that left-hand antenna *always* guards a distinct part of Labellum from the right-hand antenna.

Many of the orchids sent today are extremely valuable to me.— I would have given 5£ for the Dendrobium macrophyllum.— I *think* I understand partly Acropera: the pollinia when exposed for short time to air shrink *much* in thickness & can then be inserted into stigma; on insect flying they would rapidly be dried: I should rob your hot-house; but 2 or 3 more flowers would settle this enigma: *Is labellum uppermost*? I *think* I partly understand how rostellum must be touched to withdraw the pollinia—[4]

I shall be very curious to hear how Labellum of Bolbophyllum moves.[5] Is not Barberry & Cactus opposed to compound organs being irritable? Lateral lobes of Labellum of Listera cordata seemed covered with glands; but I did not see interest & did not observe carefully: *[*corners*]* in Spiranthes glandular.

Jamieson makes great use & thinks most highly of your Glacial-Himalayan work.—[6]

If you do really send me another Catasetum could I keep it in my greenhouse by keeping up fire night & day: we can raise it to 70°— The gardener who keeps present plant, says he has formerly found that Catasetum does not require very great Heat.—[7]

I have not seen Bonatea, but in drawings Labellum looks almost divided. Never mind about Hedyotis: Asa Gray is going to send me living *plants* of one of the same Rubiaceous group:[8] he seems wonderfully interested in the Primula case. Rhamnus lanceolatus seems another case. Can you tell me best nursery to apply to for rare trees as this Rhamnus or plants as Verbascum.

Believe me I appreciate your wonderful kindness—

Yours affectionately | C. Darwin

I have lately read heaps of letters between Mr Jamieson & Lyell;[9] & I 〈 〉

If Labellum of Bulbophyllum closes up to column when touched (in middle?) & you can tell me what angle it makes with column before being touched, I will have drawing made.—

P.S. I send Catasetum to be named.— I think I have made out mechanism of Dendrobium macrophyllum, in whatever position flower stands, unless it hangs *almost* perpendicularly downwards, with Labellum downwards. It is a frightful evil that I cannot see my way till I have examined so many specimens.

Endorsement: '/61'
Incomplete
DAR 115.2: 124

[1] Hooker's letter has not been found. It was a response to the letter to J. D. Hooker, 27 October [1861].
[2] See letter to J. D. Hooker, 27 October [1861]. The reference is to Brown 1831.
[3] For CD's description of the results of his dissection of *Catasetum*, see *Orchids*, pp. 321–3.
[4] See letters to J. D. Hooker, 15 [October 1861], 18 October [1861], and 27 October [1861]. *Acropera*, the 'opprobrium' of CD's work owing to his difficulty in working out its mechanism of pollination, is discussed in *Orchids*, pp. 203–10.
[5] See preceding letter.
[6] Hooker 1854a, 1: 242–4; 2: 116–21. See also letter to J. D. Hooker, 28 September [1861], and letter from T. F. Jamieson, 24 October 1861.
[7] The reference is to Mr Horwood, gardener to CD's neighbour George Turnbull.
[8] See letter to J. D. Hooker, 23 October [1861], and letter from Asa Gray, 11 October 1861.
[9] See the series of letters to and from Thomas Francis Jamieson, and those to and from Charles Lyell, in September and October 1861. See also Appendix IX.

To John Lindley 1 November [1861][1]

Down. | Bromley. | Kent. S.E.
Nov. 1st

My dear Lindley

Will you allow me once again to trespass on your kindness? I send in little Box by this post with a flower of a Dendrobium sent me from Kew as D. macrophyllum;[2] but it was previously sent me by an orchid grower as D. chrysanthum & is the one I asked you about. Can you tell me which it is?—

Since I wrote I have been working hard in tracing spiral ducts; & I find in *all* cases that the sides of the Labellum receive all their ducts from the two antero-lateral bundles of the ovarium, which bundles supply the 2 lower sepals & 2 stigmas (the rostellum being supplied from the posterior bundle which supplies the fertile anther & upper sepal)

You & Brown believe the Labellum is compound when it presents ridges &c;[3] but Hooker tells me he thinks I am right in inferring from the above facts that the Labellum is always an organ compounded of lower petal & 2 anthers, of outer whorl, with doubtful traces of one anther, of inner whorl.

Now this is a long preface to asking you whether the Orchid (name forgotten) which has Labellum quite similar to the two upper petals be very rare; & if not & you can spare *a single* dried flower; for I am very curious to try & trace its ducts or spiral vessels. I believe you somewhere speak of Orchid with medial spike or projection on Labellum; this would give me a good chance of tracing the one bundle of ducts which out of the 15 I have failed to discover clearly & trace.—

Forgive me, if you can, for being so troublesome & believe me

Your's sincerely obliged | Charles Darwin

Royal Botanic Gardens, Kew (Lindley letters, A–K: 195)

[1] The year is provided by reference to CD's work on orchids.

[2] See the first of the two letters to J. D. Hooker, 1 November [1861].

[3] Lindley described his views on the structure of orchids and his growing disagreement with Robert Brown's descriptions (Brown 1831) in Lindley 1853, pp. 174–5, 183–3b. See also letters to J. D. Hooker, 27 October [1861] and 1 November [1861].

To W. E. Darwin 4 November [1861][1]

Down Kent
Nov. 4th.

My dear William

I enclose Mr Hacon's note about your will;[2] you will see that it is of no importance making one at present; but of course you can do as you like. Whenever you do make a will always consult a Lawyer.— My will, as you know is equal partition; up to about £15,000, & if there should be more, all excess to be divided equally amongst the sons alone.—

I enclose a nice note from Franky.[3]

Mamma goes tomorrow to Hastings for 4 or 5 days to see Aunt Charlotte, whose health, I fear, is rapidly failing & I doubt whether she will live very long.—[4] Poor Mammy has been having bad toothach, & on Sunday Mr Engleheart pulled out one.—[5]

Skimp was pleased by your note.[6] A run-away-horse at night is no joke, & I am heartily glad you were not hurt.— (N.B people do not go at full "*tare*" along the "*rode*").— You seem to getting on well with pleasant acquaintances. The mathematicks must come hard.— I wonder in how many months you will feel up to all the account Books.—

Your letters are a great pleasure to us.—

I keep up Torquay good habits & almost every day take a pretty good walk in the country.[7] I met Montagu L. lately with arm in sling: Lenny says his manner of speaking seems altered[8]

Sir John has been laid up at Geneva by gout for a fortnight; what a time poor Lady L. must have had; but I daresay the gout will make him less cracky.—[9]

I have been doing some good work at the homologies of Orchids, by tracing their vessels, & Hooker thinks that I have made out their structure well.—

Good Night, my dear old fellow.—

Your affect^{y}. | C. Darwin

DAR 210.6

[1] The year is given by the reference to Emma Darwin's visit to Charlotte Langton (see n. 4, below).
[2] The letter from William Mackmurdo Hacon, CD's solicitor, has not been found.
[3] The letter from Francis Darwin has not been found.
[4] Charlotte Langton, Emma Darwin's eldest sister, was staying in St Leonard's-on-Sea, near Hastings, Sussex, in the hope that the sea air would improve her health (*Emma Darwin* 2: 178). According to her diary, Emma Darwin visited Charlotte from 5 to 9 November 1861. Charlotte Langton died in January 1862 at the age of sixty-five.
[5] Stephen Paul Engleheart was a surgeon in Down (*Post Office directory of the six home counties* 1862).
[6] 'Skimp' was the family nickname of Horace Darwin, aged ten.
[7] The Darwins spent July and August 1861 in Torquay, Devon.
[8] Montagu Lubbock had been seriously injured in a carriage accident in May 1861 (see letter to T. H. Huxley, 22 May [1861], and H. G. Hutchinson 1914).
[9] John William Lubbock and Harriet Lubbock were neighbours of the Darwins in Down, residing at High Elms.

To J. D. Hooker 5 [November 1861][1]

Down. | Bromley. | Kent. S.E.
5th

My dear Hooker

You will commit suicide; but do look with lens at the Bolbophyllum rhizophora (sent this morning) & touch Labellum & *white tips of upper petals which cross*, with hair or fine bristle, & see if Labellum does not close in. on column.—[2] It is a marvellous little beast.

My work must & shall soon end— God forgive me | C. Darwin

I imagine on fine day Labellum is thrown back & flower open.—

DAR 115.1: 77

[1] Dated by the relationship to the first letter to J. D. Hooker, 1 November [1861].
[2] See the first letter to J. D. Hooker, 1 November [1861].

To J. D. Hooker 6 [November 1861]

Down
6th.

My dear Hooker

There is something very odd in Rostellum of Masdevallia: the only structure I have met with at all like your wonderful Listera case—[1] I cannot, however, see any explosion.— If you have plenty of flowers & are sending anything else I sh^{d} like to see a few more; especially a rather earlier flower.

I sh[d] doubt whether there was such a shut-up flower in world; all sepals grown together except the 2 little windows for proboscis.—[2]

Ever yours | C. Darwin

Endorsement: 'Nov/61.'
DAR 115.2: 127

[1] Hooker 1854b. In *Orchids*, pp. 142–6, CD described how the rostellum of *Listera ovata*, upon being touched, instantly expels a viscid fluid that serves to attach pollen-masses to visiting insects.

[2] In *Orchids*, CD called *Masdevallia fenestrata* an 'extraordinary flower' and stated that he did not understand how insects could effect its fertilisation (*Orchids*, pp. 168, 169):

> The whole structure of the flower seemed carefully intended to prevent the withdrawal of the pollinia, as well as their subsequent insertion into the stigmatic chamber! Some new and curious contrivance has here to be made out.

To J. D. Hooker 7 November [1861]

[Down]
Nov. 7th.

My dear Hooker

Since writing this morning, I have been thinking **if** you send me more Masdevallia, the flowers had better be oldish ones instead of a bud—[1] There is something absolutely new to be made out about the action & structure of Rostellum.

C. Darwin

P.S. | I read last night your Fernando Po case. How wonderful it is! I wish to Heaven the transverse mountains were sure; it would be a superb case.[2] But confound those Indian Islands.—[3] Can these islands be a case like Madeira, according to Heer, which retains (from less severe struggle for life) for a longer period than on continent a past condition of things; ie of the Glacial Period—[4] Hurrah "Wriggler" stick to wriggling, it is a sublime art—

Endorsement: '/61'
DAR 115.2: 125

[1] See preceding letter.

[2] Hooker read a paper describing the vegetation of Clarence Peak, Fernando Po, to the Linnean Society of London on 7 March 1861 (Hooker 1862). The paper highlighted an affinity of the flora with that of East Africa and an almost total dissimilarity with the nearby Cape flora. Hooker suggested that this distributional relationship strongly favoured the possible existence of a chain of mountains crossing the unexplored area of central Africa. CD cited this information in the fourth edition of *Origin*, p. 445.

[3] CD refers to the affinity, remarked on by Hooker, between the Clarence Peak flora and the floras of Madagascar, Mauritius, and Bourbon (Réunion) in the Indian Ocean (Hooker 1862, p. 3), for which Hooker's proposed chain of mountains (see n. 2, above) could not alone account. A note pinned to CD's copy of the issue of the *Journal of the Proceedings of the Linnean Society* containing Hooker's article (DAR Library–CUL) reads: 'With respect to Indian Isl[d]. of Bourbon &c have they not retained the mundane vegetation of world during Glacial, like Madieira.—'

[4] Heer 1857. CD's abstract of Oswald Heer's discussion of the fossil plants of Madeira is in DAR 74: 3–7.

From Daniel Oliver [before 8 November 1861][1]

Royal Gardens Kew

References relative to *Orchid* structure & fertilisation. &c

On vascular bundles of Column *Link*. Bot. Zeit. 1849. 745[2]

Mohl. Nachträgliche Beobachtungen über die Sexualorgane &c. der Orchideen von R. Brown Regensburg Bot. Zeit. 1832. 673.[3]

Ménière. Fecundation of *Catasetum*. Bull. Soc. Bot. de France. i. 367[4]

Baillon, on ditto. p. 285.[5]

Guépin. on do. Ann. Soc. Linn. d'Angers. vol. i.[6]

Ménière. on Secretion of a liquid by the glandular organ of *Coryanthes*. Bull. Soc. Bot. ii. 351.[7]

Fermond. On fertilisation of *Iris* independ^t. of wind & insect-agency. Bot. Zeit. 1841. 204[8] & Journ. d. Pharmacie 1840. 751.[9]

Wiegmann. Über d. Unfruchtbarkeit bei den Bastardpflanzen. Bot. Zeit. 1839. 703.[10] (Cause of infertility to be sought only in the pollen, &c.—)

DAR 104: 225–6

CD ANNOTATIONS
8.1 *Fermond*. . . . 751. 8.2] *double scored brown crayon*
9.1 *Wiegmann*. . . . 703.] *crossed ink*

1 Dated by the fact that CD was working on *Orchids* during the latter part of 1861 (see Journal; Appendix II) and by his statement in the letter to J. D. Hooker, 8 November [1861], that he would be anxious to see Link's article in the *Botanische Zeitung* (Link 1849).
2 Link 1849.
3 Mohl trans. 1832.
4 Ménière 1854.
5 Baillon 1854.
6 Guépin 1853.
7 Ménière 1855.
8 Fermond 1841.
9 Fermond 1840.
10 Wiegmann 1839.

To J. D. Hooker 8 November [1861]

Down. | *Bromley.* | *Kent. S.E.*
Nov. 8th

I received this morning Masdevallia, which is very curious, but I cannot understand it,[1] & Cypripedium, whose ducts I will attack tomorrow. (N.B I find strong Nitric Acid most useful)[2] Your labours are over; & most truly do I thank you. No, your labours are not quite over, if you can just try irritability of Bolbophyllum *rhizophora* (? spelling?).—[3]

I have read papers on Catasetum, which do not shake me, but make me extremely anxious for the other plant.—[4]

I have made out the missing bundle of ducts proceeding from the anterior ovarian bundle: I might have looked for ever to Labellum; & I was so prepossessed by Brown's & Lindley's view that it was a chance I ever found it.[5] The enclosed gives structure of every orchid yet looked at. The missing bundle is attached to front of column, running up to beneath stigmatic cavity— So from what you say I hope I may look at homologies of Orchids as pretty safe.— I shall be curious when I get Bot. Zeitung to see what Link says on vessels of column.—[6] Farewell, with hearty thanks for all your never ending assistance.

Your affect | C. Darwin

I have sent my paper on dimorphism in Primula to Linn. Soc. I shall go up, & read it, whenever it comes on (on 21st Nov.); I hope you may be able to attend, for I do not suppose many will care a penny for the subject.—[7]

[Enclosure]

(To be torn up.—)

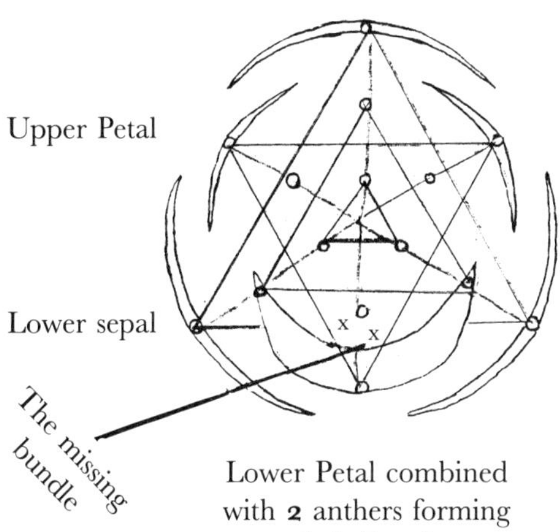

Endorsement: '/61'
DAR 115.2: 126, 129b

[1] See letters to J. D. Hooker, 6 [November 1861] and 7 November [1861].

[2] To trace the course of the spiral vessels in orchids, CD clarified the vessels by soaking longitudinal slices of orchid flowers in a solution of spirits of wine and nitric acid (see letters to J. D. Hooker, 1 November [1861] and 14 November [1861]). For a list of the species he dissected, see *Orchids*, p. 295 n.

[3] See letter to J. D. Hooker, 5 [November 1861].

[4] See second letter to J. D. Hooker, 1 November [1861].

[5] See letters to J. D. Hooker, 27 October [1861], and to John Lindley, 1 November [1861].

[6] Link 1849.

[7] CD read a paper to the Linnean Society of London on 21 November 1861 entitled 'On the two

forms, or dimorphic condition, in the species of *Primula*, and on their remarkable sexual relations' (see *Collected papers* 2: 45–63).

[8] The diagram has been reduced to 70% of its original size.

From Daniel Oliver 8 November 1861

Royal Gardens Kew
Nov. 8 61

My dear Sir

There is a work which you ought to see if not already familiar to you —"Faits pour servir à l'histoire générale de la fécondation chez les végétaux" par *Ch. Fermond*. (Rec. des Travaux de la Soc. d'Emul. pour les Sciences pharmaceutiques t. iii. 1859.) *Tirage 800 45 pp. Paris.*—[1]

Noticed in vol. vi. Bull. Soc. Bot. France p. 749 as including an exposé of all observations which have been made of recent years on Fecundation & its phenomena.

Part. 1. Du rôle que jouent les périanthes dans l'acte de la fécondation.

2. Sur quelques moyens particuliers que la nature emploie pour assurer la fécond[n]. de quelques espèces végétales.

3. Théorie mécanique de la préfloraison &c—

4. Fécondations réciproques de quelques variétés voisines.

5. Sur la fructification du Lis blanc.

Yours faithfully | Da[l]. Oliver

DAR 91: 83

[1] Fermond 1859. Charles Fermond, the chief pharmacist at the Salpetrière Hospital in Paris, reviewed in this paper the literature on plant fertilisation. CD commented on the back of his copy in the Darwin Pamphlet Collection–CUL: 'Rather poor except about involution of Iris'.

To George Bentham 9 November [1861][1]

Down. | *Bromley.* | *Kent. S.E.*
Nov. 9th

My dear Bentham

I write one line to thank you for your very kind note, & for arranging about my paper & for your Photograph.[2] The photograph hardly does you justice, though like.

Pray believe me | Yours very sincerely | Charles Darwin

P.S. As my movements are never very certain, if I do not appear at the Soc[y]. on Thursday Evening the Secretary must read my paper, but I will certainly attend if I am well.—[3]

Royal Botanic Gardens, Kew (Bentham letters: 693)

[1] The year is given by the reference to CD's paper to be read before the Linnean Society of London (see n. 3, below).

George Bentham.
(Courtesy of the Royal Botanic Gardens, Kew.)

John Lindley. Photograph by Ernest Edwards.
(Courtesy of the Royal Botanic Gardens, Kew.)

[2] Bentham's letter has not been found. A reproduction of the photograph he may have sent to CD, dated 1861, is included in this volume facing p. 334. He had been elected president of the Linnean Society earlier in the year.

[3] CD went to London on 20 November 1861, returning home on 23 November (Emma Darwin's diary). He read his paper on the two forms of *Primula* before the Linnean Society on Thursday, 21 November 1861 (see *Collected papers* 2: 45–63).

From Asa Gray 9 November 1861

Cambridge. [Massachusetts]
Nov, 9th, 1861

Dear Darwin

I asked Torrey to tell me about what he knew of diœcio-dimorphism in *Borragineæ*[1] He writes that he has observed it mostly in *Amsinckia*, but gives me no details. I have asked him for details.

In his Bot. of U.S. & Mexican Boundary, p. 140a I see he has published "The insertion of the stamens is not constant in this genus. In the same species they are sometimes placed near the base of the corolla, sometimes in the upper part of the throat."[2]

I have just had occasion to study carefully all the specimens I have of the genus *Mertensia*—of this order. I looked out for this matter. Of the 9 species examined all the specimens of 8 of them have stamens inserted on the throat, i.e. subexsert staminate. But the *style is long* also. In one species—viz. M. alpina some specimens are so also, but *others* have the stamens inserted so low down in the tube that they are wholly included—the tips of the anthers where the bases are in the common way. But in these the *style is short too*! as much shorter as the stamens are lower!

So here is a riddle for you! How varied Nature is in her ways of doing things!

Looking however now at *Amsinckia* in my herbm., I find a pencil note of Dr. Torrey's, long ago, on *Amsinckia spectabilis*.— the specimens of "504 Coulter, California.— Torrey's mem. is—"In specimens 1 & 2 the stamens are inserted on the upper part of the corolla, and the style is short. In 3, they are inserted near the base, and the style is elongated,—a kind of dioicality as in Hedyotis, Mitchella, &c." J. T.

That is all right again. Torrey is quite right about it! I fish up and send you the two sorts of flowers, by which you can verify the observation, in a good light even without dissecting the flowers.[3]

Capt. Anderson, of the Cunard Steamer "Europa",[4] has kindly consented to take any parcel for you which needs prompt delivery, & send up from Liverpool. So I went into the fields to find Houstonia for you,—supposing I could find the little seedlings which are to blossom early next spring. But not one *can I find*.—they are so tiny & concealed in the grass. Well, ⟨ ⟩ would b⟨ ⟩ by this means ⟨ ⟩ance i⟨ ⟩ ⟨ ⟩ picked up the ⟨ ⟩ ⟨su⟩bsexes! ⟨ ⟩ early spring I may get young plants over to you, under glass, late enough to select the two sorts—as they flower for some weeks, I hope to try.[5]

We are just awaiting tidings of our Naval expedition to Beaufort &c, S. Car. We only know that our *Armada* has mostly survived & weathered the gale, and is effecting a landing.[6]

I confess to have been irritated for a while by the nasty articles upon us in the Saturday Review. But now that they are said to be written by Grattan, formerly British Consul at Boston, we are quite indifferent. Grattan was never seen in good Boston society, and I am told was excluded for some disgraceful conduct or other.— Hence the spite of which we have had specimens before.[7]

Ever, dear Darwin, | Yours most cordially | Asa Gray

DAR 110 (ser. 2): 63–4

CD ANNOTATIONS
3.2 of this order.] 'Borragineæ' *added ink*
3.3 subexsert] 'subexersert' *added ink*
7.1 Capt. . . . ⟨su⟩bsexes! 7.6] *crossed blue crayon*
7.6 early . . . Darwin, 10.1] *crossed red crayon*
Top of first page: 'Bentham said some Verbenaceæ were dimorphic.'[8] *ink*

[1] John Torrey was a friend of Gray's and an expert on the flora of the United States. He and Gray had collaborated in preparing the *Flora of North America* (1838–43).

[2] Torrey [1859].

[3] CD had asked Gray for information on plant species that regularly exhibit two or more different flower forms (see letters to Asa Gray, 21 July [1861] and 16 September [1861]). In the published version of the paper read before the Linnean Society of London on 21 November 1861, CD included information from Gray's letter and referred to his examination of the dried flowers of *Amsinckia* sent by Gray (see 'On the two forms, or dimorphic condition, in the species of *Primula*, and on their remarkable sexual relations', *Journal of the Proceedings of the Linnean Society (Botany)* 6 (1862): 77–96, esp. p. 95; see also *Collected papers* 2: 45–63).

[4] James Anderson had entered Cunard's service in 1851 (*Modern English biography*).

[5] See letter from Asa Gray, 11 October 1861.

[6] The Union navy had embarked in the summer of 1861 on a large-scale operation to blockade the major southern ports and patrol the coastline of the Confederate States. Late in September, a large task force headed down the Atlantic coast toward Port Royal, South Carolina, but the fleet was scattered and several of its transports carrying ammunition and landing boats foundered during a gale on 1 November. On 7 November, however, the Union fleet was able to take control of the port, thereby securing 'the finest natural harbor on the south Atlantic coast.' (McPherson 1988, pp. 369–71).

[7] The reference is to a series of unsigned articles covering developments in the American Civil War published in the *Saturday Review of Politics, Literature, Science, and Art*. Several of the pieces were openly critical of the North, even to the point of advocating that the English and French should break the Southern blockade and re-establish commercial contact (see *Saturday Review* 12 (1861): pp. 420–1). Thomas Colley Grattan was the British consul in Boston from 1838 to 1846.

[8] This information does not occur in the extant letters from George Bentham up to 1861, but CD queried Bentham further about dimorphism in Verbenaceae in 1864 (see *Calendar* no. 4556).

To J. D. Hooker 10 November [1861]

Down Bromley Kent
Nov. 10th

My dear Hooker

For the love of God help me. I believe all my work (about a fortnight) is useless.

Look at this accursed diagram of the Butterfly orchis, which I examined after writing to you yesterday when I thought all my work done.[1] Some of The ducts of upper sepal & upper petal run to the wrong bundles or column.— I have seen no such case. This case apparently shows that not the least reliance can be placed on course of ducts. (I am sure of my facts).

There is great adhesion & extreme displacement of parts where the organs spring from top of ovarium. Asa Gray says ducts are very early developed, & it seems to me wonderful that they should pursue this course.— It may now be said that the lateral ducts in the Labellum running into the antero-lateral ovarian bundle is no argument that the Labellum consists of 3 organs blended together.—[2]

In desperation (& from the curious way the base of upper petals are soldered at basal edges) I fancied the real form of upper Sepal, upper petal, & lower sepal might be as represeted by red lines;[3] & that there had been an incredible amount of splitting of sepals & petals & subsequent fusion.— This seems a monstrous notion; but I have just looked at Bauer's drawing of allied Bonatea, & there is a degree of lobing of petals & sepals, which would account for anything.[4] Now could you spare me a dry flower out of your Herbarium of Bonatea speciosa, that I might soak & look for ducts; if I cannot explain the case of Habenaria, all my work is smashed. I was a fool ever to touch orchids.—

Your affect. | C. Darwin

[Enclosure]

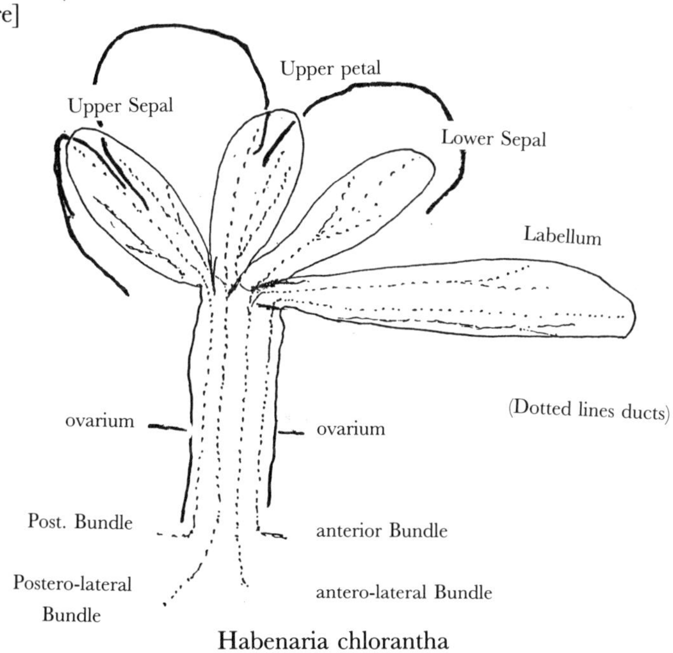

Habenaria chlorantha
Longitudinal section of ovarium.

Endorsement: '/61.'
DAR 115.2: 128, 129a

[1] See the enclosure.

[2] CD was attempting to trace the course of the spiral vessels in orchid flowers, from the ovarium to the various parts, in an attempt to establish the homologies of orchids (see letters to J. D. Hooker, 4 October [1861], 13 October [1861], 27 October [1861], and the first letter dated 1 November [1861]). In the account he gave of orchid homologies in *Orchids*, CD discussed the anomaly presented by *Habenaria* and the allied genus *Bonatea*, stating that it called into question his view that the labellum was always 'an organ compounded of one petal and two petaloid stamens' (*Orchids*, p. 303).

[3] CD drew the outermost lines, representing the 'real form' of the sepal and petal parts, in red crayon. The diagram has been reduced to 60% of its original size.

[4] Bauer 1830–8.

To Dorothy Fanny Nevill 12 November [1861][1]

Down. | *Bromley.* | *Kent. S.E.*
Nov. 12th

Madam

Dr. Lindley has told me that he thought that your Ladyship would be willing, if in your power, to assist me.—[2] I am preparing for publication a small work "on the various contrivances by which Orchids are fertilised." I much wish to examine a few more exotic forms, & if you happen to have those which I wish to see, possibly your Ladyship would be so generous as to send me two or three flowers. I am aware that it would be a remote chance that your Ladyship should possess or spare these flowers. I chiefly want any member of the great Tribe of Arethuseæ, which includes the Limodoridæ, Vanillidæ &c.

Mormodes & Cycnoches are especial desiderata, though they would be most difficult to send, as the pollen-masses move or explode when the end of the column is touched. I, also, want much Bonatea, Masdevallia & any Bolbophyllum with its lower lip or Labellum irritable. Indeed any genus with any remarkable peculiarity would be most gratefully received.—

I have much reason to apologise for thus intruding on your Ladyship; & I am far from expecting that your Ladyship will reply to this note, if you cannot assist me.—[3] I will only add that for a parcel too large for the post, my quickest address is "C. Darwin care of the Down Postman Bromley Kent". I find that orchids travel safest in tin boxes or cannisters with a little damp paper.—

With many apologies, I have the honour to remain | Your Ladyships | Obedient servant | Charles Darwin

American Philosophical Society (270)

[1] The year is given by the reference to CD's preparing the manuscript for *Orchids*, which was published in May 1862.

[2] See letter to John Lindley, 25 October [1861]. Dorothy Fanny Nevill had a notable garden at Dangstein near Midhurst, Sussex. She specialised in the cultivation of orchids, nepenthes, and other tropical plants (R. Desmond 1977).

[3] See letter to D. F. Nevill, 19 November [1861].

To George Robert Waterhouse 12 November [1861][1]

Down. | Bromley. | Kent. S.E.
Nov. 12th

My dear Waterhouse

Many thanks for the enclosed interesting letter.—[2] It always pleases me to see a clear-headed man on my side, & especially if he be bold enough to speak out, & thus render really efficient aid.— For it will require much speaking out before this side will receive fair toleration.—

Mr. Walsh has evidently been a close & good observer of Fishes.—[3] What he says about the Sea-Trout in the Loch is very curious; & it would be very desirable that some professed Ichthyologist should examine these Trout.— It would make a good case for me.— But until thus examined & pronounced on, cavillers would simply deny that the fish was a sea-trout.—

I hope to see you before long, for I shall be up on the 21st to read a paper at Linn. Soc.—[4]

Did you ever look at the skull of the Borneo Domestic Cat, which I gave you?—[5]

My dear Waterhouse | Yours very sincerely | Charles Darwin

Cleveland Health Sciences Library (Robert M. Stecher collection)

[1] Dated by CD's reference to reading a paper before the Linnean Society of London (see n. 4, below).

[2] The enclosure has not been found.

[3] Walsh has not been identified.

[4] CD read a paper on the two forms of *Primula* to the Linnean Society on Thursday, 21 November 1861 (see *Collected papers* 2: 45–63).

[5] CD had obtained specimens of cats and other domestic animals from Borneo through James Brooke, the raja of Saráwak (see *Correspondence* vol. 5, CD memorandum, [December 1855]; and vol. 6, letter to A. R. Wallace, 1 May 1857). Waterhouse had examined a South American feral cat CD brought back from the *Beagle* voyage (see *Variation* 1: 47).

To J. D. Hooker 14 November [1861]

Down. | Bromley. | Kent. S.E.
Nov. 14th.

My dear Hooker

I was very bad in bed yesterday,[1] but not too bad to be much pleased by your note & *magnificent* lot of Bonateas;[2] They have been soaked in water & are now in spirits & with a little dose of Nitric A. I have great hopes they will be clarified enough for dissection.—[3]

I was in foolish despair when I wrote, & thought all my work done for; & I cannot yet see how it is not "a great blow & discouragement",—unless indeed I do find that the lower sepal & upper petals are sometimes in allied forms deeply bifid; if they are bifid, then cohesion is so common with the orchids, that the view which I imagined & showed by red lines may be true.[4] Will it not be a very odd case that in Habenaria that the outlines of upper sepal, upper petal, & lower sepal

(& indeed of Labellum) do not one of them give the real shape of these same sepals & petals?—

You are a true friend in need.— I can hardly bear to let Bonatea soak long enough.—

Adios | C. Darwin

Endorsement: '/61'
DAR 115.2: 130

[1] Emma Darwin recorded in her diary on 13 November 1861: 'Charles in bed cold &c'.
[2] Hooker's letter has not been found.
[3] CD had written to Hooker to request specimens of the orchid *Bonatea* (see letter to J. D. Hooker, 10 November [1861]).
[4] See letter to J. D. Hooker, 10 November [1861], n. 3 and the diagram.

To W. E. Darwin 15 November [1861][1]

Down
Nov. 15

My dear William

I have no wisdom about Stock-Broker. The one whom I long trusted is dead.—[2] I have since employed Brokers through the Union Bank.—[3] Of whom the first broke, & I suspect caused the Union Bank loss; I know that just as he broke he purchased shares for me.— The second Broker employed by the Union Bank was rather discredited when the great fraud on the Union Bank was discovered.[4] I forget the name of Broker now employed; but I know nothing of him personally.—

I have forwarded several "Captain" Letters to Mr Fox.—[5] There are many resignations in the Down corps—[6]

I have had a hard days work & am not up to writing—

What a grand dinner you were at.— I rather hate such magnificence.—[7]

I hope you will be able to pay us a visit at Christmas. The Boys wish it so much.— I long to hear all sorts of trifling details about your life.—

Farewell | Your tired old | Father | C. D.

We have not horses yet & think we never shall get any! Confound the Croydon man.— I am going up to London on Wednesday to read my Primula paper.—[8]

DAR 210.6

[1] The year is given by the reference to CD's *Primula* paper (see n. 8, below).
[2] Until his death in 1853, Charles Stokes of Stokes & Hughes, London, had been CD's broker.
[3] CD banked with the Union Bank of London, with premises in Princes Street.
[4] A serious fraud on the Union Bank of London, committed by the chief cashier William George Pullinger, was discovered in April 1860. Through dealings with several Stock Exchange operators, Pullinger defrauded the bank out of £263,000 (*Banking almanac* 1861, p. 25; *The Times*, 21 April 1860).
[5] William wished to resign his commission with the Farnborough Rifle Volunteer Corps (see letters to W. E. Darwin, 22 October [1861] and [27 October 1861]). Thomas Samuel Fox was an ensign in the corps.

[6] After the immediate fear of invasion of England by France had subsided, many members of the voluntary rifle corps, who had first joined in 1859, resigned (see Beckett 1982, pp. 74–5).

[7] William had attended a dinner at the home of a Captain Vignole (see letter from W. E. Darwin, [17 November 1861]).

[8] CD went to London on Wednesday, 20 November 1861 (Emma Darwin's diary). He read a paper on the two forms of *Primula* to the Linnean Society of London on Thursday, 21 November 1861 (see *Collected papers* 2: 45–63).

To John Lindley 16 November [1861][1]

Down. | Bromley. | Kent. S.E.
Nov. 16th

My dear Lindley

I thank you warmly for your two notes & the store of specimens received this morning.[2]

I am ashamed that you have sent so many valuable specimens; for it is a chance whether I can make out for my purpose anything from dried specimens.— But I shall be *very glad* to have a try, & so again thank you cordially.— I have had a most obliging note from Mr Rucker (to whom I used your name) with promise of Catasetum & some Dendrobium.—[3] I have written also to Lady Dorothy.[4]

I know how many irons you have in the fire & will not be unreasonable & I hope give no more trouble.

Yours very truly obliged | Charles Darwin

Royal Botanic Gardens, Kew (Lindley letters, A–K: 196)

[1] The year is given by the relationship to the letter to John Lindley, 1 November [1861]. See also n. 4, below.

[2] Lindley's letters have not been found. They were written in response to the letter to John Lindley, 1 November [1861].

[3] Sigismund Rucker, who lived in West Hill, Wandsworth, is acknowledged in *Orchids*, p. 158 n., for having provided CD with specimens of these foreign orchids.

[4] Lindley had suggested that CD write to Dorothy Fanny Nevill (see letter to D. F. Nevill, 12 November [1861]).

From W. E. Darwin [17 November 1861][1]

No. 1 (not no 4) Carlton Place.
Sunday.

My Dear Father,

I was rather afraid the brokerage would turn out a failure, we shall have to choose one out of the ones Mr A. knows;[2] I am very sorry to hear that poor Aunt Charlotte is so unwell, I suppose winter coming on is very bad for asthma.[3]

If I had known the boys had been coming home last Sunday, I would have come too; I think I shall come this day fortnight or three weeks.[4] You will be astonished to hear that I am up to snuff enough to manage by myself for a short time, that is with the help of the head Clerk; Mrs A. went up to London yesterday, and Mr A. took her up, so that I had to wind up all the business yesterday and it was a very full day; he is not coming back till Tuesday, so that I shall be the cock of the walk on Monday; the head Clerk is a very nice old man, who has been there 45 years, so that in any difficulty I can always go to him. I like Mr A. better than ever and we get on perfectly together. I went up and had a quiet dinner with them on Friday, no body there but Capt. Forrest who is a very nice man, and I had a very pleasant evening.[5] I have come round to the opinion that Mrs A. is a much greater beauty than I had thought her; I can see they despise Southampton society, and know very of the town's people; she shuddered when I mentioned Capt Vignole's dinner, and said she had been there once and never means to go again.[6]

I don't think I have ever given you an account of my day here, and one day is as good as a hundred; 7.30. a.m. a very mild rap at the door; from 7.45. till 8 I generally manage to get up, I then dress up to decency pitch and go down and put the tea in the pot, and have a soliloquy before the fire what to have for breakfast. The general result is an egg, or some times a rasher, or perhaps yesterday's cold chop warmed up, but this morning a lovely kidney came up quite unexpected like. I have breakfast over by about 8.45., I then stand before the fire or have a quiet strum, till M^rs^. Pratt appears to have a solemn Confab. about dinner, and to praise yesterdays dinner or this mornings kidney, or anything to do with the establishment, of which she is mighty proud.[7] After we have made up our minds about dinner, it is about time to be off, as I have to be there by 10 minutes to 10, and it is quarter of an hour's walk; Mr A. does not come in till 10.30. I always find Mr Fall there with the keys, ready for me to open the save and strong room.[8] when it is opened and the Clerks have taken out all the books, I get out the money, and the notes, and drafts and bills &c, and then I have to look among the bills to see if any are coming due, and enter the things, drafts and bills notes powers of attorney and all sorts of things that come down from Lubbock; then theres the daily letter to write to Lubbock's to say that we have theirs, and to tell them whats what to do for us;[9] then after that there is the Ledger to be culled over, that is to see if all the business done the day before, has been put down to each separate person concerned; by this time it gets to be about 11.30, or 12. And then the times comes in, and one takes it easy, till some odds & ends want doing, of which there are generally plenty; if there is nothing to do, I finish the times or read a book, till about 1—when I generally try to do a little mathematics, but they come rather stiff after a bustling morning, and I am getting in rather a fright about my degree[10]

at about 2 or $\frac{1}{2}$ past I go out and get a turn, sometimes $\frac{1}{2}$ an hour's some times less sometimes more; before I go in I get a Sandwich or butter and roll or a bun or two, at a pastry cook's just opposite; it is by this time from 3 to 3.30. so that there is only $\frac{1}{2}$ hour more before we wind up, which is rather a long job if there is anything wrong, but we generally get away from 4.30 to 4.45: at about 5 wet or dry I go out for about a 6 mile walk, and get back for dinner at 7; I find mathematics an awful try after dinner, I sit before the fire reading a book or something, till I begin to feel sleepy, then comes the tug, am I to go to sleep or not, my general resource is to go the pianoforte and have a frantic strum, to annoyance I should guess of the widow below me; the widow and I are just Box and Cox,[11] just as I go out she comes down to breakfast, and soon after I go out my walk she comes in from hers so that I have never seen her except her back once; if the piano wakes me I do some mathematics and then read a book and so to bed; and so on; one after the other without much change.

They are going to get up Volunteer Engineers here under patronage of Sir H. James, and Mr A. gave me a hint as if he knew for a fact they were going to come and ask me to belong; but I have had enough volunteering.[12] Would it be the correct thing to go and call again on Sir H. James as he was not in; I should think not—[13] it has been a lovely day but rather cold; on Sunday I have dinner at two to suit Mrs P. after dinner I went a good long walk;

I hope you will be up to your primula paper I suppose it will be rather longer than the generality.[14]

I am your affect son | W E Darwin

am I ever to have "Orley Farm"[15]

DAR 210.16

[1] Dated by the references to CD's *Primula* paper and to Francis and George Howard Darwin being home from school (see nn. 4 and 14, below). In 1861, 17 November fell on a Sunday.

[2] See letter to W. E. Darwin, 15 November [1861]. 'M^r^ A' was William's partner in the Southampton and Hampshire Bank, George Atherley.

[3] Emma Darwin's elder sister, Charlotte Langton, had been in poor health for some time and had gone to St Leonard's-on-Sea in Sussex to recuperate (see letter to W. E. Darwin, 4 November [1861]).

[4] Francis and George Howard Darwin, who attended Clapham Grammar School, came home for the weekend on Saturday, 9 November 1861. William returned to Down House for Christmas, arriving on 24 December (Emma Darwin's diary).

[5] Captain Forrest, who lived in Winchester, was chief constable of Hampshire (*Post Office directory of Hampshire, Wiltshire, and Dorsetshire* 1859).

[6] Possibly John Vignoles, a naval commander who had retired in 1855 (*Navy list* 1861).

[7] William had taken rooms in the lodging house of Mrs William Pratt, 1 Carlton Place, Southampton (*Post Office directory of Dorsetshire, Wiltshire, and Hampshire* 1867).

[8] William perhaps refers to Phillip Carteret Fall, who had been a partner in the Southampton and Hampshire Bank before his retirement in the summer of 1861 (*Banking almanac* 1861; see also letter to W. E. Darwin, [25 May 1861]).

[9] The bank of Robarts, Lubbock & Co was the London agent for the Southampton and Hampshire Bank. It was through the intermediation of John Lubbock and his father John William Lubbock that William had learned about the vacancy in the Southampton bank. The relationship between provincial banks and their London agents was strengthened following the institution of the country clearing system in 1858. The object of the clearing system was to collect drafts payable between bankers and to expedite settlements. Country banks dispatched to London in one parcel all the cheques drawn on other country banks received in a day's trading and received in return the bulk of the cheques payable by themselves (see Matthews 1921).

[10] William left the University of Cambridge without taking a degree in order to take up the offer of a banking partnership with the Southampton and Hampshire Bank. Having kept the required number of terms (three terms for three years), he planned to enter for the mathematical tripos examination of

1862. As Francis Darwin later recalled: 'He must have gone in for the Tripos with his mathematics in a rusty condition, otherwise he might perhaps have been higher than bracketed top of the "Apostles" where his name appears.' (F. Darwin 1914, pp. 20–1).

11 In John Maddison Morton's play *Box and Cox*, John Box and James Cox are two characters who share the same apartment, one using it by day and the other by night so that they scarcely ever meet.

12 Henry James was the director of the Ordnance Survey based in Southampton. William, who was a captain in the Farnborough Rifle Volunteer Corps, was trying to resign his commission (see letters to W. E. Darwin, 22 October [1861], [27 October 1861], and 15 November [1861]).

13 CD had written a letter of introduction for William to James (see letters to W. E. Darwin, 12 October [1861] and [27 October 1861]).

14 CD read a paper on the two forms of *Primula* to the Linnean Society of London on Thursday, 21 November 1861 (see *Collected papers* 2: 45–63).

15 *Orley Farm*, by Anthony Trollope, was published in parts in *Harper's Monthly Magazine*, between May 1861 and December 1862 (Irwin 1926). It was published in book form in 1862.

To J. D. Hooker 17 November [1861][1]

Down Bromley Kent
Nov. 17th

My dear Hooker

What two *very* interesting & useful letters you have sent me.[2] You rather astound me with respect to value of grounds of generalisation in the morphology of plants.[3] It reminds me that years ago I sent you a grass to name & your answer was "it is certainly Festuca — (so & so) but it agrees *as badly with the description as most plants do*."![4] I have often laughed over this answer of a great Botanist.— All that which you say about "chorisis" is as new to me as the term itself:[5] I will be cautious; the notion seemed to me monstrous, & I should still think it so, *unless in an* **allied** *plant I can show gradation towards it*; & then surely you would admit that it was at least possible.— But I have not yet looked at Bonatea, & the longer it soaks in Spirits the better.—[6]

Lindley, from whom I asked for an orchid with simple Labellum has most kindly sent me a lot of what he marks "rare" & "rarissimo" of peloric orchids &c; but as they are dried, I know not whether they will be of use.[7] He has been most kind & has suggested my writing to Lady D. Nevill who has responded in wonderfully kind manner & has sent a lot of treasures.[8] But I must stop, otherwise by Jove I shall be transformed into a Botanist. I wish I had been one: this morphology is surprisingly interesting. Looking to your note, I may add that certainly the 15 alternating bundles of spiral vessels (mingled with odd bead-like vessels in some cases) are present in many Orchid; the inner whorl of anther-ducts are oftenest aborted.— I must keep clear of Apostasia, though I have cast many a longing look at it in Bauer.—[9]

What a very interesting case that of the Crucifer with many stamens & its relation to Papavers!—

Your note about St. Thomas is wonderfully curious: the mammals in Fernando Po show it was anciently united to mainland: but as far as I can remember I concluded on investigation that the mammals in St. Thomas were all introduced, & that the isl^d. had always been an island. Do you not think that this will bear on non-migration during glacial period. Is there not great dearth of temperate forms on Teneriffe?[10]

Pray thank heartily that living index, Oliver, for telling me of French Book, which is ordered.—[11]

I hope I may be well enough to read my own paper on Thursday, but I have been very seedy lately:[12] I see that there is paper at Royal on same night at Royal, which will more concern you on fossil plants of Bovey;[13] so that I suppose I shall not have you; but you must read my paper when published, as I shall very much like to hear what you think. It seems to me a large field for experiment.— I am now trying one on Heterocentron roseum a Melastomatous plant with 2 sorts of anthers.[14] I shall make use of my Orchid little volume in illustrating modification-of-species-doctrine; but I keep very very doubtful whether I am not doing a foolish action in publishing. How I wish you would keep to your old intention & write a book on Plants.[15]

Adios my dear old friend | C. Darwin

It strikes me as a fundamental point in value of Homologies of Ducts, in what course they are developed & I cannot find out this: when a petal is first formed, is the duct in it a prolongation upwards from a bundle below, or downwards from the petal to join a previously existing duct.— If the latter one can see how the ducts might go astray.—

(By odd chance I have just stumbled on dispute on this point in Bull. Soc. Bot. & I see both opposite views stoutly maintained.)[16]

DAR 115.2: 131

1 The year is provided by the reference to CD's *Primula* paper (see n. 12, below).

2 Hooker's letters have not been found.

3 In his discussion of spiral vessels being used as a basis for tracing the homologies of orchids, CD mentioned Hooker's assertion that he had 'never known them to speak falsely' (*Orchids*, p. 293).

4 The species was *Festuca pratensis*. See *Correspondence* vol. 5, letter to J. D. Hooker, 5 July [1855].

5 The term 'chorisis', the 'splitting of an organ into two parts, each of which is as perfect as its original' (*OED*), was proposed by Michel Félix Dunal (Dunal 1829); his explanation of the term was quoted by John Lindley (Lindley 1848, 1: 332).

6 See letters to J. D. Hooker, 10 November [1861] and 14 November [1861].

7 See letter to John Lindley, 16 November [1861].

8 See letter to D. F. Nevill, 12 November [1861].

9 Bauer 1830–8. *Apostasia* is a genus of tropical orchids indigenous to Asia, Malaysia, and Australia. In *Orchids*, CD mentioned the genus only in passing, most notably to suggest that its structural differences from most other orchids perhaps served 'to show the state of the order in ancient times, when none of the forms had become so widely differentiated from each other and from other plants' (*Orchids*, pp. 331–2).

10 See letter to J. D. Hooker, 7 November [1861]. In a note pinned to his copy of the issue of the *Journal of the Proceedings of the Linnean Society* in which Hooker 1862 was published, CD wrote: 'Hooker in letter says that Mann found at top of St. Thomas 7200 ft not one temperate form— Now I think St. Thomas no Mount Fernando Po *[here]*—' For earlier references to CD's inquiries concerning the

relationship between the faunas of Fernando Po and St Thomas (São Tomé) and the mainland fauna, see *Correspondence* vol. 6, letters from W. F. Daniell, 8 October – 7 November 1856 and 14 November 1856.

11 Fermond 1859 (see letter from Daniel Oliver, 8 November 1861).

12 CD was scheduled to read a paper on the two forms of *Primula* before the Linnean Society of London on Thursday, 21 November 1861 (see *Collected papers* 2: 45–63).

13 At a meeting of the Royal Society of London on 21 November 1861, Charles Lyell communicated two papers that described deposits found near Bovey Tracey, Devon: William Pengelly's paper on the lignites and clays (Pengelly 1862) and Oswald Heer's paper on the fossil flora (Heer 1862) (see *Proceedings of the Royal Society of London* (1861): 449–55). In 1855, Hooker had published a paper on the seed-vessels found in the Bovey Tracey coal, the first announcement of the discovery of identifiable fossils from the deposit (Hooker 1855).

14 There are notes dated 'Oct 28th. 1861—' and 'Oct 29— 1861' on *Heterocentron roseum* in DAR 205.8: 44 and 45. CD continued his experiments to test the relative fecundity of the purple- and yellow-leaved forms in 1862. He had apparently been given the plants by John Lindley.

15 In 1860, CD tried to persuade Hooker to prepare from his published materials a general work on botany (see *Correspondence* vol. 8, letters to J. D. Hooker, 8 February [1860] and 14 February [1860]).

16 The reference in the *Bulletin de la Société botanique de France* has not been traced.

To John Lindley 17 November [1861][1]

Down. | *Bromley.* | *Kent. S.E.*
Nov. 17th

My dear Lindley

I must just tell you that Lady Dorothy has written me one of the most obliging notes, I ever received, & has sent a lot of orchids now on the road![2] I used your name—

Yours very sincerely | C. Darwin

Royal Botanic Gardens, Kew (Lindley letters, A–K: 197)

1 The year is given by the reference to CD's correspondence with Dorothy Fanny Nevill (see following letter, and letter to D. F. Nevill, 12 November [1861]).

2 Lindley had suggested to CD that he write to Nevill with regard to obtaining rare foreign orchids (see letter to D. F. Nevill, 12 November [1861]).

To D. F. Nevill 19 November [1861][1]

Down. | *Bromley.* | *Kent. S.E.*
Nov. 19th

Madam

I do not know how sufficiently to thank your Ladyship for your singularly kind note & the magnificent present of Orchids.[2] I should have much enjoyed accepting your Ladyships permission to see your famous collection of Orchids; but my health is far from strong & the journey would be very fatiguing to me. The Orchids have interested me extremely & some of them have been most useful; but they were unfortunately rather too dry. If I might venture to ask your Ladyship I hope you will thank your gardener for sending me so clear a list of names.—

Only one of the sorts arrived too withered for examination (for my object) viz the Bolbophyllum. If your Ladyship could spare me one more spike, with a few *unopened* flowers on it, & have the kindness to direct it to have a little *damp* moss tied round the cut end & sent by post (in tin Box or cannister if possible), it would be of *great* use for my work. Some of the species, as your Ladyship knows, of Bolbophyllum have the little lip or Labellum within the flower irritable, so as to move when touched; I sh^d. be very truly obliged if your Ladyship could have this point ascertained,—that is if you can send me one more spike.—

If I understand your Ladyship's note, there is no chance of your being enabled to send me a Mormodes or Cycnoches, in the course of a few weeks.—

I believe that your Ladyship understands the natural enthusiasm of a naturalist over his subject, & to this alone I can trust for your Ladyship to excuse the length to which this note has run.

With the most sincere thanks, I beg leave to remain your Ladyships | Very truly obliged | Charles Darwin

I will take the liberty of doing myself the pleasure of sending to you a copy of my little book on Orchids when published.—[3]

If a single flower of Mormodes or Cycnoches should open & your Ladyship could send it; it would be necessary that it should be tied down to bottom of Box or basket, so that not even an atom of moss should fall *on the irritable top of column*; & damp moss should be tied closely round the cut end. These genera would be a priceless treasure to me.—

Cleveland Health Sciences Library (Robert M. Stecher collection)

[1] Dated by reference to the publication of *Orchids*, which appeared in May 1862.
[2] Nevill's letter has not been found. It was in response to CD's letter to her of 12 November [1861]. Nevill kept a notable garden at Dangstein, near Midhurst, Sussex, in which she especially cultivated orchids, nepenthes, and other tropical plants (R. Desmond 1977).
[3] Nevill's name appears on CD's presentation list for *Orchids* (DAR 210.18).

To J. D. Hooker 20 [November 1861][1]

Down
20th

My dear Hooker

I am very sorry to bother you, but see what Lindley says about the *Acropera luteola*.[2] What must I do? Can you tell me who named it?

Yours | C. Darwin

DAR 115.2: 132

[1] Dated by the relationship to the letter to J. D. Hooker, 25 November [1861].
[2] Hooker had sent CD flowers of an orchid he identified as *Acropera luteola*. John Lindley, however, had informed CD that he knew of 'no species so named' (*Orchids*, p. 203 and n.).

To J. D. Hooker [22 November 1861]

[Down]
Friday

My dear Hooker

I have been thinking unless you are sure that your carrier will deliver Bolbophyllum & Catasetum by 1 oclock on Thursday (to care of G. Snow Nag's Head Borough)[1] you had better send the Box per Rail addressed to "C. Darwin care of Down Postman Bromley Kent" & as soon as I heard it was at Bromley, I could send horse & cart for it.— I have friends Hot-House, who well understands Orchids & could take care of the the Catasetum—[2] No doubt Nag's Head best way if in time; & if it not in time I could send to London, for Box.— Let me know when sent to Nag's Head— Bolbophyllum will be treasure

My dear old fellow | C. D

Endorsement: 'Nov 22 /61'
DAR 115.2: 133

1 George Snow operated a weekly carrier service between Down and London, making the return trip to Down every Thursday. The 'Nag's Head' was a public house in Borough, south-east London.

2 In *Orchids*, p. 158 n., CD thanked George Henry Turnbull of The Rookery, Down, for permitting him to use his hot-house.

From William Hugh Gower 23 November 1861

Kew.
Nov. 23. 1861

To Mr. Darwin

Sir

Having mentioned to Dr. Hooker a few notes which I had made upon the fertilization of the Victoria regia in these Gardens this Summer,[1] he expresed a wish to send you word concerning it, but I fear notes so brief will be of little service to you, though from them I am led to believe that for the production of great quantities of seeds the Stigma must be impregnated with foreign pollen. 1st I noticed flowers left to impregnate themselves and they produced from 20 to 30 seeds in each Capsule, those flowers which I fertilized with their own pollen produced from 60 to 80 seeds in each Capsule, but one flower which was fertilized with the pollen of another flower though upon the same Plant, produced the immense number of 251 seeds quite perfect, but I cannot say if there would have been any perceptible difference in the Plants from these seeds and I shall not be able to come to any conclusion as they have unfortunately been mixed

I remain Sir Most | respectfully | Wm. Hugh Gower

DAR 165: 81

CD ANNOTATION

Top of first page: 'would not do to trust without capsule' *pencil*; 'Ch III'[2] *brown crayon*

[1] Gower was a foreman at the Royal Botanic Gardens, Kew. *Victoria regia* (*V. amazonica*), the giant waterlily, had flowered for the first time at Kew in 1850.

[2] The annotation refers to chapter three of CD's 'big book' on species entitled 'On the possibility of all organic beings occasionally crossing' (*Natural selection*, pp. 35–91).

To George Bentham 24 November [1861][1]

Down. | Bromley. | Kent. S.E.
Nov. 24th

My dear Bentham

I was much interested by your remarks on the Primula case, but I cannot distinctly remember to what plants you referred excepting Oxalis.[2] Could you have the kindness to give me a little more precise information so that I might endeavour to get some of the plants for experiments. Thus do you know what *species* of Oxalis are thus characterised if I knew the specific names I might perhaps find out whether they exist in nursery gardens; & so with some Labiatæ or any other orders.—[3]

I have not time for many experiments, but I should like to try a few more.—

As I am writing I will mention another subject: Mr Kippist seemed to think that if my paper was ordered to be printed it might be in the Transactions, which I should be **very** sorry for & I shd. greatly prefer, on account of quickness, it to be in the Journal.—[4]

Also he said that the Council would on application perhaps permit me to have 50 copies printed at my own expence, which I wish for to give gardeners & others who have been sending me specimens.—

Pray forgive me troubling you & grant me these favours if you can.—

Your's very sincerely | Charles Darwin

Royal Botanic Gardens, Kew (Bentham letters: 688)

[1] The year is established by the relationship to the letter to George Bentham, 26 November 1861.

[2] Bentham, president of the Linnean Society of London, had apparently made some comments following the presentation of CD's paper on dimorphism in *Primula* at the society's meeting on 21 November (see n. 4, below).

[3] CD eventually discussed the trimorphic condition of various species of *Oxalis* in *Forms of flowers*.

[4] Richard Kippist was the librarian of the Linnean Society. CD's paper, 'On the two forms, or dimorphic condition, in the species of *Primula*, and on their remarkable sexual relations', was published in the *Journal of the Proceedings of the Linnean Society (Botany)* 6 (1862): 77–96. See also *Collected papers* 2: 45–63.

To J. D. Hooker 25 November [1861][1]

Down Bromley Kent
Nov. 25th

My dear Hooker.

I fear that there is no chance of an Acropera of either species or its *ovarium* still remaining on your plants. You will remember about the mouth of the stigmatic cavity being so small; & your own remark about the stigma not being viscid, which

I noticed, but like an incredible ass did not look closely at.[2] It occurred to me that the species of Acropera may be the males of some other orchids (like Catasetum & Myanthus) & I have just looked at an ovarium in spirits & I can see no trace of ovules! Unfortunately I cut the ovarium rather short when I put 3 flowers in spirits; & this makes me want a perfect ovarium, & a flower for fresh stigmatic tissue. I sh^d like to get over this opprobrium to my work on fertilisation; of course I must look to ovaria of other unfertilised orchids & see how plain ovules are: my memory makes me think that they are plain.—[3] I have looked to Link on ducts in Orchids, & he trusted only to transverse sections & I am sure very falsely contradicts R. Brown.—[4] What you said about Brown having observed the ducts running *wrong* in Habenaria is very likely (though he ought to have said so) but then he does apparently trust to position of ducts as far as he traced them by transverse sections. I hope in a day or two to attack Bonatea.—[5]

I was much pleased & interested by your remarks on the Primula case: but I cannot remember whether you gave any additional cases of dimorphism; if you did, please tell me again. A rather tall man with upturned eyebrows, told me that Weddell says that Cinchona presents the two forms.—[6]

I went to B. Mus. & saw a few of Bate's "mimetic" butterflies & they are truly *wonderful*: He ought to have a coloured plate;[7] I told him I would give £10 towards it, but I fear a coloured plate would cost much more.— What a pity that this man sh^d. have to work for his daily bread & have only 1 or 2 hours for science; but I do not see what can be done.[8] He speaks with admiration of Wallace's talents, energy & knowledge.—[9]

Here is a good joke; I saw an extract from Lecoq. Geograph. Bot. & ordered it & hoped that it was a good sized pamphlet & my God *nine* thick volumes have arrived!—[10]

My dear old Hooker | Your affect | C. Darwin

P.S. Do you (or Oliver) know whether in *males* of *such plants*, as the **male** of Lychnis dioica, whether there are rudimentary ovules in the ovarium?[11]

I shall be so glad to see the living Bolbophyllum *rhizophora* is this spelt right? it is not in Steudel[12]

DAR 115.2: 134

[1] Dated by the reference to CD's paper on *Primula* (see n. 6, below).

[2] See letters to J. D. Hooker, 15 October [1861] and 18 October [1861]. See also letter to Daniel Oliver, 30 November [1861].

[3] See *Orchids*, pp. 206–10.

[4] Link 1849 and Brown 1831.

[5] See letters to J. D. Hooker, 10 November [1861] and 14 November [1861].

[6] Hooker apparently attended the meeting of the Linnean Society of London on 21 November 1861 at which CD read his paper 'On the two forms, or dimorphic condition, in the species of *Primula*, and on their remarkable sexual relations', *Journal of the Proceedings of the Linnean Society (Botany)* 6 (1862): 77–96 (see also *Collected papers* 2: 45–63). Hugh Algernon Weddell was a specialist on the flora of South America and had made a detailed study of *Cinchona* (see, for example, Weddell 1849).

[7] Henry Walter Bates had deposited specimens from his collection of South American butterflies at the

British Museum. His paper on insect mimetic analogies (Bates 1861b) was read before the Linnean Society of London on 21 November, the same night that CD read his paper on *Primula* (see n. 6, above). A note in DAR 205.10 (Letters) indicates that Bates showed CD his collection on 22 November 1861 (see letter from H. W. Bates, [1 December] 1861, CD note).

8 Bates worked in the family worsted hosier business in Leicester (Woodcock 1969).

9 Bates and Alfred Russel Wallace had undertaken a joint expedition to the Amazon, 1848–50.

10 Lecoq 1854–8. The work, which contains annotations by CD, is in the Darwin Library–CUL.

11 CD refers to Daniel Oliver, Hooker's colleague at the Royal Botanic Gardens, Kew.

12 The reference is to Steudel 1840–1, a copy of which is in the Darwin Library–CUL. The final two sentences of the letter were written in the margin beside '*rhizophora*'.

From George Bentham 26 November 1861

91 Victoria Street
Nov 26 /61

My dear Darwin

Your note which reached me last night crossed mine[1] I am writing to Mr Currie who as our Botanical Secretary edits the botanical part of the journal to know when your paper could appear in the Journal as we are anxious to meet the views of authors as to which of the publications they wish their papers to appear in.[2] The Council meets on Thursday in next week when the publication of yours and other papers will be decided on. The part now printing is far advanced and I almost fear there will not be room for yours till the next but we will see what can be done[3]

Your paper on hearing it read threw quite a new light on the kind of semi-dioicality which had been so much observed of late and which in the course of our examinations for Genera Plantarum[4] had become quite so common a thing that precise instances no longer attract much attention and I only mentioned a few that occurred on the spur of the moment The Labiatæ I particularly alluded to were many forms of Mentha arvensis and Thymus serpyllum and other common species of those genera which in local Floras have been distinguished as species on the character staminibus exsertis and staminibus inclusis— Kunth and others divided the Ægiphilos of tropical America into two sections on that character and we now find that each species in one section has its correspondent in the other.[5] Planchon observed that long styles & short stamens & short styles with long stamens in several Linums, and abandoned that character which had been previously made use of when he monographised the genus in Hooker's London Journal of Botany[6]—but I forget whether he remarks on the question in print although I remember conversing with him on it. So in Oxalis Jacquin distinguished some Cape species then in cultivation on that character and I think Zuccarini remarks upon its being not specific in his Monograph of Oxalis but I will look as to what are the species when I am at Kew this morning & write again—[7] In Rubiaceæ I have observed the exserted and included stamens in many species but as I say it is so common that I forget exactly which are the species— the most common in European plants that occur to me are in Mentha & Thymus as above.

I should be very anxious you would look at the case of Viola for its fertilisation is always a puzzle to me In V. canina, V. odorata, V. palustris and all of that section (but not V. tricolor V. arvensis etc) I can never find any of the perfect flowers that appear in winter and spring set their seed although both stamens and ovaries appear perfect. *After these are all gone by* a number of minute apetalous flowers in which we very seldom find any more than rudimentary stamens, go on for months setting seed in the greatest abundance— How do they get impregnated by insects when there are no flowers about for insects to get the pollen from? In most orders where this kind of dimorphism have been observed the two sets of flowers are contemporaneous— In Ononis minutissima however the apetalous and apparently stamen-less but fertile flowers appear in spring whilst the perfect (also fertile ones) come out in summer.[8] In Viola the two may be contemporaneous for a few days when one is going out and the other coming in but the majority appear at a great interval

Yours very sincerely | George Bentham

DAR III: 73–4

CD ANNOTATIONS
1.1 Your ... moment 2.5] *crossed pencil*
2.1 Your ... contemporaneous— 3.9] *crossed red crayon*
2.9 Ægiphilos] 'Aegiphilos (Labiate)' *added pencil*
2.10 Planchon ... Botany 2.13] 'Reference about Linum' *added pencil*
2.13 the genus . . . print 2.14] *double scored pencil*
2.18 Rubiaceæ] 'Rubiaceæ' *added pencil*
3.10 In ... interval 3.13] *crossed pencil*
Top of first page: 'Cleistogamy' *red crayon*

1 Letter to George Bentham, 24 November [1861]. Bentham's note has not been found.

2 Frederick Currey was the botanical secretary of the Linnean Society of London. CD had asked Bentham, who was president of the society, whether his paper on dimorphism in *Primula*, read at the meeting of 21 November 1861, could be published in the *Journal of the Proceedings of the Linnean Society* rather than in the *Transactions of the Linnean Society of London* (see letter to George Bentham, 24 November [1861]).

3 CD's paper appeared in the 1862 volume of the *Journal of the Proceedings of the Linnean Society* (see also *Collected papers* 2: 45–63).

4 Bentham refers to the major work that he and Joseph Dalton Hooker were engaged in: the production of a systematic compilation of all known vascular plant genera. The first part of the first volume of *Genera plantarum* appeared in 1862 (Bentham and Hooker 1862–83).

5 The German botanist Karl Sigismund Kunth had described many of the plants collected in South America by Alexander von Humboldt.

6 Planchon 1847–8.

7 Jacquin 1794 and Zuccarini 1825.

8 CD included this information about *Ononis minutissima* in *Forms of flowers*, p. 326.

To George Bentham 26 November [1861][1]

Down. | Bromley. | Kent. S.E.
Nov. 26th

My dear Bentham

Our notes have crossed on the road. I know it is an honour to have a paper in the Transactions & I am much obliged to you for proposing it, but I shd. *greatly* prefer to publish in Journal.[2] Nor does this apply exclusively to myself, for in old days at Geolog. Soc. I always protested against an abstract appearing when the paper itself might appear.—[3] I abominate, also, the waste of time (& it would take me a day) in making an abstract.— If the Referee on my paper should recommend it to appear in Transactions, will you be so kind as to lay my *earnest request* before the Council that it may be permitted to appear in the Journal.—

You must be very busy with your change of residence;[4] but when you are settled & have some leisure perhaps you will be so kind as to give me some cases of dimorphism like that of Primula.— Should you object to my adding them to those given me by A. Gray?[5] By the way I heard from A. Gray this morning & he gives me two very curious cases in Boriginaceæ.[6]

Pray believe me | Yours very sincerely | C. Darwin

I am much pleased that you approve of my Primula paper.—

P.S. As this note was written before your very interesting note arrived this morning, I send it off, as I have not time to write by todays post.[7]

Cordial thanks.— | In Haste | C.D.—

Royal Botanic Gardens, Kew (Bentham letters: 689)

[1] The year is provided by the relationship to the preceding letter.
[2] See letter to George Bentham, 24 November [1861]. An earlier note that CD had apparently received is missing.
[3] CD served as secretary of the Geological Society of London from 1838 to 1841.
[4] The Benthams, who had spent several months without permanent quarters, were moving into 25 Wilton Place, London (Jackson 1906, p. 190).
[5] In his paper on *Primula*, published in the *Journal of the Proceedings of the Linnean Society (Botany)* 6 (1862): 77–96, CD included information provided by Asa Gray on cases of dimorphism in Rubiaceae (see also *Collected papers* 2: 62).
[6] Letter from Asa Gray, 9 November 1861.
[7] See preceding letter.

From George Bentham 29 November 1861

Royal Gardens Kew
Nov 29/61

My dear Darwin

The Oxalis pairs distinguished on account of the proportion of the stamens and styles, and which have nevertheless been supposed to belong to one species are according to Zuccarini

O. fulgida and rubella
O. macrostylis and tubiflora
O laburnifolia and sanguinea
O. canescens and secunda
O. gracilis and reclinata
O. cuneata and cuneifolia

(Zuccar. Nachtr. Amer. Oxal. p. 19)[1] and I believe there are several others[2]
Yours very sincerely | George Bentham

DAR 109 (ser. 2): 121

CD ANNOTATION
1.4 O. fulgida ... cuneifolia 1.9] 'Oxalis' *added ink*

[1] Zuccarini 1831. See preceding letter and letter from George Bentham, 26 November 1861.
[2] CD cited Bentham concerning dimorphism in *Oxalis* at the end of his paper 'On the two forms, or dimorphic condition, in the species of *Primula*, and on their remarkable sexual relations', *Journal of the Proceedings of the Linnean Society (Botany)* 6 (1862): 77–96 (see also *Collected papers* 2: 62).

To George Bentham 30 November [1861][1]

Down. | Bromley. | Kent. S.E.
Nov. 30th

My dear Bentham

Your letter of the 26th is very valuable to me. Oxalis & Ægiphilos seem capital cases. I will follow your suggestion about violets & today had several plants put into my experimental garden: I have attended & experimented on the regular flowers of Viola: they require insect-aid for at all perfect fertility; & as insects do not freely visit them, I have always supposed that the imperfect flowers produced seeds by parthenogenesis to keep the stock up; but I will attend to this subject next summer.—[2]

Thymus is a different case from Primula: some individuals being hermaphrodite & some female by abortion of anthers. In the English Thyme this abortion to a *certain limited* extent depends on growing in dry stations (I counted hundreds of plants).[3] The garden Thyme seems to produce both forms irrespective of conditions. The female plants of the garden thyme produce nearly double the weight of seed, compared with the Hermaphrodites: the wild English female Thyme produces more than double the weight.— There is a slight difference in the stigmatic tissue in the Hermaphrodite & Female plants.— As you seem interested in dimorphic plants, I thought that perhaps you would like to hear these details.—

Since the above was written I have received your list of Oxalis, for which very many thanks.—[4]

Pray believe me— Yours sincerely & obliged | Ch. Darwin

Royal Botanic Gardens, Kew (Bentham letters: 691)

[1] Dated by the relationship to the letter from George Bentham, 26 November 1861.
[2] There are some lengthy notes describing CD's examination of the flower parts of *Viola canina*, dated May 1862, in DAR 111: 3–6.
[3] CD had studied the morphology of thyme plants during the summer of 1861 (see letter to Asa Gray, 21 July [1861]). There are notes discussing his observations in his Experimental book (DAR 157a).
[4] See preceding letter.

To Daniel Oliver 30 November [1861][1]

Down Bromley Kent
Nov. 30th

My dear Sir.

As I have lately given Hooker such a frightful amount of trouble & as you have looked at ovules of male Lychnis, will you be so kind, (the next time you have your compound microscope out) as to look at the ovules of Acropera, (sent on glass slides by this post, & to be thrown away when looked at.) which I found, after writing to Hooker lower down in the ovarium.—[2] Please first look at ovules of Maxillaria & Orchis, which resemble those of *all* the several orchids, (including ovaria of buds) which I have examined. Now see what a vast difference the ovules, (or rather mere empty membranous fringes to the placentæ) of Acropera present. These fringes *positively* lie on the same ridges or placentæ which carry the true pulpy ovules in all other orchids.— You will see them also in a transverse section.— I have examined 6 specimens of Acropera; & all the ovules on whole length of the 3 placentæ were in this condition. Why I want you so much to look at these ovules & give me your opinion is that I have never before looked at an ovule; and I have reason to believe that if I can trust these observations, I shall make out sexes of Catasetum, Myanthus &c.— Though I know nothing about ovules, the difference is most conspicuous between the placentæ of Acropera & of *all* (except Catasetum) other orchids.[3]

Besides this difference in the ovules, the stigmatic chamber of Acropera is so small that pollen-masses can hardly be forced in; the surface is but little viscous, & the stigmatic utriculi are very different from those of *every* other orchid except Catasetum in being empty.— From these several reasons I cannot avoid belief that the Acropera flowers of two species which I have examined are males.[4] Will you inform me whether all the several flowers of Acropera luteola sent me by Hooker came from *same plant*??

One other point in the ovaria of Acropera has surprised me; but unless you have carefully attended to orchids, you will perhaps not be able to express an opinion: Lindley & Link say ovarium consists of 6 carpels Brown says of three;[5] & that three of the divisions have no homological signification. Now to my surprise I see three irregular double rows of hyaline projections, running parallel to the ovules, & belonging to the three divisions of the ovarium, which do not carry ovules: you will see a few in the transverse sections & on the dissected longitudinal fragments. What can they be? it seems to me to favour Lindley's & Link's view.— According to Brown they lie on the midribs of his 3 carpellary leaves: according to Lindley, they would be, I suppose, representatives of ovules.

Pray forgive my writing at such length & believe me | Yours very sincerely | C. Darwin

Is it not a rather interesting fact that you have observed in Lychnis viz ovules in the male plant?[6]

I am glad to hear from Hooker that you are going to consider O. Heer: his Atlantis-map seems to me very wild.—[7] The view, which I give in Origin of migration by north during old warmer period seems to me much simpler & agrees better with geological facts.—[8]

P.S. I should have said that ovules of the Orchis have been in spirits: ovules of Maxillaria would have been equally plain & opake had they been in spirits; spirits make no difference to the rudimentary ovules of Acropera.— I sh^d. have imagined that the rudimentary condition of ovules & stigmatic utriculi of Acropera might have been due to culture or bad conditions, had I found any other orchids in this condition; & had not solid edges of stigmatic chamber not been so much contracted.

With respect to 6 divisions of ovarium of Orchis, are there any Monocots. with 6 carpels & only 3 pistils? I suppose not, otherwise Brown would not so have objected to Lindley's view.—

Down House (MS 10: 2)

1 The year is given by the reference to CD's work on the ovules of *Acropera*. His observations were included in *Orchids*, which was published in May 1862.
2 See letter to J. D. Hooker, 25 November [1861].
3 CD had been investigating the pollination mechanism of *Acropera* since October (see letters to J. D. Hooker, 15 October [1861], 18 October [1861], and 27 October [1861]). In *Orchids* he related how he 'had given up the whole case as inexplicable' when it occurred to him that 'although no instance of the separation of the two sexes was known in Orchids, yet that Acropera might be a male plant.' (*Orchids*, p. 206). He further suggested that several species of *Catasetum* were male forms and that *Monachanthus*, *Myanthus*, and *Cycnoches* exhibited a variety of sexual forms (*ibid*., pp. 237–47, 269).
4 CD described the results of his microscopic examination of the ovules of *Acropera*, and his reasons for concluding that the flower was a male form, in *Orchids*, pp. 207–10.
5 See Lindley 1853, p. 174, Link 1849, and Brown 1831.
6 The letter from Oliver has not been found; see, however, the letters to J. D. Hooker, 25 November [1861], and to Daniel Oliver, 7 December [1861].
7 To account for the European character of the flora of Madeira and for the presence of American forms, Oswald Heer had proposed that Madeira, the Canaries, Porto Santo, and the Azores were the remnants of a sunken land-mass, or 'Atlantis', that in the Tertiary period had been connected by a land-bridge to America and Europe (Heer 1857). CD had long been critical of attempts to explain geographical distribution by postulating former land-bridges (see especially *Correspondence* vol. 6, letters to Charles Lyell, 16 [June 1856] and 25 June [1856]). Oliver's critique of Heer's work was published in the *Natural History Review* (Oliver 1862a).
8 CD postulated that during a pre-glacial warm period, northern temperate forms lived at latitudes further north and nearer to the pole than at present, and were able to migrate freely across the almost continuous land between western Europe and eastern America via Siberia (*Origin*, pp. 369–70; see also *Correspondence* vol. 7, letter to Asa Gray, 11 August [1858]).

From Alfred Russel Wallace 30 November 1861

Sumatra, 100 miles E. of Bencoolen
Novr. 30th. 1861

My dear Mr. Darwin

On an evening in the wet season in these Central forests of Sumatra, I occupy myself in writing a few lines to you to say a few things which I may otherwise forget altogether. About a year & a half back I wrote to you from somewhere Eastward of Ceram, with more digested remarks on your book, but the letter with one to my agent Mr. Stevens never I believe reached Amboyna.[1] All I can remember of them now is to the effect that *repeated* perusals had made the whole clearer & more connected to me & the general & particular arguments clearer & more forcible than at first.

I have lent the book to two persons here in the East, neither of them with any but the vaguest & most general knowledge of or taste for Natural History but both men of much reading & with a taste for speculation & theory wh. as Bentham says is but another term for thought.[2] The first Mr. Duivenbode of Ternate, a Dutchman educated in England, read it three times through before he returned it to me, expressing himself so much pleased & interested that he wished to master the whole argument.[3] The other a merchant captain settled at Timor Delli with whom I lived 3 months kept it all the time, was constantly reading it & we made it a subject of conversation almost whenever we met, & when I was leaving he did not return it till the steamer arrived going over the recapitulations of the chapters & the conclusion to get the most of it he possibly could—[4] These humble testimonies prove I think both the attractive manner in which you have treated the subject & the clearness with which you have stated & enforced the arguments; & I trust they will be as pleasing to you as I assure you they were to me.

I met the other day on board the Dutch steamer the Geologist accompanying a Prussian voyage of discovery &c. "*Baron Richthofen*". He has been t⟨hro⟩ugh Java where he has made large collections of fossils— He ⟨is⟩ going now to leave the ship & travel in British India & afterwards go to the Amoor & overland to Europe to study & improve himself in Geology.[5] He seems a very intelligent & good naturalist. He was reading your Book wh. he had borrowed in Java, & on my asking him if he was a convert,—he smiled & said "It is very easy for a Geologist."[6] I have also seen *Dr. Schneider* who has geologized in Timor— He assures me he found many teeth of *Mastodon*,—also *terebratulæ orthoceras* & other molluscs. He had given most of his collection to the Zoologist of the same Prussian Expedition but says he has published descriptions of all the Timor fossils in the Batavian Journal of Nat. Sciences, I forget the Dutch title but you know the work no doubt—[7] They are to be published this month I think he told me. ⟨*part of page excised*⟩

I trust your great work goes on & is soon to appear.[8] I hope however you will have it copiously illustrated. I am sure it will be for the publishers interest to do so as it will I have no doubt double the circulation. There are so many things that are weak when merely mentioned or described by numbers which become clear & strong when an appeal is made to the eye. The *varieties of pigeons* the *stripes of horses*, the variations in *ants*, the formation of *honeycombs* & a hundred other things will all be better for good illustrations— They will also make the book newer & more distinct from its forerunner in the eyes of the public to whom it will otherwise appear as perhaps a mere enlargement— If this point is not decided, pray take it seriously into consideration.

I see nothing but the Athenæum so know little of what is going on among Naturalists— Huxley & Owen seem to be at open war, but I cannot glean that any one has ventured to attack you fairly on the *whole question*, or ventured to answer the whole of your argument—[9] I see by his advertisement Dr. Brees professes to do so but have no notice of his book.[10]

I have lately been very unsuccessful owing to unprecedented wet weather (the effect I suppose of the comet).[11] This is my ⟨*part of page excised*⟩

P.S. The shabby way in which your opponents ⟨*several words missing*⟩ is amusing. First comes Owen with his new interpretation of what naturalists mean by "*creation*" which it turns out is not *creation* at all, but "*the unknown manner in which species have come into existence*"!!![12] Phillips I see adopts this new interpretation which ought certainly to raise up the bench of bishops against them,[13] for what then becomes of "*special creation*" & "*special adaptation*" & "intelligent forethought visible in each special creation",—if *creation* is not creation at all, but a mere convenient expression of ignorance of the laws by which species have originated.

Has my friend Bates sent you his papers on Amazon Insects, in wh. he gives details of "variation" which are a most valuable contribution to our cause—[14] His papers also shew what mere blind work is the making of species from what may be called *chance collections*;—a species here, another there, a ♂ from one locality a ♀ from another, here a rare variety, & from another place the typical form. Of course with such materials & with localities often incorrectly indicated each succeeding *elucidator* has but added to the confusion; & the whole has formed a *chaos*, till some one collecting & observing for years over an extensive but connected district has been able to bring the whole into order.

Anor | Sumatra

Incomplete
DAR 181.1

CD ANNOTATIONS
2.2 knowledge . . . made it 2.8] *crossed ink*
3.7 "It . . . told me. 3.13] 'A. Wallace | Nov. 20/1861' *added in margin, ink*; '18'[15] *brown crayon*

[1] The letter to CD to which Wallace refers may be that dated [December? 1860] (see *Correspondence* vol. 8), of which only a part is extant. Wallace gave CD his initial comments on *Origin* in a letter, now missing, dated 16 February 1860 (see *ibid.*, letter to A. R. Wallace, 18 May 1860). Samuel Stevens acted as Wallace's London agent, handling the natural history specimens he collected (Wallace 1905, 1: 266).

[2] The reference is to Jeremy Bentham.

[3] Wallace also mentioned Mr Duivenboden in the published account of his travels in the Malay Archipelago (Wallace 1869, 2: 2).

[4] In his autobiography, Wallace identified the 'merchant captain' as Captain Hart, owner of a coffee plantation situated 'about a mile out of the town' of Dili on the island of Timor (Wallace 1905, 1: 375). For a map indicating the course of Wallace's travels in the Malay Archipelago, see Wallace 1869.

[5] Ferdinand Paul Wilhelm von Richthofen accompanied the Prussian East Asia expedition, under the leadership of Friedrich Albrecht, Graf zu Eulenburg, to Japan, China, and Thailand, 1859–62. He continued travelling in the Far East before visiting the western part of the United States (Sarjeant 1980).

[6] In his letter to Wallace, 18 May 1860 (*Correspondence* vol. 8), CD remarked that 'there are almost more Geological converts [to his theory] than of pursuers of other branches of natural science.'

[7] Schneider 1863.

[8] CD had told Wallace in his letter of 18 May 1860 (*Correspondence* vol. 8) that he was working on his 'larger work', which he planned to publish in separate volumes. Part of this material was published in *Variation* in 1868.

[9] Thomas Henry Huxley and Richard Owen were engaged in a public debate concerning the anatomical differences between human and simian brains (see R. Owen 1861a and Huxley 1861a and 1861b). Several highly charged letters from the two were published in the *Athenæum* in March and April 1861. For references to the debate in CD's correspondence, see the letters to T. H. Huxley, 1 April [1861], and to J. D. Hooker, 23 [April 1861].

[10] Bree 1860.

[11] The comet, described as one 'of extraordinary splendor', was first sighted at the end of June 1861 (*Annual register* (1861), Chronicle, pp. 99–100).

[12] Wallace refers to Owen's reference to the 'axiom of *the continuous operation of the ordained becoming of living things*' in his unsigned review of *Origin* ([R. Owen] 1860b, p. 500). See also letter to T. H. Huxley, 3 January [1861].

[13] John Phillips addressed the question of the origin of species, criticising the account provided by CD in *Origin*, in Phillips 1860. For CD's reaction to this book, see the letter to Asa Gray, 5 June [1861].

[14] Wallace and Henry Walter Bates had naturalised together in the Amazon region of South America (Marchant ed. 1916, 1: 25–30). Bates had sent CD a copy of his first paper on the insect fauna of the Amazon valley (Bates 1861a) (see letters from H. W. Bates, 18 March 1861 and 28 March 1861), and CD had heard Bates's second paper on this topic (Bates 1861b), read before the Linnean Society on 21 November 1861 (see letter to H. W. Bates, 3 December [1861] and n. 2).

[15] The annotation refers to CD's portfolio of notes on the means of dispersal of animals and plants.

From H. W. Bates [1 December] 1861[1]

King St. Leicester
31 Novem 1861

My Dear Sir

The following are the references which I promised.[2]

Volucella bombylans

V. bombylans (red-tipped abdomen)

V. plumata (Meigen) (= mystacea Fab) (white-tipped abdomen)

M. Macquart (Hist. Nat. des Diptères Vol. 1 1834, p 494) places both together, terming V. plumata "variété constante" of V. bombylans.[3] at p 479 he quotes M. de St Fargeau who observed the accouplement of the two forms:[4] and stated moreover that the two other European Volucellæ also had each a constant variety. A kind of Dimorphism?

Zeller (Entom. Zeitung, 1842 p 65) confirms the above statement from personal observation, as to V. bomb. & V. plumata.[5]

Erichson (Entom. Zeitung 1842 p 113) described all the known varieties of the species.[6] They were then 7—No 1 being the true bombylans which passed gradually into plumata through V. hæmorrhoidalis (Zetterstedt) and others.

On the meaning of the colours and markings of butterflies' wings. The article on the subject which I mentioned to you is by D^r. Rössler in the Wiener Entomologische Monatschrift 1861 p 163.[7] On reading it through carefully I find many interesting facts, which I should have done well to quote in my paper but I wrote it before D^r. R's was published. He says plainly that the colours & markings are given to deceive the enemies of the species. All the facts are from European insects. He is wrong in saying that "it is a general law that animals are clothed in the colours of their resting places." This is a kind of mistake which we are all apt to fall into.

I think the whole tenour of your book teaches what I carefully insisted on in my paper that the welfare of species is the object of all structures & forms. All species exist by virtue of some endowment enabling them to withstand adverse circumstances. If an exact mimetic dress be not required to enable a species to maintain itself, the causes which tend to produce it will not operate. There are species of the same genus of similar habits & living on the same soil; one of which is of the colour of the soil and another of a completely contrasted colour. Therefore it is not a general law that a parasitic species shall mimic its victim &c &c. the mimicry is only part of the wider law above mentioned & the exceptions are proved as they ought to be by the same wider law.

Dr Rossler remarks that the underside of the Vanessa & other butterflies are coloured to resemble the bark of trees; because as they close the wings when they alight, they become suddenly invisible. In moths (Noctuæ) which rest with wings deflexed the upper covering the lower it is the upper surface of the upper wing only which imitates bark. I will mention that in some of these moths (Catocala) the underwings are of brilliant colours; So that here we have a double dress answering two purposes in the welfare of the species. The high colour being explicable by sexual selection & the plain colour by the check caused by insectivorous animals.

The essay of Burmeister's which I mentioned as probably worth reading is called Vergangenheit & Gegenwart des Thierreichs[8]

Incomplete
DAR 205.10 (Letters)

CD ANNOTATIONS
6.1 D^r ... invisible. 6.3] *scored red crayon*
6.4 the upper covering ... bark. 6.5] *scored red crayon*
6.6 So ... animals. 6.8] *scored red crayon*
7.1 The ... Thierreichs 7.2] 'Showing earlier form general, later specialised.' *added pencil*
On cover: '*Mr Bates* | References on Dimorphism of Volucella— on Mimetic resemblance— ; 'Colours of *lower surface of wing [*interl, del pencil*] Butterflies & Moths— Sexual Selection' *ink, del pencil*; Reference to Burmeister on specialisation' *ink*; 'On under surface of wings of Vanessæ like bark | On upper surface of cut [*interl*] wings of Moth like bark— Lower wings brightly coloured (Red underwings Weir[9]—' *ink, del pencil*; 'Selection' *brown crayon, del pencil*

CD note:[10]
Nov. 22^d. 1861. Mr Bates showed me the 2 Volucellas like B. hortensis & lapidarius.— in B. Mus they are marked as distinct species; but Erichson says they blend together on continent, though distinct in

England— F. Smith has bred them ['here' *del*] one from here & has not seen variations—[11] In Amazonia there is one jet black H. Bee & *one* parasitic Bee *not Apathic [*interl*], in same region also jet black. In England some Apathic very like the Humbles they frequent—other parasitic Bees not like.—but then close resemblance not necessary, for stronger Humble-bees will take up residence occasionally with distinct species.— F. Smith showed me a brownish Volucella always parasitic in Brown Humble-bees nests.— The volucella very good case in all ways.— I saw the mimetic [*interl*] Butterflies wonderful. wonderful—

[1] The date of this letter has been given as 1 December since there are only thirty days in November.
[2] Bates had shown CD his collection of South American insects, which was being exhibited at the British Museum, during CD's recent visit to London (see CD note). CD's visit to the British Museum is also mentioned in the letter to J. D. Hooker, 25 November [1861].
[3] Macquart 1834–5.
[4] Amédée Louis Michel Lepeletier de St Fargeau was a noted French entomologist.
[5] Zeller 1842.
[6] Erichson 1842.
[7] Rössler 1861.
[8] The publication details of the essay by the German naturalist Karl Hermann Konrad Burmeister have not been traced.
[9] It is possible that CD made some of the cover annotations several years after he received the letter, during his preparation of *Descent*, in which there is a chapter on sexual selection in butterflies and moths (*Descent*, 1: 386–417). The naturalist John Jenner Weir mentioned the brightly coloured underwings and dull upperwings of certain Lepidoptera in Weir 1869–70, p. 22. CD cited Weir's article in *Descent*, 1: 395 and 417.
[10] The note is associated with the letter in DAR 205.10 (Letters).
[11] CD refers to the German entomologist Wilhelm Ferdinand Erichson and to Frederick Smith, an entomologist employed in the zoological department of the British Museum.

To J. D. Hooker 1 December [1861]

Down. | Bromley. | Kent. S.E.
Dec. 1st

My dear Hooker

The Catasetum arrived quite safe, but I have not seen it, as it is at friends hot-house, & I hear it has splendid truss of buds, which I am so very glad of, as I have thought of many experiments to try.—[1]

The Bolbyphyllum came, also, quite safe; but alas I can see no trace of irritability in Labellum.—[2]

Hearty thanks for the orchid-flowers: one of Maxillaria showed me quite a new movement in caudicle, & I was extremely glad of Acropera, which has cost you so much trouble & has so perplexed me.—[3] I send to Oliver by this post its wretched rudimentary ovules, with request that he will look at them; for I never looked at an ovule in my life before![4] Yet a blind man (to speak like an Irishman) could see the difference in the ovules of Acropera & of all the many other orchids at whose ovules I have now looked.—

Pray get your foreman G.......? to give me the precise facts about the crossing of Victoria.[5]

I am very glad to hear that you liked Bates:[6] I have seldom in my life been more struck with a man's power of mind: it is capital about the Plates for Linn. Transactions.— I do not think I could offer him the £10, but perhaps he will want it for the drawing of Plates.—[7] His explanation of these mimetic resemblances removes a difficulty which has very long perplexed me, & which never could have been understood without his facts.—[8]

When Lecoq came I was so disgusted at size, (not so much at price, which is only 3£) that I put the 9 volumes on highest shelf; but I have this evening taken down 1st & 9th vol & have cut pages & will have look & see what it is. If it does not suit me, & does suit you, you may have it for 2£: It seems full of details on range of each individual species.[9]

How little plants of Africa must be known from what you say of the Laurels![10]

Adios.— | your affect— | C. Darwin

P.S. | Have you ever attended to morphology of ovarium of Orchids, & to difference between Brown on one hand, & Lindley & Link on other, on whether it consists of 6 or 3 carpels.— In specimens on glass-slides sent to Oliver, there is row of hairs, or spine-like cells on the intermediate segments (between those which bear the ovules) which I fancy indicate that these segments are independent parts, & not mere lines of splitting, as Brown argues, to favour dissemination of seed.—[11]

But of course such a point is quite beyond my tether: no one seems to have observed these projecting spine-like cells.—

I wonder who makes abstracts for Gardener Chron: there is capital one on Primula; I am surprised much if Kippist does it so well.—[12]

Endorsement: '/61'
DAR 115.2: 135

[1] CD had asked Hooker to send him more specimens of *Catasetum* and *Bolbophyllum* (see letters to J. D. Hooker, 1 November [1861] and [22 November 1861]). In *Orchids*, George Henry Turnbull of Down is thanked for allowing CD 'free use of his hot-houses' (*Orchids*, p. 158 n.).

[2] In *Orchids*, p. 172, CD stated that one of the many 'singular properties of Orchids' was the 'irritability of the labellum in several distantly-allied forms'. He went on to say that, although *Bolbophyllum* had been described as having an irritable labellum, he had been unable to detect irritability in the specimens he examined, adding that he had not in fact seen any orchid with an irritable labellum.

[3] The movement of the pedicel of the rostellum in *Maxillaria ornithorhyncha* is described in *Orchids*, p. 191. CD had been trying to understand the pollination mechanism of *Acropera* since October (see letters to J. D. Hooker, 15 October [1861], 18 October [1861], and 27 October [1861]).

[4] See letter to Daniel Oliver, 30 November [1861].

[5] See letter from William Hugh Gower, 23 November 1861.

[6] See following letter and letter to J. D. Hooker, [9 December 1861].

[7] CD had offered Bates £10 to help defray the expense involved in including coloured plates in the published version of his paper on mimetic analogies in insects (Bates 1861b) (see letter to J. D. Hooker, 25 November [1861]). Hooker was a member of the council of the Linnean Society (L. Huxley ed. 1918, 1: 407). Bates's paper was published with two coloured plates illustrating 'mimetic analogies between various Lepidopterous insects and the Heliconidæ' (Bates 1861b, p. 564).

[8] CD included a discussion of Bates's account of mimetic resemblance in the fourth edition of *Origin* (*Origin* 4th ed., pp. 502–6).

[9] Lecoq 1854–8. See letter to J. D. Hooker, 25 November [1861].
[10] Hooker was currently describing the plants collected by Gustav Mann in the Cameroons and nearby islands (L. Huxley ed. 1918, 2: 18).
[11] See letter to Daniel Oliver, 30 November [1861].
[12] The *Gardeners' Chronicle and Agricultural Gazette*, 30 November 1861, pp. 1048–9, carried summaries of CD's paper, 'On the two forms, or dimorphic condition in the species of *Primula*, and on their remarkable sexual relations', and of Bates 1861b, both read at a meeting of the Linnean Society on 21 November 1861. Richard Kippist was librarian of the society.

To H. W. Bates 3 December [1861]

Down Bromley Kent
Dec. 3^d.

My dear Sir

I thank you for your extremely interesting letter, & valuable references,—though God knows when I shall come again to this part of my subject.—[1] One cannot of course judge of style when one merely hears a paper, but yours seemed to me very clear & good.—[2] Believe me that I estimate its value most highly. Under a general point of view, I am quite convinced (Hooker & Huxley took same view some months ago) that a philosophic view of nature can solely be driven into naturalists by treating special subjects as you have here done.— Under a special point of view I think you have solved one of the most perplexing problems which could be given to solve.[3] I am glad to hear from Hooker that Linn. Soc. will give Plates, if you can get drawings; but I suppose they might be drawn on to stone or copper.— Pray *excuse* me for again saying if ever you want £10 or £20, I shall be pleased to send it, for any aid in Natural History:—[4]

Do not complain of want of advice during your Travels; I daresay part of your great originality of views may be due to the necessity of self exertion of thought. I can understand that your reception at B. Museum would damp you; they are a very good set of men, but not the sort to appreciate your work.[5] In fact I have long thought that *too much* systematic work of description somehow blunts the faculties. The general public appreciates a good dose of reasoning, or generalisation with new & curious remarks on habits, final causes &c &c, far more than do the regular naturalists.—

I am extremely glad to hear that you have begun your Travels. (I thought your Glacial Letter admirably written);[6] I am very busy, but I shall be *truly* glad to render any aid which I can by reading your 1st Chapter or two. I do not think I shall be able to correct style,—for this reason, that after repeated trials I find I cannot correct my own style till I see the M.S. in type. Some are born with a power of good writing, like Wallace; others like myself & Lyell have to labour very hard & slowly at every sentence. I find it very good plan, when I cannot get a difficult discussion to please me, to fancy that some one comes into the room, & asks me what I am doing; & then try at once & explain to the imaginary person what it is all about.— I have done this for one paragraph to myself several times; & sometimes to M^rs.

Darwin, till I see how the subject ought to go.— It is, I think, good to read one's M.S. aloud.— But style to me is a great difficulty; yet some good judges think I have succeeded, & I say this to encourage you.— What *I think* I can do will be to tell you whether parts had better be shortened.— It is good I think to dash "in medias res", & work in later any descriptions of country or any historical details which may be necessary.— Murray likes lots of woodcuts— give some by all means of Ants.— The public appreciate Monkeys,—our poor cousins.— What sexual differences are there in monkeys? Have you kept them tame? if so about their expression.—

I fear that you will hardly read my vile handwriting, but I cannot without killing trouble, write better.—

You shall have my candid opinion on your M.S., but remember it is hard to judge from M.S.—one reads slowly & heavy parts seem much heavier.— A first rate judge thought my Journal very poor; now that it is in print, I happen to know, he likes it.—[7] I am sure you will understand why I am so egotistical.—

I was a *little* disappointed in Wallace's Book on the Amazon; hardly facts enough.—[8] On other hand in Gosse's books there is not reasoning enough to my taste.—[9] Heaven knows whether you will care to read all this scribbling—

M.S. can be sent by Book Post, if marked to be printed.— Had you not better register it?

Many thanks for Wallace's letter; he rates me much too highly & himself much too lowly.— That was an admirable paper of his in Linn. Journal.—[10] But what strikes me most about Mr Wallace is the absence of jealousy towards me: he must have a really good honest & noble disposition.[11] A far higher merit than mere intellect.

With cordial good wishes & thanks | Yours sincerely | C. Darwin

I am glad you had pleasant day with Hooker: he is an admirably good man in every sense.[12]

Endorsement: '1861'
Cleveland Health Sciences Library (Robert M. Stecher collection)

[1] Letter from H. W. Bates, [1 December] 1861.

[2] Bates's paper on the butterflies of the Amazon River basin (Bates 1861b) had been read at a meeting of the Linnean Society of London on 21 November 1861. At the same meeting CD had also read a paper, which described his study of dimorphism in *Primula* (see *Collected papers* 2: 45–63).

[3] Bates's paper described mimetic resemblances in Amazonian butterflies that could not be explained as the result purely of adaptation to similar conditions. Invoking the theory of natural selection, Bates suggested that the 'selecting agents' were insectivorous animals that destroyed those varieties of edible species that did not sufficiently resemble the inedible analogical form (Bates 1861b, pp. 511–12).

[4] See preceding letter and letter to J. D. Hooker, 25 November [1861].

[5] Members of the staff of the zoological department of the British Museum, under the direction of John Edward Gray, included the entomologist Frederick Smith, George Robert Gray (an expert on insects and birds), and Adam White (a specialist on Coleoptera and Crustacea) (British Museum (Natural History) 1904–6).

[6] Letter from H. W. Bates, 28 March 1861. Bates's account of his travels, *The naturalist on the river Amazons*, was published by John Murray in 1863.

[7] Henry Holland read CD's *Beagle* journal in manuscript and thought that it did not merit being published separately from Robert FitzRoy's general account of the voyage. See *Correspondence* vol. 1, letter to Caroline Darwin, [7 December 1836]; see also *ibid.*, letters from Emma Wedgwood to F. M. Wedgwood, 17 December 1836, and from Hensleigh Wedgwood, [20 December 1836].

[8] CD read Wallace 1853 in February 1854 (see *Correspondence* vol. 4, Appendix IV, 128: 8). There is a brief abstract of the work in DAR 205.3: 156–7.

[9] CD had read several works by Philip Henry Gosse, including Gosse 1851 and Gosse 1853 (see *Correspondence* vol. 4, Appendix IV, 128: 2 and 128: 10). There is a copy of Gosse 1851 in the Darwin Library–CUL.

[10] Alfred Russel Wallace had sent CD the manuscript of Wallace 1860, which CD communicated to the Linnean Society of London for publication (see *Correspondence* vol. 7, letter to A. R. Wallace, 9 August 1859).

[11] The reference is to Wallace's positive reaction to the arrangements made to publish his paper, which outlined the theory of natural selection, jointly with a statement of CD's theory (*Journal of the Proceedings of the Linnean Society (Zoology)* 3 (1859): 45–62). For the circumstances that led to the publication, see *Correspondence* vol. 7.

[12] Bates's description of his meeting with Joseph Dalton Hooker was presumably given in the part of his letter of [1 December] 1861 that is now missing.

To Ludwig Rütimeyer 5 December [1861][1]

Down. | Bromley. | Kent. S.E.
Dec. 5th

Dear & much respected Sir.—

I have been slowly reading through your admirable work, & this evening I came to the passage, in which you express a wish to see the skull of the wild White cattle of Chillingham Park.[2]

If you could tell me how so heavy an object as a Bull's skull could be sent to Switzerland, I would try & procure one for you.— I would write to the Earl of Tankerville, & I think (but cannot say positively) that he would grant this favour.[3] But I fear that the carriage would be too expensive. You would have to tell me clearly how it ought to be sent & directed. Of course it would be several months, before a Bull might be killed & it would take some time to partially clean the skull.— Should you wish for a few of the upper neck vertebræ?[4]

Permit me to say how much I admire your work; it seems to me the most important contribution on domestic races ever published.[5] I have been pleased to hear what you say on the Zebu; for in my M.S. I have come from habits alone, to the same conclusion with you.— Of course your evidence is of a far higher kind.—[6]

I have a volume half prepared for publication (but interrupted of late by some Botanical work) on "Variation of animals under domestication," in which I treat chiefly of Pigeons, Fowls, Rabbits &c;[7] & you may well believe how profoundly your admirable labours have interested me.—

With most sincere respect | I remain | Dear Sir | Your faithful servant | Charles Darwin

Elizabeth Rütimeyer (private collection)

[1] The year is given by the reference to Rütimeyer 1861 (see n. 2, below) and by the relationship to the letters to Ludwig Rütimeyer, 15 January [1862] and 11 February [1862] (*Correspondence* vol. 10).

[2] Rütimeyer, professor of zoology and comparative anatomy in Basel, had sent CD a presentation copy of Rütimeyer 1861, a treatise on the animal remains found during the excavation of the Swiss lake dwellings. Rütimeyer complained that his research on the origins of domestic cattle was incomplete because he had been unable to examine a skull or skeleton of one of the Chillingham cattle. This ancient breed was kept on the Chillingham Park estate of Charles Augustus Bennet, earl of Tankerville.

[3] CD asked his friend and distant relation Henry Holland, who was an acquaintance of Bennet, to ask Bennet to send a skull to Rütimeyer (see *Correspondence* vol. 10, letters from Henry Holland, [before 15 January 1862] and 15 January [1862]). It was believed that the Chillingham cattle were the closest living representatives of Britain's original wild cattle (see Hindmarsh 1839). CD himself had long been interested in the origin and habits of the Chillingham cattle (see *Notebooks*, Notebook D, p. 48). In May 1861 CD had written to Luke Hindmarsh asking him for information on the rate of increase of the Chillingham herd (see letters to Luke Hindmarsh, 3 May [1861] and 12 May [1861], and letter from William Hardy to Luke Hindmarsh, [8 May 1861]).

[4] Rütimeyer replied to CD's letter on 11 December 1861. His letter has not been found, but see the letter to Ludwig Rütimeyer, 15 January [1862] (*Correspondence* vol. 10).

[5] In Rütimeyer 1861, Rütimeyer described human and animal remains that had been found in Stone Age Swiss lake-dwellings and compared these forms with contemporary races, suggesting that his results might be relevant to the question of organic change.

[6] Zebus are humped cattle (*Bos indicus*) that inhabit tropical countries. Rütimeyer had concluded that humped and non-humped cattle represent two different species. See Rütimeyer 1861, p. 146. CD cited Rütimeyer's osteological studies several times in the discussion of cattle in *Variation* 1: 79–93.

[7] *Variation* was published in 1868.

To Daniel Oliver 7 December [1861][1]

Down. | *Bromley.* | *Kent. S.E.*
Dec 7th

My dear Sir

I am extremely much obliged for your two letters.— The quasi-ovules in Acropera certainly were not in least dried & always differed widely from those in the orchids, young & old, (*not* fertilised) of all the main divisions; so I must trust to my own judgment, & make the statement exclusively on my own authority.—[2]

Your former letter explained everything most clearly; I thought it worth while just to ask you the meaning of the spines.— I supposed that Brown was a much higher authority than Lindley & Link; & the subject does not concern me; I asked out of mere curiosity.[3] I have been interested by making out from state of ovules (according to my judgment) *& of pollen*, that Catasetum tridentatum is male Monacanthus viridis—female Myanthus barbatus—Hermaphrodite & you know they have been produced on same plants.—[4] I forgot to say in my last note, that the existence of perfect ovules in males of Lychnis dioica, well agrees with some observations which I made several years ago (but not carefully enough made for publication)[5] (I see that my facts are too doubtful to give even in letter)

Pray believe that I am truly grateful to you, so busy as you are, giving up so much time to me, & I remain | Yours very sincerely | C. Darwin

Down House (MS 10: 3)

[1] The year is given by the relationship to the letters to Daniel Oliver, 30 November [1861], and to J. D. Hooker, 1 December [1861].

[2] The letters from Oliver have not been found. CD had asked him to examine the ovules of the orchid *Acropera* (see letters to Daniel Oliver, 30 November [1861], and to J. D. Hooker, 1 December [1861]).

[3] See letter to Daniel Oliver, 30 November [1861]. CD refers to Brown 1831, Link 1849, and Lindley 1853.

[4] The description of a spike that apparently bore the flowers of the three different orchid genera mentioned in the letter was first published in Schomburgk 1837. Commenting on this and other such instances, John Lindley stated: 'Such cases shake to the foundation all our ideas of the stability of genera and species, and prepare the mind for more startling discoveries than could have been otherwise anticipated.' (Lindley 1853, p. 178).

[5] Although CD deleted this sentence (see Manuscript alterations and comments), it has been retained in order to make sense of the following sentence, which pertains to it. CD experimented on *Lychnis dioica* in 1856, using seeds that were given to him by John Stevens Henslow (see *Correspondence* vol. 6, letters to J. S. Henslow, 16 June [1856], and to J. D. Hooker, 12 April [1857]). Notes on his experiments are in DAR 157a.

To J. D. Hooker [9 December 1861]

[Down]

My dear Hooker

Lecoq is a miserable book—, dreadfully spun out, with maudlin speculations & a great dearth of precise facts: I do not believe it would be worth your having, & as here & there, miles apart, I find a reference or fact worth keeping, I will keep the monstrous work.—[1]

Bates writes he spent with you 3 or 4 of the most agreeable hours he ever spent in his life.—[2]

C. Darwin

You were, of course, quite right. Bonatea does not at all favour the splitting & subsequent fusion of sepals & petals— The vessels run in Bonatea as in Habenearia.

I cannot get out of my head that these cases throw great doubt on value of vessels in regard to homologies.[3]

[Enclosure][4]

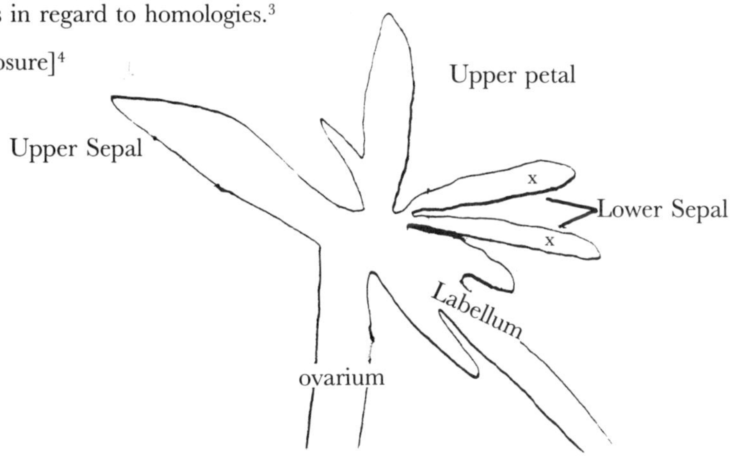

This *seems* structure of Bonatea speciosa & you will see what a little change in adhesion would *almost* convert it into Habenaria.—

Endorsement: 'Dec 9/61.'
DAR 115.2: 136, 129c

[1] CD refers to Lecoq 1854–8, which he had recently purchased. See letters to J. D. Hooker, 25 November [1861] and 1 December [1861].
[2] Bates apparently told CD about his visit with Hooker in the part of his letter of [1 December] 1861 that is now missing (see letter to H. W. Bates, 3 December [1861]).
[3] See letters to J. D. Hooker, 10 November [1861] and 14 November [1861].
[4] The diagram has been reproduced at 60% of its original size.

To Asa Gray 11 December [1861][1]

Down Bromley Kent
Dec. 11th.

My dear Gray.—

Many & cordial thanks for your two last most valuable notes.—[2] What a thing it is, that when you receive this we may be at war, & we two be bound, as good patriots, to hate each other, though I shall find this hating you very hard work.—[3] How curious it is to see two countries, just like two angry & silly men, taking so opposite a view of the same transaction! So far as I can see we rest entirely on Wilkes' acting as Judge.—[4] I fear there is no shadow of doubt we shall fight, if the two Southern rogues are not given up.[5] And what a wretched thing it will be, if we fight on side of slavery. No doubt it will be said we fight to get cotton; but I fully believe that this has not entered into the motive in the least.— Well, thank [Heaven][6] we private individuals, have nothing to do with so awful a responsibility.— Again how curious it is that you seem to think that you can conquer the south; & I never meet a soul, even those who would most wish it, who thinks it possible,—that is to conquer & retain it. I do not suppose the mass of people in your country will believe it; but I feel sure if we do go to war, it will be with the utmost reluctance by all classes, ministers of government & all.— Time will show, & it is no use writing or thinking about it.—

I called the other day on Dr. Boott & was pleased to find him pretty well & cheerful;[7] I see, by the way, he takes quite an English opinion of American affairs, though an American in heart. Buckle might write a chapter on opinion being entirely dependent on Longitude![8]

Hearty thanks for facts on dimorphism in Borragineæ.[9] What a riddle is the Mertensia: I presume that it would be impossible to get seeds.— It is very kind of you to try & send me the living plant of Houstonia. You may rely on it, I will send you a copy of my Primula paper as soon as I can get one; but I believe it will not be printed till April 1st. & therefore after my Orchid Book.[10] I care more for your & Hooker's opinion than for that of all the rest of world, & for Lyell's on geological points. Bentham & Hooker thought well of my paper when read; but no one can judge of evidence by merely hearing a paper.[11] The only thing which has interested me of late is the making out that *Catasetum tridentatum* is male, *Monacanthus viridis* female, & Myanthus barbatus the hermaphrodite, of the same species: this is clear, & accounts for all 3 forms appearing sometimes on same plant: they differ as much as or more than a peacock & pean-hen.—[12]

Bentham has given me list of species of Oxalis dimorphous like Primula & some Menthas are so he says—but I am not sure that he distinguishes such cases as Thymus.—[13] I have not yet had time to examine Amsinckia, which you sent me dried.—

If you can remember it, tell me when you write, how Dana is.—[14]

Lyell is going to publish, what will be an interesting little Book on the Geological History of man, & I believe he will touch on permanence of Species.—[15]

With respect to Design, I feel more inclined to show a white flag than to fire my usual long-range shot. I like to try & ask you a puzzling question, but when you return the compliment, I have great doubts whether it is a fair way of arguing. If anything is designed, certainly Man must be; one's "inner consciousness" (though a false guide) tells one so; yet I cannot admit that man's rudimentary mammæ; bladder drained as if he went on all four legs; & pug-nose were designed. If I was to say that I believed this, I should believe it in same incredible manner as the orthodox believe the Trinity in Unity.— You say that you are in a haze; I am in thick mud;—the orthodox would say in fetid abominable mud. I believe I am in much the same frame of mind as an old Gorilla would be in if set to learn the first book of Euclid. The old Gorilla would say it was of no manner of use; & I am much of the same mind; yet I cannot keep out of the question.

My dear Gray I have written a deal of nonsense. | Yours most cordially | C. Darwin

Gray Herbarium of Harvard University (62)

[1] The year is established by the reference to the *Trent* affair (see n. 3, below).

[2] Only one of the two letters mentioned has been found (see letter from Asa Gray, 9 November 1861).

[3] In September 1861, the Confederate government appointed two special commissioners as envoys to Europe, James Murray Mason to London and John Slidell to Paris. Having successfully passed through the Union naval blockade of the southern coast, the envoys then transferred to the British mail packet, *Trent*, in Havana on 8 November 1861. However, the Union ship *San Jacinto* intercepted the British vessel, and, under threat of force, Mason and Slidell were seized and removed. At the end of November, the British cabinet instructed their minister in Washington to demand an apology from the United States government and the surrender of the Confederate envoys to British protection. Should these demands not be met, he was told to remove himself from United States territory. The London newspapers and periodicals carried extensive reports of the diplomatic negotiations over the conflict. For a full account of what became known as the '*Trent* affair', see Ferris 1977.

[4] Charles Wilkes was captain of the thirteen-gun sloop *San Jacinto* that intercepted the *Trent*. International law allowed diplomatic dispatches to be seized as contraband of war. In Wilkes's judgment, Mason and Slidell could be considered to be the physical 'embodiment of despatches' (see McPherson 1988, p. 390). For Gray's strong reaction to the *Trent* affair, see Dupree 1959, p. 311.

[5] News that the United States government had decided to comply with British demands did not reach London until 8 January 1862.

[6] The editors have followed Francis Darwin in supplying 'Heaven' as the word needed after 'thank' (*LL* 2: 382).

[7] The American-born physician and botanist Francis Boott was a friend and correspondent of Gray (see Dupree 1959, pp. 311–12).

[8] The historian Henry Thomas Buckle promoted the use of scientific methodology as a basis for historical analysis. His 'law of climate', for example, suggested a statistical relationship between the nature of a country's climate, the productivity of its soil, and the progress of its civilisation (see Buckle 1857, 1: 38 *et seq.*). CD had read Buckle's *History of civilization in England* in 1858 (see *Correspondence* vol. 4, Appendix IV, 128: 23).

[9] See letter from Asa Gray, 9 November 1861.

[10] CD had read a paper on dimorphism in *Primula* before the Linnean Society of London on 21 November 1861. It was published in the *Journal of the Proceedings of the Linnean Society (Botany)* 6 (1862): 77–96. *Orchids* was published by John Murray in May 1862.

[11] George Bentham, president of the Linnean Society, and Joseph Dalton Hooker had attended the meeting at which CD's paper was read (see letter from George Bentham, 26 November 1861).

[12] See letter to Daniel Oliver, 7 December [1861]. On 3 April 1862, CD read a paper before the Linnean Society entitled 'On the three remarkable sexual forms of *Catasetum tridentatum*, an orchid in the possession of the Linnean Society' (*Journal of the Proceedings of the Linnean Society (Botany)* 6 (1862): 151–7; see also *Collected papers* 2: 63–9). He also described the forms in *Orchids*, p. 236.

[13] See letter from George Bentham, 29 November 1861.

[14] The American naturalist James Dwight Dana had experienced a severe breakdown of his health late in 1859.

[15] Charles Lyell published *The geological evidences of the antiquity of man* in 1863.

To H. W. Bates 15 December [1861]

Down Bromley Kent
Dec. 15

My dear Sir

I have read your chapt. with great interest.[1] I will give my opinion, whatever that may be worth, without any exaggeration. I would not shorten or omit a sentence; & I hardly remember any Travels of which I would say that. The whole seems to me *excellent*. If other chapters are as good, I believe your book will be the best of the class of Nat. History Travels.— I do not pretend to be a judge of style; for I have never attended systematically to the subject; but yours seems to me very good, just what is wanted.— I strongly suspect that you are better fitted to criticise my style than I to criticise yours.—

I have hardly anything to remark. That is a capital discussion on the effects of climate on the dress of species; very original & throws new light on subject.—[2] There is one point, I think, you had better carefully attend to; viz when referring to English insects or other animals always, when possible, to give *in addition* to scientific name a common English name—as when speaking of Vanessas, to say such as our nettle-Peacock &c— I would I think even add, when speaking of Cicindela, a word of description, such as bright green colour with spots of gold So with Amara the little oval coppery Beetles which run on hot days across our paths.— It seems to me to add greatly to interest if you can call before the Readers mind some *definite* object— Remember not one in a hundred knows the scientific names.— So with respect to the gigantic ant, I would give the length of our largest British ant R. rufa.—[3] I will scribble a few *very trifling* miscellaneous remarks.—

p. 8 & p. 40— you use "former & latter"— to my private taste this is a great bore; at p. 40 I had to read back. Macaulay, I observe, always prefers repeating the word, rather than use "latter", if not *close* to that which is referred to.[4]

p. 9. Have you goodish evidence that rocks are Cretaceous— if not, I would not use the term.—

You speak of "we"— I suppose you give Preface & there explain "we". It would add to value of your remarks, if reader from first knew that you were solely engaged in Nat. Hist pursuits.—

p. 21. what does "$4\frac{1}{5}$" of population mean?

p. 49. Would it not be well to say how many species of Lepidoptera found in Para in contrast with number in Grt. Britain

It occurs to me doubtfully, (but you would certainly be better judge) whether the History of Para (which seems to me very well done) had not better be inserted at end of chapter; as being so very distinct a subject.[5] So with Ants, it occurred to me that the little description of the Sexes & Workers ought, perhaps, to be inserted before Leaf-carrying Ant, as you allude to "worker minors", which would confound ordinary reader.— Would it not be well just to say that this Leaf-carrier has been alluded to by almost every traveller in Brazil & is so conspicuous & no explanation of habit given— I have heard it suggested gravely that each ant carried leaf as parasol!

You probably know that it is good plan, when your subject greatly changes to leave open space of a line between the two paragraphs— Have Nos of Chapter at head of Page, it is so useful for reference.— Make Index— you know of course easiest plan.—

I should think that Murray would prefer seeing *two* or three Chapters.—[6] Whenever you tell me I will write strongly to Murray & send his answer to you.[7] Remember to dispose of only 1st. Edit. Murray's usual plan is to take all risk & give certain share of profits; but all this will have to be arranged after he has considered your M.S. which will probably take him a week or two.— Perhaps he would buy the 1st. Edit. & give Bill at 6 or 9 months.

I return your M.S. by this Post.— You will see that my remarks are quite useless: but if you like to send another chapter, I will read it with pleasure. I heartily congratulate you on the probability of your producing a *popular* Book, that it will be an *excellent* one I am sure.—

My dear Sir | Yours very sincerely | Ch. Darwin

Endorsement: '1861'
Leeds University Library (Brotherton collection)

[1] CD had agreed to read the '1st Chapter or two' of the book Bates was preparing on his travels in the Amazon region of South America (see letter to H. W. Bates, 3 December [1861]). A note written on the letter indicates that Bates had sent CD '1st chapt. in M.S'.

[2] See Bates 1863, 1: 18–23.
[3] CD refers to *Formica rufa*.
[4] CD refers to the historian Thomas Babington Macaulay. CD had read volumes 3 and 4 of Macaulay 1848–61 in 1857 (see *Correspondence* vol. 4, Appendix IV, 128: 21). He had given the same advice about the use of 'former' and 'latter' to John Lubbock (see *Correspondence* vol. 6, letter to John Lubbock, 27 October [1856]).
[5] The state of Pará is in northern Brazil.
[6] CD had suggested to Bates that John Murray might be willing to publish his travels (see letter to H. W. Bates, 25 September [1861]). Murray published Bates's *The naturalist on the river Amazons* in 1863.
[7] See *Correspondence* vol. 10, letter to John Murray, 28 January [1862].

To John Innes 15 December [1861]

Down. | *Bromley.* | *Kent. S.E.*
Dec. 15th

My dear Innes.

I have got to go out & do not wish to miss a Post, so write in great Hurry, to say that we shall be delighted to have Quiz, who shall be taken great care of, & never parted with & when old & infirm shall pass from this life easily.—[1]

Most truly yours | C. Darwin

Endorsement: '61'
Cleveland Health Sciences Library (Robert M. Stecher collection)

[1] The dog 'Quiz' belonged to Innes's son, John William (see letter to John Innes, 19 December [1861]). Innes, who had served as the perpetual curate of Down since 1846, had inherited the ancestral family home, Milton Brodie, in Forres, Scotland (see Freeman 1978, p. 172, and Moore 1985, p. 469). After moving to Scotland, he assumed the name Brodie Innes.

To John Lindley 15 December [1861][1]

Down. | *Bromley.* | *Kent. S.E.*
Dec 15

My dear Lindley

Very many thanks for the Bolbophyllum & I was glad to see the curious little flower. This genus rather puzzles me.—[2]

With many thanks | Yours very sincerely | Charles Darwin

P.S. | I am so nearly ready for press, that I will not ask for anything more; unless indeed you stumbled on Mormodes in flower.—

As I am writing I will just mention that I am convinced from rudimentary state of ovules, & from state of stigma, that whole plant of Acropera luteola (& I believe A. Loddigesii) is Male.[3] Have you ever seen any form from same countries which could be females? Of course no answer expected unless you have ever observed anything to bear on this— I may add from state of ovules & of pollen, Catasetum tridentatum is Male (& never seeds according Schomburgk whom you have accidentally misquoted in Veg. K.)[4] Monacanthus viridis is female, Myanthus barbatus is the Hermaphrodite form of same species.—

Royal Botanic Gardens, Kew (Lindley letters, A–K: 198)

[1] The year is provided by the reference to the publication of *Orchids*, which appeared in May 1862.
[2] See letters to John Lindley, 16 November [1861] and 17 November [1861]. CD examined four species of *Bolbophyllum*, which are briefly described in *Orchids*, pp. 169–72.
[3] See letter to Daniel Oliver, 30 November [1861].
[4] CD refers to Schomburgk 1837, which was cited in Lindley 1853. The misquotation was Lindley's statement that Robert Hermann Schomburgk found *Catasetum* to produce seeds abundantly and that he therefore thought it to be female (Lindley 1853, p. 178).

To J. D. Hooker 18 [December 1861]

Down Bromley Kent
18th

My dear Hooker

Thanks for your note.[1] I have not written for a long time, for I always fancy, busy as you are, that my letters must be a bore. Though I like writing, & always enjoy your notes.

I can sympathise with you about fear of Scarlet Fever: to the day of my death I shall never forget all the sickening fear about the other children after our poor little Baby died of it.—[2]

The Genera Plantarum must be a tremendous work & no doubt very valuable (such a Book, odd as it may appear, would be very useful even to me) but I cannot help being rather sorry at the length of time it must take, because I cannot enter on & understand your work—[3] Will you not be puzzled when you come to the Orchids? It seems to me Orchids alone would be work for a man's life time: I cannot somehow feel satisfied with Lindley's classification: the Malaxeæ & Epidendreæ seem to me very artificially separated.[4] Not that I have seen enough to form an opinion worth anything.—

Your African plant seems to be a vegetable Ornithorhynchus,—indeed much more than that.[5] The more I read about Plants, the more I get to feel that all Phanerogams seem comparable with one class, as Lepidoptera, rather than with one Kingdom as the whole Insecta.

Thanks for your comforting sentence about the accursed ducts (accursed though they be, I should like nothing better than to work at them in other allied orders, if I had time).[6] I shall be ready for press in 3 or 4 week & have got all my woodcuts drawn.[7] I fear much that publishing separately will prove a foolish job; but I do not care much, & the work has greatly amused me.— The Catasetum has not flowered yet!

In writing to Lindley about an Orchid which he sent me, I told him a little about Acropera & in answer he suggests that Gongora may be its female:[8] he seems dreadfully busy & I feel that I have more right to kill you than to kill him; so can you send me one, or at most two dried flowers of Gongora: if you know habitat of Acropera luteola, a Gongora from same country would be best; but any true Gongora would do; if its pollen should prove as rudimentary as that of Monacanthus relatively to Catasetum, I think I could easily perceive it even in dried specimens when well soaked.—

I have picked a little out of Lecoq; but it is awful tedious hunting.[9]

Bates is getting on with his Natural History Travels in one Vol.[10] I have read 1st. Chapt. in M.S. & I think that it will be an excellent book & very well written:[11] he argues in good & new way to me, that Tropical climate has very little direct relation to the gorgeous colouring of insects (though of course he admits that Tropics have a far greater number of beautiful insects) by taking all the few genera common to Britain & Amazonia, & he finds that the species proper to the latter are not at all more beautiful. I wonder how this is in species of the same restricted genera of Plants.—

If you can remember it, thank Bentham for getting my Primula paper printed so quickly: I do enjoy getting a subject off one's hands completely.—[12] I have now got dimorphism in structure in 8 natural orders, just like Primula.— Asa Gray sent me dried flowers of capital case in Amsinckia spectabilis one of Borragineæ.[13] I suppose you do not chance to have this plant alive at Kew.—

Attend, will you please, to enclosed memorandum.—[14]

I have scribbled a long rigmarole, so good night | My dear Hooker | Your affect | C. Darwin

P.S. | I have just met in Lecoq a magnificent case (for experiment) of **Tri**morphism in Lythrum salicaria & in L. thymifolia.— Could you help me to seeds or plants?[15]

Is Mr Borrer a good-natured man.? Would it do to write to him before the Spring to ask him for some plants or seeds? Perhaps he abominates the author of the Origin.—[16]

Endorsement: 'Dec/61.'
DAR 115.2: 137

1 The letter from Hooker has not been found.

2 CD's youngest child, Charles Waring Darwin, died on 28 June 1858, at the age of eighteen months, during an epidemic of scarlet fever. See *Correspondence* vol. 7.

3 Hooker and George Bentham were preparing a systematic compilation of all known plant genera. The first part of volume 1 of *Genera plantarum* (Bentham and Hooker 1862–83) was published in 1862.

4 John Lindley separated these two tribes of orchids on the basis of a difference he perceived in the way the pollen-masses were attached to the stigma (see Lindley 1830, p. 265).

5 The reference is to an anomalous plant sent to Hooker from Angola, which he named *Welwitschia mirabilis* (see letter to J. D. Hooker, 23 October [1861], and L. Huxley ed. 1918, 2: 24–5). The species exhibits characteristics intermediate between those that distinguish what were considered to be the two main divisions of flowering plants, angiosperms and gymnosperms. *Ornithorhynchus*, the duck-billed platypus, is a monotreme, an egg-laying mammal.

6 Hooker's letter has not been found, but see the letters to J. D. Hooker, 10 November [1861] and 14 November [1861].

7 CD was preparing the manuscript of *Orchids*. The woodcuts for the work were drawn by George Brettingham Sowerby Jr.

8 The letter has not been found, but see the letter to John Lindley, 15 December [1861].

9 Lecoq 1854–8. See letters to J. D. Hooker, 25 November [1861], 1 December [1861], and [9 December 1861].

10 Henry Walter Bates was writing an account of his travels, which was published in 1863.

11 See letter to H. W. Bates, 15 December [1861].

12 CD read a paper entitled 'On the two forms, or dimorphic condition, in the species of *Primula*, and on their remarkable sexual relations' before the Linnean Society of London on 21 November 1861. It was printed in the *Journal of the Proceedings of the Linnean Society (Botany)* 6 (1862): 77–96. George Bentham was president of the society and Hooker a member of council. See letters to George Bentham, 24 November [1861] and 26 November [1861], and letter from George Bentham, 26 November 1861.

13 See letter from Asa Gray, 9 November 1861.

14 CD's memorandum has not been found.

15 Lecoq 1854–8, 6: 154–9. CD read a paper on the sexual relations of the three forms of *Lythrum salicaria* before the Linnean Society in June 1864 (see *Collected papers* 2: 106–31).

16 William Borrer was a friend of Hooker's father, William Jackson Hooker (see Allan 1967). A wealthy man with a great interest in botany, Borrer endeavoured to cultivate all the British species of hardy exotic plants that he could obtain, having at one time had as many as 6600 different species growing in his garden (*DNB*).

To John Innes 19 December [1861]

Down. | Bromley. | Kent. S.E.
Dec. 19th

My dear Innes

I hope Johnny's heart will not fail. We will take much care of Quiz. If he comes pray thank Johnny.[1] If you will see that the Dog is at London Bridge. S.E. Railway Station by one oclock, I will see that Snow, who is up as usual on Thursday, calls for him & brings him here safely.[2]

I had told Snow that he *might* be at Nag's Head this day & gave all sorts of Instructions in case he came later, after he had started.—

I suppose you will very soon be off to Scotland; most heartily do I hope you all happiness in your new career as Scotch Laird or Chieftain, as perhaps I ought to call it—[3]

Pray give my very kind remembrances to Mrs. Innes & believe me. Dear Innes | Yours very sincerely | Charles Darwin

Endorsement: '61'
Cleveland Health Sciences Library (Robert M. Stecher collection)

1 The Darwins were to keep John William Innes's dog 'Quiz' after the family moved to Scotland (see letter to John Innes, 15 December [1861]).

2 George Snow operated a carrier service between Down and London every Thursday. His point of departure from London was the Nag's Head Inn, Borough.

3 Innes, perpetual curate of Down, moved with his family to his ancestral home in Forres, Scotland in 1862.

From B. O'Neile Wilson[1] 22 December 1861

No 2 Elizabeth St | Hobartown Tasmania
Decr 22nd 1861

Sir

As in a work like yours on the Origin of Species may derive some arguments from authenticated facts and the subject is one which causes considerable interest your theory much controversy I have thought it just possible that a few facts might compensate the trouble of perusing a note from an unknown correspondent

At page 209 of the edition marked (Fith thousand)[2] you say, "If it can be shewn that instincts do vary ever so little". Again at page 243. "I have briefly endeavoured to shew that the mental qualities of our domestic animals vary"—that instincts vary *slightly* in a state of nature"

Now there are some subjects on which we colonists have far more experience than most others and especially where horses and cattle are concerned. I mean experience of a peculiar kind arising out of our commerce with great numbers of both. Instead of varying *slightly* there is the greatest possible variation in the instinct of horses and as I have frequently ridden three or four in a day and sometimes twenty or thirty in a month as these horses were many of them bred in what are colonially termed mobs and quite in a state of nature until backed and very nearly so after being ridden until tired or worked down and then turned loose perhaps for months to recruit theirs is a very natural state indeed. I speak particularly of the horses bred on the large cattle stations. Now we find every kind of instinct and mental quality in such horses. I had one an Arab that never lost his way or rather his direction if he had once gone in the direction before. The darkest night the wildest country, without path or track he would find a way and the best way to where we were going

A horse belonging to my brother in law who lived a mile from Sale in Gipp's Land the road being perfectly well beaten and defined nearly straight but open at the sides never could be got along that road after dark though travelled by him hundreds of times without losing his way. Between the two extremes I have known every gradation. Again part instinct part habit. In riding in cattle some horses will take after a stray beast or one that attempts to break from the herd with unerring sagacity another is nearly useless for this purpose another will do it if his rider requires him. Some do do it from training, others seemingly by what I call instinctive imitation. Taken in wild backed in a day or two, before the first day on which they are ridden after cattle is over one now and then will dash after a runaway and seem to feel a delight in doing so. Again the vice of the New South Wales and Victoria horses called buck jumping seems partly instinctive, it quickly becomes hereditary. The Tasmanian horses do not Buck or very rarely indeed. At least one half their descendants in Victoria do. Of our Kangaroo dogs a powerful kind of Greyhound many will come back after having killed and take their masters to the dead Kangaroo. some taught others either instinctively or following the example of an older dog, taught perhaps you will call it. Some will not shew ⟨*bottom of page excised*⟩

You civilised people have no idea of and note shades of character which if you once set an Australian squatter talking of he will give you no end of information which may be of use to supplement your own experiences on other subjects.

At page 254, you say, "So again there is reason to believe that ⟨*bottom of page excised*⟩

Incomplete
DAR 181.2

CD ANNOTATIONS
0.3 Sir . . . nature" 2.4] *crossed pencil*
3.1 Now . . . experience] *underl brown crayon*
3.4 Instead . . . variation] *underl brown crayon*
3.7 quite in a state of nature] *underl brown crayon*
3.9 I speak . . . stations. 3.10] *underl brown crayon*
3.12 The . . . way 3.13] *underl brown crayon*
4.1 A . . . him. 4.8] *scored brown crayon*
4.8 others . . . imitation. 4.9] *underl brown crayon*
4.12 buck jumping] *underl brown crayon*
4.12 it quickly becomes hereditary. 4.13] *underl brown crayon*
4.13 Buck . . . Victoria do. 4.14] *double scored brown crayon*
4.15 after having . . . shew 4.17] *scored brown crayon*
4.17 Some will not shew] 'at all.' *added ink*
5.1 You . . . subjects. 5.3] *crossed pencil*
6.1 At . . . that] *crossed ink*
Top of first page: 'B. O'Neile Wilson.—'[3] *ink, circled ink*; 'Ch X'[4] *brown crayon*

[1] The correspondent is identified by CD's annotation.
[2] The second edition of *Origin* was published in January 1860.
[3] This is the only letter from Wilson that has been found. In *Variation* 1: 83, CD cited Wilson on the fertility of species of cattle when crossed. This topic may have been discussed in the missing portion of the letter or in another communication.
[4] The annotation refers to the chapter in CD's 'big book' on species that discusses the mental powers and instincts of animals (see *Natural selection*, pp. 463–527.)

From John Innes [24 December 1861][1]

2 Coley Hill | Reading
Christmas Eve.

Dear Darwin,

We have concluded that our little party shall remain together till we are near the time of our departure, and that Quiz must celebrate my birthday here. Johnny wants his portrait taken before he goes so we will arrange to send him to you next week, ie on Thursday 2nd. Janry 1862 he shall go by the South Eastern train hence at 9.50.[2] He will arrive at the London Bridge station at 1. This will be quite time to secure his conveyance by Snow to you. I will direct him fully. He travels in a hamper—[3]

Thanks for all your kind wishes for us. I shall certainly not be anything like a chieftain, hardly to say a laird—[4] A small house and a little land hardly make any

⟨ ⟩ I retain my ⟨ ⟩ where there are to ⟨ ⟩ recollections; and ⟨ ⟩ be lightly par⟨ ⟩ ⟨ ⟩ it is by no me⟨ans⟩ ⟨ ⟩ that after a loo⟨ ⟩ ⟨ ⟩ we may return ⟨ ⟩ ⟨es⟩pecially if I can ⟨ ⟩ to get a house ⟨ ⟩ one. But I do not w⟨ant⟩ to disturb the present arrangement, it is so desirable to leave Stephens in peace.[5] It has been the greatest comfort to me ⟨ ⟩ ⟨ ⟩nt a cu⟨ ⟩ ⟨ ⟩ ⟨m⟩y best wishes ⟨ ⟩y Christmas & ⟨ ⟩ you all | Dear Darwin | ⟨ ⟩ faithfully | J. B Innes

DAR 167: 6

CD ANNOTATION
Over first paragraph: 'Dog' *brown crayon, circled brown crayon*

1 The year is given by the relationship to the letters to John Innes, 15 December [1861] and 19 December [1861].
2 See letters to John Innes, 15 December [1861] and 19 December [1861]. The dog Quiz belonged to Innes's son, John William Innes.
3 See letter to John Innes, 19 December [1861] and n. 2.
4 Innes had recently inherited the ancestral family home, Milton Brodie, in Forres, Scotland and changed his name to Brodie Innes.
5 Thomas Sellwood Stephens served as the curate of Down after Innes's departure (see Moore 1985, p. 469).

To John Lindley 24 December [1861][1]

Down. | Bromley. | Kent. S.E.
Dec. 24th

My dear Lindley

I did not thank you for your two very kind notes until I heard from Hooker.[2] All that he knows about Acropera luteola is that he "believes it was received under that name from Loddiges".[3] In structure it *closely* resembles A. Loddigesii, but is very different in colour, being all yellow. I enclose a mutilated specimen.—

The evidence from the rudimentary state of the ovules, & from the entrance into the stigmatic cavity being so contracted that the pollen-masses can rarely be inserted, & from other lesser reasons, seems sufficient to show that the Kew plant is a Male. Judging only from the contracted stigmatic cavity A. Loddigesii must be also a male.—[4] I fear there is no chance of your recognising the mutilated fragment in Spirits. I begged Hooker to send me a dried Gongora flower that I might see whether that is (according to your suggestion) the Female, as far as the state of the pollen would show; but he does not say he will, but I hope he intends to do so.—[5] I had no intention to be so troublesome to you; for I had not the most remote idea, but what A. luteola was a well-known name.

With cordial thanks | Yours very sincerely | Ch. Darwin

Royal Botanic Gardens, Kew (Lindley letters, A–K: 199)

1 The year is given by the relationship to the letter to J. D. Hooker, 18 [December 1861], and by the reference to CD's study of orchids.

[2] The letters from Lindley have not been found. Lindley had questioned the name of a species of *Acropera* sent to CD from the Royal Botanic Gardens, Kew; CD forwarded Lindley's letter to Joseph Dalton Hooker (see letter to J. D. Hooker, 20 [November 1861]).
[3] The letter from Hooker has not been found. The Loddiges family had run a nursery in Hackney, London, since 1771. The nursery was currently in the hands of Conrad Loddiges, grandson of the founder. See also letter to J. D. Hooker, 6–7 October [1861].
[4] See letter to John Lindley, 15 December [1861].
[5] See letter to J. D. Hooker, 18 [December 1861].

To J. D. Hooker 27 [December 1861]

Down
27.

My dear Hooker

Lindley has sent me a Gongora, so do not trouble yourself.—[1] He says that Acropera luteola is in no Book & that except in colour it resembles A. Loddigesii, so I will insert a note.—[2]

Will you be so kind as to write on scrap of paper & put in enclosed Envelope the name & address of whoever arranges & settles about your Flues & Heating Apparatus for Hot-Houses at Kew.— Mr C. has asked me to find out, as he wants for some experiments about Gas to get a large chamber heated very equally night & day.—[3]

I will write soon.— I write now merely to save you wasting time over Acropera & Gongora.—

Ever yours | C. Darwin

Endorsement: '/Dec. 61.'
DAR 115.2: 138

[1] See preceding letter and letter to J. D. Hooker, 18 [December 1861].
[2] The letter from John Lindley has not been found, but see the preceding letter. In *Orchids*, CD added a footnote to the description of *Acropera luteola* (*Orchids*, p. 203 n.):

> Dr. Lindley informs me that he knows of no species so named; nor is the origin of the name known at Kew. This species or variety differs in little or in nothing from A. Loddigesii, excepting in its uniform yellow colour.

[3] CD refers to his friend Edward Cresy (see letter from J. D. Hooker, [29 December 1861]).

To J. D. Hooker 28 [December 1861]

Down Bromley Kent
28th

My dear Hooker

I hope & suppose that my note was in time to stop you searching for Gongora. By Jove what a strange creature Gongora is! It has good masculine organs & cannot be female of Acropera; indeed I sh^d. not be surprised if it, also, turned out a male orchid.—[1] Thank Heaven I shall soon go to press & finish with my Hobby-horse.[2]

I see Oliver is going to lecture at Royal Institution on Northern plants; & I am very glad to hear it.[3] I hope "Atlantis" will get a good sinking.[4] These enormous continental extensions are quite an article of faith with many. Little Woodward is quite contemptuous if I hint my doubts about any island whatever having been an island within the whole recent period.—[5]

By the way tell Oliver that his note about Acropera ovules has been very useful to me; & I have had another good look; & I believe the membranous fringes are placentæ with no ovules or merest rudiments.[6]

I have written to Mr Gower to thank him about Victoria fact;[7] but further experiments would be requisite for any trust. I wrote carelessly about the value of Phanerogams; what I was thinking of was that the sub-groups seemed to blend so much more one into another than with most classes of animals.[8] I suspect Crustacea would show more differences in the extreme forms than Phanerogams; but as you say it is wild speculation. Yet it is very strange what difficulty Botanists seem to find in grouping the Families together into masses.

There is a great deal in Lecoq about colour of flowers in Latitude from Lapland to S. of Spain.[9] He shows that white flowers increase to N. but (I think) no very great difference in proportion of red & blues.—

Ever yours | C. Darwin

Endorsement: 'Dec/61'
DAR 115.2: 139

[1] See letters to Daniel Oliver, 30 November [1861], to J. D. Hooker, 18 [December 1861], and to John Lindley, 24 December [1861].

[2] *Orchids* was published by John Murray in May 1862.

[3] Daniel Oliver read a paper on the distribution of northern plants at a Friday evening lecture at the Royal Institution of Great Britain in March 1862 (Oliver 1862b). He maintained that the pattern of distribution of plants in northern Europe, North America, and Asia could be better explained by postulating that migration occurred via a former land connection in the area of the Bering Straits rather than by assuming a hypothetical Atlantic communication between Europe and America.

[4] A full account of Oliver's criticism of the 'Atlantis' hypothesis was published later in 1862 (Oliver 1862a). See also letter to Daniel Oliver, 30 November [1861] and n. 7.

[5] CD probably refers to the geologist and naturalist Samuel Pickworth Woodward.

[6] See letter to Daniel Oliver, 30 November [1861].

[7] See letter from W. H. Gower, 23 November 1861.

[8] See letter to J. D. Hooker, 18 [December 1861].

[9] Lecoq 1854–8.

To John Lindley 28 December [1861][1]

Down. | Bromley. | Kent. S.E.
Dec. 28th

My dear Lindley—

Very many thanks for information about the Acropera luteola, to which name I will attach a note.[2]

Also many thanks for the Gongora fulva— what a strange monstrous flower! It is, I am sorry to say, quite masculine in its nature & cannot be female of the A. Luteola. The pollen-masses &c are widely different from in Monacanthus viridis the female of Catasetum tridentatum. Of course, one can tell nothing safely from dry specimens, but I cannot help the suspicion that Gongora is also a male.[3]

I have been looking again very carefully at the placentæ of Acropera & I can see no escape from ranking that plant as a male.[4]

Yours very truly obliged | Charles Darwin

Royal Botanic Gardens, Kew (Lindley letters, A–K: 200)

[1] The year is provided by the relationship to a series of other letters discussing the orchids mentioned in the letter (see nn. 2–4, below).

[2] See letter to J. D. Hooker, 27 [December 1861] and n. 2.

[3] Lindley had suggested that *Gongora* might be the female of the orchid *Acropera luteola* (see letter to John Lindley, 24 December [1861]).

[4] For the basis upon which CD judged *Acropera* to be a male flower, see the letter to Daniel Oliver, 30 November [1861] and *Orchids*, pp. 206–9.

From J. D. Hooker [29 December 1861][1]

Kew
Sunday.

Dr Darwin

We employ several heaters & have written fully to Mr Cresy about them.[2]

My wicked conscience smited me about the Gongora when your letter came— I am truly shut up & confess I forgot that very trifling commission.[3] I do not give up *Acropera luteola*—& think I shall find some reference to it yet.[4]

Do you hear from Asa Gray now? & if so does he with you (as with me) avoid all allusion to politics & war. I do not know your views on this crisis I am with the popular view in this country & do confess I long to see the Yankees well drubbed by us.— I have no further sympathy with the South than that it is the weakest.—& I have no wish to see England break the blockades.[5]

Willy has returned from his Quarantine—all clear. I think he improves but is singularly backward & childish of his years.[6] he is *very* good & a charming boy in disposition & manners, so generous unselfish & conscientious, like poor dear old Henslow.—[7] he has lots of brains, but no power whatever of using them, no memory, & the most volatile disposition of any child I ever saw. The rest are all well.

I have been in bed half day with violent cold to the amazement of the household & ⟨*3 words illeg and text destroyed*⟩

Do you want *Eul*⟨*ophia*⟩ ⟨*viridis*⟩[8] Have you any inqu⟨iries⟩ ⟨ ⟩ Otago? Dr Hector ⟨ ⟩ Geolog. Surveyor ea⟨ ⟩ ⟨ ⟩ & dines with me on Fri⟨day⟩ ⟨ ⟩ send at once & I will ⟨ ⟩ for his executing any sci⟨entific⟩ commissions.[9]

I have been reading J. Haasts survey of N.W. parts of Middle Island. he alludes to extensive drift formations (Glacial) amongst the mountains.[10]

Oliver gave me enclosed reference possibly worth your consulting
Yours | J D Hooker

[Enclosure]

Lestibudois Phyllotaxie Anatomique in Ann. Sc. Nat.[11]

DAR 101: 1, 2a–c

CD ANNOTATIONS

4.1 Willy . . . saw. 4.5] 'Whateley'[12] *brown crayon*
Top of first page: 'Naudin[13] | Henslow's life'[14] *ink*
End of letter: 'Glacial' *ink, circled ink*; 'Jan.— 1862' *ink*

[1] Dated by the relationship to the letter to J. D. Hooker, 27 [December 1861]. The following Sunday was 29 December.

[2] See letter to J. D. Hooker, 27 [December 1861].

[3] See letters to J. D. Hooker, 18 [December 1861], 27 [December 1861], and 28 [December 1861].

[4] See preceding letter and letters to J. D. Hooker, 20 [November 1861] and 27 [December 1861].

[5] Hooker refers to the so-called 'Trent affair'. See letter to Asa Gray, 11 December [1861] and n. 3.

[6] Hooker's eldest child, William Henslow Hooker, was eight years old. William had been quarantined with suspected scarlet fever (see letter to J. D. Hooker, 18 [December 1861]).

[7] John Stevens Henslow, Hooker's father-in-law, had died in May 1861.

[8] *Eulophia* is a genus of orchids. See following letter.

[9] James Hector was preparing to leave for New Zealand, where he was to assume the position of geologist to the provincial government of Otago (*DNB*). In the fourth edition of *Origin*, CD cited information provided by Hector concerning evidence of past glacial action in New Zealand (*Origin* 4th ed., p. 433).

[10] Haast 1862.

[11] Lestiboudois 1848.

[12] In the letter to J. D. Hooker, 16 January [1862] (*Correspondence* vol. 10), CD related an anecdote told to him by Richard Whately, archbishop of Dublin, about a boy who, although thought a dunce throughout childhood, came to exhibit his inherited genius as an adult.

[13] Charles Victor Naudin (see following letter and n. 3).

[14] Jenyns ed. 1862.

To J. D. Hooker [30 and 31 December 1861][1]

[Down]
Monday Evening

Will you ask Mr Gower to send me 2 or 3 flowers of *Heterocentrum Mexicanum* mentioned in Gardeners Chronicle in enclosed envelope,—as I am trying experiments of H. roseum, & the sight of another species **might** be of much use to me.—[2]

Ever yours | C. Darwin

if any other Melastomad with **8 anthers** is in flower I sh^d^ be glad to see it.—

Tuesday Morning

Many thanks for note & foreign letter.—[3] I am bad today with 2 of my Boys bad—[4] But I shall indulge myself with a note to you some day soon.—

I should **rather** like to see Eulophia, but if you are not well or very busy do **not** send it—[5] Mr G. could send the Heterocentron & cause you no trouble—

DAR 115.1: 83a

[1] Dated by the relationship to the preceding letter. In 1861, 30 December fell on a Monday.
[2] William Hugh Gower was a foreman at the Royal Botanic Gardens, Kew. A notice in the *Gardeners' Chronicle and Agricultural Gazette*, 28 December 1861, pp. 1137–8, recommended *Heterocentron mexicanum* as a good winter plant. CD was investigating dimorphism in *Heterocentron* (see letter to J. D. Hooker, 17 November [1861]).
[3] See preceding letter. The 'foreign letter' has not been found; it is possible that it was from Charles Victor Naudin (see preceding letter, CD annotations).
[4] Emma Darwin recorded in her diary on 30 December 1861 that Francis and Horace Darwin were 'feverish'.
[5] See preceding letter.

From Asa Gray 31 December 1861

Dec. 31. 1861

Dear Darwin

In view of dimorphism I recommend *Valeriana* to your notice. I find that I have noticed, long ago, that the so-called *dioicism* of V. dioica, &c,—ran through the genus, and was of the same character as in Rubiaceæ, &c.[1]

I see that Weddell, in Chloris Andina, 2, p. 17, 18, describes well the dimorphism —tho' he looks upon it as if polygamous merely.[2]

Koch Fl. Germ. ed. 2. p. 369, 370, has one section 'Flores omnes conformis, stricte hermaphrodite' the other he describes dimorphism in adding,—quo sexus polygamodioicus indicatur, ut in Menthis Pulmonariis, Primulis aliisque; in specim, enim parvifloris antheræ sæpe polline carent et sexum femineum referunt."[3] Look therefore at the European Valerians where if V. dioica goes so far as to be structurally diœcious or polygamous, others are less so, and perhaps some are not dimorphous at all.

I did not intend writing to you by this mail, in acknowledgement of your very kind letter of the 11th.—[4] But while I am musing the fire burns—as the Psalmist has it.[5]

We should indeed find it hard to hate each other pro amore patriæ. Of all my English correspondents you are the only one touching upon our relations with the South & with England whose views and sentiments are perfectly satisfactory to me. And my wife—as is natural to her sex—takes a stronger line, and bids me send her *love* to Mr. Darwin, & say that his is the only Englishman whose letters do not give her a shock to read.[6]

That you should think it beyond our power to subdue the rebels is very natural. It may be true, but you would admit, if you were here that we ought to try. We do not doubt that we can, and you will also see that we do not mean that anything else shall stand in the way of it. I imagine it is now universally felt here that if we do not do it, we shall have to eat much dirt.—that the establishment of a rival power in our long southern line of the free States to be played off against us is not to be submitted to if it can be prevented at any sacrafice. God help us, indeed, if our honorable existence is to have no better safeguard than the generosity or sense of justice of more powerful nations!

As to slavery the course of things is getting to meet your views, as it is clear must be, if the South continue obstinate.[7] If they give up now they may save their institution in their own States to have the chance of abolishing it themselves in the only safe and easy way, with time and the gradual competition of white labor. But obstinate resistance will surely bring on wide sweeping manumission.

You see that we are not a going to have war at present. And it appears that the decision of our Government will be as unitedly & thoroughly sustained by the whole people, as if it had been the other way,—contrary to Mr. Russell's prediction, and to our dear friend D^r^. Boott's, who writes about our "mob" in a way he would not if he were here to see.[8] Look at an English *mob* urging up their government so that they felt obliged to back up their demands with a menacing force on our borders,—and making such a peremptory demand as you justly say "entirely on Wilkes' acting as Judge,—a matter which our Govt. would as promptly concede as yours could ask.[9]

Seemann wrote me that the general belief at the Clubs & in the city was that our Government wanted to get into war with England for an excuse to give up the South.[10] A pretty idea they must have of our wisdom and discretion! Dear Boott is firmly convinced that we have all along been trying to quarrel with England. The belief here is nearly universal the other way, and those who like England best, and perhaps the coolest and best informed men have been more and more dissatisfied as time went on.

What has caused this lamentable state of things, this complete misunderstanding? Plainly this: The secessionists in England have adroitly managed the matter and led public opinion in various lines but all in one direction, inimical to us,—and they did not think it too great a stretch to make John Bull believe that we were insane enough to want an E⟨nglish⟩ quarrel. In this they h⟨ave⟩ been ably seconded by a f⟨ew⟩ presses here—mainly by ⟨those⟩ whose loyalty is deeply s⟨uspected⟩ and whose influence is as nothing,—which is nearly as scurrilous as the Saturday Review—with no redeeming ability.—[11] And you have the result.

Will the evidence that this mail carries satisfy the English that we want to live in peace with them? But as to *good feeling*, I am afraid it is too late to expect that.

We were hurt at first, by your putting our rebels on the same footing as a government with which yours was in most amicable relations, and by the general assumption at once that we were gone past redemption.—by the failure to ⟨s⟩ee that the power had gone ⟨from⟩ the hands of those who ⟨were⟩ always making trouble ⟨for⟩ your government in some petty way or other.—&c &c &c till I think it is general⟨ly⟩ believed that the governing influence in England desires to have us a weak and divided people, and would do a good deal to secure it.[12] I am sorry to say that this is the general feeling; and this is now very much intensified.

The feelings of many are very hostile, and they would like to be strong that they might show it. Those of others—and these are mine—(who have been exceedingly fond of England, always defending her when possible), are that we must be strong to be secure and respected,—Natural selection quickly crushes out weak nations—that we have tried long enough to have intimate relations between the governments, or the people in general— Naturalists &c being enlightened people can be as intimate as they like.—but nationally let each say "God bless you, and let us see as little of each other as possible e⟨a⟩ch going our own wa⟨y.⟩

Well—enough of th⟨is.⟩

⟨Some⟩ of the representations of us ⟨in⟩ ⟨En⟩glish papers would be amusing ⟨if t⟩hey did not now do so great ⟨h⟩arm. One would think it was generally thought that there was no law and order here, nor gentlemany conduct, nor propriety of deportment among the poorer and laboring people. I wish you could come & see. As to such things, & as to intelligence, education, &c.—I have sometimes thought of the picture one could draw from individu⟨al⟩ cases. Take one—very confidentially—for I would not hurt a really good fellow by exposing his ignorance of what he might be expected to know. Here we lately had a Cambridge Graduate,—F.L.S. & godson of an English Baronet,—who in one conversation let us know most frankly, that he had no idea where Quito was, or that there were two houses of Congress in U.S.—and was puzzled to know whether Boston U.S. time was faster or slower than Greenwich!

I ⟨was⟩ much amused ⟨ ⟩ ⟨*o*⟩*ld Gorilla* learning[13] ⟨ ⟩ ⟨t⟩ell me sometime if you ⟨ ⟩ at the production of an object through natural agencies excludes ⟨ ⟩ ⟨ ⟩mption against it, I suppose you will say *Yes*.

Ever Yours | A. Gray

A happy New Year to you, and may every blessing attend you. Keep up your lively sympathies with those who have very severe trials to undergo and sacrifices to make for what they deem the cause of right, honor, & future safety. For you are not one of those who can see nothing but as it effects English interests. At least I know you will have patience with my outpourings.

I cannot write to dear Boott at present. You may send him this letter if you like. If you do let this go with it to tell him that *Winthrop* (who will not be thought prejudiced in his favor) said weeks ago that he had the very best reasons to know that the charges in England against Seward as of hostile disposition against England, were groundless.[14]

I am going to ask you to send me by *mail* the sheets of your little *Orchid* book—one by one—as soon as they come out.—I may want to write an early notice of it, and need time to ponder its teachings.[15]

Dana—his wife tells me is really getting better, and is resuming work—very gradually.[16] I send you one of the two—not the best—photographs of him—just received from his wife. If you want to ask questions, send them thro. me, & I will learn if—as I think—his family would like to have him use his powers a little.

DAR 110 (ser. 2): 65, DAR 165: 104–105

CD ANNOTATION
4.1 I did not . . . anything 6.3] *crossed ink*

[1] CD had asked Gray to assist him in his study of plants that commonly exhibit two or more different forms (see letters to Asa Gray, 21 July [1861], 16 September [1861], 17 September [1861], and 11 December [1861]). For Gray's earlier remarks about dimorphic species, see the letters from Asa Gray, [27 and 29 August] and 2 September [1861], 11 October 1861, and 9 November 1861.
[2] Weddell 1855–7.
[3] Koch 1843–4.
[4] Letter to Asa Gray, 11 December [1861].
[5] Psalms 39:3.
[6] In the wake of the *Trent* affair, Britain and the United States stood on the brink of war with one another. For CD's opinion of the situation, see the letter to Asa Gray, 11 December [1861]. Jane Loring Gray, like her husband, was a fervent supporter of the northern cause (see Dupree 1959, pp. 307–13).
[7] See letters to Asa Gray, 5 June [1861] and 21 July [1861].
[8] Gray refers to William Howard Russell, special correspondent covering the American Civil War for *The Times*. Russell had predicted that the release of the Confederate envoys would precipitate the fall of the administration in Washington (*The Times*, 10 December 1861, pp. 9–10). Gray also refers to his friend Francis Boott, an American expatriate living in England. CD mentioned Boott's opinion of events in the letter to Asa Gray, 11 December [1861]. See also Dupree 1959, pp. 311–12.
[9] As a result of the *Trent* affair, the British government, which had been planning to strengthen its defences on the Canadian border, quickly moved to implement their reinforcement (see Bourne 1961). Gray refers to Charles Wilkes, captain of the vessel that intercepted the *Trent* (see letter to Asa Gray, 11 December [1861]).
[10] The botanist Berthold Carl Seemann had recently returned to Britain from an expedition to the Fiji Islands (Seemann 1862, p. 416). This view of Union intentions was held by many in Britain (see Ferris 1977).
[11] For Gray's criticism of the articles on the American Civil War carried by the *Saturday Review*, see the letter from Asa Gray, 9 November 1861.
[12] Gray refers to Britain's declaration of neutrality with respect to the conflict in May 1861, thereby recognising the Confederacy as a belligerent power. Under international law belligerent status allowed for loans to be contracted and war materials to be purchased, and it gave the Confederacy the right to commission cruisers on the high seas with the power of search and seizure. The Union, on the other hand, regarded the activities of the Confederacy as insurrection. See McPherson 1988.
[13] CD, in the letter to Asa Gray, 11 December [1861], had compared his attempts to think about the question of design in nature with those of an 'old Gorilla . . . set to learn the first book of Euclid.'
[14] Gray may be referring to the American statesman Robert Charles Winthrop. William Henry Seward, American secretary of state, was reputed to be hostile towards Britain (see Ferris 1977).
[15] CD sent Gray sheets of the first half of *Orchids* in April 1862 (see *Correspondence* vol. 10, letter from Asa Gray, 18 May 1862). The book was published shortly afterwards, and CD sent Gray a presentation copy (see *Correspondence* vol. 10, Appendix III). Gray reviewed *Orchids* in the *American Journal of Science and Arts* 2d ser. 34 (1862): 138–44.
[16] In 1859, James Dwight Dana suffered a breakdown in his health from which he never fully recovered. His wife, Henrietta Frances Dana, was the daughter of Benjamin Silliman (*DAB*).

From Leonard Jenyns 31 December 1861

1 Darlington Place | Bath
Dec. 31. 1861.

My dear Darwin,

I do not know that I ever thanked you for your valuable contribution to my memoir of Henslow, which I received thro' Dr Hooker some months back.[1] I thought it a pity to insert it *otherwise than entire* as direct from yourself, which I accordingly did.— When however I have a proof the sheet in which it finds a place, I will take care & let you see it, according to the wish I found expressed at the end of the MS, that you may make any further corrections you desire.— I should hope this will not be long first, & unless I hear anything to the contrary, shall suppose that you are at home, & forward the proof to you the same day it gets to me, having first seen that it is correct according to what you first wrote.— If you wish to make much alteration, I should like it to be returned to me to look at,—but otherwise you may return it to the printers direct, whose address I will send with it.—

The Memoir is quite finished & out of my hands: I am afraid those friends of my lamented relative to whom it relates, & who are looking for it, will think that it has been a long time about. But like everything of the kind, when I got well into the work, one thing suggested another, & I found myself getting beyond the limits I had at first marked down. It took moreover some time looking over all his tracts & other publications, with a view to abstracting their contents. Sedgwick had promised me an analysis of his two papers on Geology of Man & Anglesea,—but alas he is too ill to do it.—[2]

It has afforded me a sort of melancholy pleasure in undertaking the task, & I have done it as well as I can,—but with what success I must leave to others to judge.—

I cannot allow you to get this on the 1st. Jan. without wishing that the New Year may bring you & your's both health & happiness.

Believe me, | My dear Darwin, | Sincerely Your's | L. Jenyns.

DAR 168: 56

1 Jenyns had been preparing a memoir of his brother-in-law, John Stevens Henslow, who died on 16 May 1861. Joseph Dalton Hooker, Henslow's son-in-law, had asked CD to contribute to the volume (see letters to J. D. Hooker, 24–5 May [1861] and 30 May [1861]). CD's brief memoir was printed in Jenyns ed. 1862, pp. 51–5 (see also *Collected papers* 2: 72–4); the text is reprinted in Appendix X.

2 Jenyns refers to Henslow 1821a and 1821b. Adam Sedgwick, professor of geology at the University of Cambridge, had been a colleague of Henslow's for many years.

APPENDIX I
Translations of letters

From Tiberius Cornelius Winkler 7 July 1861[1]

Sir!

My dissertation on the fossil fish of Oeningen, which received an award from the Dutch Society of Sciences in Harlem, has been published, and I hope the dispatch to you of a copy of this piece will convince you of the respect inspired in me by your scientific works.[2] I have had the honour of doing the Dutch translation of one of your illustrious works, your "Origin of Species",[3] and I am taking advantage of Sir Charles Lyell's kindness to offer you a copy of that translation as well.[4] I take the liberty of asking you kindly to let me know when these two works have arrived in your hands.

Accept, Sir, the assurance of my very highest esteem | T. C. Winkler

Harlem 7 July 1861
To Mr Charles Darwin.

Edinburgh University Library (Lyell 1: 6084)

[1] For a transcription of this letter in its original French, see p. 195.

[2] Winkler 1861. There is a copy of the work in the Darwin Library–Down.

[3] Winkler's translation of *Origin* into Dutch (Winkler trans. 1860) is in the Darwin Library–CUL.

[4] Winkler's letter to CD was enclosed in a letter to Charles Lyell, dated 8 July 1861 (Edinburgh University Library (Lyell 1: 6085)); Winkler asked Lyell to forward the works mentioned in the letter to CD.

From the University of Breslau 4 August 1861[1]

The undersigned faculty humbly begs to inform you, most honorable Sir, that it has conferred upon you the degree *honoris causa* of Doctor of Medicine and Surgery on the occasion of the 50th anniversary celebration of the founding of the university which took place on the 2nd. 3rd. and 4th. of August this year.[2]

The faculty thought it would lend the celebration an enhanced lustre by admitting your name which is highly honoured in science into the ranks of its honorary doctors and humbly begs to present you the diploma which was drawn up during the convocation, according to the programme published by order of the Faculty, at the University's celebration.[3]

Breslau 4th. August 1861.
The Medical Faculty of the Royal University
H. Barkow, Z.Z.D. Betschler Lebert Middeldorpf[4]

To Master of liberal arts, | Member of the Geological and Royal Society Dr. Charles Darwin | Esq in London

DAR 230

[1] For a transcription of this letter in its original German, see p. 230.

[2] The University of Breslau (present-day Wroclaw, Poland) was incorporated in 1811 out of what was formerly a Jesuit College. Breslau was the capital of the Prussian province of Silesia.

[3] The diploma was sent to CD in 1862 (see *Correspondence* vol. 10, letters from the Prussian Legation, 10 February 1862, and from H. Albertof, 13 February 1862). CD's diploma collection is in DAR 229.

[4] Hans Karl Leopold Barkow was dean of the medical faculty at Breslau and Julius Wilhelm Betschler, Hermann Lebert, and Albrecht Theodor Middeldorpf were professors in the medical faculty (see Kaufmann ed. 1911, 2: 6).

APPENDIX II
Chronology 1861

This appendix contains a transcription of Darwin's 'Journal' for the year 1861. Darwin commenced his 'Journal' in August 1838 and continued to maintain it until December 1881. In this small notebook, measuring 3 inches by $4\frac{1}{2}$ inches, Darwin recorded the periods he was away from home, the progress and publication of his work, and important events in his family life.

The version published by Sir Gavin de Beer as 'Darwin's Journal' (1959) was edited before the original 'Journal' had been found and relied upon a transcription made by an unknown copyist. The original, now in the Darwin Archive in the Cambridge University Library (DAR 158), reveals that the copyist did not clearly distinguish between the various types of entries it contains.

From 1845 onward, Darwin recorded all that pertained to his work (including his illnesses, since these accounted for time lost from work) on the left-hand pages of the 'Journal', while the periods he was away from home, and family events, were noted on the right-hand pages. In order to show clearly Darwin's deliberate separation of the types of entries he made in his 'Journal', the transcription has the left- and right-hand pages labelled, with the left-hand page of an opening preceding the right-hand page.

All alterations, interlineations, additions, and the use of a different ink or pencil have been noted. In addition, the editors have inserted relevant additional information throughout this transcription of the 'Journal' for 1861. These interpolations are enclosed in square brackets to distinguish them from Darwin's own entries.

[**Left**]

1861

March 20th. Finished Ch. III on Variation Under Domestication & began Ch. IV[1]

(In December last & Jan. prepared new & Third Edit of Origin: 2000 Copies[2]

[21 February. Went to London. Gave an account of experiments on *Drosera* at a meeting of the Philosophical Club of the Royal Society.][3]

May 16 Finished Fowls (8 weeks)[4]

— 31st Ducks[5]

July 1[6] During stay at Torquay did paper on Orchids[7] all rest of year Orchid Book

[21 November. Went to London. Read 'On the two forms, or dimorphic condition, in the species of *Primula*, and on their remarkable sexual relations' to the Linnean Society of London.][8]

[Right]

1861

[28 January. Went to London with George Howard Darwin.][9]

[9 March. Henrietta Anne Huxley and children came. Thomas Henry Huxley came on 10 March.][10]

[17 March. Thomas Henry Huxley and William Benjamin Carpenter came.][11]

The new Edit. of Origin 2000 copies. was published in April.[12]

[16–20 April. Went to London. Attended meeting of the Linnean Society. Visited Robert Chambers.][13]

July 1. Started for Torquay & returned Aug 27th. 8 weeks & a day[14]

[End of July. Visited Thomas Vernon Wollaston.][15]

[30 August. Edward Cresy and Mrs Cresy visited.][16]

[7 October. George Brettingham Sowerby Jr came.][17]

[13 October. Thomas Henry Huxley and John Lubbock called.][18]

[20–3 November. Went to London. Visited British Museum on 21 November. Called on Francis Boott.][19]

[28 November. Dined with Emma at Holwood House.][20]

[10–13 December. George Brettingham Sowerby Jr came.][21]

1 CD was preparing the manuscript of *Variation*, published in 1868. Chapter 3 discusses variation in pigs, cattle, sheep, and goats; chapter 4 treats domestic rabbits.

2 The third edition of *Origin* was published by John Murray in April 1861. Murray printed off another 2000 copies, bringing the total run of the book to date to 6250 (see letter to John Murray, 21 January [1861]).

3 Emma Darwin's diary; Philosophical Club minutes, Royal Society of London. For the text of the notice, see Appendix IV.

4 The chapter on the origin of the domestic breeds of fowl, prepared for inclusion in *Variation*, formed chapter 7.

5 CD discussed the different breeds of domestic ducks in chapter 8 of *Variation*.

6 'July 1' *interl pencil*.

7 'paper on' *interl*. At first intending to publish the results of his study of the pollination mechanisms of orchids in the *Journal of the Proceedings of the Linnean Society*, CD eventually decided in September that the subject would run to the length of a short book. Accordingly, he approached John Murray to see if he would be interested in publishing the work (see letter to John Murray, 21 September [1861]).

8 CD's paper on dimorphism in *Primula* was published in the *Journal of the Proceedings of the Linnean Society (Botany)* 6 (1862): 77–96. See also *Collected papers* 2: 45–63.

9 Emma Darwin's diary.

10 Emma Darwin's diary. Henrietta Litchfield later recalled this visit (*Emma Darwin* 2: 177–8):

> In March 1861, Mrs Huxley and her three little children came to Down for a fortnight's rest. She was seriously out of health, and unable to recover from her grief at the recent loss of her eldest little boy. My mother hardly knew her before, and this visit laid the foundation of their friendship.

The Huxleys left Down on 25 March.

[11] Emma Darwin's diary.
[12] 'of Origin . . . copies.' *interl.*
[13] Emma Darwin's diary; two letters to J. D. Hooker, 23 [April 1861].
[14] The Darwins rented a house in Hesketh Crescent, Torquay, Devon.
[15] Letter to John Lubbock, 1 [and 2] August [1861]. Thomas Vernon Wollaston had recently moved from London to Kingskerswell, a small village near Torquay.
[16] Emma Darwin's diary. Edward Cresy and his wife lived in South Darenth, a village in Kent several miles from Down (CD's Address book, Down House MS).
[17] Emma Darwin's diary. George Brettingham Sowerby Jr was preparing the illustrations for *Orchids*.
[18] Emma Darwin's diary.
[19] Emma Darwin's diary; letters to J. D. Hooker, 25 November [1861]; to H. W. Bates, 3 December [1861]; and to Asa Gray, 11 December [1861]. While at the British Museum, CD examined the collection of Amazonian insects collected by Henry Walter Bates.
[20] Emma Darwin's diary. Holwood House, 1½ miles from Down near Keston, Kent, was the seat of Robert Monsey Rolfe, Baron Cranworth. Rolfe had served as lord chancellor of England from 1852 to 1858 (*DNB*).
[21] Emma Darwin's diary.

APPENDIX III
Presentation list for Asa Gray's pamphlet on *Origin of species*

In the July 1860 issue of the American periodical *Atlantic Monthly*, the first of an unsigned, three-part series on Darwin's *Origin* appeared, the subsequent parts being printed in the August and October numbers. Darwin was greatly impressed by the July article and was pleased to learn, from Joseph Dalton Hooker, that the author was Asa Gray. In a letter to Gray of 22 July [1860] (*Correspondence* vol. 8), Darwin complimented him on the piece, stating that it was 'uncommonly pleasantly written', and would 'tell well on public.' He was likewise pleased by the August sequel. Writing to Charles Lyell on 28 August [1860], Darwin described the essay as an '*excellent* argumentative' review; he praised Gray highly: 'He is assuredly a first-rate arguer & most completely understands the subject.'

By early September 1860, Darwin was arranging to get the second article reprinted in a British periodical, sending it to the *Annals and Magazine of Natural History* and offering 'to pay for its printing, if its length prevented its insertion' (letter to Asa Gray, 26 September [1860]). It was published in the November issue of the journal.

After reading proofs of the October piece, Darwin began to consider an even 'larger scheme'. Believing that Gray's two last essays were '*far* the best Theistic essays' he had ever read (letter to Asa Gray, 26 September [1860]), and fearing that public opinion of his book was being affected by the recent publication of several works that were highly critical of *Origin*, Darwin wrote to ask Gray whether he would consider getting 'the whole three published' together as a pamphlet, adding (letter to Asa Gray, 19 October [1860]):

> It would be indispensable to have your name & title on Title-page; & very adviseable to have some remark on Title, showing its bearing on Natural Theology or Design.—

An indication of the kind of service Darwin believed the pamphlet could offer can be discerned from a letter to Asa Gray, 11 December [1860]:

> Just to exemplify use of your pamphlet. Bishop of London was asking Lyell what he thought of the Review in Quarterly, & Lyell's answer was read "Asa Gray in Atlantic".

Charles Lyell, asked for his opinion of Darwin's proposal, in turn sought the advice of John Murray on the matter. Darwin was dismayed to learn that neither thought well of the plan as a commercial venture, believing that 'it would be absolutely necessary to spend more in advertisement than the cost of publication' (letter to Asa Gray, 24 October [1860]).

Gray, however, had come to the opposite conclusion. After receiving a letter from Gray written on 16 October 1860, Darwin wrote back the same morning, stating (letter to Asa Gray, 31 October [1860]):

> I have been reflecting about getting, as you suggest, if it can be done, 200 or 250 copies of the 3 articles of the Atlantic reprinted from the Plates in America & sent here. Could it be done for 4 or 5 Pounds; I would gladly pay that.— I do not think it would be very material about the paying. But what I do consider *all* important; is that you would prefix title-page with *your name* & titles. If you did not object to this, I would post a copy to all the Scientific men I could get addresses of from the Societies. Unless your name was appended, the articles would not, *I am sure*, be read in England & it would be time & money thrown away. Their merit would not be discovered. But would it not cost you a good deal of trouble to get it done?

Gray informed Darwin of the approximate cost of reproducing the pamphlet in a letter (now missing) of 26 November 1860, to which Darwin responded in his letter of 11 December [1860]:

> Not to lose the first chance, I write to say that I will **gladly** pay half of the (about) 20£, & will distribute some copies & leave others in the agents or Trübners hands: I will put a few advertisements in newspapers & see what I can do to get it noticed in a few Periodicals.—

Gray did not communicate his final approval of the scheme until early in 1861. In a letter to Gray, 23 [January 1861], Darwin stated that he was glad Gray had decided to publish and reminded Gray that he wanted to 'bear the whole small risk of loss'. Darwin also enclosed a list of the names of those to whom he intended to send a copy of the pamphlet.

In the event, the cost of reprinting 500 copies of the pamphlet was £15. After the arrival of 250 copies at the London offices of the publisher Nicholas Trübner, Darwin set about his task. He wrote to Hooker about having the pamphlet noticed in the *Gardeners' Chronicle*, edited by Hooker's friend John Lindley (see letter to J. D. Hooker, 20 [February 1861]). Lindley devoted a full page to a review of the pamphlet. Darwin also requested the same of Thomas Henry Huxley, editor of the new series of the *Natural History Review*, and of Andrew Murray, a Scottish naturalist associated with the *Edinburgh New Philosophical Journal* (see letters to T. H. Huxley, 17 February [1861], and to Andrew Murray, 23 February [1861]). He then sent out copies to some of the country's most influential figures in the world of learning (see, for example, the letter to John Tyndall, 23 February [1861]). In a letter to Gray of 26–7 February [1861], Darwin reported:

> I have distributed 100 copies to men of science & sent some to Reviewers, Libraries &c. I am heartily rejoiced at this. I believe your pamphlet will do me & Natural Selection, right good service.

On 12 March [1861], Darwin wrote to Gray about the success of their venture and said how pleased he was to hear about the good sale of the authorised American edition of *Origin*, which Gray had superintended. Darwin again mentioned his expectation that the pamphlet would help to counter theological opposition to *Origin*:

> Your pamphlet will do the subject **very great** good, whether or no it sells.— I have had many letters about it; all full of praise— "truly admirable" says one, "& am lending my copy to one person after another". Another says, "has read nothing on the subject with anything like the satisfaction"— Another says he (ie you) "is a cunning fencer & believes in you entirely".— I wish this was half true; though I do *not* mean the cunningness, which is utterly mistaken & false.

Darwin's expenses for advertising and distributing the pamphlet apparently doubled his costs: an entry in his Account book (Down House MS) on 12 March 1862 records a payment to Trübner's of £7 13*s*. 7*d*., that probably covered promotional and mailing costs.

Lyell's and Murray's opinions notwithstanding, the remaining copies of the pamphlet apparently did sell, for by March 1862 Trübner had only 38 copies in stock. Although Darwin directed Trübner to send the £8 profits from the sale to Gray, Gray felt the money properly belonged to Darwin and returned it, also indicating that he could send more copies since the pamphlet had not sold well in the United States (see *Correspondence* vol. 10, letter from Asa Gray, 31 March 1862).

The following list of those to whom Darwin sent presentation copies of A. Gray 1861 is transcribed from two sheets in DAR 210.18. The first list is that to which Darwin referred in his letter to Gray, 23 [January 1861]; the second was obviously made subsequently, with names drawn from directories of members of the Royal Society, the Geological Society, and the Linnean Society. Not all of those listed here actually received copies, for the list includes far more than 100 names. The various ticks and crosses next to some of the names are perhaps a guide to those to whom he finally decided to send copies. All alterations, interlineations, additions, and the use of a different ink or pencil have been noted. The symbol '✗' denotes a cross marked in ink; '✘', a cross in pencil; '✓', a tick marked in ink; and '✔', a tick in brown crayon.

Following the list is a Biographical register in which those persons who are not otherwise included in the main Biographical register of the volume are identified.

Asa Gray Review

Annals of N. Hist ✗
Edin. New Phil. Journal. ✗ (A. Murray)
Athenæum ✗
Saturday Review ✗
✗ Critic
Gardener Ch ✗
Nat. Hist Review ✗
Zoologist. ✗
Geolog. Soc.— ✗
Athenæum Club. ✗
Linnean Soc. ✗
Royal Soc.—✗
Fleuriere.—✗

✓ Tyndall
✓ Woodward
✓ Dr Gray[1]
✓ H. C. Watson
✓ Hooker
✓ Bentham
✓ Carpenter
✓ Huxley[2]
✓ Lubbock
E. Darwin[3]
✓ Self
✓ Henslow
✓ Sedgwick
✓ Hopkins
B. of Oxford
B. of London[4]
✓ Phillips
✓ Herschel
✓ Whewell
✓ Lyell
✓ Rev C. Kingsley
✓ Bunbury
R. Chambers[5]
✓ Wollaston
Sir H. Holland[6]
Falconer[7]
✓ Macdonnell
✓ Harvey Dublin
More Isle of Wight[8]
✓ Boott Dr 24 Gower St[9]
✓ Symonds[10]
Laugel[11]
✓ A de Candolle
✓ Os. Heer
✓ Thomson
✓ Thwaites
Blyth.[12]
L. Horner[13]
✓ H. D. Rogers
✓ Prestwich.—
✓ Sclater
Macdonell Dublin[14]
R. Cham[15]
Lewes Physiolog[16]
✓ Adam White

Royal

✓ Allman ✘✘
✓ Acland ✔
Andrews[17]
Ansted
Dr. Arnott.[18]
✓ Duke of Argyll ✘✘
✓ Balfour ✘✘
Barclay
✓ Beckles ✘✘
✓ Bell ✘ ✔
✓ Bennett ✘✘
Bowerbank ✔
✓ Bowman ✔
✓ Brayley ✘✘
✓ Busk ✘✘
Sir J. Clark
Daubeny[19]
✓ Davidson ✘✘
Davy
✓ Egerton ✔
Everest[20]
Faraday
✓ Farr ✔
✓ Godwin-Austen ✔[21]
✓ Gosse ✘✘
✓ Forbes ✔[22]
Grove[23]
✓ Harkness ✔
✓ Hennessy ✔[24]
Kane Sir Robert
✓ Lankester ✔
✓ Latham ✔
Lawes
Moore[25]
Miers ✔
✓ Paget ✘✘
✓ Patterson ✔
✓ Playfair ✔
✓ Pritchard ✘✘
Scrope
✓ Sharpey ✘ ✔
✓ Smith Sir A. ✔
Smith, James
Smyth
Sorby
✓ Thomson Allen ✘✘
✓ Wight ✔
✓ Williamson ✔
✓ Wrottesly Ld. ✘✘
Yorke

Geolog. Cat

✓ Bigsby ✔
Buckman
Cautley
✓ Charlesworth ✔
✓ Dawes ✔
Etheridge
Evans
✓ Geikie ✘✘
Hawkins
Hull, Ed
Hutton F
James

✓ Jenyns. ✗✗
✓ Jukes ✗ ✔
McClelland
✓ McCoy ✗✗
Mallet
Manchester Bishop of ✔
Morris
✓ Mylne ✔

✓ Nicol ✗ ✔
Oldham
Ormerod
✓ Page D. ✔[26]
Percy
Phillips[27]
✓ Ramsay ✔
✓ Ruskin ✔[28]

✓ Shuttleworth ✔
✓ Salter
✓ Symonds
✓ Tennant ✔
Tremenheere[29]
Falconer
Wood

Linnean Cat.

✓ Blackwall ✔
Borrer
Brightwell
Capel
✓ Collingwood ✗✗
Couch
✓ Currey ✗✗
Edgeworth
Garner

Hanley
Hassall
✓ Hicks ✔
Hogg
Lowe
MacAndrew
✓ Munro ✗✗
✓ Rolleston ✗✗
✓ Saunders ✗ ✔

Sowerby
✓ Stainton ✗✗
Talbot[30]
Stevens Sam.
Walker
✓ Westwood ✗✗
Wight[31]
Wright ✗

(Appendix)

Denny
Salter[32]
Bishop of Manchester[33]

✓ More of Isle of Wight[34]
✓ Bishop of London[35]
Dr Gully[36]

✓ Dr Gray
✓ W. A. Wooler
✓ G. Tollet

1 'Dr' deleted in pencil.
2 '✓' *over* '✗'].
3 Deleted in ink.
4 Deleted in pencil.
5 Deleted in ink.
6 Deleted in ink.
7 Deleted in pencil.
8 Deleted in pencil.
9 Written in pencil; 'Boott' overwritten in ink.
10 Written in pencil; overwritten in ink.
11 Deleted in pencil.
12 Deleted in ink.
13 Deleted in ink.
14 Written in pencil; deleted in ink.
15 Written in pencil; deleted in pencil.
16 Written in pencil; deleted in ink.
17 Deleted in ink.
18 CD deleted a '✔' in brown crayon.
19 CD deleted a '✔' in brown crayon.
20 CD deleted a '✔' in brown crayon.

21 The '✔' was written over '✘'.
22 The '✔' was written over '✘'.
23 CD deleted a '✔' in brown crayon.
24 The '✔' was written over '✘'.
25 Although there were four members of the Royal Society in 1861 with the name Moore, CD may have meant the physician George Moore.
26 The '✔' was written over '✘'.
27 Deleted in ink.
28 The '✔' was written over '✘'.
29 CD deleted a '✔' in brown crayon.
30 CD may have meant either Christopher Rice Mansel Talbot or William Henry Fox Talbot, both of whom were fellows of the Linnean Society and the Royal Society of London.
31 Deleted in pencil.
32 Deleted in ink.
33 Written in pencil; deleted in pencil.
34 Written in pencil; 'More' overwritten in ink. The '✓' was written over '✔'.
35 Written in pencil; overwritten in ink.
36 Written in pencil.

Periodicals and societies

Annals and Magazine of Natural History
Athenæum
Athenæum Club
Critic
Edinburgh New Philosophical Journal
Gardeners' Chronicle
Geological Society of London
Linnean Society of London
Natural History Review
Royal Society of London
Saturday Review
Zoologist

Biographical register of recipients

Acland, Henry Wentworth (1815–1900). Physician. Regius professor of medicine, Oxford University. Master of Ewelme Hospital, 1857–94. Librarian of the Radcliffe scientific library, 1851–1900. FRS 1847. (*DNB*.)

Allman, George James (1812–98). Botanist and zoologist. Professor of natural history at Edinburgh University and keeper of the natural history museum, 1855–70. Specialised in marine zoology, particularly studying hydrozoa and polyzoa. FRS 1854. (R. Desmond 1977, *DNB*.)

Andrews, Thomas (1813–85). Appointed professor of chemistry at Queen's College, Belfast, in 1849. FRS 1849. (*Modern English biography*.)

Ansted, David Thomas (1814–80). Geologist. Professor of geology at King's College, London, 1840–53. Edited the *Quarterly Journal of the Geological Society of London* for many years. Consultant geologist and mining engineer. Member of the Philosophical Club of the Royal Society. FRS 1844. (*DNB*, Sarjeant 1980.)

Arnott, Neil (1788–1874). Physician and natural philosopher. Physician extraordinary to Queen Victoria. Member of the Medical Council. One of the founders and an original senate member of London University. Fellow of the Geological Society of London. FRS 1838. (*DNB*.)

Barclay, Arthur Kett (1806–69). Brewer and astronomer. Head of the brewing firm Barclay, Perkins and Co. Built an observatory at Bury hill, near Dorking, Surrey, in 1848. FRS 1852. (*Modern English biography*.)

Beckles, Samuel Husband (1814–90). Barrister. Excavated the Purbeck beds in a quarry near Swanage, Dorset, discovering the oldest known mammalian remains. FRS 1859. (*Modern English biography*, Sarjeant 1980.)

Bennett, John Joseph (1801–76). Botanist. Keeper of the department of botany, British Museum, 1859–70. Secretary of the Linnean Society of London, 1840–60. FRS 1841. (R. Desmond 1977, *DNB*.)

Bigsby, John Jeremiah (1792–1881). Geologist and physician. Army medical officer. Served in Canada, 1818–27, reporting on the geology of Upper Canada. Settled in London in 1846. Elected fellow of the Geological Society of London, 1823. FRS 1869. (*DNB*, Sarjeant 1980.)

Blackwall, John (1790–1881). Zoologist. Specialised in the study of spiders. Fellow of the Linnean Society of London. (*DNB*, Gilbert 1977.)

Bowerbank, James Scott (1797–1877). Geologist. Partner in a London distillery. Had a special interest in the fossils of the London Clay. One of the founders of the Palaeontographical Society of London in 1847. Devoted many years to the study of sponges. FRS 1842. (R. Desmond 1977, *DNB*, Sarjeant 1980.)

Bowman, William (1816–92). Ophthalmic surgeon at the Royal Ophthalmic Hospital, Moorfields, 1851–76. Member of the Philosophical Club of the Royal Society. CD consulted Bowman regarding his children's health (see *Correspondence* vol. 5). FRS 1841. (R. Desmond 1977, *DNB*, *DSB*.)

Brayley, Edward William (1802–70). Scientific writer. For many years between 1822 and 1846, he edited and contributed to the *Annals of Philosophy*, the *Zoological Journal*, and the *Philosophical Magazine*. In 1845, CD supported his (unsuccessful) application for the professorship in geology at University College London (see *Correspondence* vol. 3). FRS 1854. (*DNB*, Sarjeant 1980.)

Brightwell, Thomas (1787–1868). Solicitor in Norwich. Microscopist and botanist. Specialised on diatoms and protozoa. Fellow of the Linnean Society of London. (R. Desmond 1977.)

Buckman, James (1814–84). Geologist and agriculturalist. Professor of geology and botany at the Royal Agricultural College, Cirencester, 1848–63. Fellow of the Linnean Society of London and the Geological Society of London. (R. Desmond 1977, *DNB*, Sarjeant 1980.)

Capel, George. Clergyman. Of Middlesex. Fellow of the Linnean Society of London. (*Alum. Cantab.*, *List of the Linnean Society of London* 1860.)

Cautley, Proby Thomas (1802–71). Army engineer, palaeontologist, and geologist. Colonel in the Bengal Artillery. Constructed the Ganges canal works, 1843–5. Returned to England in 1854. Member of the council of India, 1858–68. Fellow of the Linnean Society and the Geological Society of London and member of the Philosophical Club of the Royal Society. FRS 1846. (*DNB*.)

Charlesworth, Edward (1813–93). Naturalist and palaeontologist. Assistant at the Museum of the Zoological Society, 1837–40. Curator of the Yorkshire Philosophical Society Museum, 1844–58. Edited the second series of the *Magazine of Natural History, and Journal of Zoology, Botany, Mineralogy, Geology and Meteorology*. Fellow of the Geological Society of London. (*Modern English biography*, Sarjeant 1980.)

Clark, James (1788–1870). Naval surgeon and physician. Appointed physician in ordinary to Queen Victoria in 1837. Created baronet, 1837. CD had been one of his patients (see *Correspondence* vol. 2). FRS 1832. (*DNB*.)

Couch, Jonathan (1789–1870). Naturalist. Physician in Polperro, Cornwall. Fellow of the Linnean Society of London. (*DNB*.)

Daubeny, Charles Giles Bridle (1795–1867). Professor of chemistry at Oxford University, 1822–55; of botany, 1834–67; of rural economy from 1840. Fellow of the Linnean Society of London. Member of the Philosophical Club of the Royal Society. FRS 1822. (R. Desmond 1977, *DNB*.)

Davy, John (1790–1868). Army surgeon, physiologist, and anatomist. Brother of the chemist, Humphry Davy. FRS 1814. (*DNB*.)

Dawes, Richard (1793–1867). Dean of Hereford, 1850–67. Clergyman and educational reformer. Fellow of the Geological Society of London. (*Alum. Cantab.*, *DNB*.)

Denny, Henry (1803–71). Entomologist. Curator for forty-five years of the museum of the Literary and Philosophical Society of Leeds. Associate member of the Linnean Society of London. (*DNB*, Gilbert 1977.)

Etheridge, Robert (1819–1903). Palaeontologist. Assistant palaeontologist to the Geological Survey of Great Britain, 1857; palaeontologist, 1863. Assistant at the Royal School of Mines. Co-author with Thomas Henry Huxley of a catalogue of the collection of fossils in the Museum of Practical Geology (1865). Fellow of the Geological Society of London. FRS 1871. (*DNB*, Sarjeant 1980.)

Evans, John (1823–1908). Archeologist and numismatist. Partner in the paper manufacturing firm of John Dickinson & Co., at Nash Mills, Hemel Hempstead, Hertfordshire. In 1859, he accompanied the geologist Joseph Prestwich to France to examine flint implements found in the alluvial gravels of the Somme. Fellow of the Society of Antiquaries and of the Geological Society of London. (*DNB*, Sarjeant 1980.)

Everest, George (1790–1866). Military engineer. Served in India, 1806–43. Fellow of the Geological Society and the Geographical Society of London and of the Asiatic Society of Bengal. FRS 1827. (*DNB*.)

Faraday, Michael (1791–1867). Fullerian professor of chemistry at the Royal Institution. Scientific advisor to the Trinity House from 1836. Contributed extensively to the fields of chemistry, electrochemistry, and electromagnetism. Member of the Philosophical Club of the Royal Society. FRS 1824. (*DNB*, *DSB*.)

Farr, William (1807–83). Statistician. Assistant commisioner for the censuses of 1851 and 1861. Treasurer for many years of the Statistical Society. FRS 1855. (*DNB*.)

Garner, Robert (1808–90). Surgeon and naturalist. President of the North Staffordshire Natural History Field Club. Fellow of the Linnean Society of London. (R. Desmond 1977.)

Godwin-Austen, Robert Alfred Cloyne (1808–84). Geologist. Specialised on the geology of Devon and south-eastern England. Fellow of the Geological Society of London and frequently a member of council. FRS 1849. (*DNB*, *DSB*.)

Hanley, Sylvanus Charles Thorp (1819–1900). Conchologist. Fellow of the Linnean Society of London. (*Modern English biography*.)

Harkness, Robert (1816–78). Geologist. Professor of geology at Queen's College, Cork, 1853–78. Fellow of the Royal Society of Edinburgh and of the Geological Society of London. FRS 1856. (*DNB*.)

Hassall, Arthur Hill (1817–94). Botanist and physician. Public analyst. Member of the Royal College of Surgeons. Licentiate of the Apothecaries company. Fellow of the Linnean Society of London. (R. Desmond 1977.)

Hawkins, Benjamin Waterhouse (1807–89). Sculptor and painter of zoological subjects. Fellow of the Geological Society and the Linnean Society of London. Provided plates for *Fish* and *Reptiles* (see *Correspondence* vol. 2). (*Modern English biography*.)

Hennessy, Henry (1826–1901). Physicist. Professor of physics at the Roman Catholic University, Dublin, 1855–74. FRS 1858. (*DNB*.)

Hicks, John Braxton (1823–97). Obstetrician and botanist. Fellow of the Linnean Society of London and member of the council, 1860–1. (R. Desmond 1977.)

Hogg, John (1800–69). Classical scholar and naturalist. Justice of the peace on the northern circuit. Fellow of the Linnean Society of London. FRS 1839. (R. Desmond 1977, *DNB*.)

Hull, Edward (1829–1917). Stratigrapher and geologist to the Geological Survey of Great Britain, 1850–91. Fellow of the Geological Society of London. (Sarjeant 1980.)

Kane, Robert John (1809–90). Professor of natural philosophy to the Royal Dublin Society, 1834–47. Appointed president of Queen's College, Cork in 1845. FRS 1849. (*DNB*.)

Kay-Shuttleworth, James Phillips, 1st baronet (1804–77). Physician, educationalist, and social reformer. Fellow of the Geological Society of London. (*DNB*.)

Latham, Robert Gordon (1812–88). Ethnologist, philologist, and physician. Abandoned medicine in 1849 and devoted himself to ethnology and philology. Director of the ethnology department at the Crystal Palace, 1852. FRS 1848. (*DNB*.)

Laugel, Antoine Auguste (1830–1914). French writer interested in a wide range of subjects, including history, psychology, and politics. Reviewed *Origin* in the *Revue des Deux Mondes* in 1860. (*NUC*.)

Lawes, John Bennet, 1st baronet (1814–1900). Agriculturist and chemist. Upon inheriting the family estate at Rothamsted, Lawes became interested in various aspects of agriculture, particularly the production of artificial fertilisers.

Lawes, John Bennet, cont.
Established the Rothamsted Experimental Station in 1843. FRS 1854. (R. Desmond 1977, *DNB*.)

Lee, James Prince (1804–69). Bishop of Manchester, 1848. FRS 1849. (*DNB*.)

London, Bishop of. *See* Tait, Archibald Campbell.

Lowe, Edward Joseph (1825–1900). Hybridist. Cultivated British ferns. Fellow of the Linnean Society and the Geological Society of London. (R. Desmond 1977.)

McClelland, John (1800–83). Zoologist. With the Bengal Medical Service, 1830–65. Superintendent of the Calcutta botanic garden, 1846–8. Fellow of the Linnean Society of London. (R. Desmond 1977.)

M'Coy, Frederick (1817–99). Geologist and naturalist. Professor of geology and mineralogy, Queen's College, Belfast, 1850–4. Emigrated to Australia, where he became professor of zoology and natural history at the University of Melbourne in 1854. Fellow of the Geological Society of London. (*Aust. dict. biog.*, R. Desmond 1977.)

Mallet, Robert (1810–81). Industrialist, civil engineer, and seismologist. Fellow of the Geological Society of London. FRS 1854. (*DNB*, *DSB*.)

Manchester, Bishop of. *See* Lee, James Prince.

Moore, George (1803–80). Physician and author. Resided in Hastings, Sussex. Wrote the popular book *The power of the soul over the body* (1845). FRS 1835. (*DNB*.)

Moore, John Carrick (1804–98). Scottish stratigrapher and invertebrate palaeontologist. Fellow of the Geological Society of London and member of council in 1864. FRS 1856. (Sarjeant 1980.)

Morris, John (1810–86). Geologist. Professor of geology and mineralogy, University College London, 1854–7. Identified some of the fossil shells from the *Beagle* voyage (see *Correspondence* vol. 4). Fellow of the Geological Society of London. (*DNB*, Sarjeant 1980.)

Munro, William (1818–80). Botanist. Colonel in the army. Specialist in grasses. Fellow of the Linnean Society of London. (R. Desmond 1977, *DNB*.)

Mylne, Robert William (1817–90). Architect, civil engineer, and geologist. Surveyor to the Stationer's Company. Fellow of the Geological Society of London. FRS 1860. (Sarjeant 1980.)

Ormerod, George Wareing (1810–91). Geologist. Solicitor who practised in Devon. Published a number of papers on the geology of the region. Fellow of the Geological Society of London. (*DNB*.)

Page, David (1814–79). Scientific editor to the publishing firm W. & R. Chambers, Edinburgh. Fellow of the Geological Society of London. (*Modern English biography*, Sarjeant 1980.)

Patterson, Robert (1802–72). Naturalist. Merchant in Belfast. One of the founders of the Natural History Society of Belfast and of the Ulster Society for the Prevention of Cruelty to Animals. He was also an active member of the British Association for the Advancement of Science. FRS 1859. (*DNB*.)

Percy, John (1817–89). Geologist. Professor of metallurgy at the Royal School of Mines. Fellow of the Geological Society of London. FRS 1847. (*DNB*, *DSB*.)

Playfair, Lyon (1818–98). Chemist and politician. Professor of chemistry, Edinburgh University, 1858–69. Member of the Philosophical Club of the Royal Society. FRS 1848. (*DNB*, *DSB*.)

Pritchard, Charles (1808–93). Clergyman and astronomer. Headmaster of Clapham Grammar School, 1834–62. CD sent two of his sons, George Howard Darwin and Francis Darwin, to the school. Member of the Royal Astronomical Society and a fellow of the Geological Society of London. FRS 1840. (*DNB*, *DSB*.)

Ruskin, John (1819–1900). Author, lecturer, artist, and social reformer. Held an honorary studentship at Christ Church College, Oxford, in 1858. Member of the Geological Society of London. (*DNB*.)

Saunders, William Wilson (1809–79). Entomologist and botanist. Fellow of the Linnean Society; treasurer and vice-president in 1861. FRS 1853. (R. Desmond 1977, *DNB*.)

Scrope, George Julius Poulett (1797–1876). Geologist and political economist. MP for Stroud, Gloucestershire, 1833–68. Studied the volcanic districts of Europe. Fellow of the Geological Society of London. FRS 1826. (*DNB*, *DSB*.)

Sharpey, William (1802–80). Physiologist. Professor of anatomy and physiology, University College London, 1836–74. Secretary of the Royal Society of London, 1853–72, and member of the Philosophical Club. FRS 1839. (*DNB*, *DSB*.)

Shuttleworth. *See* Kay-Shuttleworth, James Phillips.

Smith, Andrew (1797–1872). Zoologist. Spent sixteen years in the Cape Colony and became an authority on the natural history of South Africa. Director-general of the army and ordnance medical departments, 1853–8. Fellow of the Wernerian Society. Knighted, 1858. FRS 1857. (*DNB*.)

Smyth, William Henry (1788–1865). Royal Navy admiral and scientific writer. A founder of the Royal Geographical Society, 1830. Fellow of the Society of Antiquaries and of the Royal Astronomical Society. Established the Bedford Astronomical Observatory. Member of the Philosophical Club of the Royal Society. FRS 1826. (*BHGW*, *DNB*.)

Sorby, Henry Clifton (1826–1908). Naturalist and geologist who pioneered microscopic petrology. Fellow of the Geological Society of London, the Microscopical Society, and the Mineralogical Society. FRS 1857. (*DNB*, *DSB*.)

Stainton, Henry Tibbats (1822–92). Entomologist. Secretary of the Ray Society, 1861–72. Edited the *Entomologist's Annual* and the *Entomologist's Weekly Intelligencer*. Fellow of the Linnean Society and the Geological Society of London. (*DNB*.)

Tait, Archibald Campbell (1811–82). Clergyman and educationalist. Bishop of London, 1856–69. A liberal theologian, he became embroiled in the debate over *Essays and reviews* (1860), eventually yielding in his defence of the authors. FRS 1859. (*DNB*.)

Talbot, Christopher Rice Mansel (1803–90). MP for Glamorgan, 1830–85. Fellow of the Linnean Society of London. FRS 1831. (*Modern English biography*.)

Talbot, William Henry Fox (1800–77). Mathematician. Pioneer of photography. In 1841, patented the 'talbotype' photographic method of producing instantaneous photographs. Fellow of the Linnean Society of London and of the Royal Astronomical Society and a member of the Philosophical Club of the Royal Society. FRS 1831. (*DNB*.)

Tennant, James (1808–81). Mineralogist. Professor of geological mineralogy, Kings College, London, 1838–81; of geology, 1853–60. Mineralogist to Queen Victoria. Fellow of the Geological Society of London. (*DNB*.)

Thomson, Allen (1809–84). Physician. Professor of physiology at Edinburgh University, 1842–8; of anatomy, University of Glasgow, 1848–77. Fellow of the Royal Society of Edinburgh. FRS 1848. (*DNB*.)

Tollet, Georgina. Daughter of George Tollet (1767–1855), of Batley Hall, Staffordshire. Close family friend of the Darwins and Wedgwoods. Edited the manuscript of *Origin* (see *Correspondence* vol. 7). (*Burke's landed gentry* 1846.)

Tremenheere, Hugh Seymour (1804–93). Barrister, author, and social reformer. Fellow of the Geological Society of London. (*DNB*.)

Wood, Searles Valentine (1798–1880). Geologist. Practised as a solicitor in London, 1851–65. Studied the fossils of the East Anglian Crag, specialising in Eocene fossil Mollusca. Fellow of the Geological Society of London. (*DNB*, Sarjeant 1980.)

Wright, Edward Perceval (1834–1910). Naturalist. Lecturer in zoology at Trinity College, Dublin, 1858–68; professor of botany, 1865–1904. Founder and editor of the *Natural History Review*, 1854–66. Fellow of the Linnean Society of London. (R. Desmond 1977, *DNB*.)

Wrottesley, John, 2d baron (1798–1867). Barrister. President of the Royal Society, 1854–7. Member of the Philosophical Club. President of the British Association for the Advancement of Science, 1860. FRS 1841. (*DNB*.)

Yorke, Philip James (1799–1874). Chemist, mineralogist, and meteorologist. Founding member of the Chemical Society; served as vice-president in 1852, and as president, 1853–5. FRS 1849. (*DNB*.)

APPENDIX IV

Darwin's report on experiments with *Drosera* read before the Philosophical Club of the Royal Society, 21 February 1861

In July of 1860, while visiting relatives in Hartfield, Sussex, Darwin amused himself by observing the means by which the common sundew of the Sussex heaths, *Drosera rotundifolia*, was able to capture and apparently digest the unlucky flies that landed on its leaves.

Continuing to study *Drosera* for the seven weeks his family spent on holiday in Eastbourne, from 22 September to 10 November, Darwin became fascinated by the results of his early experiments. The leaves appeared to be extremely sensitive to a number of different nitrogenous substances, so much so that Darwin was almost incredulous at the results of his chemical tests. As he informed Joseph Dalton Hooker in a letter of 31 [August 1860], 'The leaves are first rate chemists & can distinguish even an incredibly small quantity of any nitrogenised substance from non-nitrogenised substances.' (*Correspondence* vol. 8).

Darwin originally intended to summarise the results of his experiments on *Drosera* in a short published paper (see *ibid.*, letters to Daniel Oliver, 17 October [1860] and 7 November [1860]). He began to write the paper late in November 1860, but then decided not to publish the work until after he had had a chance to confirm his results during the next growing season. Sharing his results with Charles Lyell, including the admission that at the moment he cared 'more about Drosera than the origin of all the species in the world', Darwin explained his reasons for not wanting to go to press immediately (*ibid.*, letter to Charles Lyell, 24 November [1860]):

> I will not publish on Drosera till next year, for I am frightened & astounded at my results.— I declare it is a certain fact, that one organ is so sensitive to touch that a weight of $\frac{1}{78,000}$ of a grain (ie seventy-eight times less weight than that, viz $\frac{1}{1000}$ of a grain, which will move the best chemical balance) suffices to cause conspicuous movement.— Is it not curious that a plant sh[d] be far more sensitive to a touch than any nerve in the human body! Yet I am perfectly sure that this is true.—

Despite his hesitation in making his results and conclusions more widely known, Darwin apparently decided to benefit from a limited sampling of scientific opinion by reporting the preliminary results of his work to the members of the Philosophical Club of the Royal Society of London. This he did at a meeting of the club on 21 February 1861. The text of the paper Darwin presented is not extant. That he did draw up a manuscript for possible publication, as he intended, can be inferred from an existing '*Index to Eastbourne paper*' in DAR 54: 1, which indicates that the paper ran to at least thirty-four pages. He performed additional experiments the following year at Bournemouth but again left the work unpublished (DAR 54: 31–49). Over the next fourteen years Darwin continued to experiment on *Drosera*, using various poisons and narcotics in consultation with chemists and physicians; he finally published his results in *Insectivorous plants* (1875). The following abstract is taken from the minutes of the Philosophical Club (Royal Society Archives) for 21 February 1861 and is transcribed in its entirety. A brief account was also published in Bonney 1919, p. 154.

Feb 21st. 1861
126th Meeting (127)

Fifteen Members present.

Mr Horner (Chair)	Dr Carpenter	Prof. Ramsay
Mr Darwin	Mr Gassiot	Dr Bence Jones
Dr W. A. Miller	Mr Grove	Mr Lubbock
General Portlock	Prof. Tyndall	Mr Partridge
Mr Stokes	Dr Frankland	Mr Prestwich

The Minutes of the last Meeting having been read, Mr Darwin gave an account of some experiments he had recently made on the *Drosera* (sun-dew) which grows abundantly on a common in his neighbourhood. The leaves of this plant are furnished with long hairs bearing glands at their summits. When a fly or other insect touches one of these glands, it pours forth a viscid secretion which attaches the insect to the hair; other hairs in the vicinity then bend towards it, and pour forth their secretion so soon as they touch it; and the insect is thus held prisoner. The hairs retain their position for three or four days, in which time the insect dies; they then return to their previous places and begin to secrete afresh. Mr Darwin expressed his belief, founded on the abundance of this plant in particular localities, that flies are destroyed by its agency *by the million* every season Having experimented for the purpose of ascertaining whether the secretion and movement were excited by contact merely or by some chemical agency, he found reason to believe the remarkable result—that the plant distinguishes *nitrogen* in the fluid or solid that touches the hairs; for whilst the contact of non-nitrogenous substances produced little or no effect, the presence of no more than 1-6000th of a grain of nitrate of ammonia caused the hairs to converge, and that of 1-64,000th of a grain of carbonate of ammonia caused a single gland to pour forth its secretion. Portions of flesh and even of hair would produce the same effect as an insect; the contact of a minute fragment of hair of the estimated weight of 1-78,000th grs. caused movement in the hairs, although it only rested in the viscid fluid of the gland.— After the action of mere chemical stimulants, the hairs returned to their normal position in a few hours.

Philosophical Club minutes, Royal Society

APPENDIX V

Correspondence between Erasmus Darwin and Josiah Wedgwood I, concerning the properties of ice

On 23 February 1861, Darwin wrote to John Tyndall, sending him the correspondence between his paternal grandfather, Erasmus Darwin, and his maternal grandfather, Josiah Wedgwood I, that discussed the phenomenon of 'congelation' in ice, a subject in which Tyndall was interested. It is likely that Darwin had mentioned these letters to Tyndall at a meeting of the Philosophical Club of the Royal Society on 21 February at which both were present. Darwin doubted whether the letters would be of any use to Tyndall but commented that it had amused him 'to see the speculations of philosophers (my two Grandfathers) nearly 80 years ago on a subject which they so little understood' (letter to John Tyndall, 23 February [1861]).

Darwin was too ungenerous to his grandfathers. In fact, Erasmus Darwin and Josiah Wedgwood I, both of whom were members of the informal scientific society, the Lunar Society of Birmingham, made original contributions on the subject of heat exchange, one of the prominent questions in the scientific research of the day.

The letters sent to Tyndall were connected with Wedgwood's attempt to design an improved pyrometer that could more accurately measure the high temperatures involved in the firing of pottery. In an attempt to calibrate the new pyrometer, he began a series of experiments late in February 1784 to relate the scale of his instrument to that of the Fahrenheit thermometer. In his experiments, he used the new crushed-ice calorimeter designed by Antoine Lavoisier and Pierre Simon Laplace and was thereby led to make observations of the phenomenon of regelation. The freezing together again of partially melted pieces of ice intrigued Wedgwood; this is recorded in his Common Place Book, where he stated: 'in my experiments the two seemingly opposite processes of nature, congelation and liquefaction, went on together, at the same instant, in the same vessel, and even in the same fragment of ice.' (Schofield 1963, p. 266).

Erasmus Darwin joined Wedgwood in reflecting on the principles underlying this phenomenon, and he eventually published a paper in which some of the theoretical views described in his letters to Wedgwood were further elaborated (E. Darwin 1788). The Scottish natural philosopher Joseph Black, another member of the Lunar Society, was also drawn into the matter. Twenty years earlier he had conducted a series of important investigations into the subject of 'latent heat' in the change of state of water. For a fuller account of these and other scientific concerns of the members of the Lunar Society, see Schofield 1963.

Tyndall had for several years been actively investigating the physical properties of glaciers, having visited the glacial regions of Switzerland repeatedly since the summer of 1856 (see Eve and Creasey 1945 and Rowlinson 1971). In 1857, Tyndall and Thomas Henry Huxley, who had accompanied him to Switzerland, read a paper on this subject before the Royal Society of London. Darwin, who was present at the meeting, expressed great interest in the topic and corresponded with Tyndall concerning the theoretical views he had propounded (see *Correspondence* vol. 6, letters to T. H. Huxley, 17 January [1857], and to John Tyndall, 4 February [1857]). In particular, Darwin was interested in Tyndall's use of the principle of 'regelation', the freezing together of ice fragments partially melted under pressure, to explain how glaciers could be moulded under pressure so that they appeared to flow as if they were viscous (see Tyndall 1860, pp. 351–2). It was his grandfathers' discussion of regelation that Darwin thought might interest Tyndall.

The letters that CD sent to Tyndall were part of a larger collection of correspondence between Erasmus Darwin and Josiah Wedgwood I that were in his possession. This material, along with papers relating to Robert Waring Darwin, has recently been deposited in the Darwin Archive (DAR 227) of Cambridge University Library by George Pember Darwin. Parts of several of the letters, transcribed from copies before the originals were discovered, have been published in King-Hele ed. 1981.

As indicated in his letter to Tyndall, 23 February [1861], Darwin numbered the four letters (in red crayon) so that he could refer to them by number. Many years later, in the biographical sketch he prepared of Erasmus Darwin in 1879 (*Erasmus Darwin*), Darwin cited passages from two of the letters in this sequence, and he marked for the typesetters the passages that were to be included.

The following letters are transcribed in their entirety. Any annotations that Darwin made on them have been noted.

Letter 1 11 March 1784[1]

Dear Sir

I admire the way in which you support your new theory of freezing steam.— you say, "Will not vapour freeze with a less degree of cold than water in the mass? instances Hoar-frost &c." Now this same et cætera, my dear friend, seems to me to be a gentleman of such consequence to your theory, that I wish he would unfold himself a little more.

I sent an account of your experiment to M^{r}. Robt. & desired him to shew it to D^{r}. Black,[2] so that I shall hope some time to hear his opinion on the very curious fact you mention, of a part of ice (during a thaw) freezing whilst you applied a heated body to another part of it.

Now in spite of your et cætera, I know no fact to ascertain that vapour will freeze with less cold than water. D^{r}. Huxham in the prolegomena to his book on air & disease, (de aere & morbis) asserts "the degree of cold at which ice is produced is not stable or certain, for I have often observed frost in march & april, when the

Spirit of Hauksbey's thermometer was at 55 only; & on the contrary, that there was no ice in december, when it descended even to 65. page. 16."[3] Now as Dr. Huxham was no philosopher, & has said other absurd things about cold, I do not believe a word of this.

Water after having been boil'd will freeze sooner I think is affirm'd by Dr. Black, but I can not recollect in what publication;[4] I suppose by the expulsion of the air in the water the particles are already got nearer together, which vicinity I suppose to be the proximate cause of freezing: althō ice will swim in water owing to the number of air-bubbles in it: which being press'd out in the act of freezing, cease to be dissolved, & assume their elastic state, & with it an increase of bulk.

I can in no way understand, why during the time you apply a heated body to one part of a piece of ice, when the air of your room was at 50, & the ice had for a day or two been in a thawing state, that a congelation should be form'd on another part of the same ice but from the following circumstances

1. There is great analogy between the laws of the propagation of heat, & those of electricity, such as the same bodies communicate them easily, as metals, & the same bodies more difficultly as glass, wax, air. they are both excitable by friction. both give light, fuse metals, *et cætera. Therefore I suppose That atmospheres of Heat of different densities, like atmospheres of electricity, will repel each other at certain distances*, like globules of quicksilver press'd against each other, and that hence by applying a heated body near one end of a cold body, the more distant end may immediately become warmer, than the end nearest to the heated body.

Mr. Canton's experiment shewing this to obtain in electricity was very agreeably conclusive. He laid a wood-scewer across a dry wine-glass, & suspended 2 light pith or cork-balls to one end of the scewer, thus,[5]

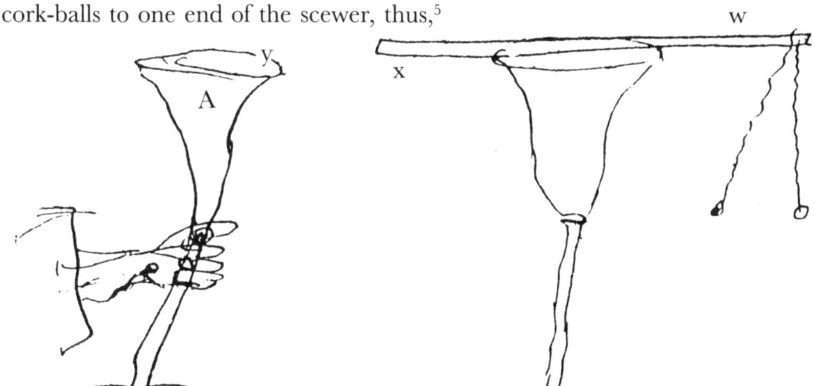

then rubbing another glass with a dry silk handckerchif He brought it nearly into contact with the other end of the scewer as x y & the cork balls diverged. He *then* touch'd the scewer with his finger at w, and the balls closed, & *then* withdrew the electrised glass, & immediately the cork-balls open'd again. Thus the greater atmosphere of the glass A, repel'd the less one of the scewer x w from x towards w.—whence w became plus, which being taken off by the finger, & the glass A withdrawn, the whole scewer was left minus, the same will succeed inversely in every respect if performed with minus electricity.[6]

Mar. 11—84

Since I wrote the above I have reconsider'd the matter, & am of opinion that steam, as it contains more of the element of heat than water, must require more absolute cold to turn it into ice, thō the same sensible cold, as is necessary to freeze water, and that the phænomenon, you have observed, depends on a circumstance which has not been attended to. When water is cool'd down to the freezing point, it's particles come so near together, as to be within the sphere of their reciprocal attractions;— what then happens?— they acceed with violence to each other, & become a solid; at the same time pressing out from between them some air, which is seen to form bubbles in ice & renders the whole mass lighter than water (on which it will swim) by this air having regain'd it's elasticity: & pressing out any saline matters, as sea-salt, or blue vitriol, which were before dissolved in it; & lastly by thus forceably acceeding together the particles of water press out also some more heat, as is seen by the rising of the thermometer immersed in such freezing water.

This last circumstance demands your nice attention, as it explains the curious fact you have observed. When the Heat is so far taken away from water, that the particles attract each other, they run together with violence, & press out some remaining Heat, which existed in their interstices. Then the contrary must also take place, when you add Heat to ice, so as to remove the particles into their reciprocal spheres of repulsion; they receed from each other violently, & thence attract more heat into their interstices; & if your piece of hot silver is become cold & has no more heat to give; or if this thawing water in this it's expansile state is in contact with other water, which is saturated with heat, it will rob it of a part, & produce freezing, if that water was but a little above 32°.

I don't know if I have express'd myself intelligibly. I shall relate an experiment I made 25 years ago, which confirms your fact. I fill'd a brewing-copper, which held about a hogshead & half, with snow; & immersed about half an ounce of water at the bottom of a glass tube in this snow, as near the center as I could guess, & then making a brisk & hasty wood-fire under it, & letting the water run off by a cock as fast as it melted, I found in a few minutes on taking out the tube that the water in it was frozen. This experiment coincides with yours, & I think can only be explain'd on the above principle.

Some blue vitriol was dissolved in a two ounce phial, & frozen in my labratory at Lichfield some years ago;[7] the phial was burst by the expansion of the ice; & after the thaw the water was dissolved away, & a thin erect column of blue vitriol was left beautifully disposed on the bottom of the broken phial. This shews that the dissolved vitriol was press'd on all sides towards the center by the freezing water, not precipitated from a want of attraction simply.

In support of the above theory I can prove from some experiments, that air when it is mechanically expanded always attracts heat from the bodies in it's vicinity, &

therefore water when expanded should do the same. But this would lengthen out my letter another sheet. I shall therefore defer it, till I have the pleasure of a personal conference with you.

Thus ice in freezing gives out heat suddenly, & in thawing gives out cold suddenly; but this last fact had not been observed (except in chemical mixtures) because when heat has been applied to thaw ice, it has been applied in too great quantities.

When shall we meet? our little boy has got the ague, & will not take bark,[8] & M^{rs}. Darwin is therefore unwilling to leave him; & begs to defer her journey to Etruria, till later in the season.—[9] Pray come this way to London, or from London.

our best compls to all yours | adieu | E Darwin

Water cool'd beneath 32°. becomes instantly ice on any small agitation, or pouring out of one vessel into another, because thus the accession of the particles to each other, & the pressing out of the air, or saline matters, & of heat, is facilitated.

'Rec^{d}. 12th March'
DAR 227 (Box 6)

CD ANNOTATIONS
1.1 I admire ... more.1.5] *scored red crayon*
2.2 some time ... it. 2.4] *double scored pencil*
3.1 Now ... bulk. 4.6] *crossed pencil*
5.1 I can ... circumstances 5.4] *double scored pencil*
5.1 I can ... body. 6.8] *enclosed in square brackets pencil*
7.1 M^{r}. Canton's ... electricity. 7.11] *crossed pencil*
8.1 Since] *after* '[' *pencil*
9.5 when ... interstices; 9.7] *double scored pencil*
10.3 & immersed ... frozen. 10.7] *scored red crayon*
11.1 Some ... simply. 11.6] *crossed pencil*
Verso of last page: '1' *brown crayon*

1 A large portion of this letter was published in *Erasmus Darwin*, pp. 98–101; it was reprinted in King-Hele ed. 1981, pp. 134–6.

2 Erasmus Darwin's son Robert Waring Darwin, aged 17, was studying medicine at Edinburgh University, where Joseph Black was professor of medicine and chemistry. Black was actively engaged in an experimental investigation of the heat transfer involved in the change of state of water (*DSB*).

3 John Huxham was a physician and writer on medical subjects. The reference is to J. Huxham 1752, 1: xvi–xvii. See also the English translation of the work, J. C. Huxham trans. 1759–67, 1: xxi–xxii. The thermometer developed by the scientific instrument maker Francis Hauksbee consisted of a glass tube filled with spirit of wine calibrated by setting the bottom bulb in freezing distilled water and taking the level of spirit as zero.

4 See Black 1775.

5 The diagram has been reduced to 65% of its original size.

6 John Canton described an experiment similar to the one mentioned by Erasmus Darwin in Canton 1753, p. 350.

7 Erasmus Darwin settled in Lichfield, Staffordshire, after qualifying as a doctor in 1756. After his second marriage in 1781, he and his family moved first to Radbourne, Derbyshire, and then in 1783 to Derby (*DNB*).

8 The 'little boy' was Edward Darwin, who was two years old (*Darwin pedigree*). The reference is probably to a medicinal preparation containing cinchona bark, which was known to have anti-febrile properties (*EB*).

[9] Elizabeth Darwin was Erasmus Darwin's second wife. Etruria, in Staffordshire, is the site of the pottery works founded by Josiah Wedgwood I in 1769; he lived in Etruria Hall (*DNB*).

Letter 2 13 March 1784[1]

Derby
Mar. 13— 84

Dear Sir

I meant to have added at the end of my long letter on your ice-experiment, that as your ice was factitious, it might possess more cold than 32°. unless it had been some time in an atmosphere of 32°. or more

2. for want of a sketch of your apparatus, I did not exactly know, where the new ice was form'd.

3. That if a part was refrozen, (all other circumstances being as I supposed them, viz. the ice having lain sometime in a thawing state previous to the experiment, or at least in at atmosphere of 32°.) I see no reason why it should not afford a correct measure of Heat. Because the new freezing, I judge, would always be exactly in proportion to the quantity thaw'd. & would therefore not affect the comparative measurement of the heats of bodies.

Thus if, by surrounding a dram of water with an artificial cold mixture, you cool it down to 32, the particles being brought within their reciprocal spheres of attraction rush together forceably, & forming ice press out a little more heat; this heat will be in proportion to the quantity frozen. So in reverse, if you surround a dram of ice with warmth, as soon as the particles receed so far as to come in to their reciprocal spheres of repulsion, they suddenly receed further, & attract more heat from the surrounding bodies, but this must be in proportion to the quantity thaw'd, & therefore will not affect the experiment as a comparative measure of the heat of two bodies. I don't know your apparatus. I think the external part of the apparatus should be surrounded with snow in its' thawing state. & the ice might be cut with a knife out of some in a thawing state, , & lastly that by means of a piece of cold blotting paper the ice might be touch'd so as instantly to take off what water might adhere to it, & then be thrown into a scale & weighed instead of weighing the water which came from it, part of which would be lost in steam.

Now as steam requires a great quantity of absolute heat to raise it, (for it consists of water & heat simply) if in your experiment any steam is raised, it should be confined so as to dissolve it's share of the remaining ice, otherwise this will be productive of great fallacy in the experiment because the heat will be carried off by the steam, which should have thaw'd more ice.

Hence the error of your proposition that steam will freeze with *less Cold* than water, seems to have arisen from it's freezing in *less time* than water, where there was cold enough, as it could be apply'd on a larger surface,—so bodies in powder are sooner dissolved, than when in a mass, but require as much of the solvend.

After having waded through two such long letters, in such bad writing, I suppose you'l be cautious how you correspond with me again on philosophical subjects.

M^{rs}. Darwin[2] says she hears your whole family are going to town in a body, like a caravan going to Mecca; & we therefore hope you will make Derby a resting place, & recruit yourselves and your camels for a few days, after having travel'd over the burning sands of Cheadle & Uttoxeter.

adieu | from yours affec^{tly}. | E Darwin

Our little boy is better, but not yet quite well of his Ague.[3] You will find three spare beds in our Caravansery.

'Rec^{d}. March 18th'
DAR 227 (Box 1, Aiii4)

CD ANNOTATION
Verso of last page: '2' *brown crayon*

[1] The last paragraph of this letter was published in Meteyard 1865–6, 2: 559, and is reprinted in King-Hele ed. 1981, p. 137.
[2] Elizabeth Darwin.
[3] See Letter 1, n. 8.

Letter 3 23 March 1784[1]

Dear Sir

I have nothing further to add upon the subject of congealation, & therefore write only to tell you how glad we shall be to see you at the time appointed, or at any time.

The expansion of ice has never been experimentally examined, as far as I know, it has generally been attributed to some elastic fluid being set at liberty & remaining in small moleculæ in the ice, & reabsorbed when it thaws. Will not a single flake of snow sink in water? will ice from salt-water sink or swim? the season is too far advanced to make these experiments.

I don't know whether my snow was below 32°. A mixture of salt & snow agitated together are said by Mussenbrook to produce artificial ice sooner if placed near a fire. but He defends a theory of frigorific particles & talks very inaccurately.[2]

I meant only to speak popularly when I talk'd of *giving out cold*.

I suspect lead will sink in melted lead, & cold glass in melted glass, & that therefore it is probable the expansion of ice may be owing to something previously dissolved in the water, which regains it's elastic state when precipitated during congealation.

In your apparatus I suspect some of the heat is lost on the cover. unless you put a plug of ice over the ignited body.

Adieu | Our Comp^{l}. to all yours | from your affect. friend | E Darwin

Derby | Mar. 23— 84

'Rec^{d}. March 25th'
DAR 227 (Box 1, Aiii4)

CD ANNOTATION
Verso of last page: '3' *brown crayon*

[1] This letter is mentioned, but not printed, in Meteyard 1865–6, 2: 470.

[2] The reference is to the Dutch physicist Peter van Musschenbroek. In his *Elements of natural philosophy*, Musschenbroek claimed that if a vessel of water was surrounded by snow mixed with salt and then the whole set-up placed over a fire, the water in the vessel would freeze as the snow mixture melted. He explained this as resulting from the 'conglatiating particles' being driven from the snow into the water in the vessel, with the addition of fire hastening the process (Musschenbroek 1744, 1: 333). His view of how ice might be formed was as follows (*ibid.*, p. 330):

> It is probable water is converted into ice, not because it is void of fire, though ice has never much fire in it; nor because it's parts are at rest, but because some subtile bodies out of the atmosphere are mingled with it, which as it were create an effervescence, expel the fire, and then fix the parts of the water by entering into the pores of the particles in the manner of nails or pegs; or are interposed between them, like glue, as water itself is the gluten of other bodies.

Letter 4 1 April 1784

Derby.
Apr. 1— 84.

Dear Sir,

I thank you for the information in your letter, thō I think the subject in respect to the expansion of ice not yet perfectly investigated. For I suppose all metallic bodies crystallize in cooling, it is even seen like a star in regulus of ♁,[1] & in the grain-like appearance in breaking bismuth & even lead. Now if the form of the crystals constitutes the cause of these expansions, these should be vacuities, visible by microscopes, since as crystallization is produced by attraction, some parts at least should be more solid than before, & should therefore press out the fluid Heat at least from those parts.

The convex or concave surfaces of cooling metals, if they contract or expand themselves, should depend on the form of the vessels which contain them, that is whether they are deeper in the middle or at the edges. The convexity of the plate of ice, when the water was let out from under it, might be owing to the contraction of the vessel. The fact you mention from de Mairan, that ice expands further (after it is form'd) by being exposed to a greater degree of cold is so unique, that I can not credit it, till it is confirm'd by other's experiments, or supported by some analogy.[2]

The detrusion or explosion of parts of icy mountains I always attributed to the water from thawing ice in the day-time filling valleys in the remaining ice, & by it's freezing there cracking the old ice by it's expansion at freezing, & not by it's expansion by an increase of cold.[3] The convexity or breaking of some ponds may be owing to the expansion of the inferiour stratum of water at the time of it's freezing, lifting up, & cracking the old ice at top, which is first form'd. This also may obtain in the northern lakes.

I rather think Sulphur also expands in congealing, by an experiment I formerly try'd; which I could then no other way account for. viz. I pour'd about 3 or 6 pounds of melted brimstone into a pot to cool, & after the surface was harden'd, a little oozing rose up through it in one place, & gradually rose into a considerable mountain.

In freezing wine & vinegar, I have observed the acid & spirit were left in cells amongst the ice, & required mashing, & laying on a sieve, but were not frozen, & hence the ice look'd flaky.

A single flake of snow swims in water.

Our little boy has had another return of his ague which has much debilitated him:[4] & to add to M^rs^. Darwin's uneasyness Miss Pole is ill of a rhumatic fever at Cambden house.[5] & if you should not favour us with your company in your going up, (as you say it is not certain) she will be much obliged to M^rs^. Wedgwood[6] to call & see Miss Pole as soon as is convenient after her arrival in town, & favour us with a line how she finds her.

M^rs^. & Miss Manwaring[7] are come to make some stay with us, but as they occupy but one bed, we can make ample room for you, & shall be truly happy to see you, but if it does not suit your convenience now, you must promise to make us a visit in your return from town.— you don't say a word about politics. NB. the wind is North West to day *therefore* I hope the frost is going. adieu M^rs^. D. joins in our best Comp^l^. to M^rs^. W. & all the W's with dear Jos

Yours affec^tly^. | E Darwin

DAR 227 (Box 6)

CD ANNOTATIONS

3.1 The . . . cold. 3.4] *double scored pencil and red crayon*

Part of letter on which address written: '(The most curious letter)' *pencil*; '4' *brown crayon*

1 The symbol is the seventeenth-century alchemical notation for antimony (Sb).

2 Jean Jacques Dortous de Mairan found that when water froze in a vessel, the ice formed a convex surface; further, he found that when the vessel was subjected to more cooling, the degree of convexity increased (see Dortous de Mairan 1749, p. 267).

3 This sentence was published in Meteyard 1865–6, 2: 470 n. 1 and reprinted in King-Hele ed. 1981, pp. 137–8. The transcription of the beginning of the sentence in Meteyard 1865–6 reads: 'The detension or expulsion of parts of icy mountains'.

4 See Letter 1, n. 8.

5 Elizabeth Pole, the elder daughter of Erasmus Darwin's wife Elizabeth by her first marriage, was 14 years old (see King-Hele ed. 1981, p. 180 n. 6).

6 Sarah Wedgwood.

7 Mrs Mainwaring had raised Erasmus's wife, Elizabeth (née Collier), who was the illegitimate daughter of Charles Colyear, the second earl of Portmore (King-Hele 1963, p. 31).

APPENDIX VI

Memorial addressed to Frederick Temple concerning *Essays and reviews*

In March 1860, a volume containing seven essays was published under the title *Essays and reviews*. The authors, six of whom were clergymen, all held influential positions within the establishment. Frederick Temple was headmaster of Rugby School; Rowland Williams the vice-principal and professor of Hebrew at St David's College, Lampeter; Baden Powell was Savilian professor of geometry at Oxford; Henry Bristow Wilson the vicar of Great Staughton, Huntingdonshire, a living presented to him by St John's College, Oxford; Mark Pattison rector of Lincoln College, Oxford; and Benjamin Jowett, Regius professor of Greek at Oxford. The only layman among the authors was Charles Wycliffe Goodwin, a fellow of Christ's College, Cambridge. The essayists wrote from the perspective of Broad Church or liberal theological opinion, and several of the pieces, particularly those of Powell, Goodwin, and Jowett, touched upon the conflict between the traditional teachings of the Church and recent findings in science. Powell, for instance, referred to Darwin's recently published *Origin of species*, calling it a 'masterly volume' and proclaiming that it was 'a work which must soon bring about an entire revolution of opinion in favour of the grand principle of the self-evolving powers of nature' (Powell 1860, p. 139).

Notices of *Essays and reviews* appeared in all the major periodicals of the day, and sales of the book were phenomenal—15,000 copies in three months and over 22,000 in a decade. But by February 1861, as the work entered its fourth edition, the simmering disapprobation it had aroused in Church circles erupted into a public forum, spearheaded by a sharp attack published in the January 1861 issue of the *Quarterly Review*. Although the review was unsigned, it was soon common knowledge that the author was Samuel Wilberforce, bishop of Oxford. Posing the rhetorical question as to why the work had excited such a degree of interest and achieved a circulation 'far greater than it would naturally have obtained by its mere literary merits', Wilberforce answered with the following reproachful attack on the essayists ([Wilberforce] 1861, pp. 249–50):

> It has been read, because to all it is new and startling—to some delightful, and to others shocking—that men holding such posts should advocate such doctrines; that the clerical head of one of our great schools, recently elected by a body of staid Conservative noblemen and country gentlemen, and a Chaplain in Ordinary to Her Majesty; two professors in our famous University of Oxford, one of whom is also tutor of one of our most distinguished colleges;

the Vice-Principal of the College at Lampeter for training the clergy of the Principality; and a country clergyman, famed in his day for special efforts on behalf of orthodoxy;—that such as these should be the putters forth of doctrines which seem at least to be altogether incompatible with the Bible and the Christian Faith as the Church of England has hitherto received it—this has been a paradox, so rare and so startling as to wake up for the time the English mind to the distasteful subject of a set of sceptical metaphysical speculations regarding many long-received fundamental truths.

Soon after the appearance of Wilberforce's review, the bishops of the Church of England, having been besieged with addresses and memorials from the rural clergy, met to decide on an appropriate response to the publication of *Essays and reviews* (see Davidson and Benham 1891, 1: 281). Their response came in the form of a letter (ostensibly in reply to a memorial submitted by a rural deanery in Dorset) signed by the archbishop of Canterbury, John Bird Sumner, and twenty-five bishops, which was published in *The Times* (16 February 1861, p. 10) and in the *Guardian* (20 February 1861). This brief letter indicated that the bishops considered many of the opinions published in *Essays and reviews* to be 'essentially at variance' with fundamental doctrines of the Church. It also announced that they were deliberating on the question, 'Whether the language in which these views are expressed is such as to make their publication an act which could be visited in the ecclesiastical courts, or to justify the synodical condemnation of the book which contains them' (see Moore 1988, p. 425).

It was the threat of 'synodical condemnation' of the essayists that prompted several London men of science to draft a memorial supporting the right to freedom of inquiry. Since there was no general editor of the volume, the memorial they drew up was symbolically addressed to the author of the first essay, Frederick Temple, headmaster of Rugby School. An informal committee was formed, with John Lubbock and William Spottiswoode as its secretaries, and the text of the printed memorial was circulated among a select group of prominent scientists late in February 1861, along with a request for them to put their name to the document.[1]

According to Lubbock's biographer, Horace G. Hutchinson, Darwin was one of the signatories of the memorial (H. G. Hutchinson 1914, 1: 58). Since, however, no other evidence has been found to corroborate this assertion, the text of the letter has been included as an appendix rather than incorporated into the main body of Darwin's correspondence for 1861.

Hutchinson also identified five other signatories of the memorial in addition to Lubbock and Spottiswoode—George Biddell Airy, George Bentham, George Busk, Leonard Horner, and Charles Lyell. Correspondence in the Lubbock papers (see n. 1, below) also indicates that Charles Spence Bate and Thomas Graham were signatories. According to James R. Moore, however, Airy and William Benjamin Carpenter 'agreed with the spirit of the memorial but may not have signed it' (Moore 1988, p. 436 n.).

Scientists who were sent copies but who decided that they could not sign the memorial included: John Frederick William Herschel, Charles Kingsley, Thomas Vernon Wollaston, and John Couch Adams. In addition, in a letter to Thomas Henry Huxley, dated 28 February 1861, Lubbock stated that Philip Lutley Sclater had agreed to sign the memorial ('but suggesting verbal alterations'), but that neither John Peter Gassiot (who had been unable to procure the book) nor George Rolleston, who described himself as 'liberal in politics, but not in religion', could support it. In a postscript, he added that Andrew Crombie Ramsay 'thinks the position of the Essayists untenable for Clergymen, so do you & many others (perhaps I too)' (Huxley papers, MS 22.61; see n. 1, below). Joseph Dalton Hooker, although sympathetic with the intent of the memorial, declined to sign it, suggesting that rather than presenting a formal memorial to the essayists, it would be better 'to give them *privately* our names in the terms of your memorial, and offer to rally round them *publicly* when the time comes for their acting, if they care to have us.' (L. Huxley ed. 1918, 2: 55–6). Apparently Huxley also decided in the end not to put his name to the document.

It is not known whether the memorial was ever actually sent to Temple. That it was being circulated was publicised in a letter from Charles Abel Heurtley (Lady Margaret professor of divinity at the University of Oxford) that was published in the 20 March 1861 issue of the *Guardian*, pp. 271–2. A week later, in the 'Ecclesiastical and University' column of the same paper, 27 March 1861, p. 294, the full text of the memorial was printed (without any names attached) together with an announcement that the memorial had been 'withdrawn from circulation'.

Whether Darwin himself ever read the articles in *Essays and reviews* is unclear. His only mention of the work comes in a letter to J. D. Hooker, 23 [April 1861], in which he recounted a recent conversation he had had with Robert Chambers. Chambers, Darwin noted, had made the interesting remark that laymen had left it to the clergy to debate the issues raised by the essayists. In August 1861, however, Darwin did make a small contribution to a fund that Lubbock and others were raising to help support the two authors, Henry Bristow Wilson and Rowland Williams, against whom charges of heresy had been brought (see letter to John Lubbock, 1 [and 2] August [1861]). For further details concerning the general controversy surrounding *Essays and reviews*, see Burgon 1870, Willey 1956, Crowther 1970, Brock and Macleod 1976, Ellis 1980, and Moore 1988.

The text of the memorial transcribed below has been taken from H. G. Hutchinson 1914, 1: 57–8, since it is the only source that lists Darwin among the signatories. There is also a printed copy of the text in the Huxley papers (MS 22.63; see n. 1, below).

To the Rev. Dr. Temple

We the undersigned have read with surprise and regret a letter in which the Archbishop of Canterbury and the other English Bishops have severely censured the volume of Articles entitled *Essays and Reviews*.

Without committing ourselves to the conclusions arrived at in the various Essays, we wish to express our sense of the value which is to be attached to enquiries conducted in a spirit so earnest and reverential, and our belief that such enquiries must tend to elicit truth, and to foster a spirit of sound religion.

Feeling as we do that the discoveries in science, and the general progress of thought, have necessitated some modification of the views generally held on theological matters, we welcome these attempts to establish religious teaching on a firmer and broader foundation.

While admitting that each writer in the *Essays and Reviews* is responsible only for the opinions expressed by himself, we address to you, as author of the first article, this expression of our sympathy and our thanks.

[1] Correspondence relating to this topic can be found in the following archives: the Thomas Henry Huxley Papers, Imperial College of Science, Technology and Medicine Archives (MSS 19.22, 22.61, and 26.217–18); and the papers of Lord Avebury (John Lubbock), Department of Manuscripts, British Library (Add. MS 498639, ff. 27–54). The Royal Society Archives holds a copy of a letter from J. F. W. Herschel to John Lubbock, 2 March 1861 (HS 13.12), the original of which is in the Avebury papers. The editors would like to thank James R. Moore for giving them access to his research materials on this topic and for sharing his expertise with them.

APPENDIX VII
Presentation copies of the third edition of *Origin*

In April 1861, John Murray issued the third edition of Darwin's *Origin of species*. Darwin had learned about the necessity for a new edition late in November 1860 (see *Correspondence* vol. 8, letter to T. H. Huxley, 22 November [1860]), and took the opportunity to make a number of corrections and additions to the text. Most of the new material had already been included in the German translation (Bronn trans. 1860) and in the American edition (*Origin* US ed.); for example, he added a 'Historical preface' listing those naturalists who had anticipated him in their published works with discussions of the possibility of species transmutation. Some of the changes incorporated into the third English edition, however, were entirely new. Darwin finished work on the new edition in January 1861.

The list of those to whom Darwin presented copies of the third edition of *Origin* is in DAR 210.18. It indicates that Darwin sent out a total of twenty-three copies. Some of those listed, such as Charles Lyell and Thomas Henry Huxley, had already received presentation copies of earlier editions (see *Correspondence* vol. 8, Appendix III).

Darwin's list is reproduced as close to its original format as possible. Darwin seems to have gone back to the list several times to add further names or address details: differences in ink quality are apparent and some entries were written in pencil. Most of the names have ticks next to them, probably indicating that copies were indeed sent. The names without ticks were subsequently deleted; Darwin presumably changed his mind about sending those individuals a copy. The names have been transcribed in a section at the end of the main list. Pertinent information about the arrangement of the original list is given in footnotes.

The list is followed by a biographical register, arranged alphabetically, identifying each recipient and giving an indication as to why Darwin would have sent them a presentation copy of *Origin*. Individuals not otherwise mentioned in the volume are only identified here and are not included in the main biographical register.

Third Edit. of Origin[1]

✓ H. W. Bates (promised)[2] King St. Leicester
✓ A. de la Fleuriere (Leader) 320 Euston Rd. NW (promised)[3]
✓ Asa Gray promised
✓ Carpenter
✓ Lyell[4]
✓ Hooker
✓ Huxley
✓ Self

✓ G. H Lewes Physiologist

✓ A. Morlot Lausanne (Soc. Vaudoise)[5]
✓ R. Colgate

✓ H C. Watson[6]
✓ Harvey
✓ Sclater (promised)[7]
✓ Wyman (*promised.*)
✓ R. Macdonnell 11 Lower Pembroke St[9] Dublin
✓ A. Murray. Edinburgh
✓ H. D. Rogers[10]
✓ E. Cresy 1 Greek St *promised*
✓ R. Chambers

✓ Hopkins?
✓ Dr Wright
✓ Geikie
Cambridge Paper[8]
✓ Rolleston

Names that were deleted
Sir H Holland
Bronn.
Fawett????[11]

1 'Third' is written above 'New', which was deleted in ink.
2 '(promised)' was written in pencil above the entry.
3 '(Leader)' was written above the entry in ink; '(promised)' was written below it in pencil.
4 This and the other left-hand entries down to and including 'Self' were written in pencil and then overwritten in ink.
5 The right-hand entries, from Morlot to Wright, were written in pencil. 'Lausanne (Soc. Vaudoise)' was written in ink.
6 CD had written Hewett Cottrell Watson's last name in pencil above that of Charles Lyell, and then subsequently deleted it in ink.
7 '(promised)' was written in pencil.
8 CD perhaps sent a copy to the *Cambridge Chronicle*. On 19 May 1860, this newspaper had carried a report of a meeting of the Cambridge Philosophical Society at which CD's theory was the subject of heated discussion. See *Correspondence* vol. 8, letter from J. S. Henslow to J. D. Hooker, 10 May 1860, and letter to J. S. Henslow, 14 May [1860].
9 '11 . . . St' was interlined.
10 This entry was written in pencil.
11 CD misspelled the name of Henry Fawcett. The name and the first two question marks were written in ink; the last two question marks were written in pencil.

Biographical register of recipients of the third edition of *Origin*

Bates, Henry Walter (1825–92). Naturalist. Undertook a joint expedition to the Amazon with Alfred Russel Wallace, 1848–50; continued to naturalise in the region until 1859. Published *The naturalist on the River Amazons* in 1863. (*DNB*, *DSB*.) CD and Bates had been corresponding since late in 1860. CD had promised to send Bates a copy of the new edition of *Origin* in his letter to H. W. Bates, 22 November [1860] (*Correspondence* vol. 8).

Carpenter, William Benjamin (1813–85). Physician and naturalist. Registrar of the University of London. (*DNB*, *DSB*.) Carpenter had written two reviews of *Origin*, and CD had sent him a presentation copy of the first edition.

Chambers, Robert (1802–71). Partner in the publishing firm W. & R. Chambers. Joint editor of *Chambers's Journal*. (*DNB*, *DSB*.) CD and Chambers not only shared an interest in geology, but CD also believed, correctly, that Chambers was the

Chambers, Robert, cont.
author of the popular evolutionary treatise *Vestiges of the natural history of creation* (London, 1844). CD visited Chambers at his north London home in April 1861 (see 'Journal'; Appendix II). CD had sent him a copy of the first edition of *Origin.*

Colgate, Robert (1816–1904). Surgeon. First practised in East Grinstead, Sussex, but by 1855 had moved to Eastbourne, where he remained for the rest of his life and became a prominent citizen. (Plarr 1930.) CD apparently made Colgate's acquaintance during the Darwins' stay in Eastbourne in the autumn of 1860, and had mentioned to him his interest in the pollination of clover by bees (see letter from Robert Colgate, 25 June 1861).

Cresy, Edward (1823–70). Principal assistant clerk at the Metropolitan Board of Works and architect to the fire brigade. (*DNB.*) Cresy, a family friend of the Darwins, had greatly assisted CD with his study of insectivorous plants in the autumn of 1860. CD had already presented Cresy with a copy of the second edition of *Origin.*

Fleuriere, A. de la. A staff writer or reviewer for the *Leader*.

Geikie, Archibald (1835–1924). Scottish geologist who worked on the glacial and volcanic history of Britain. Member of the Geological Survey of Scotland, 1855; director from 1867. (*DNB.*) CD was encouraged to learn that Geikie approved of the chapters in *Origin* on the geological record (see letters to T. W. St C. Davidson, 30 April 1861, and to George Maw, 19 July [1861]).

Gray, Asa (1810–88). American botanist. Fisher professor of natural history, Harvard University. (*DAB*, *DSB.*) Gray had been an important source of information for CD since 1855. He had helped arrange the publication of the American edition of *Origin.* His reviews and articles on *Origin* in the United States were important in swaying American opinion on the book. CD had sent Gray copies of both the first and second editions of *Origin.*

Harvey, William Henry (1811–66). Irish botanist. Colonial treasurer in Cape Town, 1836–42. Botanised in Australia, 1854–6. Keeper of the herbarium, Trinity College, Dublin, 1844; professor of botany, 1856–66. (*DNB*, *DSB.*) Harvey had described CD's algae specimens from the *Beagle* voyage. A good friend of Hooker's, Harvey initially reacted critically towards *Origin*, speaking out against it in a Dublin lecture of February 1860. He and CD subsequently engaged in a long correspondence about aspects of the theory (see *Correspondence* vol. 8).

Hooker, Joseph Dalton (1817–1911). Botanist. Assistant director of the Royal Botanic Gardens, Kew, 1855–65; director, 1865–85. (*DNB*, *DSB.*) Friend and confidant of CD. Hooker had known about the details of CD's views on transmutation since 1847 and greatly assisted the preparation of *Origin.* He applied CD's theory to explain the geographical distribution of Australian plants in the 'Introductory essay' to the *Flora Tasmaniæ* (Hooker 1859). CD had presented him a copy of the first edition of *Origin.*

Hopkins, William (1793–1866). Mathematician and geologist. A highly successful mathematics tutor at Cambridge. President of the Geological Society of

London, 1851–3. Specialised in quantitative studies of geological and geophysical questions. (*DNB*, *DSB*.) In 1860 Hopkins had written a hostile critique of the geological arguments that CD used to support his theory (Hopkins 1860); CD had considered responding to Hopkins's criticisms in the third edition of *Origin* (see *Correspondence* vol. 8, letter to Charles Lyell, 20 November [1860]).

Huxley, Thomas Henry (1825–95). Lecturer on natural history at the Royal School of Mines and naturalist to the Geological Survey of Great Britain. (*DNB*, *DSB*.) CD had sought Huxley's advice on a number of points during the preparation of *Origin*, and often stated how important Huxley's approval of and support for the theory were to him. Huxley also served as a liaison between CD and German naturalists. CD had presented Huxley with copies of the first and second editions of *Origin*, and of the German translation.

Lewes, George Henry (1817–78). Writer on literary and philosophical topics. Published an important biography of Goethe in 1855. Studied physiology and marine zoology from 1855 and published several popular works on these subjects. (*DNB*.) Lewes wrote favourably about *Origin* in an article he published in the *Cornhill Magazine* (Lewes 1860), a copy of which he may have sent to CD (see *Correspondence* vol. 8).

Lyell, Charles (1797–1875). Uniformitarian geologist. (*DNB*, *DSB*.) A long-time friend and mentor of CD, Lyell had encouraged him to begin writing his species book. After the arrival of Alfred Russel Wallace's manuscript on natural selection, Lyell had urged CD to publish a shortened version of his book. Lyell helped CD to achieve this by commenting on the manuscript and by recommending the book to John Murray for publication. CD had presented him with copies of both the first and second editions of *Origin*.

M'Donnell, Robert (1828–89). Irish surgeon. Lecturer on anatomy and physiology, Carmichael School of Medicine. Surgeon to Jervis Street hospital, Dublin, 1863; to Steevens's hospital, 1866. Medical superintendent of the Mountjoy government prison, 1857–67. (*DNB*.) M'Donnell published a paper in the *Natural History Review* in 1861 that provided evidence in favour of CD's view that classification reflected the true affinities of organisms (M'Donnell 1861).

Morlot, Charles Adolphe (1820–67). Swiss geologist in Lausanne. Member of the Société Vaudoise des sciences naturelles. Published on the geology and archaeology of Denmark and Switzerland. (*Cent cinquante ans d'histoire Vaudoise*.) There is a presentation copy of Morlot 1859, heavily annotated by CD, in the Darwin Pamphlet Collection–CUL.

Murray, Andrew (1812–78). Entomologist and botanist. Practised in the legal profession in Edinburgh until 1860. Appointed assistant secretary to the Royal Horticultural Society in 1860. Had a special interest in Coleoptera and Coniferae. Became an expert on insects harmful to crops. (*Entomologist's Monthly Magazine* 14 (1878): 215–16; Gilbert 1977.) Criticised *Origin*, particularly CD's views on the origin of blind cave animals, in an address before the Royal Society of Edinburgh in 1860 (see *Correspondence* vol. 8).

Rogers, Henry Darwin (1808–66). American naturalist. Appointed professor of geology and mineralogy, University of Pennsylvania in 1835. Regius professor of natural history, Glasgow University, 1855–66. (*DAB*.) Thomas Henry Huxley had informed CD that Rogers approved of *Origin* (see *Correspondence* vol. 8, letters to Asa Gray, 28 January [1860], and to Charles Lyell, 25 February [1860]).

Rolleston, George (1829–81). Physician and anatomist. Fellow, Royal College of Physicians, 1859, Physician to the British civil hospital at Smyrna during the Crimean War, 1855–6. Appointed Regius professor of physic at Oxford in 1857; Linacre professor of anatomy and physiology, 1860–81. (*DNB*.) Rolleston, who was favourably disposed to CD's theory, had assisted CD's study of variation by providing him with details of cases from pathological anatomy.

Sclater, Philip Lutley (1829–1913). Lawyer and ornithologist. Secretary of the Zoological Society of London, 1859–1903. FRS 1861. (*DSB*, Scherren 1905.) Sclater assisted CD in his preparation of a new edition of his *Journal of researches* in 1860, particularly by providing information about the geographical distribution of bird species in South America and the Galápagos Islands (see *Correspondence* vol. 8).

Watson, Hewett Cottrell (1804–81). Botanist with interests in phytogeography and phrenology. (*DNB*, *DSB*.) A supporter of the idea of the progressive development of plant species, Watson assisted CD's work on species, corresponding with him over a number of years about the distribution of species and varieties and the taxonomic practices of botanists (see *Correspondence* vols. 6 and 7). CD presented him with a copy of the first edition of *Origin*. Watson believed that CD's theory needed to take into account the possibility of 'convergence' of species, a criticism that CD responded to in additions to the third edition of *Origin*.

Wright, Edward Perceval (1834–1910). Irish naturalist and physician. An editor of the *Natural History Review*, 1854–66. Appointed director of the universiy museum in Dublin, 1857. Lecturer in zoology, Trinity College, Dublin, 1858–68. Lecturer in botany in the medical school of Steevens's Hospital, 1858. Graduated MD from Oxford in 1862, specialising as an oculist. Appointed professor of botany at Trinity College and keeper of the herbarium, 1869. (*DNB*.)

Wyman, Jeffries (1814–74). American comparative anatomist and ethnologist. Professor of anatomy and physiology, Hampden-Sydney Medical College, Virginia, 1843–8. Hersey professor of anatomy, Harvard University, 1847–74. Curator of the Peabody Museum of American Archaeology and Ethnology, Harvard, 1866–74. (*DAB*, *DSB*.) With Asa Gray as intermediary, Wyman and CD entered into correspondence after the publication of *Origin* (see *Correspondence* vol. 8).

Those whose names Darwin deleted:

Bronn, Heinrich Georg (1800–62). Professor of natural and technical sciences and director of the zoological collection at the University of Heidelberg. (*DSB*,

NDB.) A naturalist interested in the question of variation and development of organic life, Bronn translated *Origin* into German (1860). He had received a presentation copy of the first edition.

Fawcett, Henry (1833–84). Professor of political economy, Cambridge University, 1863. MP for Brighton, 1865–74. Blinded in a shooting accident in 1858. (*DNB.*) Fawcett wrote a favourable review of *Origin* in 1860 that particularly pleased CD because of its defence of his method of reasoning (Fawcett 1860; see *Correspondence* vol. 8).

Holland, Henry (1788–1873). Physician. Appointed physician in ordinary to Queen Victoria in 1852. President of the Royal Institution of Great Britain for many years. (*DNB*, *Physicians.*) Holland was a distant relative of the Darwins and Wedgwoods. The Darwin family often sought his medical opinion; he was consulted during Henrietta Emma Darwin's serious illness of 1860–1. CD presented him with a copy of the first edition of *Origin*.

APPENDIX VIII
The 'squib' on *Origin of species* published in *Punch*, May 1861

The 18 May 1861 issue of *Punch, or the London charivari*, p. 206, included a poem satirising evolution theory. Entitled 'Monkeyana' and signed 'Gorilla', the poem devoted two stanzas to Darwin's theory as published in *Origin of species*, particularly focusing on the implications of the theory for the question of human descent. Following verses treated of other recent scientific developments that touched upon this topic: the work of Leonard Horner, William Pengelly, and Joseph Prestwich on the antiquity of man; the debate between Richard Owen and Thomas Henry Huxley on the anatomical differences between the brains of humans and those of the higher apes; and the exhibition of Paul Belloni Du Chaillu's gorilla specimens and his descriptions, given before the Royal Geographical Society and the Royal Institution of Great Britain, of their fierce behaviour in the wild.

In a letter to Huxley of 22 May [1861], Darwin referred to the poem, stating: 'Hooker tells me that the squib in Punch was by Sir P. Egerton, which astonishes me;— I did not think it very good.—' Philip de Malpas Grey-Egerton, Conservative MP for South Cheshire, was a palaeontologist who specialised in fossil fishes. A friend and neighbour of Darwin's cousin, William Darwin Fox, Egerton was a prominent member of several London scientific societies, including the Royal Society, the Philosophical Club of the Royal Society, the Royal Geographical Society, and the Palaeontographical Society. He and Darwin had long been acquainted, having served together on the council of the British Association for the Advancement of Science in 1855 and that of the Royal Society in 1855 and 1856.

Monkeyana.

Am I satyr or man?
Pray tell me who can,
And settle my place in the scale.
A man in ape's shape,
An anthropoid ape,
Or monkey deprived of his tail?

The *Vestiges* taught,
That all came from naught
By "development," so called, "progressive;"
That insects and worms
Assume higher forms
By modification excessive.

Then DARWIN set forth.
In a book of much worth,
The importance of "Nature's selection;"
How the struggle for life
Is a laudable strife,
And results in "specific distinction."

Let pigeons and doves
Select their own loves,
And grant them a million of ages,
Then doubtless you'll find
They've altered their kind,
And changed into prophets and sages.

LEONARD HORNER relates,
That Biblical dates
The age of the world cannot trace;
That Bible tradition,
By Nile's deposition,
Is put to the right about face.

Then there's PENGELLY
Who next will tell ye
That he and his colleagues of late
Find celts and shaped stones
Mixed up with cave bones
Of contemporaneous date.

Then PRESTWICH, he pelts
With hammers and celts
All who do not believe his relation,
That the tools he exhumes
From gravelly tombs
Date before the Mosaic creation.

Then HUXLEY and OWEN,
With rivalry glowing,
With pen and ink rush to the scratch;
'Tis Brain *versus* Brain,
Till one of them's slain;
By Jove! it will be a good match!

Says OWEN, you can see
The brain of Chimpanzee
Is always exceedingly small,
With the hindermost "horn"
Of extremity shorn,
And no "Hippocampus" at all.

The Professor then tells 'em
That man's "cerebellum,"
From a vertical point you can't see;
That each "convolution"
Contains a solution,
Of "Archencephalic" degree.

Then apes have no nose,
And thumbs for great toes,
And a pelvis both narrow and slight;
They can't stand upright,
Unless to show fight,
With "DU CHAILLU," that chivalrous knight!

Next HUXLEY replies,
That OWEN he lies,
And garbles his Latin quotation;
That his facts are not new,
His mistakes not a few,
Detrimental to his reputation.

"To twice slay the slain,"
By dint of the Brain,
(Thus HUXLEY concludes his review)
Is but labour in vain,
Unproductive of gain,
And so I shall bid you "Adieu!"

Zoological Gardens, *May*, 1861. GORILLA.

APPENDIX IX
Letters between Thomas Francis Jamieson and Charles Lyell on the geology of Glen Roy, Scotland

In 1860 and 1861, in the course of preparing what he intended to be a revised edition of *Elements of geology*, Charles Lyell attended particularly to the chronological relationship between the first appearance of human artefacts in Europe and the glacial period (see K. M. Lyell ed. 1881, 2: 329–30, 337–8 and C. Lyell 1863). Discussions of this subject first entered the correspondence between Darwin and Lyell late in 1859. Although Darwin was no longer directly involved in geological investigations, he attempted to assist Lyell (see, for example, *Correspondence* vol. 8, letters from Charles Lyell, 24 November 1860 and 30 November 1860, and letters to Charles Lyell, 25 November [1860] and 4 December [1860]).

Lyell also sought assistance, early in 1861, from the geologist and agriculturalist Thomas Francis Jamieson, in particular on points concerning the geology of Scotland. Lyell had met Jamieson in London and learned that he was interested in examining the celebrated 'parallel roads' of Glen Roy, a series of terraces running parallel to each other along the sides of a glen in Lochaber, Scotland. Lyell encouraged Jamieson to visit the area 'to get more evidence on the subject' (Jamieson 1863, p. 240).

Apparently, Lyell mentioned his meeting with Jamieson in a letter (now missing) to Darwin because, two weeks later, Darwin also began to correspond with Jamieson urging him to 'take up Glen Roy' (see letter to Charles Lyell, 12 April [1861]). Darwin, as Jamieson later reported, sent him 'some useful maps and memoirs bearing upon the problem' and advised him about 'several points worthy of particular attention' (see letter to Robert Chambers, 30 April [1861], letter from T. F. Jamieson to Charles Lyell, 14 May 1861, and Jamieson 1863, p. 240). Darwin was anxious to learn whether his interpretation of the evidence, given in a paper published in 1839, as indicating that the roads were formed by the action of the sea, would be overturned by increasingly convincing accounts of the agency of glaciers. Indeed, fifteen years earlier, Darwin had almost conceded to the glacial-lake hypothesis before concluding that it still failed to account for a number of phenomena that the marine hypothesis could explain (see *Correspondence* vol. 4 and Rudwick 1974).

Darwin first became acquainted with the Glen Roy puzzle in 1834 on reading the third volume of Lyell's *Principles of geology* (C. Lyell 1830–3, 3: 131–2). Lyell commented on the parallel roads of Coquimbo, Chile, and compared them to the Glen Roy roads in Scotland. By analogy with the accounts of Glen Roy provided by

John MacCulloch (MacCulloch 1817) and Thomas Dick Lauder (Lauder 1823), the Coquimbo roads had already been interpreted as the beaches of former freshwater lakes. Lyell expressed dissatisfaction with this interpretation and instead suggested a marine origin. In May 1835, during the *Beagle* voyage, Darwin was able to visit the site (see *Correspondence* vol. 1). The continuation of the roads to the existing coastline and the abundance of marine shells in the area convinced him that Lyell was right. Upon his return to Britain, Darwin set aside time to study the Glen Roy formation and reverse the apparent analogy.

In June 1838 Darwin spent 'eight good days' in Lochaber; on 6 September 1838, he recorded in his 'Journal', 'Finished paper on Glen Roy—one of the most difficult & instructive tasks I was ever employed on' (see *Correspondence* vol. 2, Appendix II). Darwin argued that the parallel roads were the remnants of marine beaches formed by an arm of the sea during successive periods of elevation of the land ('Observations on the parallel roads of Glen Roy, and of other parts of Lochaber in Scotland, with an attempt to prove that they are of marine origin', *Philosophical Transactions of the Royal Society of London*, pt 1, pp. 39–81; *Collected papers* 1: 89–137). His account, which was refereed and recommended for publication by Adam Sedgwick, had convinced a number of prominent geologists, among them Roderick Impey Murchison, Leonard Horner, and Lyell, that the marine-beach theory was a more satisfactory explanation of the parallel roads than the prevailing diluvial lake hypothesis. Lyell, having adopted Darwin's explanation, included it in the second edition of his *Elements of geology* (1841). However, shortly afterwards, Darwin's marine theory was called into question by the Swiss naturalist Louis Agassiz, who, during a brief visit to the Lochaber region of Scotland in 1840, had been impressed by what he considered to be overwhelming evidence of former glacial action (Agassiz 1840a, 1840b, and 1842).

At this time, the prevailing orthodoxy among British geologists attributed features that would later be routinely considered 'glacial' either to the effect of the passage of debris-laiden water or to the action of great icebergs on an emerging landscape. The 'iceberg theory' accounted for extraneous deposits like erratic blocks (Darwin used the theory to explain the perched boulders above the highest of the Glen Roy roads), and for glacial striae and polished surfaces. Throughout the 1840s, Agassiz's advocation of extensive terrestrial glaciation following a great Ice Age was largely ignored and even ridiculed in Britian. Agassiz left Europe for America in 1846. By 1848, glaciologists were a beleaguered community; it had even been suggested that Agassiz had abandoned his former stance (see 'The glacial theory not abandoned by its author, professor Agassiz', *Edinburgh New Philosophical Journal* 45: 396–7). (On the reception of Agassiz's glacial theory, see North 1943, Hanson 1970, Rudwick 1970, and Rupke 1983.)

In nineteenth century geological morphology, four main approaches commanded adherence: 'cataclysmic', 'diluvial', 'marine', and 'fluvial'. Belief in a former period of sustained and extensive land-based ice action favoured none of these perspectives especially. Agassiz's formulation of the glacial theory was a cataclysmic vision of an age in which the earth was smothered under a nappe of ice, *die Eiszeit*, and, crucially, involved total organic extinction. A student of Georges Cuvier, Agassiz's attention to the agency of ice sprang from his interests in palaeontology and biology (see Agassiz 1840c; for an English translation, see Carozzi trans. 1967). Between the 1830s and 1860s the marine drift theory and the agency of icebergs enjoyed hegemony. However, the 1850s saw the action of glaciers become the subject of unprecedented attention, with explorations of Greenland and Norway adding to the interest in the agency of land-based ice. Summer excursions to the Alpine glaciers became fashionable and publishers were eager to popularise accounts of travellers' experiences (see, for example, Tyndall 1860). During the 1860s, leading geologists such as the Geological Survey's Andrew Crombie Ramsay, Joseph Beete Jukes, and Archibald Geikie, began to favour a synthesis of glacial and fluvial agencies. (On the history of geomorphology, see Davies 1969.)

Thomas Francis Jamieson had become interested in geology while engaged in work on 'improving projects' on Scottish estates. These schemes, establishing land-drainage systems for example, involved making extended networks of incisions across the surface of the land-mass; field-drains, miles of canalways, and reservoirs were cut across vast stretches of land. Jamieson was impressed by some of the natural features of the landscape, in particular by the erratic blocks that lay strewn across the countryside. His earliest contributions in geology were communicated to the Geological Society of London by Murchison (Jamieson 1859 and 1860a). At this time, Jamieson endorsed the marine drift theory as an explanation of the so-called superficial or northern drift; however, he was evidently not wholly satisfied with this position. In 1860 Jamieson travelled to the west coast of Scotland, to Argyllshire, to examine the scene of the Crinan canal 'catastrophe' (see Jamieson 1862, pp. 164–6). In the hills between the Sound of Jura and Loch Fyne three natural lakes had been artificially interconnected and enlarged to form a regulating reservoir. The uppermost lake burst its embankments and plunged through the two lower lakes; a huge body of debris-laiden water crashed into the valley 700 feet below. Jamieson visited the area with the express purpose of examining whether the passage of boulders and debris carried by the rush of such a body of water could produce scoring and polishing similar to that attributed to glaciation. He concluded that the markings left by the inundation did not resemble those associated with glacial action. He was well placed to judge: 'true glacial striæ and scores, beautifully and extensively marked' the area, running at right angles with the 'ruts made by the descending torrent'. There was no comparison between the 'scratching of poultry' effected by the debris-laiden deluge and true evidence of glaciation—the 'long, rectilinear, parallel grooves' and the characteristic glacial polish (Jamieson 1862, p. 166).

When Jamieson visited Glen Roy in August 1861, his observations confirmed the tentative view he had expressed in his letter to Lyell on 14 May 1861:

> There are so many weighty facts bearing against the theory of the marine origin of these parallel roads that I am strongly inclined to believe they may somehow or other have been caused by fresh water lakes.

Immediately upon his return from Lochaber he wrote a lengthy letter to Darwin, stating (letter from T. F. Jamieson, 3 September 1861):

> all I saw tended to impress upon me the conviction that these parallel roads have been formed along the margin of freshwater lakes and finding the marks of ice action so plain over the whole district I cannot help thinking that Agassiz hit upon the true solution of the problem when he pronounced these marks to be the effect of glacier-lakes.

In his letter to Lyell, 6 September [1861], enclosing Jamieson's first report on Glen Roy, Darwin seemed to concede defeat on the question, lamenting: 'I am smashed to atoms about Glen Roy. My paper was one long gigantic blunder from beginning to end.' Subsequent letters indicate, however, that he had not yet jettisoned his marine-beach explanation (see the letters to Charles Lyell in September and October 1861). (For a detailed account of the Glen Roy problem, and of Darwin's reluctance to relinquish his theory, see Rudwick 1974.)

Jamieson became the first geologist to amass detailed field observations in support of Agassiz's glacial explanation of the parallel roads of Glen Roy. Despite his clear conviction that the evidence fully supported the former existence of extensive glaciers in the region and of lakes formed by barriers of ice, he waited to publish the results of his study until 1863, after he had made a second extended visit to Lochaber in July 1862 (see Jamieson 1863).

By late 1861, Lyell had decided to publish the results of his protracted study of the earth's most recent geological chapters in a separate volume, the *Antiquity of man*, which appeared in 1863. In this work, he fully adopted the glacial-lake hypothesis as presented by Jamieson, abandoning his earlier support for Darwin's marine interpretation (see C. Lyell 1863, pp. 252–64).

The letters transcribed below were exchanged between Lyell and Jamieson between May and October 1861. Because Lyell sent Jamieson's letters to Darwin, they have been printed to provide information that is helpful in understanding the series of letters discussing Glen Roy that are included in this volume of the *Correspondence*. With the exception of what appear to be notes for a letter to Jamieson included in Lyell's scientific journal, all the letters are in the Lyell papers in Edinburgh University Library. Only those letters in the Lyell–Jamieson correspondence that specifically discuss the subject of Glen Roy have been included. The editors would like to thank Sandra Herbert, M. J. S. Rudwick, and James A. Secord for advice and assistance provided during the preparation of this appendix.

From T. F. Jamieson to Charles Lyell 14 May 1861

Dear Sir Charles,

I have several times been urging M^r. Smith of Jordanhill to publish a revised account of the Scottish drift shells and I am glad to learn by a letter w^h. he has sent me that he is now engaged in preparing it— I have sent him a list of those I have met with in this district, amounting to *forty* species.[1] As I believe you found some shells in elevated beds near Dundee of w^h. I have never seen any list, it has occurred to me that it would be very desirable that M^r. Smith should include some notice of them as they would connect our northern group with those of the Clyde district— I suggested to M^r. Smith that he should give separate lists of the shells found at each principal locality—as at Dalmuir, Kyles of Bute, Paisley &c. because it may turn out that they were not all strictly contemporaneous.

With regard to our so called Crag beds here it may interest you to mention that since I returned home I found in one of our sections here a fragment of what appears to be the *Nucula Cobboldiæ*— It is very provoking to get nothing but fragments but I hope some day to fall upon a spot where the shells may be found in better preservation—

M^r. Darwin has in the kindest manner possible urged me to visit Glen Roy & has sent me copies of the principal papers on that celebrated locality.[2] I am reading up the subject & hope to be able to devote a week or two in the autumn to an examination of these curious shelves. M^r. D. has also favoured me with some valuable notes as to some of the more important points to be attended to.

With regard to glacier lakes I am anxious to know whether instances occur in glacier districts of deep accumulations of water retained by a dam of *ice alone*. For in the case of some of the glacier lakes in the Alps it would appear to be the Lac-Combat *Moraines* that form the dam or barrier—and I see that in other instances the streams make a passage for themselves through below the glacier. as the Doire beneath the glacier of La Brenva. (Forbes travels.)[3]

There are so many weighty facts bearing against the theory of the marine origin of these parallel roads that I am strongly inclined to believe they may somehow or other have been caused by fresh water lakes. I should be curious to know if Prof. Rogers' examination of them has brought out any new *facts* of importance. He talked at Aberdeen about the false bedding seams always pointing *up* the glens—& that the erosion was most marked on the slopes facing down the glens.[4] I do not remember anything else of importance that he contributed to the subject on that occasion.

I am | Sir | Your very obed. servt | Thos. F. Jamieson

Ellon | Aberdeens^h. | 14 May 1861

Edinburgh University Library (MS Gen.112.2780–1)

1 The Scottish geologist James Smith of Jordanhill published a revised version of his 1839 catalogue of shells from the drift deposits of Britain and Ireland in 1862. In it he thanked Jamieson for supplying him with a list of 'Aberdeenshire Glacial Shells' (J. Smith 1862, p. vii).

2 See also letter to Robert Chambers, 30 April [1861] and Jamieson 1863, p. 240.

3 J. D. Forbes 1843, pp. 201–2.

4 Henry Darwin Rogers reported the preliminary results of his study of the parallel roads of Glen Roy at the 1859 meeting of the British Association for the Advancement of Science. A full account, delivered before the Royal Institution of Great Britain on 22 March 1861 (see following letter and n. 6), was published later in 1861 (Rogers 1861).

From T. F. Jamieson to Charles Lyell 11 September 1861

Ellon,
11. Sept 61

My Dear Sir Charles,

I have just received your long and interesting letter and shall reply to some of the points of it in detail in a day or two but in case of any little delay I am desirous of at once sending a few lines— M. Kjerulf has I find anticipated me to some extent, I agree with him in thinking the submergence theory has been pushed to the extreme.[1] After what I saw in Glen Roy nothing short of the finding of marine remains will readily convince me that the sea has occupied Lochaber since the period of these shelves to a height exceeding some 600 or at *most* 800 feet. The wonderful preservation of the deltas at the *head* of Glen Roy surpasses conception. That the submergence however which followed the great glacial epoch reached to heights of a *few hundred feet at least* can be shown in a very convincing manner in this part of Aberdeenshire for altho not a trace of marine shells of the drift period have been found at heights exceeding 200 or at most 250 feet (the statement of 350^{ft} for the Gamrie shells is quite erroneous, the shell layer there I found by Aneroid measurement to be about 130 feet above high water mark), yet along a range of open coast country we have beds of finely laminated marine clay which are of great depth (sometimes 30–60 feet) in the lower grounds & which I have traced up to the *top* of low hills of from 250 to 300 feet— Now the presence of a few feet thick of this fine muddy sediment on the *summits* of rounded eminences in our open country evidently I think requires us to suppose that there had been a sufficient depth of water covering them to keep the bottom undisturbed by the surface agitation from winds— at King Edward also the deposit containing Arctic shells ranges up to 200 or 250 feet & the character of the group indicates some depth of water at that level. So that there can be little room for doubt that the submergence reached at least a few hundred feet I should say at any rate 500 feet.

I regret that in my paper on the drift & rolled gravel I spoke so decidedly of the submergence attaining extreme heights.[2] for altho I am not satisfied that I was wrong altogether yet the *total* absence of organic remains should I think always make one hesitate in pronouncing with certainty— Nevertheless the bank of stratified drift I described at Meal Uaine in Perthshire which ranges up in one clear decided section to about 1200 feet above the sea is certainly at least an *aqueous* accumulation. In this I am confirmed by the opinion of Principal Forbes who in summer lives near it, at Pitlochry & has often examined it altho curiously enough he had never noticed the glacial striæ on the rock beneath it until I told him—[3] I am quite at a loss to conceive how this bank of drift could have been formed in a fresh water lake even by the aid of glacier agency. If then it is a marine accumulation it proves a submergence after the great glacial period to the extent of 1100 or 1200 feet.— The lower beds of it are often very finely laminated & I searched them earnestly for a long time slicing down the layers with my knife & examining closely with a lens for small fragments but deuce a crumb could I see & yet no better-suited seams of silty mud for the preservation of shells could be wanted—

I consider this bank of drift at Meal Uaine therefore my strongest proof of a great submergence subsequent to the period of chief glaciation— I hope you have seen it it is out of all sight the finest section of drift I have met with—

But It is possible that the submergence may have been greater on the East side of Scotland than on the west. The horizontality of the Lochaber lines has been tested chiefly on N & S. directions. I hoped to have been able to test the horizontality from E to W at Loch Laggan—but there is no clear line to measure from. however I saw what I thought indications of the waters of Loch Laggan having stood at greater height at the E. than at the W. end. I do not attach however any weight to this as the elements of my observations were not precise but I think the matter should be tested by levelling up to the outlet at Makoul & comparing that with the height of the old delta of the Gulban water at the W end of the Lake.

It may seem absurd in me to offer any suggestions to you on such a subject but as the tone of your letter invites it I would say that were *I* writing a new edition of a book giving a history of the drift period I should certainly not commit myself to the assertion of a submergence for Scotland beyond 500 or 600 feet for the period following the great land glaciation— The submergence *may* have been much more but, 1st the absence of all marine remains, at great heights. 2nd. The character of such remains as we do find, not indicating a profounder depth than I have stated (Ed. Forbes long ago dwelt on the fact of no deep sea corals or other proofs of profound depth of the water) occurring in the drift)—[4] 3rd. The absence of rolled shingle on the summits of *high* hills & of beds of laminated brick clay beyond heights of 300 or 400 feet— All these tend to throw a doubt on the fact of the submergence being so profound as it has been customary to suppose it.

With regard to the cross stratification or false bedding which Prof Rogers mentions in Glen Roy,[5] I kept a look out for it wherever I went but failed to find it in fact there are not, so far as I could see, any noticeable instances of false bedding.— There is merely undulating stratification arising from the form of surface on wh. the materials have been thrown down—

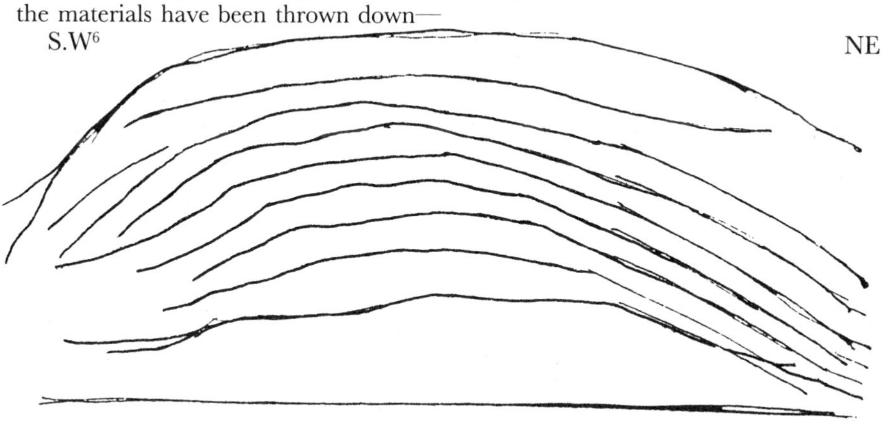

There are many long narrow mounds sloping down from the hill sides towards the bottom of the Glen, and the sections across these which have been occasionally made by the road shew beds of gravel & silt—& it so happens that in several places the cutting has been into the NE slope which shews the layers dipping up the glen but in other cases & sometimes in the same section where these layers can be followed to the S.W. slope they are seen to dip in the opposite direction, but of true false bedding (to use a paradoxical expression) I saw only one instance of wh. the

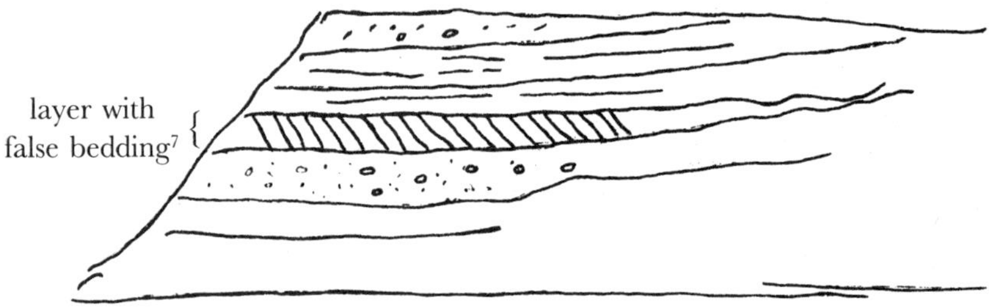

above is a sketch—and it puzzled me at first to account for it. The seam of silt shewing it was six inches to a foot thick & it was evident that the oblique lamination did not arise from layers of deposition. On slicing it up with my knife I found it arise from peroxide of iron wh. had been infiltrated somehow into the mass & had arranged itself in oblique lines as above—

It is hard to suppose that a man like Rogers could have fallen into the mistake to describing undulating layers as false bedding but after the absurd theory wh. he has brought forward for explaining the whole phenomena of these glens one can hardly wonder at what else he might do.—[8] It was indeed one of the things that particularly struck me that not only had the gravel & pebbles been left to accumulate just where the rivulets had poured them in but so little of any current had there been that even the finer sediment or silt had travelled no distance but had in many cases apparently settled down film after film just opposite the rivulet that had brought it in. It was in the upper reach of the glen between the Gap & Glen Turret that this chiefly struck me.

—With regard to your query about the Turret delta— This delta (like almost all the others & this is a remarkable thing) does not rise above the level of the lowermost Glen Roy line—at least not to any extent— We must therefore suppose that Loch Gluoy continued to exist when Loch Roy had subsided to the level of its lowest line, and in that case the outfall had to descend 320 feet or so before it fell into Loch Roy near the shepherds house in Glen Turret, & in doing so it has in some places to rush down a rapid rocky slope from wh. it could derive much debris & sediment— It was not the sediment issuing out of the lake that contributed much to the delta but the force of the large stream rushing down the channell before it joined Loch Roy.

The absence of any noticeable amount of delta matter at the levels of the two uppermost lines struck me much. I account for it by thus: these two uppermost lines represent periods of greater glacial action & consequently colder conditions of climate. so that the atmospheric moisture would fall perhaps chiefly as snow & therefore not wash down so much debris. & the level of the lake reaching higher up the hills the rivulets of course were smaller & had less force— And in the case of the Turret delta the uppermost Glen Roy line runs so far up the glen to near the head of Glen Gluoy that the outfall of Loch Gluoy had to run over but a small extent of channel & that with a very gradual slope so that it could not carry in much debris.

The case of the high lying shells on Moel Tryfan is so remarkable an exception that in thinking over it it has occurred to me Could they be *Crag*? The fact of their being accompanied by Chalk flints lends some probability to such a view. They might be Crag beds which by their position on the top of an isolated hill had escaped being swept away by the glacial action— such is the position of some curious beds of shingle wh. I consider preglacial in this district, but they are not at such a height— These shells of Moel Tryfan are described as being worn. The genera *Buccinum*, *Venus Natica* & *Turbo* to wh. they are stated to belong are all Crag forms. It might be worth while to re-examine them. I regret that when in London I forgot to see them—[9]

If we are to account for all the erratic boulders by floating ice we must push the submergence in Scotland up to 2500 or 3000 feet & *perhaps higher*. I should say at least nearly 3000 to account for those I observed on Morven— Now it seems to me that had the submergence reached to these great heights, we should find deep sea corals or beds of foraminifera or some other organic remains indicating a profound depth of water in the lower grounds, but instead of that, even in the Clyde beds, wh. are rich in shells the character of the mollusca does not point to a sea more than a few hundred feet deep— Again after submerging Scotland England & Ireland Scandinavia, the Faroe islands & North America, how is it possible to suppose an Arctic climate where there was a widespread ocean?

I shall send you some map in a few days, and shall also perhaps take the liberty of forwarding a few more notes about Lochaber and in the mean time I beg to thank you very sincerely for honouring me with your correspondence and | remain | Your most obed. servt. | Thos. F. Jamieson

Sir Charles Lyell. F R. S | &c &c

P.S. The question of a profound submergence of this country during the drift period should I think at present be left open. It would be as rash to deny it merely from the absence of marine fossils as it is to assert it without the presence of any such proof.— The skull of a walrus among the shingle at the base of Ben Muic Dhui would be very refreshing & more to the point than all the evidence we have got yet— T.F.J.

Edinburgh University Library (MS Gen.112.2786–91)

[1] Theodor Kjerulf rejected the prevalent view that striated rocks and erratic boulders in Scandinavia were the products of the agency of floating ice during the time when the submerged land was undergoing elevation; he favoured instead a land-based glacial interpretation of the phenomena (Kjerulf 1860).

[2] Jamieson 1860a.
[3] The geologist James David Forbes had been appointed principal of United College, University of St Andrews in 1859 (*DNB*).
[4] E. Forbes 1846.
[5] Rogers 1861, p. 343.
[6] The diagram has been reduced to 65% of its original size.
[7] The diagram has been reduced to 65% of its original size.
[8] Henry Darwin Rogers postulated that the parallel roads of Glen Roy had been formed by the grooving agency of floods pouring through the glens. Rogers proposed that the tsunamis, or great waves, were caused by an earthquake underneath the Atlantic Ocean (Rogers 1861).
[9] Jamieson refers to the collection of the shells of this region formed by Joshua Trimmer and described in Trimmer 1831. Trimmer's shells had been deposited at the museum of the Geological Society of London.

From Charles Lyell to T. F. Jamieson 14 September 1861

Sept. 14, 1861

Stratif^{d}. drift in Forfarsh. 800 f^{t}. & more high. Moel Tryfan, shells Norwich crag in Chillesford?

If we submerge Perthsh. 1200 ft. & raise it more than lowest shelf of Glen Roy, we have still Jamieson, difficulty of no foraminifera or deep sea corals in mud or gravel.

Glen Roy shells valuable as showing:

1st. The non-invasion of the sea to height of lowest shelf 950? f^{t}. in post glacial period.

2^{ly}. No local revolution in physical geography & surface levels within Lochaber—

Were Grampian boulders carried to the Sidlaw as Alpine erratics to the Jura by glaciers?

Wilson 1970, p. 523.

From T. F. Jamieson to Charles Lyell 14 September 1861

Ellon
Saturday Ev^{g}.
14 Sep. 61

My Dear Sir Charles,

I have sent you enclosed a map of Scotland which I colored a year or two ago to represent the submergence that would take place by depressing the country about 500 feet and shewing, as correctly as I then could, the probable outline of the new coasts that w^{d}. be thus occasioned. As 500 feet is about the upper limit to w^{h}. drift shells have been met with in Scotland, so far as I know, I have thought that this was just the thing you wanted. I have likewise pencilled the little map you sent me in accordance with the same supposition viz to shew a depression of about 500 feet.

As I have no copy of the larger map and as it w^{d}. take me some time & trouble to reconstruct one I hope you will excuse me asking you to return it some time when it has served your purpose— You c^{d}. easily reduce it I imagine to any size suitable for your intended sketch by having it photographed.

Rink's papers in the Geographical Soc^{y}. Journ. I am acquainted with— you will find them in vols 23. & 28.[1] His book however De Danske Handelsdistricter i Nordgronland I have not seen—[2] What a pity it is that no man with time money & brains will go & spend a summer or two in Greenland & tell us all about it, Some enterprising Savans shd. get up a Greenland Club. instead of always pottering away amongst the Alps.

I proceed to notice some of the queries which I did not answer in my last letter.

1. *As to the sea covering Forfarshire.*

Arctic seashells being found up to 512 feet in the Clyde district & to 250 feet in Aberdeenshire renders it probable that Mollusca must have also flourished in Forfarshire during the submergence to the heights of a few hundred feet above the present level. The fact of their being found hitherto only up to 50 feet or so is only negative evidence w^{h}. in Geology has often proved of little value— In Aberdeenshire where we do occasionally find drift shells it is surprising how very scanty they are. For instance there is a great bank of fine brick clay at Aberdeen w^{h}. has been wrought for years to a depth of from 60 to 90 feet. & yet I never cd. see or hear of a crumb of shell in it, but it is evidently marine & I believe Prof Nicol got from the workmen on one occasion some remains of star fishes but as he is not very communicative I cant give you any more account of them, but they are mentioned in the margin of his map—[3] In the same bank of clay on the South side of the river also wrought as a brick work no shells have ever to my knowledge been got but a friend of mine on one occasion got some vertebræ of a small fish—and in fact the occurrence of shells is *excessively scanty* & known only in a few localities altho the same beds have been exposed in numberless places— You talk of the multitudes of seals & walruses in Arctic seas. Now no seal remains had ever been found in the drift of the north of Scotland when one evening I got the remains of the skull of one (at least what my own palæontological endeavours made out to be such) from the bottom bed of a brickwork near this. I sent it with some other trifles to Prof. Owen enclosed in a parcel to M^{r}. Woodward, but the great anatomist never sent me any word about them.[4] There was amongst them a very curious bone got at a considerable depth in digging a well in a parish to the N of this. I got it from an old Naval surgeon who thought it to be "the right clavicular articulation of a Tortoise"— however I attach no weight to his opinion on such a point.

2nd. I am quite of the opinion that before the era of the Clyde &c beds with Arctic shells there was a period of land ice like that of Greenland, coincident probably with a considerable elevation of the country. —*this I stated in my paper* in the Geolog. Journ— vol. 16. p. 370.[5] And to this I decidedly adhere. —I myself found on the shores of Loch Fyne & also in the Kyles of Bute, beds with Arctic shells *resting upon.* glacially scored rocks—

Whether *in* **this** *country* there was a second extension of large glaciers after the submergence is a point I have not arrived at a certain conclusion about—

If we could conceive the Glen Roy lakes to have been formed by this second set of glaciers it would account for the absence of the sea in these glens since the formation of the parallel roads altho it might have overspread them during the intermediate submergence-period— This 2nd. glacier period would coincide with my 3rd. period as stated in the paper I have already alluded to, during which I supposed an upheaval sufficient to connect Britain with the Continent & also with Ireland— "Glaciers" I there said "may have still formed in favorable places but probably never regained their former extension".

Instead in looking over the sketch of the sequence of events there given the only line which "*dying I shd.* wish to blot" would be the one relating to the extent of submergence during my 2nd. period, and which I should rewrite thus:

"To this succeeded a period of submergence when the land gradually sank to a depth which has not been yet well ascertained, but proved by the occurrence of marine shells to have amounted in Scotland to at least a few hundred feet above the present sea level" &c the rest of the paragraph remaining as it is.

Your 3rd. query I answered in my last.

4th. I think if you transport the Alpine blocks to the Jura by land ice (in which I think you are quite right) it would certainly be consistent to take those of the Grampians to the Sidlaw heights by the same agency—

Query, with regard to the Sidlaw & Pentland is it altogether certain that no crystalline rocks similar to the erratic boulders occur nearer than the Grampians?

Principal Forbes[6] thought my observation important that there were no blocks from the Ben Muickdhui hills carried across the valley of the Dee to the great Quartz range on the South of that valley. This range is so lofty as to have doubtless sent glacier streams of its own to the Dee valley— I hope to be able to extend my examination to other parts of the Quartz ridge. —Another fact which I consider no less important & wh. I omitted to mention in my paper is this, that on these great granite mountains of Ben Muic Dhui, & Ben a Baird, Ben Avon &c, I found no foreign boulders—no blocks of gneiss or quartz rock or mica slate— not only was there none of these on the summits but also *none on the flanks* & lower slopes now if floating ice was the means of transporting our erratic blocks it should have brought gneiss boulders thither let the current have been from any point you choose.

Prof. MacGillivray in his Nat. hist of Braemar mentions that Mr. MacGregor of Achalater (a neighboring farm) told him he had found on the summit of Lochnagar a large piece of flint having an external white crust & internally a yellowish grey color.[7] It was broken up & used by his shepherds to light their pipes with— Other pieces also he stated had been found at various times— —If these were really flints—chalk flints how important & curious would be the fact. —MacGillivray does not seem to have seen them himself & perhaps they may have been only some variety of quartz or calcedony.

Your 6th. *query*. How much of the northern drift & boulders on the Sidlaw, Pentland & Yorkshire &c hills shd. we refer to floating ice— To this I can say little or nothing. I confess myself to be quite in the mist as to the amount of submergence— A few years ago being struck with the same considerations which seem to have impressed Kjerulf viz that of the small heights to wh. marine remains have been found[8] I made out the map I have enclosed. & searched for traces of a coast line at heights of 200—to 600 or 700 but in places wh. I thought especially adapted for shewing them, but I utterly failed to find any clear trace of a sea margin, *at any height* wh. perplexed me much. (When I say this I of course do not refer to the well marked more recent coast line of the 30 feet height on the West coast &c). I have since endeavoured to account for it by supposing the presence of coast ice had prevented appearances such as we have here viz. sandy beaches, shingle beds & cliffs—& Mr. Lamont's remarks on the angular quality of the debris along the coasts of Spitzbergen help to explain this.[9]

I do not think I have seen your account of the trains of erratic blocks in Massachussets & should like much to know where it is to be found.[10]

In reply to your 10th. or last query (wh. is the only one now that I have not referred to) I think it will be very difficult if not impossible to separate the transportation by glaciers from that by floating ice. The observations of Murchison & Verneuil as to the position of the great blocks far South in the plains of Russia & lying above beds of Arctic shells seem quite decisive as to floating ice for *that* case,[11] at least to a certain extent, for the blocks might have been carried far on their journey by the land ice of the 1st. great glacial period, & have been picked up again & floated to their present position by the coast ice of the submergence period— Might I ask if we are to be favored with a new edition of the *Principles* soon.[12]

I am | Your very obedient servt | Thos. F. Jamieson

Sir. Chas. Lyell. &c. &c

Edinburgh University Library (MS Gen.112.2792–5)

[1] Rink 1853 and 1858.

[2] Rink 1852.

[3] James Nicol was professor of natural history at the University of Aberdeen. He published a geological map of Scotland in Nicol 1844 and later had a large copy prepared of the section covering Aberdeenshire to illustrate a paper he gave to the British Association for the Advancement of Science (Nicol 1859, p. 116). This may be the map to which Jamieson refers. Nicol mentioned the occurrence of fossilised starfish in the region's boulder-clay in Nicol 1859, p. 118.

[4] Samuel Pickworth Woodward and Richard Owen were employed at the British Museum—Woodward as an assistant in the department of geology and mineralogy and Owen as superintendent of the natural history departments.

[5] Jamieson 1860a.

[6] James David Forbes.

[7] Macgillivray 1855, p. 149.

[8] Kjerulf 1860.

[9] Lamont 1860.

[10] Jamieson refers to C. Lyell 1855b.

[11] Murchison, Verneuil, and Keyserling 1845.

[12] The ninth edition of Lyell's *Principles of geology* was published in 1853; the tenth appeared in two volumes in 1867 and 1868.

From T. F. Jamieson to Charles Lyell 19 September 1861

Ellon

Friday Evening | 19. Sep. 61

My Dear Sir Charles,

I should have replied sooner to your last letters but have been occupied with some business matters that have left me no leisure for doing so.

In the first place let me thank you for the copy of your paper on the Berkshire U.S. erratics, which you have been so kind as send me—[1] It was a *perfect treat* I can assure you, the more so as in my seclusion here I had never heard of these most singular trains of blocks. It brought to my recollection that when some years ago I was engaged carrying on the reclaiming of moorlands near Birnam in Perthshire I remarked that the great boulders, which dotted so thickly the plain of drift that we were trenching, used often to run in *lines* pointing towards the hills from whence they were derived. As these large blocks cost us a great deal of trouble in clearing them off they were continually forcing themselves on my notice. I think I remember also having seen it mentioned elsewhere that a similar disposition of such boulders had been observed in other parts of Scotland.— I may perhaps take the liberty of sending you some remarks at a future time on these U-S. blocks, but in the meantime I must take up the objections or rather difficulties that M^r. Darwin has stated to the glacial lake theory for Glen Roy—[2] I think the whole of them may be surmounted.

1. "The levelled land straits with terrace like fringes above the level of the *highest* shelf"

I imagine that it is only the terrace like fringes (not the straits) that are higher than said shelf. Such fringes I conceive to be beds of stony earth due to preglacial action and which have been cut into by the outflowing stream of water from the lake— It must be borne in mind that the Glen Roy lines do not belong to the period of greatest glaciation for I found evidence that Glen Roy itself had been filled with ice before it contained a glacial lake and still more, that it had *discharged ice out at its upper end into the valley of the Spey*. I daresay I cannot expect either you or Darwin to credit this, but I hope to be able to prove it, or at any rate explain it, hereafter. I reserved these & some other extraordinary facts as I could not enter in a letter into a right acct. of them and I thought they might if unsupported by details have only had the effect of making you doubt my more sober conclusions.

2. "The beds of pebbles on very gently inclined surface between Glen Turret & Glen Gluoy. &c"

Altho I walked up Glen Turret as far as enabled me to see out towards Glen Gluoy, yet I did not get that water shed examined, but I should suppose these pebbles to have been caused by the outflow of Loch Gluoy— The large delta w^h. it has deposited at mouth of Glen Turret shews it must have considerably deepened its channel & before that was done might have run at higher levels— Darwin does not I think mean to say that these pebbles lie above the level of the Glen Gluoy line.

The greatest quantity of water worn stones that I saw due to the action of the shore lines of these lakes was on the S.W. side of the Hill of Bohuntine *at & below* the level of the lowest line, w^{h}. is the only one there marked— & standing there I could not but remark that I faced a long wide opening towards the mouth of the Caledonian Canal up w^{h}. the S.W. wind was blowing strongly as it must have done doubtless in the times of these lakes lashing the waters on the shore along this exposed slope. The pebbles were not however *well rounded*—very far less so than those at the outfall of *Makoul* where the discharge of the old lake had worn them intensely.— I also remarked that the small shingle *presently forming* along the shores of Loch Treig was more waterworn than any thing I saw along the Glen Roy terraces.

3rd. objection. as to terraced banks of detritus reaching down to the lower part of Spean valley—

—I see no difficulty at present in allowing that the sea might have occupied Glen Spean &c after the era of these lakes up to 500 or 600 feet above the present coast line, in which case these accumulations might be partly of marine origin. The marine beds of the Clyde district containing Arctic shells are clearly of posterior origin to the time of *greatest* glaciation, because I found them, myself, reposing on the top of the ice scored rocks and Smith of Jordanhill remarks it as a general fact that these beds rest on the top of the *azoic* boulder earth or till—just as Dawson describes those of Canada. Smith moreover found his Arctic shells up to a height (barometrical) of 512 feet, near Airdree—[3]

It is probable that the waters of the old glacial lake eventually found a passage out underneath the icy barrier just as the Swiss glacial lakes do where they sometimes *refill* again by the passage closing up— In this case the drainage would not be tumultuous but gradual so that the beds in the lower parts would not be swept away.

Darwin & Ramsay both call in a set of extensive glaciers after the marine drift in Wales,[4] & it is possible that the lakes of Glen Roy may have been formed in this second glacier period. I do not however quite see my way through this.

The relation of the great submergence of the drift period to these Glen Roy lakes—if *it was so profound as has been generally thought* is the thing that puzzles me most—and was a difficulty that I foresaw before going to Glen Roy & I think I mentioned it at that time in one of my letters to Mr Darwin.[5]

The following three modes of solution present themselves.

1st. These Glen Roy lakes may have been formed by the second set of glaciers w^{h}. were after the marine drift— I do not feel quite satisfied with this solution for the glaciers required are so large that it requires all my courage to stomach one set of them.

2nd. The submergence along the West coast of Scotland may not have been so great as along the East— I saw no satisfactory Evidence in Lochaber of the presence of the sea to great heights during the drift period. And the debris in the glens at the head of Loch Treig shewed no laminated silt or fine stratified sand like what I found in the high glens of Aberdeenshire—& in other parts of the West Highlands the bareness of the rocks & absence of any thick covering of drift compared with what we see on the East side of Scotland has always struck me much. Such a difference in the elevation is quite compatible I think with the horizontality of the Glen Roy lines w^h. has been tested only for short distances & chiefly in N. & S. directions. A slant sufficient to suit the purpose might be imperceptible in a length of a few miles— Still, after all this, the fact remains that marine shells have been found higher up in the Clyde district than in any part of Aberdeenshire or elsewhere on the East coast.

3^d. We may have been all mistaken about the profundity of this submergence & perhaps the sea of the drift period did not in Scotland any where reach the level of the lowest Glen Roy shelf.— —After studying Glen Roy it naturally occurred to me that the stratified aqueous accumulations I had noticed at great elevations in Aberdeenshire Perths^h. &c. being all in the hollows of glens & not on hill tops—might be fresh water beds deposited in temporary lakes or ice-dammed sheets of water. The absence of water worn shingle on *hill tops* at elevations exceeding a few hundred feet is a fact that has long struck me as remarkable on the theory of a profound submergence—

As I mentioned in a previous letter however, there are certain facts, such as my bank of drift on Meal Uaine, that seem inexplicable without a great submergence—

As to erratic boulders, I should not like to rest a theory of submergence on them *alone*, unsupported by other facts.

Darwin's case of New South Georgia I suppose extinguishes my difficulty about a glacial climate consisting in our latitudes with a widespread ocean.—[6] I may mention however, that in the Falkland islands, w^h. lie in the same lat. as Wales & between Tierra del Fuego & New S. Georgia—M^r. Vernet who lived there several years says "Snow disappears in a few hours except on the tops of mountains & ice is seldom found above an inch thick.— temp. never below 26° F. in coldest winter & general winter range 30°–50°. &c. See Geograph. Soc. Journ. vol. III.[7]"

In Tierra Del Fuego—South Shetland (& possibly New South Georgia) Sandwich land &c. the land is very mountainous reaching heights of 5000 to 6000 feet, so that they do not quite parallel the case of Britain submerged to 1500 or 2000 feet.

But I must allow that the account of Biscoe's daring run round the South Polar seas, reads very icy— (Geograph. Journ. vol III)[8] I unfortunately have not Darwin's book beside me to refer to about these parts concerning w^h. I suppose he is our chief authority—[9]

I had some other things to say but having been knocking about all day feel too drowsy to go on, | & remain | Your very obedient servant | Thos. Jamieson

Sir, Charles Lyell. | &c &c &c

Edinburgh University Library (MS Gen.112.2796–800)

[1] C. Lyell 1855b.

[2] See the letter to Charles Lyell, [15 September 1861], which Lyell forwarded to Jamieson.

[3] J. Smith 1850, and J. W. Dawson 1857 and 1860.

[4] Jamieson refers to Darwin's paper, 'Notes on the effects produced by the ancient glaciers of Caernarvonshire, and on the boulders transported by floating ice', *London, Edinburgh, and Dublin Philosophical Magazine and Journal of Science* 21 (1842): 180–8 (see also *Collected papers* 1: 163–71), and to Ramsay 1852.

[5] Jamieson's letter has not been found.

[6] See letter to Charles Lyell, [15 September 1861].

[7] Vernet 1833.

[8] Biscoe 1833.

[9] The reference is to *Journal of researches*, pp. 177–231, 273–93.

From T. F. Jamieson to Charles Lyell 20 September 1861

Ellon
20 Sept 1861

My Dear Sir Charles,

I am afraid I forgot to thank you for your kind offer to lend me the vol. of the Geograph. Soc. Journ. containing Rink's paper.[1] I however rejoice in a set of that useful publication having purchased it with some spare cash a few years ago.

On looking over your letter again I find some points I omitted to notice, & wh. I now take up.

1. The summit level of the Caledonian Canal is above 90 feet above the sea level wh. seems too high to have been submerged during the time of the 30 feet sea beach.

2. You will find some notice of the Moel Tryfane shells in Ed. Forbes's paper in the I. vol of the Geolog. Survey Memoirs.[2] I think however he examined them chiefly with the view of finding what depth of water they indicated. I have not his paper beside me but merely an extract in wh. 10 species, all recent, are mentioned— I quoted before from Trimmer's original notice in the Proceed. of Geolog. Soc. I. 331.[3]

3. As to the intermediate shelves of Glen Roy being accounted for by the more rapid growth of sward. —I was not much impressed with these intermediate shelves— they are so short & indefinite & seen in only one or two places that one is inclined to attribute them to some accidental cause on the side of Ben Erin where the three lines are beautifully marked & range along without interruption for miles I could see no trace of any intermediate line. —And from the Gap up to Glen Turret where the 3 lines are displayed with wonderful regularity on both sides of the glen I looked keenly for any hint of a higher line corresponding with that of Glen Gluoy, but not a speck belonging to such an one cd. be detected.

3rd. As to the mode of transport of the erratic blocks to the Sidlaw, I must say I incline to your original explanation in the Proceed. of the Geol. Soc. vol. III, that they were carried by land ice filling Strathmore.[4]

4. As to the peroxide of iron following the lines of oblique laminæ. I do not pretend to understand the origin of the phenomenon, but feel pretty confident that there were no such oblique lines due to aqueous deposition in the silty seam on which the circumstance occurs—

5. I have not seen the account of Dr. Hooker's case of lake beaches in the Himalaya wh. must be very interesting & am unacquainted with his Himalaya

Journal. I shall try however to procure it from the Aberdeen University library through my friend Prof. Dickie.[5]

6. I neglected to return the proofs of your two interesting diagram maps wh. I now enclose.

I have been trying to procure a copy of Hörbye's map & memoir on the glacial striæ & erosion in Norway through Williams & Norgate but they say they have failed to get it in Christiania where it is published.[6]

When in London Mr. Rupert Jones[7] shewed me it in the library of the Geol. Soc. & I should like very much to have an opportunity of studying the map & reading the memoir wh. luckily is not in Danish but in French.

—As I do not know where or how to get hold of it I have been thinking whether through your kind offices I might get a short loan of the Society's copy, which I would carefully take care of & return, if it could be transmitted by post, wh. I suppose it might be if it hasn't been bound up with any others.

I do not gather from your paper on the Massachussets erratics what is the height of the ridges above the sea shewn in the diagram.[8]

I am | Your very obedient servt | Thos. F. Jamieson.

Sir Chars. Lyell &c &c

Edinburgh University Library (MS Gen.112.2801–2)

[1] Jamieson was interested in several of Hinrich Johannes Rink's publications (see letter from T. F. Jamieson to Charles Lyell, 14 September 1861 and nn. 1 and 2).
[2] E. Forbes 1846.
[3] Trimmer 1831.
[4] Jamieson refers to C. Lyell 1840. See also Davis 1968 and Hanson 1970.
[5] The references are to Joseph Dalton Hooker and his account of his Indian travels (Hooker 1854a), and to George Dickie, professor of botany at the University of Aberdeen.
[6] Hörbye 1857. The booksellers and publishers Williams and Norgate specialised in foreign scientific literature.
[7] Thomas Rupert Jones was curator and librarian at the Geological Society of London.
[8] C. Lyell 1855b.

From T. F. Jamieson to Charles Lyell 25 September 1861

Ellon
25 Sept. 61 | Wed. evening

My Dear Sir Charles,

I have just received your letter of yesterday along with the very interesting enclosures which I herewith return—

It is very gratifying to find that in this misty subject of the drift we are getting a sight of some great land-marks— this great Greenland-state of things which looms out at the beginning is fast verging towards a certainty—in my own mind at least has become an article of faith, and forms the first act in the drama— then comes the submergence, of this also there can be no question except as to its amount—

Here then is a very decided progress— So long as this first great land-glaciation was misunderstood or overlooked, it was impossible to comprehend how the present state of things had been developed—

I should be very glad to see a good case made out for a profound submergence and the Welsh shells on Moel Tryfan are a very refreshing circumstance.

If we can get a set of glaciers after this submergence sufficient for our Glen Roy lakes it would perhaps satisfy the requirements of the case better than any other mode of explanation, for I must say I share considerably in Mr. Darwin's aversion to a *great* difference in the elevation of the W of Scotland compared with the E or with Wales.[1] And here I may point out that in my sketch of the sequence of events in the close of the Pliocene period—Geol. Journ. XVI 370. I suggested that a great elevation of the country was necessary after the marine drift wh. may have been coincident with a set of glaciers.[2] My reasons for supposing such a great elevation are.

1st. to account for the submerged forests wh. fringe the whole of Britain & Ireland

2nd. to get over the Elks & Reindeer & land shells, frogs, &c &c. to Ireland the Isle of Man, the Hebrides &c for without supposing a land connection how can the present distribution of these animals be accounted for. Again if the period during which the Reindeer, musk-ox, lemming &c roamed over England & Germany was subsequent to the marine drift, a climate suitable for glaciers is rendered not improbable—

Robert Chambers was kind enough to send me a copy of his 'Ice & Water' paper—[3] His great acquaintance with Scotland & hobby for the subject render his papers very interesting & valuable. —I think however his sketch of the succession of events must be wrong. You will see he starts with a great glaciation but refers it (altho with hesitation) to the action of *packs* & *floes* in other words he supposes the country to have been under the sea. in this I think he is entirely wrong— I have made some progress with a paper on the subject in which I shall bring forward some striking cases that I think cannot be disposed of by floating ice nor by icy debacles (Murchison's theory).[4]

His case of a terminal moraine, at Iorsa in Arran newer than the 25 feet raised beach goes very much against my digestion—& still more his supposition of a great submergence subsequent to said raised beach— This 25 to 30 feet old coast-line is I think undoubtedly contemporaneous with the raised estuary mud of the Tay & Forth forming the carses of Stirling & Gowrie & wh. in the case of the Tay (as is well shewn in Blackadder's map in the Werner Mem.)[5] extends to a similar elevation.

With regard to your map would it not answer well to make it shew by two gradations of tint. 1st. a submergence of 600 feet or so 2nd. a profound submergence to 2300 or 2500 feet—

In the Brit. Assoc. Report for 1844, Trans of Sections p. 57. you will find an interesting short notice by Oldham[6] two maps are mentioned in wh. he shewed submergence of Ireland to 500 & 1000 feet—

He also has another good notice in Brit Ass Report. 1848. p. 71–72. in wh. he states the occurrence of Arctic shells in Wicklow up to 700 feet.[7] & he says many of the ravines & cauldron shaped hollows have been denuded of drift—? Could this have been by the later glaciers.

I think your sketch of the succession of events, as good as can be given in our present state of knowledge—

I wish you could persuade Ramsay or some one of like ability to visit Lochaber—[8] for the case has been so much contested that the geological world will not readily adopt any explanation. & besides the district is so large & difficult of access & so fertile in facts that there still remains much to reward any one who will take the trouble to explore it. —It is a place where you could live for months & find something new & interesting every day—

It is certainly the case I think that fresh water lakes can accumulate well rounded pebbles on points where the wind throws the water much upon the shore— During gales the lashing of the water along such places is very considerable.

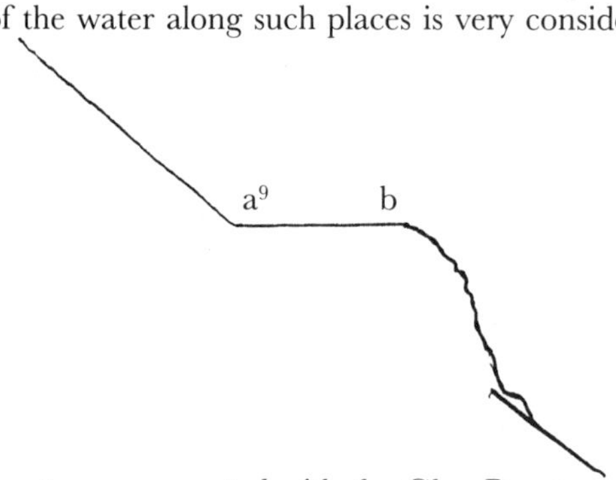

There is a circumstance connected with the Glen Roy terraces wh. struck me as very adverse to the theory of their marine origin & wh. I think I omitted to mention in my former letters on the subject. & it is this. On the flank of Ben Erin the shelves are marked with very great distinctness & precision, more especially the middle one; and I observed that from the well defined interior angle at *a* in the above sketch to the well-marked exterior angle or brink of the shelf *b* the surface is remarkably level with an inclination from A. to b not exceeding *three* feet— Now could a coast line of water rising & falling several feet every tide do this.? This part of the glen where the lines are so very neatly defined on both sides is when you stand on Ben Erin quite land-locked in all directions & very narrow so that the waters of the old lake must have been unusually calm & unruffled by winds—& hence the extreme neatness of the lines. On the face of Tom Brahn again which is exposed to the sweep of the winds coming up Glen Roy & over the hollow from Glen Gluoy. the shelves are very broadly & rudely marked. for here the water would be much more agitated & often heaped up by the wind—

With regard to the movement of pebbles down from hill tops alluded to by Mr. Darwin,[10] we have a case in this district of heaps of intensely water worn chalk flints clothing the whole top of a low ridge & wh. I believe to be the denuded relics of an expanse of chalk being confirmed in this opinion by the occurrence of a small patch of greensand on the same ridge— If I am correct in this these pebbles must have survived all the storms & rains of the drift. This ridge runs inland from near Peterhead. & if you look at the submergence map I sent you will perceive that it is at the extremity of one of the greatest expanses of low land in Scotland. I am somewhat inclined to attribute the escape of these flints & greensand & also our patch of Crag to the fact of the land ice of the first period having been unable, from their low outlying position, to carry them off—and may not such a cause also help to explain the presence of the English Crag in similar low outlying places?—

Our patch of Crag here is very handy for Hopkins used to object that no one knew how old our boulder drift of the north was seeing that it always rested on old rocks[11] & Nicol in his Geology of Scotland p. 256. thinks or rather thought, that there is "*a strong presumption*" that the boulder clay is older than the Lias![12] —Now I can shew by it, that all our drift deposits are newer than the Red Crag— This also is a little bit of progress—

I shall be delighted at all times to answer any queries I can, or in any way however slight, contribute to such local information on the districts I am acquainted with as may afford data for your reflections, & remain | Your mo. obedient serv[t] | Tho[s]. F. Jamieson

Edinburgh University Library (MS Gen.112.2803–6)

[1] See letter to Charles Lyell, 22 September [1861].
[2] Jamieson 1860a.
[3] Chambers 1861.
[4] Jamieson 1862. To account for the polished and striated rocks of Scotland, Roderick Impey Murchison advocated the view that:

> during the 'glacial epoch' Scotland probably consisted of a range of rocky hills in an icy sea, which by successive movements from different centres had thrown off dislocated sheets of ice, with much melted snow and ice, forming a detrital 'sludge' or drift which had been poured down all the lateral openings.

(R. I. Murchison 1851).
[5] Blackadder 1824.
[6] Oldham 1844.
[7] Oldham 1848.
[8] Jamieson refers to Andrew Crombie Ramsay.
[9] The diagram has been reduced to 65% of its original size.
[10] See letter to Charles Lyell, 22 September [1861].
[11] William Hopkins, president of the Geological Society of London from 1852 to 1853, published several papers on the superficial drift and the glacial theory (Hopkins 1842, 1843, 1844, and 1852).
[12] Nicol 1844.

From T. F. Jamieson to Charles Lyell 27 September 1861

Ellon
27 Sept. 1861

My Dear Sir Charles,

Many thanks for Horbye's memoir,[1] which came safe to hand yesterday. I have already sucked the juice pretty well out of it. It would appear from it that the lines of striation in Scandinavia accord with the notion of a mass of ice gliding down from the central heights of that peninsula. But there is a remarkable exception. If Horbye's account of the matter is correct it appears that near the middle of the country the striæ *ascend* out of the low ground of Sweden & run uphill right over & across the great central ridge or back bone of the peninsula! The author having traced distinct evidence of this 'burniage erratique' (*very erratique* it must be confessed) mounting over the Dörfjelde at a height of 4590 pieds norvegiens— And it seems this ascending march of the erosive agent does not depend on Horbye's observations alone but has also been noticed by M. Durocher,[2] & Horbye says the carriage of the boulders also is in the same direction— If this erratic burniage was from floating ice it implies rather an alarming amount of submergence— The district in which these ascending striæ are said to occur is not a small one but extends over a degree of latitude. But the great curiosity of it is that further *Eastwards* arrows are shewn going downhill towards the Baltic quite in the opposite direction.

—If these ascending striæ were caused by land ice it would be necessary to suppose that the relative levels of the hills in that region have been disturbed to a great degree since the period of the striation. But there is no use puzzling ones brains about it until it be seen whether it is actually the case—until it be confirmed by other observers—

The fact of the movement, however, having been otherwise so generally down the main lines of descent & radiating off in all directions from the central heights of the peninsula is strongly in favor of land ice as the cause, & in this opposite instance, *exceptio probat regulam*.

With regard to a glacier period following the marine drift you will find in Darwin's excellent paper on the Welsh glaciers (Edin Phil. Journ. vol. 33. 1842.)[3] that he 'brings' down a glacier far into Nant Francon clearing out the marine beds, & that he found the glacier markings *very plain* on the flanks of Llanberis 1000 feet above the lake, & so far as I can make out he seems to have supposed these Welsh glaciers to have been partly contemporaneous & partly subsequent to the submergence—but whether also previous does not appear. he says "From the presence of boss formed rocks low down in the valley of Nant Francon & on the shores of the lake of Llanberis (310 feet above the sea) it is evident that glaciers filled the valleys after the land had risen to nearly its present height & these glaciers must have swept the valleys clean of all rubbish left by the sea."

If there was a post marine glacier in Llanberis 1000 feet thick there might be post marine glaciers in Lochaber sufficient for the Glen Roy lakes—& by the way glaciers of such a magnitude after the submergence would have erased the lower line in Glen Spean & therefore if there were such large post marine glaciers the lower Glen Roy line could not be a sea-margin & thus arises another important objection to the marine theory of the Glen Roy lines.

These views of Darwins are adopted & confirmed by Ramsay who indeed in his last publication (on the ancient Glaciers of Wales. 1860.)[4] seems surer about these latter glaciers than he does about the earlier ones for he says p. 93. "It is probable that an earlier & greater set of glaciers preceded the Drift *but whether or not this was so*, does not interfere with *the fact* that after the re-elevation of the drift covered country, the greater glaciers ploughed the drift out of some of the larger valleys" & at p. 102. he ploughs out the drift in Nant Francon & Llanberis *as far as the lower end of Llyn Padarn* with his latter glaciers —Now if this was so it could scarcely have happened that the glaciers in Glen Treig & the adjoining Laire Glen would not have ploughed out the stratified terraces of waterworn gravel at the mouth of Loch Treig & at Inverlaire. —*ergo* these terraces of gravel formed by the lake of Lochaber when at the level of the lowest line could not have been formed by a preceding sea margin— You might put this to Ramsay & Mr Darwin if they oppose the freshwater lake theory for Glen Roy. I do not see how they can get over it if they account for the lines by the action of the margin of the drift sea.

I am | Your very obed. servt | Thos. F. Jamieson

Edinburgh University Library (MS Gen.112.2807–8)

[1] Hörbye 1857.
[2] Durocher 1846.
[3] 'Notes on the effects produced by the ancient glaciers of Caernarvonshire, and on the boulders transported by floating ice', *Edinburgh New Philosophical Journal* 33 (1842): 180–8.
[4] Ramsay 1860.

From T. F. Jamieson to Charles Lyell 5 October 1861

Ellon.
5, Oct. 1861

My Dear Sir Charles,

I am afraid I can say very little about the harmonizing of the elephantine &c remains with the glacier periods— It is only of late that Dr Falconer has introduced a distinction of the elephantine remains of this country into different species so that former observations lend us comparatively little assistance—[1] But so far as I can understand the evidence there seems to be little doubt that the Hippopotamus lived in England after both the great land glaciation & the subsequent marine drift— The clearest evidence on this head that I have seen is in a communication by T. P. Teale in the Brit. Assoc. Report. 1858. Sections p. 111,[2] where at Leeds he says they have

1. A Blue clay with transported blocks
2. A Yellow clay covering the last.
3. Above both of these what he calls *warp* containing skeleton of *Hippopotamus major* & remains of an Elephant, Ox, Deer, & Horse.

Now the *blue* clay I take to be the product of the great land glaciation the *Yellow* clay that of the marine drift the *Warp* I suppose to be contemporaneous with our old Estuary muds or carse clays of the Tay Forth & Clyde—& also probably with the *Loess* of the Continent—

The second glacier period of Darwin & Ramsay followed the marine drift but how long after that does not appear—[3] It would seem to be probably coincident with the presence of the Reindeer, Musk Ox, & Lemming in lower latitudes. & if this was subsequent to our old coast line it w^d. account for Chambers' case of the glacier moraine at Glen Iorsa thrown over upon the said old coastline & make this latter glacier period very recent indeed.[4]

M^r. Smith of Jordanhill said to me that when Edward Forbes examined the shells belonging to this old coast line w^h. are all of existing species he thought they indicated a climate rather warmer than the present—or I think he called them more of an Atlantic type. or some such term.[5]

Possibly the Hippopot. Elephas Antiquus &c. may have flourished in England during this comparatively milder period that followed the marine drift.

If I am right about the great land glaciation at the commencement, in this country it might have been coincident with a depression to the South—say at Sicily—when the Northern shells inhabited what is now the Mediterranean— Then when our Northern region was depressed these shellfish would migrate Northward & w^d. inhabit the shores of Scotland for a time on their way north *even altho the climate* here *was becoming milder*.

The Pleistocene, or as you abjure the word, the Newer Pliocene, period was evidently of vast duration & probably we shall require yet to interpolate here & there a new chapter in its history. Animals that inhabited this country before the great glaciation & drift may, like the Cyprina Islandica & other shells have retreated Southward for a time during the greatest reign of the ice to return after more genial weather came back— Besides, our great land glaciation & also the 2nd. glacier period may have arisen in a great measure from *elevation* so that low down in the valleys of the English Channel the summers may have been warm enough for Elephants to roam about & during the hotter months they may have climbed up into the higher grounds & left their remains there.

—with regard to the absence of extinct Mammalian remains in the marl lakes of Forfar, I may refer you to Prof Phillips Manual of Geology. edit. 1855. p. 430. where he mentions that with the exception of the elk there is a similar absence of extinct mammalia in the marl & peat deposits of Holderness in Yorkshire.[6]

You say that I do not explain how the ascending striæ in Scandinavia are proved.— I do not pretend to do so & in fact would fain hope that nobody else can! for if Hörbye &c are right that the striæ ascend out of the low ground of Sweden right over the top of the Dovrefield one may well stand aghast at such an enigma,[7] it will require a '*Daniel*' to read such hieroglyphics. We may sigh for the time when Sir Ja^s. Hall could make a good diluvial rush of water sweep over hill & dale & account for all the drift phenomena at one stroke.[8]

—With regard to what I said about the Glen Spean line being incompatible with such large glaciers as Darwin & Ramsay maintain to have been in Wales after the Marine drift,[9] I may explain that this Glen Spean line is the continuation of the lowest Glen Roy one & that it runs into the entrance of Loch Treig & the Laire

Glen, as you will see on referring to either Chambers' or Milne Home's map.[10] —I however ascertained that it does not occur at the upper end of Loch Treig wh. I suppose to have been at that time occupied by a glacier— Now what I meant to say was this—that if the Welsh glaciers advanced so far down the Llanberis valley as the lower end of Llyn Padarn (see Ramsay's map & p. 102 of Welsh glaciers)[11] it was inconceivable to me that further North, in Lochaber, they should not have advanced out of Glen Treig into Glen Spean & thus have ploughed out the gravel terraces that still remain there—just as they ploughed out the marine drift out of Llyn Padarn—ergo these gravel terraces of the lowest Lochaber line could not be of marine drift age— I find however that the large scale of Ramsays map misled me somewhat and the length of the said Llanberis glacier is after all not so great as I had supposed. Still I think that in the glen of the Corry Laire a glacier should have been long enough to push out the gravel terraces at Inverlaire.

You say it is singular why no ice should have been in the valley dammed up, i.e. Glen Roy—

Glen Roy is a remarkably *narrow* glen & is not connected with any wide basin for the accumulation of *névé* or snow— moreover the hills in it are all of moderate elevation so that the absence of a glacier in it is not surprising— According to Forbes (see Norway & its glaciers p. 216)[12] a sufficient continuity of mass & the disposition of snow in basins affording a large feeding surface of névé is necessary— now this is wanting in Glen Roy— Now Glen Treig itself affords a large basin & is fed by several wild mountainous glens that branch off at the upper end of the lake.

—The source of the glaciers that dammed the mouth of Glen Spean & Glen Gluoy must be looked for in Glen Arkey, Glen Nevis & the other glens that open into the mouth of the Caledonian Canal. I have no doubt a very pronounced glacier issued out of Glen Nevis & another out of Glen Arkaig & these were swelled by several other tributary streams of ice all continuing to choke up the mouth of the Caledonian Canal at Fort William & for a long way up—

—D. Archiac in his Histoire des progres de la Geologie tom II. p. 424—appears to advocate two glacier periods one at the commencement & the other at the close of his Quaternary period. & between these two he puts the *marine* till, large mammifers, & loess.[13]

With many thanks for your instructive letter I conclude in order to catch the post, | Your obed. servt. Thos F. Jamieson

Edinburgh University Library (MS Gen.112.2809–12)

[1] Falconer 1857–8.

[2] Teale 1858.

[3] The reference is to Andrew Crombie Ramsay (see preceding letter and letter from T. F. Jamieson to Charles Lyell, 19 September 1861).

[4] Chambers 1861.

[5] Jamieson refers to James Smith of Jordanhill and to Edward Forbes, who died in 1854 shortly after being appointed professor of natural history at Edinburgh University.

[6] Phillips 1855.

7 Hörbye 1857.

8 The reference is to James Hall, one of the leading 'Huttonians' (supporters of James Hutton's theory of the earth). Despite his support for Huttonian steady-state geology, Hall attributed the erratic blocks on the Jura Mountains and elsewhere to a débâcle-agency (see Hall 1815).

9 See n. 3, above.

10 Chambers 1848. None of the three published versions of David Milne-Home's paper on Glen Roy (Milne 1847a, 1847b, and 1849) includes a map of the area.

11 Ramsay 1860.

12 J. D. Forbes 1853.

13 Archiac 1847–8.

From T. F. Jamieson to Charles Lyell 14 October 1861

Ellon
14 Oct 1861

My Dear Sir Charles,

The barriers of the Glen Roy lakes if due to glaciers must have been I think mainly of ice & not of moraines altho the latter may perhaps have aided—

As to the absence of such there is a similar absence it w^d. appear from D^r. Hooker's narrative & map of the Yangma valley, in the case of several of the lakes & terraces there.[1] see also what he says Vol II. p. 119. about similar things at Mornay "the terracing is as difficult to be accounted for in this valley as in that of Yangma; both valleys being far too broad, & descending too rapidly to admit of the hypothesis of their having been blocked up in the lower part, & the upper filled with large lakes." &c.

Hooker seems to account for much of the terracing by lateral lakes formed between the side of the glacier & the flanks of the hills— see also his note at vol II. p. 121.

David Milne Home mentions some parallel roads at *Loch Tulla*.[2] & Darwin found one at *Kilfinnan*.[3] & Brewster says there are some in the *Spey* valley.[4] I have not however had an opportunity of examining any of these.

The mountain range between Glen Arkig & Glen Garry is very high one of the summits Cairn Scourach being more than 3000 feet & some others nearly as high— In the Swiss Alps, however, the largest glaciers do not proceed from the highest mountains. I have no doubt the Great Glen was filled with an ice stream fed from the lateral valleys—a glacier doubtless descended Glengarry for I saw marks of intense glacial action on a spur of a hill which runs down into the middle of the Glen at Fort Augustus, indicating a movement of ice from S.W. to N.E. this w^d. shew that the outlet to S.W. by Fort William had also been blocked up by ice, & the accumulation in the great glen had parted near its summit level at Loch Orit & gone out in two great streams one to NE the other to S.W.— You will observe that all the parallel roads viz in Kilfinnan, Glen Gluoy, & Glen Spean & Glen Roy. are all in comparatively low glens on the E. side of the Great glen. None on the W. side are as yet known— this is significant—as is also the fact of the Kilfinnan shelf opp. the summit level of the glen being the highest, Glen Gluoy next & the

Glen Roy & Glen Spean ones still lower. This is just what might be expected on the glacier theory.

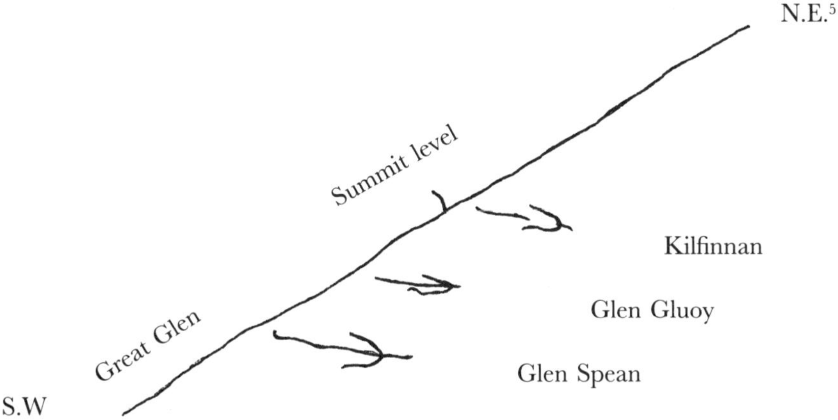

The ice being accumulated to a greatest height at the summit level afforded the highest dam at Kilfinnan—was somewhat lower at Glen Gluoy. & still lower at Glen Spean.

It is interesting to find Geikie's discovery approved of by Ramsay.—[6] I have not seen Geikie's paper.— Ramsay being so clear for a great early glaciation is also good. I hope we shall reach certainty soon on this point—

in haste for post | Your most obed. servt | Thos. F. Jamieson

Edinburgh University Library (MS Gen.112.2820–1)

1 Hooker 1854a, 1: 242; 2: 119, 121, 166.

2 Milne 1847a, p. 21; Milne 1847b, p. 360; Milne 1849, p. 415.

3 See 'Observations on the parallel roads of Glen Roy, and of other parts of Lochaber in Scotland, with an attempt to prove that they are of marine origin' (*Philosophical Transactions of the Royal Society of London* (1839), pt 1: 39–81), pp. 42–3 (see also *Collected papers* 1: 92).

4 David Brewster's observations are mentioned in Darwin's paper on Glen Roy (see n. 3, above). See also *Collected papers* 1: 93, 95.

5 The diagram has been reduced to 65% of its original size.

6 The young geologist Archibald Geikie argued that the shoreline around the Firth of Forth had risen by as much as 25 feet since Roman times (Geikie 1861). Andrew Crombie Ramsay was one of the leading geologists with the Geological Survey of Great Britain.

From T. F. Jamieson to Charles Lyell 22 October 1861

Ellon Aberdeensh.
22 Oct 1861

My Dear Sir Charles,

The occurrence of the vegetation of a genial climate in Greenland & Iceland so late as the Miocene period is a very curious & interesting fact—

I am not prepared to lessen the number of glacial periods which you enumerate. I am inclined to think that in Scotland the greater number of our large boulders have been transported by the first land glaciation—but that there were also some floated during the submergence & others carried probably by glaciers of the 2nd. period as Darwin & Ramsay suppose.[1]

It was one of the facts that struck Agassiz when he visited Scotland that so many of our larger boulders are rubbed & striated—[2] In Switzerland he says these large boulders being carried on the surface of the glaciers are seldom so marked, he therefore used this as an argument for his great crust of ice. (Edin. New. Phil Journ. vol. 33. p. 227 &c 1842)[3]

"In Britain:" he says "the ice at the time of its greatest extension seems to have covered completely great tracts of country & consequently rendered the fall of blocks on its surface if not impossible at least extremely rare; so that the great mass of the blocks was necessarily buried under the ice & was therefore subjected to all the effects of a gradual & long continued trituration just as is observed beneath the glaciers of the present day" & "Hence the majority of the detached blocks of the Scotch mountains must have been transported under the ice, & consequently rubbed, rounded, polished, & scratched." ib. 232. 233.

This I think is good—

Agassiz also assented to the agency of floating ice in some cases, for the Swedish & Norwegian blocks on the E. coast of England he allows, as you suggested, were left there by this agency.

If the change of climate was very gradual one cannot but suppose that a great Greenlandic covering of ice must have been preceded & followed by long periods of local glaciers, but it seems to be the case, in instances that I have examined, that hills of 1000 or 1500 feet which have been strongly eroded by the 1st. glaciation do not shew evidence of a subsequent scratching by local glaciers of their own.

This seems to be somewhat odd under the preceding view.

The *Scrobicularia pipirata* found in the Moray firth is says Dr. Geo Gordon "in a recently fossil state in the bed of the now freshwater loch of Spynie."[4]

If, as appears to be the case, this species has died out in the Moray Firth, the Aberdeensh. coast & the Firth of Forth (Miller)[5] it is a fact worth noting. the Spynie beds (wh. I have seen myself) appear to be raised estuary mud like that of the Ythan here, & probably belonging to the same period.

The *Scrobicularia* would appear to be here at its N. limit for MacAndrew gives its range, Britain to Mediterranean, as Mr. Jeffreys does.[6] it would therefore appear to be rather a Southern species & the change in conditions wh. has caused it to die out in all these three localities is most likely of a climatal character. Miller mentions other two companions of the *Scrobicularia* wh. have become rare or extinct in their old haunts. viz. *Thracia convexa* & *Pholas candida*. both shells of a Southern cast—[7] I find however from some M.S. notes of the late Prof. MacGillivray that he had got at least one specimen of the Scrobicularia in the Ythan estuary with the *animal* in it—[8] as he describes the animal from this specimen I conclude it was the only one he had found.

Your most obed serv | Thos F. Jamieson

Edinburgh University Library (MS Gen.112.2822–3)

[1] See *Collected papers* 1: 118–22, and Ramsay 1852 and 1860.
[2] Louis Agassiz visited the Lochaber region of Scotland in the company of William Buckland following the 1840 meeting of the British Association for the Advancement of Science, which was held in Glasgow (see Davis 1968).
[3] Agassiz 1842.
[4] Gordon 1854, p. 4308.
[5] Miller 1858, pp. 380–1.
[6] McAndrew 1853, pp. 12 and 14. The conchologist John Gwyn Jeffreys was a specialist on the Mollusca of northern seabeds.
[7] See n. 5, above.
[8] William Macgillivray was professor of natural history at Aberdeen University until his death in 1852.

From T. F. Jamieson to Charles Lyell 27 October 1861

Ellon Abberdeen^sh.
27 Oct 1861

My Dear Sir Charles

Agassiz's little map of Glen Roy is worthless,[1] he seems to have made only a short flying visit. His Ben Nevis glacier is also a mistake— it was his suggestion of glacier dams & the pointing out of the clear traces of one such at Loch Treig that form the valuable points of his theory for Glen Roy.

The most correct map of the shelves is the latest one by Chambers' & D. Milne Home. If you refer to this map in Chambers Ancient sea margins you will see how small a figure the intermediate shelves make—[2] I did notice them but was not much impressed by them. I think it is possible ledges of this sort might have been formed at some depth underneath the water without necessarily supposing that they mark a former margin of the lake— All along the W. side of GlenRoy, between Bohuntine hill & Glen Turret, where the three lines are perhaps better marked than anywhere else I could not detect a trace of any intermediate one.

I quite agree with you that the cols have determined the formation of the lines, we have now five clear lines, viz that of Kilfinnan, Glen Gluoy & 3 in Glen Roy and all coinciding with cols, this five fold proof is very strong in my opinion, and it is also significant that the lines do not coincide neatly with each col but are *slightly above* it, so as to account for some depth of water flowing out

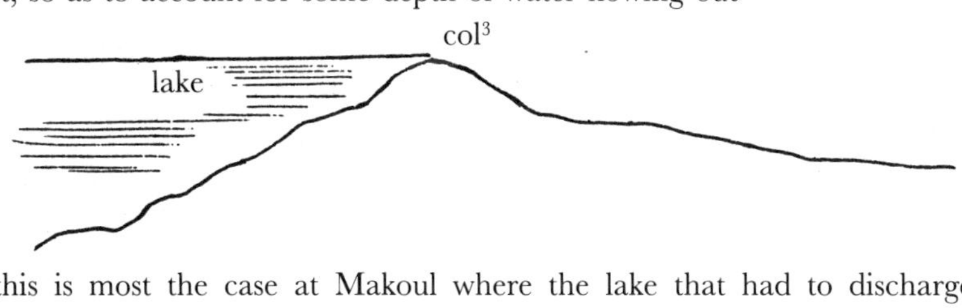

And this is most the case at Makoul where the lake that had to discharge its excess was much the longest, M^r. Milne Home says the line there is by *barometrical* measurement 21 feet above the col.—[4] doubtless the outflow had worn down the passage a little.

Of the Glen Glaster Col my notes say "on reaching the summit level of Glen Glaster I found it, by using my pocket level, to correspond with the the level of the middle line of Glen Roy, or *more accurately to be a few feet lower*."

With regard to Darwin's difficulty about the stratified matter at the mouth of Glen Spean, not having had time to examine that part I do not feel inclined to say much. But I do not feel it to be an insurmountable one. For as I hold the barrier in that quarter to have been caused by an ice stream occupying the line of the Caledonian Canal, (& that there was such I found evidence from the scoring on the rocks) it seems to me that the little glaciers & mountain torrents descending the N. flank of Ben Nevis must have carried much debris with the old lake near the present mouth of the Spean. And if, as seems likely, there was a previous period of marine drift, much stratified matter had previously existed w^h. would be ploughed through by the glacier stream of the great glen & form a side moraine across the Spean valley near its mouth w^h. would be levelled by the rushing off of the water when the lake was finally emptied. We have also the submergence of the 30 feet old coast line w^h. would perhaps invade the mouth of the Spean after the lake period—

However I have no doubt future investigation will clear up this.

With regard to the sudden termination of the lines at certain points you will observe that the marine theory skipped in a great measure that difficulty or at any rate offered no explanation w^h. will not apply to a fresh water case.

By be the bye, as you suppose the great submergence to have preceded the Glen Roy lakes. some of the matter in the bottom of Glen Roy may be marine & contain sea-shells altho the parallel roads are of fresh water origin, so that the finding of even marine fossils would not be an extinguisher to the fresh water theory!

M^r. Darwin has very kindly sent me a lot of papers bearing on Glen Roy

The best case of a glacier lake that I have seen is that described by Strachey in the Himalaya near Karakorum pass (Journ. of Geograph. Soc. vol 23 p. 55)[5]—the ice dam about 700 feet high dammed the valley he thinks nearly 50 miles back. The flood caused by the bursting was felt severely 300 miles down the valley— This is something like the thing! The barrier would appear to have been all of ice, and seems to form periodically for several floods have been caused by it.

(—In my last letter I forgot to mention that I have seen a very interesting account of the Cromlechs & primeval antiquities of the Channel islands by F. C. Lukis in the Archaeological Journal. vol I. 1846. w^h. I think would repay your looking at if you have not seen it.)[6]

There is a great deal to be done about Glen Roy yet— the area covered by the lines is so extensive & ramifying & so deficient in travelling conveniences & accomodation as to require much labour in exploring it & also loss of time. There are many gleanings still to reward fresh visitors— I do not by any means pretend to explain all the difficulties of the case. My time being limited I satisfied myself with examining some of the most crucial points, & finding these all lead to a fresh-water solution I could not but think that was the right one.

Jukes surely does not deny glacier action in Ireland—[7] (see Airy the astronomer royal on the subject. Brit. Assoc. Report. 1843 Cork transactions of sections p. 72, in a letter to Lord Rosse.)[8]

The greater part of Ireland, however, is so low, & the country itself situated so far to S.W., that probably land-ice was much less extensively developed there than in Scotland— The lands on the west side of Baffins bay opposite Greenland differ much from it in their ice condition (see the remarks of that observant voyager D^r. P. C. Sutherland on this point. Journ. of Geolog. Soc. vol IX p. 302. near foot of page)[9] the snow & ice on the land round Barrow Straits & its tributary inlets & bays all escape into the sea in running streams of water while that falling on the coasts of Davis straits makes its escape in the shape of solid ice.

—Many thanks for your explanatory notes & sketch about Hooker's lakes—the frequency of terraces in the upper Himalaya valleys being ascribed by him to glacial action[10] may not the same agency account for those terrace-like bits w^h. Darwin mentions in the high cols of Lochaber as a difficulty to the lake theory of the *roads*.

Altho Hooker found a transverse moraine barrier for his lowermost or chief lake, yet there were several cases higher up the Yangma valley of old lakes & terraces for w^h. he found no remains of a barrier & yet where a transverse barrier must have apparently once existed— at least I so understood his explanation & map—

—The foregoing pages being somewhat hurriedly written are I fear almost illegible—

I am | Your most obed ser^v | Thos F. Jamieson

Sir C. Lyell F.R.S | &c &c

The letters you sent are enclosed herewith | T.F.J.

Edinburgh University Library (MS Gen.112.2824–7)

1 See Agassiz 1842.

2 See Chambers 1848. The map was produced by Chambers, David Milne-Home, and George Steuart Mackenzie.

3 The diagram has been reduced to 75% of its original size.

4 Milne 1847a, p. 5; Milne 1847b, p. 344; Milne 1849, p. 399.

5 Strachey 1853.

6 Lukis 1846.

7 Joseph Beete Jukes was director of the Irish branch of the Geological Survey of Great Britain.

8 Airy 1843. William Parsons, the third earl of Rosse, presided over the Cork meeting of the British Association for the Advancement of Science.

9 Sutherland 1853.

10 Joseph Dalton Hooker described the glacial phenomena of the Yangma Valley in the Himalayas in Hooker 1854a. He also provided Lyell with additional information on the phenomena, referred to in C. Lyell 1863, p. 262 n.

APPENDIX X
Darwin's memoir of John Stevens Henslow[1]

I went to Cambridge early in the year 1828, and soon became acquainted, through some of my brother entomologists, with Professor Henslow, for all who cared for any branch of natural history were equally encouraged by him. Nothing could be more simple, cordial, and unpretending than the encouragement which he afforded to all young naturalists. I soon became intimate with him, for he had a remarkable power of making the young feel completely at ease with him; though we were all awe-struck with the amount of his knowledge. Before I saw him, I heard one young man sum up his attainments by simply saying that he knew everything. When I reflect how immediately we felt at perfect ease with a man older and in every way so immensely our superior, I think it was as much owing to the transparent sincerity of his character, as to his kindness of heart; and, perhaps, even still more to a highly remarkable absence in him of all self-consciousness. One perceived at once that he never thought of his own varied knowledge or clear intellect, but solely on the subject in hand. Another charm, which must have struck every one, was that his manner to old and distinguished persons and to the youngest student was exactly the same: to all he showed the same winning courtesy. He would receive with interest the most trifling observation in any branch of natural history; and however absurd a blunder one might make, he pointed it out so clearly and kindly, that one left him no way disheartened, but only determined to be more accurate the next time. In short, no man could be better formed to win the entire confidence of the young, and to encourage them in their pursuits.

His Lectures on Botany were universally popular, and as clear as daylight.[2] So popular were they, that several of the older members of the University attended successive courses. Once every week he kept open house in the evening, and all who cared for natural history attended these parties, which, by thus favouring inter-communication, did the same good in Cambridge, in a very pleasant manner, as the Scientific Societies do in London. At these parties many of the most distinguished members of the University occasionally attended; and when only a few were present, I have listened to the great men of those days, conversing on all sorts of subjects, with the most varied and brilliant powers. This was no small advantage to some of the younger men, as it stimulated their mental activity and ambition. Two or three times in each session he took excursions with his botanical class; either a long walk to the habitat of some rare plant, or in a barge down the river to the fens, or in coaches to some more distant place, as to Gamlingay, to see the wild lily of the valley, and to catch on the heath the rare natter-jack. These excursions have left a delightful impression on my mind. He was, on such occasions, in as good spirits as a boy, and laughed as heartily as a boy at the misadventures of those who chased the splendid swallow-tail butterflies across the broken and treacherous fens. He used to pause every now and then and lecture on some plant or other object; and something he could tell us on every insect, shell, or fossil collected, for he had attended to every branch of natural history. After our day's work we used to dine at some inn or house, and most jovial we then were. I believe all who joined these excursions will agree with me that they have left an enduring impression of delight on our minds.

As time passed on at Cambridge I became very intimate with Professor Henslow, and his kindness was unbounded; he continually asked me to his house, and allowed me to accompany him in his walks. He talked on all subjects, including his deep sense of religion, and was entirely open. I owe more than I can express to this excellent man. His kindness was steady: when Captain Fitzroy offered to give up part of his own cabin to any naturalist who would join the expedition in H.M.S. *Beagle*, Professor Henslow recommended me, as one who knew very little, but who, he thought, would work. I was strongly attached to natural history, and this attachment I owed, in large part, to him. During the five years' voyage, he regularly corresponded with me and guided my efforts; he received, opened, and took care of all the specimens sent home in many large boxes; but I firmly believe that, during these five years, it never once crossed his mind that he was acting towards me with unusual and generous kindness.

During the years when I associated so much with Professor Henslow, I never once saw his temper even ruffled. He never took an ill-natured view of any one's character, though very far from blind to the foibles of others. It always struck me that his mind could not be even touched by any paltry feeling of vanity, envy, or jealousy. With all this equability of temper and remarkable benevolence, there was no insipidity of character. A man must have been blind not to have perceived that beneath this placid exterior there was a vigorous and determined will. When principle came into play, no power on earth could have turned him one hair's breadth.

After the year 1842, when I left London, I saw Professor Henslow only at long intervals; but to the last, he continued in all respects the same man. I think he cared somewhat less about science, and more for his parishioners. When speaking of his allotments, his parish children, and plans of amusing and instructing them, he would always kindle up with interest and enjoyment. I remember one trifling fact which seemed to me highly characteristic of the man: in one of the bad years for the potato, I asked him how his crop had fared; but after a little talk I perceived that, in fact, he knew nothing about his own potatoes, but seemed to know exactly what sort of crop there was in the garden of almost every poor man in his parish.

In intellect, as far as I could judge, accurate powers of observation, sound sense, and cautious judgment seemed predominant. Nothing seemed to give him so much enjoyment, as drawing conclusions from minute observations. But his admirable memoir on the geology of Anglesea, shows his capacity for extended observations and broad views.[3] Reflecting over his character with gratitude and reverence, his moral attributes rise, as they should do in the highest character, in pre-eminence over his intellect.

C. Darwin.

[1] Published in *Memoir of the Rev. John Stevens Henslow* edited by Leonard Jenyns (London, 1862), pp. 51–5; reprinted in Barlow ed. 1967, pp. 221–4.

[2] On CD's attendance of Henslow's course on descriptive and physiological botany, see Sloan 1985, pp. 86 and 115 n. 34, and Walters 1981, chap. 5.

[3] Henslow 1821b.

MANUSCRIPT ALTERATIONS AND COMMENTS

The alteration notes and comments are keyed to the letter texts by paragraph and line numbers. The precise section of the letter text to which the note applies precedes the square bracket. The changes recorded are those made to the manuscript by CD or his amanuensis; readers should consult the Note on editorial policy in the front matter for details of editorial practice and intent. The following terms are used in the notes as here defined:

del	deleted
illeg	illegible
interl	interlined, i.e., inserted between existing text lines
omitted	omitted by the editors to clarify the transcription
over	written over, i.e., superimposed

To Thomas Henry Huxley 3 January [1861]
1.5 readers &] *interl*
2.3 Lecture] 'L' *over* 'l'
2.6 a] *interl*
3.8 leads] *after del illeg*

To William Darwin Fox 9 January [1861]
2.2 proofs—] '—' *over* ':'
5.2 with young] *interl*
5.2 (I forget which) 5.3] *interl*

To Joseph Dalton Hooker 15 January [1861]
1.6 thought] *interl*
2.1 I] *over* 'Do'
3.2 on] *interl*
3.5 has] *after del* 'that he'
3.5 been] *interl*
3.6 read] *above del* 'written'
3.8 I must] *interl*

To John Murray 21 January [1861]
3.1 Third Edition 3.2] *interl*
3.3 already] *interl*
4.2 end of] *interl*

To Daniel Oliver 22 January [1861]
1.2 is] *interl*
1.5 &] *interl*
1.6 at Kew] *interl*
1.7 Half-opened … subsequently.] *interl*
1.9 secreting] *interl*
1.9 to the 1.10] *interl*
1.11 lie] *after del* 'on'
1.13 to me] *interl*
1.15 enclosed] *after del* 'clo'
Enclosure
2.1 (if] *parenthesis over comma*
2.2 & Viburnum] *interl*
3.3 pistil] *after del* 'sp'

To Asa Gray 23 [January 1861]
1.3 whole] *interl*
1.3 not mis 1.4] *interl*
2.1 Price of] *interl*
Enclosure
Third column: Decandolle] *after del* 'de'; 'De' *over* 'Co'

To William Erasmus Darwin [24 January 1861]
0.1 12°oclock] *added*
1.3 just] *interl*

To Charles Lyell 2 February [1861]
1.2 (who … England)] *square brackets over parentheses in MS*
2.3 deny] *after del* 'say'

To William Bernhard Tegetmeier 3 February [1861]
2.1 males of] *interl*

2.2 in menageries] *interl*
2.4 in plumage] *interl*

To Jeffries Wyman 3 February [1861]
1.8 vibrating] *after del illeg*
1.15 prey &] '&' *over* '—'
3.1 observations on] 'on' *over* 'of'

To J. D. Hooker 4 February [1861]
2.6 & been … Woodhouse] *interl*; *caret over* ';'
2.8 the] *interl*
5.1 Ladies] 'L' *over* 'l'
6.5 Primula] *altered from* 'P.'
6.6 heads] 's' *added*
6.7 of the ammonia;] *interl*
6.8 more prolonged] *above del* 'greater'
6.8 causes] 'es' *over* 'ing'
6.9 re-] *interl*
6.10 of solid substances] *interl*
6.11 causes] 'es' *over* 'ed'
6.12 what it] *before del comma*
7.2 3 ft. 6 inches] *above del* '2. ft'
7.6 years.] *before omitted point*
8.2 is] *interl*
8.3 opponents] *after del* 'de'
8.4 maintaining] 'aining' *over* 'ing'

To John Lubbock [before 5 February 1861]
2.3 bold] *after del* 'bol'
3.5 is] *interl*

To Leonard Horner 14 February [1861]
2.2 God] *before del illeg*
2.2 parent] *interl*
2.3 should] *after del illeg*
2.4 , as] *interl after del* 'from what'
2.4 them,] *interl*
2.6 beneficial] ²'e' *over partial* 'f'
2.6 successive] *interl*

To Asa Gray 17 February [1861]
0.2 17] '7' *over illeg*
1.5 a few] *interl*
2.3 10:] '0' *over* '6'
2.5 of Origin] *interl*
3.2 at the seeds] *interl*
3.2 who … Sencio 3.3] *interl*
3.4 I] *after del* '"'
5.1 interests] ²'s' *over* 'ed'
5.3 extreme marginal] *interl*
7.2 of Bowen] 'of' *interl*
7.2 heredity.] 'i' *over* 'e'

7.5 from the] 'the' *over illeg*
11.1 Life] *interl*

To T. H. Huxley 17 February [1861]
1.1 from Trubner] *interl*
5.1 lately] *interl*

To J. D. Hooker 20 [February 1861]
2.8 Malvern] *above del* 'Malvern'

To T. H. Huxley 22 February [1861]
1.4 asking] *after del illeg*
1.4 M^rs. Huxley] *above del* 'she'

To Andrew Murray 23 February [1861]
1.4 ?] *over* '.—'
1.6 very] *interl*
3.1 Mr Crotch] *above del* '(name forgotten this minute)'
3.2 saw the] 'the' *above del* 'an'
3.3 deserted] *interl*

To John Tyndall 23 February [1861]
1.5 (my two Grandfathers) 1.6] *interl*

To John Murray 24 February [1861]
2.2 Postscript] *interl*

To W. B. Tegetmeier 25 February [1861]
3.2 (if] *parenthesis over comma*

To Asa Gray 26–7 February [1861]
1.1 the only] 'the' *interl*
1.7 I have distributed] 'I' *over illeg*
1.8 this] 'is' *over illeg*
1.10 but] 'b' *over* '&'
1.12 freit] *above del* 'cost'

To Daniel Oliver 26 February [1861]
1.6 subsequent] *interl*
1.7 resemblance] *above del* 'affinity'
4.6 Orchid] *interl*

To W. B. Tegetmeier 1 March [1861]
1.8 probably] *interl*
2.1 number,] *interl*
2.5 & broader] *interl*
4.3 discovered] *after del* 'have'

To George Rolleston 2 March [1861]
2.6 Which] 'W' *over illeg*

To John Murray 3 March [1861]
1.1 satisfactory] *interl*

To Philip Lutley Sclater 4 March [1861]
3.2 Skeletons] *after del* 'their'

To J. D. Hooker 11 March [1861]
1.3 next] *interl*

To Asa Gray 12 March [1861]
1.2 it is] *interl*
3.3 the other] 'the' *interl*
4.1 Cypripedium] 'C' *over* 'c'
7.3 revives me] *above del illeg*

To P. L. Sclater 12 [March 1861]
3.4 at first] *interl*

To Cuthbert Collingwood 14 March [1861]
1.3 is that] 'is' *over* 'it'
1.3 successful] *after del illeg*
1.3 some few] *above del* 'several'
1.6 amount] *after del* 'degree'
1.9 (but] *parenthesis over comma*; 'but' *over illeg*
2.2 me] *above del* 'you'
2.2 be] *interl*
2.9 that] *interl*
3.3 progressed] *after del illeg*
4.1 the very] 'the' *interl*
4.4 a form of] *interl*
4.5 essential] *altered from* 'essetial'
8.1 at] *interl*

To George Maw 17 March [1861]
1.4 so] *interl*
1.5 idea] *interl*
1.5 disagreeable] 'd' *over* 'D'

To J. D. Hooker 18 March [1861]
2.2 the Malay] 'the' *interl*
5.1 mundane] *added in space left by copyist*
5.6 would] *above del* 'will'
5.11 the cause] 'the' *interl*
5.12 unreasonably] *added in space left by copyist*
8.3 other] *interl*
8.3 (]*parenthesis over comma*
8.5 climate] *after del* 'new'
8.13 would] *interl*

To Leonard Horner 20 March [1861]
1.1 Address] 'A' *over* 'a'
1.6 that] *interl*
2.2 what] *after del illeg*

To W. B. Tegetmeier 22 March [1861]
1.9 S. E] *interl*
2.3 & variously 2.4] '&' *interl*
2.4 darkly] *interl*
3.7 in pencil 3.8] *interl*
3.11 each] *interl*
3.12 in my M.S.] *interl*
3.16 have corresponded] 'have' *interl*
5.1 , if you can, 5.2] *interl*
5.2 *old*] *interl*
5.4 the other] 'the' *interl*
5.6 specimens.] *interl*
5.7 the other] 'the' *added*
5.8 shorter] *above del* 'short'

To Daniel Oliver 23 March [1861]
1.4 two] *interl*
1.5 And . . . scissors 1.6] *interl*
1.6 M[r]] *over illeg*
1.10 fertilise] *after del* 'self'

To P. L. Sclater 23 March [1861]
2.4 and that] *interl above del illeg*
3.1 G] *over* 'J'
4.1 (supposing] *parenthesis over comma*

To Henry Walter Bates 26 March [1861]
2.9 the Tropics of] *interl*
3.1 temperate] *interl*
3.4 yet] 'y' *over* 'g'
5.1 so] *above del* 'not,'
6.4 the male] 'the' *interl*
6.4 several] *above del* 'many'
9.1 (Have . . . Origin?)] *square brackets in MS*

To J. D. Hooker 26 [March 1861]
1.10 on] *altered from* 'in'
2.3 away] *after del* 'for'

To J. D. Hooker 27 [March 1861]
3.2 Origin] 'O' *over* 'o'
3.2 that] *interl*
7.1 (Many . . . it—)] *square brackets in MS*

To W. B. Tegetmeier 28 March [1861]
3.2 rabbits] *above del* 'former'
6.1 The skulls . . . thanks—] *double scored*

To W. B. Tegetmeier 2 April [1861]
1.1 note] 'e' *over* 'es'

1.5 probability of] *interl*
1.5 some] *after del* 'the'
2.1 produces] 's' *over* 'd'
2.2 feared that] *before del* 'the'
2.2 a hen-tailed] 'a' *interl*
3.3 for] *interl*
3.3 You] *after del* 'So'; 'Y' *over* 'y'
3.3 must] *interl*
3.4 of clairvoyance] 'of' *interl*
5.1 young] *interl*
6.1 lately] *interl*
6.2 given by you,] *interl*
7.1 Herbert] *before del* ';'
8.1 in] *above del* 'of'

To H. W. Bates 4 April [1861]
1.7 them] *interl*
1.7 , *also*,] *interl*
2.4 their] *above del* 'any'
2.7 their] *interl*
3.6 *isolated*] *interl*
3.8 discussion 3.9] *before omitted point*
5.4 male] 'e' *over* 'es'
5.6 though] *interl*
5.9 another] *after del illeg*
5.9 fact] *after del illeg*
6.2 time] *interl*
6.4 generally] *interl*
6.4 between] *above del* 'with'

To George Busk 5 April [1861]
1.3 Esq[re]] *above del comma*
1.9 the statements] 'the' *over illeg*
3.1 The] 'T' *over* 'I'

To Samuel Birch 6 April [1861]
Enclosure
2.1 Chinese] *interl*
3.2 the feathers] 'the' *interl*

To Daniel Oliver 9 April [1861]
1.5 then] 'n' *over* 'se'
1.5 the … stamens] *interl*
1.5 (*curved … nectary*) *square brackets in MS*
1.8 again] *interl*
1.9 there is] *interl*
1.9 the hood] 'the' *interl*
2.1 is on] 'is' *interl*
2.1 the hood] 'the' *interl*

To Asa Gray 11 April [1861]
2.1 I believe] *after del illeg*
2.3 Historical] 'or' *over* 'ory'
3.2 that it is] *interl*
3.3 will have seen 3.4] *above del* 'saw'
3.4 a Review] 'a' *interl*
3.6 Remember] *above del* 'Rember'

To Charles Lyell 12 April [1861]
1.3 French] *interl*
1.4 unknown] *interl*
1.6 grand.] *full stop over comma*
1.10 the Human] 'the' *interl*
2.1 am sure] *above del* 'believe'
2.1 of] *over* 'on'
2.5 morning] *after del* 'I'
4.8 store-food] 'food' *above del* 'seed'

To W. B. Tegetmeier 14 April [1861]
2.2 (great … &c) *square brackets in MS*
4.1 (but not yet read)] *interl*
5.1 valuable] *interl*
6.4 &] *over* 'or'
9.1 opened] *after del* 'noti'
9.4 some risk] 'some' *interl*

To Oswald Heer? 20 April [1861?]
1.3 interesting] *after del illeg*

To W. E. Darwin 21 and 22 April [1861?]
0.2 21[d]] '21' *over* '22'
4.1 22[d] … splendidly] *added in pencil*

To P. L. Sclater 21 [April 1861]
1.8 antarcticus] *after del* 'Patagoni'
4.1 **seems**] *underlining in pencil*

To J. D. Hooker [2]23 [April 1861]
2.1 he is,] 'he' *over del* 'is'
2.4 in this] 'in' *interl*
2.6 from familiarity] *interl*
2.11 Edgeworth, Wallich] *interl*
2.13 dead] *interl*
3.1 dined] *after del illeg*
8.6 now] *interl*

To Daniel Oliver 1 May [1861]
3.1 The best] 'The' *above del* 'I enclose'

To P. L. Sclater 4 May [1861]
1.4 names] *altered from* 'name'
1.5 it] *after del* 'my'
2.2 (which] *parenthesis over comma*
5.3 these] 'se' *over* 'm'

To T. C. Eyton 6 May [1861]
2.2 yours (if found)] *interl above del* 'it'
3.2 other Bones] *above del* 'skeletons'
4.6 besides the skull 4.7] *interl*

To W. E. Darwin 9 May [1861]
2.7 Look … Ponys—] *added in margin*
3.4 the Boys] 'the' *above del* 'you'
6.1 Look … Envelope.] *written above printed address*

To T. C. Eyton 14 May [1861]
2.2 unreasonable] 'un' *over illeg*

To Charles William Crocker 18 May [1861]
1.8 It can] *after del* 'Yet not crossing'
1.9 castrated &] *interl*
1.11 the seedlings] 'the' *interl*
1.12 pollen of] *interl*
1.14 some very fixed vars. of] *interl*
1.16 that of] *interl*
2.4 when] *interl*

To J. D. Hooker 18 [May 1861]
1.2 to his] 'to' *interl*
1.3 how you] 'y' *over illeg*
2.1 wife] *after del* 'a'
5.2 would] *after del* 'it'

To T. H. Huxley 22 May [1861]
1.2 arrived] *after del* 'appeared'
2.3 dyspeptic] 'y' *over* 'i'

To John Frederick William Herschel 23 May [1861]
1.7 with all living productions 1.8] *interl*
1.8 that] *over* 'in'
1.11 in a highly peculiar manner] *interl*
1.15 being destroyed] 'being' *interl*
2.3 young & middle-aged] *interl*
2.6 or almost exclusively] *interl*

To J. D. Hooker 24–5 May [1861]
1.1 the Burial] *after del* 'you'
1.3 saw] *above del* 'have seen'
1.7 of] *interl*

To Bartholomew James Sulivan 24 May [1861]
2.2 me.] *full stop over comma*

To W. E. Darwin [25 May 1861]
1.3 sea-port] *interl*
2.1 £500] *after del* 'than'
2.10 only] *interl*
2.13 marry] *above del illeg*
2.15 very] *interl*
4.7 might] *after omitted point*
5.2 contrasted] *below del illeg*
5.5 to be] 'to' *interl*
7.1 John … danger 7.3] *in Emma Darwin's hand*
8.1 need] *below del illeg*
8.1 You … me.] *added in blue pencil*

To John Lubbock [25 May 1861]
1.1 about the Bank] *interl*

To Abraham Dee Bartlett 26 May [1861]
1.4 they] *interl*
1.6 freely] *after del* 'with'

To W. E. Darwin [26 May 1861]
1.2 He] 'H' *over* 'h'
1.4 his] *interl*
1.6 is.] *interl*
1.10 not] *over illeg*
1.10 a small] 'a' *interl*
2.1 for Partner] *interl*
2.2 sometime on] *interl*
2.2 leave on Friday evening &] *interl*
2.4 he supposes] *interl*
3.1 the place is] *interl*
3.2 the Robarts] 'the' *interl*
3.4 occurred.] *after del* 'have'
3.5 had not] 'not' *interl*
3.6 must] *above del* 'ought'
3.6 well] *interl*
3.8 absolutely] *interl*
3.9 you as Partner,] *after del* 'if'
3.10 interest of the] 'the' *added*
3.14 more than] *interl*
3.15 the scheme] 'the' *over* 'it'
3.15 would] *above del* 'could'
3.17 he] *above del* 'you'
3.17 you. Lubbock] *after del* 'him.'
4.3 some] *interl*
4.3 would be a] *after del* 'will'
4.3 grievous] *above del* 'grieverous'
4.3 you to] *interl*
4.5 & will never recur.] *interl*
4.10 for] *after del illeg*
4.14 consult] *above del illeg*
4.14 (& … like)] *interl*
4.16 the scheme] 'the' *above del* 'such a'
4.16 of year's trial.] *interl*

4.17 a sensible] 'a' *interl*
4.17 man—] *after del* 'a'

To Daniel Oliver 27 May [1861]
1.5 (not … flowers)] *interl*
1.6 the sides] 'the' *interl*
1.6 the corolla;] 'the' *added*
4.2 which … seed] *interl*

To W. B. Tegetmeier 28 May [1861]
1.3 scraped] 'scr' *over illeg*
3.1 of Fowls.] *interl*

To J. D. Hooker 30 May [1861]
1.4 like.] *full stop over* '—'
1.4 If] *after del* 'but I'
1.6 too trifling] 'too' *added*
7.1 We … it.] *added at top of first page in blue crayon*
8.3 minute] *interl*
8.3 *all*] *after del* 'on'
8.6 much] *interl*

To C. W. Crocker 1 June [1861]
1.7 experimented] 'ex' *over illeg*
2.5 gauze or] *interl*

To W. E. Darwin 1 [June 1861]
1.1 Lubbock.] *full stop over comma*
1.2 yet] *interl*
1.4 the copy] 'the' *over* 'a'
1.4 (*please keep this copy*)] *square brackets in MS*

To John Lubbock 1 June [1861]
1.5 time,] 't' *over illeg*
3.2 to come] *interl*

To P. L. Sclater 2 June [1861]
1.2 in Spirits 1.3] *after del* 'from'
1.5 differed] 'd' *over* 'nt'
1.7 *the general*] '*the*' *interl*
2.4 coloured] *interl*
3.2 Bartlett] 'lett' *over illeg*

To Asa Gray 5 June [1861]
1.4 read] *above del* 'wrote'
1.4 of Origin] *interl*
3.4 on] 'o' *over illeg*
3.9 comes] *after del* 'fully'
3.10 in F. Water having been] *interl*
5.4 other] *interl*
5.6 information] *after del* 'much'
6.3 the North. 6.4] 'the' *interl*

To Samuel Pickworth Woodward 5 June [1861]
3.2 Bee O. 3.3] 'O.' *interl*
6.1 St.] *after del* 'M'

To W. E. Darwin 6 [June 1861]
1.3 M^r. ——] '——' *over del illeg*
4.1 ——] *above del* 'M^r. Atherley'

To John Lubbock [8 June 1861]
1.2 & stay … inclined.] *interl*

To W. B. Tegetmeier 14 June [1861]
1.5 The … attention. 1.6] *interl*

To Bingham Sibthorpe Malden 15–16 June [1861]
4.2 & curious] *interl*
4.3 These … Habenaria. 4.4] *added*
4.4 If … flowers 4.5] *added*
5.2 (which] *parenthesis over comma*
5.2 & the … maculata 5.3] *interl*
6.4 & especially … arachnites] *interl*
8.1 some time] *interl*

To George Bentham 17 June [1861]
1.1 so] *interl*
1.3 from several curious reasons 1.4] *interl*

To J. D. Hooker 19 June [1861]
2.12 new] *interl*
3.3 my wife] *above del* 'we'
3.3 small country] *interl*
3.5 leaned for] 'for' *over* 'to'
4.5 -side] *interl*
4.6 for Linn. Soc^y] *interl*
4.10 they] *above del* 'it'

To Frederick Smith 19 June [1861]
1.2 an] *altered from* 'a'
1.5 with pollen-masses attached] *interl*

To George Bentham 22 June [1861]
2.6 Orchis] *interl*
2.9 the belief of] *interl*

To J. D. Hooker 22 June [1861]
1.2 & out of spirits 1.3] *interl*
5.1 write] *above del* 'think'
6.2 Journ] 'Jo' *over* 'Co'

To John Lubbock 24 June [1861]
1.1 had any] 'had' *interl*
1.2 of] *over* 'on'
3.2 him] *interl*

To George Gordon 6 July [1861]
1.8 *large*] *interl*

To J. D. Hooker [6 July 1861]
0.1 Hesketh] 'h' *over illeg*
1.7 & laughable] *interl*

To John Lubbock 10 July [1861]
1.5 trouble him,] 'him,' *interl*
3.3 be] *after del* 'have been'

To J. D. Hooker 13 July [1861]
1.3 but it] 'it' *interl*
3.2 of the rostellum] 'of the' *interl*
3.2 to each other] *interl*
3.4 they would] 'would' *interl*
3.4 a Cattleyea] 'a' *altered from* 'an'
3.6 Cattlyea] *interl*

To J. D. Hooker 17 [July 1861]
3.3 with some buds] *added in margin*
3.4 cut off &] *interl*
3.6 We … more] *added in margin*
4.2 spoke to] 'to' *over* 'as'

To George Maw 19 July [1861]
1.5 me] *interl*
1.7 & … such as] *interl*
1.9 the force] 'the' *interl*
2.4 much] *interl*
2.5 under] *after del* 'on'
2.6 (p. 7583)] *interl*
3.5 should] *above del* 'shall'
3.5 variations,] *comma altered from full stop*
3.5 or … Salix &c.] *added*

To Charles Lyell 20 July [1861]
1.1 the Origin] 'the' *interl*
1.4 the blind] *interl*
3.5 long] *interl*

To Asa Gray 21 July [1861]
1.1 indolent] *altered from* 'indlent'
1.2 Hybrid] *interl*
1.3 daughter's] ''s' *added*
1.3 sake] *interl*
1.4 your] *interl*
3.3 placed] *interl*
3.7 plants] *added*
4.1 the 2 forms of] *interl*
4.4 I … given.—] *added*
5.6 lately] *before del comma*
5.10 (I mean … sayings) 5.11] *interl*
6.6 a great 6.7] 'a' *interl*
10.1 We … more] *added at top of first page*

To John Lubbock 26 July [1861]
1.6 myself] *after del illeg*

To J. D. Hooker 27 July [1861]
1.6 have] *after del* 'you'
1.7 you not] *interl*
1.11 jam stick] 'stick' *interl*
1.11 also] *interl*
1.11 or anything else 1.12] *interl*
2.3 part] *interl*
2.6 almost] *interl*
3.3 (what country)] *interl*
4.2 it] *interl*

To J. D. Hooker [28 July – 10 August 1861]
2.4 a much] 'a' *above del* 'ext'
2.8 abundant] *after del* 'imp'
2.9 its] *above del* 'the'
2.12 columns of] *below del* 'pol'
3.8 but] *interl*
4.3 is sticky] *interl*
4.5 it is] *added*
5.1 the pollen] 'the' *interl*
5.3 sticky] *interl*
6.5 & Corycidæ.— 6.6] '&' *over* 'or'

To Charles Lyell [1 August 1861]
2.3 to put] 'to' *interl*
2.4 latter] *interl*
3.6 degree of] *interl*
3.7 in Hydra] 'in' *above del* 't'
3.7 in some] 'in' *interl*
3.8 which "conjugate"] 'which' *interl*
4.3 lately] *interl*
4.4 with respect to the Origin 4.5] *interl*
4.6 that God] 'that' *interl*
4.7 has] *over* 'is'
4.11 does not] *interl*
4.14 ordained] *after del* 'arr'
4.18 everything] *after del* 'th'
5.5 God foresaw] 'God' *above del* 'he'
5.7 certain changes] 'certain' *above del* 'the'

5.8 certain mountains] 'certain' *interl*
5.10 theological] *above del* 'religious'

To John Lubbock 1 [and 2] August [1861]
7.1 Goodnight] 'g' *over illeg*
8.3 tomorrow] *after del* 'on'

To John Lubbock [2 August 1861]
2.7 in Southampton] *interl*
6.2 say,] *interl*

To Charles Lyell 6 August [1861]
3.1 & old & half-] *interl*

To J. D. Hooker [11 August 1861]
1.1 & send … box 1.2] *interl*
1.6 this the] *interl*
1.7 & not] *after del* 'from'
2.2 the stigma] *interl*
3.2 to my] 'to' *over* 'as I'

To John Lubbock 9 August [1861]
1.4 hoped)] *parenthesis over comma*
3.2 much] *interl*

To Charles Lyell 13 [August 1861]
2.3 all] *interl*
2.6 or … tree, that 2.7] *interl*

To John Lubbock 14 August [1861]
1.2 friendship &] '&' *over full stop*
2.2 has] *over* 'have'

To Charles Lyell 21 August [1861]
1.3 the modification] 'the' *interl*
1.4 perhaps] *interl*
1.4 (& leaving out "Hence") 1.5] *interl*
1.6 will] *after del illeg*
1.7 in Origin] *interl*
1.8 very wide] 'very' *interl*
1.10 & important] *interl*
2.6 in Origin] *interl*
2.9 & manifest] '&' *over illeg*
2.9 previous] *interl*
2.11 this] *above del illeg*
3.4 (eheu)] *interl*
3.8 that at] 'that' *interl*
3.9 slight] *interl*
3.9 variations] *before del* 'to'
3.9 should arise] 'should' *added*
3.9 that one] 'that' *interl*
3.9 should be] *above del* 'was'
3.10 that the] 'that' *interl*
3.10 eleven should] 'should' *interl*
3.10 perish] 'h' *over* 'hed'
3.11 generations] 'g' *over comma*
4.8 formation] *interl*
4.12 on] *over illeg*
7.1 (We return Home on Monday 26th] *added at top of first page*

To J. D. Hooker 30 August [1861]
1.3 the orchids] 'the' *altered from* 'this'
1.3 orchids are] *interl above del* 'is'
1.6 nearly] *after del* 'th'
3.3 How] *interl*

To George Maw 31 August [1861]
2.8 some] *above del* 'the'

To J. O. Westwood 4 September [1861]
3.3 but] 'b' *over* 'I'

To George Rolleston 5 September [1861]
1.7 same] *interl*
2.1 variations] 's' *added*
4.1 some] *interl*

To J. D. Hooker 6 September [1861]
2.5 They] *over del* 'it'
2.5 respect &] '&' *over del* '.—'
5.2 an] *interl*

To T. F. Jamieson 6 September [1861]
2.5 myself] *after del* 'myselff.'
2.5 any case] *above del* 'it'

To Charles Lyell 6 September [1861]
1.2 was] *over illeg*

To Charles Lyell 10 September [1861]
2.1 much of the] 'the' *interl*
2.1 absence] 'abs' *over illeg*
2.2 the vast] 'the' *interl*
2.3 T. del] *after del* 'S. Amer'
2.4 afterwards] *after del* 'has lain'
4.1 at] *added*

To John Murray 10 September [1861]
3.2 as] *over* '&'

To Daniel Oliver 11 September [1861]
1.4 & I] 'I' *interl*
3.2 &] *over* 'or'

3.5 appear] *above del* 'seem'
3.5 me] *over* 'be'
3.7 filament] *above del* 'Hairs'
3.8 filament] *above del* 'Hair'
3.9 touch by a] *interl*
3.13 & sticking] '&' *below del* 'the'
3.14 single] *interl*
3.16 without exciting movement] *interl*

To Charles Lyell [15 September 1861]
2.4 the shelves] *interl*
2.12 The] 'T' *over* 't'
2.12 & Roy] *interl*
2.12 valleys] *altered from* 'valley'
2.13 ; and buttresses … Hills.—] *interl*
2.15 great] *interl*
2.17 Lochaber] *after del* 'La'
3.4 so] *over illeg*
3.5 It] *after del* 'The'
5.1 Heer] *over illeg*
5.1 opinion] 'op' *over illeg*
5.2 all] *over illeg*
5.2 Torquay] 'T' *over illeg*
5.2 Entomology] *over illeg*
5.3 Lowe] 'L' *over illeg*

To Asa Gray 16 September [1861]
3.2 it may] 'it' *after del* 'I'
4.2 our] *interl*

To Asa Gray 17 September [1861]
1.2 & Sept 2d,] *interl*
1.4 N. can] 'N' *over* 'n'
1.11 into] 'in' *added*
1.13 the commencement] 'the' *interl*
2.1 enclosed] *interl*
2.1 it] *interl*
2.2 (I … Linum) 2.4] *square brackets in MS*
2.3 of the Rubiaceæ] *interl*
2.4 in any case] *interl*
2.5 fully] *interl*
2.5 that] *above del* 'that'
2.7 Orders] 'O' *over* 'o'
2.9 such] *interl*
3.5 that the] *interl*
3.6 before] *interl*
3.8 early] *interl*
3.8 of the flower 3.9] *interl*
3.10 a little] *interl*
3.11 tips of] *interl*
3.12 necessary] *before del full stop*
3.12 for their removal. 3.13] *interl*
3.13 stuck] *above del* 'struck'
3.13 its] *interl*
3.18 of the] 'of' *interl*
4.1 & longifolia] *interl*
4.2 persons,] *added*
4.5 useless] *interl*
4.6 gets] *after del* '& is useless'
4.10 plan is] 'is' *interl*
5.2 others] *above del illeg*
6.9 yet] *after del* '&'
6.9 with] *interl*
6.9 supposed to have] *above del* 'having'
8.3 of your Ilex] *interl*
8.3 female] *after del illeg*
8.4 (ours … pollen)] *interl*

To John Murray 21 September [1861]
1.3 for the first time] *interl*
1.5 & … years] *interl*
2.2 205] *after del* '21'
2.5 that] *interl*
3.1 in order] *after del* 'to'
3.2 please] *interl*

To Charles Lyell 22 September [1861]
1.1 last] *interl*
1.5 Welch] *after del* 'Wales'
1.7 sloping] *interl*
1.13 such beds of] *interl*
1.15 to which I referred] *interl*
1.16 Some … suspicious. 1.17] *interl*
2.6 a ferruginous] 'a' *interl*
2.6 firmly] *interl*
3.1 ice-period] *after del* 'posterior'
3.1 the (marine?)] 'the' *interl*
3.2 not] *added*
3.7 (barring denudation)] *interl*
4.2 you] 'y' *over illeg*
6.1 (This … it.—)] *square brackets in MS*

To J. D. Hooker 24 September [1861]
1.1 kept] *before omitted point*
1.1 about] *interl*
1.2 as] *above del* 'until'
1.2 & until he] *above del* 'for me'
2.3 pages!!] '!!' *over full stop*
2.9 much] *interl*

To John Murray 24 September [1861]
6.3 for] *over* 'on'
6.4 an] *before omitted point*

To H. W. Bates 25 September [1861]
2.1 point!] '!' *over* ':'
2.3 Murray & it] *after del* 'to'
2.3 up to present day.—] *interl*
5.1 Travels] 'T' *over* 't'
5.1 Keep] 'ep' *over* 'p'

To J. D. Hooker 28 September [1861]
1.3 annoy] *after del* 'an'
1.4 abuse] *above del* 'abuse'
1.5 have been] *above del* 'are'
3.2 an excellent observer] *interl*
3.3 great] *interl*
3.5 spreading] *after del* 'Glacial'
6.2 or propagate for me] *interl*
6.3 wonderfull] 'l' *added*
6.5 Origin] 'O' *over* 'o'
6.6 new] *interl*
7.1 Rubiaceæ] *after interl del* 'hardy'
7.1 of the] *after del* 'not'
7.2 present] *after del* 'have'
7.7 stigma] 'a' *over* 'as'
7.8 similarly looking] *interl*

To Charles Lyell 1 October [1861]
1.6 valleys] *altered from* 'vally'
3.1 But I] 'I' *over* 'am'
3.3 a sweet] 'a' *interl*

To John Murray 3 October [1861]
2.1 cover???] '???' *over* '—'
4.2 page] *above del* 'type'
4.2 large] *after del illeg*

To J. D. Hooker 4 October [1861]
1.5 normal] *above del* 'true'
1.6 The Clinandrum] *above del* 'It'

To John Murray 5 October [1861]
2.2 ornament.] *above del* 'ornament'

To John Gould 6 October [1861]
0.2 6th.] '6' *over* '4'
1.8 all] *interl*
1.9 quite] *interl*

To J. D. Hooker 6–7 October [1861]
1.2 Do … withered. 1.3] *added blue crayon*
3.3 on the] 'the' *interl*
3.3 or] *over* '&'
3.6 to me] *interl*
3.10 !!] *over* '.—'
5.1 must] *after del* 'a'

To Daniel Oliver 8 October [1861]
1.2 well-opened flower of] *interl*
1.2 nectar is] 'is' *interl*
1.3 or cup] *interl*
1.4 base] *after del* 'ba'
1.4 cup] *above del* 'cup'
2.3 a theory] 'a' *above del* 'my'
2.3 of mine] *interl*

To J. D. Hooker 11 October [1861]
2.1 or Oliver] *interl*

To Asa Gray [after 11 October 1861]
3.3 of] 'f' *over* 'n'
3.7 distinct manner 3.8] *added before del* 'means'
3.8 by] *interl*
3.10 anthers] *after del* 'ro'
3.10 all] 'a' *over illeg*

To J. D. Hooker 13 October [1861]
3.1 Cycnoches] 'n' *interl*
4.3 touch either of] *interl*
4.5 Which is the] 'Which is' *interl*; 't' *over* 'T'
4.6 know. Touch] *full stop over semicolon*; 'T' *over* 't'
4.7 end] *interl*
4.7 I … stigma??? 4.8] *added*
4.8 I can … me. 4.9] *added blue crayon*
5.1 M^{r}] *interl*
5.3 i.e. send servant] *interl*
5.4 Perhaps … me.—] *added*
6.2 vessels;] ';' *over* '.—'
9.1 not] *above del* 'hardly get [*illeg*]'

To J. D. Hooker 15 [October 1861]
1.1 suspect,] *comma after del* 'possible'
1.2 greatly] *interl*
1.3 on the] 'the' *interl*
1.3 such] *above del* 'this'
1.3 the flower] 'the' *interl*
1.5 Try] 'r' *over* 'y'
1.7 or other] 'or' *over* '&'
2.2 use] *interl*
3.1 they have] *interl*
3.2 in] *interl*
3.3 of this movement] *interl*
4.1 Catasetum] *after del* 'If you can remember I sh^{d} be glad of **young buds** of Stanhopea or any allied genus with long caudicle, as I want to make a section of one part.'
4.4 very partially] *above del* 'partly'

4.5 rostellum] *before omitted point*
4.5 is common] 'is' *interl*
4.7 with] *interl*
4.7 foremost] *above del* 'first'
6.1 a Mormodes] 'a' *interl*
7.2 Oncidium w^{d} be best] *added*
7.4 Mr Sowerby & 7.5] *added*
9.3 do] *interl*
10.1 writing] *after del illeg*

To J. D. Hooker 18 October [1861]
1.2 structura] 'a' *below* 'a'
4.1 strange] *interl*
4.2 the] 't' *over* 'T'
4.5 depart] *after del illeg*
4.5 by this course.] *interl*
5.5 at impossibility] 'at' *interl*
5.5 ever] *interl*
5.8 points of] *interl*
6.5 potency of their] *interl*

To John Lindley 18 October [1861]
1.3 way.] *interl*

To Charles Lyell 20 October [1861]
2.1 respect] *altered from* 'respects'
2.9 (which] *parenthesis over comma*
have] 'ha' *over illeg*

To John Murray 21 October [1861]
1.7 he were] 'he' *over illeg*
2.3 tomorrow (Tuesday)] *interl*
3.1 quickly] *interl*

To W. E. Darwin 22 October [1861]
2.7 sovereign] 'ei' *over* 'ig'

To T. H. Huxley 22 October [1861]
2.1 freely.] *before omitted point*
2.3 as] *interl*

To J. D. Hooker 23 October [1861]
2.1 or allies] *interl*
2.2 hairy] *interl*
2.2 of vars.] 'of' *over* '&'
2.2 & species] *interl*
4.3 preserved] *after del* 'of'
4.5 or perhaps Rubiaceæ] *interl*
5.1 Would … does? 5.2] *double scored*
6.3 tomorrow to a] 'a' *after del caret*
6.3 neighbour's 6.4] *interl*
Enclosure 1
1 *Rather*] *interl*
2 rather] *interl*
3 genus] *interl*
4 also] *interl*
5.1 or **plant** which would be best)] *interl*
5.1 &] *over* 'or'
Enclosure 2
1 (*Malaxeæ*)] *square brackets in MS*
6 (*Epidendreæ*)] *below del* 'Vandeæ'; *square brackets in MS*
6 (Vandeæ continued)] *square brackets in MS*
10 (Vandeæ)] *square brackets in MS*
13 (Arethuseæ)] *square brackets in MS*
15 (Cypripedium.)] *square brackets in MS*

To Charles Lyell 23 [October 1861]
2.4 I can … suffice. 2.5] *interl*

To W. B. Clarke 25 October [1861]
2.1 has pleased] 'has' *interl*
3.4 Bear … exception.— 3.5] *interl*
4.5 of the] 'the' *interl*
4.6 just] *interl*
4.6 open] *above del* 'have opened'
4.8 plants] *after del* 'the'
4.8 of this Order] *interl*
5.1 dioicous] *interl*

To John Lindley 25 October [1861]
3.1 & **bud**] *interl*
3.2 it be a] *interl*
5.3 as introduction 5.4] *interl*
5.5 exotics] *interl*

To J. D. Hooker 27 October [1861]
3.2 (You … Malaxis.—3.2)] *square brackets in MS*
3.2 a wonderful] 'a' *interl*
3.5 whorl] 'l' *over* 'ls'
3.5 the two whorls of] *interl*
3.8 after what you said] *interl*
4.1 (which … it) 4.2] *square brackets in MS; double scored*
4.2 of 15] *after del* 'organ'
4.2 theoretical] *interl*
4.3 the result] *above del* 'it'
4.7 ovarian] *interl*
4.8 (only seen)] *interl*
4.9 my] *interl*
4.10 a labellum] *interl*
4.14 receives its vessels from 4.15] *interl above del* 'belongs to'
4.15 main] *added*

4.15 bundle?] '?' *over colon*
4.16 part] *after del illeg*
4.17 closely] *after del illeg*
4.19 turns] *after del* 'runs'
4.19 down it] 'it' *after del comma*
4.20 each] *after del* 'the'
4.24 as in Mormodes] *interl*
4.25 simple] *interl*
4.26 soldered] *after del comma*
6.2 course of] *interl*
8.1 plant has] *above del* 'is'

To J. D. Hooker 1 November [1861]
1.3 Labellum … attachment. 1.4] *added*
2.2 Column,] 'C' *over* 'L'

To J. D. Hooker [2]1 November [1861]
2.1 traced] *above del* 'seen'
2.2 to 2] '2' *interl*
3.2 Anthers] 'A' *over* 'a';
's' *added*
3.2 I will] *after del* 'I'
3.3 the missing] 'the' *interl*
3.3 it is] *interl*
3.4 object] *interl*
4.2 (so-called)] *interl*
4.3 or stimulus from a light touch] *interl*
4.4 & a] '&' *interl*
4.5 are] *interl*
4.6 foremost] *after del* 'first'
5.2 macrophyllum] *above del* 'chrysanthum'
5.3 when] *interl*
5.3 in thickness] *interl*
5.4 into stigma] *interl*
5.5 *Is*] '*I*' *over* '*i*'
5.7 the] *interl*
6.4 Spiranthes glandular.] *below del* 'Gl Goodyera'
8.3 has] 'h' *over* 'f'
9.1 Labellum] *interl*
9.2 living] *interl*
9.2 same Rubiaceous 9.3] *interl*
9.3 the Primula] 'the' *interl*
9.5 as this Rhamnus] *interl*
13.1 of Bulbophyllum] *interl*
14.2 of] *over illeg*

To John Lindley 1 November [1861]
1.2 with] *interl*
2.3 supply] 'y' *over* 'ies'
2.3 2 lower] '2' *interl*
4.6 bundle] *after del illeg*
4.6 failed] *before del caret*
4.6 clearly] *interl*

To W. E. Darwin 4 November [1861]
4.2 (N.B … "*rode*") 4.3] *square brackets in MS*

To J. D. Hooker 6 [November 1861]
1.2 at all] *after del* 'lik'
2.1 sepals] *interl*
3.1 C. Darwin] *above del illeg*

To J. D. Hooker 7 November [1861]
1.2 instead of a bud] *interl*
1.3 Rostellum.] *before omitted point*
3.3 these] *after del* 'it be'

To J. D. Hooker 8 November [1861]
1.2 (N.B … useful 1.3)] *square brackets in MS*
1.4 are not] 'are' *interl*
5.2 (on 21st Nov.)] *interl*; *square brackets in MS*

To J. D. Hooker 10 November [1861]
1.3 Some of] *interl*
1.4 bundles or] *interl*
1.6 (I am sure of my facts)] *square brackets in MS*
3.3 & that] *after del* 'but'; '&' *over illeg*
3.7 out of your Herbarium] *interl*
Diagram
(Dotted lines ducts)] *square brackets in MS*

To J. D. Hooker 14 November [1861]
1.2 They] 'T' *over* 't'
2.3 & upper petals] *interl*
2.3 petals] *after del illeg*
2.3 in allied forms] *interl*
2.6 that the] 'that' *interl*

To W. E. Darwin 15 November [1861]
1.4 employed] *after del* 'wa'

To John Lindley 16 November [1861]
1.1 received] *above del* 'sent'

To J. D. Hooker 17 November [1861]
1.2 with respect to value] *interl*
1.6 as new] 'as' *interl*
1.9 in Spirits] *interl*
2.1 from] 'f' *over illeg*
2.1 most kindly] *interl*
2.9 the] 't' *over* 'T'
2.9 anther-ducts] 'anther-' *interl*
4.3 on investigation] *interl*
4.4 an island] [1]'n' *added*; 'island' *after del* 'sep'

4.4 think that] 'that' *interl*
4.6 ?] *over* '.—'
6.4 my paper] *interl*
6.6 Heterocentron roseum] *interl*
6.9 in publishing.] *interl before omitted point*
9.1 (By ... maintained.) 9.2] *square brackets in MS*

To D. F. Nevill 19 November [1861]
2.8 point] *interl*
6.1 to you] *interl*
7.4 the cut] 'the' *interl*
7.4 These genera] *altered from* 'This genus'

To J. D. Hooker 20 [November 1861]
1.1 bother] 'b' *over illeg*
1.2 do?] '?' *over full stop*
1.2 it?] '?' *over full stop*

To J. D. Hooker [22 November 1861]
1.3 per Rail] *interl*
1.6 take care of the] *below del* 'forward the'

To J. D. Hooker 25 November [1861]
1.1 of either species] *interl*
1.2 your] *after del illeg*
1.5 the species of] *interl*
1.5 the males] 'the' *interl*
1.5 some other orchids] *interl*
1.6 & Myanthus] *interl*
1.8 fresh] *interl*
1.9 on fertilisation] *interl*
6.1 (or Oliver)] *interl*

To George Bentham 26 November [1861]
2.2 & have some leisure] *interl*

To Daniel Oliver 30 November [1861]
1.3 (sent ... at.) 1.4] *square brackets in MS*
1.6 ovaria of] *interl*
1.10 also] *interl*
1.11 specimens of Acropera;] *above del* 'ovaria'
1.12 ovules] *interl*
1.16 of *all*] 'of' *interl*
1.16 (except Catasetum)] *interl*
2.3 from ... Catasetum 2.4] *interl*
2.6 sent me by Hooker] *interl*
3.1 in the] *after del comma*
3.5 three irregular] *after del illeg*
3.6 belonging] *after del* 'corresponding'
3.6 divisions] *above del* 'parts'
3.8 be,] *before omitted comma*
6.2 Origin] 'O' *over* 'o'

6.3 old] *interl*
7.5 culture] *after del* 'w'
8.1 any] *interl*

To J. D. Hooker 1 December [1861]
3.5 blind] *after del* 'I'
3.5 Irishman] 'I' *over* 'i'
3.6 of all] 'of' *interl*
3.6 whose] *after del* 'which'
4.1 G.......?] *interl*
7.1 plants of] *interl*
7.1 !] *over* '.—'
9.4 or spine-like cells] *interl*
9.4 segments] *after del* 'pieces'

To H. W. Bates 3 December [1861]
3.10 for one paragraph] *interl*
3.18 their] *interl*
5.3 now that it is] *above del* 'now,'
8.2 of his] *interl*

To Ludwig Rütimeyer 5 December [1861]
1.2 wild] *interl*
3.3 M.S.] *after del illeg*
4.1 for publication] *added*

To Daniel Oliver 7 December [1861]
1.3 I must] *after omitted point*
2.7 I forgot ... publication) 2.9] *del in MS*
2.9 (I see ... letter) 2.10] *square brackets in MS*

To J. D. Hooker [9 December 1861]
1.1 Lecoq] 'q' *over* 'qu'
1.2 having, &] '&' *over* 'but'
4.2 as] *over* 'in'

To Asa Gray 11 December [1861]
1.4 How] 'H' *over* 'I'
3.6 that of] *interl*
3.12 or more than] *interl*
7.2 I like] *after del* 'When'

To H. W. Bates 15 December [1861]
1.7 criticise my] 'cri' *over* 'cir'
2.1 discussion] [1]'s' *over* 'ss'
2.4 or other animals] *interl*
2.4 scientific name 2.5] *interl*
2.7 So ... paths. 2.8] *added*
4.2 use] *after del illeg*
7.2 number in] *interl*
8.1 judge)] *parenthesis over comma*
8.2 done)] *parenthesis over comma*

8.3 a subject.] *interl*
8.7 in Brazil & is so conspicuous] *interl*
9.1 greatly] *interl*
9.3 it is] *interl*
10.1 or three] *interl*
10.3 all risk] 'all' *interl*
10.5 Perhaps … 9 months. 10.6] *added*

To John Lindley 15 December [1861]
4.2 whole plant of] *interl*

To J. D. Hooker 18 [December 1861]
3.1 Genera] 'G' *over* 'g'
3.5 Orchids alone] *above del* 'it'
3.5 for] *over* 'of'
4.2 Plants] *after del* 'the'
5.3 got] *interl*
6.3 you] *over* 'him'
6.4 me] *interl*
8.4 (though] *parenthesis over comma*
8.5 few] *interl*
8.7 species] *after del* 'pl'
9.4 spectabilis] *interl*
9.5 alive] *added*

12.2 in L.] 'in' *interl*

To John Innes 19 December [1861]
1.1 Quiz] *above del* 'him'
1.2 Johnny] *above del* 'him'

To John Lindley 24 December [1861]
2.1 rudimentary] *interl*
2.3 seems] *after del* 'I certainly'

To J. D. Hooker 27 [December 1861]
1.1 Lindley] 'L' *over* 'C'

To J. D. Hooker 28 [December 1861]
1.2 good] *after del* 'a'
1.5 at Royal Institution] *interl*
3.4 most] *interl*
3.5 more] *after del* 'f'

To John Lindley 28 December [1861]
2.2 A.] *interl*

To J. D. Hooker [30 and 31 December 1861]
1.2 mentioned … Chronicle] *interl*
5.1 but] *after del* 'if'

BIBLIOGRAPHY

The following bibliography contains all the books and papers referred to in this volume by author–date reference or by short title. Short titles are used for some standard reference works (e.g., *DNB, OED*), for CD's books, and for editions of his letters and manuscripts (e.g., *Descent, LL, Notebooks*). Works referred to by short titles are listed in alphabetical order according to the title; those given author–date references occur in alphabetical order according to the author's surname.

Abbott, Charles Compton. 1861. Notes on the birds of the Falkland Islands. *Ibis* 3: 149–67.

ADB: *Allgemeine deutsche Biographie*. Under the auspices of the Historical Commission of the Royal Academy of Sciences. 56 vols. Leipzig: Duncker and Humblot. 1875–1912.

Agassiz, Jean Louis Rodolphe. 1840a. Discovery of the former existence of glaciers in Scotland, especially in the Highlands. *Scotsman* 24: 7 October 1840.

——. 1840b. On glaciers, and the evidence of their having once existed in Scotland, Ireland, and England. *Proceedings of the Geological Society of London* 3 (1838–42): 327–32.

——. 1840c. *Etudes sur les glaciers*. Neuchâtel.

——. 1842. The glacial theory and its recent progress. *Edinburgh New Philosophical Journal* 33: 217–83.

——. 1857–62. *Contributions to the natural history of the United States of America*. 4 vols. Boston and London.

——. 1859. *An essay on classification*. London. [Reprint of vol. 1, pt 1, of *Contributions to the natural history of the United States of America*. 4 vols. 1857–62.]

——. 1860. On the Origin of species. *American Journal of Science and Arts* 2d ser. 30: 142–54. [Reprinted in *Annals and Magazine of Natural History* 3d ser. 6 (1860): 219–32.]

Airy, George Biddell. 1843. Letter from the astronomer royal to the earl of Rosse. *Report of the 13th meeting of the British Association for the Advancement of Science held at Cork*, Transactions of the sections, p. 62.

Allan, Mea. 1967. *The Hookers of Kew, 1785–1911*. London: Michael Joseph.

Allen, David Elliston. 1986. *The botanists. A history of the Botanical Society of the British Isles through a hundred and fifty years*. Winchester: St Paul's Bibliographies.

Alum. Cantab.: *Alumni Cantabrigienses. A biographical list of all known students, graduates and holders of office at the University of Cambridge, from the earliest times to 1900. Part II. From 1752 to 1900*. Compiled by J. A. Venn. 6 vols. Cambridge: Cambridge University Press. 1940–54.

Alum. Oxon.: *Alumni Oxonienses: the members of the University of Oxford, 1715–1886*. By Joseph Foster. 4 vols. Oxford. 1888.

Annual register (1861): *The annual register, or a view of the history and politics of the year 1861*. London. 1862.

Anon. 1869. *Memoir of W. H. Harvey*. London.

Archiac, Etienne Jules Adolphe Desmier de Saint-Simon. 1847–8. *Histoire des progrès de la géologie de 1834 à 1845*. 2 vols. Paris.

Army list: *The army list*. London. 1815–81.

Aust. dict. biog.: *Australian dictionary of biography: 1788–1850; 1851–1890*. Edited by Douglas Pike and Bede Nairn. 6 vols. Melbourne: Melbourne University Press. 1966–76.

Autobiography: *The autobiography of Charles Darwin 1809–1882. With the original omissions restored*. Edited and with appendix and notes by Nora Barlow. London: Collins. 1958.

Azara, Félix d'. 1801. *Essais sur l'histoire naturelle des quadrupèdes de la province du Paraguay*. 2 vols. Paris.

Babington, Charles Cardale. 1851. *Manual of British botany, containing the flowering plants and ferns arranged according to the natural orders*. 3d ed. London.

Bachman, John. 1850. An investigation of the cases of hybridity in animals on record, considered in reference to the unity of the human species. *Charleston Medical Journal* 5: 168–97.

Baillon, Ernest Henri. 1854. Sur le mode de fécondation du *Catasetum luridum* (Lindley). *Bulletin de la Société Botanique de France* 1: 285–8.

Banking almanac: *The banking almanac, directory, year book and diary*. Edited by D. Morier Evans. London. 1845–1919.

Barlow, Nora, ed. 1933. *Charles Darwin's diary of the voyage of H.M.S. "Beagle"*. Cambridge: Cambridge University Press.

——, ed. 1963. Darwin's ornithological notes. *Bulletin of the British Museum (Natural History). Historical series* 2 (1959–63): 203–78.

——, ed. 1967. *Darwin and Henslow. The growth of an idea*. London: Bentham-Moxon Trust/John Murray.

Barrett, Paul H. 1973. Darwin's gigantic blunder. *Journal of Geological Education* 21: 19–28.

Bartlett, Edward, ed. 1900. *Bartlett's life among wild beasts in the 'zoo' . . . by A. D. Bartlett*. London: Chapman and Hall.

Bartley, Mary. 1992. Darwin and domestication: studies on inheritance. *Journal of the History of Biology* 25: 307–33.

Bates, Henry Walter. 1861a. Contributions to an insect fauna of the Amazon valley. Diurnal Lepidoptera. [Read 5 March and 24 November 1860.] *Transactions of the Entomological Society of London* n.s. 5 (1858–61): 223–8, 335–61.

——. 1861b. Contributions to an insect fauna of the Amazon valley. Lepidoptera: Heliconidæ. [Read 21 November 1861.] *Transactions of the Linnean Society of London* 23 (1862): 495–566.

Bates, Henry Walter. 1863. *The naturalist on the river Amazons. A record of adventures, habits of animals, sketches of Brazilian Indian life, and aspects of nature under the equator, during eleven years of travel*. 2 vols. London.

Bates, Henry Walter and Smith, Frederick. 1855. Descriptions of some species of Brazilian ants belonging to the genera *Pseudomyrma*, *Eciton* and *Myrmica* (with observations on their economy). [Read 1 January 1855.] *Transactions of the Entomological Society of London* n.s. 3 (1854–6): 156–69.

Bauer, Franz Andreas. 1830–8. *Illustrations of orchidaceous plants ... with notes and prefatory remarks by John Lindley*. London.

Bechstein, Johann Matthäus. [1789–95.] *Gemeinnützige Naturgeschichte Deutschlands nach allen drey Reichen*. 4 vols. Leipzig.

Beckett, Ian F. W. 1982. *Riflemen form. A study of the Rifle Volunteer Movement, 1859–1908*. Aldershot: The Ogilby Trusts.

Bénézit, Emmanuel. 1976. *Dictionnaire critique et documentaire des peintres, sculpteurs, dessignateurs et graveurs de tous les temps et de tous les pays, par un groupe d'écrivains spécialistes français et étrangers*. Revised edition. 10 vols. Paris: Librairie Gründ.

Bennett, John Hughes. 1841. *Treatise on the oleum jecoris aselli, or cod liver oil as a therapeutic agent in certain forms of gout, rheumatism, and scrofula; with cases*. London.

Bentham, George. 1858. *Handbook of British flora; a description of the flowering plants and ferns indigenous to, or naturalized in, the British Isles*. London.

——. 1861a. On the species and genera of plants, considered with reference to their practical application to systematic botany. *Natural History Review* n.s. 1: 133–51.

——. 1861b. *Outlines of elementary botany, as introductory to local floras*. London.

Bentham, George and Hooker, Joseph Dalton. 1862–83. *Genera plantarum ad exemplaria imprimis in herbariis kewensibus servata definita*. 3 vols. London.

Berg, Carlos. 1895. Carlos Germán Conrado Burmeister. Reseña Biográfica. *Anales del Museo Nacional de Buenos Aires* 2d ser. 1: 316–57.

BHGW: *Biographisch-literarisches Handwörterbuch zur Geschichte der exacten Wissenschaften*. By Johann Christian Poggendorff. Vols. 1–4. Leipzig: Johann Ambrosius Barth. 1863–1904. Vol. 5. Leipzig and Berlin: Verlag Chemie. 1926.

Bibby, Harold Cyril. 1959. *T. H. Huxley, scientist, humanist and educator*. London: Watts.

Billings, Elkanah. 1859a. On some new genera and species of Brachiopoda from the Silurian and Devonian rocks of Canada. *Canadian Naturalist* 4: 131–5.

——. 1859b. Description of a new genus of Brachiopoda, and on the genus Cyrtodonta. *Canadian Naturalist* 4: 301–3.

Birds: Pt 3 of *The zoology of the voyage of H.M.S. Beagle*. By John Gould. Edited and superintended by Charles Darwin. London. 1838–41.

Biscoe, John. 1833. Recent discoveries in the Antarctic Ocean. From the log-book of the brig Tula, commanded by Mr. John Biscoe, R.N. *Journal of the Royal Geographical Society of London* 3: 105–12.

BLA: *Biographisches Lexikon der hervorragenden Aerzte aller Zeiten und Völker*. Edited by A. Wernich, E. Gurlt, and August Hirsch. 6 vols. Vienna and Leipzig: Urban & Schwarzenberg. 1884–8.

Black, Joseph. 1775. On the supposed effect of boiling water in disposing it to freeze more readily. *Philosophical Transactions of the Royal Society of London* 65 (1775): 124–8.

Blackadder, Alexander. 1824. On the superficial strata of the Forth district. [Read 15 May 1824.] *Memoirs of the Wernerian Natural History Society* 5 (1823–4): 424–39.

Blainville, Henri Marie Ducrotay de. 1839–64. *Ostéographie, ou description iconographique comparée du squelette et du système dentaire des mammifères récents et fossiles, pour servir de base à la zoologie et à la géologie*. 4 vols. Paris.

BLC: *The British Library general catalogue of printed books to 1975*. 360 vols. and supplement (6 vols.). London. 1979–88.

BLKO: *Biographisches Lexikon des Kaiserthums Oesterreich, enthaltend die Lebensskizzen der denkwürdigen Personen, welche seit 1750 in den österreichischen Kronländern geboren wurden oder darin gelebt und gewirkt haben*. By Constant von Wurzbach. 60 vols. Vienna. 1856–90.

Blyth, Edward. 1859. On the different animals known as wild asses. *Journal of the Asiatic Society of Bengal* 28: 229–53. [Reprinted in *Annals and Magazine of Natural History* 3d ser. 6 (1860): 233–54.]

BNB: *Biographie nationale publiée par l'Académie royale des sciences, des lettres et des beaux-arts de Belgique*. 28 vols. Brussels. 1866–1944.

Bonney, Thomas George. 1919. *Annals of the Philosophical Club of the Royal Society, written from its minute books*. London: Macmillan and Co.

Bourne, K. H. 1961. British preparations for war with the North. *English Historical Review* 76: 600–32.

[Bowen, Francis.] 1860a. Review of *Origin of species*. *North American Review* 90: 474–506.

——. 1860b. Remarks on the latest form of the development theory. [Read 27 March, 10 April, and 1 May 1860.] *Memoirs of the American Academy of Arts and Sciences* n.s. 8 (1861): 97–122.

——. 1861. Observations of the supposed hereditability of peculiar traits of bodily and mental organization, and especially of mental disease. [Read 8 January 1861.] *Proceedings of the American Academy of Arts and Sciences* 5 (1860–2): 102–10.

Bowler, Peter J. 1983. *The eclipse of Darwinism. Anti-Darwinian evolution theories in the decades around 1900*. Baltimore and London: Johns Hopkins University Press.

Bravais, Auguste. [1843?] Sur les lignes d'ancien niveau de la mer dans le Finmark. In vol. 1 of *Voyages en Scandinavie, en Laponie, au Spitzberg et aux Feröe, pendant les années 1838, 1839 et 1840, sur la corvette la Recherche . . . Géographie physique*, edited by Paul Gaimard. 2 vols. Paris.

——. 1845. On the lines of ancient level of the sea in Finmark. *Quarterly Journal of the Geological Society of London* 1: 534–49.

Bree, Charles Robert. 1860. *Species not transmutable, nor the result of secondary causes. Being a critical examination of Mr. Darwin's work entitled "Origin and variation of species"*. London and Edinburgh.

British Museum (Natural History). 1904–6. *The history of the collections contained in the natural history departments of the British Museum*. 2 vols. London.

Brock, William H. and Macleod, Roy M. 1976. The scientists' declaration: reflections on science and belief in the wake of *Essays and reviews*. *British Journal of the History of Science* 9: 39–66.

Brogan, Hugh, ed. 1975. *"The Times" reports. The American Civil War. Extracts from "The Times", 1860–1865*. London: Times Books.

Bronn, Heinrich Georg. 1858. *Untersuchungen über die Entwickelungs-Gesetze der organischen Welt während der Bildungs-Zeit unserer Erd-Oberfläche. Eine von der Französischen Akademie im Jahre 1857 gekrönte Preisschrift ... Natura doceri*. Stuttgart.

——, trans. 1860. *Charles Darwin über die Entstehung der Arten im Thier- und Pflanzen-Reich durch natürliche Züchtung*. Stuttgart.

Brown, Robert. 1810. *Prodromus floræ Novæ Hollandiæ et Insulæ Van-Diemen*. Vol. 1. [No more published.] London.

——. 1831. On the organs and mode of fecundation in Orchideæ and Asclepiadeæ. [Read 1 and 15 November 1831.] *Transactions of the Linnean Society of London* 16 (1833): 685–745.

——. 1832. Additional observations on the mode of fecundation in Orchideæ. [Read 5 June 1832.] *Transactions of the Linnean Society of London* 16 (1833): 739–45.

Brown-Séquard, Charles Edouard. 1859–60. Hereditary transmission of an epileptiform affection accidentally produced. *Proceedings of the Royal Society of London* 10: 297–8.

Buckle, Henry Thomas. 1857. *History of civilization in England*. 2 vols. London.

Buckley, Samuel Botsford. 1860. The cutting ant of Texas. *Proceedings of the Academy of Natural Sciences of Philadelphia*, pp. 233–6.

Bulhof, Ilse N. 1974. The Netherlands. In *The comparative reception of Darwin*, edited by T. F. Glick. Austin and London: University of Texas Press.

Bunbury, Frances Joanna, ed. 1891–3. *Memorials of Sir C. J. F. Bunbury, Bart*. *Early life* 1 vol.; *Middle life* vols. 1–3; *Later life* vols. 1–5. Mildenhall.

Burgon, John William. 1870. *Dr. Temple's "explanation" examined*. Oxford and London.

Burke's landed gentry: *Burke's genealogical and heraldic history of the landed gentry*. By John Burke. 1–18 editions. London: Burke's Peerage Ltd. 1836–1972.

Burke's peerage: *Burke's peerage and baronetage*. 1–105 editions. London: Burke's Peerage (Genealogical Books). 1826–1980.

Burkhalter, Lois Wood. 1965. *Gideon Lincecum, 1793–1874. A biography*. Austin: University of Texas Press.

BWN: *Biographisch Woordenboek der Nederlanden, bevattende Levensbeschrijvingen van zoodanige Personen, die zich op eenigerlei wijze in ons Vaderland hebben vermaard gemaakt.* Edited by K. J. R. van Harderwijk and G. D. J. Schotel. 20 vols. Haarlem. 1852–77.

Calendar: *A calendar of the correspondence of Charles Darwin, 1821–1882*. Edited by Frederick Burkhardt and Sydney Smith. New York and London: Garland. 1985.

Cambridge history of China: *The Cambridge history of China*. Vol. 7: *The Ming dynasty, 1368–1644*. Pt 1. Edited by Frederick W. Mote and Denis Twitchett. Cambridge: Cambridge University Press. 1988.

Campbell, George Douglas. 1860. Presidential address. *Proceedings of the Royal Society of Edinburgh* 4 (1857–62): 350–77.

Canton, John. 1753. Electrical experiments, with an attempt to account for their several phænomena; together with some observations on thunder-clouds. *Philosophical Transactions of the Royal Society of London* 48: 350–8.

Carozzi, Albert Victor, trans. 1967. *Studies on glaciers, preceded by the Discourse of Neuchâtel*. By Louis Agassiz. New York: Hafner.

Carpenter, William Benjamin. 1860. Review of *Origin of species*. *British and Foreign Medico-Chirurgical Review* 25: 367–404.

Catalogue of graduates, University of Dublin: *A catalogue of graduates who have proceeded to degrees in the University of Dublin, from the earliest recorded commencements to July, 1866, with supplement to December 16, 1868*. Dublin. 1869.

CDEL: *A critical dictionary of English literature and British and American authors, living and deceased*. By S. Austin Allibone. 3 vols. Philadelphia and London. 1859–71. Supplement by John Foster Kirk. 2 vols. Philadelphia and London. 1891.

Cent cinquante ans d'histoire Vaudoise: *Cent cinquante ans d'histoire Vaudois, 1803–1953*. Published by the Société Vaudoise d'Histoire et d'Archéologie. Lausanne: Payot. 1953.

Chambers, Robert. 1848. *Ancient sea margins*. Edinburgh.

——. 1853. On glacial phenomena in Scotland and parts of England. *Edinburgh New Philosophical Journal* 54 (1852–3): 229–81.

——. 1861. Ice and water: a review of the superficial formation. In *Edinburgh papers*. London and Edinburgh. 1859–61.

Chancellor, Gordon, DiMauro, Angelo, Ingle, Ray, and King, Gillian. 1988. Charles Darwin's *Beagle* collections in the Oxford University Museum. *Archives of Natural History* 15: 197–231.

Clarke, William Branwhite. 1860. *Researches in the southern gold fields of New South Wales*. Sydney.

——. 1861. *Recent geological discoveries in Australasia*. 2d ed. Sydney.

Clergy list: *The clergy list . . . containing an alphabetical list of the clergy*. London. 1841–.

Collected papers: *The collected papers of Charles Darwin*. Edited by Paul H. Barrett. 2 vols. Chicago and London: University of Chicago Press. 1977.

Collingwood, Cuthbert. 1860a. On homomorphism; or, organic representative form. *Proceedings of the Literary and Philosophical Society of Liverpool* 14 (1859–60): 181–216.

——. 1860b. On Agassiz's views of Darwin's theory of species. *Proceedings of the Literary and Philosophical Society of Liverpool* 15 (1860–1): 81–99.

——. 1860c. On recurrent animal form, and its significance in systematic zoology. *Annals and Magazine of Natural History* 3d ser. 6: 81–91.

Collins, William Wilkie. 1860. *The woman in white*. 3 vols. London.

Complete peerage: *The complete peerage of England Scotland Ireland Great Britain and the United Kingdom extant extinct or dormant*. By George Edward Cokayne. New edition, revised and much enlarged. Edited by the Hon. Vicary Gibbs and others.

12 vols. London: The St Catherine Press. 1910–59.

Comte, Isidore Auguste Marie François Xavier. 1830–42. *Cours de philosophie positive*. 6 vols. Paris.

Coral reefs: *The structure and distribution of coral reefs. Being the first part of the geology of the voyage of the Beagle, under the command of Capt. FitzRoy, R.N. during the years 1832 to 1836*. By Charles Darwin. London. 1842.

Correspondence: *The correspondence of Charles Darwin*. Edited by Frederick Burkhardt, Sydney Smith, and others. Cambridge: Cambridge University Press. 1985–.

Corsi, Pietro and Weindling, Paul J. 1985. Darwinism in Germany, France and Italy. In *The Darwinian heritage*, edited by David Kohn. Princeton: Princeton University Press in association with Nova Pacifica.

[Cowell, F. R.] 1975. *The Athenaeum Club and social life in London, 1824–1974*. London: Heinemann.

[Crawfurd, John.] 1859. Review of *Origin of species*. *Examiner*, 3 December 1859, pp. 772–3.

——. [1860a.] On the relation of the domestic animals to civilization. London: privately printed.

——. 1860b. The history of the horse, and his comparative value for military and other purposes. [Read 27 January 1860.] *Journal of the Royal United Service Institution, Whitehall Yard* 4 (1861): 10–24.

——. 1863. On Sir Charles Lyell's 'Antiquity of man' and on Prof. Huxley's 'Evidence as to man's place in nature'. [Read 14 April 1863.] *Transactions of the Ethnological Society of London* n.s. 3 (1864–5): 58–70.

Crisp, Edwards. 1861. On some points relating to the habits and anatomy of the oceanic and of the freshwater ducks, and also of the hare (Lepus timidus) and of the rabbit (L. cuniculus), in relation to the question of hybridism. *Proceedings of the Zoological Society of London*, pp. 82–7.

Crockford's: *The clerical directory, a biographical and statistical book of reference for facts relating to the clergy and the Church*. London. 1858. Continued as *Crockford's clerical directory etc*. London. 1860–.

Cross and self fertilisation: *The effects of cross and self fertilisation in the vegetable kingdom*. By Charles Darwin. London. 1876.

Crowther, M. A. 1970. *Church embattled: religious controversy in mid-Victorian England*. Newton Abbot, Devon: David & Charles. Hamden, Conn.: Archon Books.

Czermak, Johann Nepomuk. 1857. Ueber den schallerzeugenden Apparat von Crotalus. *Zeitschrift für wissenschaftliche Zoologie* 8: 294–301.

DAB: *Dictionary of American biography*. Under the auspices of the American Council of Learned Societies. 20 vols., index, and 8 supplements. New York: Charles Scribner's Sons. London: Oxford University Press. 1928–88.

Dana, James Dwight. 1848. *Zoophytes*. Vol. 8 of *United States Exploring Expedition. During the years 1838, 1839, 1840, 1841, 1842. Under the command of Charles Wilkes, U.S.N.* Philadelphia.

Darwin pedigree: *Pedigree of the family of Darwin*. Compiled by H. Farnham Burke.

Privately printed. 1888. [Reprinted in facsimile in Freeman, Richard Broke, *Darwin pedigrees*. London: printed for the author. 1984.]

Darwin, Erasmus. 1788. Frigorific experiments on the mechanical expansion of air. *Philosophical Transactions of the Royal Society of London* 78: 43–52.

——. 1789–91. *The botanic garden; a poem, in two parts*. Pt 1. *Containing the economy of vegetation*. London. 1791. Pt 2. *The loves of the plants. With philosophical notes*. Lichfield. 1789.

——. 1794–6. *Zoonomia; or, the laws of organic life*. 2 vols. London.

Darwin, Francis. 1914. William Erasmus Darwin, 1839–1914. *Christ's College Magazine* 29: 16–23.

Davidson, Randall Thomas and Benham, William. 1891. *Life of Archibald Campbell Tait, archbishop of Canterbury*. 2 vols. London.

Davidson, Thomas William St Clair. 1848. Mémoire sur les Brachiopodes du système silurien supérieur d'Angleterre. *Bulletin de la Société Géologique de France* 2d ser. 5 (1847–8): 309–53.

——. 1851–86. *British fossil Brachiopoda*. 6 vols. and 2 vols. plates. London.

——. 1861. On British Carboniferous Brachiopoda. *Geologist* 4: 41–59.

——. 1862. On some Carboniferous brachiopods collected in India. *Quarterly Journal of the Geological Society of London* 18: 25–35.

Davies, Gordon Leslie. 1968. The tour of the British Isles made by Louis Agassiz in 1840. *Annals of Science* 24: 131–46.

——. 1969. *The earth in decay; a history of British geomorphology, 1578–1878*. London: Macdonald Technical and Scientific.

Dawson, John William. 1857. On the Newer Pliocene and Post-Pliocene deposits of the vicinity of Montreal, with notices of fossils recently discovered in them. *Canadian Naturalist* 2: 401–26.

——. 1860. Notice of the Tertiary fossils from Labrador, Maine, &c., and remarks on the climate of Canada in the New Pliocene or Pleistocene period. *Canadian Naturalist* 5: 100–20.

Dawson, Warren R. 1946. *The Huxley papers. A descriptive catalogue of the correspondence manuscripts and miscellaneous papers of . . . Thomas Henry Huxley*. London: Macmillan and Co.

DBF: *Dictionnaire de biographie Française*. Under the direction of J. Balteau, M. Barroux, M. Prevost, and others. 17 vols. and 4 fascicles of vol. 18 (A–Jumelle). Paris: Librairie Letouzey et Ané. 1933–92.

DBL: *Dansk biografisk leksikon*. Founded by Carl Frederick Bricka. 3d ed. Edited by Sv. Cedergreen Bech. 16 vols. Copenhagen: Gyldendal. 1979–84.

de Beer, Gavin, ed. 1959. Darwin's journal. *Bulletin of the British Museum (Natural History). Historical Series* 2: 3–21.

Delamer, Eugene Sebastian, *pseud*. (Edmund Saul Dixon.) 1854. *Pigeons and rabbits, in their wild, domestic, and captive states*. London.

Descent: *The descent of man, and selection in relation to sex*. By Charles Darwin. 2 vols. London. 1871.

Desmond, Adrian. 1984a. Robert E. Grant: the social predicament of a pre-Darwinian transmutationist. *Journal of the History of Biology* 17: 189–223.

——. 1984b. Robert E. Grant's later views on organic development: the Swiney lectures on "palaeozoology", 1853–1857. *Archives of Natural History* 11: 395–413.

Desmond, Ray. 1977. *Dictionary of British and Irish botanists and horticulturists, including plant collectors and botanical artists*. 3d ed. London: Taylor and Francis.

Dictionary of Ming biography: *Dictionary of Ming biography 1368–1644*. Edited by L. Carrington Goodrich and Chaoying Fang. 2 vols. New York and London: Columbia University Press. 1976.

Dictionnaire universel des contemporains: *Dictionnaire universel des contemporains contenant toutes les personnes notables de la France et des pays étrangers*. Edited by G. Vapereau. Paris. 1858. 3d ed., 1865. 4th ed., 1870. 5th ed., 1880.

DNB: *Dictionary of national biography*. Edited by Leslie Stephen and Sidney Lee. 63 vols. and 2 supplements (6 vols.). London: Smith, Elder, and Co. 1885–1912. *Dictionary of national biography 1912–85*. Edited by H. W. C. Davis, J. R. H. Weaver, and others. 8 vols. London: Oxford University Press. 1927–90.

DNZB: *A dictionary of New Zealand biography*. Edited by G. H. Scholefield. 2 vols. Wellington: Department of Internal Affairs, New Zealand. 1940.

Don, David. 1837. On the modifications of æstivation observable in certain plants, formerly referred to the genus *Cinchona*. *Transactions of the Linnean Society of London* 17: 139–43.

Dortous de Mairan, Jean Jacques. 1749. *Dissertation sur la glace*. Paris.

DSAB: *Dictionary of South African biography*. Edited by W. J. de Kock, D. W. Krüger, and C. J. Beyers. Vol. 1. Cape Town: Nasionale Boekhandel Beperk. Vols. 2–3. Cape Town: Tafelberg Publishers. Vol. 4. Durban and Pretoria: Butterworth & Co. 1968–81.

DSB: *Dictionary of scientific biography*. Edited by Charles Coulston Gillispie. 14 vols., supplement, and index. New York: Charles Scribner's Sons. 1970–80.

Du Chaillu, Paul Belloni. 1861. *Explorations & adventures in equatorial Africa; with accounts of the manners and customs of the people, and of the chace of the* Gorilla, *crocodile, leopard, elephant, hippopotamus, and other animals*. London.

Dufour, Léon. 1833. Recherches anatomiques et physiologiques sur les hémiptères, accompagnées de considérations relatives à l'histoire naturelle et à la classification de ces insectes. *Mémoires présentés par divers Savans a l'Académie Royale des Sciences de l'Institut de France. Sciences mathématiques et physiques* 4: 129–461.

Duméril, André Marie Constant and Bibron, Gabriel. 1834–54. *Erpétologie générale, ou histoire naturelle complète des reptiles*. 9 vols. in 10 and atlas. Paris.

Dunal, Michel Félix. 1829. *Considérations sur la nature et les rapports de quelques uns des organes de la fleur*. Montpellier.

Duncan, Peter Martin. 1866–72. *A monograph of the British fossil corals. Second series. Being a supplement to the 'Monograph of the British fossil corals,' by MM. Milne-Edwards and Jules Haime*. London.

Dupree, Anderson Hunter. 1951. Some letters from Charles Darwin to Jeffries Wyman. *Isis* 42: 104–10.

——. 1959. *Asa Gray, 1810–1888*. Cambridge, Mass.: Belknap Press of Harvard University.

Durocher, Joseph Marie Elisabeth. 1846. Observations sur les phénomènes d'érosions et les dépôts de transport de la Scandinavie. *Comptes Rendus Hebdomadaires des Séances de l'Académie des Sciences* 23: 206–10.

East-India register and army list: *The East-India register and army list*. Compiled from the official returns, by permission of the Honourable the East-India Company. 16 vols. London. 1845–60.

EB: *Encyclopædia Britannica*. 11th ed. 29 vols. Cambridge: Cambridge University Press. 1910–11.

Elliot, Clark A. 1979. *Biographical dictionary of American science: the seventeenth through the nineteenth centuries*. Westport, Conn. and London: Greenwood Press.

Ellis, Ieuan. 1980. *Seven against Christ. A study of 'Essays and reviews'*. Leiden: E. J. Brill.

Emma Darwin: *Emma Darwin: a century of family letters 1792–1896*. Edited by Henrietta Litchfield. Revised edition. 2 vols. London: John Murray. 1915.

ENZ: *An encyclopaedia of New Zealand*. Edited by A. H. McLintock. 3 vols. Wellington, New Zealand: R. E. Owen. 1966.

Erasmus Darwin: *Erasmus Darwin*. By Ernst Krause. Translated from the German by W. S. Dallas, with a preliminary notice by Charles Darwin. London. 1879.

Erichson, Wilhelm Ferdinand. 1842. Ueber Volucella bombylans und plumata. *Entomologische Zeitung* 3: 113–15.

Essays and reviews. 1860. London.

Eve, A. S. and Creasey, C. H. 1945. *Life and work of John Tyndall*. London: Macmillan and Co.

Falconer, Hugh. 1857–8. On the species of mastodon and elephant occurring in the fossil state in Great Britain. Pt 1. Mastodon. Pt 2. Elephas. *Quarterly Journal of the Geological Society of London* 13 (1857): 307–60; 14 (1858): 81–4.

——. 1860. On the ossiferous caves of the peninsula of Gower, in Glamorganshire, South Wales. *Quarterly Journal of the Geological Society of London* 16: 487–91.

Fallén, Carl Frederik. 1814–27. *Diptera Sveciæ*. 2 vols. in 1. Lund.

Farley, John. 1982. *Gametes & spores: ideas about sexual reproduction, 1750–1914*. Baltimore and London: The Johns Hopkins University Press.

Fawcett, Henry. 1860. A popular exposition of Mr. Darwin on the Origin of species. *Macmillan's Magazine* 3 (1860–1): 81–92.

——. 1861. On the method of Mr. Darwin in his treatise on the origin of species. *Report of the 31st meeting of the British Association for the Advancement of Science held at Manchester*, Transactions of the sections, pp. 141–3.

Ferguson, George. 1854. *Ferguson's illustrated series of rare and prize poultry, including comprehensive essays upon all classes of domestic fowl*. London.

Fermond, Charles. 1840. Du rôle que jouent les périanthes dans l'acte de la fécondation. *Journal de Pharmacie et des Sciences Accessoires* 26: 751–7.

Fermond, Charles. 1841. Über die Rolle, welche die Blüthendecken bei dem Acte der Befruchtung spielen. *Flora oder allgemeine botanische Zeitung* 24 (1841): 204–8.

——. 1859. Faits pour servir à l'histoire générale de la fécondation chez les végétaux. Vol. 3 of *Recueil des travaux de la Société d'émulation pour les sciences pharmaceutiques*.

Ferris, Norman B. 1977. *The* Trent *affair; a diplomatic crisis*. Knoxville: The University of Tennessee Press.

Fish: Pt IV of *The zoology of the voyage of H.M.S. Beagle*. By Leonard Jenyns. Edited and superintended by Charles Darwin. London. 1840–2.

Flett, John Smith. 1937. *The first hundred years of the Geological Survey of Great Britain*. London: His Majesty's Stationery Office.

Forbes, Edward. 1846. On the connexion between the distribution of the existing fauna and flora of the British Isles and the geological changes which have affected their area, especially during the epoch of the Northern Drift. *Memoirs of the Geological Survey of Great Britain, and of the Museum of Practical Geology in London* 1: 336–432.

Forbes, James David. 1843. *Travels through the Alps of Savoy and other parts of the Pennine chain with observations on the phenomena of glaciers*. Edinburgh.

——. 1853. *Norway and its glaciers: visited in 1851*. Edinburgh.

Forms of flowers: *The different forms of flowers on plants of the same species*. By Charles Darwin. London. 1877.

Freeman, Richard Broke. 1977. *The works of Charles Darwin; an annotated bibliographical handlist*. 2d ed. Folkestone, Kent: William Dawson and Sons. Hamden, Conn.: Archon Books, Shoe String Press.

——. 1978. *Charles Darwin; a companion*. Folkestone, Kent: William Dawson and Sons. Hamden, Conn.: Archon Books, Shoe String Press.

Freke, Henry. 1860. Observations upon Mr. Darwin's recently published work—"On the origin of species by means of natural selection". Dublin: privately printed.

——. 1861. *On the origin of species by means of organic affinity*. London, Dublin, and Edinburgh.

Gage, Andrew Thomas. 1938. *History of the Linnean Society of London*. London: Linnean Society of London.

Gärtner, Karl Friedrich von. 1844. *Beiträge zur Kenntniss der Befruchtung der vollkommeneren Gewächse*. Pt 1: *Versuche und Beobachtungen über die Befruchtungsorgane der vollkommeneren Gewächse und über die natürliche und künstliche Befruchtung durch den eigenen Pollen*. Stuttgart.

——. 1849. *Versuche und Beobachtungen über die Bastarderzeugung im Pflanzenreich. Mit Hinweisung auf die ähnlichen Erscheinungen im Thierreiche*. Stuttgart.

Geikie, Archibald. 1861. On a rise of the coast of the Firth of Forth within the historical period. *Edinburgh New Philosophical Journal* n.s. 14: 102–12.

Gesner, Konrad. 1558. *De piscium & aquatilium animantium natura*. Vol. 4 of *Historiae animalium*. 1551–8. Tiguri.

Ghiselin, Michael T. 1969. *The triumph of the Darwinian method*. Berkeley and Los Angeles: University of California Press.

——. 1984. Foreword. In *The various contrivances by which orchids are fertilised by insects*, by Charles Darwin. Facsimile reprint of the 2d ed. (1877). Chicago: University of Chicago Press.

Gilbert, Pamela. 1977. *A compendium of the biographical literature on deceased entomologists*. London: British Museum (Natural History).

[Glennie, John David.] 1855. "The whole duty of man:" popular error. *Notes and Queries* 11: 489–90.

Goodwin, Charles Wycliffe. 1860. On the Mosaic cosmogony. In *Essays and reviews*. London.

Gordon, George. 1854. A list of the Mollusca hitherto found in the province of Moray and in the Moray Firth. *Zoologist* 12: 4300–18, 4421–35, 4453–62.

Gosse, Philip Henry. 1840. *The Canadian naturalist. A series of conversations on the natural history of lower Canada*. London.

——. 1851. *A naturalist's sojourn in Jamaica*. Assisted by Richard Hill. London.

——. 1853. *A naturalist's ramble on the Devonshire coast*. London.

Gould, John. 1861. *An introduction to the Trochilidæ or family of humming-birds*. London: privately printed.

Grande encyclopédie: *La grande encyclopédie inventaire raisonné des sciences, des lettres et des arts*. Edited by F. Camille Dreyfuss and others. 31 vols. Paris. 1886–1902.

Grant, Robert Edmond. 1826. On the structure and nature of the Spongilla friabilis. *Edinburgh Philosophical Journal* 14: 270–84.

——. 1861. *Tabular view of the primary divisions of the animal kingdom, intended to serve as an outline of an elementary course of recent zoology (cainozoology), or the natural history of existing animals*. London.

Gratiolet, Louis Pierre. 1854. *Mémoire sur les plis cérébraux de l'homme et des primates*. Paris.

Gray, Asa. 1848–9. *Genera floræ Americæ Boreali-Orientalis illustrata; the genera of the plants of the United States*. 2 vols. Vol. 1: Boston, New York, and London; vol. 2: New York.

——. 1856. *Manual of the botany of the northern United States . . . including Virginia, Kentucky, and all east of the Mississippi: arranged according to the natural system*. 2d ed. New York.

——. 1858. Note on the coiling of tendrils in plants. *Proceedings of the American Academy of Arts and Sciences* 4: 98–9.

——. 1858–9. Diagnostic characters of new species of phænogamous plants, collected in Japan by Charles Wright, botanist of the U.S. North Pacific Exploring Expedition . . . With observations upon the relations of the Japanese flora to that of North America, and of other parts of the northern temperate zone. With an appendix by N. J. Andersson. [Read 14 December 1858 and 11 January 1859.] *Memoirs of the American Academy of Arts and Sciences* n.s. 6 (1857–9): 377–452.

[Gray, Asa.] 1860a. Darwin on the origin of species. *Atlantic Monthly* 6: 109–16, 229–39; Darwin and his reviewers. 6: 406–25.

——. 1860b. Review of Darwin's theory on the origin of species by means of natural selection. *American Journal of Science and Arts* 2d ser. 29: 153–84.

——. 1860c. Discussion between two readers of Darwin's treatise on the origin of species, upon its natural theology. *American Journal of Science and Arts* 2d ser. 30: 226–39.

——. 1861a. *Natural selection not inconsistent with natural theology. A free examination of Darwin's treatise on the origin of species, and of its American reviewers*. Boston and London.

——. 1861b. Review of *Life on earth*, by John Phillips. *American Journal of Science and Arts* 2d ser. 31: 444–9.

——. 1862. Review of *Orchids*, by Charles Darwin. *American Journal of Science and Arts* 2d ser. 34: 138–44, 420–29.

Gray, George Robert. 1849. Notice of two examples of the genus Gallus. *Proceedings of the Zoological Society of London* 17: 62–3.

Green, J. Reynolds. 1914. *A history of botany in the United Kingdom from the earliest times to the end of the 19th century*. London and Toronto: J. M. Dent & Sons. New York: E. P. Dutton & Co.

Greene, Joseph Reay. 1859–61. *A manual of the animal kingdom*. Pt 1: Protozoa. Pt 2: Cœlenterata. London.

GSE: *Great Soviet encyclopedia*. Translation of the 3d ed. of *Bol'shaia Sovetskaia entsiklopediia*. (Moscow. 1970–9.) 31 vols. New York: Macmillan. London: Collier Macmillan. 1973–82.

Guépin, Jean Baptiste Pierre. 1853. Note sur l'hybridation des Orchidées. *Annales de la Société Linnéenne d'Angers* 1: 52–82.

Günther, Albert Charles Lewis Gotthilf. 1853. *Die Fische des Neckars, untersucht und beschreiben von A. Günther*. Stuttgart.

——. 1855. Beiträge zur Kenntniss unserer Süsswasserfische. *Archiv für Naturgeschichte* 21: 197–212.

——. 1859–70. *Catalogue of acanthopterygian fishes in the collection of the British Museum*. 8 vols. London.

Gunther, Albert E. 1975. *A century of zoology at the British Museum through the lives of two keepers, 1815–1914*. Folkestone, Kent: William Dawson and Sons.

Haast, John Francis Julius von. 1862. *Notes on the geology of the province of Canterbury, New Zealand*. [Reprinted from the *New Zealand Government Gazette*, vol. 9.] Christchurch, New Zealand.

Hall, James. 1815. On the revolutions of the earth's surface. *Transactions of the Royal Society of Edinburgh* 7: 139–211.

Hall, James. 1860. Descriptions of new species of fossils from the Silurian rocks of Nova Scotia. *Canadian Naturalist* 5: 144–59.

Hanson, Bert. 1970. The early history of the glacial theory in British geology. *Journal of Glaciology* 9: 135–41.

Harcourt, Edward William Vernon. 1855. Notes on the ornithology of Madeira. *Annals and Magazine of Natural History* 2d ser. 15: 430–8.

Hartmann, Georg Leonhard. 1827. *Helvetische Ichthyologie, oder ausführliche Naturgeschichte in der Schweiz sich vorfindenden Fische*. Zürich.

Harvey, Joy. 1987. 'Strangers to each other': male and female relationships in the life and work of Clémence Royer. In *Uneasy careers and intimate lives: women in science, 1789–1979*, edited by Pnina G. Abir-Am and Dorinda Outram. New Brunswick and London: Rutgers University Press.

Harvey, William Henry. 1861. On "omne vivum ab ovo", or, the natural evolution of organic species considered. *Dublin Hospital Gazette*, 15 May 1861.

Heer, Oswald. 1857. Ueber die fossilen Pflanzen von St. Jorge in Madeira. [Read 5 November 1855.] *Neue Denkschriften der allgemeinen schweizerischen Gesellschaft für die gesammten Naturwissenschaften* 15: 1–40.

——. 1860. *Untersuchungen über das Klima und die Vegetationsverhältnisse des Tertiärlandes*. Winterthur.

——. 1862. On the fossil flora of Bovey Tracey. *Philosophical Transactions of the Royal Society of London* 152: 1039–86.

Henslow, John Stevens. 1821a. Supplementary observations to Dr. Berger's account of the Isle of Man. [Read 7 April 1820.] *Transactions of the Geological Society* 5: 482–505

——. 1821b. Geological description of Anglesea. [Read 26 November 1821.] *Transactions of the Cambridge Philosophical Society* 1 (1822): 359–452.

Herbert, William. 1837. *Amaryllidaceae; preceded by an attempt to arrange the monocotyledonous orders, and followed by a treatise on cross-bred vegetables, and supplement*. London.

——. 1847. On hybridization amongst vegetables. *Journal of the Horticultural Society of London* 2: 1–28, 81–107.

Herschel, John Frederick William. 1831. *A preliminary discourse on the study of natural philosophy*. In Dionysius Lardner's *Cabinet cyclopædia*. London.

——. 1861. *Physical geography. From the Encyclopædia Britannica*. Edinburgh.

Hewitt, Ethel M. 1912. Industries. Introduction to vol. 5 of *The Victoria history of the counties of England: Hampshire and the Isle of Wight*, edited by William Page. Westminster: Archibald Constable and Company.

Higgins, Henry Hugh. 1860. On Darwin's theory of the "Origin of species". *Proceedings of the Literary and Philosophical Society of Liverpool* 15 (1860–1): 42–9.

Hindmarsh, Luke. 1839. On the wild cattle of Chillingham Park. *Annals of Natural History* 2: 274–84.

Historical register of the University of Cambridge: *The historical register of the University of Cambridge . . . to the year 1910*. Edited by J. R. Tanner. Cambridge: Cambridge University Press. 1917.

Hodge, M. J. S. 1985. Darwin as a lifelong generation theorist. In *The Darwinian heritage*, edited by David Kohn. Princeton: Princeton University Press in association with Nova Pacifica.

Hoeven, Jan van der. 1856–8. *Handbook of zoology*. Translated from the second Dutch edition by William Clark. 2 vols. London.

Hofacker, J. D. 1828. *Über die Eigenschaften welche sich bei Menschen und Thieren von den Eltern auf die Nachkommen vererben, mit besonderer Rücksicht auf die Pferdezucht, mit Beiträgen von F. Notter*. Tübingen.

Hooker, Joseph Dalton. 1849. *The rhododendrons of Sikkim-Himalaya; being an account ... of the rhododendrons recently discovered in the mountains of eastern Himalaya ... by Joseph Dalton Hooker*. Edited by William Jackson Hooker. London.

——. 1854a. *Himalayan journals; or, notes of a naturalist in Bengal, the Sikkim and Nepal Himalayas, the Khasia Mountains, &c.* 2 vols. London.

——. 1854b. On the functions and structure of the rostellum of *Listera ovata*. *Philosophical Transactions of the Royal Society of London* 144: 259–63.

——. 1855. On some small seed-vessels (*Folliculites minutulus*, Bronn) from the Bovey Tracey coal. *Journal of the Geological Society of London* 6: 566–70.

——. 1859. *On the flora of Australia, its origin, affinities, and distribution; being an introductory essay to the Flora of Tasmania*. [Reprinted from the *Botany of the Antarctic expedition*, pt 3, *Flora Tasmania*, vol. 1.] London.

——. 1861. Outlines of the distribution of Arctic plants. [Read 21 June 1860.] *Transactions of the Linnean Society of London* 23 (1862): 251–348.

——. 1862. On the vegetation of Clarence peak, Fernando Po; with descriptions of the plants collected by Mr. Gustav Mann on the higher parts of that mountain. [Read 7 March 1861.] *Journal of the Proceedings of the Linnean Society of London (Botany)* 6: 1–23.

——. 1863. On *Welwitschia*, a new genus of *Gnetaceæ*. *Transactions of the Linnean Society of London* 24 (1863–4): 1–48.

Hooker, Joseph Dalton and Thomson, Thomas. 1855. *Flora Indica: being a systematic account of the plants of British India, together with observations on the structure and affinities of their natural orders and genera*. London.

Hopkins, William. 1842. On the elevation and denudation of the district of the lakes of Cumberland and Westmoreland. *Proceedings of the Geological Society of London* 3 (1838–42): 757–66.

——. 1843. On the motion of glaciers. [Read 1 May and 11 December 1843.] *Transactions of the Cambridge Philosophical Society* 8 (1849): 50–74, 159–69.

——. 1844. On the transport of erratic blocks. [Read 29 April 1844.] *Transactions of the Cambridge Philosophical Society* 8 (1849): 220–40.

——. 1852. Anniversary address of the president. *Quarterly Journal of the Geological Society of London* 8: xxiv–lxxx.

——. 1860. Physical theories and the phenomena of life. *Frazer's Magazine* 61: 739–52; 62: 74–90. [Reprinted in Hull 1973, pp. 229–72.]

Hörbye, Jens Carl. 1857. *Observations sur les phénomènes d'érosion en Norvège ... Programme de l'Université pour le 1er semestre, 1857*. Christiania.

Horner, Leonard. 1861. Anniversary address of the president. *Quarterly Journal of the Geological Society of London* 17: xxxi–lxxii.

Hull, David Lee. 1973. *Darwin and his critics: the reception of Darwin's theory of evolution by the scientific community*. Cambridge, Mass.: Harvard University Press.

Humboldt, Alexander von. 1814–29. *Personal narrative of travels to the equinoctial regions of the New Continent, during the years 1799–1804, by Alexander de Humboldt and Aimé Bonpland . . . translated into English by Helen Maria Williams*. 7 vols. London.

Hutchinson, Horace Gordon. 1914. *Life of Sir John Lubbock, Lord Avebury*. 2 vols. London: Macmillan and Co.

Hutchinson, William Nelson. 1850. *Dog breaking. The most expeditious, certain, and easy method; whether great excellence or only mediocrity is required*. 2d ed. London.

Hutton, Frederick Wollaston. 1861. Some remarks on Mr. Darwin's theory. *Geologist* 4: 132–6, 183–8.

——. 1887. *Darwinism. A lecture*. Christchurch.

Huxham, John. 1752. *Observationes de aëre et morbis epidemicis*. 2 vols. London and Plymouth.

Huxham, John Corham, trans. 1759–67. *Observations on the air and epidemic diseases*. 2 vols. London and Plymouth.

Huxley, Leonard, ed. 1900. *Life and letters of Thomas Henry Huxley*. 2 vols. London: Macmillan and Co.

——, ed. 1918. *Life and letters of Sir Joseph Dalton Hooker*. 2 vols. London: John Murray.

[Huxley, Thomas Henry.] 1860. Darwin on the origin of species. *Westminster Review* n.s. 17: 541–70.

——. 1861a. On the zoological relations of man with the lower animals. *Natural History Review* n.s. 1: 67–84.

——. 1861b. On the brain of *Ateles paniscus*. *Proceedings of the Zoological Society of London*, pp. 247–60.

[——.] 1861c. Review of vol. 3 of *Contributions to the natural history of the United States of America*, by Louis Agassiz. *Natural History Review* n.s. 1: 433–44.

IBN: *Index bio-bibliographicus notorum hominum*. Pt C: *Corpus alphabeticum sectio generalis*. Edited by Jean-Pierre Lobies and Otto and Wolfram Zeller. 60 vols. (A–Durau, Pierre). Osnabrück: Biblio Verlag. 1974–92.

Insectivorous plants: *Insectivorous plants*. By Charles Darwin. London. 1875.

Irwin, Mary Leslie. 1926. *Anthony Trollope: a bibliography*. New York: Burt Franklin.

Jackson, B. Daydon. 1906. *George Bentham*. London: J. M. Dent & Co. New York: E. P. Dutton & Co.

Jacquin, Nikolaus Josef. 1794. *Oxalis. Monographia, iconibus illustrata*. Vienna.

Jamieson, Thomas Francis. 1859. On an outlier of *Lias* in Aberdeenshire. *Quarterly Journal of the Geological Society of London* 15: 131–3.

——. 1860a. On the drift and rolled gravel of the north of Scotland. *Quarterly Journal of the Geological Society of London* 16: 347–71.

——. 1860b. On the occurrence of Crag strata beneath the Boulder Clay in Aberdeenshire. *Quarterly Journal of the Geological Society of London* 16: 371–3.

——. 1862. On the *ice-worn rocks* of Scotland. *Quarterly Journal of the Geological Society of London* 18: 164–84.

Jamieson, Thomas Francis. 1863. On the parallel roads of Glen Roy, and their place in the history of the Glacial period. *Quarterly Journal of the Geological Society of London* 19: 235–59.

Jenyns, Leonard, ed. 1862. *Memoir of the Rev. John Stevens Henslow, late rector of Hitcham, and professor of botany in the University of Cambridge*. London.

Jervis, James. 1945. *W. B. Clarke: "The father of Australian geology"*. Sydney: Royal Australian Historical Society.

Jesperson, P. Helveg. 1948. Charles Darwin and Dr. Grant. *Lychnos* (1948–9): 159–67.

Johnston, Ivan M. 1943. Publication dates for the botanical parts of the Pacific Railway reports. *Journal of the Arnold Arboretum, Harvard University* 24: 237–42.

Jones, William, trans. 1796. *Institutes of Hindu law: or, the ordinances of Menu, according to the gloss of Cullúca. Comprising the Indian system of duties, religious and civil*. Calcutta and London.

Journal and remarks: Vol. 3 of *Narrative of the surveying voyages of His Majesty's Ships Adventure and Beagle*. By Charles Darwin. Edited by Robert FitzRoy. London. 1839. [Separately published as *Journal of researches*.]

Journal of researches: *Journal of researches into the geology and natural history of the various countries visited by H.M.S. Beagle, under the command of Captain FitzRoy, R.N., from 1832 to 1836*. By Charles Darwin. London. 1839.

Journal of researches 2d ed.: *Journal of researches into the natural history and geology of the countries visited during the voyage of H.M.S. Beagle round the world, under the command of Captain FitzRoy, R.N.* Second edition, corrected, with additions. By Charles Darwin. London. 1845.

Journal of researches (1860): *Journal of researches into the natural history and geology of the countries visited during the voyage of H.M.S. Beagle around the world, under the command of Capt. FitzRoy R.N.* By Charles Darwin. Reprint edition. London. 1860.

Jussieu, Adrien Henri Laurent de. 1843. *Monographie de la famille malpighiacées*. Paris.

Kaufmann, Georg, ed. 1911. *Festschrift zur Feier des hundertjährigen Bestehens der Universität Breslau*. 2 vols. Breslau: Ferdinand Hirt.

Kerr, Robert. 1811–24. *A general history and collection of voyages and travels . . . forming a complete history of the origin and progress of navigation, discovery, and commerce, by sea and land, from the earliest ages to the present time*. 18 vols. Edinburgh.

Keynes, Richard Darwin, ed. 1979. *The 'Beagle' record: selections from the original pictorial records and written accounts of the voyage of H.M.S. 'Beagle'*. Cambridge: Cambridge University Press.

King-Hele, Desmond. 1963. *Erasmus Darwin*. London: Macmillan & Co.

——, ed. 1981. *The letters of Erasmus Darwin*. Cambridge: Cambridge University Press.

Kirby, William and Spence, William. 1815–26. *An introduction to entomology*. 4 vols. London.

Kjerulf, Theodor. 1860. Ueber das Friktions-Phänomen. *Zeitschrift der Deutschen geologischen Gesellschaft* 12: 389–408.

Koch, Wilhelm Daniel Joseph. 1843–4. *Synopsis florae Germanicae et Helveticae*. 2d ed. 2 pts. Frankfurt am Main.

Kohn, David. 1985. Darwin's principle of divergence as internal dialogue. In *The Darwinian heritage*, edited by David Kohn. Princeton: Princeton University Press in association with Nova Pacifica.

Köstlin, Otto. 1860. *Ueber die Unveränderlichkeit der organischen Species*. Stuttgart.

Kurr, Johann Gottlob von. 1833. *Untersuchungen über die Bedeutung der Nektarien in den Blumen*. Stuttgart.

Lamont, James. 1860. Notes about Spitzbergen in 1859. [Read 28 March 1860.] *Quarterly Journal of the Geological Society of London* 16: 428–44.

——. 1861. *Season with the sea-horses; or, sporting adventures in the northern sea*. London.

Lauder, Thomas Dick. 1823. On the parallel roads of Lochaber. [Read 2 March 1818.] *Transactions of the Royal Society of Edinburgh* 9: 1–64.

Lecoq, Henri. 1854–8. *Etudes sur la géographie botanique de l'Europe, et en particulier sur la végétation du plateau central de la France*. 9 vols. Paris.

Le Couteur, John. 1836. *On the varieties, properties and classification of wheat*. London.

Leidy, Joseph. 1869. The extinct mammalian fauna of Dakota and Nebraska, including an account of some allied forms from other localities, together with a synopsis of the mammalian remains of North America. *Journal of the Academy of Natural Sciences of Philadelphia* 2d ser. 7: 23–472.

Lestiboudois, Gaspard Thémistocle. 1848. Phyllotaxie anatomique ou recherches sur les causes organiques. *Annales des Sciences Naturelles (Botanique)* 3d ser. 2: 15–105.

Leuckart, Rudolf. 1852. *Vesicula Prostatica*. In vol. 4 of *The cyclopædia of anatomy and physiology*, edited by Robert Bentley Todd. 5 vols. 1835–59. London.

Lewes, George Henry. 1860. Studies in animal life. *Cornhill Magazine* 1: 61–74, 198–207, 283–95, 438–47, 598–607, 682–90.

Leydig, Franz von. 1850. Zur Anatomie der männlichen Geschlechtsorgane und Analdrüsen der Säugethiere. *Zeitschrift für wissenschaftliche Zoologie* 2: 1–57.

Lindley, John. 1830. *An introduction to the natural system of botany: or, a systematic view of the organisation, natural affinities, and geographical distribution, of the whole vegetable kingdom*. London.

——. 1846. *The vegetable kingdom*. London.

——. 1848. *An introduction to botany*. 4th ed. 2 vols. London.

——. 1853. *The vegetable kingdom*. 3d ed. London.

Link, Heinrich Friedrich. 1849. Bemerkungen über den Bau der Orchideen, besonders der Vandeen. *Botanische Zeitung* 7: 745–50.

List of the Geological Society of London: *List of the [members of the] Geological Society of London*. London. 1864.

List of the Linnean Society of London: *Lists of the [members of the] Linnean Society of London 1805–1915*. 8 vols. London.

List of officers of the army: *A list of the officers of the army and of the corps of Royal Marines, on full, retired, and half-pay; with an index*. London. 1779–1869.

Living Cirripedia (1851): *A monograph of the sub-class Cirripedia, with figures of all the species. The Lepadidæ; or, pedunculated cirripedes*. By Charles Darwin. London. 1851.

Living Cirripedia (1854): *A monograph of the sub-class Cirripedia, with figures of all the species. The Balanidæ (or sessile cirripedes); the Verrucidæ, etc*. By Charles Darwin. London. 1854.

LL: *The life and letters of Charles Darwin, including an autobiographical chapter*. Edited by Francis Darwin. 3 vols. London. 1887.

Loiseleur Deslongchamps, Jean Louis Auguste. 1842–3. *Considérations sur les céréales et principalement sur les froments*. 2 pts. Paris.

London and provincial medical directory: *The London and provincial medical directory, and general medical register*. London. 1848–60.

London & provincial medical directory: *The London & provincial medical directory, inclusive of the medical directory for Scotland, and the medical directory for Ireland, and general medical register*. London. 1861–9.

Lubbock, John. 1857. An account of the two methods of reproduction in *Daphnia*, and of the structure of the ephippium. *Philosophical Transactions of the Royal Society of London* 147: 79–100.

——. 1859. On the ova and pseudova of insects. *Philosophical Transactions of the Royal Society of London* 149: 341–69.

——. 1860. On some *Entomostraca* collected by Captain Toynbee. [Read 7 June 1860.] *Transactions of the Linnean Society of London* 23 (1860–2): 173–91.

——. 1861a. On Sphærularia bombi. *Natural History Review* n.s. 1: 44–57.

——. 1861b. Notes on the generative organs, and on the formation of the egg in the Annulosa. [Read 21 February 1861.] *Philosophical Transactions of the Royal Society of London* 151: 595–626.

——. 1861c. The Kjökkenmöddings: recent geologico-archæological researches in Denmark. *Natural History Review* n.s. 1: 489–504.

——. 1862. Notes on the Thysanura. [Read 6 February and 19 June 1862.] *Transactions of the Linnean Society of London* 23 (1860–2): 429–48, 589–601.

Lukis, F. C. 1846. Cromlech du Tus. Guernsey. *Journal of the British Archaeological Association* 1: 25–9.

Lurie, Edward. 1960. *Louis Agassiz: a life in science*. Chicago: University of Chicago Press.

Lyell, Charles. 1830–3. *Principles of geology, being an attempt to explain the former changes of the earth's surface, by reference to causes now in operation*. 3 vols. London.

——. 1840. On the geological evidence of the former existence of glaciers in Forfarshire. *Proceedings of the Geological Society of London* 3 (1838–42): 337–45.

——. 1841. *Elements of geology*. 2d ed. 2 vols. London.

——. 1847. *Principles of geology; or, the modern changes of the earth and its inhabitants considered as illustrative of geology*. 7th ed. London.

——. 1853. *Principles of geology; or, the modern changes of the earth and its inhabitants considered as illustrative of geology*. 9th ed. London.

——. 1855a. *A manual of elementary geology: or, the ancient changes of the earth and its inhabitants as illustrated by geological monuments*. 5th ed. London.

Lyell, Charles. 1855b. On certain trains of erratic blocks on the western borders of Massachusetts, United States. [Read 27 April 1855.] *Notices of the Proceedings of the Meetings of the Members of the Royal Institution* 2 (1854–8): 86–97.

——. 1863. *The geological evidences of the antiquity of man; with remarks on theories of the origin of species by variation*. London.

——. 1865. *Elements of geology; or, the ancient changes of the earth and its inhabitants as illustrated by geological monuments*. 6th ed. London.

——. 1867–8. *Principles of geology or the modern changes of the earth and its inhabitants considered as illustrative of geology*. 10th ed. 2 vols. London.

Lyell, Katharine Murray, ed. 1881. *Life, letters and journals of Sir Charles Lyell, Bart*. 2 vols. London.

——, ed. 1890. *Memoir of Leonard Horner, consisting of letters to his family and from some of his friends*. 2 vols. London: privately printed.

McAndrew, Robert. 1853. On the geographical distribution of testaceous Mollusca in the north-east Atlantic and neighbouring seas. [Read 31 October 1853.] *Proceedings of the Literary and Philosophical Society of Liverpool* 8 (1853–4): 8–56.

Macaulay, Thomas Babington. 1848–61. *The history of England, from the accession of James II*. 5 vols. London.

MacCulloch, John. 1817. On the parallel roads of Glen Roy. *Transactions of the Geological Society* 4: 314–92.

M'Donnell, Robert. 1861. On an organ in the skate which appears to be the homologue of the electrical organ of the torpedo. *Natural History Review* n.s. 1: 57–60.

Macgillivray, William. 1837–52. *History of British birds, indigenous and migratory*. 5 vols. London.

——. 1855. *The natural history of Dee Side and Braemar*. Edited by Edwin Lankester. London: privately printed.

McPherson, James M. 1988. *Battle cry of freedom. The Civil War era*. New York and Oxford: Oxford University Press.

Macquart, Justin Pierre Marie. 1834–5. *Histoire naturelle des insectes. Diptères*. 2 vols. Paris.

Marchant, James, ed. 1916. *Alfred Russel Wallace; letters and reminiscences*. 2 vols. London: Cassell and Co.

Marginalia: *Charles Darwin's marginalia*. Edited by Mario A. di Gregorio, with the assistance of N. W. Gill. Vol. I. New York: Garland Publishing. 1990.

Martin, William Charles Linnaeus. 1855. *The poultry yard: comprising the management of fowls, for use and exhibition*. New edition revised by Elizabeth Watts. London and New York.

Matthew, Patrick. 1831. *On naval timber and arboriculture*. London and Edinburgh.

Matthews, Philip W. 1921. *The bankers' clearing house: what it is and what it does*. London: Sir Isaac Pitman & Sons.

Maw, George. 1861a. Review of *Origin of species* and other works. *Zoologist* 19: 7577–611.

——. 1861b. The pavements of Uriconium. *Journal of the British Archaeological Association* 17: 100–10.

Max Müller, Friedrich. 1856. Comparative mythology. In vol. 2 of *Oxford essays, contributed by members of the University*. 4 vols. 1855–8. London.

Medical directory for Ireland: *The medical directory and general medical register for Ireland*. London. 1852–60.

Meding, Karl Heinrich. 1861. *Goethe als Naturforscher in Beziehung zur Gegenwart*. Dresden.

Ménière, Prosper. 1854. Note sur la fécondation des orchidées. *Bulletin de la Société Botanique de France* 1: 367–71.

——. 1855. Note sur la sécrétion d'un liquide abondant par l'organe glanduleux des *Coryanthes*. *Bulletin de la Société Botanique de France* 2: 351–5.

Meteyard, Eliza. 1865–6. *The life of Josiah Wedgwood*. 2 vols. London.

Mill, John Stuart. 1862. *A system of logic, ratiocinative and inductive, being a connected view of the principles of evidence, and the methods of scientific investigation*. 5th ed. 2 vols. London.

Miller, Hugh. 1858. *The Old Red Sandstone; or, new walks in an old field. To which is appended a series of geological papers, read before the Royal Physical Society of Edinburgh*. 7th ed. Edinburgh.

Milne, David. 1847a. *On the parallel roads of Lochaber; with remarks on the change of relative levels of sea and land in Scotland, and on the detrital deposits in that country*. Edinburgh: privately printed.

——. 1847b. On the parallel roads of Lochaber, with remarks on the change of relative levels of sea and land in Scotland. *Edinburgh New Philosophical Journal* 43: 339–64.

——. 1849. On the parallel roads of Lochaber, with remarks on the change of relative levels of sea and land in Scotland, and on the detrital deposits in that country. [Read 1 March and 5 April 1847.] *Transactions of the Royal Society of Edinburgh* 16: 395–418.

Mineka, Francis E. and Lindley, Dwight N., eds. 1972. *The later letters of John Stuart Mill, 1849–1873*. Vol. 15 of *The collected works of John Stuart Mill*, edited by F. E. L. Priestley and J. M. Robson. 33 vols. Toronto: University of Toronto Press. London: Routledge & Kegan Paul. 1963–91.

ML: *More letters of Charles Darwin: a record of his work in a series of hitherto unpublished letters*. Edited by Francis Darwin and Albert Charles Seward. 2 vols. London: John Murray. 1903.

Modern English biography: *Modern English biography containing many thousand concise memoirs of persons who have died since the year 1850*. By Frederic Boase. 3 vols. and supplement (3 vols.). Truro: printed for the author. 1892–1921.

Mohl, Hugo von, trans. 1832. Nachträgliche Beobachtungen über die Sexualorgane und die Befruchtungsweise der Orchideen und Asclepiadeen von Robert

Brown. *Flora oder allgemeine botanische Zeitung* 15 (1832): 673–6.

Montgomery, William M. 1974. Germany. In *The comparative reception of Darwinism*, edited by T. F. Glick. Austin: University of Texas Press.

Moon, Harold Philip. 1976. *Henry Walter Bates ... explorer, scientist and Darwinian*. Leicester: Leicestershire Museums, Art Galleries & Records Service.

Moore, James R. 1985. Darwin of Down: the evolutionist as squarson-naturalist. In *The Darwinian heritage*, edited by David Kohn. Princeton: Princeton University Press in association with Nova Pacifica.

——, ed. 1988. *Sources*. Vol. 3 of *Religion in Victorian Britain*. Manchester and New York: Manchester University Press in association with the Open University.

Moquin-Tandon, Horace Bénédict Alfred. 1841. *Eléments de tératologie végétale, ou histoire abrégée des anomalies de l'organisation dans les végétaux*. Paris.

Morlot, Charles Adolphe. 1859. Etudes géologico-archéologiques en Danemark et en Suisse. [Read January 1859.] *Société Vaudoise des Sciences Naturelles* 6 (1860): 259–329.

Morrell, Jack and Thackray, Arnold. 1981. *Gentlemen of science. The early years of the British Association for the Advancement of Science*. Oxford: Clarendon Press.

Morren, Charles François Antoine. 1836. On the agency of insects in causing sterility in flowers by the removal of the masculine organs, observed amongst the Asclepiadeæ. *Transactions of the Entomological Society of London* 1: Proceedings, pp. xliv–xlv.

——. 1839. On the production of vanilla in Europe. *Annals and Magazine of Natural History* 3: 1–9.

Morton, Leslie T. 1970. *A medical bibliography (Garrison and Morton): an annotated check-list of texts illustrating the history of medicine*. 3d ed. London: André Deutsch.

Morton, Samuel George. 1847. Hybridity in animals, considered in reference to the question of the unity of the human species. *American Journal of Science and Arts* 2d ser. 3: 39–50, 203–12.

——. 1850a. Letter to the Rev. John Bachman, D.D., on the question of hybridity in animals, considered in reference to the unity of the human species. *Charleston Medical Journal* 5: 328–44.

——. 1850b. Additional observations on hybridity in animals, and on some collateral subjects; being a reply to the objections of the Rev. John Bachman, D.D. *Charleston Medical Journal* 5: 755–805.

Murchison, Charles, ed. 1868. *Palæontological memoirs and notes of the late Hugh Falconer ... With a biographical sketch of the author*. 2 vols. London.

Murchison, Roderick Impey. 1843. Anniversary address of the president. *Proceedings of the Geological Society of London* 4 (1843–5): 65–151.

——. 1851. On the scratched and polished rocks of Scotland. *Report of the 21st meeting of the British Association for the Advancement of Science held at Ipswich*, Transactions of the sections, pp. 66–7.

Murchison, Roderick Impey, Verneuil, Edouard, and Keyserling, Alexander. 1845. *The geology of Russia in Europe and the Ural Mountains*. 2 vols. London.

Murray, Andrew. 1860. On Mr. Darwin's theory of the origin of species. [Read 20 February 1860.] *Proceedings of the Royal Society of Edinburgh* 4 (1857–62): 274–91.

Musschenbroek, Peter van. 1744. *Elements of natural philosophy*. Translated from the Dutch by John Colson. 2 vols. London.

Nathusius, Hermann von. 1860. *Die Racen des Schweines. Eine zoologische Kritik und Andeutungen über systematische Behandlung der Hausthier-Racen*. Berlin.

Natural selection: *Charles Darwin's Natural selection; being the second part of his big species book written from 1856 to 1858*. Cambridge: Cambridge University Press. 1975.

Naudin, Charles Victor. 1852. Considérations philosophiques sur l'espèce et la variété. *Revue Horticole* 4th ser. 1: 102–9.

Navy list: *The navy list*. London. 1815–70.

NBL: *Norsk biografisk leksikon*. Edited by Edvard Bull, Anders Krogvig, Gerhard Gran, and others. 19 vols. Oslo: H. Aschehoug & Co. 1923–83.

NBU: *Nouvelle biographie universelle*. Edited by Jean Chrétien Ferdinand Hoefer. 46 vols. Paris. 1852–66.

NDB: *Neue deutsche Biographie*. Under the auspices of the Historical Commission of the Bavarian Academy of Sciences. 16 vols. (A–Melanchthon). Berlin: Duncker and Humblot. 1953–90.

NDBA: *Nuevo Diccionario Biografico Argentino (1750–1930)*. Edited by Vicente Osvaldo Cutolo. 7 vols. Buenos Aires: Editorial Elche. 1968–85.

Neale, Edward Vansittart. 1861. On typical selection, as a means of removing the difficulties attending the doctrine of the origin of species by natural selection. *Proceedings of the Zoological Society of London*, pp. 1–11. [Reprinted in *Annals and Magazine of Natural History* 3d ser. 7: 330–40.]

Newman, Henry Wenman. 1851. On the habits of the Bombinatrices. *Transactions of the Entomological Society of London* n.s. 1: 86–92, 109–12, 116–18.

Nicol, James. 1844. *Guide to the geology of Scotland; containing an account of the character, distribution, and more interesting appearances of its rocks and minerals*. Edinburgh and London.

——. 1859. On the geological structure of the vicinity of Aberdeen and the north-east of Scotland. *Report of the 29th meeting of the British Association for the Advancement of Science held at Aberdeen*, Transactions of the sections, pp. 116–19.

NNBW: *Nieuw Nederlandsch biografisch woordenboek*. Edited by P. C. Molhuysen, P. J. Blok, and K. H. Kossmann. 10 vols. Leiden: A. W. Sijthoff. 1911–37.

North, F. J. 1943. Centenary of the glacial theory. *Proceedings of the Geologists' Association* 54 (1942–3): 1–28.

Notebooks: *Charles Darwin's notebooks, 1836–1844. Geology, transmutation of species, metaphysical enquiries*. Transcribed and edited by Paul H. Barrett, Peter J. Gautrey, Sandra Herbert, David Kohn, and Sydney Smith. Cambridge: Cambridge University Press for the British Museum (Natural History). 1987.

NUC: *The national union catalog*. Pre-1956 imprints. 685 vols. and supplement (68 vols.). London and Chicago: Mansell. 1968–81.

OBL: *Österreichisches biographisches Lexikon 1815–1950*. Edited by Leo Santifaller, Eva

Obermayer-Marnach, and others. 9 vols. and 3 fascicles of vol. 10 (A–Schmid). Vienna: Österreichischer Akademie der Wissenschaften. 1957–92.

OED: *The Oxford English dictionary*. 2d ed. Prepared by J. A. Simpson and E. S. C. Weiner. First edited by James A. H. Murray, Henry Bradley, W. A. Craigie, and C. T. Onions. Combined with *A supplement to the Oxford English dictionary*, edited by R. W. Burchfield. 20 vols. Oxford: Clarendon Press. 1989.

Oldham, Thomas. 1844. On the occurrence of marine shells in the gravel of Ireland. *Report of the 14th meeting of the British Association for the Advancement of Science held at York*, Transactions of the sections, p. 57.

——. 1848. On the geology of the county of Wicklow. *Report of the 18th meeting of the British Association for the Advancement of Science held at Swansea*, Transactions of the sections, pp. 71–2.

——. 1858. On some additions to the knowledge of the Cretaceous rocks of India. *Journal of the Asiatic Society of Bengal* 27: 112–24.

Oliver, Daniel. 1860. On *Sycopsis*. [Read 15 March 1860.] *Transactions of the Linnean Society of London* 23 (1860–2): 83–8.

——. 1861. Review of literature on phanerogamic plants. *Natural History Review* n.s. 1: 85–115.

——. 1862a. The Atlantis hypothesis in its botanical aspect. *Natural History Review* n.s. 2: 149–70.

——. 1862b. On the distribution of northern plants. *Notices of the Proceedings at the Meetings of the Members of the Royal Institution of Great Britain* 3 (1858–62): 431–3.

Olmsted, Frederick Law. 1856. *A journey in the seaboard slave states, with remarks on their economy*. New York.

——. 1857. *A journey through Texas; or, a saddle-trip on the southwestern frontier: with a statistical appendix*. New York.

——. 1860. *A journey in the back country*. 2 vols. New York and London.

Orchids: *On the various contrivances by which British and foreign orchids are fertilised by insects, and the good effects of intercrossing*. By Charles Darwin. London. 1862.

Orchids 2d ed.: *On the various contrivances by which British and foreign orchids are fertilised by insects, and the good effects of intercrossing*. Revised, etc. By Charles Darwin. London. 1877.

Origin: *On the origin of species by means of natural selection, or the preservation of favoured races in the struggle for life*. By Charles Darwin. London. 1859.

Origin 2d ed.: *On the origin of species by means of natural selection, or the preservation of favoured races in the struggle for life*. By Charles Darwin. London. 1860.

Origin 3d ed.: *On the origin of species by means of natural selection, or the preservation of favoured races in the struggle for life*. With additions and corrections. By Charles Darwin. London. 1861.

Origin 4th ed.: *On the origin of species by means of natural selection, or the preservation of favoured races in the struggle for life*. With additions and corrections. By Charles Darwin. London. 1866.

Origin US ed.: *On the origin of species by means of natural selection, or the preservation of favoured races in the struggle for life*. A new edition, revised and augmented by the author. By Charles Darwin. New York. 1860.

Owen, Richard. 1846. *A history of British fossil mammals, and birds*. London.

——. 1857. On the characters, principles of division, and primary groups of the class Mammalia. *Journal of the Proceedings of the Linnean Society of London* 2 (1858): 1–37.

——. 1858. Presidential address. *Report of the 28th meeting of the British Association for the Advancement of Science held at Leeds*, pp. xlix–cx

——. 1859. *On the classification and geographical distribution of Mammalia, being the lecture on Sir Robert Reade's foundation, delivered before the University of Cambridge, in the Senate-House, May 10, 1859. To which is added an appendix "On the gorilla," and "On the extinction and transmutation of species."* London.

——. 1860a. *Palæontology or a systematic summary of extinct animals and their geological relations*. Edinburgh.

[——.] 1860b. Review of *Origin of species* and other works. *Edinburgh Review, or critical journal* 111: 487–532.

——. 1861a. On the cerebral characters of man and the ape. *Annals and Magazine of Natural History* 3d ser. 7: 456–8.

——. 1861b. On a national museum of natural history. *Athenæum*, 27 July 1861, pp. 118–20; 3 August 1861, pp. 153–5; 10 August 1861, pp. 187–9.

Owen, Richard Startin, ed. 1894. *The life of Richard Owen . . . with the scientific portions revised by C. Davies Sherborn. Also an essay on Owen's position in anatomical science by the Right Hon. T. H. Huxley, F.R.S.* 2 vols. London.

Pallas, Pyotr Simon. 1767–80. *Spicilegia zoologica quibus novae imprimis et obscurae animalium species iconibus, descriptionibus atque commentariis illustrantur.* 14 pts in 2 vols. Berlin.

Pancaldi, Giuliano. 1985. Darwin's intellectual development. In *The Darwinian heritage*, edited by David Kohn. Princeton: Princeton University Press in association with Nova Pacifica.

Peckham, Morse, ed. 1959. *The Origin of species by Charles Darwin; a variorum text.* Philadelphia: University of Pennsylvania Press.

Pelzeln, August von. 1861. *Bemerkungen gegen Darwin's Theorie vom Ursprung der Spezies.* Vienna.

Pengelly, William. 1862. The lignites and clays of Bovey Tracey, Devonshire. *Philosophical Transactions of the Royal Society of London* 152: 1019–38.

Pennant, Thomas. 1771. *A tour in Scotland*. Chester.

Phillips, John. 1855. *Manual of geology: practical and theoretical*. London and Glasgow.

——. 1860. *Life on earth, its origin and succession*. Cambridge and London.

Physicians: *The roll of the Royal College of Physicians of London*. Vol. 3: 1801–25. By William Munk. 2d ed., revised and enlarged. London. 1878. Vol. 4: *Munk's roll, vol. 4: Lives of the fellows of the Royal College of Physicians of London 1826–1925*. Compiled by G. H. Brown. London: Royal College of Physicians. 1955.

Pickering, Charles. 1850. *The races of man; and their geographical distribution. New edition.*

To which is prefixed, an analytical synopsis of the natural history of man, by John Charles Hall. London.

Planchon, Jules Emile. 1847–8. Sur la famille des Linées. *London Journal of Botany* 6 (1847): 588–603; 7 (1848): 165–86, 473–501, 507–28.

Plarr, Victor Gustave. 1930. *Plarr's lives of the fellows of the Royal College of Surgeons of England*. Revised by Sir D'Arcy Power. 2 vols. London: Simpkin Marshall.

Porter, Duncan M. 1985. The *Beagle* collector and his collections. In Kohn, David, ed., *The Darwinian heritage*. Princeton: Princeton University Press in association with Nova Pacifica.

Post Office directory of the county of Durham: *The Post Office directory of the county of Durham, and the principal towns and adjacent places in Northumberland, Cumberland & Westmorland*. London. 1873–.

Post Office directory of Hampshire, Wiltshire, and Dorsetshire: *Post Office directory of Hampshire, Wiltshire, and Dorsetshire*. London. 1848–1939.

Post Office directory of Lancashire: *Post Office directory of Lancashire*. London. 1858–.

Post Office directory of Northumberland: *Post Office directory of Westmorland, Cumberland, Northumberland, and Durham*. London. 1858.

Post Office directory of the six home counties: *Post Office directory of the six home counties, viz., Essex, Herts, Kent, Middlesex, Surrey and Sussex*. London. 1845–.

Post Office London directory: *Post Office London directory*. London. 1802–.

Post Office London suburban directory: *The Post Office London suburban directory*. London. 1860–1918.

Powell, Baden. 1860. On the study of the evidences of Christianity. In *Essays and reviews*. London.

Quatrefage de Bréau, Jean Louis Armand de. 1861. *Unité de l'espèce humaine*. Paris.

Ramsay, Andrew Crombie. 1852. On the superficial accumulations and surface-markings of North Wales. [Read 26 March 1851.] *Quarterly Journal of the Geological Society of London* 8: 371–6.

——. 1860. *The old glaciers of Switzerland and North Wales*. London.

Rapp, Wilhelm Ludwig von. 1854. *Die Fische des Bodensees, untersucht und beschreiben von W. v. Rapp*. Stuttgart.

Rehbock, Philip F. 1979. The early dredgers: 'naturalizing' in British seas, 1830–1850. *Journal of the History of Biology* 12: 293–368.

Reichenbach, Heinrich Gustav. 1852. *De pollinis Orchidearum genesi ac structura, et de Orchideis in artem ac systema redigendis*. Leipzig.

Reinikka, Merle A. 1972. *A history of the orchid*. Coral Gables, Florida: University of Miami Press.

Reptiles: Part V of *The zoology of the voyage of H.M.S. Beagle*. By Thomas Bell. Edited and superintended by Charles Darwin. London. 1842–3.

Richardson, Edmund William. 1916. *A veteran naturalist; being the life and work of W. B. Tegetmeier*. London: Witherby & Co.

Richardson, John. 1861. *The polar regions*. Edinburgh.

Rink, Hinrich Johannes. 1852. *De danske Handelsdistrikter i Nordgrønland, deres geographiske*

Beskaffenhed og produktive Erhvervskilder. 2 vols. Copenhagen.

Rink, Hinrich Johannes. 1853. On the large continental ice of Greenland, and the origin of icebergs in the Arctic seas. [Read 9 May 1853.] *Journal of the Royal Geographical Society* 23: 145–54.

——. 1858. On the supposed discovery, by Dr. E. K. Kane, U.S.N., of the north coast of Greenland, and of an Open Polar Sea, &c.; as described in "Arctic explorations in the years 1853, 1854, 1855." [Read 12 April 1858.] *Journal of the Royal Geographical Society* 28: 272–87.

Rogers, Henry Darwin. 1861. On the origin of the parallel roads of Lochaber (Glen Roy), Scotland. [Read 22 March 1861.] *Notices of the Proceedings at the Meetings of the Members of the Royal Institution of Great Britain* 3 (1858–62): 341–5.

Rolleston, George. 1861a. On the affinities of the brain of the orang utang. *Natural History Review* n.s. 1: 201–17.

——. 1861b. On correlations of growth, with a special example from the anatomy of a porpoise. *Natural History Review* n.s. 1: 484–8.

——. 1861c. On the homologies of the lobes of the liver in Mammalia. *Report of the 31st meeting of the British Association for the Advancement of Science held at Manchester*, Transactions of the sections, pp. 174–5.

Rolleston, George and Robertson, Charles. 1859. On the aquiferous and oviductal systems in the lamellibranchiate mollusks. [Read 3 February 1859.] *Proceedings of the Royal Society of London* 9 (1857–9): 633–4.

——. 1862. On the aquiferous and oviducal system in the lamellibranchiate mollusks. *Annals and Magazine of Natural History* 3d ser. 10: 446–58.

[Rorison, Gilbert.] 1861. *The three barriers: notes on Mr. Darwin's "Origin of species"*. Edinburgh and London.

Rössler, Adolf. 1861. Gedanken über die Bedeutung der Malerei auf den Schmetterlingsflügeln. *Wiener Entomologische Monatschrift* 5: 163–6.

Rothschild, Nathaniel Mayer Victor. 1965. *A classification of living animals*. 2d ed. London.

Roulin, François Désiré. 1835. Recherches sur quelques changemens observés dans les animaux domestiques transportés de l'ancien dans le nouveau continent. *Mémoires présentés par divers Savans a l'Académie Royale des Sciences de l'Institut de France. Sciences mathématiques et physiques* 6: 319–52.

Rowlinson, J. S. 1971. The theory of glaciers. *Notes and records of the Royal Society of London* 26: 189–204.

Royal Society catalogue of scientific papers: *Royal Society catalogue of scientific papers (1800–1900). Compiled and published by the Royal Society of London*. 19 vols. and index (3 vols.). London and Cambridge. 1867–1925.

Royal Society record: *The record of the Royal Society of London for the promotion of natural knowledge*. 4th ed. London: Royal Society. 1940.

Rudwick, M. J. S. 1970. The glacial theory. *History of Science* 8: 137–57.

——. 1974. Darwin and Glen Roy: A "great failure" in scientific method? *Studies in History and Philosophy of Science* 5: 97–185.

Rugby School register: *Rugby School register*. Vol. 1: from April 1675 to October 1857. Revised and annotated by Godfrey A. Solly. Rugby: George Over. 1933.

Rupke, Nicolaas A. 1983. *The great chain of history: William Buckland and the English school of geology (1814–1849)*. Oxford: Clarendon.

——. 1988. The road to Albertopolis: Richard Owen (1804–92) and the founding of the British Museum of Natural History. In *Science, politics and the public good: essays in honour of Margaret Gowing*, edited by Nicolaas A. Rupke. Basingstoke and London: Macmillan Press.

Ruschenberger, W. S. W. 1892. A sketch of the life of Joseph Leidy, M.D., LL.D. *Proceedings of the American Philosophical Society* 30: 135–84.

Ruse, Michael. 1975. Darwin's debt to philosophy: an examination of the influence of the philosophical ideas of John F. W. Herschel and William Whewell on the development of Charles Darwin's theory of evolution. *Studies in History and Philosophy of Science* 6: 159–81.

Rütimeyer, Ludwig. 1861. *Die Fauna der Pfahlbauten in der Schweiz. Untersuchungen über die Geschichte der wilden und der Haus-Säugethiere von Mittel-Europa*. Basel.

Sarjeant, William A. S. 1980. *Geologists and the history of geology: an international bibliography from the origins to 1978*. 5 vols. and supplement (2 vols., 1987). London: Macmillan.

SBL: *Svenskt biografiskt lexikon*. Edited by Bertil Boëthius, Erik Grill, Birgitta Lager-Kromnow, and Göran Nilzén. 26 vols. and 2 fascicles of vol. 27 (A–Olsson). Stockholm: Albert Bonnier and P. A. Norstedt. 1918–93.

Scherren, Henry. 1905. *The Zoological Society of London. A sketch of its foundation and development and the story of its farm, museum, gardens, menagerie and library*. London, Paris, New York, and Melbourne: Cassell & Co.

Schleicher, August. [1861.] Die Darwin'sche Theorie und die Thier- und Pflanzenwelt. *Zeitschrift für die Landwirtschaft*, vol. 15.

Schneider, C. F. A. 1863. Bijdrage tot de geologische kennis van Timor. *Natuurkundig Tijdschrift voor Nederlandsch-Indië* 25 (1863): 87–98, 107

Schofield, Robert E. 1963. *The Lunar Society of Birmingham. A social history of provincial science and industry in eighteenth-century England*. Oxford: Clarendon Press.

Scholes, Percy A. 1970. *The Oxford companion to music*. 10th ed. Edited by John Owen Ward. London, New York, and Toronto: Oxford University Press.

Schomburgk, Robert Hermann. 1837. On the identity of three supposed genera of orchideous epiphytes. *Transactions of the Linnean Society of London* 17: 551–2.

Sclater, Philip Lutley. 1860. Remarks on the geographical distribution of recent terrestrial Vertebrata. *Report of the 30th meeting of the British Association for the Advancement of Science held at Oxford*, Transactions of the sections, pp. 121–2.

——. 1860–1. Catalogue of the birds of the Falkland Islands. *Proceedings of the Zoological Society of London* 28 (1860): 383–91; 29 (1861): 45–7.

——. 1862. *Catalogue of a collection of American birds belonging to Philip Lutley Sclater, M.A., Ph.D., F.R.S.* London.

[Sedgwick, Adam.] 1860. Objections to Mr. Darwin's theory of the origin of species.

Spectator, 24 March 1860, pp. 285–6. [Reprinted with revisions in the *Spectator*, 7 April 1860, pp. 334–5.]

Sedgwick, William. 1861. On sexual limitation in hereditary disease. *British and Foreign Medico-Chirurgical Review* n.s. 27: 477–89; 28: 198–214.

Seemann, Berthold Carl. 1862. *Viti: an account of a government mission to the Vitian or Fijian islands in the years 1860–61*. Cambridge.

Sidebotham, Joseph. 1847. Note on raising Primulas from seed. *Phytologist* 2 (1845–7): 887.

——. 1849. Experiments on the specific identity of the cowslip and primrose. *Phytologist* 3 (1848–9): 703–5.

Sloan, Phillip R. 1985. Darwin's invertebrate program, 1826–1836: preconditions for transformism. In *The Darwinian heritage*, edited by David Kohn. Princeton: Princeton University Press in association with Nova Pacifica.

Smith, Audrey Z. 1986. *A history of the Hope entomological collections in the University Museum Oxford; with lists of archives and collections*. Oxford: Clarendon Press.

Smith, James. 1839. On the last changes in the relative levels of the land and sea in the British Islands. *Memoirs of the Wernerian Society* 8: 49–113.

——. 1850. On the occurrence of marine shells in the stratified beds below the till. *Quarterly Journal of the Geological Society of London* 6: 386–8.

——. 1862. *Researches in newer Pliocene and post-Tertiary geology*. Glasgow.

South America: *Geological observations on South America. Being the third part of the geology of the Beagle, under the command of Capt. Fitzroy, R.N. during the years 1832 to 1836*. By Charles Darwin. London. 1846.

Spencer, Herbert. 1860–2. *First principles*. London.

——. 1904. *An autobiography*. 2 vols. London.

Sprengel, Christian Konrad. 1793. *Das entdeckte Geheimniss der Natur im Bau und in der Befruchtung der Blumen*. Berlin.

Stenton, Michael. 1976. *Who's who of British members of Parliament, 1832–1885*. Brighton: Harvester Press.

Stephen, Leslie. 1885. *Life of Henry Fawcett*. London.

Steudel, Ernst Gottlieb. 1840–1. *Nomenclatur botanicus seu: synonymia plantarum universalis, enumerans ordine alphabetico nomina atque synonyma, tum generica tum specifica, etc*. 2d ed. Stuttgart and Tübingen.

Stewart, Joyce, ed. 1992. *Orchids at Kew*. London: Her Majesty's Stationery Office.

Strachey, Henry. 1853. Physical geography of western Tibet. *Journal of the Royal Geographical Society* 23: 1–69.

Stur, Dionys. 1860. *Beiträge zu einer Monographie des Genus Astrantia*. Vienna.

Sutherland, Peter Cormack. 1853. On the geological and glacial phenomena of the coasts of Davis' Strait and Baffin's Bay. *Quarterly Journal of the Geological Society of London* 9: 296–312.

Swinhoe, Robert. 1861a. Notes on the birds observed about Talien Bay (North China), from June 21 to July 25, 1860. *Ibis* 3: 251–62.

——. 1861b. Notes on ornithology taken between Takoo and Peking, in the neighbourhood of the Peiho River, Province of Chelee, North China, from August to December, 1860. *Ibis* 3: 323–45.

Symonds, William Samuel. 1861. *Old bones; or, notes for young naturalists*. London.

Taxonomic literature: *Taxonomic literature. A selective guide to botanical publications and collections with dates, commentaries and types*. 2d ed. By Frans A. Stafleu and Richard S. Cowan. 7 vols. Utrecht: Bohn, Scheltema & Holkema. The Hague: dr. W. Junk b.v., Publishers. 1976–88.

Taylor, Tom, ed. 1860. *Autobiographical recollections. By the late Charles Robert Leslie, R.A.* 2 vols. in 1. London.

Teale, Thomas Pridgin. 1858. On the superficial deposits of the valley of the Aire at Leeds. *Report of the 28th meeting of the British Association for the Advancement of Science held at Leeds*, Transactions of the sections, pp. 111–13.

Tegetmeier, William Bernhard, ed. 1856. On the very remarkable peculiarities existing in the skulls of the feather-crested variety of the domestic fowl, now known as the Polish. *Proceedings of the Zoological Society of London* 24: 366–8.

——, ed. 1856–7. *The poultry book: including pigeons and rabbits; comprising the characteristics, management, breeding and medical treatment of poultry, being the results of personal observation and practice of the Rev. W. Wingfield, G. W. Johnson, Esq*. Re-arranged and edited by W. B. Tegetmeier. London.

——. 1858. On the formation of the cells of bees. *Report of the 28th meeting of the British Association for the Advancement of Science held at Leeds*, Transactions of the sections, pp. 132–3.

——. 1859. Cells of the honey bee. *Transactions of the Entomological Society of London* n.s. 5 (1858–61): Proceedings, pp. 34–5.

——. 1861. Undescribed variations in plumage. *The Field, the Farm, the Garden, the Country Gentleman's Newspaper* 17: 313.

Thwaites, George Kendrick. 1847. On conjugation in the Diatomaceæ. *Report of the 17th meeting of the British Association for the Advancement of Science held at Oxford*, Transactions of the sections, p. 87.

Topham, Jonathan Richard. 1993. 'An infinite variety of arguments': the *Bridgewater treatises* and British natural theology in the 1830s. (PhD thesis: University of Lancaster).

Torrey, John. 1843. *A flora of the state of New York, comprising full descriptions of all the indigenous and naturalized plants hitherto discovered in the state; with remarks on their economical and medicinal properties*. Pt 2 of *Natural History of New York*. 2 vols. New York and Boston. 1842–94.

——. 1857. *Descriptions of the general botanical collections*. In vol. 4 of *Reports of explorations and surveys, to ascertain the most practicable and economical route for a railroad from the Mississippi River to the Pacific Ocean. Made under the direction of the Secretary of War, in 1853–4*. 12 vols. Washington. 1855–60.

——. [1859.] Botany of the boundary. In *Report on the United States and Mexican boundary survey, made under the direction of the Secretary of the Interior, by William H. Emory*. Washington, DC.

Torrey, John and Gray, Asa. 1838–43. *A flora of North America: containing abridged descriptions of all the known indigenous and naturalized plants growing north of Mexico: arranged according to the natural system*. 2 vols. New York and London.

Trautschold, Hermann Adolfovich. 1861. Übergänge und Zwischenvarietäten. [Reprinted from the Moscow University Bulletin, no. 4, 1860.] Moscow: privately printed.

Tree, Isabella. 1991. *The ruling passion of John Gould: a biography of the bird man*. London: Barrie & Jenkins.

Trimmer, Joshua. 1831. On the diluvial deposits of Caernarvonshire, between the Snowdon chain of hills and the Menai Strait, and on the discovery of marine shells in diluvial sand and gravel on the summit of Moel Tryfane, near Caernarvon, 1000ft. above the level of the sea. [Read 8 June 1831.] *Proceedings of the Geological Society of London* 1 (1826–33): 331–2.

Trollope, Anthony. 1861. *Framley parsonage*. 3 vols. London.

——. 1862. *Orley Farm*. 2 vols. London.

Turrill, W. B. 1959. *The Royal Botanic Gardens, Kew. Past and present*. London: Herbert Jenkins.

——. 1963. *Joseph Dalton Hooker: botanist, explorer, and administrator*. London: Thomas Nelson and Sons.

Tyndall, John. 1860. *The glaciers of the Alps. Being a narrative of excursions and ascents, an account of the origin and phenomena of glaciers, and an exposition of the physical principles to which they are related*. London.

Variation: *The variation of animals and plants under domestication*. By Charles Darwin. 2 vols. London. 1868.

Vaucaire, Michel. 1930. *Paul Du Chaillu: gorilla hunter*. New York and London: Harper & Brothers.

Vernet, Luis Elie. 1833. Account of East Falkland Island. *Journal of the Royal Geographical Society of London* 3: 94–9.

Victoria history of the county of Kent: *The Victoria history of the county of Kent*. Edited by William Page. 3 vols. London: Archibald Constable and St Catherine Press. 1908–32.

Volcanic islands: *Geological observations on the volcanic islands visited during the voyage of H.M.S. Beagle, together with some brief notices of the geology of Australia and the Cape of Good Hope. Being the second part of the geology of the voyage of the Beagle, under the command of Capt. FitzRoy, R.N. during the years 1832 to 1836*. By Charles Darwin. London. 1844.

Wagner, Rudolph. 1860. Louis Agassiz's Prinzipien der Classification der organischen Körper insbesondere der Thiere mit Rücksicht auf Darwin's Ansichten im Auszuge dargestellt und besprochen. *Göttingische gelehrte Anzeigen* (1860) pt 2: 761–800.

Walford, Edward, ed. 1862. *Men of the time; a biographical dictionary of eminent living characters of both sexes. A new edition, thoroughly revised and brought down to the present time*. London.

Wallace, Alfred Russel. 1853. *A narrative of travels on the Amazon and Rio Negro*. London.

——. 1860. On the zoological geography of the Malay Archipelago. *Proceedings of the Linnean Society of London (Zoology)* 4: 172–84.

——. 1869. *The Malay archipelago; the land of the orang-utan, and the bird of paradise. A narrative of travel, with studies of man and nature*. 2 vols. London.

——. 1905. *My life: a record of events and opinions*. 2 vols. London: Chapman & Hall.

Walters, S. M. 1981. *The shaping of Cambridge botany: a short history of whole-plant botany in Cambridge from the time of Ray into the present century*. Cambridge: Cambridge University Press.

Watson, Hewett Cottrell. 1847–59. *Cybele Britannica; or, British plants, and their geographical relations*. 4 vols. London.

——. 1860. *Part first of a supplement to the Cybele Britannica*. London.

Weddell, Hugh Algernon. 1849. *Histoire naturelle des quinquinas ou monographie du genre Cinchona suivie d'une description du genre Cascarilla et de quelques autres plantes de la même tribu*. Paris.

——. 1855–7. *Chloris Andina. Essai d'une flore de la région alpine des Cordillères de l'Amérique du Sud*. 2 vols. Paris.

Wedgwood, Barbara and Wedgwood, Hensleigh. 1980. *The Wedgwood circle, 1730–1897: four generations of a family and their friends*. London: Studio Vista.

[Wedgwood, Frances Julia.] 1860–1. The boundaries of science, a dialogue. *Macmillan's Magazine* 2 (1860): 134–8; 4 (1861): 237–47.

Weir, John Jenner. 1869–70. On insects and insectivorous birds; and especially on the relation between the colour and the edibility of *Lepidoptera* and their larvæ. *Transactions of the Entomological Society of London* (1869): 21–6; (1870): 337–9.

Wellesley index: *The Wellesley index to Victorian periodicals 1824–1900*. Edited by Walter E. Houghton and others. 5 vols. Toronto: University of Toronto Press. London: Routledge and Kegan Paul. 1966–89.

Wiegmann, Arend J. Friedrich. 1839. [Über die Unfruchtbarkeit bei den Bastardpflanzen.] *Flora oder allgemeine botanische Zeitung* 22 (1839): 703.

Wight, Robert and Arnott, George Arnott Walker. 1834. *Prodromus floræ peninsulæ Indiæ orientalis: containing abridged descriptions of the plants found in the peninsula of British India, arranged according to the natural system*. London.

[Wilberforce, Samuel.] 1860. On the Origin of species. *Quarterly Review* 108: 225–64.

[——.] 1861. Review of *Essays and reviews*. *Quarterly Review* 109: 248–305.

Willey, Basil. 1956. *More nineteenth-century studies: a group of honest doubters*. London: Chatto and Windus.

Wilson, Leonard Gilchrist, ed. 1970. *Sir Charles Lyell's scientific journals on the species question*. New Haven and London: Yale University Press.

Winkler, Tiberius Cornelius, trans. 1860. *Het ontstaan der soorten van dieren en planten door middel van de natuurkeus, of het bewaard blijven van bevoorregte rassen in den strijd des levens*. [*Origin*], by Charles Darwin. Translated from the English. Haarlem.

——. 1861. Description de quelques nouvelles espèces de poissons fossiles des calcaires d'eau douce d'Oeningen. *Natuurkundige Verhandelingen van de Hollandsche*

Maatschappij der Wetenschappen te Haarlem 14 (1861): 1–64.

Woodcock, George. 1969. *Henry Walter Bates: naturalist of the Amazons*. London: Faber and Faber.

Woodward, Samuel Pickworth. 1851–6. *A manual of the Mollusca; or, a rudimentary treatise of recent and fossil shells*. 3 pts. London.

Wrangel, Ferdinand Petrovich. 1840. *Narrative of an expedition to the Polar Sea, in the years 1820, 1821, 1822, & 1823. Commanded by Lieutenant, now Admiral, Ferdinand von Wrangell, of the Russian imperial navy*. [Translated from the German edition by Elizabeth Juliana Sabine.] Edited by Edward Sabine. London.

WWW: *Who was who*. A companion to *Who's who*. 6th ed. 7 vols. and cumulated index, 1897–1980. London: Adam & Charles Black. 1988–9.

Wyman, Jeffries. 1859. On some unusual modes of gestation. *American Journal of Science and Arts* 2d ser. 27: 5–13.

——. 1861. On the mode of formation of the rattle of the rattlesnake. *Boston Society of Natural History* 8 (1861–2): 121.

Zeller, Philipp Christoph. 1842. Beitrag zur Kenntniss der Volucella plumata und bombylans. *Entomologische Zeitung* 3: 65–7.

Zuccarini, Joseph Gerhard. 1825. *Monographie der amerikanischen Oxalis-arten*. Munich.

——. 1831. *Nachtrag zu Monographie der amerikanischen Oxalis-arten*. Munich.

BIOGRAPHICAL REGISTER AND INDEX TO CORRESPONDENTS

This list includes all correspondents and all persons mentioned in the letters and notes whom the editors have been able to identify. Dates of letters to and from correspondents are given in chronological order. Letters to the correspondents are listed in roman type; letters from correspondents in italic type; third-party letters are listed in italic type with the name of the recipient given in square brackets. Following the register, a list of biographical sources referred to in the entries is given. These works are also listed in the main bibliography.

Abbott, Charles Compton. Commanded the detachment of troops stationed in Stanley, Falkland Islands, 1858–60. Wrote on the ornithology of the island group. (Abbott 1861, *List of officers of the army* 1861.)

Adams, John Couch (1819–92). Astronomer and mathematician. Co-discoverer of the planet Neptune. Fellow and tutor, St John's College, Cambridge, 1843–52; fellow, Pembroke College from 1853. Appointed Lowndean professor of astronomy and geometry, Cambridge University, in 1858; director of the Cambridge Observatory, 1861. FRS 1849. (*DNB*, *DSB*.)

Agassiz, Jean Louis Rodolphe (Louis) (1807–73). Swiss geologist and zoologist. Professor of natural history, Neuchâtel, 1832–46. Emigrated to the United States in 1846. Professor of natural history, Harvard University, 1847–73. Established the Museum of Comparative Zoology, Harvard, in 1859. Foreign member, Royal Society, 1838. (*DAB*, *DSB*.)

Airy, George Biddell (1801–92). Plumian professor of astronomy and director of the Cambridge Observatory, 1828–35. Astronomer royal, 1835–81. FRS 1836. (*DNB*, *DSB*.)

Anderson, James (1824–93). Entered service for the Cunard company in 1851 and commanded fourteen different steamers of their fleet. Commanded the steamer *Great Eastern* during the laying of the Atlantic telegraph, 1865–6. Managing director of Eastern Telegraph Company, 1866–93. Knighted, 1866. (*Modern English biography*.)

Appleton, D. & Co. (publishers). New York publishing firm founded in 1838 by Daniel Appleton (1785–1849) in partnership with his son, William Henry Appleton (1814–99). (*DAB*.)

Archiac, Etienne Jules Adolphe Desmier de Saint-Simon, Vicomte d' (1802–68). French army officer, geologist, palaeontologist, and historian of geology. An active member of the Société géologique de France. Professor of palaeontology, Muséum d'histoire naturelle, 1861. (*DSB*, Sarjeant 1980.)

Argyll, Duke of. *See* Campbell, George Douglas.

Arnott, George Arnott Walker (1799–1868). Botanist. Worked with William Jackson Hooker in Glasgow, 1830–40. Regius professor of botany, Glasgow, 1845–68. (R. Desmond 1977, *DNB*.)

Atherley, George (1819–83). Partner in the Southampton and Hampshire Bank, Southampton. (*Alum. Cantab.*, *Banking almanac* 1861.)

Azara, Félix d' (1746–1811). Spanish explorer and army officer. Sent to South America to survey Spanish and Portuguese territories. Published works on the fauna of Paraguay. (*NBU*.)

Babington, Charles Cardale (1808–95). Botanist and archaeologist. Greatly involved in the natural history activities at Cambridge for more than forty years. A founding member of the Entomological Society in Cambridge and the Cambridge Antiquarian Society. Professor of botany, Cambridge University, 1861–95. An expert on plant taxonomy. FRS 1851. (*DNB*, *DSB*.)

Bachman, John (1790–1874). American clergyman and naturalist. A close friend and collaborator of John James Audubon. (*DAB*, Elliot 1979.)

Bagg, William. Artist with premises at 63 Gower Street, Bedford Square, London. (*Post Office London directory* 1846.)

Baillon, Ernest Henri (1827–95). French botanist. Professor of medical natural history and of hygiene and natural history applied to industry, École centrale des arts et manufactures, Paris, 1868–95. (*DBF*.)

Baily, John. Poulterer and dealer in live birds at 113 Mount Street, Berkeley Square, London. Author of works on the management of domestic and game fowl. (Freeman 1978, *Post Office London directory* 1860.)

Bain, Alexander (1810–77). Inventor. Pioneered the development of high-speed telegraphy with his invention of the automatic chemical telegraph in 1843. (*DNB*.)

Baird, Spencer Fullerton (1823–87). American vertebrate zoologist. Professor of natural history, Dickinson College, Pennsylvania, 1846–50. Assistant secretary, Smithsonian Institution, 1850–78; secretary, 1878–87. Instrumental in augmenting the zoological and anthropological collections of the Smithsonian's National Museum. (*DAB*, *DSB*.)

Baker, Samuel C. and Charles N. Dealers in ornamental poultry and live wild fowl in Chelsea, London. (*Post Office London directory* 1858.)

Balfour, John Hutton (1808–84). Physician and botanist. Professor of botany, Glasgow University, 1841–5. Professor of botany and Regius keeper of the Royal Botanic Garden, Edinburgh, 1845–79. FRS 1856. (*DNB*, *DSB*.)

Ball, James. Oilman. Of Duke Street, Grosvenor Square West. (*Post Office London directory* 1861.)

Baly, Joseph Sugar (1817–90). Physician and entomologist. Surgeon at Warwick, 1868–90. A leading authority on phytophagous Coleoptera. (*Modern English biography*.)

Banks, Joseph (1743–1820). Naturalist and patron of science. Accompanied James Cook, as naturalist, on his circumnavigation of the globe, 1768–71. President of the Royal Society, 1778–1820. Persuaded King George III to develop the Royal Botanic Gardens, Kew as a botanical research centre. FRS 1766. (*DNB*, *DSB*.)

Barkow, Hans Karl Leopold (1798–1873). Professor of anatomy, University of Breslau, 1835–73. Director of the Anatomy Institute, Breslau, from 1845. (*BLA*.)

Bartlett, Abraham Dee (1812–97). Taxidermist in London, 1834–52. Superintendent of the natural history department of the Crystal Palace, 1852–9; of the Zoological Society's gardens, Regent's Park, 1859–97. (*Modern English biography*.)
26 May [1861]

Bate, Charles Spence (1819–89). Dentist and scientific writer. Practised dentistry in Swansea, 1841–51, then in Plymouth. Secretary of the Plymouth Institution, 1854–60; president, 1861–2 and 1869–70. Studied the Crustacea. FRS 1861. (*DNB*.)

Bates, Henry Walter (1825–92). Naturalist. Undertook a joint expedition to the Amazon with Alfred Russel Wallace, 1848–50; continued to explore the area alone until 1859. Provided the first comprehensive scientific explanation for the phenomenon subsequently known as Batesian mimicry. Published *The naturalist on the River Amazons* in 1863. Assistant secretary, Royal Geographical Society, 1864–92. FRS 1881. (*DNB*, *DSB*.)
18 March 1861, 26 March [1861], *28 March 1861*, 4 April [1861], *[before 25 September 1861]*, 25 September [1861], *30 September 1861*, 2 October 1861, *[1 December] 1861*, 3 December [1861], 15 December [1861]

Bauer, Franz Andreas (1758–1840). German artist who visited England in 1788. Was pursuaded to remain there by Joseph Banks, who employed him as botanical painter for the Royal Botanic Gardens, Kew, for the sum of £300 per annum for life. Noted for his detailed botanical drawings of orchids published in his *Illustrations of orchidaceous plants* (1830–8). (Reinikka 1972.)

Baxter, William Walmisley (b. 1829). Chemist and druggist of Bromley, Kent. (*Post Office directory of the six home counties* 1859.)
22 January 1861

Beale, Lionel Smith (1828–1906). Physician and microscopist. Professor of physiology and general and morbid anatomy, King's College, London, 1853–69. Professor of pathological anatomy, 1869; of medicine, 1876–96. FRS 1857. (R. Desmond 1977, *DNB*.)
26 October [1861]

Beaton, Donald (1802–63). Scottish gardener interested in hybridisation. (R. Desmond 1977.)

Bechstein, Johann Matthäus (1757–1822). German forestry scientist and ornithologist. Teacher at the Salzmann educational establishment, Schnepfenthal, 1784–94. Director of the Forestry Academy, Dreißigacker bei Meiningen, 1801–22. (*NDB*.)

Bell, Thomas (1792–1880). Dental surgeon at Guy's Hospital, London, 1817–61. Appointed professor of zoology, King's College, London in 1836. President of the Ray Society, 1843–59. One of the secretaries of the Royal Society, 1848–53. President of the Linnean Society, 1853–61. Described the reptiles from the *Beagle* voyage. FRS 1828. (R. Desmond 1977, *DNB*.)

Bence-Jones, Henry. *See* Jones, Henry Bence.

Bennet, Charles Augustus, 6th earl of Tankerville (1810–99). Styled Lord Ossulston, 1822–59. Tory MP for North Northumberland, 1832–59. Succeeded as 6th earl in 1859. Lord Steward of the Household to Queen Victoria, 1867–8. (*Burke's peerage* 1980, *Modern English biography*, Stenton 1976.)

Bennett, John Hughes (1812–75). English physician who visited Paris and Germany and learned there of the beneficial effects of cod-liver oil. Published a book in 1841 in which he pointed out the value of the oil for medicinal purposes. (Morton 1970, p. 223.)

Bennett, Thomas (1814–98). Of Oslo. Author of an English–Norwegian phrase-book and of a handbook for travellers in Norway, both of which went through many editions. (*BLC*, *NUC*.)

Bentham, George (1800–84). Botanist. Pursued his botanical studies at the Royal Botanic Gardens, Kew. President of the Linnean Society, 1861–74. Published *Genera plantarum* (1862–83) with Joseph Dalton Hooker. FRS 1862. (*DNB*, *DSB*.)
17 June [1861], 22 June [1861], 9 November [1861], 24 November [1861], *26 November 1861*, 26 November [1861], *29 November 1861*, 30 November [1861]

Bentham, Jeremy (1748–1832). Writer on jurisprudence. (*DNB*.)

Berens, Edward Osmond. Lieutenant in the Rifle Volunteer Corps of Farnborough, Kent. (*Army list* 1861.)

Berthoud, Samuel Henri (1804–91). French writer. Wrote for *La Mode*, *Revue des deux mondes*, and *La Presse*. Became one of the editors of *La Patrie* in 1848. Wrote a number of novels, books on natural history, and scientific articles. (*DBF*.)

Betschler, Julius Wilhelm (1796–1865). Extraordinary professor and director of the institute of obstetrics, Breslau, 1827; ordinary professor, 1832–65. (*BLA*.)

Bibron, Gabriel (1806–48). French zoologist who described the reptiles collected by Alcide Charles Victor Dessalines d'Orbigny in South America. (*DBF*.)

Bigelow, John Milton (1804–78). American surgeon and botanist. Formed the botanical collections for the Pacific railroad surveys of 1853 and 1854. (*Taxonomic literature*.)

Billings, Elkanah (1820–76). Canadian naturalist who worked with the Geological Survey of Canada, 1856–76. Founder of the journal *Canadian Naturalist*. (Sarjeant 1980.)

Birch, Samuel (1813–85). Egyptologist and archaeologist. Assistant keeper of the department of antiquities in the British Museum, 1844–61; keeper of the oriental, British, and medieval antiquities, 1861–85. Translator of classical Chinese texts. (*DNB*.)

Birch, Samuel, cont.
6 April [1861], 21 April [1861]

Biscoe, John (1794–1843). Antarctic explorer. (*IBN*.)

Black, Joseph (1728–99). Scottish physician and chemist. Discovered the phenomenon of latent heat. Professor of medicine, University of Glasgow, 1756–66. Professor of medicine and chemistry, University of Edinburgh, 1766–97. (*DNB*, *DSB*.)

Blackadder, Alexander. Civil engineer. Of Stirling, Scotland.

Blainville, Henri Marie Ducrotay de (1777–1850). French anatomist and zoologist. Appointed professor of comparative anatomy, Muséum d'histoire naturelle, Paris, in 1832. Foreign member, Royal Society, 1832. (*DBF*, *DSB*.)

Blyth, Edward (1810–73). Owned an unsuccessful druggist's business in Tooting. Wrote zoological articles and edited zoological works before being appointed curator of the museum of the Asiatic Society of Bengal in 1841. He retired from this post in 1862 and returned to England, where he continued to write on zoology and on the question of the origin of species. (*DNB*, *DSB*.)

Bond, Frederick (1811–89). Entomologist and ornithologist. Elected member of the Entomological Society in 1841; of the Zoological Society, 1854. Collector of British Lepidoptera. (*Entomologist's Monthly Magazine* 25 (1888–9): 384, Gilbert 1977.)

Bonham-Carter, Henry (1827–1921). Called to the bar, 1853, but gave up practice in 1861. Managing director of the Guardian Fire and Life Assurance Company, 1861–1902. Collaborated with Florence Nightingale, his cousin, in the organisation of the Nightingale Training School for Hospital Nurses. (*Alum. Cantab.*, *Burke's landed gentry* 1952.)

Boott, Francis (1792–1863). Physician and botanist. Had a successful medical practice in London, 1825–32. Lecturer on botany at the Webb Street School of Medicine, London, 1825. Secretary, Linnean Society, 1832–9; treasurer, 1856–61. (R. Desmond 1977, *DNB*.)

Borrer, William (1781–1862). Botanist who studied, in particular, the genera *Salix*, *Rubus*, and *Rosa*. FRS 1835. (R. Desmond 1977, *DNB*.)

Bowen, Francis (1811–90). Philosopher. Alvord professor of natural religion, moral philosophy, and civil polity, Harvard College, 1853–89. (*DAB*.)

Bowring, John Charles (1821–93). Entomologist. For some time, partner in the Chinese commercial firm Jardine, Matheson & Co. Presented his collection of Coleoptera to the British Museum. (*Entomologist's Monthly Magazine* 29 (1893): 169, Gilbert 1977.)

Boys, John Cowell (1824–89). Born in Sussex. Served on the survey staff of the New Zealand Company, 1842–5. Returned to England and completed his professional qualifications before going to New Zealand again in 1849. Later, owing to ill health, gave up field surveying and became inspector of surveys. Noted breeder of Romney sheep. Advised the introduction of the hedge sparrow to combat the caterpillar in Canterbury, New Zealand. (*DNZB*.)

Bravais, Auguste (1811–63). French naval officer, botanist, and physicist. Travelled with the Commission scientifique du Nord in Norwegian Lapland, 1838–9, and published an account of the expedition. Professor of physics at the École polytechnique, 1845–57. (*DBF*, *DSB*.)

Bravard, Pierre Joseph Auguste (1803–61). French geologist and palaeontologist. Emigrated to Buenos Ayres in 1852. Inspector general of mines in Argentina and director of the national museum in Paraná. Died in an earthquake that struck Mendoza, Argentina, in 1861. (*DBF*, *EB*, *NDBA*.)

Bree, Charles Robert (1811–86). Zoologist and medic. Senior physician to the Essex and Colchester Hospital. Opposed CD's views. (*CDEL*, Freeman 1978.)

Brent, Bernard Philip (d. 1867). Bird fancier and author. Studied pigeon breeding in France and Germany. (*CDEL*.)

29 May 1861, *15 June 1861*

Breslau, University of

4 August 1861

Brewster, David (1781–1868). Scottish physicist who specialised in optics. Invented the kaleidoscope, 1816. Edited the *Edinburgh encyclopædia*, 1807–29. Assisted in the establishment of the British Association for the Advancement of Science in 1831. Knighted, 1832. Principal of the United College of St Salvatore and St Leonard, University of St Andrews, 1838–59; of University of Edinburgh, 1859–68. FRS 1815. (*DNB*, *DSB*.)

Brodie Innes, John. *See* Innes, John.

Brodie Innes, John William. *See* Innes, John William.

Brooke, James (1803–68). Raja of Saráwak, Borneo, 1841–63. Appointed British commissioner and consul-general of Borneo in 1847. (*DNB*.)

Bronn, Heinrich Georg (1800–62). German palaeontologist. Ordinary professor of natural science at Heidelberg, 1833. Translated and superintended the first German edition of *Origin* (1860) and of *Orchids* (1862). (*DSB*, *NDB*.)

Brown, Robert (1773–1858). Botanist. Naturalist on a voyage to Australia commanded by Matthew Flinders (*DNB*), 1801–5; published descriptions of the plants he collected. Librarian to Joseph Banks, 1810–20. Keeper of the botanical collections, British Museum, 1827–58. FRS 1811. (*DNB*, *DSB*.)

Brown-Séquard, Charles Edouard (1817–94). French physiologist. Practised medicine in France and America before accepting a professorship at Virginia Medical College, Richmond, in 1854. Returned to Paris in 1855, but moved to England in 1858. Physician, National Hospital for the Paralysed and Epileptics, 1860–3. Professor of physiology and pathology, Harvard Medical School, 1864–7. Professor of medicine, Collège de France, 1878–94. Conducted pioneering research in neurology and endocrinology. FRS 1861. (*DBF*, *DSB*.)

Buckland, William (1784–1856). Reader in geology at Oxford University, 1818–49. President of the Geological Society, 1824–5 and 1840–1. Dean of Westminster from 1845. Author of the Bridgewater treatise on geology (1836). FRS 1818. (*DNB*, *DSB*.)

Buckle, Henry Thomas (1821–62). Historian of civilisation. (*DNB*.)

Buckley, Samuel Botsford (1809–83). American botanist and field naturalist. Collected plants extensively in the southern states. Worked on his homestead farm, 1843–55. Assistant geologist and naturalist, Texas Geological Survey, 1860–1. Chief examiner, statistical department, United States Sanitary Commission, 1862–5. State geologist, Texas, 1865–7 and 1874–7. (*DAB*.)

Bunbury, Charles James Fox, 8th baronet (1809–86). Botanist, palaeobotanist. Collected plants in South America, 1833–4; in South Africa, 1838–9. Accompanied Charles Lyell to Madeira in 1853. Married Frances Joanna Horner, Charles Lyell's sister-in-law, in 1844. FRS 1851. (R. Desmond 1977, Sarjeant 1980.)

Burmeister, Karl Hermann Konrad (1807–92). German zoologist, ethnographer, and geologist. Published extensively on entomology. Professor of zoology, University of Halle, 1837–61. Travelled in Brazil, 1850–2; in Argentina, 1857–60. Director of the Museo Nacional in Buenos Aires, 1861–80. (Berg 1895, *NBU*, Sarjeant 1980.)

Busk, George (1807–86). Naval surgeon and naturalist who served on the hospital ship at Greenwich, 1832–55. Member of several scientific societies and president of the Microscopical Society, 1848–9. Hunterian professor of comparative anatomy, Royal College of Surgeons, 1856–9. Specialised in palaeontology and in the study of Bryozoa. FRS 1850. (*DNB*, *DSB*.)
5 April [1861]

Campbell, George Douglas, 8th duke of Argyll (1823–1900). Whig politician and author of works on science, religion, and politics. A defender of the concept of design in nature. President of the Royal Society of Edinburgh, 1861. Postmaster general, 1855–8. FRS 1851. (*DNB*.)

Candolle, Alphonse de (1806–93). Swiss botanist whose home was a centre of botanical activity. Professor of botany and director of the botanic gardens, Geneva, 1835–50. After 1850, concentrated on his own research. Son of Augustin Pyramus de Candolle. Foreign member, Royal Society, 1869. (*DSB*.)

Canterbury, Archbishop of. *See* Sumner, John Bird.

Canton, John (1718–78). Teacher. Experimented on various aspects of electricity and magnetism. FRS 1749. (*DSB*.)

Carpenter, William Benjamin (1813–85). Physician and naturalist. Fullerian professor of physiology at the Royal Institution, 1844–56; professor of forensic medicine at University College London; physiology lecturer, London Hospital, 1845–56. Registrar of the University of London, 1856–79. FRS 1844. (*DNB*, *DSB*.)

Carter, Henry Bonham. *See* Bonham-Carter, Henry.

Chaillu, Paul Belloni Du. *See* Du Chaillu, Paul Belloni.

Chambers, Robert (1802–71). Publisher, writer, and geologist. Partner, with his brother William Chambers, in the Edinburgh publishing company W. & R. Chambers. Joint editor of *Chambers's Journal* from 1832. Anonymous author of *Vestiges of the natural history of creation* (1844). (*DNB*, *DSB*.)
30 April [1861]

Clarke, William Branwhite (1798–1878). Divine and geologist. Emigrated to Australia in 1839, where he carried out extensive geological surveys. Discovered gold in New South Wales in 1841. Minister in Willoughby, New South Wales, 1847–70. FRS 1876. (*DNB*, *DSB*.)
[August 1861], 25 October [1861]

Clowes, William & Sons (printers). William Clowes (1807–83), eldest son of William Clowes (1779–1847), joined his father's printing business in 1823 and the name of the firm was changed to William Clowes & Sons in 1846. Printed the offical catalogue of the Great Exhibition of 1851. Introduced improvements in type-music printing. Printer to John Murray. (*DNB*.)

Colburn, Henry (d. 1855). London publisher who published the first edition of CD's *Journal of researches* (1839). (*DNB*.)

Colgate, Robert (1816–1904). Surgeon. First practised in East Grinstead, Sussex, moving before 1855 to Eastbourne, where he remained for the rest of his life. Appointed medical officer to the first district of Eastbourne Union, and Admiralty Surgeon and agent to the Eastbourne division of the coastguard. Became a prominent citizen of Eastbourne. (Plarr 1930.)
25 June 1861

Collingwood, Cuthbert (1826–1908). Naturalist and writer on religious topics. Lecturer on botany to the Royal Infirmary Medical School, Liverpool, 1858–66. Also lectured on biology at the Liverpool School of Science. Served as surgeon and naturalist on HMS *Rifleman* and HMS *Serpent* on voyages of exploration in the China seas, 1866–7; carried out research in marine zoology. On his return to Liverpool, became senior physician of the Northern Hospital and took a leading part in the intellectual life of the city. (R. Desmond 1977, *DNB*.)
14 March [1861]

Collins, William Wilkie (1824–89). Novelist. His best known novels include *The woman in white* (1860) and *The moonstone: a romance* (1868). (*DNB*.)

Colyear, Charles, 2d earl of Portmore (1700–85). MP for Andover, 1727–30. When young, known as 'Beau Colyear'. He was a great patron of the turf and a conspicuous member of London society. (*Complete peerage*, *DNB*.)

Comte, Isidore Auguste Marie François Xavier (Auguste) (1798–1857). French philosopher. Private secretary to Claude Henri de Rouvroy, comte de Saint-Simon, 1817–23. Founded the Association Polytechnique, a group devoted to education of the working classes, in 1830; and the Société Positiviste, devoted to the promulgation of the 'Cult of Humanity' in 1848. Adopted the term 'positivism' for his philosophy. (*DSB*.)

Cossham, Mr. Bailiff to Henry Wenman Newman, Thornbury Park, Gloucestershire.

Covington, Syms (1816?–61). Became CD's servant on the *Beagle* in 1833 and remained with him as assistant, secretary, and servant until 1839. Emigrated to Australia in 1839. (Freeman 1978.)

Cranworth, Baron. *See* Rolfe, Robert Monsey.

Crawfurd, John (1783–1863). Orientalist. Held several civil and political posts in Java, India, Siam, and Cochin China. Returned to England in 1827 and promoted the study of Indo-China. FRS 1818. (*DNB*.)
25 March [1861], 7 April 1861

Cresy, Edward (1823–70). Principal assistant clerk at the Metropolitan Board of Works and architect to the fire brigade. (*DNB*.)
28 May [1861]

Crisp, Edwards (1806–82). Physician and medical scientist. General practitioner in London from 1830. Practised as a physician in London, 1849–82. Wrote a number of books on medical subjects; published many papers on medicine and zoology. Built up a museum of comparative and pathological anatomy of over 5000 specimens. (*Lancet* (1882), pt 2: 1010–11, *Modern English biography*.)

Crocker, Charles William (1832–68). Foreman of the propagating department, Royal Botanic Gardens, Kew, 1857–62. Spent time in Germany as a gardener at the summer residence of Victoria, Crown Princess of Prussia, near Berlin. (R. Desmond 1977, *Journal of Horticulture* n.s. 14 (1868): 206–7.)
18 May [1861], 1 June [1861]

Crotch, William Duppa (1832?–1903). Entomologist who specialised in the Lepidoptera, Coleoptera, and Hemiptera. Travelled with his brother George Robert Crotch on several collecting expeditions. Married a Swede and lived for many years in Sweden. (*Entomologist's Monthly Magazine* 39 (1903): 256, Gilbert 1977.)
25 January 1861

Crowe, John Rice (1792–1877). British consul-general in Norway, 1843–75. Knighted, 1875. (*Modern English biography*.)

Currey, Frederick (1819–81). Mycologist. Called to the bar, 1844. Secretary, Linnean Society, 1860–80; vice-president and treasurer, 1880–1. One of the editors of the *Natural History Review*, 1861. FRS 1858. (*Alum. Cantab.*, *DNB*.)

Cuvier, Georges (1769–1832). French systematist, comparative anatomist, palaeontologist, and administrator. Professor of natural history, Collège de France, 1800–32; professor of comparative anatomy, Muséum d'histoire naturelle, 1802–32. Permanent secretary of the Académie des sciences from 1803. Foreign member, Royal Society, 1806. (*DBF*, *DSB*.)

Czermak, Johann Nepomuk (1828–73). Czech physiologist who established a private laboratory in his home in Prague in 1860 after leaving the University of Budapest. Professor of zoology, Jena, 1865–70; Leipzig, 1870–3. (*ADB*, *NDB*.)

Dana, Henrietta Frances. Daughter of Benjamin Silliman. Married James Dwight Dana in 1844. (*DAB*.)

Dana, James Dwight (1813–95). American geologist and zoologist. Naturalist with the United States Exploring Expedition to the Pacific, 1838–42; wrote reports on the geology, zoophytes, and Crustacea. An editor of the *American Journal of Science and Arts* from 1846. Professor of geology, Yale University, 1856–90. Foreign member, Royal Society, 1884. (*DAB*, *DSB*.)

Darwin, Anne Elizabeth (Annie) (1841–51). CD's eldest daughter. (*Darwin pedigree*.)

Darwin, Caroline. *See* Wedgwood, Caroline Sarah.

Darwin, Charles Waring (1856–8). Youngest child of CD. Died of scarlet fever. (*Darwin pedigree*.)

Darwin, Charlotte Maria Cooper (1827–85). Remote cousin of CD. Wife of Francis Rhodes Darwin. Cousin of William Darwin Fox. (*Darwin pedigree*, Freeman 1978.)

Darwin, Edward (1782–1829). Army officer. Eldest son of Erasmus and Elizabeth Collier Darwin. (*Darwin pedigree*.)

Darwin, Elizabeth (1747–1832). Illegitimate daughter of Charles Colyear, the second earl of Portmore. First married Edward Sacheveral Chandos-Pole. Became Erasmus Darwin's second wife in 1781. (*Darwin pedigree*.)

Darwin, Emily Catherine (1810–66). CD's sister. Resided at The Mount, Shrewsbury, until she married Charles Langton in 1863. (*Darwin pedigree*.)

Darwin, Emma (1808–96). Youngest daughter of Josiah Wedgwood II. Married CD, her cousin, in 1839. (*Emma Darwin*.)

[June 1861]

Darwin, Erasmus (1731–1802). CD's grandfather. Physician, botanist, and poet. Advanced an evolutionary theory similar to that subsequently expounded by Lamarck. FRS 1761. (*DNB*, *DSB*.)

Darwin, Erasmus Alvey (1804–81). CD's brother. Attended Shrewsbury School, 1815–22. Matriculated Christ's College, Cambridge, 1822; M.B., 1828. At Edinburgh University, 1825–6. Qualified but never practised as a physician. Lived in London from 1829 to his death. (*Alum. Cantab.*)

Darwin, Francis (Frank) (1848–1925). CD's son. BA, Trinity College, Cambridge, 1870. Collaborated with CD on several botanical projects, 1875–82. Lecturer in botany, Cambridge University, 1884; reader, 1888–1904. Edited CD's letters. FRS 1882. (*DNB*, *DSB*.)

Darwin, Francis Rhodes (1825–1920). Assumed the name Darwin in 1850 after inheriting the estate of his brother-in-law, Robert Alvey Darwin. Resided at Creskeld Hall, Yorkshire, and Elston Hall, Nottinghamshire. JP for Nottinghamshire and Yorkshire, a role in which he was very active. (*Alum. Cantab.*)

Darwin, George Howard (1845–1912). CD's son. BA, Trinity College, Cambridge, 1868; fellow, 1868–78. Called to the bar in 1872 but did not practise law. Plumian professor of astronomy and experimental philosophy, Cambridge University, 1883–1912. FRS 1879. (*DNB*, *DSB*.)

Darwin, Henrietta Emma (Etty) (1843–1927). CD's daughter. Married Richard Buckley Litchfield (*Alum. Cantab.*) in 1871. (*Burke's landed gentry* 1952.)

Darwin, Horace (1851–1928). CD's son. BA, Trinity College, Cambridge, 1874. Apprenticed to an engineering firm in Kent before returning to Cambridge in 1875 to design and make scientific instruments. Formed the Cambridge Scientific

Darwin, Horace, cont.
Instrument Company in 1885 and served as chairman. Mayor of Cambridge, 1896–7. FRS 1903. (*Alum. Cantab.*, *DNB.*)

Darwin, Leonard (1850–1943). CD's son. Attended the Royal Military Academy, Woolwich. Entered the Royal Engineers in 1871; major, 1889. Served on several scientific expeditions, including those for the observation of the transit of Venus in 1874 and 1882. Instructor in chemistry, School of Military Engineering, Chatham, 1877–82. Staff, Intelligence Department, War Office, 1885–90. Liberal Unionist MP, Lichfield division of Staffordshire, 1892–5. President, Royal Geographical Society, 1908–11; Eugenics Education Society, 1911–28. Chairman, Bedford College, London University, 1913–20. (Sarjeant 1980, *WWW*.)

Darwin, Robert Alvey (1826–47). Of Elston Hall, Nottinghamshire. (*Darwin pedigree*.)

Darwin, Robert Waring (1766–1848). CD's father. Physician. MD, Leiden, 1785. Had a large practice in Shrewsbury and resided at The Mount, which he built *c.* 1796–8. Son of Erasmus Darwin and his first wife, Mary Howard. Married Susannah, daughter of Josiah Wedgwood I, in 1796. FRS 1788. (Freeman 1978.)

Darwin, Susan Elizabeth (1803–66). CD's sister. Lived at The Mount, Shrewsbury, until her death. (*Darwin pedigree*.)

Darwin, William Alvey (1726–83). Brother of Erasmus Darwin. Grandfather of William Darwin Fox. (*Darwin pedigree*.)

Darwin, William Erasmus (1839–1914). CD's oldest child. Attended Rugby School. BA, Christ's College, Cambridge, 1862. Banker in Southampton. Keen amateur photographer. (*Alum. Cantab.*)

[24 January 1861], 17 February [1861], 21 and 22 April [1861?], 9 May [1861], [25 May 1861], [26 May 1861], 1 [June 1861], 6 [June 1861], 12 October [1861], 17 [October 1861], 22 October [1861], [27 October 1861], 4 November [1861], 15 November [1861], *[17 November 1861]*

Davidson, Thomas William St Clair (1817–85). Artist and palaeontologist. Fellow of the Geological Society of London. Expert on fossil brachiopods. FRS 1857. (*DNB*, Sarjeant 1980.)

26 April 1861, 30 April 1861, *3 May 1861*

Dawson, John William (1820–99). Canadian geologist and educationist. Investigated the geology of maritime provinces with Charles Lyell in 1846 and 1852. Superintendent of education for Nova Scotia, 1850. Appointed principal and professor of geology at McGill University in 1855. FRS 1862. (*DNB*, *DSB*.)

Delamer, Eugene Sebastian. *See* Dixon, Edmund Saul.

Derby, Earl of. *See* Stanley, Edward George Geoffrey Smith.

Dick, Thomas Lauder. *See* Lauder, Thomas Dick.

Dickie, George (1812–82). Scottish botanist. Professor of natural history, Belfast, 1849–60. Professor of botany, Aberdeen, 1860–77. Specialised in the study of marine algae and described many of the collections sent to the Royal Botanic Gardens, Kew. FRS 1881. (R. Desmond 1977, *DNB*.)

[5 July 1861]

Dixon, Edmund Saul (1809–93). Rector of Intwood with Keswick, Norfolk, 1842–93. Author of books on the history and management of poultry. From 1854, published under the pseudonym Eugene Sebastian Delamer. (*Alum. Cantab.*, *Modern English biography*.)

Don, David (1800–41). Professor of botany, King's College, London, 1836–41. Librarian, Linnean Society, 1822–41. (R. Desmond 1977, *DNB*.)

Dortous de Mairan, Jean Jacques (1678–1771). French physicist. Studied wide variety of topics, including heat, light, sound, motion, the shape of the earth, and the aurora. Made *pensionnaire géomètre*, Academy of Sciences, Paris, in 1746. Foreign member, Royal Society. (*DBF*, *DSB*.)

Drummond, James (1784–1863). Botanist. Emigrated to Australia in 1829. Superintendent of the government gardens, Western Australia. (*Aust. dict. biog.*, *DNB*.)

Du Chaillu, Paul Belloni (1835–1903). Explorer. Undertook an exploring expedition into Central Africa, 1856–60. Collected rare birds and animals, bringing back to America the first gorillas ever seen there. His account of his journey, *Explorations and adventures in equatorial Africa*, caused great controversy among scientists, some of whom were sceptical about his discoveries. However, later travellers confirmed many of his findings. Revisited Africa, 1863–5. After spending several years writing and lecturing in New York, in 1871 he went to study the people of Sweden and Norway. (*DAB*.)

Dufour, Léon Jean Marie (1780–1865). French entomologist. (Gilbert 1977.)

Duméril, André Marie Constant (1774–1860). French entomologist. Professor of anatomy, Paris, 1801–12. Became professor of herpetology and ichthyology at the Jardin des plantes, Paris, in 1813. (*DBF*, Gilbert 1977.)

Dunal, Michel Félix (1789–1856). French botanist. Professor of botany, Montpellier, from 1829. (*DBF*.)

Duncan, Peter Martin (1821–91). Physician, botanist, zoologist, invertebrate palaeontologist, and stratigrapher. Practised medicine in Colchester, 1848–60; then in Blackheath. Devoted much of his spare time to the study of corals. Appointed professor of geology, King's College, London, in 1870. Shortly afterwards, appointed professor of geology at Cooper's Hill College. Made a special study of living and fossil corals and echinids, also taking much interest in sponges and protozoa. FRS 1868. (*DNB*, Sarjeant 1980.)

18 July [1861]

Durocher, Joseph Marie Elisabeth (1817–60). Mining engineer. Accompanied voyage around Scandinavia under the auspices of the Commission scientifique du Nord, 1838–40. Appointed assistant professor of geology in the science faculty at Rennes in 1841. Studied and theorised on the glacial phenomena of Scandinavia. (*BHGW*, *DBF*.)

Earle, Augustus (1793–1838). Artist and traveller. In America, 1818–24; in Australia and New Zealand, 1825–8. Returned to England in 1830. Joined HMS *Beagle* as artist in 1831. Left at Montevideo due to ill health; his position was taken by Conrad Martens. (*Aust. dict. biog.*, Bénézit 1976.)

Edgeworth, Michael Pakenham (1812–81). Irish botanist who collected plants in Aden, India, and Ceylon. Commissioner for the settlement of the Punjab, 1850. (R. Desmond 1977, *DNB*.)

Edmondston, Laurence (1795–1879). Physician and naturalist. Had a medical practice in Unst, the most northerly of the Shetland Islands. Was an accomplished chemist, archaeologist, linguist, and musician. Made numerous additions to the list of British birds. (*DNB*.)

Egerton, Francis Henry, 8th earl of Bridgewater (1756–1829). Scholar and patron of learning. His literary works, many on his notable ancestors, were mostly printed for private circulation. In his will, he bequeathed 8000 guineas for a work on 'The goodness of God as manifested in the Creation'. The disposal of the money was left to the president of the Royal Society, who divided it among eight persons who were each to write a suitable essay. The resulting essays have since become known as the 'Bridgewater treatises'. FRS 1781. (*DNB*.)

Egerton, Philip de Malpas Grey- (1806–81). Of Oulton Park, Cheshire. Tory MP for South Cheshire, 1835–68. Palaeontologist who specialised in fossil fish. FRS 1831. (*DNB*, Sarjeant 1980.)

Ellice, Russell. Partner in the London bank, Robarts, Lubbock & Co. (*Banking almanac* 1861.)

Engleheart, Stephen Paul. Surgeon in Down, Kent. (*Post Office directory of the six home counties* 1862.)

Erichson, Wilhelm Ferdinand (1808–49). German entomologist. Extraordinary professor and assistant at the zoological museum, University of Berlin. Edited, after 1842, the *Archiv für Naturgeschichte*. (Gilbert 1977.)

Erskine, Thomas (1788–1864). Judge. Chief judge of the Court of Review in Bankruptcy, 1831–42. Judge of the Court of Common Pleas, 1839–44. Retired in 1844 owing to poor health. A friend and supporter of Charles Kingsley. (*DNB*.)

Euclid. Greek mathematician of the third century B.C. Author of *Elements*, a study of plane geometry comprising thirteen books. (*EB*.)

Eulenburg, Friedrich Albrecht, Graf zu (1815–81). Prussian statesman. Entered the diplomatic service in 1852. Led the Prussian East Asia expedition, 1859–62, during which he settled trade deals with Japan, China, and Siam. In 1862, appointed minister of the interior. (*NDB*.)

Eyton, Thomas Campbell (1809–80). Shropshire naturalist. Friend and Cambridge contemporary of CD. Author of several works on natural history. On coming into possession of the family estate at Eyton, Shropshire, in 1855, he built a museum in which he formed a fine collection of skins and skeletons of the birds of Europe. (*DNB*.)

3 May [1861], 6 May [1861], 14 May [1861]

Fairbeard, William and Julius. Nurserymen at Teynham, Sittingbourne, Kent. (*Post Office directory of the six home counties* 1855.)

Falconer, Hugh (1808–65). Palaeontologist and botanist. Superintendent of the botanic garden, Saharanpur, India, 1832–42. Superintended the arrangement of Indian fossils for the British Museum, 1844. Superintendent of the Calcutta botanic garden and professor of botany, Calcutta Medical College, 1848–55. FRS 1845. (*DNB*, *DSB*.)

23 June 1861, [24 June 1861]

Fall, Phillip Carteret. Partner in the Southampton and Hampshire Bank, Southampton until 1861. (*Banking almanac* 1861.)

Fallén, Carl Frederik (1764–1830). Swedish entomologist. Appointed professor of natural history at Lund University in 1812. (*SBL*.)

Fawcett, Henry (1833–84). Statesman. BA, Cambridge, 1856. Blinded in a shooting accident in 1858. Appointed professor of political economy, Cambridge University, 1863. MP for Brighton, 1865–74. Became MP for Hackney and postmaster general in 1880. FRS 1882. (*DNB*.)

16 July [1861], 20 July 1861, 18 September [1861]

Felton, Cornelius Conway (1807–62). American philologist. Eliot professor of Greek literature, Harvard, 1834–60. President of Harvard University, 1860–2. (*DAB*.)

Ferguson, George. Author of a work on fancy poultry. (*CDEL*.)

Fermond, Charles (1810 – *c.* 1875). French pharmacist and botanist. Head pharmacist at the Salpetrière hospital, Paris, 1837. His botanical studies encompassed the germination of plants, their movements, flowering, fertilisation, hybridisation, and the production of monstrosities. (*DBF*.)

Fleurière, A. de la. Staff writer or reviewer for the *Leader*.

Field

[before 27 April 1861], [before 4 May 1861], [before 25 May 1861], [before 15 June 1861]

Fitch, Walter Hood (1817–92). Botanical artist at the Royal Botanic Gardens, Kew, from 1841. Produced over 10,000 published drawings. Among the works he illustrated was Joseph Dalton Hooker's *Illustrations of Himalayan Plants* (1855). (Gilbert 1977, *Proceedings of the Linnean Society of London* (1891–2): 68.)

FitzRoy, Robert (1805–65). Naval officer, hydrographer, and meteorologist. Commander of HMS *Beagle*, 1828–36. Tory MP for Durham, 1841–3. Governor of New Zealand, 1843–5. Superintendent of the dockyard at Woolwich, 1848–50. Chief of the meteorological department of the Board of Trade, 1854. Vice-admiral, 1863. FRS 1851. (*DNB*, *DSB*.)

Forbes, Edward (1815–54). Zoologist, botanist, and palaeontologist. Naturalist on board HMS *Beacon*, 1841–2. Appointed professor of botany, King's College, London, and curator of the museum of the Geological Society in 1842. Palaeontologist with the Geological Survey, 1844–54. Professor of natural history, Edinburgh University, 1854. FRS 1845. (*DNB*, *DSB*.)

Forbes, James David (1809–68). Professor of natural philosophy, Edinburgh University, 1833–60. Secretary of the Royal Society of Edinburgh, 1840–51. Principal of the United College, University of St Andrews, 1860–8. FRS 1832. (*DNB*, *DSB*.)

Forchhammer, Johan Georg (1794–1865). Danish geologist and chemist. Became a lecturer in geology at the University of Copenhagen in 1821; professor of mineralogy and geology, 1831. Professor of chemistry and mineralogy, Polytechnic Institute, and manager of one of its two chemical laboratories. Secretary, Royal Danish Academy of Sciences, 1851–65. (*DBL*, *DSB*.)

Forrest, Captain. Of West Hill, Winchester. Chief constable of Hampshire. (*Post Office directory of Hampshire, Wiltshire, and Dorsetshire* 1859.)

Fox, Eliza Ann (1801–86). Sister of William Darwin Fox. (*Darwin pedigree*.)

Fox, Harriet Emma. Daughter of Harriet and William Darwin Fox. (*Darwin pedigree*.)

Fox, John Woodham. Of Grinstead Green, Farnborough, Kent. (*Post Office directory of the six home counties* 1859.)

Fox, Thomas Samuel. Ensign in the Farnborough Rifle Volunteer Corps. (*Army list* 1861.)

Fox, William Darwin (1805–80). Clergyman. CD's second cousin. A close friend of CD's at Cambridge who shared his enthusiasm for entomology. Maintained an active interest in natural history throughout his life and provided CD with much information. Rector of Delamere, Cheshire, 1838–73. Spent the last years of his life at Sandown, Isle of Wight. (*Alum. Cantab.*)

9 January [1861], 8 July [1861]

Frankland, Edward (1825–99). Professor of chemistry at Owens College, Manchester, 1851–7. Lecturer in chemistry, St Bartholomew's Hospital, London, 1857–64. FRS 1853. (*DNB*, *DSB*.)

Freke, Henry. Irish physician. Resident medical officer, Dr Steevens's Hospital, Dublin. Physician, Fever Hospital, Kilmainham. Lecturer on clinical medicine, and later, professor of physic, Dr Steevens's Hospital Medical College. (*Catalogue of graduates, University of Dublin*; *Medical directory for Ireland* 1860.)

Gardeners' Chronicle

[before 5 January 1861], [before 9 February 1861], [before 15 June 1861], [before 14 September 1861], [before 14 September 1861]

Gärtner, Karl Friedrich von (1772–1850). German physician and botanist. Practised medicine in Calw, Germany, from 1802. Contracted an eye ailment in the course of his microscopical investigations that led him to discontinue them and cease medical practice. From about 1824, he devoted his time to the study of plant hybridisation. (*DSB*, *NDB*.)

Gassiot, John Peter (1797–1877). Wine merchant and scientific writer. Chairman of Kew Observatory. Carried out electrical experiments. FRS 1840. (*DNB*, *DSB*.)

Geikie, Archibald (1835–1924). Scottish geologist. Appointed member of the Scottish branch of the Geological Survey in 1855. Director of the Geological Survey of Scotland, 1867–82. Director-general of the Geological Survey of Great Britain,

1882–1901. Murchison professor of geology and mineralogy, Edinburgh University, 1871–81. FRS 1865. (*DNB*, *DSB*.)

Geoffroy Saint-Hilaire, Isidore (1805–61). French zoologist. Succeeded his father, Étienne Geoffroy Saint-Hilaire, as professor at the Muséum d'histoire naturelle in 1841. Continued his father's work in teratology. Became professor of zoology at Paris in 1850. (*DBF*, *DSB*.)

Gescheidt, Ludovic Anton (b. 1808). Physician and ophthalmologist. (*BLA*, *NUC*.)

Gesner, Konrad (1516–65). Swiss classical scholar who was the author of several treatises on natural history and philology. Interested in systematic botanical and zoological classification, being one of the first botanists to realise the importance of floral structures as a means of establishing a systematic key to the classification of plant life. (*DSB*.)

Glennie, John David, Jr (1825–1903). Clergyman. Curate of Postling, Kent, 1849–51; of Lyminge, 1851–3. Diocesan Inspector, London, 1853–7. HM Inspector for London and Middlesex, 1857–8; for Norfolk, Suffolk, and Essex, 1858–60. Curate of Blore Ray, Staffordshire, 1861–9. Vicar of Croxton, near Eccleshall, 1869–1903. (*Alum. Cantab.*)

6 April 1861

Glover, Thomas. Of Smedley Hill, Cheetham, near Manchester. Breeder of cactuses. (*Post Office directory of Lancashire* 1858.)

Goodwin, Charles Wycliffe (1817–78). Egyptologist and lawyer. Called to the bar, 1848. Contributed to the publications of the Cambridge Antiquarian Society. In 1860 wrote one of the articles in *Essays and reviews*, to which he was the only lay contributor. Appointed assistant judge in the supreme court for China and Japan in 1865. Died in Shanghai. (*DNB*.)

Gordon, George (1801–93). Scottish botanist and clergyman. Minister of Birnie, near Elgin, 1832–89. (Gilbert 1977.)

17 June [1861], 22 June [1861], 4 July [1861], 6 July [1861]

Gosse, Philip Henry (1810–88). Naturalist, traveller, and writer. Studied marine invertebrates. FRS 1856. (*DNB*.)

Gould, Charles (1834–95). Son of John Gould. Emigrated to Tasmania and was employed by the Tasmanian Geological Survey. (*Aust. dict. biog*, Tree 1991, p. 209.)

6 October [1861]

Gould, John (1804–81). Ornithologist and artist. Taxidermist to the Zoological Society of London, 1826–81. Travelled in Australia, 1838–40. Described the birds collected on the *Beagle* and *Sulphur* expeditions. Published forty-one illustrated folio volumes. FRS 1843. (*DNB*, *DSB*.)

6 October [1861]

Gower, William Hugh (1835–94). Foreman, Royal Botanic Gardens, Kew, until 1865. Employed by various London nurseries. Principally known for his knowledge of orchids. (R. Desmond 1977.)

23 November 1861

Graham, Thomas (1805–69). Chemist. Professor of chemistry at the Andersonian Institution, Glasgow, 1830–7; University College London, 1837–55. Master of the mint, 1855–69. FRS 1836. (*DNB*, *DSB*.)

Grant, Robert Edmond (1793–1874). Scottish physician and zoologist. An early advocate of the transmutation of species. Befriended CD in Edinburgh. Professor of comparative anatomy and zoology, University College London, 1827–74. FRS 1836. (*DNB*, Sarjeant 1980.)

16 May 1861

Gratiolet, Louis Pierre (1815–65). French anatomist and anthropologist. Laboratory assistant to Henri de Blainville at the Muséum d'histoire naturelle, Paris, 1842–50. Lectured on anatomy at the museum as deputy to Blainville, 1844–50. Continued as laboratory assistant until 1853, when he was put in charge of anatomical studies at the museum. Deputy professor of zoology, faculty of science, Paris, 1862–3; professor, 1863–5. (*DSB*.)

Grattan, Thomas Colley (1792–1864). Writer and statesman. Began his career as a writer after settling in Bordeaux. His *High-ways and by-ways* (1823) made his name both in England and on the Continent. Resided for some time in Belgium and, partly owing to the patronage of King Leopold, was appointed British consul to the state of Massachusetts in 1839. After a successful period in office, he resigned the consulship in 1846 and returned to England where he resumed his literary work. (*DNB*.)

Gray, Asa (1810–88). American botanist. Fisher professor of natural history, Harvard University, 1842–73. Devoted much time to the Harvard botanic garden and herbarium. Foreign member, Royal Society, 1873. (*DAB*, *DSB*.)

23 [January 1861], 17 February [1861], 26–7 February [1861], 12 March [1861], 11 April [1861], 5 June [1861], 21 July [1861], *[27 and 29 August] and 2 September [1861]*, 16 September [1861], 17 September [1861], *11 October 1861*, [after 11 October 1861], *9 November 1861*, 11 December [1861], *31 December 1861*

Gray, George Robert (1808–72). Zoologist; an expert on insects and birds. Assistant in the zoological department of the British Museum, 1831–72. Brother of John Edward Gray. FRS 1865. (*DNB*, Gilbert 1977.)

Gray, Jane Loring. Wife of Asa Gray. (Dupree 1959.)

Gray, John Edward (1800–75). Naturalist. Assistant keeper of the zoological collections of the British Museum, 1824; keeper, 1840–74. FRS 1832. (R. Desmond 1977, *DNB*.)

Greene, Joseph Reay. Irish naturalist. One of the editors of the *Natural History Review*. Studied Protozoa and coelenterates. (*Catalogue of graduates, University of Dublin*; L. Huxley ed. 1900.)

24 September [1861]

Grey-Egerton, Philip de Malpas. *See* Egerton, Philip de Malpas Grey-.

Grove, William Robert (1811–96). Judge and physical scientist. Professor of experimental philosophy, London Institution, 1841–6. An active member of the Royal

Society; treasurer and chairman of the executive committee of the Philosophical Club, 1847. Became a member of the Queen's Counsel in 1853. Defended William Palmer, the 'Rugeley poisoner', in 1856. FRS 1840. (*DNB*, *DSB*.)

Guépin, Jean Baptiste Pierre (1778–1858). French botanist. (*NUC*.)

Guillaumin et cie. French publishers. Founded in 1833 by Gilbert Urbain Guillaumin (1801–64) initially to publish works on political economy and business. (*DBF*, *NUC*.)

Gully, James Manby (1808–83). Physician. Practised medicine in London, 1830–42. In 1842, he set up a hydropathic establishment in Malvern and was a successful practitioner of the water treatment until his retirement in 1872. His reputation suffered when, in 1876, his affair with Mrs Bravo, who was accused of poisoning her husband, was revealed. (*DNB*, *Modern English biography*.)

Günther, Albrecht Carl Ludwig (Albert Charles Lewis) Gotthilf (1830–1914). German-born zoologist. Began his association with the British Museum in 1857 when he was charged with completing the museum's catalogues of amphibia, reptiles, and fish. Officially joined the staff of the museum in 1862. Assistant keeper of the zoological department, 1872–5; keeper, 1875–95. Edited the *Record of Zoological Literature*, 1865–9. FRS 1867. (*DNB*, *NDB*.)
14 June 1861

Haast, John Francis Julius (Julius) von (1824–87). German geologist who emigrated to New Zealand in 1858. Appointed provincial geologist, 1861. Conducted the first geological survey of Canterbury province, 1861–8. Knighted, 1886. FRS 1867. (*DNB*, *DSB*, *ENZ*, Sarjeant 1980.)

Hacon, William Mackmurdo (d. 1885). CD's solicitor. Member of the firm Rowland & Hacon, with premises in Fenchurch Street, London. (CD's Address book (Down House MS), Freeman 1978.)

Hall, James, 4th baronet (1761–1832). Geologist and chemist. Succeeded to the baronetcy in 1776. Travelled in Europe, 1783–6, where he met several European scientists and became actively interested in geology and chemistry. Was a strong supporter of James Hutton, carrying out experimental work to counter criticisms of Hutton's *Theory of the earth* (1795). MP for Michael, Cornwall, 1807–12. Became president of the Royal Society of Edinburgh in 1812. FRS 1806. (*DNB*, *DSB*.)

Hall, James (1811–98). American geologist and palaeontologist. Palaeontologist to the New York State geological survey from 1843. Also state geologist of Iowa, 1855–8; of Wisconsin, 1857–60. Appointed director of New York State Museum in 1866; made state geologist in 1893. (*DAB*, *DSB*.)

Harcourt, Edward William Vernon (1825–91). An active member of the artillery volunteer movement from 1862. Sheriff of Oxfordshire, 1875. MP for Oxfordshire, 1878–85; for South Oxfordshire, 1885–6. Studied the natural history of Madeira. (*Modern English biography*.)

Hardinge, Charles Stewart, 2d viscount (1822–94). An accident in youth meant Hardinge had to use an artificial leg and was prevented from following his intended profession in the army. Accompanied his father, Henry Hardinge, when

Hardinge, Charles Stewart, cont.
he was appointed governor-general of India, as his private secretary, 1844–8. MP for Downpatrick, 1851–6. Under-secretary for war, 1858–9. Appointed trustee of the National Portrait Gallery in 1868; chairman of the trustees, 1876–94. (*DNB*.)

Hardy, Charles (1803–85). Clergyman. Curate of the Subdeanery Church, Chichester, 1827–32. Vicar of North and South Hayling, Hampshire, 1832–80. (*Alum. Cantab.*, *Crockford's*.)
4 September [1861]

Hardy, William. Bailiff to Charles Augustus Bennet, earl of Tankerville, and steward of the wild cattle of Chillingham Castle, Northumberland. (*Post Office directory of Northumberland* 1858.)
[8 May 1861] [To Luke Hindmarsh]

Hartmann, Georg Leonhard (1764–1828). Swiss zoologist and ichthyologist. (*NUC*.)

Harvey, William Henry (1811–66). Irish botanist. Colonial treasurer in Cape Town, 1836–42. Keeper of the herbarium, Trinity College, Dublin, from 1844; professor of botany, 1856–66. (*DNB*, *DSB*.)

Hauksbee, Francis (*c.* 1666–1713). Experimental natural philosopher and scientific instrument maker. Demonstrator and curator of instruments, Royal Society of London, 1704–13. Known for his improved model of the air pump. (*DSB*.)

Hector, James (1834–1907). Scottish geologist and naturalist. Assistant to Edward Forbes, professor of natural history at Edinburgh University, 1854. Surgeon and geologist to the British North American Exploring Expedition, 1857–60. In 1861, Hector was appointed geologist to the provincial government in Otago, New Zealand, becoming director in 1865. Director of the meteorological and weather department of the New Zealand Institute and of the colonial museum and botanical gardens, 1866–1903. FRS 1866. (*DNB*.)

Heer, Oswald (1809–83). Swiss botanist, palaeontologist, and entomologist. An expert on Tertiary flora. Lecturer in botany, University of Zürich, 1834–5; undertook direction of botanic garden in 1834; associate professor, 1835–52; professor of botany and entomology, 1852–83. Great traveller and mountain climber. He contracted tuberculosis in 1849, but, after spending the winter of 1850–1 on Madeira, his condition improved; however, the illness returned in 1871, preventing further mountain excursions. (*DSB*, *NDB*.)
20 April [1861?]

Henry, Joseph (1797–1878). American physicist. Professor of mathematics and natural philosophy, Albany Academy, New York, 1826–32. Professor of natural philosophy, College of New Jersey at Princeton, 1832–46. Became the first secretary and director of the Smithsonian Institution in 1846. Elected member of the American Philosophical Society in 1835; helped to organise the American Association for the Advancement of Science. (*DAB*.)

Henslow, John Stevens (1796–1861). Clergyman, botanist, and mineralogist. Professor of mineralogy, Cambridge University, 1822–7; professor of botany, 1825–61.

Extended and remodelled the Cambridge botanic garden. Curate of Little St Mary's Church, Cambridge, 1824–32; vicar of Cholsey-cum-Moulsford, Berkshire, 1832–7; rector of Hitcham, Suffolk, 1837–61. CD's teacher and friend. (*DNB*, *DSB*.)

Herbert, William (1778–1847). Naturalist, classical scholar, linguist, politician, and clergyman. Noted for his work on plant hybridisation. Rector of Spofforth, Yorkshire, 1814–40. Dean of Manchester, 1840–7. (*DNB*, *DSB*.)

Herschel, John Frederick William (1792–1871). Astronomer, mathematician, chemist, and philosopher. Member of many learned societies. Carried out astronomical observations at the Cape of Good Hope, 1834–8. Master of the mint, 1850–5, retiring after a bout of illness. Created baronet, 1838. FRS 1813. (*DNB*, *DSB*.)
23 May [1861]

Heurtley, Charles Abel (1806–95). Clergyman and academic. Curate, Wardington, Oxfordshire, 1831–40. Rector, Fenny Compton, Warwickshire, 1840–72. Lady Margaret professor of Divinity, Oxford, 1853–95. (*Alum. Cantab.*, *DNB*.)

Hewitt, Edward. Judge of poultry exhibitions. Resided in Sparkbrook, Birmingham. (Tegetmeier ed. 1856–7, p. 89.)

Higgins, Henry Hugh (1814–93). Clergyman and geologist. Inspector of National Schools, Liverpool, 1842–8. Chaplain of Rainhill Asylum, Liverpool, 1853–86. President of the Naturalist's Field Club, Liverpool, 1861–81. (*Alum. Cantab.*, *Crockford's*.)

Hindmarsh, Luke. A tanner and timber merchant in Alnwick, Northumberland. (*Post Office directory of Northumberland* 1858.)
3 May [1861], 12 May [1861]

Hoeven, Jan van der (1801–68). Dutch comparative anatomist and anthropologist. Lecturer on zoology, Leiden, 1826–68. An adherent of Cuvier's theory of catastrophism; believed in creation rather than evolution, attempting in his works to show how the infinite wisdom of the Creator is reflected in the harmony of the world's creatures. (*BWN*, *DSB*.)

Hofacker, J. D. German author of work on heredity in man and animals (1828).

Hogg, Robert (1818–97). Horticulturist and pomologist. Gained practical horticultural training with Peter Lawson and Sons, Edinburgh, and Mr Ronalds, Brentford, after which he studied in Paris, Belgium, and Germany. Became a partner with Gray and Adams, nurserymen of Brompton Park, London in 1845; the business was dissolved in 1851. Co-editor of the *Cottage Gardener* (*Journal of Horticulture*), 1855–95. (R. Desmond 1977.)

Holland, Henry (1788–1873). Physician. Distant relative of the Darwins and Wedgwoods. Travelled extensively. Physician in ordinary to Queen Victoria, 1852. President of the Royal Institution for many years. Created baronet, 1853. FRS 1815. (*DNB*, *Physicians*.)

Hooker, Brian Harvey Hodgson (b. 1860). Fifth child of Frances Harriet and Joseph Dalton Hooker. (Allan 1967.)

Hooker, Charles Paget (1855–1933). Son of Frances and Joseph Dalton Hooker. (Allan 1967.)

Hooker, Frances Harriet (1825–74). Daughter of John Stevens Henslow. Married Joseph Dalton Hooker in 1851. (R. Desmond 1977.)

Hooker, Harriet Anne (1854–1945). Daughter of Frances and Joseph Dalton Hooker. (Allan 1967.)

Hooker, Joseph Dalton (1817–1911). Botanist. Accompanied James Clark Ross on his Antarctic expedition, 1839–43, and published the botanical results of the voyage. Appointed palaeobotanist to the Geological Survey of Great Britain in 1846. Travelled in the Himalayas, 1847–50. Assistant director, Royal Botanic Gardens, Kew, 1855–65; director, 1865–85. Worked chiefly on taxonomy and plant geography. Son of William Jackson Hooker. Friend and confidant of CD. FRS 1847. (*DNB*, *DSB*.)

15 January [1861], 4 February [1861], 8 [February 1861], 20 [February 1861], 11 March [1861], 18 March [1861], 26 [March 1861], 27 [March 1861], 4 April [1861], 11 April [1861], 14 [April 1861], 23 [April 1861], 23 [April 1861], 14 May [1861], 18 [May 1861], 24–5 May [1861], 30 May [1861], 19 June [1861], 22 June [1861], [6 July 1861], 13 July [1861], 17 [July 1861], 27 July [1861], [28 July – 10 August 1861], [11 August 1861], 13 [August 1861], 30 August [1861], 6 September [1861], 24 September [1861], *[28 September 1861]*, 28 September [1861], 4 October [1861], 6–7 October [1861], 11 October [1861], 13 October [1861], 15 [October 1861], 18 October [1861], 23 October [1861], 27 October [1861], 1 November [1861], 1 November [1861], 5 [November 1861], 6 [November 1861], 7 November [1861], 8 November [1861], 10 November [1861], 14 November [1861], 17 November [1861], 20 [November 1861], [22 November 1861], 25 November [1861], 1 December [1861], [9 December 1861], 18 [December 1861], 27 [December 1861], 28 [December 1861], *[29 December 1861]*, [30 and 31 December 1861]

Hooker, Maria Elizabeth (1857–63). Daughter of Joseph Dalton Hooker and Frances Harriet Hooker. (Allan 1967.)

Hooker, William Henslow (1853–1942). Oldest child of Frances Harriet and Joseph Dalton Hooker. (Allan 1967.)

Hooker, William Jackson (1785–1865). Botanist. Regius professor of botany, Glasgow University, 1820. Established the Royal Botanic Gardens at Kew in 1841 and served as first director. Father of Joseph Dalton Hooker. FRS 1812. (*DNB*, *DSB*.)

Hope, Frederick William (1797–1862). Entomologist and clergyman. Gave his collection of insects to Oxford University and founded a professorship of zoology, 1849. FRS 1834. (*DNB*, Gilbert 1977.)

Hopkins, William (1793–1866). Mathematician and geologist. A highly successful mathematics tutor at Cambridge. President of the Geological Society, 1851–3.

Specialised in quantitative studies of geological and geophysical questions. FRS 1837. (*DNB*, *DSB*.)

Hørbye, Jens Carl (1815–1906). Norwegian geologist and forester. Scientific officer, mineralogy, University of Kråkstad, 1844–57. Head forester, county of Hedemarkens, 1860–75; of Akershus and Smålenenes district, 1875–89. (*NBL*.)

Horner, Anne Susan (1789–1862). Married Leonard Horner in 1806. (Freeman 1978.)

Horner, Leonard (1785–1864). Geologist and educationist. Warden of University College London, 1827–31. Inspector of factories, 1833–56. A promoter of science-based education at all social levels. President of the Geological Society, 1845–7, 1860–2. Father-in-law of Charles Lyell and Charles James Fox Bunbury. FRS 1813. (*DNB*, *DSB*.)
14 February [1861], 20 March [1861]

Horwood, Mr. Gardener to George Henry Turnbull of Down, Kent.

Humboldt, Friedrich Wilhelm Heinrich Alexander (Alexander) von (1769–1859). Eminent Prussian naturalist and traveller. Official in the Prussian mining service, 1792–6. Explored equatorial South America, 1799–1804. Travelled in Siberia, 1829. Foreign member, Royal Society, 1815. (*DSB*, *NDB*.)

Hunt, James (1833–69). Speech therapist and anthropologist. Honorary secretary of the Ethnological Society of London, 1859–62. Founder and first president of the Anthropological Society of London, 1863. (*DSB*.)
28 May [1861]

Hunt, Thornton Leigh (1810–73). Journalist. Contributed regularly to the *Spectator*, 1840–60; also wrote for other newspapers and for magazines. In 1850, helped his friend George Henry Lewes to establish the *Leader*. In 1855, joined the staff of the *Daily Telegraph*, writing mainly on political subjects. (*DNB*.)

Hutchinson, William Nelson (1803–95). Joined the army in 1820 and rose steadily through the ranks. Colonel in the Duke of Wellington's West Riding regiment, 1863–95. Patented inventions relating to the steering of balloons and to steam engines. Author of a book on dog-breaking (1848). (*Modern English biography*.)

Hutton, Frederick Wollaston (1836–1905). Geologist and army officer. Commissioned in Royal Welsh Fusiliers, 1835; captain, 1862. Appointed assistant geologist to the New Zealand geological survey in 1871; provincial geologist of Otago, 1873. Professor of natural science, University of Otago, 1877–90. Professor of biology and geology, Canterbury University College of the University of New Zealand at Christchurch, 1890–3. Curator of the Canterbury Museum, 1893–5. (*DNB*, *ENZ*.)
20 April 1861

Hutton, James (1726–97). Scottish natural philosopher and geologist. Propounded a uniformitarian view of geological history in his *Theory of the earth* (1795). (*DNB*, *DSB*.)

Hutton, Mrs Lewis. Of Sydney, New South Wales, Australia.

Huxham, John (1692–1768). Physician. Practised medicine in Plymouth. Author of *An essay on fevers and their various kinds* (1755), which established his reputation as a physician and writer on medical subjects. FRS 1739. (*DNB*.)

Huxham, John Corham. Son of John Huxham. Edited several of his father's works. (*Alum. Oxon.*, *DNB*.)

Huxley, Henrietta Anne (Nettie) (1825–1915). Lived in Australia, where she met Thomas Henry Huxley in 1847. Came to England and married Huxley in 1855. (Freeman 1978.)

Huxley, Jessie Oriana (b. 1858). Daughter of Henrietta Anne and Thomas Henry Huxley. (Bibby 1959.)

Huxley, Leonard (1860–1933). Son of Thomas Henry Huxley. Biographer. Assistant editor of the *Cornhill Magazine*, 1901–16; editor, 1916–33. (*DNB*.)

Huxley, Mariane (b. 1859). Daughter of Henrietta Anne and Thomas Henry Huxley. (Bibby 1959.)

Huxley, Noel (1856–60). Oldest child of Henrietta Anne and Thomas Henry Huxley. Died of scarlet fever. (L. Huxley ed. 1900.)

Huxley, Thomas Henry (1825–95). Assistant-surgeon in HMS *Rattlesnake*, 1846–50, during which time he investigated Hydrozoa and other marine invertebrates. Lecturer on natural history, Royal School of Mines, 1854; professor, 1857. Appointed naturalist to the Geological Survey of Great Britain in 1855. Served on several Royal commissions. Hunterian professor of comparative anatomy, Royal College of Surgeons, 1863–9. Fullerian professor of physiology, Royal Institution, 1863–7. President of the Royal Society, 1883–5. FRS 1851. (*DNB*, *DSB*.)

3 January [1861], 17 February [1861], 22 February [1861], 1 April [1861], 22 May [1861], 22 October [1861], 31 October [1861]

Innes, John (1817–94). Curate of Farnborough, Kent, 1842. Perpetual curate of Down, 1846; vicar, 1860–9. Left Down in 1862 after inheriting property in Scotland; changed his surname to Brodie Innes *c.* 1860. Priest in charge of Milton Brodie Mission from 1861. Chaplain to the bishop of Moray, 1861–80. (*Crockford's*, Freeman 1978.)

[before 6 April 1861], 15 December [1861], 19 December [1861], *[24 December 1861]*

Innes, John William (1848–1923). Son of John Innes. Barrister and novelist. Called to the bar, 1876. Advocate of the Scottish bar, 1888. Interested in antiquarian research, romance, demonology, witchcraft, and criminology. (*Alum. Cantab.*, Freeman 1978).

Jacquin, Nikolaus Josef (1727–1817). Dutch botanist. Studied medicine in Paris before moving to Vienna to continue his studies with the financial help of a family friend. Sent to West Indies and South America by Francis I for the purpose of enlarging the imperial natural history collections, 1754–9. Appointed professor of chemistry and mineralogy at the Mining School in Schemnitz, Hungary, in 1763. Professor of botany and chemistry, University of Vienna, 1768–96. Created baron, 1806. (*BWN*, *DSB*, *NBU*.)

James, Henry (1803–77). Lieutenant, Royal Engineers, 1831; captain, 1846; colonel, 1857. Joined the staff of the Ordnance Survey in 1827. Local superintendent, Geological Survey of Ireland, 1843–6. Transferred to admiralty employment in 1846, but returned to the Ordnance Survey in 1850; director-general, 1854–75. Director of the topographical and statistical department of the War Office, 1857–70. (*DNB*.)

Jamieson, Thomas Francis (1829–1913). Scottish agriculturalist of Ellon, Aberdeenshire. Factor on the Ellon estate for many years; later took the farm of Mains, Waterton. Appointed Fordyce lecturer on agricultural research, University of Aberdeen in 1862. Carried out notable researches on Scottish Quaternary geology and geomorphology. Became a fellow of the Geological Society in 1862. (*Geological Magazine* 50 (1913): 332–3, Sarjeant 1980.)
13 June 1861, *3 September 1861*, 6 September [1861], *24 October 1861*

Jeffreys, John Gwyn (1809–85). Conchologist, zoologist, and solicitor. Conducted extensive deep-sea dredging researches. FRS 1840. (*DNB*, *DSB*.)

Jenyns, Leonard (1800–93). Naturalist and clergyman. Brother-in-law of John Stevens Henslow. Vicar of Swaffham Bulbeck, Cambridgeshire, 1828–49. Moved near Bath in 1850 and settled in Bath itself in 1860. Founded the Bath Natural History and Antiquarian Field Club in 1855. Member of many scientific societies. Described the *Beagle* fish specimens. Adopted the name Blomefield in 1871. (R. Desmond 1977, *DNB*.)
31 December 1861

Johnson, George William (1802–86). Writer on gardening. Professor of moral and political economy, Calcutta, 1839–42. Founded the *Cottage Gardener* in 1848. (R. Desmond 1977, *DNB*.)

Jones, Henry Bence (1814–73). Physician and chemist. Physician to St George's Hospital, 1846–62. Acquired a large and remunerative practice. In his chemical studies, devoted himself especially to the application of chemistry to pathology and medicine. FRS 1846. (*DNB*.)
23 June [1861]

Jones, Thomas Rupert (1819–1911). Geologist and palaeontologist. Assistant-secretary, Geological Society of London, 1851–62. Lecturer on geology, Royal Military College, 1858; professor, 1862. Afterwards, professor at the Staff College, Sandhurst; retired, 1880. An authority on Phyllopoda, Ostracoda, and Foraminifera. (*DNB*.)

Jones, William (1746–94). Orientalist and jurist. Judge of the high court in Calcutta, 1783–94. Founded the Asiatic Society of Bengal in 1784. FRS 1772. (*DNB*.)

Journal of Horticulture
[before 14 May 1861], [17 May 1861], [before 18 June 1861],
[before 9 July 1861], [before 10 September 1861], [before 22 October 1861]

Jowett, Benjamin (1817–93). Master of Balliol College and Regius professor of Greek, Oxford University, 1855–93. Author of an article in *Essays and reviews* (1860). (*DNB*.)

Jukes, Joseph Beete (1811–69). On the staff of the Geological Survey of Newfoundland, 1839–40. Naturalist aboard HMS *Fly*, 1842–6. Worked in North Wales for the Geological Survey of Great Britain, 1846–50. Director of the Irish branch of the Geological Survey, 1850–69. FRS 1853. (*DNB*, *DSB*.)

Jussieu, Adrien Henri Laurent de (1797–1853). French botanist. In 1826, succeeded his father, Antoine de Jussieu, as professor of botany at the Muséum d'histoire naturelle, Paris, where he helped to build up a large herbarium. (*DSB*, *NBU*.)

Kerr, Robert (1755–1813). Scientific writer and translator. Entered the medical profession as surgeon to the Edinburgh Foundling Hospital, but left to manage a paper mill. After the failure of this venture, occupied himself with historical and biographical work. (*DNB*.)

Key, Henry Cooper (1819–79). Rector of Stretton-Sugwas, Herefordshire, 1846–79. (*Alum. Oxon.*, *Clergy list*.)

Keyserling, Alexander Andreevich (1815–91). Russian geologist, palaeontologist, and botanist. Accompanied Roderick Impey Murchison on his expedition to Russia in 1841. Official handling special missions in the Mining Department of the Russian government, 1841–50. Thereafter, travelled very little until 1860 when he made several journeys to the Pyrenees with Edouard de Verneuil. (*DSB*, Sarjeant 1980.)

Kingsley, Charles (1819–75). Author and clergyman. Lecturer on English literature, Queen's College, London, 1848–9. Professor of modern history at Cambridge University, 1860–9. Rector of Eversley, Hampshire, 1844–75. Chaplain to the Queen, 1859–75. (*Alum. Cantab.*, *DNB*.)

Kippist, Richard (1812–82). Botanist. Librarian of the Linnean Society of London, 1842–81. Specialist on Australian plants. (R. Desmond 1977, *DNB*.)
7 January [1861], 13 February [1861?]

Kirby, William (1759–1850). Clergyman and entomologist. Vicar of Barham, Suffolk, 1797–1850. One of the first fellows of the Linnean Society, 1788. FRS 1818. (*Alum. Cantab.*, *DNB*.)

Kjerulf, Theodor (1852–88). Lecturer in mineralogy and geology, University of Christiania, Norway, 1858; professor, 1866. (*BHGW*, *NBL*.)

Koch, Wilhelm Daniel Joseph (1771–1849). German physician and botanist. Professor of botany, Erlangen University, 1824. (*ADB*.)

Köstlin, Otto (1818–84). Physician and teacher. Professor of natural sciences at the grammar school in Stuttgart, 1846–81. (*ADB*, *BLA*.)

Kunth, Karl Sigismund (1788–1850). German botanist. Between 1813 and 1829, worked in Paris on the plants collected by Alexander von Humboldt on his travels in South America. Appointed professor of botany at Berlin in 1829. (*ADB*.)

Kurr, Johann Gottlob von (1798–1870). Professor of botany and mineralogy at the polytechnic institute in Stuttgart from 1838. Appointed director of the establishment in 1841. Continued to practise medicine until 1850. (*ADB*, *BHGW*.)

Lamarck, Jean Baptiste Pierre Antoine de Monet de (1744–1829). French naturalist. Held various botanical positions at the Jardin du Roi, 1788–93. Appointed professor of zoology, Muséum d'histoire naturelle, in 1793. Believed in spontaneous generation and the progressive development of animal types; propounded a theory of transmutation. (*DSB*.)

Lamb, William, 2d Viscount Melbourne (1779–1848). Politician. Called to the bar, 1804, but gave up the bar for politics in 1805. MP for Leominster, 1806–7; for Portarlington, 1807–12; for Northampton, 1816–19; for Hertfordshire, 1819–25; for Newport, Isle of Wight, 1827–9. Irish secretary, 1827–8. On the death of his father in 1829, took his seat in the House of Lords. Appointed home secretary in 1830. Prime minister, 1834; 1835–41. (*DNB*.)

Lamont, James (1828–1913). Arctic and African traveller. MP for Buteshire, Scotland, 1865–8. Fellow of the Geological Society of London. (Freeman 1978, *Rugby School register*.)
25 February [1861]

Langton, Charlotte (1797–1862). Emma Darwin's sister. Married Charles Langton in 1832. Resided at Maer, Staffordshire, 1840–6; at Hartfield Grove, Sussex, 1847–61. (*Emma Darwin*.)

Lankester, Edwin (1814–74). Physician, lecturer, and author. An active supporter of the British Association for the Advancement of Science. Professor of natural history, New College, London, 1850. FRS 1845. (*DNB*.)

Laplace, Pierre Simon, marquis de (1749–1827). French mathematician, physicist, and cosmologist. (*DSB*.)

Latter, Miss. Governess at Down House from April 1859 until February 1860. (CD's Account book (Down House MS).)

Lauder, Thomas Dick, 7th baronet (1784–1848). Scottish writer. Author of novels, works on Scottish topography, and several scientific papers. Appointed secretary to the Board of Scottish Manufactures, later united to the Board of White Herring Fishery, in 1839. (*DNB*, Sarjeant 1980.)

Lavoisier, Antoine Laurent (1743–94). French scientist and public servant. Foremost a chemist, he discovered the role of oxygen in chemical reactions. Had a lifelong interest in geology, was a pioneer in scientific agriculture, and a financier of ability. Executed on the guillotine during the 'Reign of Terror'. (*DSB*.)

Lebert, Hermann (1818–78). Professor of the medical clinic, Zürich, 1853–9; Breslau, 1859–74. (*BLA*.)

Lecoq, Henri (1802–71). French naturalist and vulcanologist. Professor of natural history, University of Clermont-Ferrand, and director of the town's botanic garden, 1826–54. Taught at the Preparatory School of Medicine and Pharmacy, Clermont-Ferrand, from 1840; professor, science faculty, from 1854. Published widely on botany, agriculture, and meteorology. (*Grande encyclopédie*, Sarjeant 1980.)

Le Couteur, John (*fl.* 1830s–60s). Colonel, Jersey Militia. President, Jersey

Le Couteur, John, cont.
Agricultural Society, 1834–9. Author of a well-known work on wheat (1836). FRS 1843. (R. Desmond 1977.)

Leidy, Joseph (1823–91). American physician and naturalist. Demonstrator of anatomy, Franklin Medical College, 1846. Visited Europe in 1848 and 1850. Professor of anatomy, University of Pennsylvania, 1853–91. Curator of the Academy of Natural Sciences, Philadelphia. Published extensively on the fossil remains of extinct vertebrates of North America. (*DAB*, *DSB*.)
4 March [1861]

Lepeletier de Saint Fargeau, Amédée Louis Michel (1770–1845). French entomologist with a particular interest in the Hymenoptera. Elected president of the Société entomologique de France in 1833. (*Annales de la Société Entomologique de France* 2d ser. 4 (1846): 193–5, Gilbert 1977.)

Leslie, Charles Robert (1794–1859). Painter. Born in London of American parents. Taken to America in 1800 where he was educated and began to show a decided talent for drawing. As his family could not afford to pay for his training as an artist he became apprenticed to a Philadelphia publishers in 1808. A subscription having been raised on his behalf, he travelled to England in 1811 to become a student of the Royal Academy. He remained in England as a painter for the rest of his life. (*DNB*.)

Lestiboudois, Gaspard Thémistocle (1797–1876). French politician and naturalist. Sometime teacher of anatomy and physiology at the science faculty, Paris. (*Grande encyclopédie*.)

Leuckart, Karl Georg Friedrich Rudolf (Rudolf) (1822–98). German zoologist who carried out important work in morphology and parasitology. Lecturer in zoology, University of Göttingen, 1847–50. Extraordinary professor of zoology, University of Giessen, 1850–69. Appointed professor of zoology in Leipzig in 1869 and dean of the philosophical faculty in 1873. (*DSB*, *NDB*.)

Lewes, George Henry (1817–78). Writer. Studied physiology from 1855 and published several works on the subject. Common law husband of Mary Ann Evans (George Eliot). (*DNB*.)

Leydig, Franz von (1821–1908). German zoologist. Lecturer in physiology, University of Würzburg, 1849–55; extraordinary professor, 1855–7. Professor of zoology at Tübingen, 1857–75; of comparative anatomy, Bonn, 1875–87. Foreign member, Royal Society, 1901. (*DSB*, *NDB*.)

Lightbody, George (d. 1872). Of Falkirk, Scotland. Grew auriculas and raised numerous varieties. (R. Desmond 1977.)

Lincecum, Gideon (1793–1874). American physician and naturalist. Studied medicine privately; taught in Georgia. Was for several years an Indian trader in the Choctaw country of central Mississippi. Practised medicine in Cotton Gin Port and Columbus, Mississippi, 1830–48. Settled in Texas in 1848. Studied the behaviour of agricultural ants. (*DAB*.)
4 March 1861

Lincoln, Abraham (1809–65). Republican president of the United States of America during the time of the American Civil War, from 1861 until his assassination in 1865. (*DAB*.)

Lind, Johanna Maria (Jenny) (1820–87). Opera singer who achieved world-wide popular acclaim. (*DNB*.)

Lindley, John (1799–1865). Botanist and horticulturist. Assistant secretary to the Horticultural Society, 1822–41; vice-secretary, 1841–58. Professor of botany at London University (later University College London), 1828–60. Editor of the *Gardeners' Chronicle* from 1841. FRS 1828. (*DNB*, *DSB*.)
18 October [1861], 25 October [1861], 1 November [1861],
16 November [1861], 17 November [1861], 15 December [1861],
24 December [1861], 28 December [1861]

Link, Heinrich Friedrich (1767–1851). Professor of zoology, botany, and chemistry, University of Rostock, 1792–1811; of chemistry and botany, University of Breslau, 1811–15. In 1815, appointed professor of botany and director of the botanic garden at the University of Berlin. Had a strong interest in the philosophical foundations of the natural sciences. (*DSB*, *NDB*.)

Li Shih-chen (1518–93). Chinese naturalist and pharmacologist. The compiler of China's most important pharmacopoeia, the *Pen ts'ao kang-mu*. (*Dictionary of Ming biography*.)

Litchfield, Henrietta Emma. *See* Darwin, Henrietta Emma.

Loddiges, Conrad (d. 1865). Nurseryman in Hackney, London. Ran a nursery founded in 1771 by his grandfather, Conrad Loddiges (*c*. 1739–1826), and then directed by his father, George Loddiges (1784–1846). (R. Desmond 1977.)

Loiseleur Deslongchamps, Jean Louis Auguste (1775–1849). French botanist and physician. Studied the medicinal qualities of plants. (*NBU*.)

Longfellow, Frances Elizabeth (d. 1861). Second wife of Henry Wadsworth Longfellow. Sister of Mary Appleton, the wife of Robert Mackintosh (brother of Fanny Mackintosh Wedgwood, CD's sister-in-law). Was burned to death in her home when her dress caught fire. (*DAB*.)

Longfellow, Henry Wadsworth (1807–82). American poet. Married Frances Elizabeth Appleton, the sister-in-law of Robert Mackintosh, in 1843. (*DAB*.)

Lovegrove, Charles. Church-warden at Down church. (Freeman 1978.)
9 July [1861?]

Low, Sampson (1797–1886). Publisher. Senior partner in the publishing firm Sampson Low, Marston & Co., with offices in Ludgate Hill, London. Publisher of the *Publisher's Circular*. (*DNB*, *Modern English biography*.)

Low, Sampson, Jr (1822–71). Publisher. In business with his father, Sampson Low, 1847–71. (*Modern English biography*.)

Lowe, Richard Thomas (1802–74). Clergyman and botanist. English chaplain on Madeira, 1832–54. Rector of Lea, Lincolnshire, 1854–74. Published a flora of Madeira (1857–72). Drowned on a journey to Madeira. (R. Desmond 1977, *DNB*.)

Lubbock, Ellen Frances (d. 1879). Married John Lubbock in 1856. (*Burke's peerage* 1970.)

Lubbock, Harriet (1810–73). Married John William Lubbock in 1833. (Freeman 1978.)

Lubbock, John, 4th baronet and 1st Baron Avebury (1834–1913). Banker, politician, and naturalist. Son of John William Lubbock and a neighbour of CD's in Down. Studied entomology and anthropology. An active supporter of CD's theory of natural selection. Succeeded his father to become 4th baronet in 1865. Created Baron Avebury in 1900. FRS 1858. (*DNB*, *DSB*.)
[before 5 February 1861], [25 May 1861], 1 June [1861], 6 June [1861], [8 June 1861], 24 June [1861], 10 July [1861], 14 July [1861], 26 July [1861], 1 [and 2] August [1861], [2 August 1861], 9 August [1861], 10 August [1861], 14 August [1861]

Lubbock, John William, 3d baronet (1803–65). Astronomer, mathematician, and banker. First vice-chancellor of London University, 1837–42. CD's neighbour at Down. Succeeded to the baronetcy, 1840. FRS 1829. (*DNB*, *DSB*.)

Lubbock, Montagu (b. 1842). Younger brother of John Lubbock. (*Burke's peerage* 1970.)

Lubbock, Nevile (1839–1914). Brother of John Lubbock. President of the West India Committee, 1909–14. (*Burke's peerage* 1970.)

Ludwig, Camilla. Governess to the Darwin children. Joined the household in 1860. Translated German works for CD. (CD's Account book (Down House MS).)

Lukis, Frederick Collings. Of Guernsey. Surgeon, first regiment, Guernsey Royal Militia. Fellow of the Royal Society of Antiquaries. (*London & provincial medical directory* 1861.)

Lutwidge, Charles R. Fletcher. Lieutenant in the Tunbridge Wells corps of the Kent Rifle Volunteers. (*Army list* 1861.)

Luxford, George (1807–54). Printer and botanist. Lecturer in botany, St Thomas's Hospital, 1846–51. Edited the *Phytologist*, 1841–54. (R. Desmond 1977.)

Lyell, Charles (1797–1875). Uniformitarian geologist whose *Principles of geology* (1830–3) and *Elements of geology* (1838) appeared in many editions. Professor of geology, King's College, London, 1831–3. President of the Geological Society, 1834–6, 1849–50. Travelled widely and published accounts of his trips to the United States. Knighted, 1848; created baronet, 1864. Scientific mentor and friend of CD. FRS 1826. (*DNB*, *DSB*.)
2 February [1861], 12 April [1861], 20 July [1861], [1 August 1861], 6 August [1861], 13 [August 1861], 21 August [1861], 6 September [1861], 10 September [1861], [15 September 1861], 22 September [1861], *30 September 1861*, 1 October [1861], 20 October [1861], 23 [October 1861]

Lyell, Katharine Murray (1817–1915). Daughter of Leonard Horner. Married Henry Lyell, brother of Charles Lyell, in 1848. Collected plants in India. Edited

Life, letters and journals of Sir Charles Lyell (1881) and memoirs of Charles James Fox Bunbury and Leonard Horner. (R. Desmond 1977.)

Lyell, Leonard, 1st baron (1850–1926). Eldest son of Henry Lyell, Charles Lyell's brother, and Katharine Murray Lyell. MP for Orkney and Shetland, 1885–1900. Created baronet, 1894; peer, 1914. (*Burke's peerage* 1980.)

Lyell, Mary Elizabeth (1808–73). Eldest child of Leonard Horner. Married Charles Lyell in 1832. (Freeman 1978.)

McAndrew, Robert (1802–73). Liverpool merchant, yachtsman, and naturalist. Member of the dredging committee of the zoological section of the British Association for the Advancement of Science, 1844. FRS 1853. (Rehbock 1979.)

Macaulay, Thomas Babington (1800–59). Historian, writer, and politician. Called to the bar, 1826. Rector of Glasgow University, 1848. Took the title Baron Macaulay of Rothley in 1857. (*DNB*.)

MacCulloch, John (1773–1835). Physician, chemist, and geologist. President of the Geological Society, 1816–17. Commissioned to prepare a geological map of Scotland, 1826. FRS 1820. (*DNB*, *DSB*.)

M'Donnell, Robert (1828–89). Irish surgeon. His name was variously spelt M'Donnell or McDonnell. Served as a surgeon in the Crimean War, 1855. Demonstrator of anatomy, Carmichael School of Medicine, 1856; afterwards lecturer on anatomy and physiology. Surgeon to Jervis Street Hospital, Dublin, 1863; to Steevens's Hospital, 1866. Medical superintendent of the Mountjoy government prison, 1857–67. Elected president, Royal College of Surgeons of Ireland, 1877; president, Academy of Medicine in Ireland, 1885–8. FRS 1865. (*DNB*.)

Macgillivray, William (1796–1852). Conservator of the museum of the Royal College of Surgeons, Edinburgh, 1831–41. Appointed professor of natural history, Aberdeen University, in 1841. (*DNB*.)

Mackay, Charles (1814–89). Poet and journalist. Editor, *Illustrated London News*, 1852–9. Established the *London Review and Weekly Journal of Politics, Literature, Art and Science* in 1860; however, the journal was not a success. Special correspondent of *The Times* in New York, 1862–5, during the American Civil War. (*DNB*.)

Mackenzie, George Steuart, 7th baronet (1780–1848). Mineralogist. Succeeded to the baronetcy in 1796. Undertook geological excursions to Iceland and the Faroe Islands. From 1826 to 1848, contributed numerous papers to the discussion of the origin of the 'parallel roads' of Glen Roy. Fellow, later president, of the physical class of the Royal Society of Edinburgh. FRS 1815. (*DNB*.)

Mackintosh, Mary (Molly). American-born wife of Robert Mackintosh. (Wedgwood and Wedgwood 1980.)

Mackintosh, Robert (1806–64). Brother of Frances Mackintosh Wedgwood, CD's sister-in-law. (Wedgwood and Wedgwood 1980.)

Macquart, Justin Pierre Marie (1778–1855). French entomologist. (Gilbert 1977, *NBU*.)

Mainwaring, Mrs (née Collier). Reared Elizabeth Collier Darwin, who was the natural daughter of the second earl of Portmore. (King-Hele 1963.)

Mairan, Jean Jacques Dortous de. *See* Dortous de Mairan, Jean Jacques.

Malden, Bingham Sibthorpe (1830–1906). Curate of Clayton, Sussex, 1854–8; of St George-the-Martyr, Canterbury, 1858–63; of Pattingham, Staffordshire, 1863–70. Vicar of Sheldwich, Kent, 1870–1906. (*Alum. Cantab.*)
15–16 June [1861]

Manchester, dean of. *See* Herbert, William.

Mann, Gustav (1836–1916). German-born explorer and botanist. Gardener to the Royal Botanic Gardens, Kew. Appointed botanist to the Niger Expedition, under the direction of William Balfour Baikie, 1859–62. (R. Desmond 1977, *Taxonomic literature*.)

Mansel, Henry Longueville (1820–71). Metaphysician. Tutor, St John's College, Oxford, from 1843. Elected reader in moral and metaphysical theology, Magdalen College, Oxford, in 1855; 'professor fellow' of St John's, 1864. Appointed professor of ecclesiastical history, Oxford, in 1866. (*DNB*.)

Mantell, Walter Baldock Durrant (1820–95). Geologist and naturalist. Emigrated to New Zealand, 1840. Made important studies of the *Dinornis* beds. Appointed Commissioner of Crown Lands, Otago, in 1851. Supported Maori claims for land. Entered politics in the late 1850s and was elected for Wallace in 1861. Member of the Legislative Council, 1866–95. (*DNZB*, Sarjeant 1980.)

Marshall, William (1815–90). Solicitor of Ely. (R. Desmond 1977.)

Martens, Conrad (1801–78). Landscape painter. Joined HMS *Beagle* in Montevideo in 1833 and served as draughtsman until 1834. Settled in Australia in 1835. (*Aust. dict. biog.*, Bénézit 1976.)

Martin, William Charles Linnaeus (1798–1864). Writer on natural history. Superintendent of the museum of the Zoological Society of London, 1830–8. Wrote many books on animals and birds. (*DNB*, *Modern English biography*.)

Martineau, James (1805–1900). Unitarian divine, writer, and lecturer. Ordained in 1828. Had ministries in Dublin and Liverpool. Retaining his congregational charge in Liverpool, he was appointed professor of mental and moral philosophy and political economy at the Manchester New College in 1840. Became professor of mental, moral, and religious philosophy in 1857 by which time the college had moved to University Hall, Gordon Square, London. Returned to the pulpit in 1859 as minister of Little Portland Street Chapel, London. Became principal of Manchester New College in 1869. Gave up preaching owing to ill health in 1872. (*DNB*.)

Mason, James Murray (1798–1871). American politician and diplomat. United States senator from 1847 to 1861, when his state, Virginia, seceded from the Union. Appointed Confederate diplomatic commissioner to England in 1861; his detention by the United States Navy on his departure nearly caused a war between Britain and the United States. (*DNB*.)

Masson, V. et fils. French publishing company owned by Victor Masson (1807–79). Published elegant editions of scientific works. (*Dictionnaire universel des contemporains*.)

Masters, Maxwell Tylden (1833–1907). Physician. Lecturer, St George's Hospital, 1855–68. Editor, *Gardeners' Chronicle*, 1865–1907. FRS 1870. (R. Desmond 1977.)

Matthew, Patrick (1790–1874). Author of works on political and agricultural subjects. Advanced a theory of natural selection in the 1830s. (R. Desmond 1977.)

Maurice, John Frederick Denison (1805–72). Divine. Chaplain of Guy's Hospital, London, 1836–46. Professor of modern history and English literature, King's College, London, 1840–6; of ecclesiastical history, 1846–53. Chaplain of Lincoln's Inn, 1846–60. Perpetual curate of the chapel of St Peter's, Vere Street, London, 1860–6. Knightbridge professor of moral philosophy, Cambridge, 1866–72. Vicar of St Edward's, Cambridge, 1870–2. (*Alum. Cantab.*, *DNB.*)

Maw, George (1832–1912). Tile manufacturer and amateur geologist, botanist, and antiquarian. Wrote on the geology of western England and north Wales. Travelled to Morocco and Algeria with Joseph Dalton Hooker in 1871 and on his own in 1873, writing on the geology of these countries. (R. Desmond 1977, Sarjeant 1980.)

15 March 1861, 17 March [1861], 13 July [1861], 19 July [1861],
27 August [1861], 31 August [1861]

Max Müller, Friedrich (1823–1900). German orientalist and philologist. Published an edition of the 'Rigveda', the most important of the sacred books of the Brahmans (1849–73). Moved to Paris, 1845. Settled in Oxford in 1848 after fleeing the revolution in France. Deputy Taylorian professor of modern European languages, Oxford, 1850–4; professor, 1854–68; professor of comparative philology, 1868–75. Curator of the Bodleian library, 1856–63 and 1881–94. (*DNB.*)

Meding, Karl Heinrich (1791 – *c.* 1862). Physician. Anatomist at the Academy of Medicine and Surgery, Dresden. Had a large and popular practice as physician and surgeon. Worked for a time as a school doctor in Meissen. (*BLA.*)

Melbourne, Lord. *See* Lamb, William.

Mendelssohn-Bartholdy, Jacob Ludwig Felix (1809–47). Composer. Travelled extensively in his early twenties, becoming a great favourite in England. Town director of music in Düsseldorf, 1833–5. Director of the Conservatory, Leipzig, and conductor of the Gewandhaus Concerts, 1845–7. (*ADB*, Scholes 1970.)

Ménière, Prosper (1799–1862). French anatomist. (*Grande encyclopédie.*)

Meteyard, Eliza (1816–79). Author. Biographer of the Wedgwood family. (*DNB.*)

Middeldorpf, Albrecht Theodor (1824–68). Professor of surgery and ophthalmology and director of the outpatient clinic at the University of Breslau. (*BLA.*)

Miers, John (1789–1879). Botanist and engineer. Travelled and worked in South America, 1819–38. Author of many papers describing South American plants. FRS 1843. (R. Desmond 1977, *DNB.*)

Mill, John Stuart (1806–73). Philosopher and political economist. (*DNB*, *DSB.*)

Miller, Hugh (1802–56). Scottish quarryman, poet, ecclesiastical journalist, and geologist. Examined the stratigraphy of parts of Great Britain, in particular

Miller, Hugh, cont.
the Edinburgh district and the Hebrides. Author of popular works on geology. (*Modern English biography*, Sarjeant 1980.)

Miller, William Allen (1817–70). Chemist and astronomer. Professor of chemistry, King's College, London, from 1845. Pioneered the use of spectroscopic analysis in chemistry. Developed a telescope to give spectra of celestial objects, thereby procuring detailed information on stellar chemistry. FRS 1845. (*DNB*, *DSB*.)

Milne, David. *See* Milne-Home, David.

Milne-Home, David (1805–90). Scottish advocate and geologist. Studied earthquakes and the parallel roads of Glen Roy. Founder of the Scottish Meteorological Society. (*DNB*, *Modern English biography*, Sarjeant 1980.)

Mohl, Hugo von (1805–72). German biologist. Professor of physiology, Bern, 1832–5; of botany, Tübingen, 1835–72. Known for his work on the microscopic anatomy of plants and for his contributions to knowledge of the plant cell. (*DSB*.)

Moquin-Tandon, Horace Bénédict Alfred (Alfred) (1804–63). French botanist and naturalist. Professor of zoology, Marseille, 1829–33. Professor of botany and director of the Jardin des plantes, Toulouse, 1833–53. Professor of natural history and director of the Jardin des plantes, faculty of medicine, Paris, 1853–63. (*Dictionnaire universel des contemporains*, *NBU*.)

More, Alexander Goodman (1830–95). Naturalist. Lived on the Isle of Wight before settling in Ireland in the 1860s. Studied the flora of Ireland, publishing his *Cybele Hibernica* in 1866. In his zoological work he concentrated mainly on birds. Assistant in the Dublin Natural History Museum, 1867–81; curator, 1881–7. (R. Desmond 1977, Gilbert 1977, *Natural Science* 6 (1895): 351.)
8 March 1861, 2 June 1861, 4 June 1861, 17 June 1861, 7 July 1861, 17 July 1861, 19 July 1861, 13 August 1861, 23 September 1861, 1 October 1861

Morlot, Charles Adolphe (1820–67). Swiss stratigrapher and archaeologist. Appointed professor of geology and mineralogy, Lausanne, in 1851. At one time, conservator of the Bern Archaeological Museum. Wrote on the Tertiary and Quaternary geology of Austria, Switzerland, and Denmark. Originator of the term 'Quaternary'. (*ADB*, Sarjeant 1980.)

Morren, Charles François Antoine (1807–58). Belgian botanist and horticultural writer. Taught physics at the industrial school in Gant, 1831–5; studied medicine and qualified in 1835. Extraordinary professor of botany, University of Liège, 1835–7; ordinary professor, 1837–54. (*BNB*, Gilbert 1977.)

Morton, John Maddison (1811–91). Dramatist. Prolific author of farcical plays. (*DNB*.)

Morton, Samuel George (1799–1851). American physician and ethnologist, known for his research on human skulls. Professor of anatomy in the medical department, Pennsylvania College, 1839–43. (*DAB*, *DSB*.)

Mueller, Ferdinand Jakob Heinrich von (1825–96). German-born explorer and botanist who emigrated to Australia in 1847. Government botanist in Victoria,

1853. Botanist to the North-west Australia Expedition, 1855–7. Director of the botanic garden, Melbourne, 1857–73. FRS 1861. (*Aust. dict. biog.*, R. Desmond 1977.)

Murchison, Charles (1830–79). Physician. Assistant physician, King's College Hospital, London, 1856–60; at Middlesex Hospital, 1860–6; full physician, Middlesex, 1866–71. Physician with the London Fever Hospital from 1861. A keen amateur geologist, he was a good friend of Hugh Falconer's; edited Falconer's *Palaeontological memoirs* in 1868. FRS 1866. (*DNB*.)

Murchison, Roderick Impey (1792–1871). Geologist noted for his work on the Silurian system. A leading figure in the Geological Society, the British Association for the Advancement of Science, and the Royal Geographical Society. Appointed director-general of the Geological Survey of Great Britain in 1855. Knighted, 1846; created baronet, 1866. FRS 1826. (*DNB*, *DSB*.)

Murray, Andrew (1812–78). Entomologist and botanist. Practised in the legal profession in Edinburgh until 1860 when he was appointed assistant secretary to the Royal Horticultural Society, London. Became an expert on insects harmful to crops. In entomology, specialised on the Coleoptera; in botany, on the Coniferae. (*Entomologist's Monthly Magazine* 14 (1878): 215–16, Gilbert 1977.)

23 February [1861], *3 March 1861*, 3 March [1861]

Murray, John (1808–92). Publisher and author of guide-books. CD's publisher from 1845. (*DNB*.)

21 January [1861], 24 February [1861], 3 March [1861], 30 April [1861], 3 May [1861], *7 May 1861*, 8 May [1861], 10 September [1861], 21 September [1861], *23 September 1861*, 24 September [1861], 3 October [1861], 5 October [1861], 21 October [1861]

Musschenbroek, Peter van (1692–1761). Dutch physicist. Professor of natural philosophy and mathematics, Utrecht, 1723–40; also professor of astronomy from 1832. Professor at Leiden, 1840–61. (*DSB*.)

Namgyé, Cho-phoe. Raja of Sikkim, 1790–1861. (*EB*.)

Nathusius, Hermann Engelhard von (1809–79). Studied zoology with Johannes Müller in Berlin, 1827–9. Author of works on various botanical and zoological topics, notably a book on pigs, *Ueber die Rassen des Schweines* (1860). Senior civil servant and advisor in the Prussian Ministry of Agriculture from 1869. (*ADB*.)

Naudin, Charles Victor (1815–99). French botanist. Experimented widely on plants, particularly on acclimatisation and hybridity. Published a theory of transmutation based on hybridisation. (*DSB*.)

Neale, Edward Vansittart (1810–92). Christian socialist. Called to the bar, 1837. Keenly interested in social reform. Founded the first London co-operative stores. Established the Central Co-operative Agency in 1851, to promote the cause of co-operation, which he believed would improve the condition of the working classes. Devoted his wealth and energies to the co-operative movement throughout his life. (*DNB*.)

Nevill, Dorothy Fanny (1826–1913). Had a notable garden at Dangstein in Sussex,

Nevill, Dorothy Fanny, cont.
where she cultivated orchids, nepenthes, and other tropical plants. (R. Desmond 1977.)
12 November [1861], 19 November [1861]

Newman, Edward (1801–76). Naturalist. Founder of the Entomological Club (later the Entomological Society), 1826. Natural history editor of the *Field*, 1858–76. (R. Desmond 1977, *DNB*, Gilbert 1977.)

Newman, Henry Wenman (1788–1865). Of Thornbury Park, Gloucestershire. Lieutenant-colonel commandant in the Royal South Gloucestershire Militia, 1854–60. (*Modern English biography*.)
[before 22 October 1861]

Nicol, James (1810–79). Scottish mineralogist and stratigrapher. Professor of geology, Queen's College, Cork, 1849–53. Professor of natural history, University of Aberdeen, 1853–79. (*Modern English biography*.)

Nightingale, Florence (1820–1910). Reformer of hospital nursing. Nursed British troops during the Crimean War, 1854–6. Established a training school for nurses at St Thomas's Hospital, London, in 1860. (*DNB*.)

Nightingale, William Edward (1794–1874). Father of Florence Nightingale. Owned two estates, one at Lea Hurst in Derbyshire and the other at Embley Park, Hampshire. (*DNB*.)

Norman, Ebenezer. Schoolmaster in Down whom CD employed as a copyist. (Freeman 1978.)

Oldham, Thomas (1816–78). Irish geologist. Chief geological assistant to the Geological Survey of Ireland, 1839–50. Professor of geology, Trinity College, Dublin, 1845–51. In 1849, discovered in the rocks of Bray Head, County Wicklow, the fossils or organic marks, named after him, Oldhamia. Director, Geological Survey of India, 1850–76. (*Modern English biography*, Sarjeant 1980.)

Oliver, Daniel (1830–1916). Assistant in the herbarium of the Royal Botanic Gardens, Kew, 1858; keeper, 1864–90. Professor of botany, University College London, 1861–88. (R. Desmond 1977.)
22 January [1861], 26 February [1861], 23 March [1861], 1 April [1861], 4 April [1861], 9 April [1861], 1 May [1861], 27 May [1861], 11 September [1861], 8 October [1861], *[after 8 October 1861]*, *[before 8 November 1861]*, *8 November 1861*, 30 November [1861], 7 December [1861]

Olmsted, Frederick Law (1822–1903). American landscape architect. Farmed on Staten Island, New York for several years; also pursued literary interests and travelled widely. Wrote extensively on slavery and the economic and social conditions of the American South. Appointed superintendent of Central Park, New York City, 1857; architect in chief, 1858. Superintendent, Frémont Mariposa mining estates, California, 1863–5. Reappointed landscape architect, Central Park, 1865, until his removal in 1878. Thereafter worked independently. (*DAB*.)

Ossulston, Lord. *See* Bennet, Charles Augustus.

Ottinge, J. D. *See* Glennie, John David, Jr.

Overton, Samuel Charlesworth (b. 1837). Curate of Hackness, Yorkshire, 1861–3; of Stretton, Cheshire, 1864–7. (*Crockford's*.)

Owen, Richard (1804–92). Comparative anatomist. Assistant-conservator of the Hunterian Museum, Royal College of Surgeons, 1827; Hunterian professor, 1836–56. Superintendent of the natural history departments, British Museum, 1856–84. Described the *Beagle* fossil mammal specimens. FRS 1834. (*DNB*, *DSB*.)

Oxenden, George Chichester (1797–1865). Author of satiric verse and parodies. Lived at the family seat at Broome Park, near Canterbury, Kent. (*Alum. Cantab.*, *Burke's peerage* 1895.)

Oxford, Bishop of. *See* Wilberforce, Samuel.

Paget, James (1814–99). Assistant surgeon at St Bartholomew's Hospital, London, 1847; surgeon, 1861–71. Arris and Gale professor of anatomy and surgery at the Royal College of Surgeons, 1847–52. Lectured on physiology in the medical school, St Bartholomew's, 1859–61; on surgery, 1865–9. Appointed surgeon-extraordinary to Queen Victoria, 1858; serjeant-surgeon, 1877. Created baronet, 1871. FRS 1851. (*DNB*.)

Pallas, Pyotr Simon (1741–1811). German naturalist and geographer who travelled extensively in the Russian empire. Foreign member, Royal Society, 1763. (*ADB*, *DSB*.)

Parker, Robert (*fl.* 1850s–80s). In partnership with Benjamin Samuel Williams in a nursery in Holloway, London, 1854–61. Had a nursery in Tooting, Surrey. (R. Desmond 1977.)

Parslow, Joseph (1809/10–98). CD's manservant at 12 Upper Gower Street *c.* 1840–2 and butler at Down House until 1875. (Freeman 1978.)

Parsons, William, 3d earl of Rosse (1800–67). Astronomer. Constructed a telescope with a 72-inch disc at Birr Castle, Ireland, in 1845. Made important observations on nebulae. President of the Royal Society, 1848–54. Appointed chancellor of the University of Dublin in 1862. FRS 1831. (*DNB*, *DSB*.)

Partridge, Richard (1805–73). Surgeon, King's College Hospital, London, 1840–70. Appointed professor of anatomy at the Royal Academy in 1858. FRS 1837. (*DNB*, Plarr 1930.)

Pattison, Mark (1813–84). Academic and author. Appointed tutor at Lincoln College, Oxford, in 1843; rector, 1861. Contributed to *Essays and reviews* (1860). (*DNB*.)

Pelzel von Pelzeln, August von (1825–91). Austrian zoologist. Became a trainee in the Royal Zoological Museum in Vienna in 1851. Curator of the bird collection, 1857; of the mammal and bird collections, 1869. Founding member of the Zoological-botanical and Ornithological societies of Vienna. (*OBL*.)

Pengelly, William (1812–94). Geologist. An expert on the geology of Devon; explored the plant-bearing deposit at Bovey Tracey and several caves including Brixham Cave and Kent's Hole at Torquay. FRS 1863. (*DNB*, Sarjeant 1980.)

Pennant, Thomas (1726–98). Traveller and naturalist. Encouraged Gilbert White to begin writing the series of letters that became *The natural history and antiquities of Selbourne* (1789); many of the letters are to Pennant. Pennant's writings emphasise the goodness of nature, which he considered a reflection of a sanctified creation. Supported the classificatory views of John Ray and later those of Linnaeus. (*DNB*, *DSB*.)

Phillips, John (1800–74). Geologist. Keeper of the museum of the Yorkshire Philosophical Society, 1825–40. Assistant secretary of the British Association for the Advancement of Science, 1832–62. Professor of geology, King's College, London, 1834–40. Palaeontologist to the Geological Survey of Great Britain, 1840–4. Deputy reader in geology, Oxford University, 1853; professor, 1860–74. FRS 1834. (*DNB*, *DSB*.)

Pickering, Charles (1805–78). American physician and naturalist. Curator of the Academy of Natural Sciences, Philadelphia, 1833–7. Chief zoologist to the United States Exploring Expedition, 1838–42. Published on the geographical distribution of plants and animals. (*DAB*.)

Planchon, Jules Emile (1823–88). French botanist. Keeper of the herbarium, Royal Botanic Gardens, Kew, 1844–9. Professor, Institute of Horticulture, Gand, Belgium, 1849–51. Professor, School of Medicine and Pharmacy, Nancy, 1851–3. Professor of botany, faculty of sciences, Montpellier, 1853–81, and director of the school of pharmacy. In 1881, became professor of botany in the faculty of medicine. (*Grande encyclopédie*.)

Pole, Elizabeth (1770–1821). Daughter of Elizabeth Darwin by her first marriage. (King-Hele ed. 1981.)

Poole, Skeffington. Entered the East India Company's service in 1820. Served in the first regiment light cavalry of the Bombay Military Establishment. Retired with the rank of lieutenant-colonel in 1850. (*East-India register and army list* 1853.)

Portlock, Joseph Ellison (1794–1864). Major-general in the Royal Engineers. Engaged on the geological and productive economical sections of the Irish survey, 1824–43. FRS 1837. (*DNB*.)

Portmore, Lord. *See* Colyear, Charles, 2d earl of Portmore.

Powell, Baden (1796–1860). Writer on theological topics and active in university reform. Vicar of Plumstead, Kent, 1821–7. Savilian professor of geometry at Oxford, 1827–60. Author of an article in *Essays and reviews* (1860). FRS 1824. (*DNB*, *DSB*.)

Pratt, Mrs William. Keeper of a lodging house at 1 Carlton Place, Southampton (*Post Office directory of Hampshire, Wiltshire, and Dorsetshire* 1867.)

Prestwich, Joseph (1812–96). Geologist. Proprietor of the family wine business in London. An expert on the Tertiary geology of Europe. Prominent in studies of early mankind. Professor of geology at Oxford, 1874–88. FRS 1853. (*DNB*, *DSB*.)

Pugh, Miss. Governess of the Darwin children, 1856–7. Later became insane and was confined to an asylum. (Freeman 1978.)

Pullinger, William George. Chief cashier with the Union Bank of London. Exposed in 1860 for having embezzled £263,000 from the bank and was sentenced to twenty years' imprisonment. (*Banking almanac* 1861.)

Quatrefages de Bréau, Jean Louis Armand de (1810–92). French zoologist and anthropologist. Undertook a study of invertebrates and, in 1844, accompanied Henri Milne-Edwards on a zoological excursion off the coast of Sicily. Awarded the chair of natural history at the Lycée Henri IV, Paris, 1850; professor of the natural history of man, Muséum d'histoire naturelle, 1855. Foreign member, Royal Society, 1879. (*DSB*.)
25 April [1861]

Quekett, Isabella Mary Anne (d. 1872). Married John Thomas Quekett in 1846. (*DNB*.)

Quekett, John Thomas (1815–61). Histologist and microscopist. Appointed assistant conservator of the Hunterian Museum in 1843; conservator, 1856. Appointed demonstrator of minute anatomy, Royal College of Surgeons, in 1844, his title being changed to professor of histology in 1852. President, Microscopical Society of London, 1860–1. FRS 1860. (*DNB*.)

Ramsay, Andrew Crombie (1814–91). Geologist with the Geological Survey of Great Britain, 1841; senior director for England and Wales, 1862; director-general, 1871–81. Professor of geology, University College London, 1847–52; lecturer on geology at the Government School of Mines, 1852–71. Knighted, 1881. FRS 1849. (*DNB*, *DSB*.)

Rapp, Wilhelm Ludwig von (1794–1868). Physician and naturalist. Practised medicine in Stuttgart, 1818–19. Extraordinary professor of anatomy, Tübingen, 1819–28; ordinary professor from 1828. Also taught zoology and undertook scientific journeys throughout Europe, publishing the results of his studies. Retired from academic teaching in 1858 and practised as a physician until his death. (*ADB*.)

Raymond, E. French entomologist with a particular interest in the Coleoptera.

Rede, Robert (d. 1519). Chief justice of the common pleas, 1509. Endowed the Rede lectures at Cambridge University. (*DNB*.)

Reeve, Henry (1813–95). Man of letters well known in European literary and artistic circles. Register of the judicial committee of the privy council, 1843–87. Worked for *The Times*, 1840–55. Editor of the *Edinburgh Review*, 1855–95. (*DNB*.)

Reichenbach, Heinrich Gustav (1824–89). German botanist. Appointed extraordinary professor of botany, Leipzig, in 1855. Director of the botanic gardens in Hamburg, 1863–89. Wrote extensively on orchids. (*ADB*.)

Richardson, John (1787–1865). Arctic explorer and naturalist. After serving as assistant surgeon and surgeon on various of Her Majesty's ships, acted as surgeon and naturalist on John Franklin's polar expeditions, 1819–22; 1825–7. Surgeon to Chatham division of marines, 1824–38. Appointed physician to the Royal Hospital, Haslar, in 1838. Conducted search expedition for John Franklin, 1847–9.

Richardson, John, cont.
Retired to Grasmere where he accomplished much literary work. FRS 1825. Knighted, 1846. (*DNB*.)

Richthofen, Ferdinand Paul Wilhelm von, Freiherr (1833–1905). German geologist, geographer, and explorer. Accompanied the Eulenberg expedition to Japan, China, and Thailand, 1860. Continued travelling alone in Indonesia, the Philippines, and Burma, and then explored the western part of the United States. Explored China, 1868–72. Professor at the University of Bonn, 1879–83; of Leipzig, 1883–6; and Berlin, from 1886. (Sarjeant 1980.)

Rink, Hinrich Johannes (1819–93). Danish naturalist. Explored Greenland, 1848–51; member of a Danish commission to investigate trade with Greenland, 1851. Resided in Greenland from 1853. (*DBL*.)

Robarts, Abraham John (1838–1926). Partner in the banking firm Robarts, Lubbock & Co. (*WWW*.)

Robertson, Charles Alexander Lockhart (d. 1897). Physician. Assistant staff surgeon, Military lunatic asylum, Great Yarmouth, 1845–50. Appointed Linacre demonstrator in anatomy, Oxford, in 1860. Medical superintendent, Sussex County lunatic asylum, Hayward's Heath, 1859–70. Lord chancellor's visitor in lunacy, 1870–96. (*Alum. Cantab.*, *Modern English biography*.)

Rogers, Henry Darwin (1808–66). American geologist. Professor of geology and mineralogy, University of Pennsylvania, 1835. Director of the geological survey of New Jersey, 1835; of Pennsylvania, 1836. Regius professor of natural history, Glasgow University, 1857–66. FRS 1858. (*DAB*, *DSB*.)

Rolfe, Robert Monsey, Baron Cranworth (1790–1868). Lord chancellor, 1853–8, 1865–6. Created Baron Cranworth, 1850. Lived at Holwood House, near Keston, Kent. (*DNB*.)

Rolleston, George (1829–81). Physician. A physician to the British civil hospital at Smyrna, 1855–6. Appointed Regius professor of physic, Oxford, in 1857. Linacre professor of anatomy and physiology, Oxford, 1860–81. FRS 1862. (*DNB*.)
2 March [1861], 7 March [1861], *16 April 1861*, *1 September 1861*,
5 September [1861]

Rorison, Gilbert. Clergyman in Peterhead, Aberdeenshire.

Rössler, Adolf (1814–85). German entomologist. Judicial official in Wiesbaden. (*NUC*.)

Roulin, François Désiré (1796–1874). French naturalist. Spent the years 1821–8 in Central America and became an authority on its natural history. Returned to France in 1828 where he worked as a librarian and edited the proceedings of the Academy of Sciences in Paris. (*Dictionnaire universel des contemporains*.)

Rowland, David. Solicitor. Member of the firm Rowland & Hacon, Fenchurch Street, London. (CD's Address book, Down House MS.)

Royer, Clémence Auguste (1830–1902). French author and economist. Studied natural science and philosophy in Switzerland. In Lausanne in 1859, founded a course on logic aimed at women. Translated *Origin* into French in 1862. (*Dictionnaire universel des contemporains*.)

Rucker, Sigismund. East and West India broker with premises in Great Tower Street, City, London. Resided at West Hill, Wandsworth. (*Post Office London directory* 1861, *Post Office London suburban directory* 1860.)

Russell, William Howard (1820–1907). War correspondent who worked primarily for *The Times*. Covered the Crimean War, the Mutiny in India, the American Civil War (from March 1861 to April 1862), the Prussian–Austrian War, the Franco–Prussian War, and the Zulu War in South Africa. (*DNB*.)

Rütimeyer, Ludwig (1825–95). Palaeozoologist and geographer. Ordinary professor of zoology and comparative anatomy, University of Basel, 1855; rector, 1865; professor in the medical and philosophical faculties, 1874–93. Made significant contributions to the natural history and evolutionary palaeontology of ungulate mammals. (*DSB*.)

5 December [1861]

Sabine, Edward (1788–1883). Geophysicist and army officer. General secretary of the British Association for the Advancement of Science, 1838–59. FRS 1818. Foreign secretary of the Royal Society, 1845–50; treasurer, 1850–61; president, 1861–71. (*DNB*, *DSB*.)

Salter, John William (1820–69). Geologist and palaeontologist. Apprenticed to James de Carle Sowerby in 1835. Assistant to Adam Sedgwick, 1842–6. Assistant to Edward Forbes on the Geological Survey of Great Britain, 1846; palaeontologist to the survey, 1854–63, where his duties included the preparation of specimens for exhibition in the Museum of Practical Geology. (*DNB*.)

Salvin, Osbert (1835–98). Naturalist. Between 1857 and 1874, made many trips abroad, particularly to Central America, collecting natural history specimens. Strickland curator of ornithology, University of Cambridge, 1874–82. (*Alum. Cantab*.)

Sanders, Henry Martyn (1831–1902). Curate of Bexley, Kent, 1856–7; of Skidby, Yorkshire, 1857–9. Vicar of Skidby, 1859–75; of Sutton-on-the-Forest, 1875–1902. (*Alum. Cantab*.)

Schlagintweit, Adolph (1829–57). German geologist and explorer. Lecturer in geology, University of Munich, 1853–4. Accompanied his brothers on a scientific expedition to India and Tibet, during which he was put to death by the amir of Kashgar. (*ADB*, *EB*.)

Schlagintweit, Hermann Rudolph Alfred (1826–82). German geographer and explorer. Lecturer in physical geography at the University of Berlin, 1851–4. Undertook with his brothers a scientific expedition to India and Tibet, 1854–7. (*ADB*, *BHGW*.)

Schlagintweit, Robert (1833–85). German traveller and explorer. With his brothers Adolph and Hermann, undertook a scientific expedition, sponsored by the East India Company and supported by the Prussian government, to India and

Schlagintweit, Robert, cont.
Tibet, 1854–7. Appointed extraordinary professor of geography, Giessen, in 1864. (*ADB*, *BHGW*.)

Schleicher, August (1821–68). Philologist. Published many works on the science of language. Related CD's theory of evolution to the study of language. (*ADB*.)

Schneider, C. F. A. Studied the geology of the Dutch East Indies.

Schomburgk, Robert Hermann (1804–65). German-born explorer and naturalist who spent much of his time in British Guiana. Knighted, 1844. FRS 1859. (*DNB*.)

Sclater, Philip Lutley (1829–1913). Lawyer and ornithologist. Founder of *Ibis*, 1858; editor, 1858–65 and 1878–1912. Secretary of the Zoological Society of London, 1859–1903. FRS 1861. (*DSB*, Scherren 1905.)
4 March [1861], 12 [March 1861], 23 March [1861], *17 April 1861*,
21 [April 1861], 4 May [1861], 2 June [1861]

Scott, Hercules R. (*fl.* 1830s). Of Edinburgh. Advocate. (R. Desmond 1977.)

Scott, John (1836–80). Foreman in the propagating department of Edinburgh Botanic Garden for seven years. Through CD's patronage, was able to begin a professional career as head of the herbarium department, Botanic Garden, Calcutta. Published papers on the propagation and irritability of *Drosera* and *Dionaea* and experimented on the fertilisation of orchids. (R. Desmond 1977, *Transactions of the Botanical Society of Edinburgh* 14 (1883): 160–1.)

Sebright, John Saunders, 7th baronet (1767–1846). Whig politician and agriculturist. Published works on animal breeding. (*DNB*.)

Sedgwick, Adam (1785–1873). Geologist and clergyman. Woodwardian professor of geology, Cambridge University, 1818–73. Canon of Norwich, 1834–73. FRS 1821. (*DNB*, *DSB*.)

Sedgwick, William. Surgeon to the St Marylebone Provident Dispensary. (*London & provincial medical directory* 1861.)

Seemann, Berthold Carl (1825–71). German-born traveller and botanist. Went to the Royal Botanic Gardens, Kew, 1844, to learn about botanical collecting. Naturalist to HMS *Herald*, 1847–51. Commissioned in 1860 to report on the Fiji Islands before their cession and published a botanical catalogue of the islands. Travelled in Venezuela, 1864; in Nicaragua, 1866–7. (R. Desmond 1977, *DNB*.)

Seward, William Henry (1801–72). American statesman. Secretary of state under President Abraham Lincoln. (*DAB*.)

Sidebotham, Joseph (1824–85). Calico printer in Manchester. Botanist and entomologist. One of the founders of the Manchester Field Naturalists' Society. Collected diatoms. (R. Desmond 1977, Gilbert 1977.)

Silliman, Benjamin (1779–1864). American chemist, geologist, and mineralogist. Professor of chemistry and natural history, Yale University, 1802–53. Founder and first editor of the *American Journal of Science and Arts*, 1818. (*DAB*, *DSB*.)

Sinclair, Andrew (1796–1861). Naval surgeon and botanist. Surgeon on board HMS *Sulphur*, 1834–42. Collected plants in Australia, New Zealand, Mexico, and Central America. Colonial secretary in New Zealand, 1844–56. (*DNB*, *DNZB*.)

Skipworth, Mrs. Of Caistor, Lincolnshire.

Skipworth, Septimus Patrick (1839–1917). Member of Christ's College, Cambridge, 1856–61. Sometime surveyor and inspector for Caistor, Lincolnshire, until 1893. Friend of William Erasmus Darwin. (*Alum. Cantab.*)

Slidell, John (1793–1871). American politician and diplomat. United States senator, 1853–61. Appointed Confederate diplomatic commissioner to France in 1861; his detention with James Murray Mason by the United States Navy caused a tense diplomatic situation to develop between the United States and Britain. (*DAB*.)

Smith, Frederick (1805–79). Entomologist in the zoological department of the British Museum from 1849. Specialised in the Hymenoptera. (*Entomologist* 12 (1879): 89–92, Gilbert 1977.)
19 June [1861]

Smith, James (1782–1867). Known as Smith of Jordanhill. Scottish antiquarian, numismatist, and geologist. Made major early contributions to Quaternary geology; also wrote on the geology of Madeira, Gibraltar, and Portugal. Partner in the firm of West India merchants, Leitch and Smith. FRS 1830. (*DNB*, Sarjeant 1980.)

Smith, John (1798–1888). Gardener, Royal Botanic Gardens, Edinburgh, 1818; Kew, 1822. Curator, Kew, 1842–64. (R. Desmond 1977.)

Smith, John. Farmer, Court Farm, Down, Kent. (*Post Office directory of the six home counties* 1862.)

Snow, George (d. 1885). Carrier between Down and London. (Down churchyard, *Post Office directory of the six home counties* 1862.)

Soemmering, Samuel Thomas (1755–1830). German anatomist. Professor of anatomy and surgery, Collegium Carolinum, Kassel, 1779–84. Professor of anatomy and physiology, University of Mainz, 1784–97. Thereafter, practised medicine in Frankfurt before becoming a resident member of the Bavarian Academy of Sciences in Munich in 1805. Returned to Frankfurt in 1820 where he practised medicine and studied sunspots. (*ADB*, *DSB*.)

Sowerby, George Brettingham, Jr (1812–84). Conchologist and artist. Illustrated numerous works on fossils. Prepared the illustrations for *Orchids*. (*DNB*, Sarjeant 1980.)

Spence, William (1783–1860). Entomologist. A founder of the Entomological Society, 1833; president, 1847. FRS 1834. (*DNB*, Gilbert 1977.)

Spencer, Herbert (1820–1903). Writer. Civil engineer on the railways, 1837–41 and 1844–6. Supported the complete suffrage movement and in 1844 became sub-editor of the *Pilot*, a newspaper devoted to the movement. Sub-editor of the *Economist*, 1848–53. From 1852, author of papers on evolution and numerous works on philosophy and the social sciences. (*DNB*, *DSB*.)

Spottiswoode, William (1825–83). Mathematician and physicist. Succeeded his father as Queen's printer in 1846. Throughout his life pursued mathematical studies in which he supplied new proofs of known theorems and also did important original work, notably producing a series of memoirs on the contact of curves and surfaces. FRS 1853; president, 1878–83. (*DNB*.)

Sprengel, Christian Konrad (1750–1816). German botanist who studied the insect-aided fertilisation of flowers. Rector of the Great Lutheran Town School, Spandau, where he taught languages and natural science, 1780–94. Moved to Berlin where he worked as a private tutor. Published his major work on the fertilisation of flowers in 1793. (*ADB*, *DSB*.)

Stanhope, Philip Henry, 5th earl (1805–75). Historian. Known as Viscount Mahon from 1816 until 1855 when he succeeded to the earldom. Tory MP for Wootton Bassett, 1830–2; for Hertford, 1832–3 and 1835–52. President of the Society of Arts, 1846–75. Resided at the family seat in Chevening, Kent. FRS 1827. (*DNB*.)

Stanley, Edward George Geoffrey Smith, 14th earl of Derby (1799–1869). Scholar, orator, statesman, and minister. MP for North Lancashire, 1832–44. Colonial Secretary, 1833–4 and 1841–4. Prime minister, 1852, 1858–9, 1866–8. Appointed chancellor of Oxford University in 1852. Supported parliamentary reform. (*Burke's peerage* 1980, *DNB*.)

Steenstrup, Johannes Japetus Smith (Japetus) (1813–97). Danish zoologist. Professor of zoology and director of the Zoology Museum, University of Copenhagen, 1846–85. Foreign member, Royal Society, 1863. (*DBL*, *DSB*.)

Stephen, Leslie (1832–1904). Philosopher and man of letters. Fellow, Trinity Hall, Cambridge, 1854–67; tutor, 1856–62. Left Cambridge for London in 1864 to pursue a literary career. Contributed articles to newspapers and journals on a wide variety of subjects. Editor of the *Cornhill Magazine*, 1871–82. First editor of the *Dictionary of national biography* from 1882. Knighted, 1902. (*DNB*.)

Stephens, Thomas Sellwood (b. 1825). Clergyman. Curate of Wanstead, Essex, 1853–9; of Down, Kent, 1859–67. Rector of St Erme, Cornwall, 1867–1904. (*Alum. Oxon.*, *Crockford's*.)

Steudel, Ernst Gottlieb (1783–1856). Swiss physician and botanist. Began medical practice in Esslingen, Switzerland, in 1806. Senior consultant, 1828–56. (*ADB*.)

Stevens, Samuel. Natural history agent and collector of British Coleoptera and Lepidoptera. Active member of the Zoological Society of London. Alfred Russel Wallace's agent, 1848–63. (Wallace 1905, 1: 266.)

Stokes, Charles (1783–1853). Stockbroker in London; collector of coins and drawings, naturalist, and geologist. Collected fossil wood. FRS 1821. (R. Desmond 1977, *Modern English biography*, Sarjeant 1980.)

Stokes, George Gabriel, 1st baronet (1819–1903). Mathematician and physicist. Lucasian professor of mathematics, Cambridge University, 1849–1903. Secretary of the Royal Society, 1854–85; president, 1885–90. FRS 1851. (*DNB*, *DSB*.)

Strachey, Henry (1816–1912). Lieutenant-colonel in the Bengal army. Member of the Tibetan Boundary Commission, 1848. Surveyed the Upper Indus Valley. Collected plants at Pagong Lake in eastern Ladakh, 1846. (R. Desmond 1977.)

Stur, Dénes Rudolf Jossef (Dionys) (1827–93). Hungarian stratigrapher, palaeobotanist, and vertebrate palaeontologist. Joined the staff of the Imperial-Royal Geological Survey of Vienna in 1850; director, 1885. (*BHGW*, Sarjeant 1980, *Quarterly Journal of the Geological Society of London* (1894), Proceedings, pp. 55–6.)

Sulivan, Bartholomew James (1810–90). Naval officer and hydrographer. Lieutenant in HMS *Beagle*, 1831–6. Surveyed the Falkland Islands in HMS *Arrow*, 1838–9. Commander of HMS *Philomel*, 1842–6. Resided in the Falkland Islands, 1848–51. Commanded HMS *Lightning* in the Baltic, 1854–5. Naval officer of the marine department of the Board of Trade, 1856–65. Admiral, 1877. (*DNB*.)
24 May [1861]

Sumner, John Bird (1780–1862). Held the second prebendal stall in Durham cathedral, 1827–48. Bishop of Chester, 1828–48. Archbishop of Canterbury, 1848–62. Believed that the evidences of both geology and the Scriptures supported the existence of the Creator. (*DNB*.)

Sutherland, Peter Cormack (1822–1900). Scottish physician, naturalist, and explorer. Undertook expeditions to the west coast of Africa and to Greenland. Accompanied an expedition in search of John Franklin, 1850. Government geologist and surveyor-general in Natal, South Africa, 1853–86. (R. Desmond 1977, *DSAB*.)

Swale, William (1816–75). Gardener from Norfolk who emigrated to Christchurch, New Zealand, and became a prominent nurseryman.

Swinhoe, Robert (1836–77). Official in the British foreign service. Attached to the British consulate in Amoy, China, 1861. British consul, Taiwan, 1865; Ning-po, 1873–5. Collected plants and animals in Formosa. FRS 1876. (R. Desmond 1977, *Modern English biography*.)

Symonds, William Samuel (1818–87). Geologist, author, and clergyman. Curate of Offenham, near Evesham, 1843–5. Became rector of Pendock, Worcestershire, in 1845. Devoted his spare time to studying local archaeology and geology and that of Wales. In the 1870s, visited Auvergne and the Ardèche to search for traces of ancient glaciers. (*DNB*, Sarjeant 1980.)
26 March [1861]

Tankerville, Earl of. *See* Bennet, Charles Augustus.

Taylor, Tom (1817–80). Dramatist and editor of *Punch*. Professor of English literature and the English language, London University, 1845–7. Called to the bar, 1846. Assistant secretary, Board of Health, 1850–4; promoted secretary in 1854. Secretary, sanitary department, until 1871. A successful playwright who provided over seventy plays to the principal London theatres in thirty-five years. Active member of the staff of *Punch* until 1874. (*DNB*.)

Teale, Thomas Pridgin (1800–67). Surgeon. Senior surgeon, Leeds Public Dispensary, 1824–33; Leeds General Infirmary, 1833–64. Consulting surgeon, Leeds General Infirmary, 1864–7. Lecturer on anatomy and physiology, Leeds School of Medicine. (*Modern English biography*, Plarr 1930.)

Tegetmeier, William Bernhard (1816–1912). Editor, journalist, lecturer, and naturalist. Pigeon fancier and an expert on fowls. Secretary of the Apiarian Society of London. (E. W. Richardson 1916.)

3 February [1861], 25 February [1861], 1 March [1861], 22 March [1861], 28 March [1861], 2 April [1861], 14 April [1861], *4 May [1861]*, *[28 May 1861]*, 28 May [1861], 14 June [1861], 29 August [1861]

Temple, Frederick (1821–1902). Fellow and tutor of Balliol College, Oxford, 1842–8. Headmaster of Rugby School, 1857–69. Bishop of Exeter, 1869–85; of London, 1885–96. Archbishop of Canterbury, 1896–1902. Author of 'The education of the world' published in *Essays and reviews* (1860). (*DNB*.)

Tethmosis III. Pharaoh of Egypt from *circa* 1500 B.C. (*EB*.)

Thomson, Thomas (1817–78). East India Company surgeon and botanist in India. Collaborated with Joseph Dalton Hooker on various publications and joined him on his Himalayan expedition in 1849. Curator of the Asiatic Society's museum, Calcutta, 1840. Superintendent of the Calcutta botanic garden and professor of botany at the Calcutta Medical College, 1854–61. FRS 1855. (R. Desmond 1977, *DNB*.)

Thorley, Emily. Sister of C. A. Thorley, the governess at Down House between 1850 and 1856. Acted as governess to the Darwin children at Ilkley in October 1859.

Thwaites, George Henry Kendrick (1811–82). Botanist and entomologist. Superintendent of the Peradeniya botanic gardens, Ceylon, 1849; director, 1857–80. FRS 1865. (R. Desmond 1977, *DNB*.)

Ticknor and Fields. Boston publishing company, founded in 1832 by William Davis Ticknor (1810–64). Soon joined by James Thomas Fields (1817–81), the company took the name of Ticknor and Fields in 1854. Published the *Atlantic Monthly* as well as the works of many leading British and American writers. (*DAB*.)

Todd, Robert Bentley (1809–60). Physician, physiologist, and medical editor. Professor of physiology at King's College, London, 1836–53. FRS 1838. (*DNB*.)

Torrey, John (1796–1873). American botanist and chemist. Collected and described plants from various government explorations. Professor of chemistry at the College of Physicians and Surgeons, New York City, 1827–54. Assayer of the United States Mint in New York, 1853–73. (*DAB*, *DSB*.)

Town, Daniel. Carpenter in Down. (*Post Office directory of the six home counties* 1862.)

Trautschold, Hermann Adolfovich (1817–1902). Russian geologist, mineralogist, and palaeontologist. Attached to the Petrovsky Academy in Moscow. Travelled extensively in Russia and the Caucasus. Specialist on the Jurassic stratigraphy and invertebrates of Russia. (Sarjeant 1980.)

Trevelyan, Walter Calverley, 6th baronet (1797–1879). Of Wallington, Northumberland. Discovered *Romulea parviflora*, a member of the Iris family. (*Alum. Oxon.*, R. Desmond 1977.)

Trimmer, Joshua (1795–1857). Geologist. Employed on the Geological Survey of England, 1846–54. Studied the relationship between geology and agriculture. (*DNB*, Flett 1937, Sarjeant 1980.)

Trollope, Anthony (1815–82). Novelist and post-office official. His numerous novels include the series known as the 'Barsetshire Chronicles'. His writing is noted for its humour and detailed human observation. (*DNB*.)

Trübner, Johann Nicolaus (Nicholas) (1817–84). German-born publisher. Worked for the Longman publishing company, 1843–51. Entered into partnership with Thomas Delf and, later, David Nutt and established a successful publishing business. His main interests were philology, philosophy, religion, and oriental studies. Studied Sanskrit and Hebrew. (*ADB*, *DNB*.)

Turnbull, George Henry. Resided at the Rookery, Down. (*Post Office directory of the six home counties* 1862.)

Tyndall, John (1820–93). Irish physicist, lecturer, and populariser of science. Professor of natural philosophy at the Royal Institution, London, 1853; superintendent of the Royal Institution, 1867–87. FRS 1852. (*DNB*, *DSB*.)

23 February [1861], 24 July [1861]

University of Breslau

4 August 1861

Ussher, James (1581–1656). Professor of divinity at Dublin, 1607–20/1. Vice-chancellor of Trinity College, Dublin, 1614. Appointed archbishop of Armagh in 1824/5. His version of biblical chronology was the standard adopted in editions of the English Bible. (*DNB*.)

Van Voorst, John (1804–98). Publisher. Employed by the Longman publishing company, 1826–33. Publisher with premises in Paternoster Row, London, 1833–86. Published many works on natural history. (*Modern English biography*.)

Veitch, James (1815–69). In partnership with his father James Veitch (1792–1863) as the owners of a nursery in Chelsea. (R. Desmond 1977.)

Vernet, Luis Elie (1791–1871). German-born Argentinian pioneer, merchant, and civil servant. A major party in the dispute between the Argentinians, the Americans, and the British over the control of the Falkland Islands (Malvinas), 1819–33. Formed a settlement on East Falkland Island and resided there for several years under an authority from the government of Buenos Aires. (*NDBA*, Vernet 1833.)

Verneuil, Philippe Edouard Poulletier (Edouard) de (1805–73). French geologist and palaeontologist. Studied, in particular, Palaeozoic stratigraphy and fossils. Travelled in Europe, 1835–8. Toured European Russia with Roderick Impey Murchison, 1840–1. Carried out geological studies in Spain, 1849–62. President of the Geological Society of France, 1840, 1853, and 1867. (*DSB*, Sarjeant 1980.)

Victoria Adelaide Mary Louise, Princess Royal of Great Britian and German Empress (1840–1901). Eldest child of Queen Victoria and Prince Albert. Married Prince Frederick William of Prussia in 1858. (*DNB*.)

Vignoles, John. Naval commander who retired from active service in 1855. (*Navy list* 1861.)

Wagner, Rudolph (1805–64). German anatomist, physiologist, and anthropologist. Appointed ordinary professor of physiology, comparative anatomy, and natural history at Göttingen in 1840. (*ADB*, *DSB*.)

Walford, Edward (1823–97). Compiler. Ordained priest in 1847. Edited *Once a Week*, 1859–65; *Gentleman's Magazine*, 1866–8. Engaged in the publication of several biographical and genealogical works including *County families of the United Kingdom* (1860) and *Windsor peerage* (1890). In later life, established a reputation as an antiquary. (*DNB*.)

13 June [1861]

Walker, Francis (1809–74). Entomologist who specialised in the small parasitic Hymenoptera, the chalcids, and was also an authority on the Diptera. Described the Chalcididae CD collected on the *Beagle* voyage. Catalogued a number of the insect collections in the British Museum in a manner that was subsequently criticised. (*Entomologist's Monthly Magazine* 11 (1874): 140–1, Gilbert 1977.)

1 February 1861

Wallace, Alfred Russel (1823–1913). Collector in the Amazon, 1848–52; in the Malay Archipelago, 1854–62. Independently formulated a theory of natural selection in 1858. Lecturer and author of works on protective coloration, mimicry, and zoogeography. Became a spiritualist in 1865 and was considered a controversial figure because of his views on spiritualism, vaccination, land nationalisation, women's rights, and socialism. FRS 1893. (*DNB*, *DSB*.)

30 November 1861

Wallich, George Charles (1815–99). Surgeon and zoologist. Served with the Indian army, 1838–57. Joined HMS *Bulldog* as naturalist to the North Atlantic telegraphic expedition in 1860; discovered evidence of animal life at great oceanic depths and published his findings. Also published on the Protozoa. (*DNB*, *DSB*.)

Wang Ch'i (*fl.* 1565–1614). Chinese official and author. Compiled several encyclopaedic works valuable for their treatment of the Ming period. (*Dictionary of Ming biography*.)

Ward, Nathaniel Bagshaw (1791–1868). Botanist and physician. Inherited his father's medical practice in Whitechapel, London. Invented the 'Wardian' case for the transport of plant specimens. Examiner in botany for the Society of Apothecaries, 1836–54. Took a great interest in the Chelsea Botanical Garden. An original member of the Botanical Society of Edinburgh and a founding member of the Microscopical Society. (R. Desmond 1977, *DNB*.)

Waterhouse, George Robert (1810–88). Naturalist. A founder of the Entomological Society, 1833. Curator of the Zoological Society, 1836–43. Joined the mineralogical and geological branch of the natural history department of the

British Museum in 1843; keeper, geology department, 1857–80. Described CD's mammalian and entomological specimens from the *Beagle* voyage. (*DNB*, Gilbert 1977.)
  12 November [1861]

Watson, Hewett Cottrell (1804–81). Botanist, phytogeographer, and phrenologist. Published various guides to the distribution of British plants. Edited the *Phrenological Journal*, 1837–40. Thereafter, returned to the study of botany to which he devoted the rest of his life. Collected plants in the Azores, 1842. An early supporter of the idea of the progressive development of plant species. (*DNB*, *DSB*.)
  [17 July 1861], *24 July 1861*, *[after 24 July 1861]*,
  20 September 1861

Watts, Elizabeth. Breeder of domesticated fowls. Author of works on cookery, gardening, and poultry keeping. For some years, editor of the poultry department in the *Field* and editor of the *Poultry Chronicle*. (*NUC*.)

Weddell, Hugh Algernon (1819–77). British-born botanist and physician. MD, Paris, 1841. Worked in France. Specialist on South American flora. (R. Desmond 1977.)

Wedgwood, Alfred Allen (Tim) (1842–92). Son of Hensleigh and Frances Mackintosh Wedgwood. (Freeman 1978.)

Wedgwood, Caroline Sarah (1800–88). CD's sister. Married Josiah Wedgwood III, her cousin, in 1837. (*Darwin pedigree*.)

Wedgwood, Frances Julia (Snow) (1833–1913). Novelist, biographer, historian, and literary critic. Daughter of Hensleigh and Frances Mackintosh Wedgwood. (*Burke's peerage* 1980, Wedgwood and Wedgwood 1980.)
  11 July [1861]

Wedgwood, Frances Mackintosh (Fanny Hensleigh) (1800–89). Second child of James Mackintosh and Catherine Allen. Married Hensleigh Wedgwood in 1832. (Freeman 1978.)

Wedgwood, Frances Mosley (Fanny Frank) (1808–74). Married Francis Wedgwood in 1832. (*Burke's peerage* 1980.)

Wedgwood, Francis (Frank) (1800–88). Master-potter and partner in the Wedgwood pottery works at Etruria, Staffordshire, until 1876. Emma Darwin's brother. Married Frances Mosley in 1832. (*Alum. Cantab.*)

Wedgwood, Godfrey (1833–1905). Son of Francis and Frances Mosley Wedgwood. Joined the family pottery works in Staffordshire. (Freeman 1978, Wedgwood and Wedgwood 1980.)

Wedgwood, Hensleigh (1803–91). Philologist and barrister. Fellow, Christ's College, Cambridge, 1829–30. Metropolitan police magistrate at Lambeth, 1832–7. Registrar of metropolitan carriages, 1838–49. An original member of the Philological Society, 1842. Published the first volume of his *Dictionary of English etymology* in 1859. Emma Darwin's brother. Married Frances Mackintosh in 1832. (*DNB*.)

Wedgwood, Hope Elizabeth (Dot) (1844–1934). Daughter of Hensleigh and Frances Mackintosh Wedgwood. Second wife of Godfrey Wedgwood. (Freeman 1978.)

Wedgwood, James Mackintosh (Bro) (1834–64). Eldest son of Hensleigh and Frances Mackintosh Wedgwood. Held a position in the Colonial Office. (Freeman 1978, Wedgwood and Wedgwood 1980.)

Wedgwood, Josiah, I (1730–95). Master-potter. Founded the Wedgwood pottery works at Etruria, Staffordshire. Grandfather of CD and Emma Darwin. Greatly interested in experimental chemistry. Contributed several papers on the measurement of high temperature to the Royal Society's *Philosophical Transactions*. Actively associated with scientists and scientific societies. FRS 1783. (*DNB*, *DSB*.)

Wedgwood, Sarah (1734–1815). Married Josiah Wedgwood I in 1764. (*Emma Darwin*.)

Weir, John Jenner (1822–94). Naturalist and customs official. (R. Desmond 1977, Freeman 1978.)

Welwitsch, Friedrich (1807–72). Austrian botanist and traveller. Lived in Portugal and studied the flora, 1839–53. Botanised in West Africa, 1853–61. The resources in Lisbon for dealing with his large plant collections proved insufficient and Welwitsch had to move to London. A fire in his house in 1872 damaged his collection. This led to a nervous breakdown and eventually his death. (*ADB*, *BLKO*.)

Westwood, John Obadiah (1805–93). Entomologist and palaeographer. An active member of the Entomological Society of London. Hope professor of zoology, Oxford University, 1861–93. Entomological referee for the *Gardeners' Chronicle*. Awarded the Royal Medal of the Royal Society in 1855. (*DNB*, Gilbert 1977.)
15 August [1861], 4 September [1861], *26 September 1861*

Whately, Richard (1787–1863). Principal of St Alban Hall, Oxford, 1825–9. Professor of political economy, Oxford, 1829–31. Archbishop of Dublin, 1831–63. (*DNB*.)

Whewell, William (1794–1866). Mathematician and historian and philosopher of science. Master of Trinity College, Cambridge, 1841–66. Professor of moral philosophy, Cambridge University, 1838–55. FRS 1820. (*DNB*, *DSB*.)

White, Adam (1817–79). Naturalist. Official in the zoological department of the British Museum, 1835–63. Described some of the insect specimens collected on the *Beagle* voyage. (*DNB*, Gilbert 1977.)

Wiegmann, Arend J. Friedrich (1771–1853). Pharmacist in Braunschweig, Germany. (*BHGW*.)

Wight, Robert (1796–1872). Botanist. Assistant surgeon to the East India Company in Madras, 1819–26; naturalist, 1826–8; garrison surgeon, 1828–31; surgeon, 1834–6; superintendent of cotton cultivation, 1836–53. Published numerous works on the botany of India. FRS 1855. (R. Desmond 1977, *DNB*.)

Wilberforce, Samuel (1805–73). Clergyman. Rector of Brighstone, Isle of Wight, 1830–40; of Alverstoke, Hampshire, 1840–3. Appointed dean of Westminster in

1845. Chaplain to Prince Albert, 1841. Bishop of Oxford, 1845–69. Opposed CD's theory of evolution. FRS 1845. (*DNB*, Sarjeant 1980.)

Wilkes, Charles (1798–1877). Naval officer and explorer. Commanded the United States Exploring Expedition, 1838–42. From 1843 until 1861, on special service, mainly in Washington, preparing for publication the information collected by the expedition. Took command of the *San Jacinto* in 1861 and sparked off the 'Trent affair' by forcibly removing two Confederate diplomatic commissioners from their ship and detaining them. (*DAB*.)

Wilkinson, George Pearson. Perpetual curate of Thornley, Wolsingham, Durham, from 1857. (*Crockford's*.)

Willdenow, Karl Ludwig (1765–1812). German botanist and apothecary in Berlin. Gave up his apothecary's shop when he became professor of natural history at the Royal College of Medicine and Surgery, Berlin, in 1798. Appointed chief botanist of the Berlin Academy of Science and curator of the botanical gardens in 1801. Professor of botany, University of Berlin, 1809–12. (*ADB*, *DSB*.)

Williams, Benjamin Samuel (1824–90). Gardener to the Warner family at Hoddeston. Partner with Robert Parker in a nursery at Seven Sisters Road, Holloway, London, 1854–61; subsequently moved to Victoria and Paradise Nurseries, Upper Holloway. A specialist on orchids, he was the author of the popular work *The orchid grower's manual* (1852), which was the first successful manual giving instructions for the cultivation of orchids. (R. Desmond 1977, Reinikka 1972, Stewart ed. 1992.)

Williams, Edward Augustus. Surgeon residing in Bromley High Street. Member of the Royal College of Surgeons of England, 1824. (*London and provincial medical directory* 1860, *Post Office directory of the six home counties* 1859.)

Williams and Norgate. London booksellers and publishers who specialised in foreign scientific literature. The business was a partnership between Edmund Sydney Williams (1817–91) and Frederick Norgate. Publishers of the *Natural History Review*. (*Modern English biography*.)

Williams, Rowland (1817–70). Divine. Classical tutor, King's College, Cambridge, 1843–51. Vice-principal and professor of Hebrew, St David's Theological College, Lampeter, 1850–62. Vicar of Broad Chalke with Bower Chalke, Wiltshire, 1858–70. His contribution to *Essays and reviews* (1860) led to his being prosecuted for heterodoxy in the Court of Arches in 1861; however, the judgement against him was subsequently reversed on appeal. (*Alum. Cantab.*, *DNB*.)

Williamson, William Crawford (1816–95). Surgeon and naturalist. Surgeon to the Chorlton-on-Medlock dispensary, Manchester, 1842–68; to the Manchester Institute for Diseases of the Ear, 1855–70. Professor of natural history, anatomy, and physiology, Owens College, Manchester, 1851–91. Began a comprehensive study of the plants of the coal measures in 1858. FRS 1854. (*DNB*, *DSB*.)

Wilson, Benjamin O'Neile (*c.* 1806–86). Newspaper editor and proprietor in Hobart, Tasmania. Editor of the *Hobart Town Advertiser* from 1845. Usher of the

Wilson, Benjamin O'Neile, cont.
Black Rod in the Legislative Council of the Tasmanian Parliament, 1876–9. (State Archives, University of Tasmania Library.)
22 December 1861

Wilson, Henry Bristow (1803–88). Divine. Fellow, St John's College, Oxford, 1825–50. Rawlinsonian professor of Anglo-Saxon, Oxford, 1839–44. Vicar, Great Staughton, Huntingdonshire, 1850–88. Contributed an article to *Essays and reviews* (1860) as a result of which he was prosecuted for heresy in the Court of Arches. His case was tried together with that of Rowland Williams; he was found guilty of three of the articles brought against him, but, as with Williams, the decision was reversed on appeal. (*DNB*.)

Wilson, Horace Hayman (1786–1860). Orientalist. Assay-master at the Calcutta mint, 1811–19. Appointed secretary of the Asiatic Society of Bengal in 1811. Boden professor of Sanskrit, Oxford University, 1832–60. Librarian to the East India Company, 1836–60. Director of the Royal Asiatic Society of London, 1837–60. FRS 1834. (*DNB*, *Modern English biography*.)

Winkler, Tiberius Cornelius (1822–97). Dutch geologist and palaeontologist. Curator of the Teylor Museum, Haarlem, from 1864. Translated *Origin* into Dutch (1860). (*NNBW*, Sarjeant 1980.)
7 July 1861

Winthrop, Robert Charles (1809–94). American statesman. Representative and senator from Massachusetts. (*DAB*.)

Wollaston, Thomas Vernon (1822–78). Entomologist and conchologist. Resided in Cambridge, then London, until, for health reasons, he moved to Devon. Passed many winters in Madeira where he collected insects and shells. Had a particular interest in the Coleoptera. (*DNB*, Gilbert 1977.)

Wolstenholme, Joseph (1829–91). Mathematician. Fellow of Christ's College, Cambridge, 1852–69. Professor of mathematics at the Royal Indian Engineering College, Cooper's Hill, London, 1871–89. (*Alum. Cantab.*, *DNB*.)

Wood, George. Farmer. Of Down, Kent. (*Post Office directory of the six home counties* 1862.)

Wood, Harriet Charlotte (d. 1878). Daughter of Western Wood of North Cray Place, Kent. Married Nevile Lubbock in 1861. (*Burke's peerage* 1980.)

Woodhouse, Alfred J. Dentist at 1 Hanover Square, London. The Darwin family dentist. (CD's Account book (Down House MS).)

Woodward, Samuel Pickworth (1821–65). Naturalist. Sub-curator, Geological Society, 1839–45. Appointed professor of geology and natural history at the Royal Agricultural College, Cirencester, 1845. First-class assistant in the department of geology and mineralogy, British Museum, 1848–65. (*DNB*, Sarjeant 1980.)
5 June [1861]

Wooler, William A. Landowner who resided at Sadberge Hall near Darlington, County Durham (*Post Office directory of the county of Durham* 1873.)
4 February 1861

Worsaae, Jens Jacob (1821–85). Danish archaeologist. Inspector of ancient monuments, Denmark. Appointed professor of archaeology at the University of Copenhagen in 1854. Director, National Museum, Copenhagen, 1865–85. (*DSB*.)

Wrangel, Ferdinand Petrovich (1796–1870). Russian navigator who explored northern waters. Director of the Russian American Company, 1840–9. Naval minister, 1855–7. (*GSE*.)

Wright, Chauncey (1830–75). American mathematician and philosopher. In 1852, became a calculator for the newly established *American Ephemeris and Nautical Almanac*, for which he devised new methods of calculation. Recording secretary, American Academy of Arts and Sciences, Boston, 1863–70. Published the first of a series of philosophical essays in the *North American Review* in 1864. (*DAB*.)

Wyman, Jeffries (1814–74). American anatomist and ethnologist. Curator of the Lowell Institute, Boston, 1840; lecturer, 1840–1 and 1848–9. Travelled in Europe, 1841–3. Professor of anatomy and physiology, Hampden-Sydney Medical College, Virginia, 1843–8. Hersey professor of anatomy, Harvard University, 1847–74. Collected in South America, 1857–8. Curator of the Peabody Museum of American Archaeology and Ethnology, Harvard, 1866–74. (*DAB*, *DSB*.)
8 January [1861], 3 February [1861]

Yarrell, William (1784–1856). Zoologist. Newspaper agent and bookseller in London. Author of standard works on British birds and fishes. An original member of the Zoological Society of London from its formation in 1826. (*DNB*.)

Zarco, J. Gonzales (*fl.* 1420). Portuguese navigator. Travelled on the *Ceuta* expedition of 1415. Navigated the coast of Africa with Tristan Vaz in 1417. Claimed Madeira for the Portuguese crown in 1419 and was granted part of the island in recognition of his service. (*NBU*, *Variation* 1: 112 n. 23.)

Zeller, Philipp Christoph (1808–93). German entomologist. Teacher in Glogau. (Gilbert 1977.)

Zuccarini, Joseph Gerhard (1797–1848). German botanist. Professor of applied botany in the economics faculty at the University of Munich, 1826–48. Described some of the plants collected in Brazil by J. B. von Spix and K. F. P. von Martius (1817–20); of Japan by P. F. von Siebold (1824–30). (*ADB*.)

Bibliography of biographical sources

Abbott, Charles Compton. 1861. Notes on the birds of the Falkland Islands. *Ibis* 3: 149–67.

ADB: *Allgemeine deutsche Biographie*. Under the auspices of the Historical Commission of the Royal Academy of Sciences. 56 vols. Leipzig: Duncker and Humblot. 1875–1912.

Allan, Mea. 1967. *The Hookers of Kew, 1785–1911*. London: Michael Joseph.

Alum. Cantab.: *Alumni Cantabrigienses. A biographical list of all known students, graduates and holders of office at the University of Cambridge, from the earliest times to 1900. Part II.*

From 1752 to 1900. Compiled by J. A. Venn. 6 vols. Cambridge: Cambridge University Press. 1940–54.

Alum. Oxon.: *Alumni Oxonienses: the members of the University of Oxford, 1715–1886*. By Joseph Foster. 4 vols. Oxford. 1888.

Army list: *The army list*. London. 1815–81.

Aust. dict. biog.: *Australian dictionary of biography: 1788–1850; 1851–1890*. Edited by Douglas Pike and Bede Nairn. 6 vols. Melbourne: Melbourne University Press. 1966–76.

Banking almanac: *The banking almanac, directory, year book and diary*. Edited by D. Morier Evans. London. 1845–1919.

Bénézit, Emmanuel. 1976. *Dictionnaire critique et documentaire des peintres, sculpteurs, dessinateurs et graveurs de tous les temps et de tous les pays, par un groupe d'écrivains spécialistes français et étrangers*. Revised edition. 10 vols. Paris: Librairie Gründ.

Berg, Carlos. 1895. Carlos Germán Conrado Burmeister. Reseña Biográfica. *Anales del Museo Nacional de Buenos Aires* 2d ser. 1: 316–57.

BHGW: *Biographisch-literarisches Handwörterbuch zur Geschichte der exacten Wissenschaften*. By Johann Christian Poggendorff. Vols. 1–4, Leipzig: Johann Ambrosius Barth. 1863–1904. Vol. 5, Leipzig and Berlin: Verlag Chemie, G.m.b.H. 1926.

Bibby, Harold Cyril. 1959. *T. H. Huxley, scientist, humanist and educator*. London: Watts.

BLA: *Biographisches Lexikon der hervorragenden Aerzte aller Zeiten und Völker*. Edited by A. Wernich, E. Gurlt, and August Hirsch. 6 vols. Vienna and Leipzig: Urban & Schwarzenberg. 1884–8.

BLC: *The British Library general catalogue of printed books to 1975*. 360 vols. and supplement (6 vols.). London. 1979–88.

BLKO: *Biographisches Lexikon des Kaiserthums Oesterreich, enthaltend die Lebensskizzen der denkwürdigen Personen, welche seit 1750 in den österreichischen Kronländern geboren wurden oder darin gelebt und gewirkt haben*. By Constant von Wurzbach. 60 vols. Vienna. 1856–90.

BNB: *Biographie nationale publiée par l'Académie royale des sciences, des lettres et des beaux-arts de Belgique*. 28 vols. Brussels. 1866–1944.

Burke's landed gentry: *Burke's genealogical and heraldic history of the landed gentry*. By John Burke. 1–18 editions. London: Burke's Peerage Ltd. 1836–1972.

Burke's peerage: *Burke's peerage and baronetage*. 1–105 editions. London: Burke's Peerage (Genealogical Books). 1826–1980.

BWN: *Biographisch Woordenboek der Nederlanden, bevattende Levensbeschrijvingen van zoodanige Personen, die zich op eenigerlei wijze in ons Vaderland hebben vermaard gemaakt.* Edited by K. J. R. van Harderwijk and G. D. J. Schotel. 20 vols. Haarlem. 1852–77.

Catalogue of graduates, University of Dublin: *A catalogue of graduates who have proceeded to degrees in the University of Dublin, from the earliest recorded commencements to July, 1866, with supplement to December 16, 1868*. Dublin. 1869.

CDEL: *A critical dictionary of English literature and British and American authors, living and deceased*. By S. Austin Allibone. 3 vols. Philadelphia and London. 1859–71.

Supplement by John Foster Kirk. 2 vols. Philadelphia and London. 1891.

Clergy list: *The clergy list . . . containing an alphabetical list of the clergy*. London. 1841–.

Complete peerage: *The complete peerage of England Scotland Ireland Great Britain and the United Kingdom extant extinct or dormant*. By George Edward Cokayne. New edition, revised and much enlarged. Edited by the Hon. Vicary Gibbs and others. 12 vols. London: The St Catherine Press. 1910–59.

Crockford's: *The clerical directory, a biographical and statistical book of reference for facts relating to the clergy and the Church*. London. 1858. Continued as *Crockford's clerical directory etc*. London. 1860–.

DAB: *Dictionary of American biography*. Under the auspices of the American Council of Learned Societies. 20 vols., index, and 7 supplements. New York: Charles Scribner's Sons. London: Oxford University Press. 1928–81.

Darwin pedigree: *Pedigree of the family of Darwin*. Compiled by H. Farnham Burke. Privately printed. 1888. Reprinted in Freeman, Richard Broke, *Darwin pedigrees*. London: printed for the author. 1984.

DBF: *Dictionnaire de biographie Française*. Under the direction of J. Balteau, M. Barroux, M. Prevost, and others. 17 vols. and 4 fascicles of vol. 18 (A–Jumelle). Paris: Librairie Letouzey et Ané. 1933–92.

DBL: *Dansk biografisk leksikon*. Founded by Carl Frederick Bricka. 3d ed. Edited by Sv. Cedergreen Bech. 16 vols. Copenhagen: Gyldendal. 1979–84.

Desmond, Ray. 1977. *Dictionary of British and Irish botanists and horticulturists, including plant collectors and botanical artists*. 3d ed. London: Taylor and Francis.

Dictionary of Ming biography: *Dictionary of Ming biography 1368–1644*. Edited by L. Carrington Goodrich and Chaoying Fang. New York and London: Columbia University Press. 1976.

Dictionnaire universel des contemporains: *Dictionnaire universel des contemporains contenant toutes les personnes notables de la France et des pays étrangers*. Edited by G. Vapereau. Paris. 1858. 3d ed., 1865. 4th ed., 1870. 5th ed., 1880.

DNB: *Dictionary of national biography*. Edited by Leslie Stephen and Sidney Lee. 63 vols. and 2 supplements (6 vols.). London: Smith, Elder, and Co. 1885–1912. *Dictionary of national biography 1912–85*. Edited by H. W. C. Davis, J. R. H. Weaver, and others. 8 vols. London: Oxford University Press. 1927–90.

DNZB: *A dictionary of New Zealand biography*. Edited by G. H. Scholefield. 2 vols. Wellington: Department of Internal Affairs, New Zealand. 1940.

DSAB: *Dictionary of South African biography*. Edited by W. J. de Kock, D. W. Krüger, and C. J. Beyers. Vol. 1, Cape Town: Nasionale Boekhandel Beperk. Vols. 2–3, Cape Town: Tafelberg Publishers. Vol. 4, Durban and Pretoria: Butterworth & Co. 1968–81.

DSB: *Dictionary of scientific biography*. Edited by Charles Coulston Gillispie. 14 vols., supplement, and index. New York: Charles Scribner's Sons. 1970–80.

Dupree, Anderson Hunter. 1959. *Asa Gray, 1810–1888*. Cambridge, Mass.: The Belknap Press of Harvard University.

East-India register and army list: *The East-India register and army list*. Compiled from the official returns, by permission of the Honourable the East-India Company. 16 vols. London. 1845–60.

EB: *Encyclopædia Britannica*. 11th ed. 29 vols. Cambridge: Cambridge University Press. 1910–11.

Elliot, Clark A. 1979. *Biographical dictionary of American science: the seventeenth through the nineteenth centuries*. Westport, Conn. and London: Greenwood Press.

Emma Darwin: *Emma Darwin: a century of family letters, 1792–1896*. Edited by Henrietta Litchfield. Revised edition. 2 vols. London: John Murray. 1915.

ENZ: *An encyclopaedia of New Zealand*. Edited by A. H. McLintock. 3 vols. Wellington, New Zealand: R. E. Owen. 1966.

Flett, John Smith. 1937. *The first hundred years of the Geological Survey of Great Britain*. London: His Majesty's Stationery Office.

Freeman, Richard Broke. 1978. *Charles Darwin: a companion*. Folkestone: W. Dawson & Sons. Hamden, Conn.: Archon Books, Shoe String Press.

Gilbert, Pamela. 1977. *A compendium of the biographical literature on deceased entomologists*. London: British Museum (Natural History).

Grande encyclopédie: *La grande encyclopédie inventaire raisonné des sciences, des lettres et des arts*. Edited by F. Camille Dreyfuss and others. 31 vols. Paris. 1886–1902.

GSE: *Great Soviet encyclopedia*. [Translation of 3d ed. of *Bol'shaia Sovetskaia entsiklopediia*. Moscow. 1970–9.] 31 vols. New York: Macmillan. London: Collier Macmillan. 1973–82.

Huxley, Leonard, ed. 1900. *Life and letters of Thomas Henry Huxley*. 2 vols. London: Macmillan and Co.

IBN: *Index bio-bibliographicus notorum hominum*. Pt C: *Corpus alphabeticum sectio generalis*. Edited by Jean-Pierre Lobies and Otto and Wolfram Zeller. 60 vols. (A–Durau, Pierre). Osnabrück: Biblio Verlag. 1974–92.

King-Hele, Desmond. 1963. *Erasmus Darwin*. London: Macmillan & Co.

——, ed. 1981. *The letters of Erasmus Darwin*. Cambridge: Cambridge University Press.

List of officers of the army: *A list of officers of the army and of the corps of Royal Marines, on full, retired, and half-pay; with an index*. London. 1779–1869.

London and provincial medical directory: *The London and provincial medical directory, and general medical register*. London. 1848–60.

London & provincial medical directory: *The London & provincial medical directory, inclusive of the medical directory for Scotland, and the medical directory for Ireland, and general medical register*. London. 1861–9.

Medical directory for Ireland: *The medical directory and general medical register for Ireland*. London. 1852–60.

Modern English biography: *Modern English biography containing many thousand concise memoirs of persons who have died since the year 1850*. By Frederic Boase. 3 vols. and supplement (3 vols.). Truro: printed for the author. 1892–1921.

Moore, James R. 1985. Darwin of Down: the evolutionist as squarson-naturalist. In *The Darwinian heritage*, edited by David Kohn. Princeton: Princeton University Press in association with Nova Pacifica.

Morton, Leslie T. 1970. *A medical bibliography (Garrison and Morton): an annotated check-list of texts illustrating the history of medicine*. 3d ed. London: André Deutsch.

Navy list: *The navy list*. London. 1815–70.

NBL: *Norsk biografisk leksikon*. Edited by Edvard Bull, Anders Krogvig, Gerhard Gran, and others. 19 vols. Oslo: H. Aschehoug & Co. 1923–83.

NBU: *Nouvelle biographie universelle*. Edited by Jean Chrétien Ferdinand Hoefer. 46 vols. Paris. 1852–66.

NDB: *Neue deutsche Biographie*. Under the auspices of the Historical Commission of the Bavarian Academy of Sciences. 16 vols. (A–Melanchthon). Berlin: Duncker and Humblot. 1953–90.

NDBA: *Nuevo Diccionario Biografico Argentino (1750–1930)*. Edited by Vicente Osvaldo Cutolo. 7 vols. Buenos Aires: Editorial Elche. 1968–85.

NNBW: *Nieuw Nederlandsch biografisch woordenboek*. Edited by P. C. Molhuysen, P. J. Blok, and K. H. Kossmann. 10 vols. Leiden: A. W. Sijthoff. 1911–37.

NUC: *The national union catalog*. Pre-1956 imprints. 685 vols. and supplement (68 vols.). London and Chicago: Mansell. 1968–81.

OBL: *Österreichisches biographisches lexikon 1815–1950*. Edited by Leo Santifaller, Eva Obermayer-Marnach, and others. 9 vols. and 3 fascicles of vol. 10 (A–Schmid). Vienna: Österreichischer Akademie der Wissenschaften. 1957–92.

Physicians: *The roll of the Royal College of Physicians of London*. Vol. 3: 1801–25. By William Munk. 2d ed., revised and enlarged. London. 1878. Vol. 4: *Munk's roll, vol. 4: Lives of the fellows of the Royal College of Physicians of London 1826–1925*. Compiled by G. H. Brown. London: Royal College of Physicians. 1955.

Plarr, Victor Gustave. 1930. *Plarr's lives of the fellows of the Royal College of Surgeons of England*. Revised by Sir D'Arcy Power. 2 vols. London: Simpkin Marshall.

Post Office directory of the county of Durham: *The Post Office directory of the county of Durham, and the principal towns and adjacent places in Northumberland, Cumberland & Westmorland*. London. 1873–.

Post Office directory of Hampshire, Wiltshire, and Dorsetshire: *Post Office directory of Hampshire, Wiltshire, and Dorsetshire*. London. 1848–1939.

Post Office directory of Lancashire: *Post Office directory of Lancashire*. London. 1858–.

Post Office directory of Northumberland: *Post Office directory of Westmorland, Cumberland, Northumberland, and Durham*. London. 1858.

Post Office directory of the six home counties: *Post Office directory of the six home counties, viz., Essex, Herts, Kent, Middlesex, Surrey and Sussex*. London. 1845–.

Post Office London directory: *Post Office London directory*. London. 1802–.

Post Office London suburban directory: *The Post Office London suburban directory*. London. 1860–1918.

Rehbock, Philip F. 1979. The early dredgers: 'naturalizing' in British seas, 1830–1850. *Journal of the History of Biology* 12: 293–368.

Reinikka, Merle A. 1972. *A history of the orchid*. Coral Gables, Florida: University of Miami Press.

Richardson, Edmund William. 1916. *A veteran naturalist; being the life and work of W. B. Tegetmeier*. London: Witherby & Co.

Rugby School register: *Rugby School register*. Vol. 1: from April 1675 to October 1857. Revised and annotated by Godfrey A. Solly. Rugby: George Over. 1933.

Sarjeant, William A. S. 1980. *Geologists and the history of geology: an international bibliography from the origins to 1978*. 5 vols. and supplement (2 vols., 1987). London: Macmillan.

SBL: *Svenskt biografiskt lexikon*. Edited by Bertil Boëthius, Erik Grill, Birgitta Lager-Kromnow, and Göran Nilzén. 26 vols. and 2 fascicles of vol. 27 (A–Olsson). Stockholm: Albert Bonnier and P. A. Norstedt. 1918–93.

Scherren, Henry. 1905. *The Zoological Society of London. A sketch of its foundation and development and the story of its farm, museum, gardens, menagerie and library*. London, Paris, New York, and Melbourne: Cassell & Co.

Scholes, Percy A. 1970. *The Oxford companion to music*. 10th ed. Edited by John Owen Ward. London, New York, and Toronto: Oxford University Press.

Stenton, Michael. 1976. *Who's who of British members of Parliament, 1832–1885*. Brighton: Harvester Press.

Stewart, Joyce, ed. 1992. *Orchids at Kew*. London: HMSO.

Taxonomic literature: *Taxonomic literature. A selective guide to botanical publications and collections with dates, commentaries and types*. 2d ed. By Frans A. Stafleu and Richard S. Cowan. 7 vols. Utrecht: Bohn, Scheltema & Holkema. The Hague: dr. W. Junk b.v., Publishers. 1976–88.

Tegetmeier, William Bernhard, ed. 1856–7. *The poultry book: including pigeons and rabbits; comprising the characteristics, management, breeding and medical treatment of poultry, being the results of personal observation and practice of the Rev. W. Wingfield, G. W. Johnson, Esq*. Re-arranged and edited by W. B. Tegetmeier. London.

Tree, Isabella. 1991. *The ruling passion of John Gould: a biography of the bird man*. London: Barrie & Jenkins.

Variation: *The variation of animals and plants under domestication*. By Charles Darwin. 2 vols. London. 1868.

Vernet, Luis Elie. 1833. Account of East Falkland Island. *Journal of the Royal Geographical Society of London* 3: 94–9.

Wallace, Alfred Russel. 1905. *My life: a record of events and opinions*. 2 vols. London: Chapman & Hall.

Wedgwood, Barbara and Wedgwood, Hensleigh. 1980. *The Wedgwood circle, 1730–1897: four generations of a family and their friends*. London: Studio Vista.

WWW: *Who was who*. A companion to *Who's who*. 6th ed. 7 vols. and cumulated index, 1897–1980. London: Adam & Charles Black. 1988–9.

INDEX

The dates of letters to and from Darwin's correspondents are listed in the Biographical register and index to correspondents and are not repeated here. Darwin's works are indexed under the short titles used throughout this volume and listed in the bibliography.

Table o

Robert Darwin = Elizabeth Hill
1682–1754 1702–97

William Alvey Darwin = Jane Brown
1726–83 1746–1835

Elizabeth Collier Chandos-Pole 1747–1832 =[2] Erasmus Darwin 1731–1802 =[1] Mary Ho 174

Samuel Fox = Anne
1765–1851 1777–1859

Jane Harriett Ryle = Francis Sacheverel
1794–1866 1786–1859

Robert 1766–

Samuel Tertius Galton = Frances Anne Violetta
1783–1844 1783–1874

Elizabeth Ann (Bessy) b.1808

Henry Parker = Maria
1788–1856 1798–18

Mary Ann = Samuel Ellis Bristowe
1800–29 1800–55

Lucy Harriot 1809–48

Eliza 1801–86

Milicent Adèle 1810–83

Susa Eliza 180

Emma 1803–85

Emma Sophia b.1811

William Darwin = Harriet Fletcher
1805–80 1799–1842

Darwin b.1814

Erasmus A 1804

Frances Jane b.1806

Erasmus b.1815

Julia b.1809

Francis 1822–1911

Emily Cathe 1810